Lecture Notes in Networks and Systems

Volume 229

Series Editor

Janusz Kacprzyk, Systems Research Institute, Polish Academy of Sciences, Warsaw, Poland

Advisory Editors

Fernando Gomide, Department of Computer Engineering and Automation—DCA, School of Electrical and Computer Engineering—FEEC, University of Campinas—UNICAMP, São Paulo, Brazil

Okyay Kaynak, Department of Electrical and Electronic Engineering, Bogazici University, Istanbul, Turkey

Derong Liu, Department of Electrical and Computer Engineering, University of Illinois at Chicago, Chicago, USA; Institute of Automation, Chinese Academy of Sciences, Beijing, China

Witold Pedrycz, Department of Electrical and Computer Engineering, University of Alberta, Alberta, Canada; Systems Research Institute, Polish Academy of Sciences, Warsaw, Poland

Marios M. Polycarpou, Department of Electrical and Computer Engineering, KIOS Research Center for Intelligent Systems and Networks, University of Cyprus, Nicosia, Cyprus

Imre J. Rudas, Óbuda University, Budapest, Hungary

Jun Wang, Department of Computer Science, City University of Hong Kong, Kowloon, Hong Kong

AF896369

The series "Lecture Notes in Networks and Systems" publishes the latest developments in Networks and Systems—quickly, informally and with high quality. Original research reported in proceedings and post-proceedings represents the core of LNNS.

Volumes published in LNNS embrace all aspects and subfields of, as well as new challenges in, Networks and Systems.

The series contains proceedings and edited volumes in systems and networks, spanning the areas of Cyber-Physical Systems, Autonomous Systems, Sensor Networks, Control Systems, Energy Systems, Automotive Systems, Biological Systems, Vehicular Networking and Connected Vehicles, Aerospace Systems, Automation, Manufacturing, Smart Grids, Nonlinear Systems, Power Systems, Robotics, Social Systems, Economic Systems and other. Of particular value to both the contributors and the readership are the short publication timeframe and the world-wide distribution and exposure which enable both a wide and rapid dissemination of research output.

The series covers the theory, applications, and perspectives on the state of the art and future developments relevant to systems and networks, decision making, control, complex processes and related areas, as embedded in the fields of interdisciplinary and applied sciences, engineering, computer science, physics, economics, social, and life sciences, as well as the paradigms and methodologies behind them.

Indexed by SCOPUS, INSPEC, WTI Frankfurt eG, zbMATH, SCImago.

All books published in the series are submitted for consideration in Web of Science.

More information about this series at http://www.springer.com/series/15179

Radek Silhavy
Editor

Artificial Intelligence in Intelligent Systems

Proceedings of 10th Computer Science On-line Conference 2021, Vol. 2

Springer

Editor
Radek Silhavy
Faculty of Applied Informatics
Tomas Bata University in Zlín
Zlín, Czech Republic

ISSN 2367-3370 ISSN 2367-3389 (electronic)
Lecture Notes in Networks and Systems
ISBN 978-3-030-77444-8 ISBN 978-3-030-77445-5 (eBook)
https://doi.org/10.1007/978-3-030-77445-5

© The Editor(s) (if applicable) and The Author(s), under exclusive license
to Springer Nature Switzerland AG 2021, corrected publication 2022
This work is subject to copyright. All rights are solely and exclusively licensed by the Publisher, whether the whole or part of the material is concerned, specifically the rights of translation, reprinting, reuse of illustrations, recitation, broadcasting, reproduction on microfilms or in any other physical way, and transmission or information storage and retrieval, electronic adaptation, computer software, or by similar or dissimilar methodology now known or hereafter developed.
The use of general descriptive names, registered names, trademarks, service marks, etc. in this publication does not imply, even in the absence of a specific statement, that such names are exempt from the relevant protective laws and regulations and therefore free for general use.
The publisher, the authors and the editors are safe to assume that the advice and information in this book are believed to be true and accurate at the date of publication. Neither the publisher nor the authors or the editors give a warranty, expressed or implied, with respect to the material contained herein or for any errors or omissions that may have been made. The publisher remains neutral with regard to jurisdictional claims in published maps and institutional affiliations.

This Springer imprint is published by the registered company Springer Nature Switzerland AG
The registered company address is: Gewerbestrasse 11, 6330 Cham, Switzerland

Preface

Artificial Intelligence and Intelligent Systems papers are presented in this proceedings. This proceedings is a Vol. 2 of the Computer Science On-line Conference 2021.

This book constitutes the refereed proceedings of the Artificial Intelligence in Intelligent Systems section of the 10th Computer Science On-line Conference 2021 (CSOC 2021), held online in April 2021.

CSOC 2021 has received (all sections) more than 390 submissions from more than 30 countries. More than 68% of accepted submissions were received from Europe, 19% from Asia, 5% from Africa, 5% from America and 3% from Australia.

CSOC 2021 intends to provide an international forum to discuss the latest high-quality research results in all areas related to computer science.

Computer Science On-line Conference is held online, and modern communication technology, which is broadly used, improves the traditional concept of scientific conferences. It brings equal opportunity to participate for all researchers around the world.

I believe that you find the following proceedings exciting and valuable for your research work.

April 2021 Radek Silhavy

Organization

Program Committee

Program Committee Chairs

Petr Silhavy	Tomas Bata University in Zlin, Faculty of Applied Informatics, Czech Republic
Radek Silhavy	Tomas Bata University in Zlin, Faculty of Applied Informatics, Czech Republic
Zdenka Prokopova	Tomas Bata University in Zlin, Faculty of Applied Informatics, Czech Republic
Roman Senkerik	Tomas Bata University in Zlin, Faculty of Applied Informatics, Czech Republic
Roman Prokop	Tomas Bata University in Zlin, Faculty of Applied Informatics, Czech Republic
Viacheslav Zelentsov	Doctor of Engineering Sciences, Chief Researcher of St. Petersburg Institute for Informatics and Automation of Russian Academy of Sciences (SPIIRAS)
Roman Tsarev	Department of Informatics, Siberian Federal University, Krasnoyarsk, Russia

Program Committee Members

Boguslaw Cyganek	Department of Computer Science, University of Science and Technology, Krakow, Poland
Krzysztof Okarma	Faculty of Electrical Engineering, West Pomeranian University of Technology, Szczecin, Poland

Monika Bakosova	Institute of Information Engineering, Automation and Mathematics, Slovak University of Technology, Bratislava, Slovak Republic
Pavel Vaclavek	Faculty of Electrical Engineering and Communication, Brno University of Technology, Brno, Czech Republic
Miroslaw Ochodek	Faculty of Computing, Poznan University of Technology, Poznan, Poland
Olga Brovkina	Global Change Research Centre Academy of Science of the Czech Republic, Brno, Czech Republic and Mendel University of Brno, Czech Republic
Elarbi Badidi	College of Information Technology, United Arab Emirates University, Al Ain, United Arab Emirates
Luis Alberto Morales Rosales	Head of the Master Program in Computer Science, Superior Technological Institute of Misantla, Mexico
Mariana Lobato Baes	Research-Professor, Superior Technological of Libres, Mexico
Abdessattar Chaâri	Laboratory of Sciences and Techniques of Automatic Control and Computer Engineering, University of Sfax, Tunisian Republic
Gopal Sakarkar	Shri. Ramdeobaba College of Engineering and Management, Republic of India
V. V. Krishna Maddinala	GD Rungta College of Engineering and Technology, Republic of India
Anand N. Khobragade (Scientist)	Maharashtra Remote Sensing Applications Centre, Republic of India
Abdallah Handoura	Computer and Communication Laboratory, Telecom Bretagne, France

Technical Program Committee Members

Ivo Bukovsky, Czech Republic
Maciej Majewski, Poland
Miroslaw Ochodek, Poland
Bronislav Chramcov, Czech Republic
Eric Afful Dazie, Ghana
Michal Bliznak, Czech Republic
Donald Davendra, Czech Republic
Radim Farana, Czech Republic
Martin Kotyrba, Czech Republic
Erik Kral, Czech Republic
David Malanik, Czech Republic
Michal Pluhacek, Czech Republic
Zdenka Prokopova, Czech Republic
Martin Sysel, Czech Republic
Roman Senkerik, Czech Republic
Petr Silhavy, Czech Republic
Radek Silhavy, Czech Republic
Jiri Vojtesek, Czech Republic

Eva Volna, Czech Republic
Janez Brest, Slovenia
Ales Zamuda, Slovenia
Roman Prokop, Czech Republic
Boguslaw Cyganek, Poland
Krzysztof Okarma, Poland
Monika Bakosova, Slovak Republic
Pavel Vaclavek, Czech Republic
Olga Brovkina, Czech Republic
Elarbi Badidi, United Arab Emirates

Organizing Committee Chair

Radek Silhavy Tomas Bata University in Zlin, Faculty of Applied Informatics, Czech Republic, email: radek@silhavy.cz

Conference Organizer (Production)

Silhavy s.r.o.
Website: https://www.openpublish.eu
Email: csoc@openpublish.eu

Conference Website, Call for Papers

https://www.openpublish.eu

Contents

Online Teaching as an Important Means of Technogenic Society Development During the Pandemic

Irina V. Batunova(✉), Ekaterina I. Lobyneva, Albina Yu. Nikolaeva, and Svetlana V. Kokorina

Siberian Federal University, 79 Svobodny pr., 660041 Krasnoyarsk, Russia

Abstract. The article is devoted to the problems that have arisen during the students' education of non-language institutes to distance learning due to the pandemic of 2020. The relevance of the research topic is due, on the one hand, to modern requirements for the digitalization of education in general and higher vocational education in particular cases. On the other hand, the organisation to distance learning of students at the Siberian Federal University carried out with the use of various digital resources, which had a number of positive results; it was accompanied by a number of difficulties, the analysis of some of these problems are presented in this article. In the article you also can find the most effective digital methods of teaching students that were identified, taking into account direction of the student and acquisition of unqualified competencies in accordance with the discipline program, all these findings were based on the analysis fulfilled by the authors. Other digital resources considered in the article by the authors can be applied additionally, according to their advantages and disadvantages. The practical significance of the article lay in the justification of the feasibility of combining the resources under consideration in the organization of distant study of students in the course "Foreign language" as an example of the work organized during the pandemic. #CSOC1120

Keywords: Pandemic · Online teaching · Digital resources

1 Introduction

The process of updating to the online learning is actively developing in higher education in nowadays. Online teaching of the foreign languages is one of the most pressing issues and trends in non-language universities. The unprecedented situation with the coronavirus pandemic since March 2020 forced all educational systems to switch to online format, which caused many problems, spurred it online "… transformation educational process at all its levels" (Distance learning: analysis, opinions, judgments of Shmurygin O. IN. The exemplary process in the conditions of a pandemic//Vocational education and the labor market. 2020. - № 2 - pp. 51–52) [1]. At the Siberian Federal University teachers were ready for this situation, as they have long actively developed their own electronic training courses and teaching materials, introduced it online test programs, combined it

© The Author(s), under exclusive license to Springer Nature Switzerland AG 2021
R. Silhavy (Ed.): CSOC 2021, LNNS 229, pp. 1–7, 2021.
https://doi.org/10.1007/978-3-030-77445-5_1

online lectures, interactive practical tasks, webinars, teleconferences with international participation and integrated data development into classroom sessions using MOODLE platform. The organization of independent work of students, who performed it online tests, drew up glossaries, recorded their own kinds of videos in the studied language, and presented their scientific works online. Particularly these resources were used by part-time students, masters, postgraduates and teachers who improve their skills in the process [2].

2 Methods

According to the order of the Ministry of Science and Higher Education of the Russian Federation of 14.03.2020 №397 "On the organization of educational activities in organizations implementing educational programs higher education and related additional professional programs, in the conditions of prevention of the spread of coronovurus infection in the territory of the Russian Federation", the Siberian Federal University (SibFU) has introduced the procedure for organizing intermediate certification in distance format.

The presence of effective introduction of online education in SibFU gave the opportunity to provide continuous distance learning during the isolation. Distance learning at the university, built on the basis of the using modern, digital technologies, with almost completely independent work of students, has become one of the possible and affordable forms of training in the spread of a new disease on the world scale (Ponomarenk E.P., Ivanova A.L. Problems of the organization of reflexive teaching of a foreign language remotely in the period of a pandemic/World of Pedagogy and Psychology: international scientific and practical journal. 2020. 04 (45) [2]. Technical capabilities of the university allowed to disconnect access to available resources as much as possible.

Simultaneously with the use of videoconferencing services and online meetings for the implementation of separate procedures within the framework of intermediate certification activities (issuance and verification of training assignments, testing, surveys, document exchange, etc.) the following services of the University's EIOS were also used:

- system "E-courses" (https://e.sfu-kras.ru/);
- corporate social network service "My SibFU". (The use of these services is carried out in accordance with the Regulations on the implementation of e-learning and distance educational technologies in SibFU (http://about.sfu-kras.ru/node/9739).

The shifting to online learning in the university took place in a very short time, teachers had to analyze a lot of resources, identify the effectiveness of used during the pandemic digital resources. Therefore, the purpose of the research is to analyze, systematize and describe the results of the use of digital resources used during the pandemic, identify their advantages and disadvantages. Experienced training in the discipline "Foreign Language" was organized with students of the engineering and construction and architectural and design schools of the first and second year education, masters and postgraduate students during 4 months. Taking into account the analysis, online training was based on an electronic course in the MOODLE system, and both ready and cut are considered during the course of changes in training organizations, video broadcasting

Table 1. Analysis of digital resources used

Digital resources	Advantages	Disadvantages
MOODLE	- change the training content according to the discipline program - test the students - view the statistics of the done works and the quality of their fulfillment - monitor the work of teachers and students by the department of the education at the university	- there were difficulties with access to the course - the inability to download large files - no mobile application giving uninterrupted access to the system
ZOOM and Skype	- providing it with a line of contact with students - the possibility of holding a full-fledged classroom on the training schedule - doing tests and examinations online - the opportunity to record practical lessons, tests and exams for the report on the work done	- intermittent communication interference (hovering, signal interruption) - lack of electricity or failure of technical equipment - the inability of some students to enter to the video conference on the schedule of classes - limit of the number of participants who can simultaneously communicate in real time on Skype - time limit for free communication in ZOOM
Viber and WhatsApp	- convenient for organizing the educational process - ongoing contact of teachers and students - an easiest type of communication for students - the ability to find out for yourself any questions very quickly	- for teachers communication through these applications was round-the-clock, uninterrupted
Server "My SibFU"	- the possibility of rapid communication between teachers and students - consultations of students - attachments of completed tasks - downloads of different files of different sizes - Tracking the work of teachers and students	- possible communication interruptions due to site load
e-mail	- the possibility of communication between teachers and students - providing a group or individual work with a student with general information	- the huge number of letters on the email it is not always possible to respond to letters immediately - not all students have business communication skills

(*continued*)

Table 1. (*continued*)

Digital resources	Advantages	Disadvantages
VK	- the presence of students accounts and the usability of this resource for them - the ability to post information, photos, audio and video files of various formats	- lack of possibility of placing quality tests - lack of control of students' work

using the ZOOM platform, video calls to Skype, Viber, WhatsApp, working in small groups in the chat "My SibFU", e-mail, VK [4].

These digital resources were certainly useful in providing online learning with the students, they could use those resources that they had access to or that were more convenient to them. However, if you analyze the resources used, you can conclude that everyone has its pros and cons (Table 1).

3 Results

According to the detailed table data we can affirm that e-courses in the MOODLE system—on the one hand, give a wide opportunity to change educational content in accordance with the discipline program, to conduct intermediate testing of students, view the statistics of the performed works and their quality, monitor the work of teachers and students by the educational department of the university. On the other hand, some students had difficulty finding the necessary course, there were difficulties with access to the course could not download large files, had no special mobile application giving uninterrupted access to the system. Organizing videoconferencing on the platform ZOOM and Skype- the advantage was the provision of online contact with students and the possibility of conducting a full-fledged classroom on the academic schedule, conducting scores and exams he line. Possibility to provide a record of classes, credits and examinations to the training department for the report of the work done. As some disadvantages we can note the intermittent communication interference (hovering, signal interruption), lack of electricity or failure of technical equipment, inability of some students to connect to video Conferences on the schedule of classes, a limit on the number of participants who can simultaneously communicate in real time on Skype, and a limit on the time of free communication in ZOOM. If we speak about Viber and WhatsApp, these applications were very convenient for organizing the educational process, fast contact of teachers and students, comfortable for students, as these applications for they could find out for themselves any questions very quickly. For teachers, communication through these applications was round-the-clock. Working in the chat "My SibFU" was a very effective resource, giving the possibility of rapid communication between teachers and students, consultations of students, attachment of completed tasks, upload different files of different sizes, track the work of teachers and students [5, 6].

Don't forget about email, because it is the possibility of communication between teachers and students, providing a group or individual student with general information.

Of the huge number of letters in the mail, it was not always possible to respond to letters immediately. Speaking about social network, we decided to choose VK—the advantage was the availability of accounts for students and the habitability of this resource for them, the ability to place information, photos, audio and video files of various formats and sizes. The lack of possibility of placing quality tests, as well as control by the educational department of work of teachers and students.

The university's work in the context of distance education during coronovurus infection has shown that our professional resources and technologies are easily adapted to the tasks of education; it clearly demonstrated the opportunities of online education. The educational process in SibFU has been completely transferred to online form, with the retention of the number of audit hours and planned at the beginning of the academic year.

4 Discussions

After analyzing the above digital resources used in the process of students' learning, the authors emphasize the greatest efficiency of e-courses created by university's teachers for each school. Within the framework of the educational process, these online resources were thematically completed, structured by the author, which was presented to students through the Internet. Multimedia facilities provided the opportunity to organize practical classes on various types of resources and to study the course "Foreign language" in the context of a pandemic without reducing educational material with the help of other digital resources. The courses developed by the authors on the MOODLE platform, accessed by the login and password of the student and the teacher of the SibFU, were presented with a variety of materials strictly consistent with the textbook, on which students were trained to the pandemic period, for example:

- lectures and explanations on the topics of the course;
- additional materials (methodological recommendations, glossaries, articles and texts for translation and doing of corresponding text tasks for individual work of students);
- practical tasks for work on the topics of the course;
- control tasks and tests with the possibility of viewing correct answers, which made in this period tests not just control, but also training, the student could see not the same evaluation for the done work, but also the teacher's comment explaining the mistakes made by the student in the test;
- video tutorials for independent study and analysis on the system of evaluation score;
- a forum with students' questions and teacher' answers, links to self-created videos, audio and other materials for viewing and evaluation by the teacher and other students.

Open access to all materials presented in the course for all students of technical directions gave the opportunity to successfully implement students' training in SibFU during the pandemic.

In this situation, these courses allowed to fulfill the programmatic requirements for each direction: comprehensively to form and develop the communicative competence of students technical directions in the whole variety of its components (linguistic, sociolinguistic, socio-cultural, strategic, discursive, educational and cognitive);

communicative-cognitive skills to search and select the necessary material for learning, to generalize, classify, analyze and synthesize the information received; communicative skills to represent and discuss the results of work with the results of completed tasks in the course, required by the teacher; the ability to use the resources of electronic courses with direct links to the Internet self-education in order to familiarize with the cultural and historical heritage of different countries and peoples in the professional branch of students in the studied language, as well as to submit reports in language conferences of various levels, to discuss student scientific work; the ability to use e-course resources to mark their information and educational interests and needs (E. M. Rodionova, A. A. Hanzhina Internet resources in teaching foreign language file: ///C: /users/home/downloads/internet-resurs-v-obuchenii-inostrannomu-yaz-ku.pdf) [7].

5 Conclusion

Thus, it can be concluded that during the pandemic in SibFU, the organisation of e-learning, which occurred in the shortest possible time, involved all basic and additional digital resources. This gave the opportunity to conduct the training of students effectively, without getting knocked out of the classroom schedule and reducing school hours. Positive moments of training for teachers can be considered: increased interest in the use of digital resources in the educational process in higher school, surge of activity (albeit forced) to create and adjust electronic courses for students, masters and postgraduates of non-language specialties, awareness of the need to improve the skills of teachers in the field of digital technology.

For students, the advantages were immersion in the familiar virtual environment for young people and, "as a result, the removal of the psychological barrier in situations of live, albeit online, communication" (Sidorova L.V. Krupskaya Yu.V. Pros, Cons and Prospects of Online Education//Vocational Education and the Labor Market. -2020.-№2.-p. 87–88) [8].

Negative points can be considered the difficulty of controlling students' work in time of online classes, especially in case of communication failures, inability for some students to open certain files, computer breakdowns, and overabundance of communication. At the same time there was an increase in the load in teaching work. In addition to working out all the auditor hours planned at the beginning of the year for teachers, they were littered with a mass of student work, on which they spent a lot of time not taken into account in the load, and were also in contact with students of their groups practically abnormal time. Therefore, in the future in such situations, in case of implementation of distance learning, it is not possible to focus on the number of hours planned for the admitter for contact work at traditional forms of learning [9, 10]. The priority should be to acquire and master students' learning results with their further assessment and validation, rather than mechanically adhering to the class schedule. It is necessary to plan time for consultations of students and teachers and individual consultations with the teacher in case of misunderstanding or difficulties when working students with educational materials online and ability to discuss problems on the content and form of tasks in any additional digital format convenient for the student: Whatsapp, Viber, VK, email, workgroup chats, forums with teachers and other participants, etc. In electronic courses, it is necessary to structure the material so that it is convenient for students to perceive—theoretical material needs to be crushed in small portions and in a variety of forms, e.g. audio, video, text information so that students have more opportunities to learn it. It is also necessary to develop training exercises to secure material and corresponding test tasks for intermediate certification. Another important factor is the availability of free access to the e-learning course from home devices, as well as the provision of technical support in case of problems.

In conclusion, we would like to note that the situation with the pandemic has started irreversible processes of qualitative change of educational system using digital resources in Russia in general and also in SibFU. Of course, such important elements of auditor contact work with students as practice, work in language laboratories, dialogue between all participants of the educational process cannot be replaced nothing, but the system of higher vocational education should acquire a form in which there would be a combination of auditor work with high-quality modern digital equipment that would give the ability to evaporate the advantage of digital resources.

References

1. Shmurygin, O.I.N.: Distance learning: analysis, opinions, judgments of the exemplary process in the conditions of a pandemic. Vocat. Educ. Labor Market **2**, 51–52 (2020)
2. Sharples, M., Taylor, J., Vavoula, G.A.: VS Verlag fur Sozialwissenschaften. Theory of learning for the mobile age. In: Medienbildung in Neuen Kuturraumen, pp. 87–99 (2010)
3. Ponomarenk, E.P., Ivanova, A.L.: Problems of the organization of reflexive teaching of a foreign language remotely in the period of a pandemic. World of Pedagogy and Psychology: International Scientific and Practical Journal. https://scipress.ru/pedagogy/articles/problemy-organizatsii-refleksivnogo-obucheniya-inostrannomu-yazyku-distantsionno-v-period-pandemii.html. Accessed 18 Aug 2020
4. Batunova, I.V., Lobinova, E.I., Nikolaeva, A.Yu.: Organization of contact work of students and teachers at a technical higher school through the moodle system. Int. Res. J. **5**(95) 3, 61–64 (2020)
5. Vonog, V.V., Yarutskaya, L.V.: Learning individual's development as function of productive assessment. Asian EFL J. **6**(20), 429–449 (2018)
6. Batunova, I.V., Lobinova, E.I., Nikolaeva, A.U.: Primenenie effektivnyx metodov obucheniya na zanyatiyax po inostrannomu yazyku v neyazykovyx texnicheskix vuzax (Application of effective teaching methods in foreign language classes in non-linguistic (technical) universities). Ekaterinburg **1**(67, Part 4), 11–14 (2018)
7. Rodionova, E.M., Hanzhina, A.A.: Internet resources in teaching foreign language. file: ///C: /users/home/downloads/internet-resurs v obuchenii inostrannomu-yaz-ku pdf. Accessed 15 Aug 2020
8. Sidorova, L.V., Krupskaya, Yu.V.: Pros, cons and prospects of online education. Vocat. Educ. Labor Market **2**, 87–88 (2020)
9. Herrington, J., Reeves, T.C.: A Guide to Authentic E-Learning. Routledge (2009). http://murdoch.edu.au. Accessed 8 Aug 2020
10. Cochrane, M., Shelly, S., Kirey, A.: Turn your phone on: using Android devices to collect scientific data. In: Education Research Highlights in Mathematics, Science and Technology, Ames, IA, pp. 89–95. International Society for Research in Education and Science (ISRES) (2019)

Application of Principal Component Analysis Based Support Vector Machine for Long-Run Economic Growth Forecasting

Moses Olaifa(✉) and Tranos Zuva

Department of ICT, Faculty of Applied and Computer Science,
Vaal University of Technology, Andries Potgieter Blvd,
Vanderbijlpark 1900, South Africa
tranosz@vut.ac.za

Abstract. This study aims at building a Support Vector Machine model based on Principal Component Analysis to forecast the direction of economic growth. Evaluation of the generated model is based on multiple economic indicators that have direct influence on economic performance. The performance of the improved SVM is compared with the SVM and regression analysis (RA). Results show that the PCA based SVM out performed the SVM and regression analysis with higher forecast accuracy. In addition, the proposed PCA based SVM method produced smaller error and more accurate forecast for longer time-steps.

Keywords: Economic growth · Support Vector Machine · Economic indicators · Forecasting

1 Introduction

Forecasting of economic growth has attracted a great deal of interest from the academia and practice across the world for a long time. Work done in previous studies [1–4] show that, persistent challenges in this area have always revolved around how to improve accuracy of forecast and what influential economic parameters to consider for the forecast.

Sokolov-Mladenovi et al. [5] proposed an improved artificial neural network using back propagation and extreme learning machine to predict GDP trend based on economic trade parameters for a single time-step ahead. The economic trade parameters used include; trade in services, exports of goods and services, imports of goods and services, trade and merchandise trade. However, the prediction only investigated trade factors which exclude other influential indicators such as inflation, government consumption expenditure, household consumption expenditure, gross capital formation among others, which also contribute to determining the economic growth of different countries.Beyond trade parameters, different studies have shown that other economic indicators such as inflation [6,7], government consumption expenditure [8–10], household consumption

© The Author(s), under exclusive license to Springer Nature Switzerland AG 2021
R. Silhavy (Ed.): CSOC 2021, LNNS 229, pp. 8–16, 2021.
https://doi.org/10.1007/978-3-030-77445-5_2

expenditure [11–13], capital formation [14–16] and many more can affect the state of the economy growth. The composition and dimensionality of the data set play major roles in producing accurate forecast [17,18].

Furthermore, the model evaluation was performed for a period of 1-time-step. Such short-run forecasting may not provide significant foresight to help make adequate economic decision. Most models will probably perform well in short time-steps. However, the level of forecasting accuracy may degrade as time-steps increases. Hence the need for methods that can not only incorporate high dimensional data features into forecasting but also produce high level accuracy in forecasting higher time-steps or long-run. The approach proposed in our work enables the contribution of various economic indicators to forecasting the trend in economic growth and produce lower error for long-run prediction.

In this study, an improved Support Vector Machine for economic growth forecasting is introduced. The SVM is a machine learning algorithm that finds its basis in the statistical theory proposed by Vapnik [19]. SVMs have been successfully applied to different problems areas such as regression problems [20, 21], image retrieval [22,24,25], fault detection [26,27] and intrusion detection [28, 29]. According to this study, the proposed Support Vector Machine based on Principal Component Analysis is constructed for economic growth forecast for various indicators. Experimental results of this study shows a prediction accuracy of the improved SVM over regression analysis and ordinary SVM.

2 Study Methodology

2.1 Principal Component Analysis

One of the major challenges in classification problems is the reduction of high dimensional data. Principal Component Analysis (PCA) is a classical statistical correlation analysis technique for addressing this problem. This technique can be used to determine the relationship between input data space and the reduced dimensional data. With the use of PCA for feature fusion or extraction, classifiers can produce more accurate predictions.

Given a data matrix $X = [x_1, x_2.x_3, ..., x_n]^T_m$, $e_1 \geq e_2 \geq ... \geq e_k \geq 0$, with the mean $E(X) = \mu(X)$ and co-variance $cov(v, X)$, then we assume there exist eigenvalues $\lambda_1 \geq \lambda_2 \geq ... \geq \lambda_k$ of the co-variance matrix and unit orthogonal eigenvectors $P_i = [v_{1i}, v_{2i}, ..., v_{qi}]$. Therefore, according to the cummulative contribute rate $R(t) = \sum_{i=1}^{t} e_i / \sum_{m=1}^{n} e_i$, the generated principal component PC_i is given as $PC_i = XP_i$.

The purpose of the PCA in this study is to infuse economic indicators to generate more useful Principal Components for the models rather than filtering out these indicators and selecting few as done in previous studies.

2.2 Support Vector Machine

Support Vector Machine (SVM) is a supervised machine algorithm that can be applied to solve both classification and regression problems. It performs a non-linear mapping from an input space to a higher dimensional feature space. According to [23], SVM objective is to minimize upper bound for generalization errors. If presented with a training data set given as $\{x_i, y_i\}^n$, where $x_i \in R^q$ is the input vector with d attributes, y_i is the target value and n is the size of the data points, the regression model of the SVM can be constructed using Eq. 1:

$$f(x) = \omega^T \dot{\phi}(x) + b \tag{1}$$

where $f(x)$ is regression estimate function built from the learning of the sample data points, ω is the weight vector, $\phi(x)$ is a non-linear mapping function from the input space to high dimensional space and b is the bias term. The SVM approximates the functions given in Eq. 1 and 2 below:

$$R = \frac{1}{2}||(\omega)||^2 + C\frac{1}{n}\sum_{i=1}^{n} |y_i - f(x_i)|_\varepsilon \tag{2}$$

where $||(\omega)||^2$ is the vector of feature weights used to constrain the model capacity. C is the regularized constant used to compute the trade-off between the regularization term and empirical error. To determine parameters ω and the bias b, the regularized function in Eq. 2 must be minimized as

$$minimize\left\{\frac{1}{2}||(\omega)||^2 + C\sum_{i=1}^{n}(\xi_i + \xi_i^*)\right\} \tag{3}$$

$$\begin{cases} \omega^T \dot{\phi}(x) + b - y_i \leq \varepsilon + \xi_i^* & i = 1, 2, ..., n \\ y_i - \omega^T \dot{\phi}(x) + b \leq \varepsilon + \xi_i & i = 1, 2, ..., n \\ \xi_i, \xi_i^* \leq 0 & i = 1, 2, ..., n \end{cases} \tag{4}$$

Equation 3 introduces positive stack variables ξ_i and ξ_i^* denoting the upper and lower excess deviation that penalize training errors by the loss function over the error tolerance. Equation 4 is generally computed in two different forms that include using Lagrangian multipliers and enforcing the Karush-Kuhn-Tucker (KKT) optimality condition. Input vectors with nonzero Lagrangian multipliers under the KKT condition support the structure of the estimator. These input vectors are called support vectors. A comprehensive discussion on algorithms that can be used to solve the SVM dual optimization problem is available in [30]. The SVM model uses different kernel functions that include polynomial, exponential radial basis function (ERBF), radial basis function (RBF), sigmoid and linear kernels.

3 Data Description

The data set used for evaluating the prediction of our models represents the GDP growth of South Africa over a period of 39 years. Each GDP is derived from 13 economic indicators including inflation, consumer prices; imports of goods and services; export of goods and services; gross capital formation; central government total expenditure and net spending among others. The data set contains 1 output and 13 input variables. The data set consists of GDP data in 1 year interval that spans from 1920 to 2019.

The data set is sourced from the Humanitarian Data Exchange for African Development Bank Group, the data provides update on African Economic Outlook update.

4 Models Performance Indicators

The performances of the algorithms considered in this study are evaluated based on different statistical measure criteria. The prediction accuracy of the three models are evaluated according to three different indicators, Mean Absolute Error (MAE), Means Square Error (MSE) and Root Mean Square Error (RMSE). These three are used because they are in the same units as the response variable. The MAE computes the sum of absolute value of the difference between the predicted and actual value. This metric is useful in situations where errors increases linearly over time. The RMSE which is a measure of fit, determines the spread of the prediction errors between the models' predicted values and the actual values. These three indicators are computed using the equations given below:

$$MAE = \frac{\sum_{i=1}^{n}(p_i - q_i)}{n} \tag{5}$$

where p_i is the predicted growth, q_i is the actual economic growth and n is the number of data points.

$$MSE = \frac{1}{n}\sum_{i=1}^{n}(p_i - q_i)^2 \tag{6}$$

By computing the square root of MSE, we get the RMSE given in Eq. 7

$$RMSE = \sqrt{\frac{\sum_{i=1}^{n}(p_i - q_i)^2}{n}} \tag{7}$$

5 Experiment and Results Discussion

In this section, we discuss the experimental setup, performance assessment of the PCA based SVM and the performance comparison of the proposed technique with Regression analysis and SVM. The focus of the performance evaluation is on the forecast accuracy of these models over 9 time-step. These three algorithms are used for prediction on the normalized economic data set. Both the SVM and PCA based SVM chose the polynomial kernel with a complexity parameter c lower than 10.0. From the trial experiment, selecting any c value greater than 10.0 yields a more accurate prediction for short run forecasting but the prediction error increases as the time unit increases. In view of the long term prediction objective of this study, a c-value less than 10.0 is more realistic for better results.

Table 1 presents the forecast accuracy result on the training data set with RA, SVM and PCA-SVM algorithms respectively. Each of the algorithms performed well on the training data with the PCA-SVM presenting the least prediction error or higher accuracy; an indication of a very small deviation between the predicted and actual growth. With a minimum and maximum RMSE of 0.5136 and 0.5702 respectively, the PCA-SVM out performs the other models. The RA and SVM present a close performance on the training data with RA producing a minimum and maximum RMSE of 1.4772 and 1.7044, while the SVM produces 1.5777 and 1.806 as minimum and maximum RMSE respectively. This shows a slightly better performance by the RA on training data. Furthermore, the three models did not show much difference in prediction accuracy with increase in time step. Each of the models recorded similar performances in MAE indicator with PCA-SVM leading by an average of 0.406 MAE error (minimum 0.383 and maximum 0.4299) over a 9-step time period, followed by RA and finally SVM with MAE averages of 1.1790 (minimum 1.0324 and maximum 1.2511) and 1.3138 (minimum 1.0567 and maximum 1.4198) respectively. Generally, none of the models present any significant difference in accuracy over long time steps.

Table 1. Training data set forecasting result

	RA			SVM			PCA-SVM		
	MAE	MSE	RMSE	MAE	MSE	RMSE	MAE	MSE	RMSE
T-01	1.2489	2.4819	1.5754	1.0567	2.4891	1.5777	0.383	0.3135	0.5599
T-02	1.2511	2.7879	1.6697	1.28	2.9529	1.7184	0.4053	0.3045	0.5518
T-03	1.1949	2.7692	1.6641	1.3644	3.0433	1.7445	0.4279	0.3242	0.5694
T-04	1.1753	2.9050	1.7044	1.4198	3.2616	1.806	0.4299	0.3251	0.5702
T-05	1.1812	2.8170	1.6784	1.3875	3.0878	1.7572	0.4091	0.2670	0.5167
T-06	1.2047	2.8487	1.6878	1.389	3.1198	1.7663	0.4068	0.2727	0.5222
T-07	1.1904	2.8043	1.6746	1.3705	3.1375	1.7713	0.4031	0.2669	0.5166
T-08	1.132	2.7156	1.6479	1.3225	3.0496	1.7463	0.395	0.2638	0.5136
T-09	1.0324	2.1821	1.4772	1.2334	2.6002	1.6125	0.3939	0.2722	0.5217

Forecasting performances on different percentages of test data are presented in Tables 2, 3 and 4. The first test result in Table 2 is based on 30% test data. PCA-SVM produces the best result with an average MAE and RMSE of 0.9264 and 0.9838 over 9-time steps. In terms of the time-steps; The first time step produces a RMSE of 1.1516 and MAE of 1.0229; Time-step-5 produces a RMSE of 0.9633 and MAE of 0.9366; and time-step-9 produces a RMSE of 0.8363 and MAE of 0.8224. The SVM produces an average MAE and RMSE of 11.1104 and 11.6582 respectively, while RA model produces an average MAE and RMSE of 41.0492 and 45.6826.

In time-step 1, RA showed a better prediction accuracy for lower time-step. However, the prediction error grows exponentially as the time step increases. Table 2 shows the RA with a MAE and RMSE of 3.0834 and 3.3304 respectively at time-step-1 but increased to 101.5046 and 103.7415 at time-step-9. The SVM showed low MAE and RMSE values of 6.4412 and 7.2100 at time-step-1, but this values linearly increased and showed 13.0916 and 13.1539 respectively at time-step 9.

The proposed model showed consistently high forecast accuracy for both short and long time-steps. it is obvious in the results shown in Tables 3 and 4. Both the RA and SVM showed fairly good prediction results at short time steps, however the models displayed significant increase in prediction errors as the time-steps get larger. Generally, the improved SVM shows a better growth forecasting performance than the RA and SVM.

Table 2. 0.3 Test data

	RA			SVM			PCA-SVM		
	MAE	MSE	RMSE	MAE	MSE	RMSE	MAE	MSE	RMSE
T-01	3.0834	11.0916	3.3304	6.4412	51.9841	7.2100	1.0229	1.3262	1.1516
T-02	6.6013	48.2122	6.9435	10.3278	126.0792	11.2285	1.3705	2.1509	1.4666
T-03	11.5551	157.3645	12.5445	11.8365	161.8340	12.7214	1.1148	1.3905	1.1792
T-04	18.9566	440.3912	20.9855	11.6790	158.6769	12.5967	0.9259	0.9063	0.9520
T-05	30.0063	1142.2980	33.7979	11.5171	148.2379	12.1753	0.9366	0.9279	0.9633
T-06	44.1608	2493.8138	49.9381	11.3710	138.0649	11.7501	0.7907	0.6675	0.8170
T-07	65.2621	5762.6469	75.9121	11.5199	138.1307	11.7529	0.6661	0.4903	0.7002
T-08	88.3123	10805.5193	103.9496	12.2097	152.1473	12.3348	0.6880	0.6211	0.7881
T-09	101.5046	10762.2988	103.7415	13.0916	173.0251	13.1539	0.8224	0.6994	0.8363

Table 3. 0.2 Test data

	RA			SVM			PCA-SVM		
	MAE	MSE	RMSE	MAE	MSE	RMSE	MAE	MSE	RMSE
T-01	3.4962	17.9852	4.2409	6.5122	45.0040	6.7085	0.7683	0.6699	0.8185
T-02	5.8288	42.3007	6.5039	10.2176	108.6077	10.4215	0.8155	0.7663	0.8754
T-03	6.5806	48.0000	6.9282	11.0186	125.9557	11.2230	0.7437	0.6121	0.7824
T-04	5.2494	35.8023	5.9835	10.0392	104.4995	10.2225	0.6861	0.5243	0.7241
T-05	5.4325	37.0922	6.1394	7.7678	62.0960	7.8801	0.8436	0.7230	0.8503
T-06	6.0161	47.1214	6.8645	6.8821	48.0887	6.9346	0.2927	0.1247	0.3531
T-07	18.6700	359.7243	18.9664	7.5718	58.2810	7.6342	0.1997	0.0587	0.2423
T-08	29.1414	849.2212	29.1414	9.6676	93.4025	9.6676	0.0743	0.0055	0.0743

Table 4. 0.5 Test data

	RA			SVM			PCA-SVM		
	MAE	MSE	RMSE	MAE	MSE	RMSE	MAE	MSE	RMSE
T-01	10.4333	225.1320	15.0044	10.9691	192.0719	13.8590	0.9674	1.2850	1.1336
T-02	10.9603	237.5883	15.4139	16.6939	445.8306	21.1147	1.2406	1.9642	1.4015
T-03	11.3952	250.8137	15.8371	19.9705	637.5676	25.2501	0.9529	1.2526	1.1192
T-04	9.3855	208.2047	14.4293	23.2638	858.4138	29.2987	1.0486	1.3995	1.1830
T-05	9.9132	221.9415	14.8977	17.8532	479.2246	21.8912	1.3258	2.5799	1.6062
T-06	10.3405	236.6921	15.3848	14.1329	269.3734	16.4126	1.4177	2.8285	1.6818
T-07	21.0159	1887.7200	43.4479	12.6920	196.1204	14.0043	1.3211	3.5797	1.8920
T-08	22.7206	2045.0022	45.2217	11.0988	134.1752	11.5834	2.4354	8.1174	2.8491
T-09	24.1136	2225.9430	47.1799	14.9726	240.0640	15.4940	2.6508	8.1447	2.8539

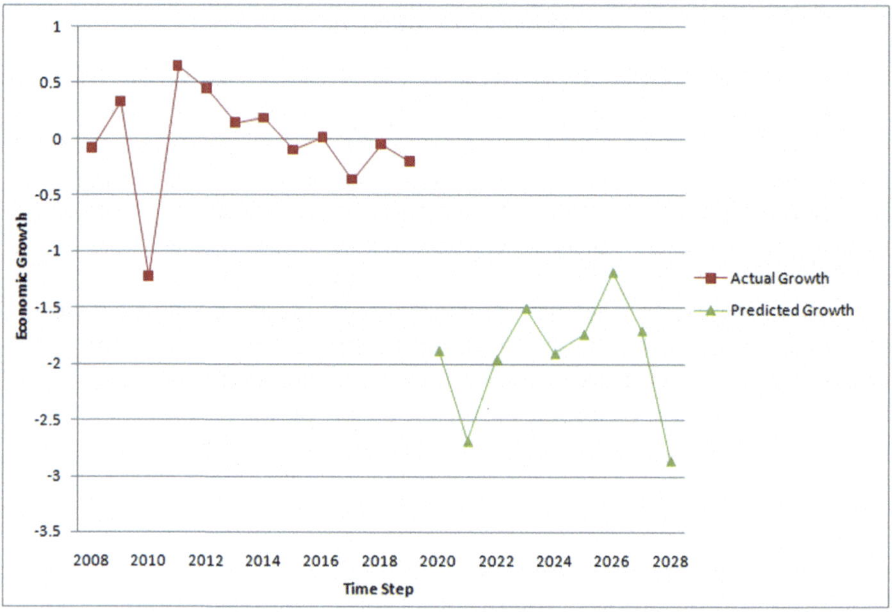

Fig. 1. PCASVM 9-Time-step forecast on 30% of test data

Figure 1 shows the economic growth forecast generated by the proposed enhanced SVM over a 9-year period between 2020 and 2028. The figure shows a drop in growth in 2020 and further drop in 2021. However, there is a rise in the growth in following 4 years.

6 Conclusion

In this study, the RA, SVM and PCASVM were built for the purpose of forecasting economic growth using multiple economic indicators that contributes to GDP. In order to achieve this goal, South Africa - one of the countries in sub-sahara Africa was chosen as a case study. This choice gives the opportunity to study Africa's economic data which has received less attention in previous research in this area. Moreover, unlike previous studies that focused on few indicators, the available data presents twelve economic indicators used to test the performances of the models in forecasting economic growth. Obtained results shows that RA, SVM and PCASVM models can be used for growth in short time steps. However, for long time steps, the PCASVM shows very low MAE and RMSE values compared to RA and SVM. This can translate to a better and consistent accuracy by PCASVM over the RA and SVM. In future work, further investigation will be necessary to improve accuracy for longer-run forecasting considering larger input data.

References

1. Feng, L., Zhang, J.: Application of artificial neural networks in tendency forecasting of economic growth. Econ. Model. **40**, 76–80 (2014)
2. Kordanuli, B., Barjaktarović, L., Jeremić, L., Alizamir, M.: Appraisal of artificial neural network for forecasting of economic parameters. Physica A: Stat. Mech. Appl. **465**, 515–519 (2017)
3. Milačić, L., Jović, S., Vujović, T., Miljković, J.: Application of artificial neural network with extreme learning machine for economic growth estimation. Physica A: Stat. Mech. Appl. **465**, 285–288 (2017)
4. Tümer, A.E., Akkuş, A.: Forecasting gross domestic product per capita using artificial neural networks with non-economical parameters. Physica A **512**, 468–473 (2018)
5. Sokolov-Mladenović, S., Milovančević, M., Mladenović, I., Alizamir, M.: Economic growth forecasting by artificial neural network with extreme learning machine based on trade, import and export parameters. Comput. Hum. Behav. **65**, 43–45 (2016)
6. Umaru, A., Anono, A.Z.: Effect of inflation on the growth and development of the Nigerian economy (an empirical analysis). Int. J. Bus. Soc. Sci. **3**(10) (2012)
7. Saymeh, A.A., Orabi, M.M.: The effect of interest rate, inflation rate, GDP, on real economic growth rate in Jordan. Asian Econ. Finan. Rev. **3**(3), 341 (2013)
8. Fouladi, M.: The impact of government expenditure on GDP, employment and private investment a CGE model approach. Iran. Econ. Rev. **15**(27), 53–76 (2010)
9. Halkos, G.E., Paizanos, E.: The effect of government expenditure on the environment: an empirical investigation. Ecol. Econ. **91**, 48–56 (2013)
10. Hajamini, M., Falahi, M.A.: The nonlinear impact of government consumption expenditure on economic growth: evidence from low and low-middle income countries. Cogent Econ. Finan. **2**(1), 948122 (2014)
11. Gerstberger, C., Yaneva, D.: Analysis of EU-27 household final consumption expenditure–Baltic countries and Greece still suffering most from the economic and financial crisis. Stat. Focus **2** (2013)

12. Tapsin, G., Hepsag, A.: An analysis of household consumption expenditures in EA-18. Eur. Sci. J. **10**(16) (2014)
13. Bonsu, C.O., Muzindutsi, P.F.: Macroeconomic determinants of household consumption expenditure in Ghana: a multivariate cointegration approach. Int. J. Econ. Finan. Issues **7**(4) (2017)
14. Hartwig, J.: Is health capital formation good for long-term economic growth?–Panel Granger-causality evidence for OECD countries. J. Macroecon. **32**(1), 314–325 (2010)
15. Adhikary, B.K.: FDI, trade openness, capital formation, and economic growth in Bangladesh: a linkage analysis. Int. J. Bus. Manag. **6**(1), 16 (2011)
16. Zheng, X., Wang, R., Wood, R., Wang, C., Hertwich, E.G.: High sensitivity of metal footprint to national GDP in part explained by capital formation. Nat. Geosci. **11**(4), 269–273 (2018)
17. Boivin, J., Ng, S.: Are more data always better for factor analysis? J. Econometrics **132**(1), 169–194 (2006)
18. Caggiano, G., Kapetanios, G., Labhard, V.: Are more data always better for factor analysis? Results for the euro area, the six largest euro area countries and the UK. J. Forecast. **30**(8), 736–752 (2011)
19. Cortes, C., Vapnik, V.: Support-vector networks. Mach. Learn. **20**(3), 273–97 (1995)
20. Shao, Y.H., Zhang, C.H., Yang, Z.M., Jing, L., Deng, N.Y.: An ε-twin support vector machine for regression. Neural Comput. Appl. **23**(1), 175–185 (2013)
21. Hemmati-Sarapardeh, A., Shokrollahi, A., Tatar, A., Gharagheizi, F., Mohammadi, A.H., Naseri, A.: Reservoir oil viscosity determination using a rigorous approach. Fuel **116**, 39–48 (2014)
22. Rao, C., Kumar, S.S., Mohan, B.C.: Content based image retrieval using exact Legendre moments and support vector machine. arXiv preprint arXiv:1005.5437, May 2010
23. Meyer, D., Wien, F.T.: Support vector machines. The interface to LibSVM in package e1071 28, August 2015
24. Kanimozhi, T., Latha, K.: An integrated approach to region based image retrieval using firefly algorithm and support vector machine. Neurocomputing **151**, 1099–1111 (2015)
25. Sharif, U., Mehmood, Z., Mahmood, T., Javid, M.A., Rehman, A., Saba, T.: Scene analysis and search using local features and support vector machine for effective content-based image retrieval. Artif. Intell. Rev. **52**(2), 901–925 (2019)
26. Banerjee, T.P., Das, S.: Multi-sensor data fusion using support vector machine for motor fault detection. Inf. Sci. **217**, 96–107 (2012)
27. Yin, S., Gao, X., Karimi, H.R., Zhu, X.: Study on support vector machine-based fault detection in Tennessee Eastman process. In: Abstract and Applied Analysis, vol. 2014. Hindawi, April 2014
28. Ahmad, I., Basheri, M., Iqbal, M.J., Rahim, A.: Performance comparison of support vector machine, random forest, and extreme learning machine for intrusion detection. IEEE Access **6**, 33789–33795 (2018)
29. Shams, E.A., Rizaner, A.: A novel support vector machine based intrusion detection system for mobile ad hoc networks. Wirel. Netw. **24**(5), 1821–1829 (2018)
30. Shevade, S.K., Keerthi, S.S., Bhattacharyya, C., Murthy, K.R.: Improvements to the SMO algorithm for SVM regression. IEEE Trans. Neural Netw. **11**(5), 1188–1193 (2020)

Experimental Study of Speed Parameters and Resource Intensity of Programming Languages for Embedded Systems

A. E. Averiyanihin(✉), D. O. Andronov, S. A. Melikyan, and I. I. Skalchenkov

Bauman Moscow State Technical University, Moscow, Russian Federation

Abstract. This article is concerned with an experimental study of speed parameters shown by the C and C++ programming languages. After so much time since the release of these languages, it is still not clear whether the C++ programming language is inferior in efficiency compared to C, its ancestor. The paper is primarily focused on writing functionally identical data structures in both languages and subsequent testing of the speed parameters and resource intensity of the programs. It briefly considers basic algorithms of data structure various representations—graphs. The article describes an experiment was on numerous input data, with the subsequent read-out of the parameters. Based on the data obtained, diagrams were plotted, illustrating the comparison of these languages, conclusions were drawn from the work performed. #CSOC1120

Keywords: Graphs · Compilers · Programming · Algorithms and data structures · Performance · Memory consumption · Testing · Matrices · Arrays · Adjacency lists

1 Introduction

The paper analyses the results of experimental studies on speed parameters and convenience of the C and C++ languages in software development. The speed parameters evaluation is based on the implementation of algorithms for working with graph structures. There are several basic approaches for implementation of graphs [1]. Each of them is represented in the computer memory in a different way, as a result of which there is a discrepancy in speed parameters of each implementation on different kinds of tasks [2]. Implementing the same algorithms in two different languages, we can compare the effectiveness of these languages by analysing the results of testing the speed parameters of the test algorithm, as well as the speed and efficiency of memory allocation.

The C programming language, developed in 1969–1973 by Dennis Ritchie, from the very beginning implied closeness to assembler language machine instructions [3]. Through this, the language is as simple and effective as possible. This contributed to spreading the language widely across all available platforms. Many operating systems have been written in C, but the most striking example is UNIX, for the development

© The Author(s), under exclusive license to Springer Nature Switzerland AG 2021
R. Silhavy (Ed.): CSOC 2021, LNNS 229, pp. 17–33, 2021.
https://doi.org/10.1007/978-3-030-77445-5_3

of which Ritchie created C. Currently, this language is most often used in embedded development and for writing low-level parts of operating systems [4]. For example, most of the Linux kernel is written in C [5]. Also, most programs requiring maximum speed parameters (servers, databases, software for space and rocket systems, etc.) are developed in C. But recently, a tendency to move to higher-level programming languages in such projects can be noticed [6].

One of them is the C++ programming language—the direct heir to C. It encapsulates all the advantages of its ancestor inside itself and adds many new features that not only facilitate development but also reduce the amount of code required. The main features of C++ are the paradigm of object-oriented- and meta-programming [7]. Because of new features, as well as a rich standard library of the language, C++ code is more expressive and capacious. On top of everything else, C++ programs are safer when using standard language and library features [8].

The development of systems in the context of the industry digital transformation raises the question—which of these languages to use and in what cases? Is it worth it to completely abandon one in favour of the other, and whether any of these languages give advantages in any of the parameters? We will try to answer these questions with arguments.

2 Literature Review

We will carry out a comparative analysis of programming languages based on the implementation of algorithms and data structures from graph theory [9]. Adequate evaluation requires the implementation of functionally identical data structures, as well as the development of the basic algorithms required for their performance [10]. In work, we chose graph as the data structure, because its implementation uses the widest range of basic data structures [11–21]. Also, graphs are widely used in various fields:

- artificial intelligence [10];
- path construction [10];
- printed circuit board development [8];
- compilation of program source codes [3].

Various applications, which process multiple choices to construct a path to achieve the final goal, use some elements of graph theory [7]. The comparison method consists in using the GoogleTest framework. Identical tests written for the corresponding languages are launched, measurements are made in such parameters as the execution time of certain algorithms, the application's memory usage during program execution. The testing is based on the breadth-first search [1]. There is also a variable 'count', which indicates the number of vertices in the graph. Changing this variable will change the amount of allocated memory, as well as the execution time of tests. After carrying out the proper number of experiments, testing is considered as completed. Then the results are processed, and conclusions are drawn about the difference in speed parameters and resource consumption.

In order to develop functions that work with a graph, it is necessary to simulate the system operation based on certain algorithms. We introduce the following assumptions and provisions to cope with the task:

- a graph is an object consisting of 2 sets: points and lines, which are in some relation to each other;
- an edge is a line segment connecting two vertices of the graph;
- a directed graph is a graph in which all edges have a direction;
- adjacent vertices are vertices that are connected by an edge;
- an adjacency matrix is a square matrix of size m*m, each element of which takes on the value 1 if the vertex xi and xj are adjacent and 0 otherwise;
- lists of adjacent vertices—each vertex of the graph corresponds to a list consisting of "neighbours" of this vertex;
- breadth-first search is a graph traversal when vertices are traversed in order of increasing distance from the starting vertex;
- a queue is a data structure from which the element that was first added is removed first;
- a list is a data structure consisting of elements linked by pointers;
- a dense graph is a graph in which the number of edges is close to the maximum possible.

We formulate the test task as follows:

- to represent graphs in the form of an adjacency matrix, adjacency list, array of edges and the main functions of working with them;
- to implement a system for testing the resulting structures using the breadth-first search;
- to evaluate the speed of memory allocation, filling and algorithm execution for each implementation.

3 Methods

We implement three types of graph structures. All implementations have a common interface for getting and adding data, implemented by the following methods:

GetNextVertices/*type*_graph_get_next_vertices—getting a list of available vertices from the given one;
GetPrevVertices/*type*_graph_get_prev_vertices—getting a list of vertices from which it is possible to get to this vertex;
AddEdge/*type*_graph_add_edge—adding an edge between the 'from' and 'to' vertices (graph-oriented);
VerticesCount/*Type_graph*-> count—getting the number of vertices.

Further, the following designations are used: V is the number of vertices, E is the number of edges, subsequent vertices are vertices directly accessible from the desired one, previous vertices are the vertices directly leading to the desired one.

Graph implementation as an adjacency matrix (Matrix) (see Fig. 1).

$$A = \begin{pmatrix} 0 & 0 & 1 & 0 & 0 \\ 1 & 0 & 0 & 0 & 0 \\ 0 & 0 & 0 & 1 & 0 \\ 0 & 0 & 0 & 0 & 1 \\ 0 & 0 & 1 & 1 & 0 \end{pmatrix}_i$$

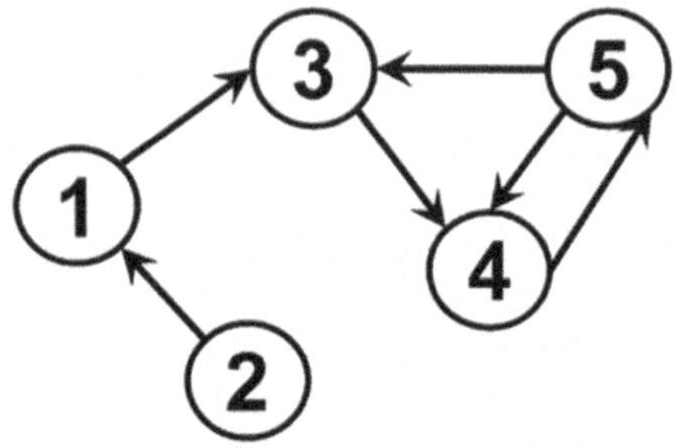

Fig. 1. Graph representation as a matrix.

This structure is a matrix A with the dimension of V*V, in which the presence of a connection between two vertices i and j is denoted as 1, and the absence is denoted as 0.

The advantages of this method are:

- determination of an edge in the graph as O(1);
- efficient storage of dense graphs.

Disadvantages of an adjacency matrix:

- it requires O(|V2|) of memory;
- it takes a relatively long time to get a list of subsequent vertices since it is necessary to go through the entire line of the required vertex.

Implementation of the graph in the form of adjacency lists (List) (see Fig. 2).

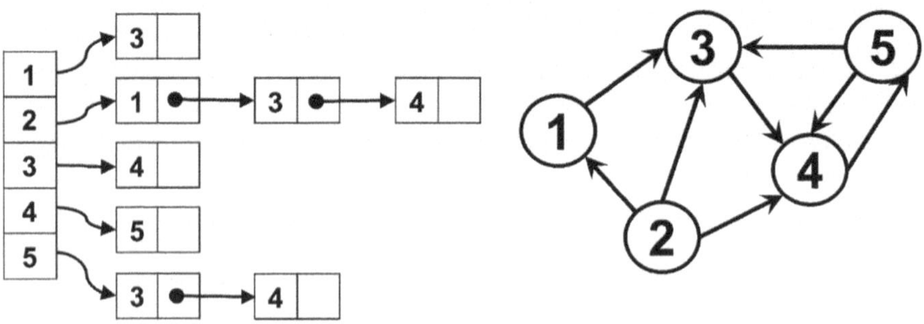

Fig. 2. Graph representation as a list.

This structure is a list of vertices of size V, in which, for each vertex, a list of subsequent vertices is stored.

The advantages of this method are:

- it requires O(|V + E|) of memory, therefore, sparse graphs are stored effectively;
- the fastest of the three ways to get a list of subsequent vertices, since the necessary information is already stored in the graph representation.

Disadvantages of an adjacency matrix:

- determination of an edge in the graph as O(E);
- a slower search for previous vertices, since it is necessary to go through the lists of all vertices to find the desired one. The situation improves when replacing the data structure of the vertex list from an array to a binary tree or hash table since the search time for a vertex decreases from O(n) to O(log n)/O(1) on average. To filter external radiation, modulation of the transmitted signal and cascaded filters are used on the receiver side of the device.

Graph implementation as an array of edges (Arc) (Fig. 3):

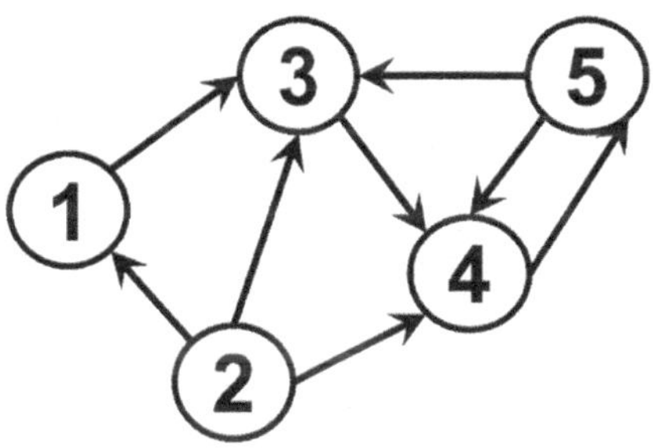

1		2		3		4		5		6		7		8	
1	3	2	1	3	4	5	3	4	5	2	3	2	4	5	4

Fig. 3. Graph representation as an array of edges.

This structure is an array of edges, where each edge is represented by two vertices it connects.

The advantages of this method are:

- it requires O(|E|) of memory, therefore, sparse graphs are stored most effectively;
- it enables instant access to the number of edges;
- linear data storage enables to use sorting and binary search algorithms.

Disadvantages of an adjacency matrix:

- the slowest search for subsequent/previous vertices, since it is necessary to go through the entire length of the array and check all the edges, provided that the array is not sorted;
- it is mostly ineffective when storing dense graphs.

3.1 Code Writing and the Difference Between C and C++ Implementations

At this stage, we demonstrate the differences in writing C and C++ code using the depth-first search.

When implementing graphs in C++, there was used one of OOP paradigms—inheritance. Thanks to this, it became possible to reduce the amount of duplicated code, as well as increase the usability of these structures. The graph general interface is shown in Fig. 4.

```
struct IGraph {
    virtual ~IGraph() = default;

    virtual void AddEdge(unsigned int from, unsigned int to) = 0;

    [[nodiscard]] virtual unsigned int VerticesCount() const = 0;

    virtual std::vector<unsigned int>
    GetNextVertices(unsigned int vertex) const = 0;
    virtual std::vector<unsigned int>
    GetPrevVertices(unsigned int vertex) const = 0;
};
```

Fig. 4. The graph interface.

Further, descendant classes implementing the features of their own kind of graph were inherited from this interface. An example implementation of a graph on adjacency lists is shown below. Figure 5 shows a header file that presents the implementation of the descendant class. Figure 6 shows the file with the implementation code for this class methods.

```
#include <vector>

#include "interface.hpp"

#ifndef GRAPH_ARCGRAPH_HPP
#define GRAPH_ARCGRAPH_HPP

class ArcGraph : public IGraph {
  public:
    explicit ArcGraph(unsigned int count);
    explicit ArcGraph(const IGraph &graph);
    ~ArcGraph() override;

    [[nodiscard]] unsigned int VerticesCount() const override;

    [[nodiscard]] std::vector<unsigned int>
    GetNextVertices(unsigned int vertex) const override;
    [[nodiscard]] std::vector<unsigned int>
    GetPrevVertices(unsigned int vertex) const override;

    void AddEdge(unsigned int from, unsigned int to) override;

  private:
    std::vector<std::pair<unsigned int, unsigned int>> arcs;
    unsigned int count;
};

#endif // GRAPH_ARCGRAPH_HPP
```

Fig. 5. Header file of graph on adjacency lists for C++.

In the C language, due to the lack of an object-oriented paradigm for each type of graph, we had to implement our own structure and write our own breadth-first search method. A general view of the implementation in C is shown in Fig. 7.

Figure 8 below shows the implementation of these methods for a graph on adjacency lists.

The implementation of the remaining classes and data structures, as well as tests, can be found on the remote repository with the source code: https://github.com/Sermelyan/KNIRS.

```
#include <cassert>
#include <vector>

#include "ArcGraph.hpp"

ArcGraph::ArcGraph(unsigned int count) : arcs(count * 2), count(count) {}

ArcGraph::ArcGraph(const IGraph &graph) : ArcGraph(graph.VerticesCount()) {
    for (int i = 0; i < graph.VerticesCount(); i++)
        for (const auto &j : graph.GetNextVertices(i))
            arcs.emplace_back(i, j);
}

ArcGraph::~ArcGraph() = default;

void ArcGraph::AddEdge(unsigned int from, unsigned int to) {
    assert(from >= 0 && from < count);
    assert(to >= 0 && to < count);
    arcs.emplace_back(from, to);
}

unsigned int ArcGraph::VerticesCount() const { return count; }

std::vector<unsigned int> ArcGraph::GetNextVertices(unsigned int vertex) const {
    assert(vertex >= 0 && vertex < count);
    std::vector<unsigned int> result;
    for (const auto &arc : arcs)
        if (arc.first == vertex)
            result.push_back(arc.second);
    return result;
}

std::vector<unsigned int> ArcGraph::GetPrevVertices(unsigned int vertex) const {
    assert(vertex >= 0 && vertex < count);
    std::vector<unsigned int> result;
    for (const auto &arc : arcs)
        if (arc.second == vertex)
            result.push_back(arc.first);
    return result;
}
```

Fig. 6. Implementation of graph methods on adjacency lists for C++.

```
typedef type_graph {
    Data_type *data;
    unsigned int count;
} Type_graph;

Type_graph *type_graph_create(unsigned int count);
void type_graph_delete(Type_graph *ths);

void type_graph_add_edge(Type_graph *ths, unsigned int from, unsigned int to);

Vector *type_graph_get_next_vertices(Type_graph *ths, unsigned int vertex);
Vector *type_graph_get_prev_vertices(Type_graph *ths, unsigned int vertex);
```

Fig. 7. General structure of a graph made using C.

```
#include "arc_graph.h"

Arc_graph *arc_graph_create(unsigned int count) {
    Arc_graph *temp = (Arc_graph *) calloc(1, sizeof(Arc_graph));
    if (!temp) {
        perror("Can't allocate memory for graph!\n");
        return NULL;
    }
    temp->count = count;
    temp->arcs = create();
    if (!temp->arcs) {
        perror("Can't allocate memory for graph array!\n");
        arc_graph_delete(temp);
        return NULL;
    }
    return temp;
}

void arc_graph_delete(Arc_graph *ths) {
    delete_vec(ths->arcs);
    free(ths);
}

inline void arc_graph_add_edge(Arc_graph *ths, unsigned int from, unsigned int to)
    assert(from >= 0 && from < ths->count);
    assert(to >= 0 && to < ths->count);
    push_back(ths->arcs, from);
    push_back(ths->arcs, to);
}

inline Vector *arc_graph_get_next_vertices(Arc_graph *ths, unsigned int vertex) {
    assert(vertex >= 0 && vertex < ths->count);
    Vector *result = create();
    for (unsigned int i = 0; i < ths->arcs->size; i += 2)
        if (get_at(ths->arcs, i) == vertex)
            push_back(result, get_at(ths->arcs,i + 1));
    return result;
}

Vector *arc_graph_get_prev_vertices(Arc_graph *ths, unsigned int vertex) {
    assert(vertex >= 0 && vertex < ths->count);
    Vector *result = create();
    for (unsigned int i = 1; i < ths->arcs->size; i += 2)
        if (get_at(ths->arcs, i) == vertex)
            push_back(result, get_at(ths->arcs,i - 1));
    return result;
}
```

Fig. 8. Implementation of graph methods on adjacency lists for C.

In order to test the speed parameters of various types of graphs, we implement a breadth-first search function. The block diagram of its operation is shown in the figure below (Fig. 9).

In the language, the graph traversal method is implemented for each type of graph and differs only in the accepted type and function of obtaining vertices. By inheriting from a common interface, one method implementation can be used for all descendant classes in C++.

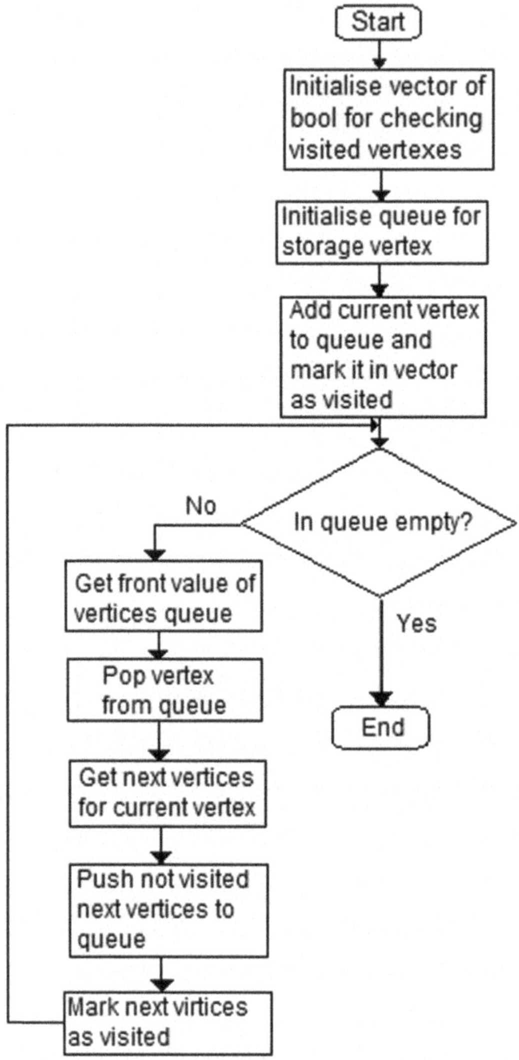

Fig. 9. Breadth-first search block diagram.

4 Result

For the experiment, we used a computer-based on processors i7-7700k@4.8 GHz/Ryzen 2700X@4.3 GHz with 16 GB RAM. The GCC 9.2.1 Compiler. The experiment consists in an implementation of a test program developed in C and C++. IDE JetBrainsCLion 2019.3 x64 is used for code development. Libraries used: STL (C++), GoogleTest (C++), self-written Vector (C).

4.1 General Testing Methodology

1. Generating a certain number of vertices and random edges between them.
2. Memory allocation for specific graph structure.
3. Filling the graph with data that was generated in the first stage.
4. Executing the breadth-first search algorithm for a certain number of vertices.
5. Resource deallocation.

The elapsed time is measured on stages 2, 3 and 4.

Also, after stage 2, it is possible to measure utilised memory.

Testing was carried out on two different computers. This is necessary to demonstrate the platform-independence of differences between C and C++. So, if the differences between two identical tests on platforms of approximately the same performance do not differ by more than 10%, then the test results can be considered reliable.

4.2 Speed Parameter Testing

These tests are necessary in order to find out which language generates faster and more optimised programs. It will also identify the most appropriate graph storage structure for working with the breadth-first search algorithm.

In this series, 3 graphs of different implementations are created and filled with the same data. Then, for each of the graphs, a breadth-first search is started for each vertex of the graph. This procedure is repeated for a different number of vertices. The results of testing the graph in matrix form are presented in Figs. 10 and 11 on AMD and INTEL platforms, respectively. Further, for all diagrams of this stage, the solid line is the implementation in C++, the dashed line is in C.

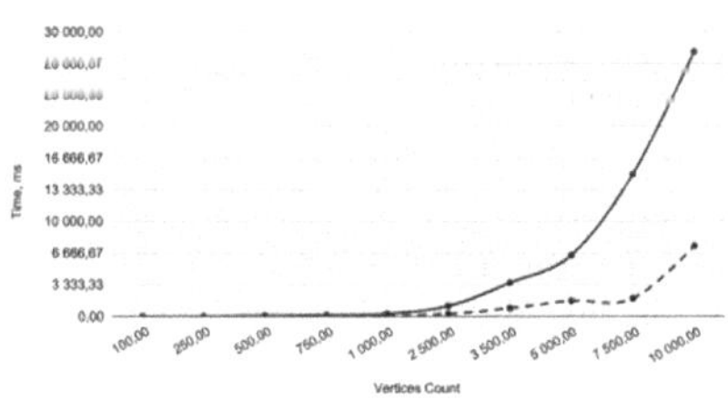

Fig. 10. Results of testing graph in matrix form on AMD platform.

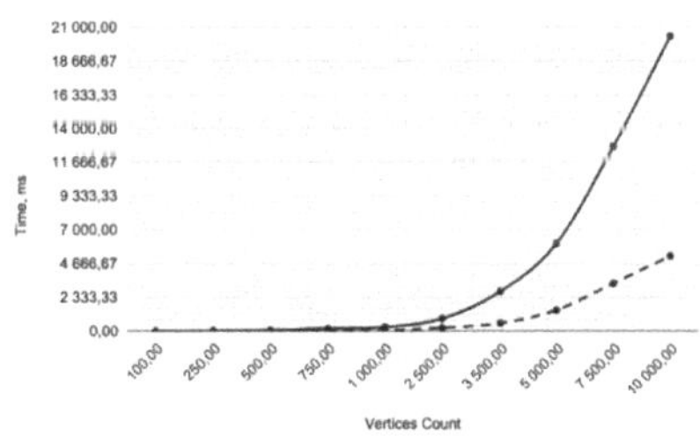

Fig. 11. Results of testing graph in matrix form on INTEL platform

The results of testing the graph in the form of adjacency list are presented in Figs. 12 and 13 on AMD and INTEL platforms, respectively.

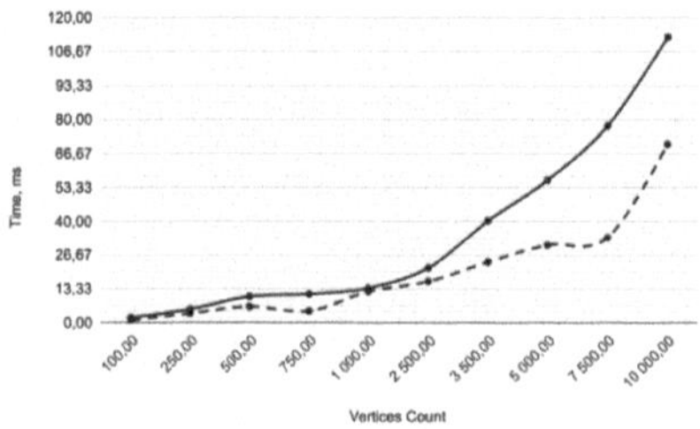

Fig. 12. Results of testing the graph in the form of adjacency list on AMD platform.

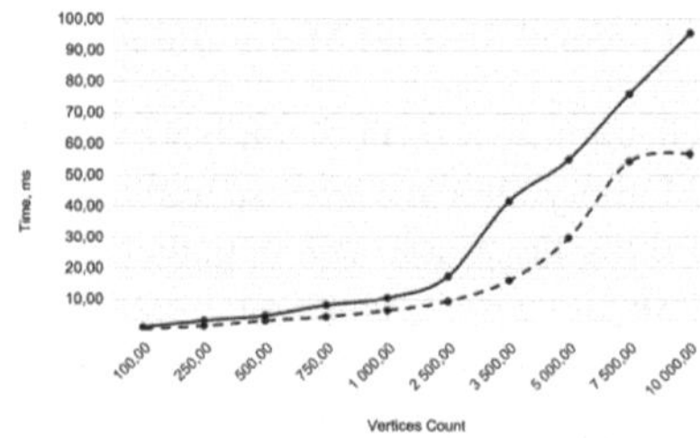

Fig. 13. Results of testing the graph in the form of adjacency list on INTEL platform.

The results of testing the graph in the form of edge array are presented in Figs. 14 and 15 on AMD and INTEL platforms, respectively.

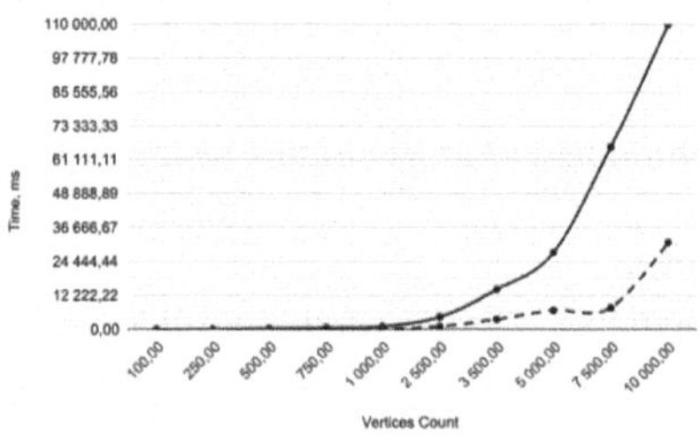

Fig. 14. Results of testing the graph in the form of edge array on AMD platform.

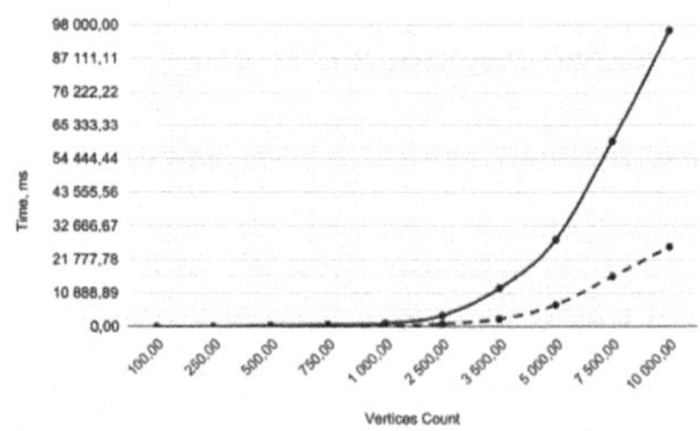

Fig. 15. Results of testing the graph in the form of edge array on INTEL platform.

As you can see from the diagrams presented above, C has a speed advantage of up to 3 times. It also may be noticed that ListGraph is the most suitable implementation for working with the breadth-first search algorithm.

As expected, the difference between the two platforms is no more than 10%, which indicates the consistency of our tests.

4.3 Testing for Resource Intensity and Memory Operating Speed

This series of tests is needed in order to find out which language works best with allocating and deallocating memory. A feature of these tests is separate test operations for each type of graphs.

In this experiment, for each of the implementations, we took readings of the memory usage required to store the graph, the time of allocating this memory, and also the execution time. For convenience, the results are presented in the form of diagrams.

The first test was carried out for ListGraph and ArcGraph. The dependence of memory used on the number of vertices (same for both platforms) is shown in Fig. 16. Then, the test was carried out for MatrixGraph. The results are shown in Fig. 17.

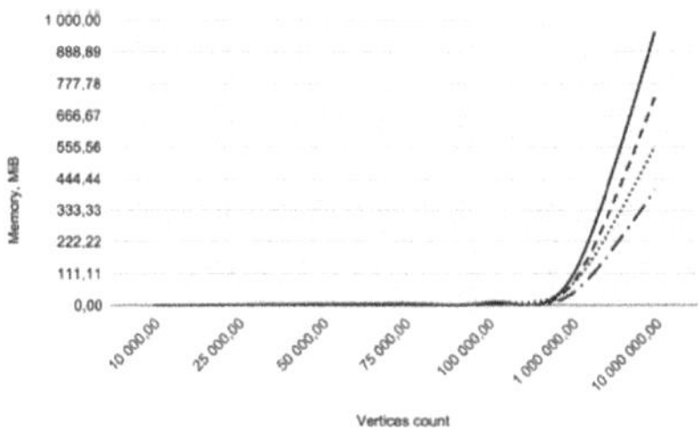

Fig. 16. Dependence of memory used on the number of vertices for the ListGraph and ArcGraph. • The solid line shows the diagrams of testing the List structure implemented in C. • The dashed line shows the diagrams of testing the List structure implemented in C++. • The dotted line shows the diagrams of testing the Arc structure implemented in C++. • The dash-dotted line shows the diagrams of testing the Arc structure implemented in C.

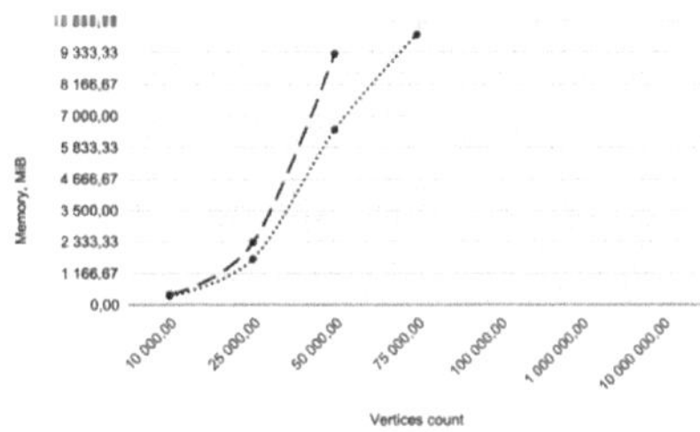

Fig. 17. Dependence of memory used on the number of vertices for MatrixGraph. • The dashed line shows the diagrams of testing the Matrix structure implemented in C++. • The dotted line shows the diagrams of testing the Matrix structure implemented in C.

In Figs. 16 and 17 we can see that C language consumes less memory that does not correspond to theoretical calculations. For example, for MatrixGraph with 50,000 vertices and unsigned type (4 bytes) it should take:

$$4 * 50{,}000 * 50{,}000 = 10{,}000{,}000{,}000\,\text{bytes} \approx 9.31\,\text{GiB}$$

Which is observed for C++. However, somehow the consumption is only 6.5 GiB in C.

The next test was to measure the time spent on allocating memory needed to implement graphs (Alloctime) depending on the number of vertices in the graph.

Results (Figs. 18 and 19):

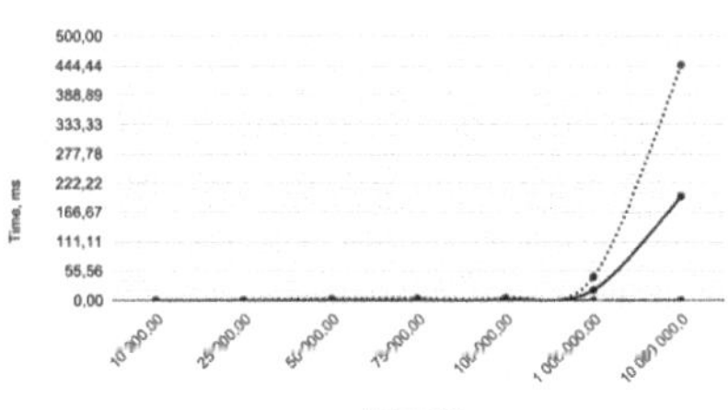

Fig. 18. Dependence of the time spent on allocating memory on the number of vertices for ListGraph and ArcGraph for AMD platform.

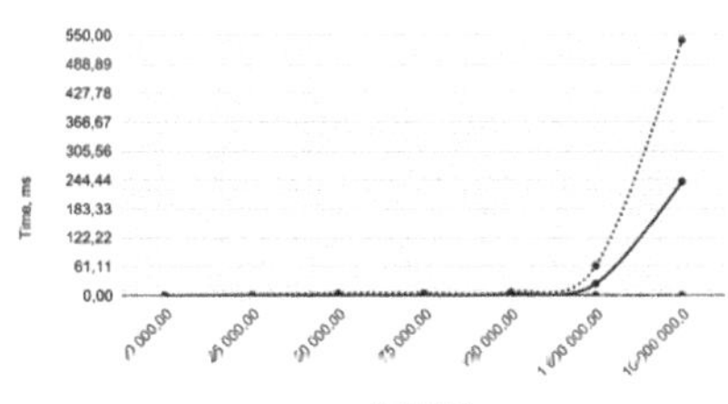

Fig. 19. Dependence of the time spent on allocating memory on the number of vertices for ListGraph and ArcGraph for INTEL platform.

- The solid line shows the diagrams of testing the List structure implemented in C++.
- The dashed line shows the diagrams of testing the Arc structure implemented in C++.
- The dotted line shows the diagrams of testing the List structure implemented in C.

- The dash-dotted line shows the diagrams of testing the Arc structure implemented in C.

The memory allocation time for a graph on edge arrays always tends to zero, since this implementation consumes the least amount of memory (Figs. 20 and 21).

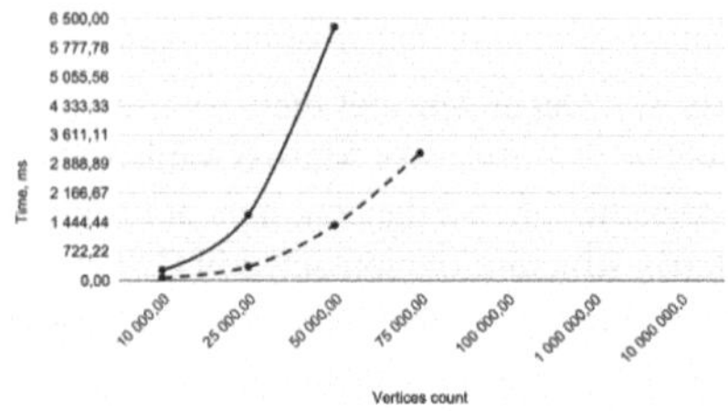

Fig. 20. Dependence of the time spent on allocating memory on the number of vertices for MatrixGraph for AMD platform.

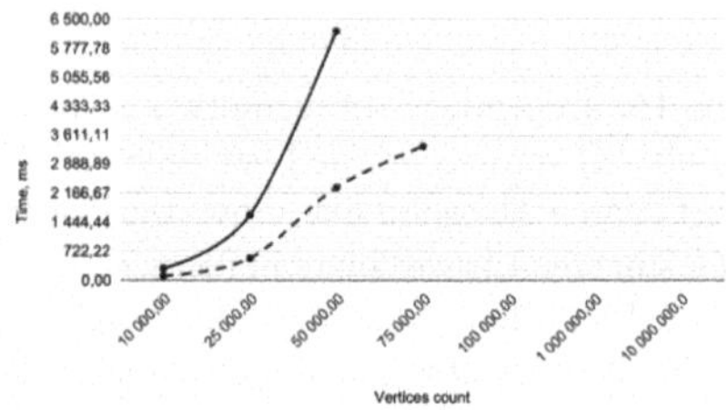

Fig. 21. Dependence of the time spent on allocating memory on the number of vertices for MatrixGraph for INTEL platform.

- The solid line shows the diagrams of testing the Matrix structure implemented in C++.
- The dashed line shows the diagrams of testing the Matrix structure implemented in C.

The next test was to measure the time spent to fill all the necessary connections of the graph (Filltime) depending on the number of vertices in the graph (Figs. 22 and 23).

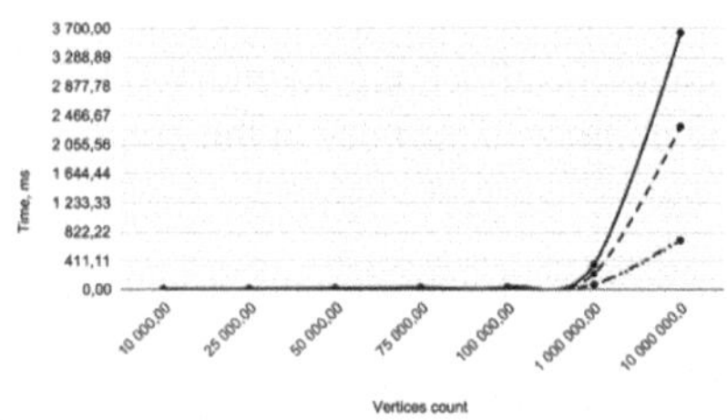

Fig. 22. Dependence of the time spent to fill all the necessary connections of the graph on the number of vertices in the graph for ListGraph and ArcGraph on AMD platform.

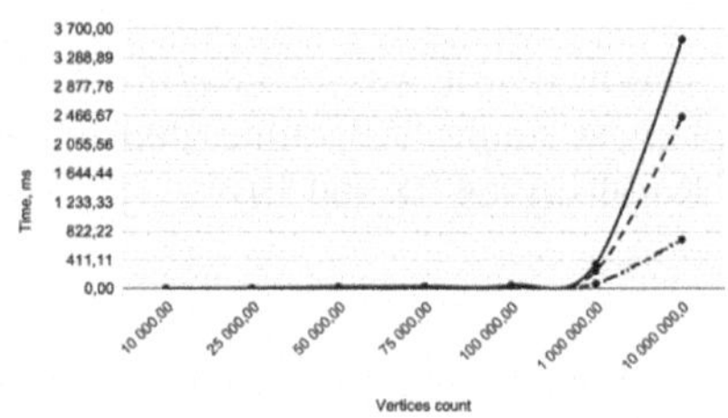

Fig. 23. Dependence of the time spent to fill all the necessary connections of the graph on the number of vertices in the graph for ListGraph and ArcGraph on INTEL platform.

- The solid line shows the diagrams of testing the List structure implemented in C++.
- The dashed line shows the diagrams of testing the Arc structure implemented in C+.
- The dotted line shows the diagrams of testing the List structure implemented in C.
- The dash-dotted line shows the diagrams of testing the Arc structure implemented in C (Fig. 24).

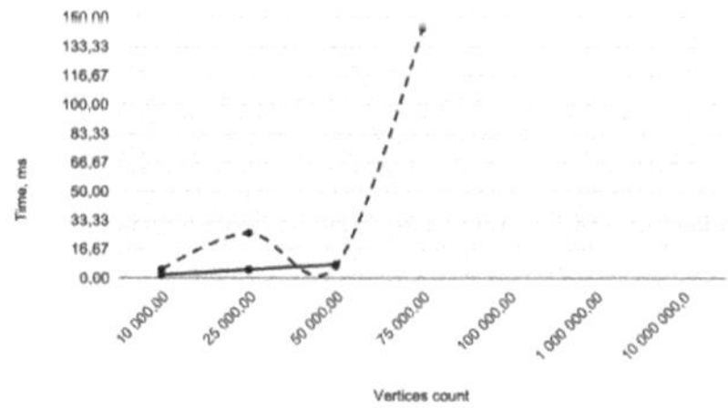

Fig. 24. Dependence of the time spent to fill all the necessary connections of the graph on the number of vertices in the graph for MatrixGraph on INTEL platfo.

- The solid line shows the diagrams of testing the Matrix structure implemented in C.
- The dashed line shows the diagrams of testing the Matrix structure implemented in C++.

As you can see from the diagrams presented above, C has a speed advantage of up to 3 times. It also may be noticed that ListGraph is the most suitable implementation for working with the breadth-first search algorithm.

As expected, the difference between the two platforms is no more than 10%, which indicates the consistency of our tests.

5 Discussion

5.1 Speed Parameter Analysis

From the first test data, it can be clearly seen that the C language gives a big performance gain, especially in tasks requiring a rough approach (direct enumeration of data). This can be achieved by better optimisation of the assembler code generated by the compiler. In addition, C lacks additional abstractions that would decrease the speed of data access and complicate optimisation at the compiler level.

However, for a graph on adjacency lists, gain of C language is less than 2 times. In this case, the implementation in C++ is more convenient for further use.

5.2 Memory Operating Analysis

If we compare languages in terms of resource intensity and memory operating speed, then on the basis of experimental data we can see that the C language, on average, is 4 times faster in filling the graph. Considering the process of memory allocation, we can see that, for the C language, is takes twice as long, which does not coincide with theoretical expectations, the reason for which may be the use of non optimal structures. According to theoretical expectations, memory usage should also be approximately the same but, for not entirely clear reasons, in our experiment, the amount of memory used in C is 1.5 times less.

Thus, if the issue is about the maximum speed of the program, then the C language is the right choice in this case. But it should be understood that highly qualified specialists are needed to achieve the set goal since even small mistakes when operating memory or ignorance of some language subtleties can lead to catastrophic performance losses. If speed parameters are important, but the priority is also the convenience of maintaining and reusing the code, as well as the comparative ease of development when a code base grows to giant scale, C++ is a preferable choice.

5.3 Further Prospects

In further research, it is necessary to find out the reason for such a large difference between the memory consumption by programs made in C and C++. Since this trend cannot be an accidental error, as it is observed on different platforms and even on different compilers (the situation is similar on MacOS/Clang), the study of this problem becomes doubly interesting.

It is also interesting to study not languages but graph implementations themselves using various algorithms. For example, List showed the best speed parameters in the breadth-first search algorithm, thanks to the immediate return of subsequent vertices. But getting the previous vertices of the graph on the adjacency lists is much slower and requires traversing all the vertices. Therefore, algorithms using this data will probably work faster on other types of graphs. So, a list of algorithms and best graph implementations can be made.

6 Conclusion

Analysing the results, it can be stated that the C language is not yet outdated and maybe well applied in modern projects that require maximum speed parameters, for example, in system development. Since many years of optimisation of the C compiler, and the simplicity of the language enable creating programs with maximum speed parameters. However, in large projects, it is better to use C++ with using C or even Assembler inserts where maximum performance is needed. Modern language paradigms provide ample opportunities to reuse code and use all modern trends in programming.

Also, during the experiments, it was found that the most optimal type of graph implementation for its further processing by the breadth-first search algorithm is a graph on adjacency lists. This structure provides the most optimal speed and memory consumption characteristics of those studied.

Acknowledgements. Some results were obtained in the framework of the State task #0705-2020-0041 "Component's digital transformation methods' fundamental research for micro- and nanosystems".

References

1. Manber, U.: Algorithms: A Creative Approach
2. Vlasov, A.I., Shakhnov, V.A., Filin, S.S., Krivoshein, A.I.: Sustainable energy systems in the digital economy: concept of smart machines. Entrepreneurship Sustain. Issues **6**(4), 1975–1986 (2019)

3. Ritchie, D.M., Kernighan, B.W.: The C Programming Language
4. Yuldashev, M.N., Vlasov, A.I.: Mathematical model of the general problem of state classification in wireless sensor networks. In: IOP Conference Series: Materials Science and Engineering, p. 012002 (2019)
5. Zhalnin, V.P., Zakharova, A.S., Uzenkov, D.A., Vlasov, A.I., Krivoshein, A.I., Filin, S.S.: Configuration-making algorithm for the smart machine controller based on the internet of things concept. Int. Rev. Electr. Eng. **14**(5), 375–384 (2019)
6. Shakhnov, V.A., Kurnosenko, A.E.: Simulation of digital production of electronics as part of the Industry 4.0 concept. In: Akberdina, V.V. (ed.) The Anthology: Digital Transformation of Industry: Trends, Management, Strategies. Materials of the 1st International Scientific and Practical Conference, pp. 585–594 (2019)
7. Kleinberg, J., Tardos, E.: Algorithm Design
8. Khalzev, S.E., Vlasov, A.I., Shakhnov, V.A.: Visual methods of high-level system design for digital hardware components. In: Journal of Physics: Conference Series, p. 42024 (2020)
9. Cormen, T.H., Leiserson, C.E.: Introduction to Algorithms
10. Prudius, A.A., Karpunin, A.A., Vlasov, A.I.: Analysis of machine learning methods to improve efficiency of big data processing in Industry 4.0. In: Journal of Physics: Conference Series, vol. 1333, p. 032065. https://doi.org/10.1088/1742-6596/1333/3/032065
11. Averyanikhin, A.E., Vlasov, A.I., Malevany, A.Yu.: The concept of test environment management in synchronous programming. In: Proceedings of the International Symposium: Reliability and Quality, vol. 2, pp. 135–142 (2019)
12. Demin, A.A., Vlasov, A.I.: Visual methods of formalization of knowledge in the conditions of the synchronous technologies of system engineering. In: ACM International Conference Proceeding Series 13. Cep. CEE-SECR 2017 - Proceedings of the 13th Central and Eastern European Software Engineering Conference in Russia, p. 3166098 (2017)
13. Vlasov, A., Yuldashev, M.: Performance analysis of algorithms for energy-efficient data transfer in wireless sensor networks 2019. In: International Conference on Industrial Engineering, Applications and Manufacturing, ICIEAM 2019, p. 8743087 (2019)
14. Vlasov, A.I., Muraviev, K.A., Prudius, A.A., Uzenkov, D.A.: Load balancing in big data processing systems. Int. Rev. Autom. Control **12**(1), 42–47 (2019)
15. Nazarov, A.V., Shakhnov, V.A., Vlasov, A.I., Novikov, A.N.: Effective algorithm for routing integral structures with twolayer switching. In: Journal of Physics: Conference Series, vol. 1015, p. 032098 (2018)
16. Vlasov, A.I., Grigoriev, P.V., Krivoshein, A.I., Shakhnov, V.A., Filin, S.S., Migalin, V.S.: Smart management of technologies: predictive maintenance of industrial equipment using wireless sensor networks. Entrepreneurship Sustain. Issues **6**(2), 489–502 (2018)
17. Vlasov, A.I., Juravleva, L.V., Shakhnov, V.A.: Knowledge-based model for formal representation of complex system visual models. In: Arai, K., Kapoor, S., Bhatia, R. (eds.) IntelliSys 2020. AISC, vol. 1251, pp. 618–632. Springer, Cham (2021). https://doi.org/10.1007/978-3-030-55187-2_45
18. Chernov, I.E., Kondakov, N.A., Perederin, K.D., Zhalnin, V.P., Shakhnov, V.A.: Use of the "Via-in-Pad" method to ensure the high density layout of the conductive pattern when designing multilayer switching structures. Int. J. Adv. Sci. Eng. Inf. Technol. (2020)
19. Blass, A., Gurevich, Y.: Algorithms: A Quest for Absolute Definitions
20. Skiena, S.S.: The Algorithm Design Manual
21. Levitin, A.: Introduction to the Design and Analysis of Algorithms

Classification of Cardiotocography Data for Fetal Health Using Feature Selection Techniques

Imsal Rafique(✉), Mudasir Dilawar, Amina Umer, and Muhammad Ahmad Hassan

University of Engineering and Technology, Lahore, Pakistan
imsal.rafique@se.uol.edu.pk, mudasir.dilawar@ue.edu.pk, ahmed.hasan@kics.edu.pk

Abstract. Cardiotocography is used to record the information of fetal health through fetal heart rate and uterine contractions during pregnancy. Cardiotocography is widely used technique by doctors to evaluate the health of newborn and maternal health, particularly in an emergency condition. Mostly, maternity period complications restrict fetus from proper growth and can cause impairment or death. Safe maternity period, by predicting risk levels before the occasion of complications, encourage proper fetal growth. Prediction of fetal health state from a set of pre-classified patterns knowledge is important in developing a predictive classifier model using data processing algorithms. The main goal of this research is to build a classification model using machine learning algorithms with feature selection techniques that predict whether the fetal is healthy or unhealthy.

Keywords: Classification · Data mining · Feature selection techniques · Imbalance data · Over-sampling

1 Introduction

Data mining refers to extracting useful information and patterns from a large sized dataset. Our world full of information has a huge amount of raw data available. This large amount of data can be used to make effective strategies and decisions if we transform it into a useful information. Data mining's main goal is to automatically analyze the large sized data and extract useful information or pattern out of it. This data mining extracted useful information can be helpful in many areas such as Education, Healthcare, Business market analysis, Government strategies, etc. In recent days data mining played an important role to make decisions and helped to get advantage over competitors.

Data mining is also called knowledge discovery and is about to discover the unknown patterns in the data. Data mining is a huge concept and involves different steps. First step is pre- processing in which raw data is converted into

© The Author(s), under exclusive license to Springer Nature Switzerland AG 2021
R. Silhavy (Ed.): CSOC 2021, LNNS 229, pp. 34–44, 2021.
https://doi.org/10.1007/978-3-030-77445-5_4

understandable format. Pre-processing includes data cleaning, data integration, data transformation, data reduction and data discretization. Second step is to apply data mining technique which includes classification, Association rule mining, clustering, regression and others. Next step includes the evaluation of the model on the basis of accuracy of results. In data mining, pre-processing is an important part that solves the problems of high dimensionality, class imbalance and noisy data. In case of high dimensional data, data mining techniques are time consuming and sometimes decrease the performance of the classifier. Imbalanced dataset also decreases the performance of the model by making it biased towards a specific class. Class imbalance means that the dataset has majority or minority of one or more classes and consequences in biased prediction.

Class imbalance problem can be solved using Synthetic Minority Oversampling Technique (SMOTE). SMOTE creates the synthetic samples to balance the dataset. Performance of classification model also depends on the number of features being used for classification. Dataset may include irrelevant features and these irrelevant features increase the time and complexity of model results in performance decrements. Performance of classification model can be increased by selecting and using most relevant subset of features. Feature selection techniques help in this regard to improve the accuracy of classification model by selecting relevant subset of features.

Different feature selection techniques include Filter, Wrapper and embedded methods. Feature selection techniques select the feature subset which plays important role in classification of data and these selected features can be used for classification. Filter method ranks and selects the subset of features which are highly correlated with the target variable on the basis of their scores. Wrapper method uses a predictive model to select subset of features. Different subsets of features are used to get results and performance of subset features are compared with each other. Then subset of features which gives best results over specific classification model is selected for further model training. Embedded methods are the combination of filter methods and wrapper methods. Embedded methods mostly use regularization to select the best subset of features on which model gives the best accuracy.

In current research work, subset of filter methods and wrapper methods subset are compared, and work is tested over UCI machine learning repository cardiotocography (CTG) dataset.

2 Related Work

In this chapter, different techniques used for the classification of fetal health are discussed in detail. Researchers have used different machine learning algorithms for the classification of fetal health but feature selection algorithms along with data balancing techniques are not used for the classification of fetal health over CTG dataset. According to the literature survey, imbalance dataset leads towards the biased prediction in classification of fetal health dataset.

Literature survey shows that subset of features selected from feature selection techniques can improve performance of machine learning algorithm applied on the CTG dataset. Different Machine Learning techniques like data mining, SMOTE, Feature selection algorithm are also discussed in detailed.

Chamidah and Wasito used hybrid K-SVM and SVM algorithm to classify cardiotocography data. Feature extraction algorithm is used for pre-processing which is helpful for better classification. Authors proved that the KSVM provides better accuracy that the SVM [4]. Classification of cardiotocography is done using Weka. Suspicious and pathological class are identified successfully in this research using various classification algorithms. Random Forest algorithm provided best result for the classification of cardiotocography [3]. Feature selection algorithms are implemented with various machine learning techniques. Authors proved that the use of feature selection algorithms can reduce the time and space complexity. The results are evaluated using heart disease dataset of UCI machine learning repository [18]. A feature selection approach is developed and evaluated over different medical data sets. The authors proved that the dimensions of the data set can be reduced with the developed algorithm named M-Cluster feature selection technique. Experiment results show better performance over existing methods [15]. Comparison of different classification algorithms for classification of cardiotocography data is done by the authors. According to the authors, ANN model provides the best result over other classification algorithm but suggested that using hybrid model of classification algorithm can produce better results [10].

In this research performance of classification algorithms are evaluated over multiple healthcare datasets. Evaluation of performance is evaluated over twelve medical data sets and five classification algorithms. Author also suggest that the feature selection algorithms are useful to reduce the dimensionality of dataset which boost the performance of model [12]. An ANN model for fetal health monitor is proposes by the authors. Recordings of cardiotocography is used to train ANN model and the work is carried out in MATLAB tool. The proposed model of ANN successfully identifies the fetal health [14]. In this paper feature selection algorithm PCA and AdaBoost is used to classify the fetal state on the cardiotocography dataset. Author proves that the hybrid model of PCA and AdaBoost provide the better results. Machine learning algorithms are used to classify caesarean sections over cardiotocography patterns. Author extracted the important feature using feature selection algorithm and balance the data set. Random forest and SVM provides highest results [20]. Machine learning algorithms are used to classify caesarean sections over cardiotocography patterns. Author extracted the important feature using feature selection algorithm and also balance the data set. Random forest and SVM provides highest results [7]. In this research paper decision tree modeling is used to find the unknown patterns for cardiotocography data. Decision tree is commonly used data mining algorithm and has potential to successfully find the unknown patterns for fetal health [11].

Trained activation function approach was used to determine optimal activation function in artificial neural networks that is an important issue. Each neuron was trained using linear regression for activation function. Training is done on the training dataset which is composed the sums of inputs of every neuron within the hidden layer and preferred outputs. This approach was used in random weight artificial neural network (RWN) and tested on different datasets. Obtain results shows that RWN that uses activation function perform better than RWN that uses simple activation function [6]. eCTG an electronic procedure has been proposed which ex- tract digital signals from Digital CTG images by scanning CTG paper report. CTU-UHB Intrapartum CTG database is used which contains 552 CTG recordings of digital signals. Digital FHR and UC signals from digital CTG images is extracted accurately using Matlab and submitted to eTCG procedure. In this study Wavelet, transform technique is used to measure Fetal Electrocardiogram (FECG) characteristics for finding out the fetal heart rate and stress. MATLAB simulation is used to extract the fetal heart rate and suggested the implementation using hardware processors of proposed work [17].

Fetal electronic monitoring system is developed to monitor the fetal heart rate. To improve the accuracy of classification fetal heart record is divided into ten d-window. Through feature selection features are extracted and these features are given as input to classification algorithms like SVM, MP and CNN. The result shows that the CNN provide the higher accuracy [13]. Fetal distress and hypoxia are a major case for caesarean section during pregnancy. In this study Different machine learning algorithms (ANN, SVM, RF) are used to classify fetal health using PH value that are facing oxygen problems over Clinical data from Physionet. Selected features extracted for data is given as input to each model. According to the results neural network gives high accuracy [1].

Electro Hysterographic Signal (EHG) are studies which are collected from 20 patients who can face preterm labor. EHG signals are analysis using feature selection algorithm and machine learning algorithms to trace pattern for preterm labor. According to the results, SVM provides the best result [19]. In this paper e-Health application is used to help doctors and patients to monitor the health of fetal complications. Clinical data of 96 pregnant women are used to train the classification model. Decision forest produces the best result to predict the heath of fetal. The model is also implemented on real life user that gives the higher accuracy than traditional model [2]. Class imbalance is an important issue in medical data set for classification. In this paper different sampling techniques are discussed, and performance is measured using Artificial Neural Network with sampling techniques. Over sampling and combi-sampling predict the classes more efficiently than under sampling [16].

In literature survey different classification algorithms are used for the classi fication of cardiotocography data. In this research paper, Artificial Neural Networks (ANN) and Support Vector Machine (SVM) algorithms are used. Due to massive growth of data a large amount of data is being generated has following problems associated with it.

2.1 Mislabelled Data

Mislabelled data can lead to less accuracy and affect the results of training models. To check a large amount of data whether it is correctly labelled or not in training data is difficult. The percentage of mislabeled data increases as data increases.

2.2 Missing Values

In every real dataset, missing values are a common problem and can lead to inaccurate model creation. Many methods deal with missing values in data mining that includes ignoring, exclude or replace missing records.

2.3 Noisy Data

Noisy data is unstructured data that has no meaning and cannot be understood by the machine. Noisy data can affect the results and increase the amount of storage. Noisy data can be handled by many techniques including regression, clustering, etc.

2.4 High Dimensionality

Training a data model on a dataset with many dimensions usually requires enormous time and space complexity. It also often leads to overfitting. Not all the features we have available to us are relevant to our problem. If we reduce the dimensions, we can reduce the noise and find those that are surprising, very closely related. Thus, we can apply small and simple algorithms easily. We can divide dimensionality reduction into two categories such as feature selection and extraction [9]. Feature selection is all about finding most relevant features to a problem. Feature extraction means finding new features after transforming the data from a high dimensionality space to a lower dimensional space. This problem can be solved using principal component analysis (PCA) and feature selection techniques.

2.5 Imbalance Data

The percentage of the classes in the total sample plays an important role. However, there are scenarios where you may be dealing with imbalanced dataset by imbalanced mean there is a presence of a minority class. In such a case model try to fit the major class and provide a biased prediction at the same time it gives a false sense of accuracy. This research uses the SMOTE [5] to address the imbalance class, and oversampling technique is used to balance the imbalanced the class. The synthetic minority oversampling technique creates new synthetic observations or new data points. Firstly, every data point is plotted, and it then identifies the feature vector and its nearest neighbor. Then it takes the difference between the two simply. Finally multiplies the difference with the random number between the “0” and “1”. The next step is to identify a new point on the line segment by adding the random number to the feature vector and then repeat the process for identified the feature vectors.

3 Methodology

Cardiotocography (CTG) data from UCI Machine Learning repository [8] is used for evaluation and development of fetal health prediction model during pregnancy. CTG provides the detail health visualization of fetal health and problems can be solved earlier. CTG data is collected from three expert obstetricians and labels are assigned by them. Automatic SISPORTO 2.0 Software is used to collect the UCI CTG dataset.

The dataset contains 2126 measurements of fetal heart rate signals that belong to different classes. CTG data set has 21 attributes which contain discrete, continuous and nominal values. The classification is of three class problems as normal, suspicious and Pathological. The details of the data set is given in Table 1.

Table 1. Dataset description.

S. no	Attribute name	Description
1	LB	Baseline value
2	AC	Accelerations
3	FM	Fetal movement
4	UC	Uterine contractions
5	DL	Light decelerations
6	DS	Severe decelerations
7	DP	Prolonged decelerations
8	ASTV	Percentage of time with abnormal short-term variability
9	MSTV	Mean value of short-term variability
10	ALTV	Percentage of time with abnormal long-term variability
11	MLTV	Mean value of long term variability
12	Width	Histogram width
13	Min	Low freq. of the histogram
14	Max	High freq. of the histogram
15	Nmax	Number of histogram peaks
16	Mode	Histogram mode
17	Mean	Histogram mean
18	Median	Histogram median
19	Variance	Histogram variance
20	Tendency	Histogram tendency
21	Class label	Normal, suspect, pathological

Prior to classification the main problem to be solved is to balance the dataset as the dataset is suffering from imbalance problem. From total 2126 records

'Normal' class has 1665, 'suspect' class has 295 records and 'pathological' class has 176 records. It is clear that the normal class is not balanced with suspect and pathological class. To balance the class SMOTE algorithm is used to balance the classes. So, samples of suspect and pathological class are created equal to normal class 1665 with equal ratio 1:1:1. Now each class has 1665 samples of records.

The intension of proposed methodology is to improve the accuracy of classification model by reducing the feature subset. From a dataset consisting of 'M' features, it is supposed to consider the most relevant features for better classification and analyzing features in case of high dimensional data can decrease the performance. To solve this problem in this research paper, Correlation feature selection (CFS), Chi-square and Recursive feature selection techniques are used for the section top 'N' features for the improved classification results. In this research, top 10 features are selected for the classification of fetal health. Finally following classification algorithms (ANN) and (SVM) are used for the classification of fetal health for the derived best subset of features and experiment is carried out in python programming language. Methodology of our system is given in (Fig. 1).

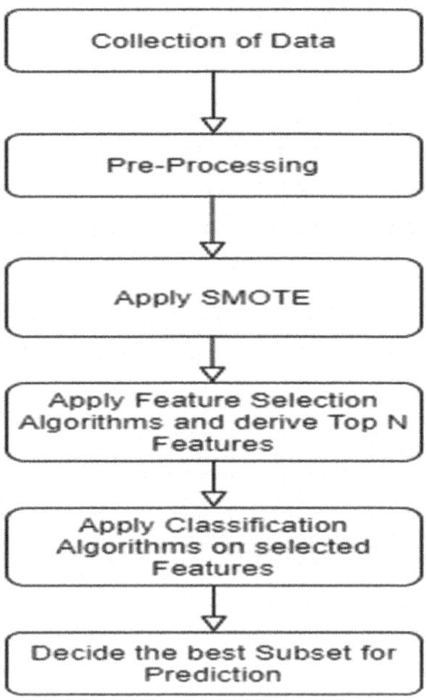

Fig. 1. Methodology

4 Experiments and Results

In this section obtained results with possible discussion is articulated. As discussed earlier, three different feature selection techniques are applied on the CTG dataset. Subset of features selected, best for classification models, using feature selection techniques as CFS, CHI and RFE are shown in Table 2.

Table 2. Results of feature selection techniques.

Feature selection techniques	Selected attribute names
Correlation Feature Selection (CFS)	Mean, MSTV, Width, Min, Median, Mode, Variance, DL, Nmax
Chi-Square (CHI)	AC, DL, Variance, ALTV, MSTV, Width, Mean, Nmax, Max
Recursive Feature Elimination (RFE)	ALTV, ASTV, UC, Median, MLTV, LB, MSTV, AC, Mode, NSP

Feature subsets obtained by applying feature selection techniques are used in different classification algorithms to decide the best feature set on the basis of accuracy of the algorithms. Classification algorithms applied on selected feature subsets are Naïve Bayesian, K-NN, SVM and ANN. During the training phase of a classification algorithms, loss function is checked. When the value of loss function stops decreasing, it indicates that the model has stopped learning. At this point the training of the model is stopped. The loss function value of the ANN is shown in (Fig. 2). Training phases also includes a testing, n fold testing, to ensure that model is learning rather than memorizing the training samples. So the loss function is calculated for both train and test samples in training phase as shown in (Fig. 2).

Decrease in loss function value in training process indicates increase in accuracy of model in correctly classifying the samples. During training process, accuracy value can also be checked along with loss value to ensure models learning. Accuracy values w.r.t training epochs are shown in (Fig. 3). As discussed earlier, train and test accuracy values are to ensure that model is learning rather than memorizing, (Fig. 3) includes train and test accuracy.

After the training phase of the model, final testing is done to check the model accuracy on unseen data, sample which are not used in training process. For better understanding of number of samples correctly and falsely classified, confusion matrix of the ANN model is shown in Table 3. Accuracy of the model can be calculated by dividing the sum of the diagonal values with the total number of samples. Thus, accuracy of ANN model is calculated as follows:

$$1407/1490 = 94.9\%$$

Comparison of classification models on the basis of accuracy for different set of feature sets selected by feature selection techniques is given in Table 4.

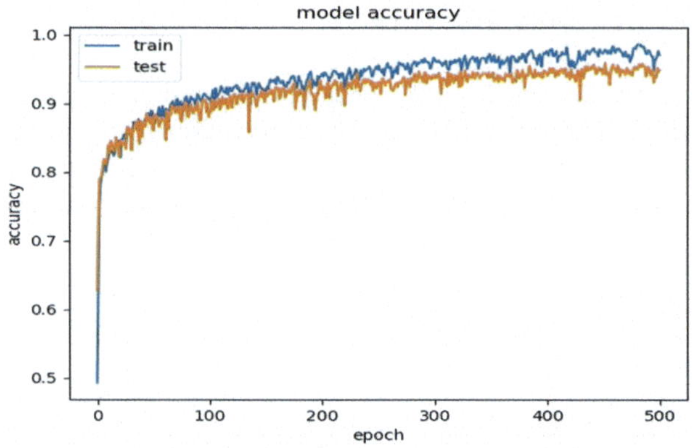

Fig. 2. Loss function of ANN

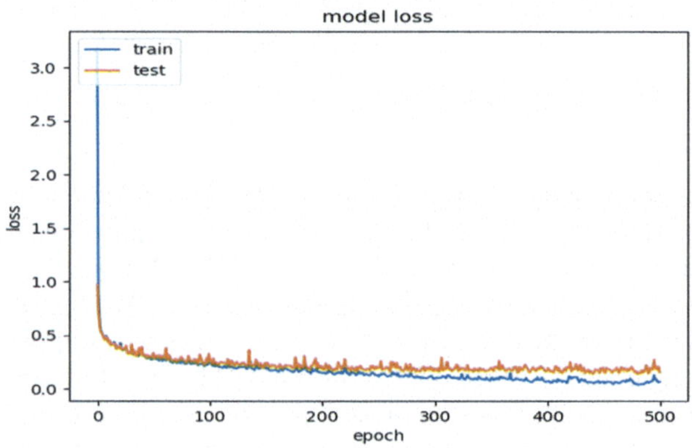

Fig. 3. Loss function of ANN

Table 3. Confusion matrix of ANN.

	Normal	Suspect	Pathological
Normal	480	2	4
Suspect	22	465	6
Pathological	46	3	462

Results show that the using features selection techniques before classification models can help in improvement of performance. ANN shows best performance as compared to other three algorithms. Results also show that using SMOTE technique improves the results obtained for classification compared to recent research work. The comparison of ANN results using SMOTE in current research with other sampling techniques applied in recent research is provided in Table 5.

Table 4. Comparison of classification mode.

	Naïve Bayes (%)	K-NN (%)	SVM (%)	ANN (%)
Without FS	59	62	78	88
CFS	66	63	82	90
Chi-square	64	67	86	93
RFE	61	64	89	94

Table 5. Comparison of class accuracy (%) with recent research.

	Marcia Amstelvina Saul et al. [16]				Current Research			
	Normal	Suspect	Pathological	Acc. (%)	Normal	Suspect	Pathological	Acc. (%)
Normal	90	7	4	89	480	2	4	98
Suspect	9	84	7	84	22	465	6	94
Pathological	0	10	90	90	46	3	462	90

5 Conclusion

Current research provides comparison of results of different classification algorithms along with usage of feature selection techniques to predict Cardiotography dataset state. Cardiotography dataset used in current research has the problem of imbalanced data due to which classification model predictions become biased toward class with more samples. The problem of imbalanced data is solved using Synthetic Minority Oversampling Technique (SMOTE). After using SMOTE feature selection techniques are used for the selection of top features set. According to the results in currents research, feature selection techniques improved the accuracy of classification algorithms. Current research also proves that using SMOTE technique along with Feature selection techniques to predict Cardiotography dataset class provides better results compared to sampling techniques used in recent researches.

In future work, hybrid algorithms can be implemented for the better prediction of classes. Dataset enrichment by collecting large sized dataset and further features can also help in improvement of prediction of fetal health state.

References

1. Abbas, R., Hussain, A.J., Al-Jumeily, D., Baker, T., Khattak, A.: Classification of foetal distress and hypoxia using machine learning approaches. In: International Conference on Intelligent Computing, pp. 767–776. Springer (2018)
2. Akbulut, A., Ertugrul, E., Topcu, V.: Fetal health status prediction based on maternal clinical history using machine learning techniques. Comput. Methods Programs Biomed. **163**, 87–100 (2018)
3. Bhatnagar, D., Maheshwari, P.: Classification of cardiotocography data with WEKA. Int. J. Comput. Sci. Netw.-IJCSN **5**(2) (2016)

4. Chamidah, N., Wasito, I.: Fetal state classification from cardiotocography based on feature extraction using hybrid k-means and support vector machine. In: 2015 International Conference on Advanced Computer Science and Information Systems (ICACSIS), pp. 37–41. IEEE (2015)
5. Chawla, N.V., Bowyer, K.W., Hall, L.O., Philip Kegelmeyer, W.: Smote: synthetic minority over-sampling technique. J. Artificial Intell. Res. **16**, 321–357 (2002)
6. Ertuğrul, Ö.F.: A novel type of activation function in artificial neural networks: trained activation function. Neural Netw. **99**, 148–157 (2018)
7. Fergus, P., Selvaraj, M., Chalmers, C.: Machine learning ensemble modelling to classify caesarean section and vaginal delivery types using cardiotocography traces. Comput. Biol. Med. **93**, 7–16 (2018)
8. Frank, A.: UCI Machine Learning Repository (2010). http://archive.ics.uci.edu/ml
9. Witten Ian, H., Eibe, F.: Data mining: practical machine learning tools and techniques (2005)
10. Jagannathan, D., Phil, M.: Cardiotocography-a comparative study between support vector machine and decision tree algorithms. Int. J. Trend Res. Dev. **4**(1) (2018)
11. Kamath, R., Kamat, R.: Modeling fetal morphologic patterns through cardiotocography data: decision tree-based approach. J. Pharmacy Res. **12**(10) (2018)
12. Kedia, A., Narsaria, M., Goswami, S., Taparia, J.: Empirical study to evaluate the performance of classification algorithms on healthcare datasets. World **5**(1), 1–11 (2017)
13. Li, J., et al.: Automatic classification of fetal heart rate based on convolutional neural network. IEEE Internet Things J. **6**(2), 1394–1401 (2018)
14. Mazumdar, S., Choudhary, R., Swetapadma, A.: An innovative method for fetal health monitoring based on artificial neural network using cardiotocography measurements. In: 2017 Third International Conference on Research in Computational Intelligence and Communication Networks (ICRCICN), pp. 265–268. IEEE (2017)
15. Potharaju, S.P., Sreedevi, M.: A novel m-cluster of feature selection approach based on symmetrical uncertainty for increasing classification accuracy of medical datasets. J. Eng. Sci. Technol. Rev. **10**(6) (2017)
16. Saul, M.A., Rostami, S.: A comparison of re-sampling techniques for pattern classification in imbalanced data-sets. In: UK Workshop on Computational Intelligence, pp. 240–251. Springer (2018)
17. Sbrollini, A., et al.: eCTG: an automatic procedure to extract digital cardiotocographic signals from digital images. Comput. Methods Programs Biomed. **156**, 133–139 (2018)
18. Silwattananusarn, T., Kanarkard, W., Tuamsuk, K.: Enhanced classification accuracy for cardiotocogram data with ensemble feature selection and classifier ensemble. arXiv preprint arXiv:2010.14051 (2020)
19. Tamilselvi, R., Parisa Beham, M., Merline, A., Roomi, S.M.M., Saravanan, B., Ruba, T.: Detection of fetal stress from maternal abdominal electrocardiogram signal. J. Comput. Sci. Eng. **6**(4), 65–70 (2018)
20. Zhang, Y., Zhao, Z.: Fetal state assessment based on cardiotocography parameters using PCA and adaboost. In: 2017 10th International Congress on Image and Signal Processing, BioMedical Engineering and Informatics (CISP-BMEI), pp. 1–6. IEEE (2017)

Intersection as an Event- and Agent-Based System

Daniel Kuchár(✉) and Peter Schreiber

Institute of Applied Informatics, Automation and Mechatronics, Faculty of Materials Science and Technology in Trnava, Slovak University of Technology in Bratislava, Trnava, Slovakia
{daniel.kuchar,peter.schreiber}@stuba.sk

Abstract. The paper presents the capabilities of Matlab/Simulink environment, namely the toolboxes Stateflow and SimEvents, of modelling the event- and agent-based systems in the shape of a symmetrical intersection at a micro-scale level. This simple but complex system deals with a random occurrence of the agents (the traffic participants), their own behaviour patterns, given road traffic regulations and a discretization of the intersection space model. The goal consists in proposing of a control algorithm, simulating of its function at a given model abstraction, and finally, generating a programming code for a PLC. With this kind of an approach, it is possible to work within one development environment with the systems which are impossible to be solved analytically and whose real testing is time and effort consuming.

Keywords: Intersection · Event-based · Agent-based · Stateflow · SimEvents

1 Introduction

1.1 Systems in Matlab/Simulink

There are various systems, such as physical or social. From a particular point of view on a given system, it is possible to choose its model while there are no exact rules how to do it [1]. The next choice of a programming environment and the methods how to use the model are conditional on this decision. Discrete-Event Systems [2] and System Dynamics [3] are two commonly used techniques for dealing with majority of systems. In case of complex systems, there is a need for a different technique called Agent-Based Modelling and Simulation [4]. Matlab/Simulink offers the following possibilities for working with event-based and agent-based systems [35]. Stateflow provides a graphical language that includes state transition diagrams, flow charts, etc. for describing how Simulink models react to input signals, events, and time-based conditions [36]. SimEvents can be used to model any event-driven process for optimizing its various performance characteristics [37] by means of hybrid system models, agent-based models, state charts, and process flows [38].

© The Author(s), under exclusive license to Springer Nature Switzerland AG 2021
R. Silhavy (Ed.): CSOC 2021, LNNS 229, pp. 45–54, 2021.
https://doi.org/10.1007/978-3-030-77445-5_5

1.2 Related Work in Transportation

The basic terminology of Traffic Dynamics simulations could be found in [5]. Bungartz [6] informs about two Road Traffic simulations: a microscopic (a cellular automaton as a short road segment) and a macroscopic (individual quantities depicted as continuous flows). [33] deals with a road segment hybrid model: a continuous part of a car flow and a discrete part of driver's behaviour and the traffic lights. Besides more Urban Traffic models, [7, pp. 391, pp. 402] presents an intersection model with different features (only cars without pedestrians, four directions without turning, a car never stops inside the intersection, all of the cars take the same time for passing through the intersection) in CD++ environment.

1.3 Our Goal

We intended to use the capabilities of Matlab/Simulink environment for modelling and simulating of a textbook discrete-event and agent-based system. The system in shape of a symmetrical intersection is designed at a micro-scale level. The goal was to develop a control algorithm with respect to a foreseen traffic participants behaviour pattern, their number, and an applied level of a model abstraction. Having checked the relevance of the algorithm, its programming code in.scl format for Siemens PLC was generated. The proposed model could be extended or altered at any time, of course with different simulation goal in mind.

2 Methods

The textbook example of an intersection modelling and simulation was divided in four main components: intersection, entities, road traffic regulations, and traffic lights control algorithm.

2.1 HW and SW Configuration

Windows 10 Pro, Version 1809, 64bit
MATLAB Version: 9.7.0.1190202 (R2019b)
Simulink Version 10.0 (R2019b)
SimEvents Version 5.7 (R2019b)
Stateflow Version 10.1 (R2019b)
TIA Portal Version V15.1
Simatic S7-1200 CPU 1214C DC/DC/DC, firmware V2.2.0.

2.2 Intersection

Figure 1 shows the model of a symmetrical intersection with four entries Left, Bottom, Top, and Right. The space of Bottom entry was discretized with the following principle: the left turning lane was split into three yellow fields A1, A2, and A3, the straight lane was created with the blue field B1 and four violet fields B2-B5, and finally, the right

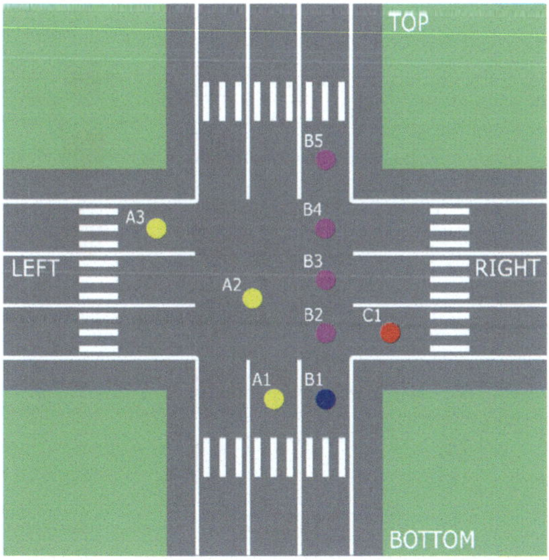

Fig. 1. Intersection

turning lane had a form of the blue field B1 and the orange field C1. B1 field is common for the straight and the right turning lanes. The others entries were solved analogically.

The pedestrians of Bottom zebra crossing were shaped by SimEvents diagram in the form illustrated on Fig. 2. The entity generator generated the pedestrians who waited in an infinite queue. Green Light (OK_Bottom_P_GR signal) opened the gate EN which allowed up to 30 pedestrians to enter the zebra. The number of pedestrians was indicated by Bottom_P_ZebraO signal. The others zebras were solved analogically and the parameters of pedestrians were different for each of the zebras.

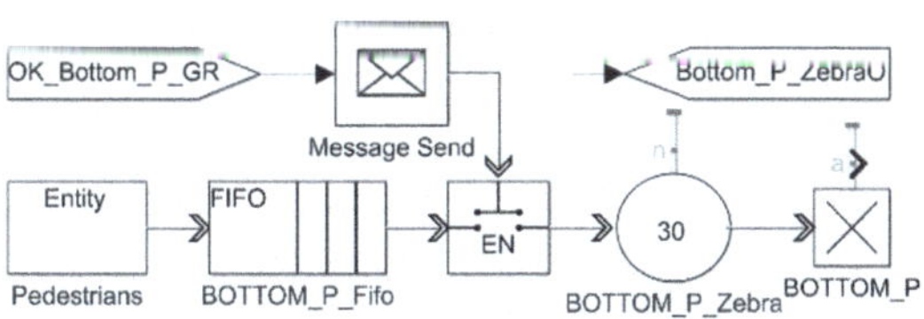

Fig. 2. Zebra crossing Bottom

The cars of Bottom entry were shaped by SimEvents diagram in the form depicted on Fig. 3. They waited before the zebra in two infinite queues. Green Lights OK_Bottom_GR_L (lane L for turning left) and OK_Bottom_GR_SR (lane SR for passing straight and turning right) opened the gates M1 and M2. The coloured lanes depicted an analogy of the coloured fields from Fig. 1. A car could enter a next field (server) on condition that a gate was opened (Green Light or a rule of the road traffic regulations) and the next field was empty. For that reason, the particular servers were connected serial and their Service Numbers equaled to one (no overtaking cars).

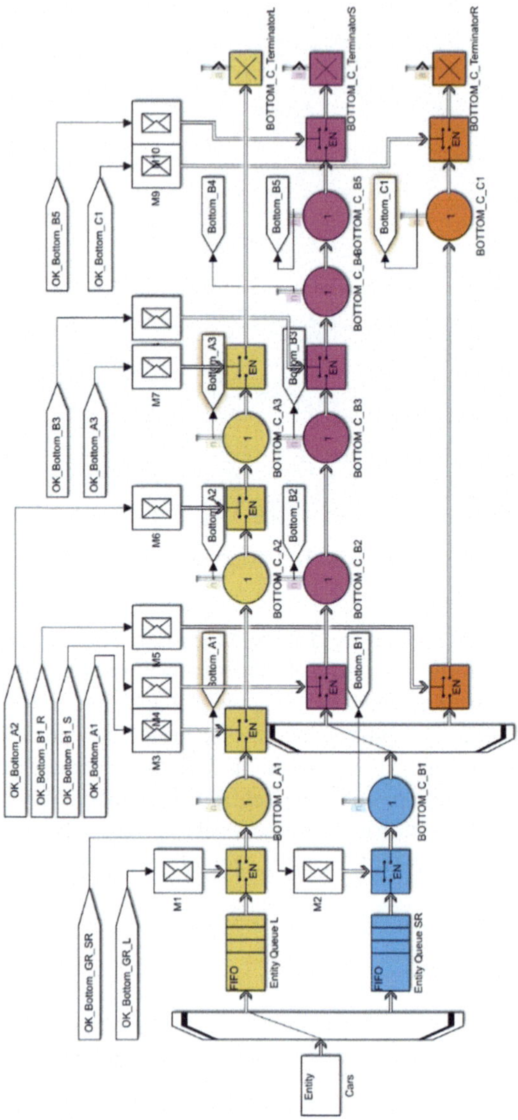

Fig. 3. Cars of Bottom entry

2.3 Entities

Pedestrians entity was generated at random moments e.g. (0,22)sec and with ServiceTime random value e.g. (5,15)sec which represented a time spent by a pedestrian on the zebra. Cars entity was generated at random moments e.g. (0,5)sec or regular moments e.g. 1 s with the random values Queue and ServiceTime e.g. (0.2, 0.9)sec. Queue value determined whether a car advanced in L lane (left) or SR lane (straight or right) and consequently passed straight or turned right. Analogically, ServiceTime defined a time spent by a car in the server.

2.4 Road Traffic Regulations

The road traffic regulations set conditions when a car could enter a next field. Figure 4 depicts an example of these conditions by means of the standard Simulink blocks. A car absence on the fields Top_B1, Top_B2, Right_B2, Right_B3, and Right_B4 allowed a car in the field Bottom_A2 to enter the field Bottom_A3. Some fields had no conditions or different conditions e.g. no possibility to pass through a zebra with nonzero pedestrian occupancy.

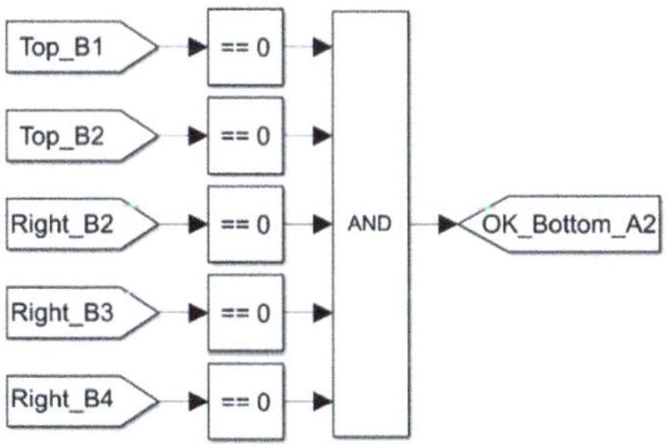

Fig. 4. Road traffic regulations

2.5 Traffic Lights Control Algorithm

The traffic light control realised by means of StateFlow is depicted on Fig. 5.

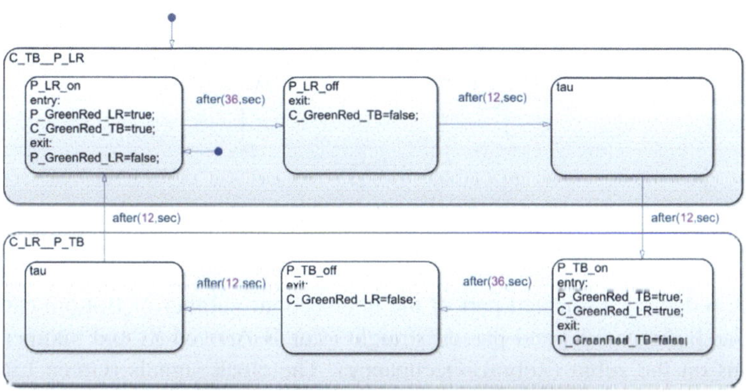

Fig. 5. Traffic lights control algorithm

The temporal logic with a given pedestrians and cars profile was verified by simulations. P_LR_on state stands for Green Light of Left and Right pedestrians (at the same time for Top and Bottom cars). P_LR_off state stands for Red Light of the pedestrians but Green Light of the cars mentioned above. The state Tau means Red Light for all the traffic participants in order to empty the intersection (possible slowness of some pedestrians and cars). The other three states operated analogically.

It was necessary to decide which part of the code would be generated for a PLC and choose the output variables types, too. Subsequently, the procedure from [34, pp. 1–3, pp. 1–14] or in [39] followed. Having imported a.scl code into the Projet tree and sequentially generated the Function blocks, those were at disposal in the folder Program blocks. The input ssMethodType had to be served by means of OB100 Startup block for a correct initialization of the generated block. More about programming the PLC S7-1200 could be found in [8].

3 Results

During the intersection modelling (Sect. 2.2), the entities attributes setting (Sect. 2.3) and the road traffic regulations setting (Sect. 2.4), it is necessary to avoid blocking the intersection model operation – deadlock. In a similar way, it is important to deal correctly with random numbers [46] and know the nature of the stochastic processes (entities arriving in the real environment) for achieving an authenticity of the model entities behaviour.

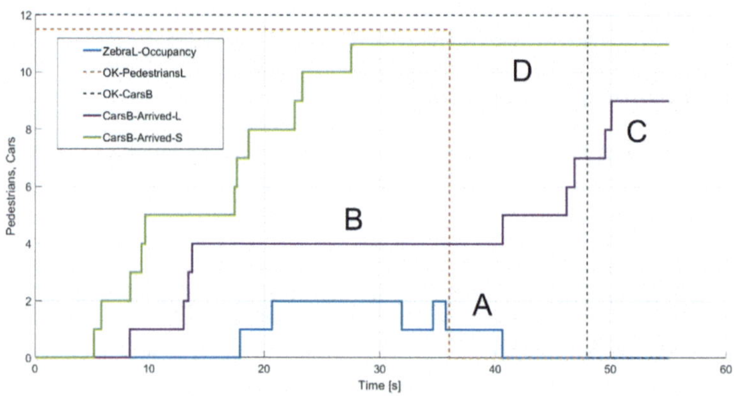

Fig. 6. Intersection agents behaviour

Figure 6 depicts a selected part of the intersection: number of Bottom cars turned to left (CarsB-Arrived-L) and passed straight (CarsB-Arrived-S) and number of Left pedestrians on the zebra (ZebraL-Occupancy). The clock signals (Green Lights) for Bottom cars (OK-CarsB) and Left pedestrians (OK-PedestriansL) are indicated by the dashed lines. The time of 60 s represents the state C_TB__P_LR from Fig. 5. Our simulations were performed in the longer time spans, of course. The difference between

CarsB-Arrived-S and CarsB-Arrived-L increases little by little because turning left one should obey opposite direction and pedestrians rights of way, too.

Description of the situations:

A – when Left pedestrians Green Light has changed off, those are emptying the zebra (the state P_LR_off from Fig. 5)
B – Left pedestrians are blocking Bottom cars while turning left
C – Bottom cars while turning left are emptying the intersection when Green Light has changed off (the state Tau in the state C_TB__P_LR from Fig. 5)
D – Bottom cars while heading straight are blocked by Bottom cars heading right which must wait for Right pedestrians (the fields B1 and C1 in Fig. 1 and Fig. 3).

4 Conclusion

4.1 Basic Terms

There are three basic terms in simulation terminology: system, model and simulation. A **system** consists of the objects with mutual relationships separated from universe [9]. A closer look at the topic showes a detailed segmentation into system of systems [10], complex system [11], integrated system [12], multi-agent system [13], etc. A **model** represents a set of essential system features in an appropriate form. Those can be experimentally verified in order to find out the system properties [14]. While modelling, it is important to know: goal, essential features, simulation data accessibility, and application area [15]. A good model is simple, valid, and robust [16]. There are various steps during the model creation process: real system -> conceptual model (basic ideas) -> formal model (after aplying a formalim [7]) -> simulation model (written program in a SW environment) [17, pp. 17, 7 pp. 12]. For each non-primitive system exists a group of its models which describes it from different points of view (model hierarchies) [18] e.g.: level of details (spatial complexity), characteristic sizes (multi-scale modelling), characteristic time (temporal complexity). A **simulation** is an applied methodology which describes the system behaviour by means of its model [19]. A simulation engine is an SW application with the model and a simulation infrastructure represents all the HW and SW used during the simulation [20].

We chose the modelling at the micro-scale level with a simple discretization of cars space. Analogous principle of discretization could be applied for pedestrians zebra in order to facilitate car passing. The control algorithm of the intersection consists exclusively of temporal logic – in case of need, it is possible to add the inputs from the process: sensing the presence of the waiting cars or a request for Green Light for the pedestrians.

4.2 Modelling and Simulation

According to [9], the modelling and simulation tools could be described as graphical, algebraic and formal-language based. As an alternative SW environment to Matlab/Simulink could serve e.g. Anylogic [21, 41], CD++ [7, 42], Compose and Activate

[22, 43], NetLogo [44], Repast [45]. [1] informs about the key **components** common for all of discrete-event systems: clock, events list, random-number generators, statistics, ending condition. There are **common errors** during a simulation [23] e.g. inappropriate level of detail, improper selection of the programming language, invalid models, improperly selected initial conditions, short run times, poor random number generators, etc. A typical agent-based modelling [4] consists of three components: a set of agents with their attributes; a set of agent relationships and methods of interaction; the agents' environment. A **topology** represents a way how and with whom the appropriate agents interacts: cellular automaton depicts a regular cell structure while a network model is interconnected in an irregular manner [11].

The model validity could be verified with a sufficiently long time span for detecting a potential deadlock. The decisive influence belongs to the implementation of the road traffic regulations. In our case, we used only positional rules (presence of the other cars on the appropriate fields) but the velocity aspect (there is an other car on particular fields but moves relatively slowly) could be added, too.

4.3 Further Development

The model could be extended with a **3D visualisation** and a **collision detection** in Simulink [24]. Simulink 3D Animation allows to work with two standards: VRML97 and X3D [25]. In case of need for a more demanding graphics, the environment Blender offers collision detection under the name Collision Physics [26, pp. 246] and the scripting language Python could be used [27, pp. xviii], too. In the same manner, Collision [28, pp. 161] and the sensor implementation [29, pp. 48] in Unity environment could be mentioned. Among the scripting languages, there are Boo, C# or JavaScript [30, pp. 334]. One important thing has to be said, the space model is changing from a discretized into a continuous one, but of course, the goal of the simulation is completely different.

Under condition of leaving behind the purpose of an intersection modelling for a moment, a discrete-event system could be used for another role, too. Stateflow with the conception of finite state machines [31, pp. 390] enables you to design and develop **supervisory control** and hybrid systems [36]. Valavanis [32, pp. 871] informs about **mode switching problem** for multiple control modes in case of a UAV (taking-off, hovering, path-following, landing, etc.).

Acknowledgements. This publication is the result of implementation of the KEGA Project 0009STU-4/2018: "The innovation of the subject Intelligent Control Methods at the Faculty of Materials Science and Technology of Slovak University of Technology".

References

1. Chaturvedi, D.K.: Modeling and Simulation of Systems Using MATLAB and Simulink. CRC Press Taylor & Francis Group, Boca Raton (2010)
2. Cassandras, C., Lafortune, S.: Introduction to Discrete Event Systems, 2nd edn. Springer, New York (2008)

3. Dukkipati, R.V.: Solving Engineering System Dynamics Problems with MATLAB. New Age International (P) Ltd Publishers, New Delhi (2007)
4. Taylor, S.J.: Agent-Based Modeling and Simulation. Palgrave Macmillan, New York (2014)
5. Ferrara, A., Sacone, S., Siri, S.: Freeway Traffic Modelling and Control. Springer, Cham (2018)
6. Bungartz, H.J., Zimmer, S., Buchholz, M.: Modeling and Simulation An Application-Oriented Introduction. Springer, Heidelberg (2014)
7. Wainer, G.A.: Discrete-Event Modeling and Simulation A Practitioner's Approach. CRC Press Taylor & Francis Group, Boca Raton (2009)
8. Kamel, K., Kamel, E.: Programmable Logic Controllers Industrial Control. McGraw-Hill Education, New York (2014)
9. Hrúz, B., Zhou, M.C.: Modeling and Control of Discrete-event Dynamic Systems with Petri Nets and Other Tool. Springer, London (2007)
10. Rainey, L.B., Tolk, A.: Modeling and Simulation Support for System of Systems Engineering Applications. Wiley, Hoboken (2015)
11. Sayama, H.: Introduction to the Modeling and Analysis of Complex Systems. Open SUNY Textbooks, New York (2015)
12. Murray-Smith, D.: Modelling and Simulation of Integrated Systems in Engineering Issues of Methodology, Quality, Testing and Application. Woodhead Publishing Limited, Sawston (2012)
13. Vamvoudakis, K.G., Jagannathan, S.: Control of Complex Systems Theory and Applications. Elsevier Inc., Kidlington (2016)
14. Roffel, B., Betlem, B.: Process Dynamics and Control Modeling for Control and Prediction. Wiley, Chichester (2006)
15. DiStefano, J., III.: Dynamic Systems Biology Modeling and Simulation. Elsevier Inc., London (2013)
16. Igelnik, B.: Computational Modeling and Simulation of Intellect: Current State and Future Perspectives. IGI Global, Hershey (2011)
17. Liu, J.: Radial Basis Function (RBF) Neural Network Control for Mechanical Systems Design, Analysis and Matlab Simulation. Tsinghua University Press, Beijing and Springer, Berlin (2013)
18. Hangos, K., Cameron, I.: Process Modelling and Model Analysis. Academic Press, London (2001)
19. Sokolowski, J.A., Banks, C.M.: Modeling and Simulation Fundamentals Theoretical Underpinnings and Practical Domains. Wiley, Hoboken (2010)
20. Siegfried, R.: Modeling and Simulation of Complex Systems A Framework for Efficient Agent-Based Modeling and Simulation. Springer, Wiesbaden (2014)
21. Borshchev, A.: The Big Book of Simulation Modeling Multimethod Modeling with Anylogic 6, AnyLogic North America (2013)
22. Campbell, S.L., Nikoukhah, R.: Modeling and Simulation with Compose and Activate. Springer, Cham (2018)
23. Guizani, M., Rayes, A., Khan, B.: Network Modeling and Simulation a Practical Perspective. Wiley, Chichester (2010)
24. Khaled, N.: Virtual Reality and Animation for MATLAB and Simulink Users. Springer, London (2012)
25. Brutzman, D., Daly, L.: X3D: Extensible 3D Graphics for Web Authors. Elsevier Inc., San Francisco (2007)
26. Guevarra, E.T.: Modeling and Animation Using Blender: Blender 2.80: The Rise of Eevee. Apress, Laguna (2020)
27. Conlan, C.: The Blender Python API Precision 3D Modeling and Add-on Development. Apress, Bethesda (2017)

28. Tristem, B., Geig, M.: SamsTeachYourself Unity Game Development in 24 Hours, 2nd edn. Pearson Education, Indiana (2016)
29. Barrera, R., Kyaw, A.S., Swe, T.N.: Unity 2017 Game AI Programming, 3rd edn. Packt Publishing, Birmingham (2018)
30. Watkins, A.: Creating Games with Unity and Maya. Elsevier Inc., Burlington (2011)
31. Xue, D., Chen, Y.Q.: System Simulation Techniques with MATLAB and Simulink. Wiley, Chichester (2014)
32. Valavanis, K.P., Vachtsevanos, G.J. (eds.): Handbook of Unmanned Aerial Vehicles. Springer, Dordrecht (2015). https://doi.org/10.1007/978-90-481-9707-1
33. Zhang, Y., Cassandras, Ch.G., Li, W.: A simevents model for hybrid traffic simulation. In: Proceedings of the 2017 Winter Simulation Conference (2017)
34. Simulink PLC Coder User's Guide (R2019b), MathWorks, Inc., Natick (2019)
35. https://www.mathworks.com/products.html?s_tid=gn_ps. Accessed 22 Mar 2020
36. https://www.mathworks.com/products/stateflow.html. Accessed 22 Mar 2020
37. https://www.mathworks.com/products/simevents.html. Accessed 22 Mar 2020
38. https://www.mathworks.com/solutions/discrete-event-simulation.html. Accessed 22 Mar 2020
39. https://www.youtube.com/watch?v=E1ppmAuf47o. Accessed 22 Mar 2020
40. https://www.mathworks.com Documentation: Integrate Generated Code with Custom Code. Accessed 22 Mar 2020
41. https://www.anylogic.com. Accessed 22 Mar 2020
42. http://cell-devs.sce.carleton.ca/mediawiki/index.php/Main_Page. Accessed 22 Mar 2020
43. https://www.altair.com/mbd2019/. Accessed 22 Mar 2020
44. http://ccl.northwestern.edu/netlogo/. Accessed 22 Mar 2020
45. https://repast.github.io/index.html. Accessed 22 Mar 2020
46. https://www.mathworks.com Documentation: Event Action Languages and Random Number Generation. Accessed 22 Mar 2020

Comparison of UAV Landing Site Classifications with Deep Neural Networks

Daniel Kuchár(✉) and Peter Schreiber

Institute of Applied Informatics, Automation and Mechatronics, Faculty of Materials Science and Technology in Trnava, Slovak University of Technology in Bratislava, Trnava, Slovakia
{daniel.kuchar,peter.schreiber}@stuba.sk

Abstract. The paper compares classification results of three deep neural networks: AlexNet, VGG16 and VGG19 for ten classes of a UAV landing site with respect to accuracy, memory requirements, and a typical running time. The overall results offer an insight into typical false friends among particular classes of landing site and a zooming technique for better understanding of a surrounding context in a landing situation. The algorithm presents just one of alternatives how to find out if the appropriate terrain is suitable for landing or not by means of an optical way. Nowadays, this kind of sensing is generally present on every single piece of UAV in the shape of a camera which could be easily adopted to the purpose of an automatic landing. All above, when an emergency of a loss of the controlling signal arises. The process of the training set preparation is described, too.

Keywords: UAV landing site · Deep learning · AlexNet · VGG16 · VGG19

1 Introduction

The commercial UAVs [1, 2] are able to independently take off from a suitable surface; remember the homing point at the height of seven meter if it is clearly differentiable from the surroundings and return to the homing point on condition that the situation has not significantly changed.

Moreover, they are able to recognize whether any arbitrary terrain is suitable for landing. If yes, they land. If not, they hover and wait for the next operator command. The algorithm for taking the decision is a subject of know-how and generally not known. The paper is offering one possible solution to this kind of algorithm, together with its limitations.

Nowadays, the visual information serves as a primary source for overall perception. For the image processing, it is possible to use the artificial intelligence by means of:

1. machine learning with feature extraction – this process is time consuming.
2. deep learning – this process leads quickly to the results with satisfying quality. On the other hand, it is time and computational power demanding for the full training or transfer learning. [3] informs about training times and computational power needed for some deep neural netwoks, [4–6] present the principles of the convolutional neural networks.

© The Author(s), under exclusive license to Springer Nature Switzerland AG 2021
R. Silhavy (Ed.): CSOC 2021, LNNS 229, pp. 55–63, 2021.
https://doi.org/10.1007/978-3-030-77445-5_6

The relations and links among artificial intelligence, machine learning and deep learning are explained in [7].

2 Methods

2.1 HW and SW Configuration

We used the following HW and SW equipment for the comparison:

Windows 10 Pro, Version 1809, 64 bit
i7-7820HQ, SSD, 64 GB RAM
GPU: NVidia Quadro P3000 6 GB RAM
Matlab 9.2.0.556344 (R2017a)
Neural Network Toolbox Model for AlexNet Network version 17.1.0
Neural Network Toolbox Model for VGG-16 Network version 17.1.0
Neural Network Toolbox Model for VGG-19 Network version 17.1.0.

2.2 Deep Neural Networks

We used three types of deep neural networks for classification of a UAV landing site [8]: AlexNet [9], VGG16, and VGG19.

2.3 Classes of Landing Site

In Fig. 1 ten classes of landing site are shown - the first line from the left to the right: asphalty, concrete, brizolit, pavement, and needles. The second line from the left to the right: shingle, leaves, sand, grass, and soil.

Within each of a class many variations of colours and shapes of a depicted feature were presented.

Fig. 1. Ten classes of landing site

2.4 Training Sets

For the high quality training process with deep neural networks AlexNet, VGG16, and VGG19 we needed the sets of appropriate images. Although there are free sets of images on the Internet at disposal [10–12] – those were practically not suitable because of their size, topical content, conditions of the taking moment (height of the taking, angle, various seasons), or resolutions.

Therefore, we took own images with a camera with 3216 × 4288 (Y × X format) pixels resolution from a height circa one meter. Following the next selection programmable process, we obtained two sets of images SetAlex and SetVGG. The normalized input for AlexNet are images with resolution 227 × 227 pixels, and for VGG images with resolution 224 × 224 pixels. Both sets were not equal because of the difference between AlexNet and VGG input image requirements. This quantity difference was supposed not to be significant due to the high element numbers of both sets.

We used the rotations 90°, 180°, 270° and zoom, techniques of data augmentation [13], for increasing the number of the normalised images. Regarding the zoom technique, we took a sample from an original image with higher resolution and downsampled it to the appropriate resolution of a normalised image. As a result, we obtained a wider context and suppressed some disturbing elements (stub of a cigarette). Naturally, with this procedure we lost some details from the original sample. Therefore, the zoom technique is not generally suitable for any surface – these aspects are known as image pyramids [8, 14, 15].

Table 1 depicts the numbers of each training set together with the comments about their important features (crack in the concrete, tuft of grass in the asphalty, stubs of cigarettes, wet spots, leaves and branches on the surface, presence of a unique season or weather – influence on the leaf colour, etc.). Some of these extraneous elements (stub of cigarette) are normally present and did not constitue a dominant impact on the background character of the image. Regarding the others (unique season or weather), we took them into account and in case of need it would be necessary to supply the rest of missing aspects.

2.5 Classification of the Test Set

An image with resolution 3216 × 4288 pixels was split into a matrix of small images, which came into the classification process independetly (green area), except for the first and last row and the first and last column (red area). Those were not used because of 3 × 3 zoom. The small yellow sample WP with 227 × 227 pixels served for the algorithm AlexNet and after having been cut to 224 × 224 pixels for VGG nets, too. The blue sample WPZoom depicted 3 × 3 zoom. In case of need, the green area could be shifted by the coordinations Xorigin and Yorigin. The classification is depicted in Fig. 2.

We chose (apart from those within the training sets) ten images for each of ten surface classes. Each of them represented 192 normalised images for a classification process – altogether 19200 normalised images. Majority category from the classification was regarded as right (a general content of the images for a classification is supposed, too). All of rest were regarded as false. Number of right classified images over the total number of the images is accuracy.

Table 1. Normalised sets SetAlex (227 × 227), SetVGG (224 × 224)

Category	SetAlex	SetVGG	Comment
	Number	Number	
Asphalty	16 896	14 336	Cracks, grass, polygons, wet spots
Concrete	13 728	11 648	Cracks, gravel, wet spots, sole
Brizolit	9 272	9 956	Litter: stub, leaf, branch
Pavement	20 640	23 220	Wet/dry, colour/mono, grass
Needles	11 088	9 408	Grass, litter: leaf, cone, branch
Shingle	11 712	12 576	Grass, litter: leaf, stub
Leaves	10 176	11 448	Autumn, grass, needles, maple+oak
Sand	13 248	14 904	Sole, colours, litter: leaf, papers
Grass	19 536	16 576	Green/dry, types, litter
Soil	12 688	13 624	Black/grey, litter: stone, grass

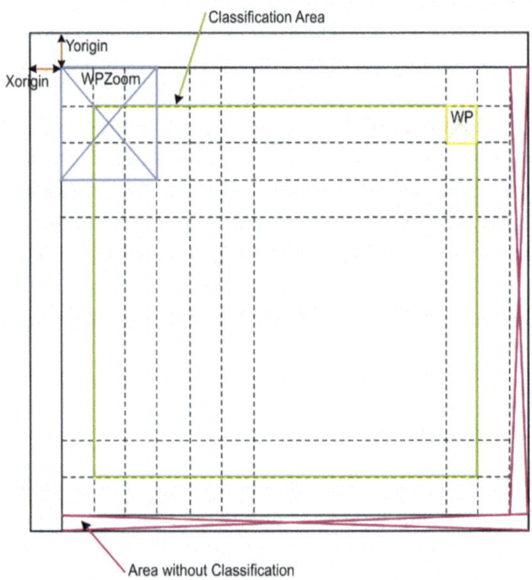

Fig. 2. Classification algorithm

3 Results

Overall classification results for AlexNet, VGG16, and VGG19 are summarised in Table 2 both for zoom 1 × 1 and for zoom 3 × 3, too. Table 3 states the memory requirements and a typical running time for a classification of one image with 3216 × 4288 pixels resolution.

Table 2. Accuracy of AlexNet, VGG16, and VGG19

	Zoom	Number of false classifications	Accuracy
AlexNet	1 × 1	563	97.07%
	3 × 3	115	99.40%
VGG16	1 × 1	810	95.78%
	3 × 3	206	98.93%
VGG19	1 × 1	730	96.20%
	3 × 3	257	98.66%

Table 3. Memory requirements + running time for an image (3216 × 4288 p.) classification

Type	Size [MB]	Running time [s]
AlexNet	244	~9
VGG16	554	~16
VGG19	575	~17

Detailed classification results are shown in Table 4 and Table 5 in the form of triplets: AlexNet, VGG16, and VGG19. The left column states for a ground-truth surface category, while in the first row there are false categories.

Table 4. False classifications in the form of triple AlexNet, VGG16, VGG19

	Zoom	Asphalty	Concrete	Brizolit	Pavement	Needles
Asphalty	1 × 1		4,11,8	0,1,2	16,11,28	1,16,3
	3 × 3		4,0,0	0,6,4	14,4,14	1,0,0
Concrete	1 × 1	1,44,34			3,12,0	
	3 × 3	0,14,21				
Brizolit	1 × 1					
	3 × 3	1,0,0			1,0,0	
Pavement	1 × 1	40,116,95	86,98,112			0,3,1
	3 × 3	0,16,3	23,10,3			

(*continued*)

Table 4. (*continued*)

	Zoom	Asphalty	Concrete	Brizolit	Pavement	Needles
Needles	1 × 1	0,2,1			1,0,0	
	3 × 3					
Shingle	1 × 1	2,0,1	1,0,2	0,1,0	0,0,1	0,4,1
	3 × 3		1,0,0			
Leaves	1 × 1		2,0,0		2,1,0	4,3,2
	3 × 3					
Sand	1 × 1	8,77,49	16,13,7		67,14,13	10,0,2
	3 × 3		3,0,0		21,0,3	0,0,2
Grass	1 × 1		0,0,1		1,0,1	9,17,23
	3 × 3					
Soil	1 × 1	0,0,1	1,0,0	1,0,0	0,5,1	5,1,7
	3 × 3					

Table 5. False classifications in the form of triple AlexNet, VGG16, VGG19

	Zoom	Shingle	Leaves	Sand	Grass	Soil
Asphalty	1 × 1		1,3,6	0,1,0	4,1,2	4,1,0
	3 × 3					
Concrete	1 × 1	1,9,4	1,3,0	1,1,3		3,25,2
	3 × 3					
Brizolit	1 × 1					
	3 × 3					
Pavement	1 × 1	1,0,0	93,105,126	63,65,83	14,16,7	15,7,3
	3 × 3		37,110,173	3,17,18	2,11,2	0,2,0
Needles	1 × 1		1,11,11		5,7,6	4,8,11
	3 × 3		0,0,8		0,9,6	1,4,0
Shingle	1 × 1		10,7,15	0,0,1	1,1,1	
	3 × 3					
Leaves	1 × 1	5,2,1		3,14,5	6,13,4	

(*continued*)

Table 5. (*continued*)

	Zoom	Shingle	Leaves	Sand	Grass	Soil
	3 × 3				1,0,0	
Sand	1 × 1		10,3,11			11,19,12
	3 × 3		0,1,0			0,2,0
Grass	1 × 1		1,0,3			16,15,13
	3 × 3					2,0,0
Soil	1 × 1	3,0,3		0,0,1	2,3,2	
	3 × 3					

4 Conclusion

4.1 On Results

From Table 2 it is obvious that 3 × 3 zoom generally improved the accuracy of classification although not in every single case. At the same time, not all of normalised images stated as false were actually false – a tuft of grass in the asphalty without its surroundings is still just a tuft of grass and only with zoom covering its ambient could be right classified as an asphalty. All of three deep neural networks had comparable results with 3 × 3 zoom. In favour of AlexNet spoke its half requirements on the memory and the running time. Another deep neural network candidates suitable for the purpose of this paper are presented in [16, 17] and [8].

From the viewpoint of the deep neural networks running, some time optimization is possible (during the program, a storing of the normalized images into the computer memory had been running which was not necessary), but the times are good enough for the user to have an idea about the possibility to deploy these algorithms in a real situation online. Next time savings could arise with the camera resolution reduction or with the compiled version of a program since Matlab is an interpreter.

It is not possible to assume that an image from a height of one meter could have the same informative value as from the height of 20 or 50 m (a common flying altitude of the commercial UAVs [1, 2]). A loss of details occurs and there is absolutely different content in the image. Therefore, with assumption of two basic flying altitudes, two distinct algorithms would be necessary.

4.2 Indoor-Outdoor, Invariance

All of training set images were taken in outdoor environment in good lighting conditions. [18, 19] state that it could be necessary to take various particularities of outdoor and indoor environments into account such as reflections, shadows, contrast changes, etc.

The deep neural networks could be used for classifying the individual objects (threats) in the scene, too. In this case, as much as possible affine [20–23] and colour invariances should be considered. This can be achieved with an appropriately prepared training set.

Regarding the colour invariance, there are more options for colour models [24], some of them with just one intensity channel. In case of poor lighting conditions, it is possible to use night vision [25] which also leads into the greyscale option. The structure of the deep neural network is then simpler.

In the present methodology, the UAV ability to recognize the geometrical profile of a terrain formed by various obstacles, slopes, or roughness and estimate its intrinsic attributes in the form of the surface stability or strength is absent. In the same way, unfavourable conditions of direct optic visibility (fog, smoke) cause the problem. For solving these aspects of a landing process, next sensor equipment and means of artificial intelligence are needed.

4.3 Bio-inspired Approach

Contrary to the classical electronic artificial intelligence approach there is another one provided by the nature. [26] states that the hardware and software combination made by human being still lags behind when even a small 'non-intelligent' bug is able to excel the most powerful computers while performing its daily routines such as own sensory data processing, visual perception, or motor control. As a logical continuation could serve Tau methodology [27] which mimics behaviour of animals or insect at near-ground manoeuvres (landing) exclusively by visual and/or vestibular information disregarding own height position or velocity.

Acknowledgements. This publication is the result of implementation of the KEGA Project 0009STU-4/2018: "The innovation of the subject Intelligent Control Methods at the Faculty of Materials Science and Technology of Slovak University of Technology".

References

1. DJI Phantom 4 PRO/PRO+ User Manual V1.4 2017.10 (2017)
2. DJI Mavic 2 PRO/ZOOM User Manual V1.8 2019.01 (2019)
3. https://adeshpande3.github.io/adeshpande3.github.io/A-Beginner's-Guide-To-Understanding-Convolutional-Neural-Networks/. Accessed 3 Apr 2020
4. Buduma, N., Lacascio, N.: Fundamentals of Deep Learning. O'Reilly, Sebastopol (2017)
5. Goodfellow, I., Bengio, Y., Courville, A.: Deep Learning. Massachusetts Institute of Technology, London (2016)
6. Patterson, J., Gibson, A.: Deep Learning: A Practitioner's Approach. O'Reilly, Sebastopol (2014)
7. Kim, P.: MATLAB Deep Learning: With Machine Learning, Neural Networks and Artificial Intelligence. Apress (2017)
8. Krig, S.: Computer Vision Metrics. Springer, Heidelberg (2016)
9. Krizhevsky, A., Sutskever, I., Hinton, G.E.: ImageNet Classification with Deep Convolutional Neural Networks, pp. 1–9 (2012)
10. https://data.vision.ee.ethz.ch/cvl/rrothe/imdb-wiki. Accessed 3 Apr 2020
11. http://ufldl.stanford.edu/housenumbers. Accessed 3 Apr 2020
12. http://www.vision.caltech.edu/Image_Datasets/Caltech101. Accessed 3 Apr 2020

13. https://medium.com/nanonets/how-to-use-deep-learning-when-you-have-limited-data-part-2-data-augmentation-c26971dc8ced. Accessed 3 Apr 2020
14. Beyerer, J., León, F.P., Frese, Ch.: Machine Vision. Springer, Heidelberg (2016)
15. Sonka, M., Hlavac, V., Boyle, R.: Image Processing, Analysis, and Machine Vision. Cengage Learning, Stamford (2015)
16. Gong, S., Liu, C., Ji, Y.: Advanced Image and Video Processing. Springer, Cham (2019)
17. Khan, S., Rahmani, H., Shah, S.A.: A Guide to Convolutional Neural Networks for Computer Vision. Morgan & Claypool (2018)
18. Bhanu, B., Lee, S.: Genetic Learning for Adaptive Image Segmentation. Springer, New York (1994)
19. Chen, Ch., Ren, Y., Kuo, C.C.: Big Visual Data Analysis Scene Classification and Geometric Labeling. Springer, Singapore (2016)
20. Costa, L.F.: Shape Classification and Analysis Theory and Practice. Taylor & Francis Group, LLC, Boca Raton (2009)
21. Flusser, J., Suk, T., Zitová, B.: 2D and 3D Image Analysis by Moments. Wiley, Chichester (2017)
22. Flusser, J., Suk, T., Zitová, B.: Moments and Moment Invariants in Pattern Recognition. Wiley, Chichester (2009)
23. Reiss, T.H. (ed.): Recognizing Planar Objects Using Invariant Image Features. Springer, Heidelberg (1993). https://doi.org/10.1007/BFb0017553
24. Sangwine, S.J., Horne, R.E.: The Colour Image Processing Handbook. Chapman & Hall, Cambridge (1998)
25. Bai, L., Han, J., Yue, J.: Night Vision Processing and Understanding. Springer, Singapore (2019)
26. Cristóbal, G., Perrinet, L., Keil, M.: Biologically-inspired Computer Vision Fundamentals and Applications. Wiley-VCH Verlag GmbH & Co., Weinheim (2016)
27. Becerra, V.: Autonomous Control of Unmanned Aerial Vehicles. MDPI, Basel (2019)

An Approach to Optimize Multi-objective Problems Using Hybrid Genetic Algorithms Supported by Initial Centroid Selection Optimization Enhanced K-Means Based Selection Operator

Ahmed Maghawry[1(✉)], Rania Hodhod[2], Yasser Omar[1], and Mohamed Kholief[1]

[1] College of Computing and Information Technology, Arab Academy for Science, Technology and Maritime Transport (AASTMT), Alexandria, Egypt

[2] TSYS School of Computer Science, Columbus State University, Columbus, GA, USA

hodhod_rania@columbusstate.edu

Abstract. The process of optimization is approached as a searching problem, where an optimization algorithm attempts to find the best possible solution to a given objective function within a permissible search domain. Such problems are complicated since we attempt to find the best possible solution to a given objective function. The problem becomes harder when there is more than one objective function that can be defined as multi-objective optimization problems. In such problems, the algorithm attempts to optimize more than one objective function. Furthermore, the problem becomes worse when these objectives are contradicting. Evolutionary algorithms are used to solve such problems including genetic algorithms (GAs). Hybridizing genetic algorithms is also utilized to overcome the sub optimal solution tendency of basic genetic algorithms. In this paper, an enhanced hybrid genetic algorithm is introduced with an advanced selection operator mechanism based on the K-means clustering algorithm that is also supported by the initial centroid selection optimization to ensure the best possible selection process. The proposed algorithm was tested against 4 benchmark multi-objective optimization algorithms where it succeeded to maximize the balance between search space exploration performed by the GA and search space exploitation performed by the PSO, that was reflected in the optimization ability of the algorithm. The enhanced ICSO/K-means selection operator also succeeded to enhance the optimization ability of the proposed algorithm by assuring fair distribution of the selected individuals from each generation.

Keywords: Genetic algorithms · Particle swarm optimization · Hybrid genetic algorithm · Multi-objective optimization · K-means

© The Author(s), under exclusive license to Springer Nature Switzerland AG 2021
R. Silhavy (Ed.): CSOC 2021, LNNS 229, pp. 64–87, 2021.
https://doi.org/10.1007/978-3-030-77445-5_7

1 Introduction

1.1 Evolutionary Computation

In the field of computer science, evolutionary computation is the sub field of artificial intelligence and soft computing studying the family of algorithms that tackles the problems of global optimization derived by the concepts of biological evolution inspired from nature. Such algorithms are referred to as evolutionary-based algorithms (EAs), and they mimic the biological process of evolution. An EA makes use of operators inspired from biological evolution such as breeding, crossover, mutation, and selection. EAs are population-based algorithms in which each individual represents a candidate solution to an optimization problem. To determine feasibility and efficiency of a candidate solution EAs utilize one or more fitness functions that measures the quality of a candidate solution. The process of evolution in an EA takes place by repeating the evolutionary operators mentioned above [1].

1.2 Multi-objective Optimization Problems

The main target of solving optimization problems is to find the best possible solution from among a permissible search domain. In such problems, an optimization algorithm is proposed to traverse the search domain to find the best solution. Single-objective optimization problems are such problems where the algorithm attempts to optimize one objective function to find the global minimum/maximum according to the target of optimization. Multi-objective optimization problems are such problems that contain more than one optimization functions where the algorithm attempts to optimize more than one objective function while traversing through the search domain to find a solution or a solution set that would achieve the optimization target considering all the given objective functions [1, 2].

In computer science, there are various multi-objective benchmark optimization functions for testing purposes, however in a real life scenario a single objective optimization problem would be "an attempt to find the best satellite design that can perform weather monitoring tasks, in such a problem, the optimization algorithm will focus only on finding the best satellite design that when manufactured will produce a satellite that can perform weather monitoring tasks regardless of any other feature. On the other hand, a real life scenario of a multi-objective optimization problem would be "an attempt to find the best satellite design that when manufactured will produce a satellite that can perform weather, geographical and military tasks and to have a cheap cost all at the same time".

From this scenario, it is very hard to design such a satellite that can perform all these tasks and at the same time will have a cheap cost, which introduces the challenging problem of optimizing a multi-objective problems especially when there are conflicting objectives. The rest of the paper includes a background in the coming section, followed by challenges section that discusses the challenges facing this research and then a section for the proposed model architecture followed by the results then a section discussing the results and finally the conclusion.

2 Background

2.1 Genetic Algorithms

Genetic algorithms belong to the family of evolutionary algorithms. A GA is considered a search heuristic that mimics the process of natural evolution where the fittest individuals are selected to produce the offspring of the new generation [3]. A genetic algorithm is a mimetic technique that targets optimization problems. A candidate solution is referred to as a chromosome also can be referred to as an individual. Such individual consists of genes that can also be referred to as features. Each gene defines one feature of the candidate solution. The chromosome structure is a derived from the problem encoding. Problem encoding is a process where the algorithm designer encodes the targeted optimization problem in a way that will enable the GA to operate on. An objective function is then designed to measure the fitness of a candidate solution or a chromosome to specify how good a chromosome is as a candidate solution to the targeted optimization problem. After a successful problem encoding the chromosome is designed, an initial randomly generated population of individuals is created [4].

After then, an iterative procedure commences to evaluate the fitness of each chromosome and searches for the best possible solution in the form of a chromosome that will achieve the objective function's goal. Based on the evaluation results, one or more chromosomes will survive to the next generation, while the rejected chromosomes will be dismissed. And finally, the survived individuals will go through the mating process by applying crossover and mutation genetic operators. The crossover operator commences between two chromosomes where both share genes to form new chromosome according to a previously defined probability (Pc). Then mutation follows for one or more genes with predefined probability of mutation (Pm). Once fitness evaluation, crossover and mutation steps are completed, a new generation (offspring) is now ready to replace the old population and become the main population where the iterative process will start all over again to create a new generation and so on until a termination criterion is reached [5]. The basic flow of a genetic algorithm is shown in Fig. 1.

2.2 Particle Swarm Optimization

Sharing information among members of a fish school leads to intelligent behavior, from this general belief, computer scientists introduced the particle swarm optimization algorithm (PSO). The PSO is inspired from the instinctual behavior of fish school members that synchronize with each other with great precision to perform intelligent behavior without colliding. This behavior has been studied in various researches [6, 7]. The particle swarm optimization algorithm is considered a member of the wide category of swarm intelligence techniques [8] that was introduced in 1995 as a method for optimization as it mimics the social behavior of swarms.

The PSO was also applied with success in various optimization problems like function optimization and training of neural networks [9]. The PSO is considered a computationally inexpensive optimization algorithm as its system requirements are low [10]. The PSO is a population-based algorithm that searches for the optimal value of a targeted objective function. A population of candidate solutions are the backbone of the PSO that

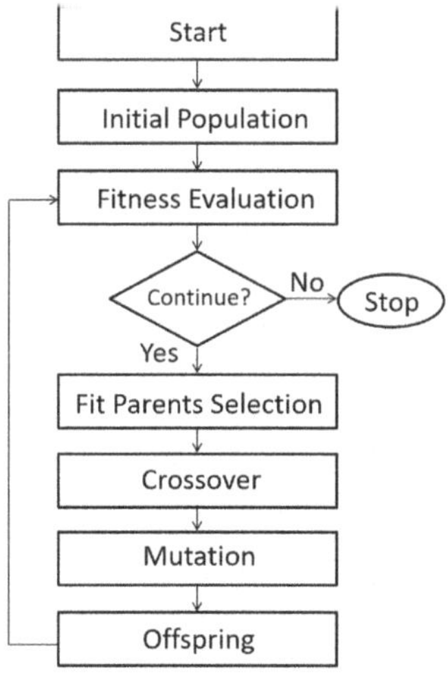

Fig. 1. Genetic algorithm operation steps

consists of particles where each particle is approached as a possible solution. Particles in the population are metaphor of bird in bird flocks or fish in fish schools. Particles are given initial random values and can traverse a permissible search domain.

During operation, each particle updates its own position and velocity depending on the best result reached so far as well as the best result reached so far by the entire swarm member particles. The continuous updating process will eventually guide all swarm members toward the area in the search space that have the best possible result. First, a population of particles is generated randomly and is initialized from a permissible range of values. Then all velocities of all swarm members are updated according to Eq. 1:

$$\vec{v}_i = w\vec{v}_i + c_1 R_1(\vec{p}_{i,best} - \vec{p}_i) + c_2 R_2(\vec{g}_{i,best} - \vec{p}_i) \tag{1}$$

Where $\vec{p}_i$ and $\vec{v}_i$ are the position and velocity of the particle i; $\vec{p}_{i,best}$ is the personal best of the particle i and $\vec{g}_{i,best}$ is the best value reached so far for the objective function from the entire population's perspective; w is the parameter that control the movement dynamics of a particle; R_1 and R_2 both are random variables with permissible domain of [0, 1]; c_1 and c_2 both are factors that manipulate the weighting of the corresponding term. The random variables grant PSO the possibility to perform random searching, while c_1 and c_2 are weighting factors that refer to the tradeoff between search space exploration and search space exploitation. As the updating process takes place, $\vec{v}_i$ is checked and maintained within the acceptable domain to eliminate stray random walking. Then PSO updates the position of its particles according to Eq. 2:

$$\vec{p}_i = \vec{p}_i + v_i \tag{2}$$

As particles positions are successfully updated, $\vec{p}_i$ is checked to be restricted to the permissible domain of values. Then the saved personal best and global best $\vec{p}_{i,best}$ and $\vec{g}_{i,best}$ are updated according to Eqs. 3 and 4:

$$\vec{p}_{i,best} = \vec{p}_i \quad \text{if } f(\vec{p}_i) > f(\vec{p}_{i,best}) \tag{3}$$

$$\vec{g}_{i,best} = \vec{g}_i \quad \text{if } f(\vec{g}_i) > f(\vec{g}_{i,best}) \tag{4}$$

$f(\vec{x})$ is the objective function. Finally, the algorithm loops through from step two to the step four until a termination criterion is reached. Then the algorithm presents the values of $\vec{g}_{i,best}$ and $f(\vec{g}_{i,best})$ as the solution to the optimization problem. Figure 2 presents the particle swarm optimization algorithm.

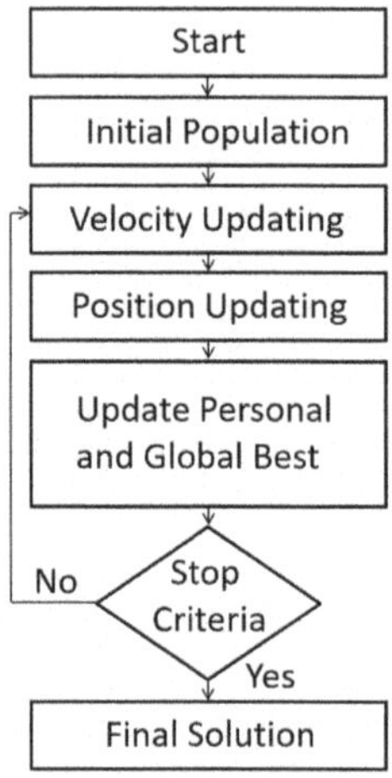

Fig. 2. Particle swarm optimization operation steps.

2.3 Hybrid Genetic Algorithms

Genetic algorithms (GAs) are a population based meta-heuristic search techniques. They simulate the process of natural evolution as they utilize the concepts of natural selection and genetic operators to solve optimization problems. The concept of genetic algorithms was first presented by Holland [10] and is discussed further with examples in [11]. Theoretically, GA's performance depends on balancing search space exploration as well as search space exploitation [1]. However, in reality there will be problems because Holland assumed that the population size is infinite, and the fitness function accurately reflects the suitability of a solution and the interactions between the genes are very small [10].

On the other hand, practically, the population size is finite which influences both performance as well as the sampling ability of the algorithm. Hybridization can eliminate most of the obstacles faced because of a finite population size, it is also responsible for the genetic drift problem [12] by introducing new genes. It may also accelerate the search process toward the global optimum [13]. The approaches in [14] have proposed that hybridization is a possible way to build efficient genetic algorithms.

2.4 The K-Means Clustering Algorithm

K-means is a partitioning based and non-hierarchical clustering algorithm [17]. It is one of the widely used algorithms that has been utilized in various science and technology fields [18, 19]. It is commonly used because it can be applied on different types of data. The K-means algorithm initially starts with several targeted numeric objects X and an integer number k. Then the algorithm starts an effort to partition all member objects of X into k groups/clusters while pushing the sum of squared errors to the minimum [20].

The algorithm randomly initializes k cluster centers, then attempts to allocate each member of X to its closest center according to the square of the Euclidean distance from the cluster [21]. Then the value of each center is updated by calculating the mean value of each cluster. The updating process results from the change of membership of cluster members [22]. Then the algorithm loops through to update cluster centers to reassign members of each cluster until no more changes in the membership is achieved. To following equation is used to calculate how near a data vector is to a cluster's center:

$$d\left(z_p,\ a_j\right) = \sqrt{\sum\nolimits_{k=1}^{d} \left(z_{pk} - a_{jk}\right)^2} \tag{5}$$

However, the final result of the k-means clustering algorithm depends on the quality of the randomly generated initial clusters centers. On the other hand, various techniques were proposed to overcome this problem [42].

2.5 Multi-objective Optimization Benchmark Functions

Benchmark functions are usually used to evaluate the efficiency of optimization algorithms [25]. Using such functions facilitates efficiency comparison between different optimization algorithms. Several multi-objective optimization benchmark functions are used in this research to evaluate the proposed methodology.

1) Binh and Korn function [36].
   Functions:

$$f_1(x, y) = 4x^2 + 4y^2 \tag{6}$$

$$f_2(x, y) = (x - 5)^2 + (y - 5)^2 \tag{7}$$

   Constrains:

$$g_1(x, y) = (x - 5)^2 + y^2 \le 25 \tag{8}$$

$$g_2(x, y) = (x - 8)^2 + (y + 3)^2 \ge 7.7 \tag{9}$$

   Search domain:

$$0 \le x \le 5.0 \le y \le 3 \tag{10}$$

2) Chakong and Haimes function [39].
Functions:

$$f_1(x, y) = 2 + (x - 2)^2 + (y - 1)^2 \tag{11}$$

$$f_2(x, y) = 9x + (y - 1)^2 \tag{12}$$

Constrains:

$$g_1(x, y) = x^2 + y^2 \leq 225 \tag{13}$$

$$g_2(x, y) = x - 3y + 10 \leq 0 \tag{14}$$

Search domain:

$$-20 \leq x, y \leq 20 \tag{15}$$

3) Constr-Ex Problem [37].
Functions:

$$f_1(x, y) = x \tag{16}$$

$$f_2(x) = \frac{1 + y}{x} \tag{17}$$

Constrains:

$$g_1(x, y) = y + 9x \ \geq 6 \tag{18}$$

$$g_2(x, y) = -y + 9x \ \geq 1 \tag{19}$$

Search domain:

$$0.1 \leq x \leq 1 \tag{20}$$

$$0 \leq y \leq 5 \tag{21}$$

4) Poloni's two objective [38].
Functions:

$$f_1(x, y) = \left[1 + (A_1 - B_1(x, y))^2 + (A_2 - B_2(x, y))^2\right] \tag{22}$$

$$f_2(x) = (x + 3)^2 + \ (y + 1)^2 \tag{23}$$

$$A_1 = 0.5 \sin(1) - 2 \cos(1) + \sin(2) - 1.5 \cos(2) \tag{24}$$

$$A_2 = 1.5 \sin(1) - \cos(1) + 2 \sin(1) - 0.5 \cos(2) \tag{25}$$

$$B_1(x, y) = 0.5 \sin(x) - 2 \cos(x) + \sin(y) - 1.5 \cos(y) \tag{26}$$

$$B_2(x, y) = 1.5 \sin(x) - \cos(x) + 2 \sin(y) - 0.5 \cos(y) \tag{27}$$

Search domain:

$$-\pi \leq x, y \leq \pi \tag{28}$$

3 Challenges

3.1 Challenges Facing Genetic Algorithms

A genetic algorithm is theoretically designed to achieve the balance between exploring and exploiting a search space. To explore a search space, the GA traverses through it looking for new solutions, while exploitation a search space is another concept where the GA attempts to benefit from all possible opportunities to get the best out of the searching process. It was assumed that a GA can achieve this perfect balance as assumed in [23] that "the population size is infinite and the fitness function accurately reflects the suitability of a solution and the gene interactions are minimum". However, practically, population size is finite which affects the performance as well as the sampling ability of the genetic algorithm. GA's behavior is also highly affected by the fitness function that chooses fit individuals to survive to the next generation, while dismissing individuals that do not pass the fitness evaluation, even if they have good genes. This is because the GA searches for high-quality chromosomes not high-quality genes. This behavior punishes individuals that may not pass the fitness evaluation but may contain good genes that can take the searching process to promising places in the search space.

3.2 Multi-objective Optimization Challenges

Optimization algorithms were utilized in a several fields including image processing [24, 25], industry [26–28], and manufacturing [29, 30]. Such algorithms also face a significant challenge in applied science [31–33], especially when they deal with multi-objective optimization problems (MOPs) [34, 35]. In such problems, the algorithm has to simultaneously consider all the objectives in the optimization process. A multi-objective optimization problem can formally be described as follows:

$$Minimize \quad F(x) = \left[f1(x), f2(x), \ldots, fm(x)\right]^T \tag{29}$$

$$s.t \quad \boldsymbol{x} = (\boldsymbol{X}_1, \boldsymbol{X}_2, \ \ldots, \boldsymbol{X}_n)^T \ \in \Omega \tag{30}$$

where $\boldsymbol{x}$ is the decision variable vector and Ω is the search space and $\boldsymbol{Rm}$ is the objective vector space. $\boldsymbol{F}(\boldsymbol{x})$ represents the objective vector with $\boldsymbol{m}$ real value objective functions. In multi-objective optimization problems, the existence of a relation between the two optimization functions will make it difficult to a single point in the search space to minimize/maximize all objective functions simultaneously. Hence, a multi-objective optimization problem is solved by searching for several promising points in the search domain where their objective functions evaluation values achieve a balanced minimum/maximum optimization value.

On the other hand, multi-objective evolutionary algorithms (MOEAs) utilize the evolution process to search for solutions to a multi-objective optimization problem. During this process, the algorithm performs many calculations as all individuals are evaluated in all generations. In some fields, the computation cost of an algorithm is critical. Therefore, it would be optimal to reduce the number of fitness evaluations and maximize the quality of solutions.

4 Proposed Model Architecture

4.1 Proposed Algorithm

The proposed algorithm utilizes GA search as the main search technique that will target a multi-objective benchmark function for optimization. The algorithm also utilizes PSO to perform the proposed concept of the rehabilitation of individuals rejected by the GA. Furthermore, the proposed algorithm K-means clustering to enhance the selection process by ensuring fair distribution in both selected to survive and selected for rehabilitation individuals. We assume that there is a chance that the rejected chromosomes contain good genes that may lead the searching process to places in the search domain where promising results could be found. Accordingly, the rejected chromosomes are transferred to the PSO algorithm for rehabilitation, where all the rejected non-fit chromosomes will form the population of the PSO where they will communicate with each other to update both their velocity and position to reach the best possible outcome from these non-fit chromosomes. The proposed model is as shown in Fig. 3.

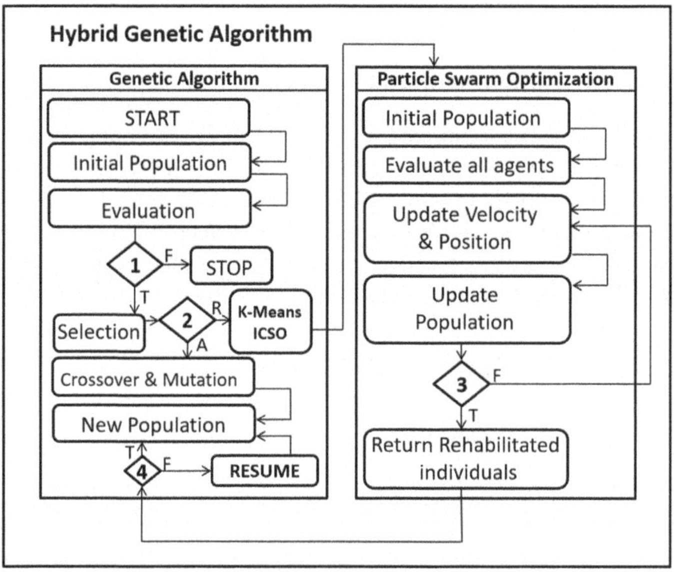

Fig. 3. Proposed model

In Fig. 3, condition one evaluates a chromosome and checks the stopping criterion which is the maximum number of generations; if such criterion is reached, the algorithm will stop (F), otherwise it will continue (T). Condition two is considered as a two-way selection mechanism, where on one side selects the fittest chromosomes to survive to next generation (A). On the other side, and with the support of the enhanced k-means + ICSO algorithm, it splits the rejected chromosomes into k clusters where k is the number of optimization functions in a given multi-objective optimization problem to fairly transfer a balanced group of rejected individuals to the PSO for rehabilitation

(R). Condition 3 checks for the PSO stopping criteria, that is also a maximum number of iterations, where it will either continue looping through the PSO (F) or return the rehabilitated individuals into the GA's population (T). Condition 4 either injects the incoming individuals into the upcoming population of the GA (T) or force the GA to resume (F) if the GA has already stopped for an additional pre-configured number of iterations (for example 5000) iterations.

Proposed algorithm (main part):
Input: BenchMarkFunction(X, Y): The targeted benchmark function, IterMax: maximum number of iterations.
Output: List <Individual> ParetoSet.
Variables: Individual: Array of X and Y, GAPopulation: List of individual, Pm: probability of mutation, Pc: probability of crossover, i: individual#, j: iteration#.
Process:

```
Begin
        Initialize GAPopulation = GetRandomPopulation()
        Foreach (Individual INDi  in GAPopulation)
            Calculate: BenchMarkFunction(INDi.X, INDi.Y)
        End Foreach
        Foreach (Individual INDi  in GAPopulation)
            If(GetRandom() >= Pc)
                CrossOver(INDi)
            End If
            Foreach( Gene Gj in INDi)
                If(GetRandom() > Pm)
                    Mutate(Gj)
                End If
            End Foreach
        Separate(GAPopulation, out Selected, out Rejected)
        Rehabiliatate = EKMeansSelection(Rejected)
        OffSpring = Selected
        If(Online)
            NewThread(PSORehabilitation(Rehabiliatate))
        Else
            PSORehabilitation(Rehabiliatate)
        Output = GetParetoSet(OffSpring)
        End Foreach
        If(MaxGen)
            Return Output
        Else
            Continue
END
```

This is the initial and main component of the proposed algorithm. GetRandomPopulation() is a subroutine used to generate initial individuals with random values of X and Y. BenchMarkFunction() is a subroutine delegating the targeted multi-objective benchmak test function. GetRandom() a function used to get random value to be compared against Pm and Pc. CrossOver (Individual) is a subroutine that applies a fixed point crossover on a targeted individual. Mutate (Gene) is a subroutine that applies the mutation process on a targeted gene. Separate (GAPopulation, out Selected, out rejected) is a subroutine that scans the current population and returns the selected individuals and the rejected individuals. EKMeansSelection (Rejected) is a subroutine that applies clustering where K = the chromosome length to assure a balanced distribution of the rejected individuals that are selected for rehabilitation. PSO Rehabilitation (Rehabilitate) is a subroutine that applies the rehabilitation process according to the condition of the main GA either online or offline. GetParetoSet (OffSpring) is a subroutine that gets the solution pareto set from the evaluated offspring.

Proposed algorithm (PSO):
Input: List <Individual>: rejected individuals from GA.
Output: Rehabilitated Individual.
Variables: V: Velocity, P: Position, i: individual#, j: iteration#. GB: global best, PB: personal best.
Process:

```
Begin
        Initialize PSO.Population = Input
         Foreach (Individual INDi  in Input)
            Calculate: V(INDij) &  PB( INDij)
            If(PB(INDij) < GB)
              GB = INDij
            Else
             If(j =Iteration Limit)
              Output =  GB & break
             Else
              Calculate: P(INDij) & V(INDij)
              i++ & j++
              Continue
         End Foreach
        Return Output
END
```

The algorithm above describes the process of rehabilitation using the PSO algorithm where Output represents the global best achieved by all particles.

1) *Problem Encoding and solution Decoding.*
   The proposed methodology targets a set of benchmark multi-objective optimization test functions with objective functions that require two inputs. As a result, the problem encoding will produce a chromosome structure that consists of two genes (one for each input) as shown in Fig. 4.

I1	I2

Fig. 4. Chromosome structure

- *Encoding:*
  All four targeted benchmark functions share the same specification of having two inputs I1 and I2 with two objective functions F1 and F2. Both I1 and I2 stand for (X, Y) coordinates of a point in a constrained search space, where the proposed methodology is expected to find the point with the minimum value for both objective functions in the search space of each targeted benchmark function. The proposed algorithm searches for I1 and I2 and for each I1 and I2 the value of both objective functions is calculated. As a result, we encode a candidate solution of each targeted benchmark function in the form of a chromosome of I1 and I2 as shown in Fig. 4.
- *Decoding:*
  As the test subjects of our research are multi-objective optimization benchmark test functions, it will not be of real business value to translate the results of these benchmark test functions by decoding the I1 and I2 value, as the encoded input of the targeted benchmark test functions represent X and Y coordinates where the proposed algorithm is supposed to find I1 and I2 that will minimize both F1 and F2, furthermore, the targeted multi-objective benchmark test functions were designed in such a way that F1 and F2 are conflicting such that, generally minimizing F1 will maximize F2 and vice versa. As a result, decoding the chromosome of I1 and I2 will lead to nothing but the X value and Y coordinates values that the proposed algorithm is searching for in each search space of the targeted benchmark test functions.

2) *Population Specifications*
   A group of n candidate solutions is randomly generated to initialize the algorithm. n is a pre-determined value indicating the size of a population. During the execution phase, the algorithm is tested against different values of n to test the effect of changing population size on the algorithm's output.
3) *Genetic Operators*
   The genetic operators are a primary part of the proposed methodology as they are utilized to mimic the process of natural evolution. Genetic operators are crossover, mutation, evaluation, and selection. As soon as the problem encoding is done properly, the algorithm can then apply these operators on chromosomes searching for the best possible solution.
   a. Crossover
   The chromosome structure is the same in the entire benchmark test functions used in this research; as a result, a single point crossover will be used in all test cases so that individuals may share genes as a form of knowledge sharing toward the search for the optimal result.
   b. Mutation
   The proposed algorithm applies gene mutation in a chromosome according to the probability of mutation Pm. However, in a constrained search domain, the algorithm is designed to mutate the genes within a permissible range of values.

c. Evaluation
For each generation, the proposed algorithm evaluates all candidate solutions to assign a fitness score for each one. In this research, we focus on multi-objective optimization problems. Each of the targeted benchmark test functions have two objective functions. As a result, the fitness of each individual will be represented by two values each of them represents one of the fitness functions.
d. Selection
In this research, elitism selection is utilized where a set of the top performing individuals is selected to survive to the next generation. Chromosomes that are fitter can replace the elite individuals in order, else, such group survives through all generations unchanged as far as no better individuals are found. Furthermore, rejected individuals are clustered by the K-means algorithm supported by ICSO according to their fitness evaluation. Such step assures a more balanced distribution of individuals passed to PSO for rehabilitation.

4) *K-Means Clustering*
The K-means clustering algorithm is utilized to support the individuals filtering process in the selection phase of the proposed algorithms. It is viewed as an enhanced selection technique that gathers all individuals that did not pass the fitness evaluation criterion, cluster them in to $K = 2$ clusters, then passes a balanced group of rejected individuals to the PSO while assuring the existence of all unique individuals without losing lone individuals in an offspring as shown in Fig. 5.
Figure 5 represents hypothetical case of individuals with values to show the selection mechanism of the K-Means based selection method. After the algorithm rejects individuals in a generation, their evaluation values are clustered by the K-Means algorithm to produce K clusters where K = the chromosome length. According to the clustering result, the algorithm selects a group of individuals for rehabilitation asserting the selection of at least one individual from each cluster, hence, asserting the existence of all unique chromosomes. The PSO will then operate on them and return a better individual.

5) *ICSO*
Furthermore, the K-Means operation mechanism was also targeted for enhancement. The K-Means final clustering result is very sensitive to the selection of initial centroids. Initial centroid selection optimization ICSO enables the K-Means clustering algorithm to select optimized initial centroids that will be balanced to assure best possible clustering results. Figure 6 shows the operation of ICSO [42].

6) *Stopping Criterion*
The stopping criterion is set to a maximum number of generations that was set to only 100 unlike the 10000 in our previous research.

7) *PSO Integration*
Individuals that failed to pass the fitness function are then clustered and passed to the PSO algorithm for rehabilitation. The passed individuals will act as particles in a swarm where they will communicate updating both their velocity and position. As soon as all particles agree on the best solutions, this solution is passed back to the GA's offspring replacing the least fit offspring member(s) in case the algorithm did not terminate. In case the algorithm has already terminated then the PSO will force it to continue and if no better result is reached for a predefined number of iterations,

the algorithm will finally terminate introducing a group of solutions called a pareto set.

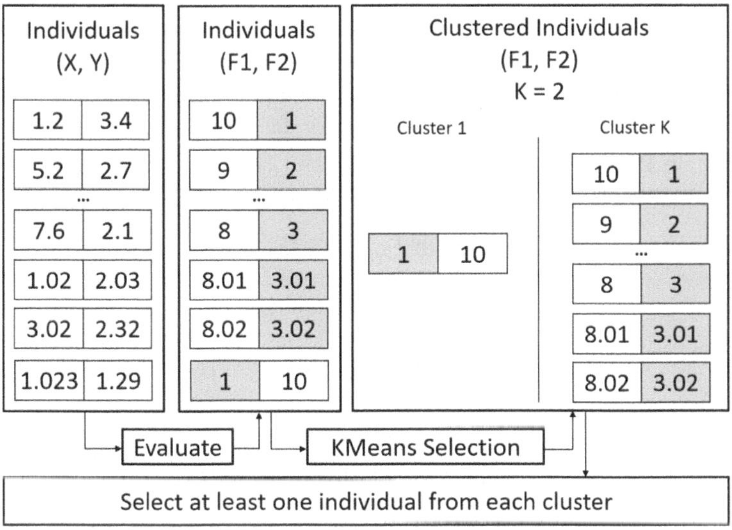

Fig. 5. KMeans selection.

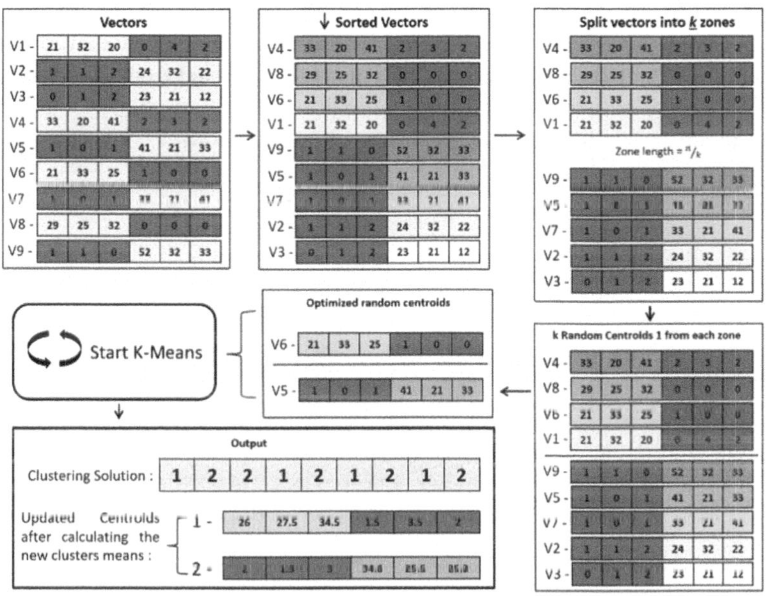

Fig. 6. Initial centroid selection optimization [42].

5 Results

The proposed algorithm was tested against four benchmark functions: Binh and Korn [35], Chakong and Haimes [38], Constr-Ex Problem [36], and Poloni's two objective functions [37]. The algorithm execution took place on a 2.6 GHz Intel Core i7 vPro machine with 16 GB of RAM and a magnetic HDD (10000 rpm) on a 64-bit OS. Test cases executed in this research are as described in Table 1 where the used algorithm is configured as either the genetic algorithm (GA) or the hybrid genetic algorithm (HGA). PS specifies the population size.

Table 1. Test cases.

#	Benchmark function	Algorithm	PS
1	Binh and Korn	HGA	1M
2	Chakong and Haimes	HGA	1M
3	Constr-Ex Problem	HGA	1M
4	Poloni's Two Objective	HGA	1M

All test cases have been executed and the results are noted in detail in Tables 2, 3, 4 and 5. Each table(s) introduces the pareto set as extracted from each test case. A pareto set is a group of the highest performing solutions to a multi-objective optimization problem. In this research the pareto set size is set to 7 solutions.

Table 2. Pareto set of test case 1.

#	X	Y	F1	F2	ITR
1	4.85E−06	4.85E−06	1.87989E−10	49.99990305	45
2	6.97E−05	5.78E−05	3.27784E−08	49.99872531	3
3	5.78E−05	0.000489582	9.7213E−07	49.99452635	2
4	6.97E−05	0.002153637	1.8572E−05	49.97777164	1
5	4.985165417	2.999851323	135.4039288	4.000814794	10
6	4.992365807	2.999851323	135.6912972	4.00065301	18
7	4.999969306	2.999851323	135.9952041	4.00059473	28

Table 3. Pareto set of test case 2.

#	X	Y	F1	F2	ITR
1	1.114051282	3.704687307	10.10023856	2.711128107	69
2	1.113019791	3.704687307	10.10206732	2.701844688	5
3	1.056422424	3.685484861	10.10216758	2.295972879	3
4	1.135449303	3.711985687	10.10231427	2.864177362	1
5	0.015713349	14.97881503	201.344663	−195.2658494	4
6	0.015746523	14.98199905	201.4335592	−195.3545786	12
7	3.36E−05	14.99748025	201.929319	−195.9291509	72

Table 4. Pareto set of test case 3.

#	X	Y	F1	F2	ITR
1	0.389387389	2.504181518	0.389387389	8.999216766	97
2	0.389441145	2.504181518	0.389441145	8.997974572	6
3	0.389456027	2.504181518	0.389456027	8.997630726	1
4	0.999819395	0.000593872	0.999819395	1.000774617	8
5	0.999976984	0.000593872	0.999976984	1.000616902	7
6	0.999992802	0.000593872	0.999992802	1.000601074	3
7	0.999995596	4.651956E−06	0.999995596	1.000009056	84

Table 5. Pareto set of test case 4.

#	X	Y	F1	F2	ITR
1	2.022739664	0.730736764	1.000000002	28.22336347	56
2	2.022638336	0.730791302	1.000000021	28.22253438	3
3	2.022767003	0.730791302	1.000000034	28.22382689	2
4	1.000713567	1.996122669	1.00001216	24.98246009	1
5	0.000506802	5.35E−05	38.16911445	10.00314802	7
6	2.78E−06	5.35E−05	38.1777455	10.00012363	42
7	2.78E−06	2.78E−06	38.1790504	10.00002224	47

Figure 7 shows a comparison between the minimization ability of the previous model and the enhanced model regarding Binh&Korn Function 1.

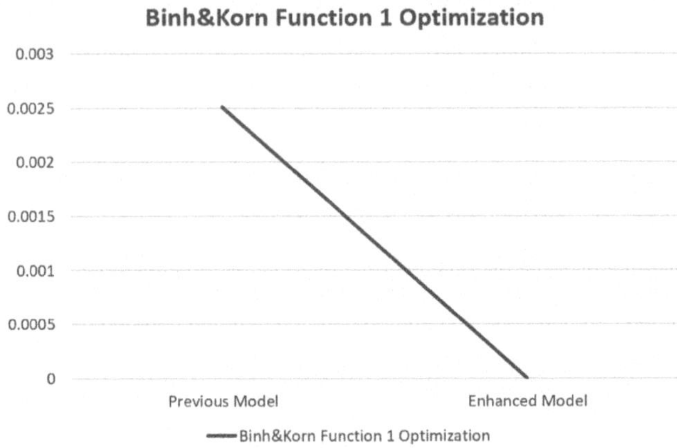

Fig. 7. Binh&Korn F1 Optimization Comparison.

Figure 8 shows a comparison between the minimization ability of the previous model and the enhanced model regarding Binh&Korn Function 2.

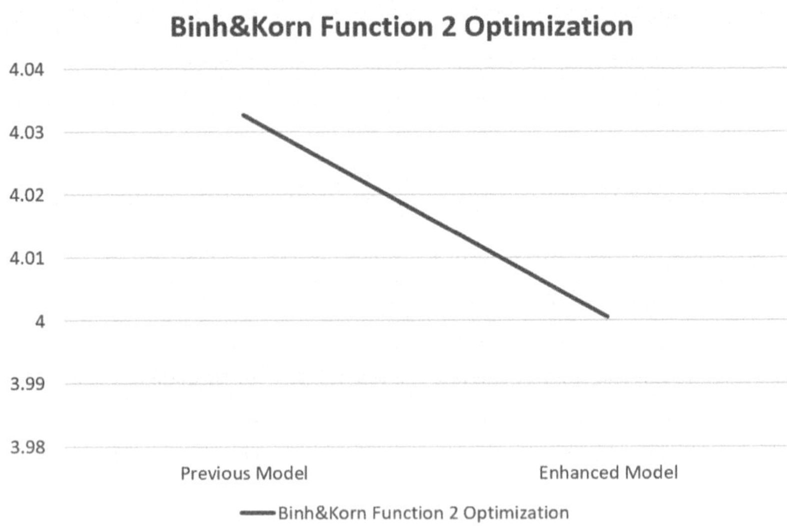

Fig. 8. Binh&Korn F2 optimization comparison.

Figure 9 shows a comparison between the minimization ability of the previous model and the enhanced model regarding Chakong&Haimes Function 1.

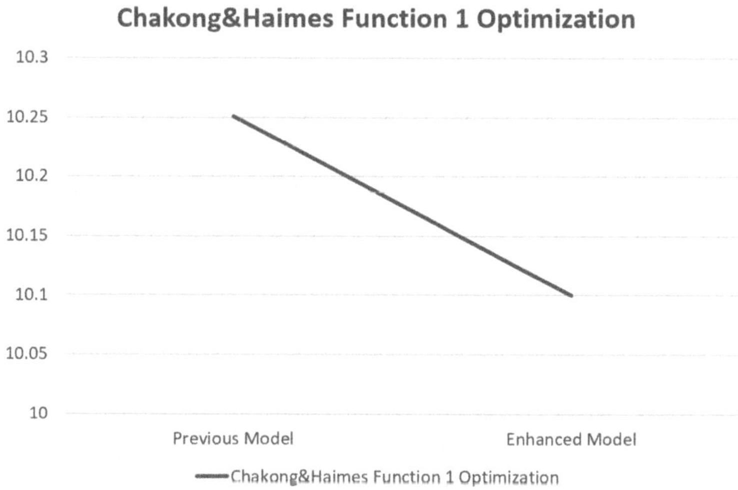

Fig. 9. Chakong&Haimes F1 optimization comparison.

Figure 10 shows a comparison between the minimization ability of the previous model and the enhanced model regarding Chakong&Haimes Function 2.

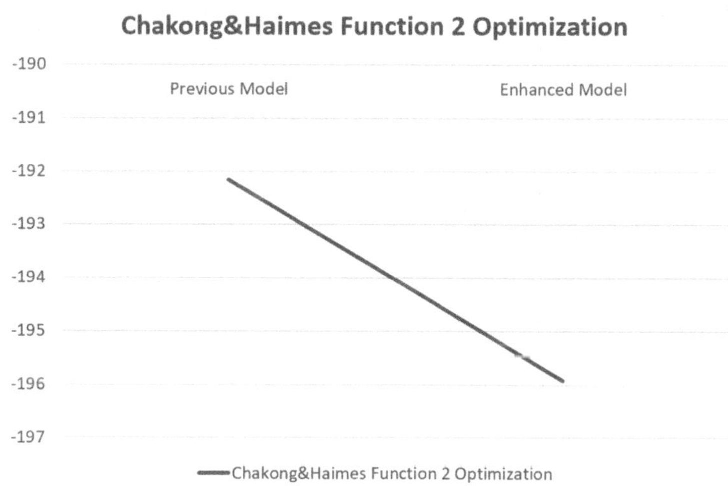

Fig. 10. Chakong&Haimes F2 optimization comparison.

Figure 11 shows a comparison between the minimization ability of the previous model and the enhanced model regarding ConstrEx Function 1.

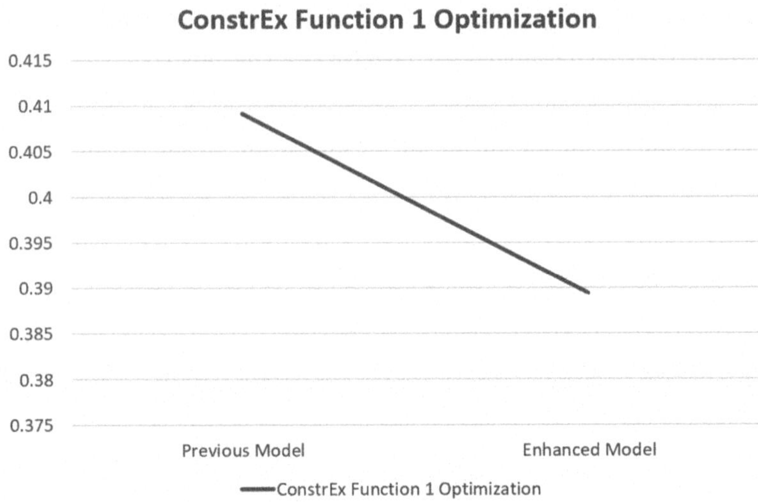

Fig. 11. Constr-Ex F1 optimization comparison.

Figure 12 shows a comparison between the minimization ability of the previous model and the enhanced model regarding ConstrEx Function 2.

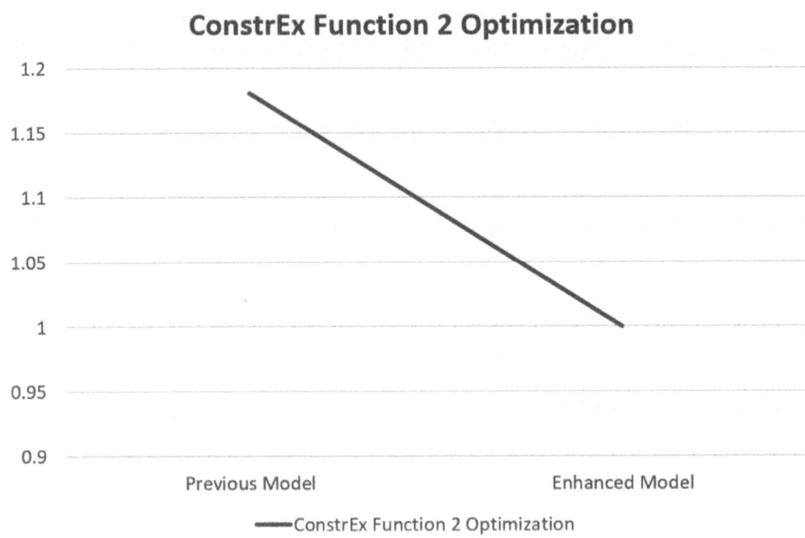

Fig. 12. Constr-Ex F2 optimization comparison.

Figure 13 shows a comparison between the minimization ability of the previous model and the enhanced model regarding Poloni's Two Objective Function 1.

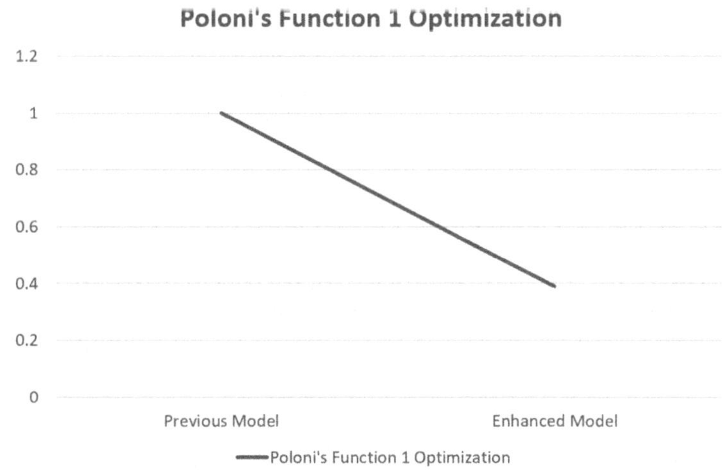

Fig. 13. Poloni's F1 optimization comparison.

Figure 14 shows a comparison between the minimization ability of the previous model and the enhanced model regarding Poloni's Two Objective Function 2.

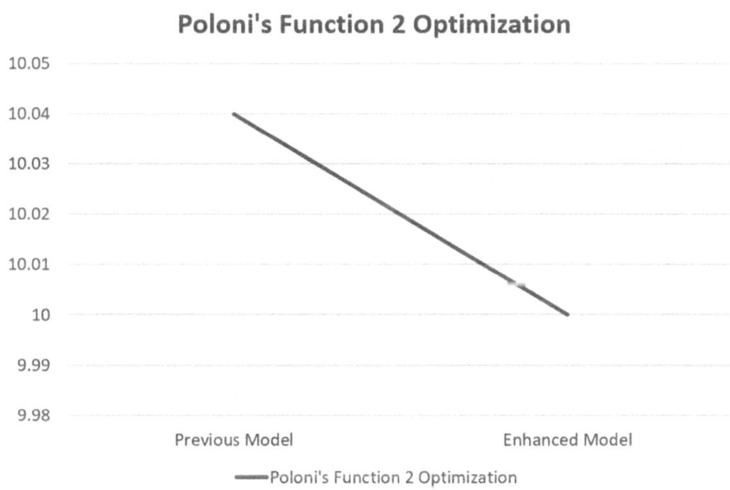

Fig. 14. Poloni's F2 optimization comparison.

6 Discussion of Results

The proposed enhanced HGA was tested against four benchmark functions and compared against previous results [43] that resulted in a total of 4 test cases. All benchmark functions used in this research share the same characteristic of having 2 objective functions F1 and F2, where each was targeted with the enhanced HGA with 100 iterations and 1 million population size.

6.1 Results Analysis

From the previous research [43] it was obvious that the population size has a great impact in the optimization ability of the proposed model there. Also, the initial centroid selection phase of the K-means based selection operator suffered from sub-optimal clustering tendency because the overall clustering result highly depended on the random selection of initial centroids. As a result, in this research the K-means selection operator was modified by ICSO [42] and the population size was increased to 1 million individual per generation and to balance the performance, the maximum number of iterations was decreased to only 100 iteration. Results from test cases 1, 2, 3 and 4 when compared to the best results achieved by the previous model [43] showed great enhancement in the optimization ability of the enhanced model.

6.2 Complexity

The proposed enhanced hybrid algorithm utilizes two evolutionary algorithms as well as the enhanced K-Means clustering based selection with overall 5 loop structures as assumed below:

(n): GA maximum number of iterations.
(x): GA Population size.
(y): K-Means maximum number of iterations.
(c): ICSO optimization.
(z): PSO maximum number of iterations.
(m): PSO population size.
The combined complexity will be as follows:

$$O(n) \times \left(O(x) + O\left(cy^2\right)\right) + O(z) \times O(m) \tag{31}$$

That can be further calculated as:

$$O(nx) + O\left(ncy^2\right) + O(zm) \tag{32}$$

Where $O(nx)$ is the the GA's maximum number of iterations by the process of genetic operators, while $O\left(ncy^2\right)$ is the GA's maximum number of generations by the K-Means process by the ICSO operations and finally $O(zm)$ is the PSO's population size by the PSO's velocity and position process. As a result, better optimization was successfully achieved despite a trade off in performance and resource consumption.

6.3 Contributions

The proposed technique utilizes a genetic algorithm that targets search space exploration supported by the K-Means algorithm to enhance the selection mechanism as described in Sect. 4 – 4. The particle swarm optimization algorithm was also utilized to target the rejected individuals of each generation to fulfill the concept of the rehabilitation of rejected individuals to maximize the utilization of all individuals in each generation. To test the effect of each component of the proposed hybrid algorithm, the hybrid algorithm was tested against 4 benchmark functions under several configurations that resulted in 16 different test cases where their results were finely analyzed in Sect. 6 – 1.

From this analysis it was observed that the proposed K-Means based selection mechanism enhanced the optimization ability of the genetic algorithm however added more computational cost to the GA. It was also observed that the PSO further enhanced the optimization ability of the hybrid algorithm, however, also added to the computational complexity of the hybrid algorithm. From Sect. 6-2 it was also observed that the overall optimization process is enhanced significantly, however, and as a tradeoff to this enhancement, more computational cost was added compared to a basic GA.

7 Conclusion

In this research, an enhanced hybrid genetic algorithm was proposed to target multi-objective optimization problems. The enhanced hybrid genetic algorithm utilized the PSO in addition to the ICSO supported K-means algorithm to solve multi-objective optimization problems. Four benchmark multi-objective optimization test functions were used to evaluate the proposed enhanced HGA. Practically, the GA was utilized to achieve search space exploration supported by the ICSO enhanced K-Means algorithm to further enhance the selection operator of the GA. On the other hand, the PSO was utilized to achieve search space exploitation. The proposed enhanced model was put to test against four benchmark multi-objective optimization functions with fixed maximum iterations of 100 and population size of 1 million individual per generation the lead to 4 different test cases as shown in Table 1. All 4 test cases where executed and the results where noted in Tables 2, 3, 4 and 5. From the results we can conclude that the proposed enhanced HGA has successfully achieved a better optimization results by minimizing the objective functions in all test cases that achieved values lower than the previous model [43].

Compliance with Ethical Standards. This article does not contain any studies with human participants or animals performed by any of the authors.

Conflict of Interest. Author Ahmed Maghawry declares that he has no conflict of interest. Author Rania Hodhod declares that she has no conflict of interest. Author Yasser Omar declares that he has no conflict of interest. Author Mohamed Kholief declares that he has no conflict of interest.

References

1. Luo, J., He, F., Yong, J.: An efficient and robust bat algorithm with fusion of opposition-based learning and whale optimization algorithm. Intell. Data Anal. **24**(3), 581–606 (2020)
2. Li, H.-R., He, F.-Z., Yan, X.: IBEA-SVM: an indicatorased evolutionary algorithm based on preelection with classification guided by SVM. Appl. Math.-A J. Chin. Univ. **34**(1), 1–26 (2019). https://doi.org/10.1007/s11766-019-3706-1

3. Durairaj, M., Dhanavel, C.: A survey on cloud service scheduling using genetic algorithm. Int. J. Comput. Sci. Eng. **6**(6), 1201–1207 (2018)
4. Kaur, G., Aggarwal, S.: A survey of genetic algorithm for association rule mining. Int. J. Comput. Appl. **67**(20), 25–28 (2013)
5. Chen, J., Zhang, D., Liu, D., Pan, Z.: A network selection algorithm based on improved genetic algorithm. In: 2018 IEEE 18th International Conference on Communication Technology (ICCT) (2018)
6. Heppener, F., Grenander, U.: A stochastic nonlinear model for coordinate bird flocks. In: Krasner, S. (ed.) The Ubiquity of Chaos. AAAS Publications, USA (1990)
7. Reynolds, C.W.: Flocks, herds and schools: a distributed behavioral model. ACM SIGGRAPH Comput. Graph. **21**(4), 25–34 (1987)
8. Prado, R.P., García-Galán, S., Yuste, A., Munoz-Expósito, J.E.: Genetic fuzzy rule-based meta-scheduler for grid computing. In: Proceedings of the 4th International Workshop on Genetic and Evolutionary Fuzzy Systems. IEEE Computational Intelligence Society, Piscataway (2010)
9. Rakitianskaia, A., Engelbrecht, A.: Weight regularisation in particle swarm optimisation neural network training. In: 2014 IEEE Symposium on Swarm Intelligence (2014)
10. Holland, J.: Adaptation in Natural and Artificial Systems. The University of Michigan (1975)
11. De Jong, K.: An analysis of the behavior of a class of genetic adaptive systems, Doctoral Dissertation. The University of Michigan, Ann Arbor (1975)
12. Goldberg, D.E.: Genetic Algorithms in Search, Optimization, and Machine Learning. Addison-Wesley, Boston (1989)
13. Beasley, D., Bull, D.R., Martin, R.: An overview of genetic algorithms: part 1, fundamentals, University Computing, vol. 15, pp. 58–69 (1993)
14. Asoh, H., Mühlenbein, H.: On the mean convergence time of evolutionary algorithms without selection and mutation. In: Davidor, Y., Schwefel, H.-P., Männer, R. (eds.) PPSN 1994. LNCS, vol. 866, pp. 88–97. Springer, Heidelberg (1994). https://doi.org/10.1007/3-540-58484-6_253
15. Hart, W.E.: Adaptive global optimization with local search. Doctoral Dissertation. University of California, San Diego (1994)
16. Goldberg, D.: The race, the hurdle, and the sweet spot: lessons from genetic algorithms for the automation of design innovation and creativity. In: Evolutionary Design by Computers, pp. 105–118. Morgan Kaufmann (1999)
17. Survey report on K-means clustering algorithm. Int. J. Mod. Trends Eng. Res. **4**, 218–221 (2017). https://doi.org/10.21884/ijmter.2017.4143.lgjzd
18. Xu, R., Wunschii, D.: Survey of clustering algorithms. IEEE Trans. Neural Netw. **16**, 645–678 (2005). https://doi.org/10.1109/tnn.2005.845141
19. An Introduction to Classification and Clustering. Cluster Analysis Wiley Series in Probability and Statistics, pp. 1–13 (2011). https://doi.org/10.1002/9780470977811.ch1
20. Hamerly, G., Drake, J.: Accelerating Lloyd's algorithm for k-means clustering. In: Celebi, M.E. (ed.) Partitional Clustering Algorithms, pp. 41–78. Springer, Cham (2015). https://doi.org/10.1007/978-3-319-09259-1_2
21. Shrivastava, P., Kavita, P., Singh, S., Shukla, M.: Comparative analysis in between the k-means algorithm, k-means using with Gaussian mixture model and fuzzy c means algorithm. Commun. Comput. Syst. (2016). https://doi.org/10.1201/9781315364094-186
22. Lei, X.-F.: An efficient clustering algorithm based on local optimality of K-means. J. Softw. **19**, 1683–1692 (2008). https://doi.org/10.3724/sp.j.1001.2008.01683
23. Beasley, D., Bull, D.R., Martin, R.: An overview of genetic algorithms: part 1, fundamentals. University Computing, vol. 15, pp. 58–69 (1993)
24. Chen, X., He, F., Yu, H.: A matting method based on full feature coverage, multimedia tools and applications. https://doi.org/10.1007/s11042-018-66901.

25. Zitzler, E., Künzli, S.: Indicator-based selection in multiobjective search. In: Yao, X., et al. (eds.) PPSN 2004. LNCS, vol. 3242, pp. 832–842. Springer, Heidelberg (2004). https://doi.org/10.1007/978-3-540-30217-9_84
26. Li, W., Mcmahon, C.: A simulated annealing-based optimization approach for integrated process planning and scheduling. Int. J. Comput. Integr. Manuf. **20**(1), 80–95 (2007)
27. Lin, X., Zhang, Q., Kwongs, S.: A decomposition based multiobjective evolutionary algorithm with classification. In: IEEE Congress on Evolutionary Computation, pp 3292–3299. IEEE (2016)
28. Zhu, H., Zhou, M.: Role-based collaboration and its kernel mechanisms. IEEE Trans Syst. Man Cybern. Part C: Appl. Rev. **36**(4), 578–589 (2006)
29. Gui, W.-H., Zhang, H.-N.: Asymptotic properties and expectation-maximization algorithm for maximum likelihood estimates of the parameters from Weibull-Logarithmic model. Appl. Math.-A J. Chin. Univ. **31**(4), 425–438 (2016). https://doi.org/10.1007/s11766-016-3391-2
30. Zhang, D.J., He, F.Z., Han, S.H., et al.: Quantitative optimization of interoperability during feature-based data exchange. Integr. Comput.-Aided Eng. **23**(1), 31–50 (2016)
31. Kim, W., Xiong, S., Liang, Z.: Effect of loading symbol of online video on perception of waiting time. Int. J. Human–Comput. Interact. **33**(12), 1001–1009 (2017)
32. Ni, B., He, F.-Z., Pan, Y.-T., Yuan, Z.-Y.: Using shapes correlation for active contour segmentation of uterine fibroid ultrasound images in computer-aided therapy. Appl. Math.-A J. Chin. Univ. **31**(1), 37–52 (2016). https://doi.org/10.1007/s11766-016-3340-0
33. Qian, T., Zhang, L.-M.: Mathematical theory of signal analysis vs. complex analysis method of harmonic analysis. Appl. Math.-A J. Chin. Universities **28**(4), 505–530 (2013). https://doi.org/10.1007/s11766-013-3225-4
34. Bandaru, S., Ng, A., Deb, K.: On the performance of classification algorithms for learning pareto-dominance relations. In: Evolutionary Computation (CEC), pp 1139–1146. IEEE (2014)
35. Coello, C.: Twenty years of evolutionary multi-objective optimization: a historical view of the field. IEEE Comput. Intell. Mag. **1**(1), 28–36 (2006)
36. Binh, T., Korn, U.: MOBES: a multiobjective evolution strategy for constrained optimization problems. In: Proceedings of the Third International Conference on Genetic Algorithms, Czech Republic, pp. 176–182 (1997)
37. Han, Z., Ning, C., Wei, Y.: MOPSO for BIM: a multi-objective optimization tool using particle swarm optimization algorithm on a BIM based visual programming platform. "Hello, Culture!". In: 18th International Conference, CAAD Futures, pp. 39–51 (2019)
38. Poloni, C.: Hybrid GA for multiobjective aerodynamic shape optimization. In: Winter, G., Periaux, J., Galan, M., Cuesta, P. (eds.) Genetic Algorithms in Engineering and Computer Science, pp. 397–414. Wiley, New York (1997)
39. Chankong, V., Haimes, Y.Y.: Multiobjective Decision Making: Theory and Methodology. North Holland, New York (1983)
40. Nopiah, Z., Khairir, M., Abdullah, S., Baharin, M., Arifin, A.: Time complexity analysis of the genetic algorithm clustering method, pp. 171–176 (2010)
41. Li, H., He, F., Liang, Y., Quan, Q.: A dividing-based many-objective evolutionary algorithm for large-scale feature selection. Soft. Comput. **24**, 1–20 (2019). https://doi.org/10.1007/s00500-019-04324-5
42. Maghawry, A.M., Omar, Y., Badr, A.: Initial centroid selection optimization for k-means with genetic algorithm to enhance clustering of transcribed arabic broadcast news documents. In: Silhavy, R., Silhavy, P., Prokopova, Z. (eds.) CoMeSySo 2017. AISC, vol. 662, pp. 86–101. Springer, Cham (2018). https://doi.org/10.1007/978-3-319-67621-0_8
43. Maghawry, A., Hodhod, R., Omar, Y., Kholief, M.: An approach for optimizing multi-objective problems using hybrid genetic algorithms. Soft. Comput. **25**(1), 389–405 (2020). https://doi.org/10.1007/s00500-020-05149-3

Numerical Modelling of the Impact of Dredging on Stability of Oum Rabiâ Estuary (Morocco) Using SWAN Model

Hassan Ayad[1](✉), Mustapha Mouhid[1], Dalila Loudyi[2], Laila Mouakkir[1], and Mohamed Chagdali[1]

[1] Polymer Physics and Critical Phenomena Laboratory, Faculty of Sciences Ben M'sik, University Hassan II, P.O. BOX 7955, Casablanca, Morocco
ayadhassan01@gmail.com

[2] Faculty of Sciences and Technologies Mohammedia, Hassan II University of Casablanca, Casablanca, Morocco

Abstract. The excess of sediments at the Oum Rabiâ estuary impacts the closures of the mouth of the river which directly influences the socio-economic activities in the region. For this reason, dredging is a key solution to avoid the excess of sediments. Further, the numerical modeling of wave propagation is necessary to study the wave's characteristics before operating in the dredging. The goal of this paper is revolved around discussing the impact of dredging on waves characteristic, which are done at the Oum Rabiâ estuary located on the Moroccan Atlantic coast near the city of Azemmour. The impact of the seabed depression on swell characteristics is analyzed using SWAN model to check the impact of the sea depression on waves characteristics when the depth of the depression is greater than the limit depth of action of the waves (closing depth), and also in the case of the depth of the dredging trench is close to the height of the waves.

Keywords: Dredging · Oum Rabiâ · Waves propagations · Atlantic coast · SWAN model

1 Introduction

The overflow of sediments (sand, silt, pebble…) is always impacting the closures of mouths of the rivers, which influence directly the environment and many socio-economic activities related to the existence and the life of the rivers [1, 2]. The extraction of marine sediments by dredging is a major key solution to avoid the excess of sediments. Further, the process of dredging needs a deep numerical study and modeling to avoid problems that can impact the environment, SWAN model is a numerical code developed to define the propagation of the swell in the coastal zone where the dredging will be done and makes it possible to highlight the phenomena to be taken into account when the swell is transferred from the sea to the coast and when it is propagated on the extraction zones.

The original version of this chapter was revised: The author name "Dalila Loudy" has been changed to "Dalila Loudyi". The correction to this chapter is available at https://doi.org/10.1007/978-3-030-77445-5_68

© The Author(s), under exclusive license to Springer Nature Switzerland AG 2021, corrected publication 2022
R. Silhavy (Ed.): CSOC 2021, LNNS 229, pp. 88–94, 2021.
https://doi.org/10.1007/978-3-030-77445-5_8

The dredging is an important environmental and economic activity in coastal and riverine systems around the world and can be carried out for the purposes of land reclamation, remediation of pollution, mining, and navigation [3, 4]. Dredging may influence the physical environment by modifying the bathymetry, altering current velocities and wave conditions [5] which influencing the sedimentary system and can cause erosion under seagrass beds [6]. Furthermore, dredging and resuspension can also influence the consistency of water by the release of pollutants [7].

A multiple geospatial dredging optimization methods, known as Dredging Management Unit-DMU, was developed by the US Army Corps of Engineering.[8], this was developed in a paper published by Wasserman [9]. In the preparation of the dredging activities, the technique was applied in such a way that the dredging areas and volumes were previously specified. Other approaches to dredging operations management procedures have been developed, applying Multi-Criteria Decision Analysis -MCDA [10–13].

The motivation of this work is revolved around studying the process of kinematics modification close to the bottom, and dredging trench on the stability of the coastline. Furthermore, this study focuses on analyzing the impact of a dredging trench on the propagation of the swell using the numerical third-generation wave model (SWAN) for the Azemmour site at the Atlantic coast of Morocco. This study requires knowledge of the swell in the coastal zone as well as its characteristics (Height, direction and wavelength) that are already available in our laboratory.

2 Problem Statement

The exact position of dredging is at the Oum Rabiâ estuary is located on the Moroccan Atlantic coast near the city of Azemmour. Its coordinates WGS84 are: Latitude 33° 17 16.01″ N Longitude 8° 20 31.99″ W (Fig. 1).

The development of the coastline between 1982 and 2016 shows a closure of the mouth of the river, this situation justifies the existence of a dredging project at the mouth to improve the circulation of water masses between the river and the ocean (Fig. 1). The motivation of our work is to analyze the impact of this dredging project on the stability of the coastline using the numerical third-generation wave model (SWAN).

3 Description of the Numerical Model

In this section, we describe the numerical model used for transporting data from offshore to the coast. This operation is done using SWAN (SWAN III. VERSION 41.20) model (Simulating Waves Nearshore) developed at the Delft University of Technology (The Netherlands) [14]. For sake of description, SWAN model is a third-generation spectral digital model that provides information of waves propagations by exploring data of wind, bathymetry and current at the zone studied. In the SWAN model [15], the dynamic spectrum balance equation is adopted based on the theory of linear and random surface gravity waves, the random waves are presented in two-dimensional dynamic spectral density rather than two-dimensional energy spectral density. Further, the dynamic spectral density $N(\sigma,\theta)$ is the ratio of the spectral energy density $E(\sigma, \theta)$ to the intrinsic

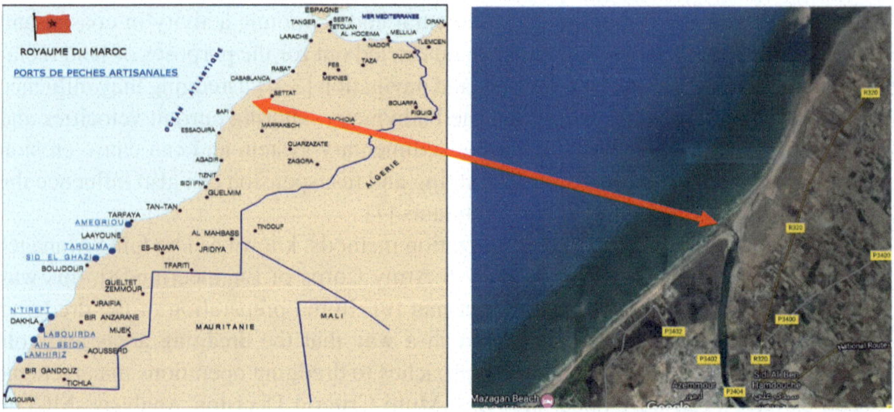

(a)

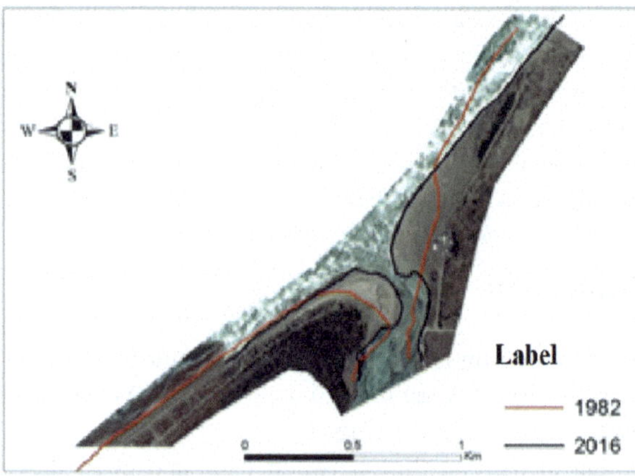

(b)

Fig. 1. (a) Satellite view of the Oum Rabiâ river mouth, (b) The Development of the coastline between 1982 and 2016.

representative wave frequencies. The equation is describing the wave motion within the SWAN model is written as:

$$\frac{\partial N}{\partial t} + \frac{\partial}{\partial x}(C_x N) + \frac{\partial}{\partial y}\left(C_y N\right) + \frac{\partial}{\partial \sigma}(C_\sigma N) + \frac{\partial}{\partial \theta}\left(C_{y\theta} N\right) = \frac{S_{en}}{\sigma} \quad (1)$$

where

$N(\sigma, \theta)$: Wave action density (=ρg.E (ρ, θ)/σ[J.s/m^2]);
$E(\sigma, \theta)$: spectral variance;
σ: relative angular frequency (relative to current)[Hz];
C_x, C_y, C_θ, C_ρ: propagation speeds according to *x*, *y* and θ and angular frequency;
S_{en}: term of energy sources [J/m^2].

where x and y are horizontal Cartesian coordinates, t is time, θ is the propagation direction of each wave component, C_x, C_y, C_σ, and C_θ are respectively the propagation velocity in x-space, y-space, σ-space, and θ-space, and Stott represents the energy source.

Our application in this coastal zone, we have used the following discretization directional resolution $\Delta\theta = 15° - 10°$, Frequency interval $F_{min} = 0.04$ Hz and $F_{max} = 1.00$, the implicit "Upwind" scheme was adopted. This scheme is implanted in the SWAN code in the SWAN model because of its following characteristics: Unconditionally stable; allows the use of fairly large time steps. Details of these processes can be found in the SWAN technical documentation manual [16].

4 Results and Discussion

To better analyze the impact of dredging on the equilibrium state of the coastline, we present in the Fig. 2 the impact of depression on the heights of the waves. Thereafter, we describe in the Fig. 3 the analysis of the heights of waves at the upstream and downstream of trench for −2 m and −5 m of depth. Next, we plot in the Fig. 4 the profiles of friction velocity at the upstream and downstream of the trench for 2 m of depth as versus of x-coordinates.

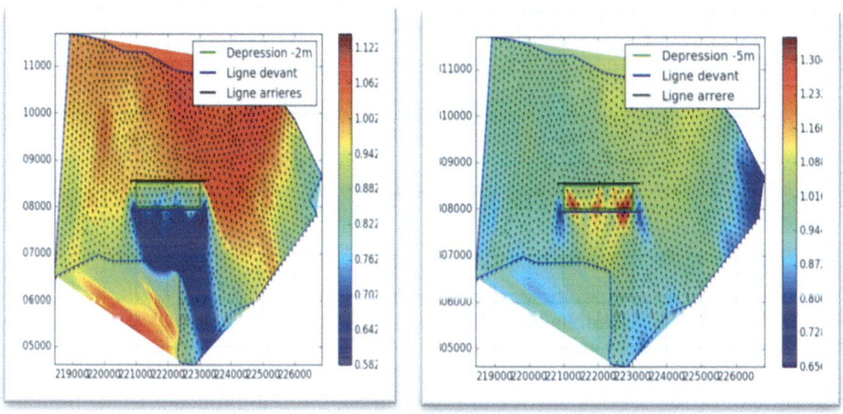

(a) (b)

Fig. 2. Impact of depression on wave propagations: (a) 2 m of depth, (a) 5 m of depth.

The Fig. 3 shows that swell characteristics are changed as a function of the depth of the trench. In the case of 2 m of depth, we figure out that the waves are higher at the upstream of the trench, as we increase the depth into 5 m, the waves are higher at the downstream. Furthermore, we observe also that the intensity of the acceleration at the corners depends on the depth. As long as the depth is greater than 2 m, the effect of the acceleration of the corners is greater. In general, the results prove that heights of waves are more important when the depth of the trench is greater than 2 m. In the Fig. 4, the evolution of the friction velocity at the bottom is evaluated because of its direct link

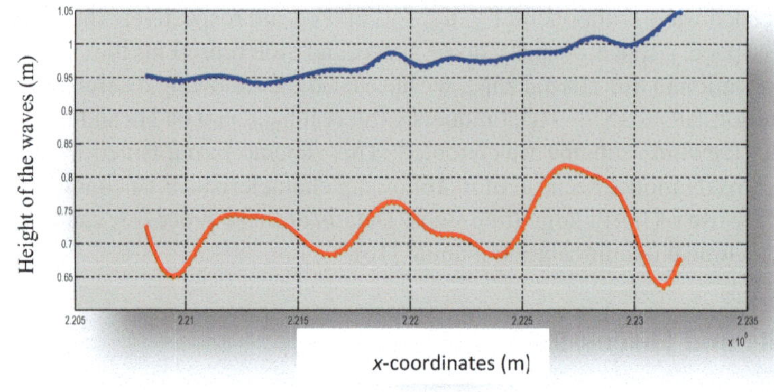

(a)

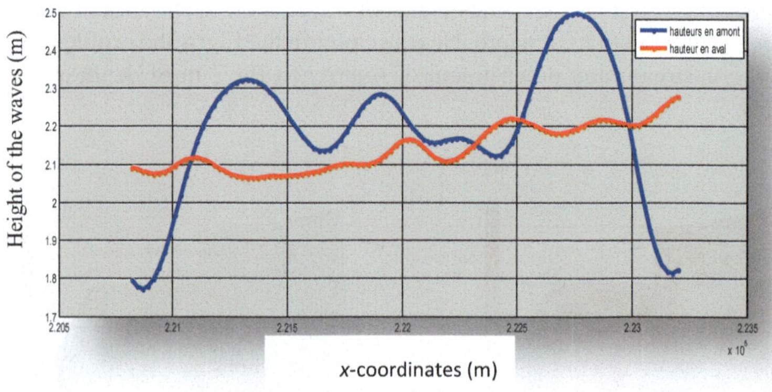

(b)

Fig. 3. Height of the waves upstream and downstream of the trench for (a) 2 m of depth of the trench, (b) 5 m of depth of the trench.

with the rate of erosion. For sake of clarity, when the swell approaches to the trench, we observe that there is a significant decrease in the friction velocity at the upstream and a slight increase at the downstream, the friction velocity on the bottom is diminished by 18% with a difference of 1 to 2% compared to the current case. This analysis using SWAN numerical model allowed us to conclude that the depression must remain in the order of 2 m to keep the stability of the coastline.

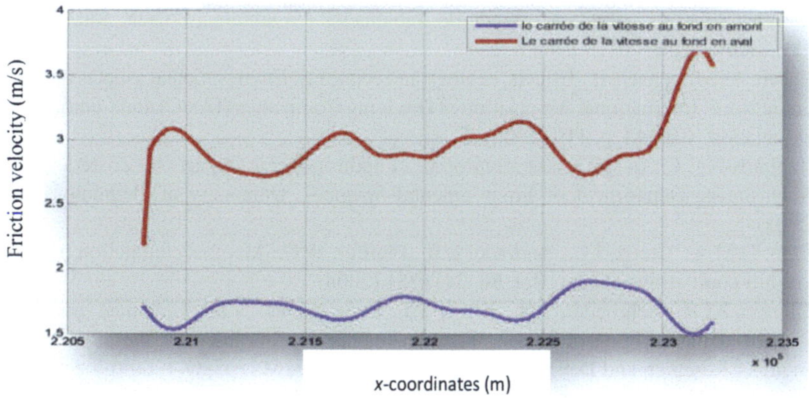

Fig. 4. Friction velocity profiles at the upstream and downstream of the trench for 2 m of depth as versus of *x*-coordinates.

5 Conclusion

This paper is revolved around analyzing the effect of dredging on waves characteristics. The dredging is performed at the Oum Rabiâ estuary situated on the Moroccan Atlantic coast near the city of Azemmour. The influence of the seabed depression on swell characteristics is studied using SWAN model, the outcomes demonstrated in this study are proved that the influence of the sea depression on waves characteristics depends on the bathymetry, the depth of the depression, and the direction of the incident waves, when the depth of the depression is greater than the maximum depth of operation of the waves, the swell does not interfere with the bottom. The depression would then not change the characteristics of the waves. When the depth of the dredging trench is similar to the height of the waves, we find that the features of the waves depend on the depression's depth. Furthermore, the friction velocity allowed as to figure out the stability zone of the coastline. Lastly, the numerical model SWAN is proven that is an efficient numerical tool that can provides accurate analysis before operating on the dredging. As a perspective, we endeavor to include further numerical methods of analysis for analyzing the stability of the coastlines and the marine structures.

References

1. Molisani, M.M., Marins, R.V., Machado, W., Paraquetti, H.H.M., Bidone, E.D., Lacerda, L.D.: Environmental changes in Sepetiba Bay, SE Brazil. Reg. Environ. Chang. **4**, 17–27 (2004). https://doi.org/10.1007/s10113-003-0060-9
2. Sturve, J., et al.: Effects of dredging in Göteborg harbor, Sweden, assessed by biomarkers in eelpout (Zoarces viviparus). Environ. Toxicol. Chem. **24**(8), 1951–1961 (2005)
3. Bray, N., Cohen, M.: Dredging for Development: International Association of Dredging Companies – International Association of Ports and Harbors (2010)
4. Freitas, M.B., Rodrigues, S.C.A.: Consequences of territorial removal from artisanal fishermen in Sepetiba Bay: a regard over health issues of workers and environment (As consequências do processo de desterritorialização da pesca artesanal na Baía de Sepetiba (RJ, Brasil):

um olhar sobre as questões de saúde do trabalhador e o ambiente). Cien. Saude Colet. **19**, 4001–4009 (2014)
5. Jensen, A., Mogensen, B.: Effects, ecology and economy. Environmental aspects of dredging - Guide No. 6. International Association of Dredging Companies (IADC) and Central Dredging Association (CEDA), p. 119 (2000)
6. MacInnis-Ng, C.: In situ stress monitoring of anthropogenic impacts on Zostera capricorni. Ph.D. thesis, Department of Environmental Sciences, University of Technology, Sydney (2003)
7. Filho, G.M.A., Creed, J.C., Andrade, L.R., Pfeiffer, W.C.: Metal accumulation by Halodule wrightii populations. Aquat. Bot. **80**, 241–251 (2004)
8. Palermo, M.R., Schroeder, P.R., Estes, T.J., Francingues, N.R.: Technical guidelines for environmental dredging of contaminated sediments. Environmental Laboratory U.S. Army Engineer Research and Development Center, Vicksburg, MS, p. 304 (2008)
9. Wasserman, J.C., Barros, S.R., Lima, G.B.A.: Planning dredging services in contaminated sediments for balanced environmental and investment costs. J. Environ. Manag. **121**, 48–56 (2013)
10. Jeong, A., Kim, S., Kim, M., Jung, K.: Development of optimization model for river dredging management using MCDA. In: Kim, J.H., Kim, H.S., Yoo, D.G., Jung, D., Song, C.G. (eds.), 12th International Conference on Hydroinformatics. vol. 154, pp. 369–373. Elsevier Science Bv, Amsterdam (2016)
11. Linkov, I., et al.: Comparative assessment of several multi-criteria decision analysis tools for management of contaminated sediments. In: Linkov, I., Kiker, G.A., Wenning, R.J. (eds.) Environmental Security in Harbors and Coastal Areas. NSSS, pp. 195–215. Springer, Dordrecht (2007). https://doi.org/10.1007/978-1-4020-5802-8_14
12. Manap, N., Voulvoulis, N.: Environmental management for dredging sediments – the requirement of developing nations. J. Environ. Manag. **147**, 338–348 (2015)
13. Yeh, C.H., Deng, H., Pan, H.: Multi-criteria analysis for dredger dispatching under uncertainty. J. Oper. Res. Soc. **50**, 35–43 (1999)
14. Booij, N., Ris, R.C., Holthuijsen, L.H.: A third-generation wave model for coastal Regions 1. Model description and validation. J. Geophys. Res. **104**, 7649–7666 (1999)
15. Mouhid, M., Mouakkir, L., Mordane, S., Loukili, M., Chagdali, M., El Bouni, B.: Stability study of a protection structure by stacking GSC geosynthetics: application to the Port of Corisco (Equatorial Guinea). In: 4th Computational Methods in Systems and Software 2020. AISC, vol. 1294, pp. 225–241. Springer, Cham (2020). https://doi.org/10.1007/978-3-030-63322-6_18
16. SWAN team: SWAN User Manual, SWAN Cycle III version 41.20AB. Delft University of Technology, p. 133 (2018). http://www.swan.tudelft.nl

New Method of Verifying Cryptographic Protocols, Based on the Process Model

Andrew M. Mironov(✉)

Innopolis University, Universitetskaya Str 1, 420500 Innopolis, Republic of Tatarstan, Russian Federation

Abstract. A cryptographic protocol (CP) is a distributed algorithm designed to provide a secure communication in an insecure environment. CPs are used, for example, in electronic payments, electronic voting procedures, database access systems, etc. Errors in the CPs can lead to great financial and social damage, therefore it is necessary to use mathematical methods to justify the correctness and safety of the CPs. In this paper, a new mathematical model of a CP is introduced, which allows one to describe both the CPs and their properties. It is shown how, on the basis of this model, it is possible to solve the problems of verification of CPs.

Keywords: Cryptographic protocols · Process model · Verification

1 Introduction

A **cryptographic protocol (CP)** is a distributed algorithm that describes the order in which messages are exchanged between agents. Examples of such agents are computer systems, bank cards, people, etc. To ensure security properties of a CP (such as, for example, the secrecy of transmitted data), cryptographic transformations (encryption, electronic signature, hash functions, etc.) can be used in the CP. Many CP vulnerabilities are related not with poor cryptographic qualities of the cryptographic primitives used in them, but with logical errors in protocols. For example, a logical error was found in the CP for logging into a Google portal that allows a user to identify himself only once and then get access to various applications (such as Gmail or Google Calendar), allowing a dishonest service provider to impersonate any of its users. There are many other examples of CPs (see for example [1–5]), which have been used for a long time in security-critical systems, but then it was discovered that these CPs contain vulnerabilities of the following type:

- participants of these CPs can receive distorted messages (or even lose them) as a result of interception, deleting or distorting of transmitted messages by the adversary, which violates the integrity property,

This research has been financially supported by the Ministry of Digital Development, Communications and Mass Media of the Russian Federation and Russian Venture Company (Agreement No. 004/20 dd. 20.03.2020, IGK 0000000007119P190002).

© The Author(s), under exclusive license to Springer Nature Switzerland AG 2021
R. Silhavy (Ed.): CSOC 2021, LNNS 229, pp. 95–109, 2021.
https://doi.org/10.1007/978-3-030-77445-5_9

- the adversary can discover a secret information contained in the intercepted messages as a result of erroneous or malicious actions of CP participants.

Vulnerabilities were also detected in one of the most well-known CPs **Kerberos** [6]. The absence of vulnerabilities in the patched version of Kerberos was justified in [4]. There are many other examples of CP vulnerabilities used to authentication for cell phone providers, ATM cash withdrawals, e-passports, electronic elections, etc. All of the above examples justify the fact that an informal analysis of the required properties is not enough for CPs used in the security critical systems, it is necessary

- to construct a **mathematical model** of the analyzed CPs,
- describe properties of analyzed CPs in the form of a mathematical objects called **specifications** of these CPs, and
- to construct proofs of statements that the analyzed CPs meet (or do not meet) the specifications, the procedure for constructing such proofs is called **verification** of the analyzed CPs.

In this work, a new mathematical model of CPs is constructed. In terms of this model it is possible to express such properties of correctness of CPs as, for example, integrity of transmitted messages (i.e., justification of the following property of the analyzed CPs: messages sent by one participant of a CP to another participant of this CP, reach the recipient in an undistorted form).

Historically, first formal approach for CP verifying was the BAN-logic of Burrows M., Abadi M., and Needham R., [7]. This approach has very large limitations, in particular, it does not allow considering the case of unlimited generation of sessions of the analyzed protocol. A more popular approach to CP verification is the strand spaces formalism developed by Joshua D. Guttman, Jonathan C. Herzog, F. Javier Thayer Fabrega, [8–10]. One of the CP verification formalisms is the approach associated with the use of Horn clauses and Constraint Systems, developed in the works of Abadi, Blanchet, Cortier and other specialists [11]. Among other CP models, the most popular ones are logic models (see for example [7,13,15]). These models make it possible to reduce the problems of CP verification to the problems of constructing proofs of theorems that CPs under analysis meet their specifications. Algebraic and logical approaches to CP verification are also considered in [12,14,16,17].

2 Sequential and Distributed Processes

In this paper, we outline the concepts of sequential and distributed processes. These concepts are basic mathematical objects for building a CP process model. This model is a development of the **Calculus of Cryptographic Protocols** of Abadi-Gordon (**SPI-calculus**, [16]). It can serve as a theoretical basis for a new method for verifications of CPs, where CP verification means the construction of a mathematical proof that an analyzed CP has the desired properties. Examples of such properties are **integrity** and **secrecy** properties. In the process model described in this text, CPs and their formal specifications are represented as distributed processes.

One of the most important advantages of the proposed CP process model is the low complexity of proofs of CP correctness. In particular, this model eliminates the need to build the sets of all reachable states of the analyzed CPs. This provides an important advantage when analyzing sets of states of the analyzed CP in the case when sets of these states are potentially unlimited. Another important advantage of the proposed CP model is the high degree of automation of solving the CP verification problem based on this model.

2.1 Auxiliary Concepts

We assume that there are given sets $Types$, Con, Var and Fun. The elements of these sets are called **types**, **constants**, **variables**, and **function symbols (FS)**, respectively. Each element x of Con, Var and Fun is associates with some type $\tau(x) \in Types$, and if $x \in Fun$, then $\tau(x)$ has the form $(\tau_1, \ldots, \tau_n) \to \tau$, where $\tau_1, \ldots, \tau_n, \tau \in Types$.

The concept of a **term** is defined inductively. Each term e is associated with a type $\tau(e) \in Types$. The definition of a term is as follows:

- $\forall\, x \in Con \cup Var$ x is a term of the type $\tau(x)$,
- if $f \in Fun$, $e_1, \ldots, e_n$ are terms, and $\tau(f) = (\tau(e_1), \ldots, \tau(e_n)) \to \tau$, then the notation $f(e_1, \ldots, e_n)$ is the term of the type τ.

We will use the following notations:

- Tm denotes the set of all terms,
- $\forall\, e \in Tm$ Var_e denotes the set of all variables, occurred in e,
- $\forall\, X \subseteq Var$ $Tm(X)$ denotes the set $\{e \in Tm \mid Var_e \subseteq X\}$,
- $\forall\, E \subseteq Tm, \forall\, \tau \in Types$ E^τ denotes the set $\{e \in E \mid \tau(e) = \tau\}$.

$\forall\, e, e' \in Tm$ e is called a **subterm** of e', if either $e = e'$, or e' has the form $f(e_1, \ldots, e_n)$, where $f \in Fun$, and $\exists\, i \in \{1, \ldots, n\}$: e is a subterm of e_i.

The notation $e \subseteq e'$, where $e, e' \in Tm$, means that e is a subterm of e'.

Below, for each function of the form $\varphi : E \to E'$, where $E, E' \subseteq Tm$, we will assume that $\forall\, e \in E$ $\tau(\varphi(e)) = \tau(e)$.

We shall assume that $Types$ has the following types:

- type **A**, terms of this type are called **agents**,
- type **C**, terms of this type are called **channels**, they denote communication channels used by agents for communication with each other by sending messages,
- type **K**, terms of this type are called **keys**, they denote cryptographic keys, that agents can use to encrypt or decrypt messages,
- type **M**, terms of this type are called **messages**, they denote messages, that agents can send to each other in the work flow,
- type **P**, terms of this type are called **processes**.

The notations *Agents*, *Channels*, *Keys* and *Processes* denote the sets of all agents, channels, keys, and processes, respectively.

We will use the following conventions and notations:

- *Channels* has a constant denoted by $\circ$, and called an **open channel**,
- an occurrence of a key k in a term e is said to be **hidden**, if this occurrence is first occurrence of k in a subterm of the form $k(e') \subseteq e$,
- type **M** includes any other types from *Types*, i.e. a term of any type is also a term of type **M**,
- set Var contains **shared variables**, each such variable has the form $x_{P_1 \ldots P_n}$, where $P_1, \ldots, P_n$ are different constants of the type **P**.

We will assume that Fun contains the FSs *encrypt* and *decrypt* of type $(\mathbf{K}, \mathbf{M}) \to \mathbf{M}$. Terms of the form $encrypt(k, e)$ and $decrypt(k, e)$ denote messages received by encrypting (and decrypting, respectively) the message e on the key k. Terms of the form $encrypt(k, e)$ and will be denoted by the notations of the form $k(e)$, this terms are called **encrypted messages**.

An **expression** is a notation of one of the following forms:

- any set of terms $E \subseteq Tm$,
- X_P, where $P \in Processes$,
- M_c, where $c \in Channels$,
- $k^{-1}(E)$, where $k \in Keys$, and E is an expression,
- $E \cap E'$, $E \cup E'$, $\neg E$, where E, E' are expressions.

The set of all expressions is denoted by $Expr$. $\forall\, E \in Expr$ the notation Var_E denotes the set of all variables occurred in E.

An **elementary formula (EF)** is a notation of one of the following forms:

- $E = E'$, $E \subseteq E'$, $E \supseteq E'$, where $E, E' \in Expr$,
- $x \perp P$, $x \perp C$, where $x \in Var$, $P \in Processes$, $C \subseteq Channels$,
- $k \perp_{\mathbf{K}} P$, $k \perp_{\mathbf{K}} C$, where $k \in Keys$, $P \in Processes$, $C \subseteq Channels$.

A **formula** is a set of EFs. The set of all formulas is denoted by the notation Fm. $\forall\, \beta \in Fm$ the notation Var_β denotes the set of all variables, occurred in β.

A **binding** is a function of the form $\theta : Var \to Tm$.

We say that a binding θ binds the variable $x \in Var$ with the term $\theta(x)$.

We will use the following notations:

- the set of all bindings is denoted by the symbol Θ,
- id denotes identical binding: $\forall\, x \in Var \;\; id(x) = x$,
- $\forall\, X \subseteq Var$ notation Θ_X denotes the set

$$\{\theta \in \Theta \mid \forall\, x \in Var \setminus X \;\; \theta(x) = x\},$$

- a binding $\theta \in \Theta$ can be denoted by the notations

$$x \mapsto \theta(x) \quad \text{or} \quad (\theta(x_1)/x_1, \ldots, \theta(x_n)/x_n)$$

(second notation is used when $\theta \in \Theta_{\{x_1, \ldots, x_n\}}$),
- $\forall\, \theta \in \Theta$, $\forall\, e \in Tm$ the notation e^θ denotes a term derived from e by replacing $\forall\, x \in Var_e$ each occurrence of x in e by the term $\theta(x)$,
- $\forall\, \theta \in \Theta$, $\forall\, E \subseteq Tm$ the notation E^θ denotes the set $\{e^\theta \mid e \in E\}$,
- $\forall\, \theta, \theta' \in \Theta$ the notation $\theta\theta'$ denotes the binding $x \mapsto (x^\theta)^{\theta'}$.

2.2 Sequential Processes

An **action** is a notation of one of the following forms:

$$c!e, \quad c?e, \quad e := e', \quad \text{where } c \in Channels,\ e, e' \in Tm,$$

which are called **sending** message e to channel c, **receiving** message e from channel c, and **assignment**, respectively.

Actions of the form $c!e$ and $c?e$ are called **external** actions, and actions of the form $e := e'$ are called **internal** actions.

The set of all actions is denoted by the notation Act. $\forall\, \alpha \in Act$ the set of all variables occurred in α, is denoted by the notation Var_α.

If $\theta \in \Theta$ and $\alpha \in Act$, then the notation α^θ denotes an action $c^\theta!e^\theta$, $c^\theta?e^\theta$ and $e^\theta := (e')^\theta$, if $\alpha = c!e$, $c?e$ and $e := e'$, respectively.

In some cases, actions can be written in brackets, i.e., for example, instead of $c!e$, the notation $(c!e)$ might be used, etc.

A **sequential process (SP)** is a triple $(P, X, \bar{X})$, whose components have the following meaning:

- P is a graph with a selected node (called an **initial** node, and denoted by P^0), each edge of which is labeled by an action $\alpha \in Act$,
- $X \subseteq Var \cup Con$ is a set of **initialized variables** and constants, $\circ \in X$,
- $\bar{X} \subseteq X \cap Var$ is a set of **hidden variables**, these variables denote secret keys, hidden channels, and objects with unique values called **nonces**.

A SP is a formal description of the behavior of a dynamic system, which works by sequentially performing actions related to sending/receiving messages and initializing uninitialized variables. For each SP $(P, X, \bar{X})$

- this SP can be abbreviated by the same symbol P as the corresponding graph, the set of nodes of the graph P also is denoted by P,
- notations X_P, $\bar{X}_P$ denote the corresponding components of the SP P,
- Var_P denotes the set of all variables occurred in P,
- if P has no edges and $X_P = \emptyset$, then P is denoted by $\mathbf{0}$.

Each SP is associated with a constant from $Processes$, called a **name** of this process. In order to simplify notations, we will denote the names of processes with the same notations that denote the processes themselves.

Actions of the form $\circ!e$ and $\circ?e$ will be shortened as $!e$ and$?e$ respectively.

The **adversary process** is a SP P_* with the following features:

- the SP graph P_* has a single node,
- $Con \subseteq X_{P_*}$, $\forall\, \tau \in Types$ the sets $\bar{X}_{P_*}$ and $X_{P_*} \setminus \bar{X}_{P_*}$ have a countable set of variables of the type τ,
- $\forall\, \alpha \in Act$ graph P_* has an edge labeled by α.

Let P be a SP. A **state** of P is a 4-tuple $s = (v, \alpha, X, \theta)$, where $v \in P$ is a **current node**, $\alpha \in \{init\} \sqcup Act$ is a **current action**, $X \subseteq Var$ is a **current set of initialized variables**, and $\theta \in \Theta$ is a **current binding**.

Components of s are denoted by v_s, α_s, X_s, and θ_s, respectively.

A state of the SP P is said to be **initial**, and is denoted by $\odot$, if it has the form $(P^0, init, X_P, id)$.

An **execution** of a SP P can be understood as a walk through the graph P, starting from P^0, with the execution of actions that are labels of traversed edges. Each step of an execution of P is associated with a state of P, called a **current state** at this step (a current state at first step is $\odot$), and a **current channels state**, which is a family of sets $M = \{M_c \subseteq Tm \mid c \in Channels\}$.

If a current step of the execution of P is not a final step, then the following actions are performed at this step:

- the current state s on this step is changed on a state s', which will be a current state at the next step of the execution: if s has the form (v, α, X, θ), then there is selected an edge of P outgoing from v, whose label α' meets one of the following conditions:

$$\left.\begin{array}{l} \text{(a) } \alpha' = c!e,\ c^\theta \in X^\theta, e \in Tm(X) \\ \text{(b) } \alpha' = c?e,\ c^\theta \in X^\theta,\ \exists \hat{\theta} \in \Theta_{Var\setminus X} : (e^{\hat{\theta}})^\theta \in M_{c^\theta} \\ \text{(c) } \alpha' = (e := e'),\ e' \in Tm(X),\ \exists \hat{\theta} \in \Theta_{Var\setminus X} : (e^{\hat{\theta}})^\theta = (e')^\theta \end{array}\right\} \quad (1)$$

  and components of $s' = (v', \alpha', X', \theta')$ have the following form: v' is the end of the selected edge, α' is the label of the selected edge, and
  - if (a) in (1) holds, then $X' = X$, $\theta' = \theta$,
  - if (b) or (c) in (1) holds, then $X' = X \cup Var_e$, $\theta' = \hat{\theta}\theta$, and
- a replacement of the channels state M with the channels state M', which will be the current channels state at the next step of the execution: M' either is equal to M, or is obtained by adding terms to the sets from M, and
  - this adding can be performed by P as well as those SPs that use shared channels with P, and
  - if (a) in (1) holds, then one of such addings is that P adds the term e^θ to the set M_{c^θ}.

We will say that s' is obtained by a **transition** from s, and denote this by the notation $s \to s'$. During each execution of each SP P the variables from Var_P have the following features: $\forall x \in Var_P$

1. if $x \notin X_P$, then at the initial step of each execution of P the variable x is not initialized, i.e. there is no value associated with x,
2. if $x \in \bar{X}_P$ and x is not a shared variable, then at first step of each execution *Exec* of P this variable is associated with a **unique value**, i.e. such a value that differs from values associated with other initialized variables at *Exec*, and from values associated with initialized variables at any execution $Exec' \neq Exec$ of any SP,
3. if a variable from $\bar{X}_P$ is shared and has the form $x_{P_1 \ldots P_n}$, then
   - $P_1, \ldots, P_n$ is a list of names of all SPs, executed together with P (and P is one of the SPs in this list), which have the variable $x_{P_1 \ldots P_n}$ among his hidden variables, and

- at the initial moment of each joint execution of SPs from the list $P_1, \ldots, P_n$ variable $x_{P_1 \ldots P_n}$ is initialized in all these SPs with the same value, which is unique, i.e. has the properties described in the point 2.

2.3 Prefix Actions and Renamings

A **refined action** is a triple $\tilde{\alpha} = (\alpha, \hat{X}, \bar{X})$, where $\alpha \in Act$, and $\hat{X}, \bar{X}$ are disjoint subsets of the set Var_α. We will denote the refined action $\tilde{\alpha} = (\alpha, \hat{X}, \bar{X})$ by the notation obtained from the notation of the action α by replacing each variable $x \in Var_\alpha$ to $\hat{x}$ or $\bar{x}$, if $x \in \hat{X}$ or $x \in \bar{X}$, respectively.

Let $\tilde{\alpha} = (\alpha, \hat{X}, \bar{X})$ be a refined action and P be a SP. An operation of a **prefix action** maps the pair $(\tilde{\alpha}, P)$ to a SP $\tilde{\alpha}.P$, having the following components:

- a graph of the SP $\tilde{\alpha}.P$ is obtained by adding a new node v to P, which will be an initial node in $\tilde{\alpha}.P$, and an edge $v \xrightarrow{\alpha} P^0$,
- $X_{\tilde{\alpha}.P} = (X_P \cup Var_\alpha) \setminus \hat{X}, \quad \bar{X}_{\tilde{\alpha}.P} = \bar{X}_P \cup \bar{X}$.

Below we will omit the symbol $\sim$ in the notations of the refined actions.

A renaming is a partial injective function $\zeta : Var \to Var$, where for each shared variable $x_{P_1 \ldots P_n} \in Dom(\zeta)$ the variable $\zeta(x_{P_1 \ldots P_n})$ has the form $y_{P_1 \ldots P_n}$.

For each renaming ζ, each term e and each SP P the notations e^ζ and P^ζ denote a term or a SP respectively, obtained from e or P by replacing $\forall x \in Dom(\zeta)$ of each occurrence of x by $\zeta(x)$.

If $Var_P \subseteq Dom(\zeta)$, then the SPs P and P^ζ are assumed to be the same.

2.4 Distributed Processes

Let $P_I = \{P_i \mid i \in I\}$ be a family of SPs. $\forall i \in I$ let $\tilde{X}_{P_i}$ be a set of variables from Var_{P_i}, which either do not belong to X_{P_i}, or belong to $\bar{X}_{P_i}$ and are not shared. We shall assume that for each family of SPs P_I under consideration the sets $\tilde{X}_{P_i}$ are disjoint (if this is not the case, then we rename accordingly variables in SPs from the family P_I).

A **distributed process (DP)** corresponding to the family P_I is an object denoted by the notation $\prod_{i \in I} P_i$. A DP is a model of a distributed algorithm, components of which are SPs from the family P_I, interacting by transmitting messages through channels. The meaning of a DP is explained in Sect. 2.4.

If P is a DP of the form $\prod_{i \in I} P_i$, then

- $Var_P = \bigcup_{i \in I} Var_{P_i}$, $X_P = \bigcup_{i \in I} X_{P_i}$, $\bar{X}_P = \bigcup_{i \in I} \bar{X}_{P_i}$,
- if ζ is a renaming, then P^ζ denotes the DP $\prod_{i \in I} P_i^\zeta$, if $Var_P \subseteq Dom(\zeta)$, then P and P^ζ are assumed to be the same,
- P can be denoted by the notation $(P_1, \ldots, P_n)$, if $I = \{1, \ldots, n\}$

A **state** a DP $P = \prod_{i \in I} P_i$ is a pair S of the following objects:

- a set $\{s^S_{P_i} \mid i \in I\}$ of states of SPs from P_I,
- a **channel state**: $M^S = \{M^S_c \subseteq Tm \mid c \in Channels\}$.

A state S of DP P is said to be **initial**, and is denoted by $\odot$, if

$$\forall i \in I \ s^S_{P_i} = \odot, \quad \forall c \in Channels \ M^S_c = \emptyset.$$

If S is a state of the DP $P = \prod_{i \in I} P_i$, and $i \in I$, then

- $v^S_{P_i}$, $\alpha^S_{P_i}$, $X^S_{P_i}$, $\theta^S_{P_i}$ denote the corresponding components of the state $s^S_{P_i}$,
- V^S denotes the set $\{v^S_{P_i} \mid i \in I\}$,
- θ^S denotes a binding, such that $\forall i \in I$, $\forall x \in X^S_{P_i}$ $\theta^S(x) = \theta^S_{P_i}(x)$.

An **execution** of a DP $P \prod_{i \in I} P_i$ can be understood as non-deterministic interleaving of executions of SPs from P_I. At each step of an execution of P at most one SP from P_I performs its current action, and other SPs from P_I are in the waiting status. An execution of a DP P can be formally defined as a generation of a sequence of states of this DP (starting with $\odot$), in which each state S that is not terminal, is associated with the next state S' by a **transition relation**, which means the following: $\exists\, i \in I$:

$$\begin{array}{l} s^S_{P_i} \to s^{S'}_{P_i}, \quad \forall i' \in I \setminus \{i\} \quad s^{S'}_{P_{i'}} = s^S_{P_{i'}}, \quad \text{and if } s^{S'}_{P_i} = (v, \alpha, X, \theta), \text{then} \\ \text{if } \alpha = c!e, \text{then} \left\{ \begin{array}{l} M^{S'}_{c^\theta} = M^S_{c^\theta} \cup \{e^\theta\}, \\ M^{S'}_{c'} = M^S_{c'} \text{ when } c' \neq c^\theta \end{array} \right\}, \text{otherwise } M^{S'} = M^S. \end{array} \tag{2}$$

For each states S, S' of DP P

- $S \to S'$ means that S is related with S' by a transition relation,
- $S \xrightarrow{\alpha_{P_i}} S'$ means that $S \to S'$, and (2) holds,
- $S \Rightarrow S'$ means that either $S = S'$, or there is a sequence $S_0, \ldots, S_n$ of states of P, such that $S_0 = S$, $S_n = S'$, $\forall i = 0, \ldots, n-1$ $S_i \to S_{i+1}$.

A state S of P is said to be **reachable**, if $\odot \Rightarrow S$.
The set of reachable states of P is denoted by Σ_P.

2.5 Schemes of Distributed Processes

Let P be a DP of the form $\prod_{i \in I}$, and $\forall i \in I$ SP P_i has the form $\alpha_1. \ \ldots \ \alpha_n.P'_i$. The sequence of actions $\alpha_1 \ldots \alpha_n$ and SP P'_i will be called a **prefix** and a **postfix** of SP P_i, respectively. If each external actions in the prefix of P_i is a sending (receiving) a message to (from) a SP $P_j \in P_I$, and the action of SP P_j corresponding to the receiving (sending) this message is in the prefix of P_j, then these dependencies between actions can be expressed as a **scheme** of DP P, which has the following form:

- each SP $P_i \in P_I$ is represented by a **thread**, i.e. by a vertical line, on which there are marked points corresponding to nodes of the graph P_i belonging to the prefix of P_i (the upper point of the thread corresponds to P^0_i), and
  - near each such point it might be specified an identifier of the corresponding node, near the upper point of the thread a name of SP P_i is specified, if $P'_i \neq \mathbf{0}$, then the postfix name P'_i is specified near the bottom point,

- the segments connecting the neighboring points of the thread correspond to edges of P_i related to the prefix of P_i, there are the specified labels of the corresponding edges beside these segments,

– for each segment O of the thread connecting neighboring points, if the corresponding action is sending a message, then there is an arrow in the scheme, such that the start of this arrow lies on the segment O, and the end of this arrow lies on the segment O', the label of which is an action of the corresponding SP $P_j \in P_I$ to receive this message.

An example of a DP is a **Wide-Mouth Frog (WMF)** cryptographic protocol, consisting of three SPs named A, B and T (where T is a trusted intermediary). This DP is a model of a transmission of encrypted message $k_{AB}(x)$ from A to B through open channel $\circ$ with use of T, with whom A and B communicate through open channel $\circ$. SP A creates the secret key k_{AB}, sends B this key in an encrypted form using T, and then sends B encrypted message $k_{AB}(x)$. It is assumed that A and T (B and T) have a shared secret key k_{AT} (k_{BT}), on which they can encrypt and decrypt messages using a symmetric encryption system, and only A and T (B and T) know secret key k_{AT} (k_{BT}). This DP works as follows.

– A creates a secret key k_{AB} (at first only A knows this key) and sends T encrypted message $k_{AT}(k_{AB})$ through $\circ$, then A sends B encrypted message $k_{AB}(x)$ through $\circ$,
– T receives a message from A, decrypts this message, then encrypts the extracted key k_{AB} with the key k_{BT}, and sends B encrypted message $k_{BT}(k_{AB})$ through $\circ$,
– B extracts key k_{AB} from the message received from T, and then uses this key to extract message x from the message received from A, writes x to variable y, and then behaves in the same way as SP P.

SPs A, B and T are defined as follows:

$$\begin{array}{lll} A = \alpha_1.\alpha_2.\mathbf{0}, \text{ where } \alpha_1 = !\bar{k}_{AT}(\bar{k}_{AB}), & \alpha_2 = !\bar{k}_{AB}(x), \\ T = \gamma_1.\gamma_2.\mathbf{0}, \text{ where } \gamma_1 = ?\bar{k}_{AT}(\hat{u}), & \gamma_2 = !\bar{k}_{BT}(u), \\ B = \beta_1.\beta_2.P, \text{ where } \beta_1 = ?\bar{k}_{BT}(\hat{v}), & \beta_2 = ?v(\hat{y}). \end{array} \quad (3)$$

A scheme of DP (A, B, T) has the following form:

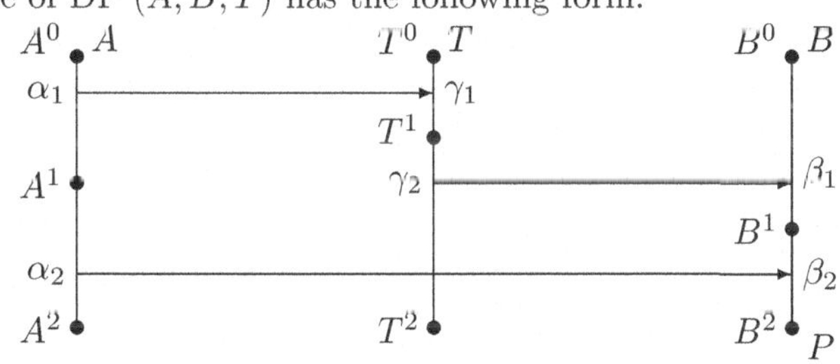

(4)

2.6 Transition Graphs of Distributed Processes

Let P be a DP of the form $\prod_{i \in I} P_i$.

A **transition graph (TG)** of DP P is a graph G_P such that

- a set of nodes of G_P is the Cartesian product of the sets of nodes of graphs from P_I, i.e. each node of G_P is a family of nodes $V = \{v_i \mid i \in I\}$, where $\forall\, i \in I \;\; v_i \in P_i$,
- each edge of G_P has the form $\{v_i \mid i \in I\} \xrightarrow{\alpha_{P_i}} \{v'_i \mid i \in I\}$, where P_i has the edge $v_i \xrightarrow{\alpha} v'_i$ and $\forall\, i' \in I \setminus \{i\} \;\; v_{i'} = v'_{i'}$.

The node $\{P_i^0 \mid i \in I\} \in G_P$ is said to be an **initial** node of G_P, and is denoted by G_P^0. An edge $V \xrightarrow{\alpha_{P_i}} V'$ is said to be a **realizable** edge, if $\exists\, S, S' \in \Sigma_P$: $V = V^S$ and $V' = V^{S'}$.

For each DP P the graph G_{P^*} can be considered as a completion of the graph G_P with cyclic edges corresponding to the actions of P_*.

If DP P has the form $(P_1, \ldots, P_n)$, then the following conventions will be used in a graphical representation of G_P: each node $V = \{v_i \mid i = 1, \ldots, n\}$ of G_P is represented by an oval, there is a list $v_1 \ldots v_n$ of components of V inside this oval, an initial node G_P^0 is represented by a double oval.

For example, a TG for a DP described by scheme (4) has the following form:

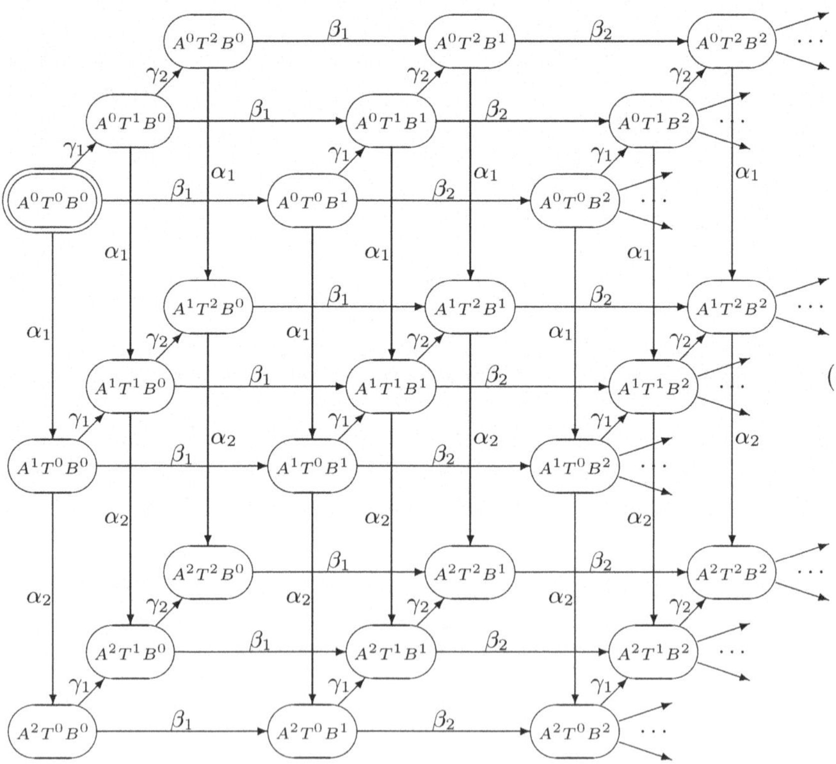

(5)

2.7 Values of Expressions and Formulas in States of Distributed Processes

Let there are given the DP $P = \prod_{i \in I} P_i$, the state $S \in \Sigma_P$, the expression $E \in Expr$, and the formula $\beta \in Fm$. The notation E^S denotes a subset of Tm, called a **value of the expression** E **in the state** S, and defined as follows:

- if $E \subseteq Tm$, then $E^S = E^{\theta^S}$,
- if $E = X_P$, then $E^S = (X_P^S)^{\theta^S}$,
- if $E = M_c$, then $E^S = M^S_{c^{\theta^S}}$,
- if $E = k^{-1}(E')$, then $E^S = \{e \in Tm \mid \exists e' \in (E')^S : k^{\theta^S}(e) \subseteq e'\}$,
- $(E \cap E')^S = E^S \cap (E')^S$, $(E \cup E')^S = E^S \cup (E')^S$, $(\neg E)^S = Tm \setminus E^S$.

The notation $S \models \beta$ denotes the statement β **holds in** S, which is true iff one of the following cases holds:

- • $\beta = (E = E')$, $(E \subseteq E')$, or $(E \supseteq E')$, where $E, E' \in Expr$, and
  • $E^S = (E')^S$, $E^S \subseteq (E')^S$, or $E^S \supseteq (E')^S$, respectively,
- • $\beta = (x \perp P_i)$, where $x \in Var$, $i \in I$, and
  • $\forall e \in (X^S_{P_i})^{\theta^S}$ $x \notin Var_e$,
- • $\beta = (x \perp C)$, where $x \in Var$, $C \subseteq Channels$, and
  • $\forall c \in C$, $\forall e \in M^S_c$ $x \notin Var_e$,
- • $\beta = (k \perp_{\mathbf{K}} P_i)$, where $k \in Keys$, $i \in I$, and
  • $\forall e \in (X^S_{P_i})^{\theta^S}$ each occurrence of k in e is hidden,
- • $\beta = (k \perp_{\mathbf{K}} C)$, where $k \in Keys$, $C \subseteq Channels$, and
  • $\forall c \in C$, $\forall e \in M^S_c$ each occurrence of k in e is hidden,
- $\beta = \{\beta_i \mid i \in I\}$ is a family of EFs, $\forall i \in I$ $S \models \beta_i$.

Below we prove theorems that some formulas have the same values in states related by a transition relation.

Theorem 1. Let $P = \prod_{i \in I} P_i$ be a DP and $S, S' \in \Sigma_P$ be states such that

$$\exists i \in I : S \xrightarrow{\alpha_{P_i}} S'.$$

Then the implication $S \models \beta \Rightarrow S' \models \beta$ holds, where β is a formula of one of the following forms:

1. $\beta = \{x \perp P_i, x \perp Channels\}$, where $x \in X_P$,
2. $\beta = \{k \perp_{\mathbf{K}} P_i, k \perp_{\mathbf{K}} Channels\}$, where $k \in X^{\mathbf{K}}_{P_i}$.

Theorem 2. Let $P = \prod_{i \in I} P_i$ be a DP, and $S, S' \in \Sigma_P$ be states such that

$$\exists i \in I : S \xrightarrow{\alpha_{P_i}} S'.$$

Then the implication $S \models \beta \Rightarrow S' \models \beta$ holds, where β is a formula of one of the following two forms:

$$\{c \perp P_i, c \perp Channels, M_c = E\}, \text{ where } c \in X^{\mathbf{C}}_P, \tag{6}$$

$$\left\{\begin{array}{l} k \perp_{\mathbf{K}} P_i,\ k \perp_{\mathbf{K}} Channels, \\ k^{-1}(M_c) \subseteq E \quad (\forall c \in Channels), \\ k^{-1}(X_{P_i}) \subseteq E \end{array}\right\}, \text{where } k \in X_P^{\mathbf{K}}, E \subseteq Tm(X_P). \tag{7}$$

Theorem 3. Formula (7) in theorem 2 can be replaced by a formula

$$\left\{\begin{array}{l} k \perp_{\mathbf{K}} P_i,\ k \perp_{\mathbf{K}} Channels, \\ k^{-1}(M_{c_0}) = E \\ k^{-1}(M_c) \subseteq E \quad (\forall c \in Channels), \\ k^{-1}(X_{P_i}) \subseteq E \end{array}\right\}, \text{ where } c_0 \in X_P^{\mathbf{C}},\ k \in X_P^{\mathbf{K}}, E \subseteq Tm(X_P).$$

2.8 Marking of a Transition Graph

Let P be a DP of the form $\prod_{i \in I} P_i$. A **marking** of the TG G_P is a pair

$$(G, \{\beta_V \in Fm \mid V \in G\}) \tag{8}$$

where G is a subset of the set of nodes of G_P, such that $G_P^0 \in G$, and $\forall V \in G$, if G_P has an edge of the form $V' \to V$, then $V' \in G$.

Marking (8) is said to be **correct**, if $G_P^0 \models \beta_{G_P^0}$, and $\forall S, S' \in \Sigma_P$, if $S \to S'$ and $V^S, V^{S'} \in G$, then the implication $S \models \beta_{V^S} \;\Rightarrow\; S' \models \beta_{V^{S'}}$ holds.

It was noted in Sect. 2.6 that for each DP P the graph G_{P^*} is a completion of the graph G_P with cyclic edges corresponding to actions of the adversary P_*. Therefore, for each DP P, any marking of the TG G_P can also be considered as a marking of the corresponding TG G_{P^*}.

Below, a marking of any TG G_P is said to be correct, if it is a correct marking (in the sense of the above definition) of the corresponding TG G_{P^*}.

An example of a correct marking for TG (5), where actions $\alpha_i, \beta_i, \gamma_i$ ($i = 1, 2$) are defined according to (3), has the form:

- $G = \left\{\begin{array}{l} A^0T^0B^0, A^1T^0B^0, A^2T^0B^0, A^1T^1B^0, A^2T^1B^0, \\ A^1T^2B^0, A^2T^2B^0, A^1T^2B^1, A^2T^2B^1, A^2T^2B^2 \end{array}\right\}$
- • $\beta_{A^0T^0B^0} \stackrel{\text{def}}{=} \left\{\begin{array}{l} k_{AT}^{-1}(M_\circ) = k_{BT}^{-1}(M_\circ) = k_{AB}^{-1}(M_\circ) = \emptyset \\ \{k_{AT}, k_{BT}, k_{AB}\} \perp_{\mathbf{K}} P_* \\ \{k_{AT}, k_{BT}, k_{AB}\} \perp_{\mathbf{K}} Channels \end{array}\right\}$,
  - • $\beta_{A^1T^0B^0} \stackrel{\text{def}}{=} \left\{\begin{array}{l} k_{AT}^{-1}(M_\circ) = \{k_{AB}\}, k_{BT}^{-1}(M_\circ) = k_{AB}^{-1}(M_\circ) = \emptyset \\ \{k_{AT}, k_{BT}, k_{AB}\} \perp_{\mathbf{K}} P_* \\ \{k_{AT}, k_{BT}, k_{AB}\} \perp_{\mathbf{K}} Channels \end{array}\right\}$,
  - • $\beta_{A^2T^0B^0} \stackrel{\text{def}}{=} \left\{\begin{array}{l} k_{AT}^{-1}(M_\circ) = \{k_{AB}\}, k_{BT}^{-1}(M_\circ) = \emptyset, k_{AB}^{-1}(M_\circ) = \{x\} \\ \{k_{AT}, k_{BT}, k_{AB}\} \perp_{\mathbf{K}} P_* \\ \{k_{AT}, k_{BT}, k_{AB}\} \perp_{\mathbf{K}} Channels \end{array}\right\}$,
  - • $\beta_{A^1T^1B^0} \stackrel{\text{def}}{=} \left\{\begin{array}{l} u = k_{AB}, k_{AT}^{-1}(M_\circ) = \{k_{AB}\}, k_{BT}^{-1}(M_\circ) = k_{AB}^{-1}(M_\circ) = \emptyset \\ \{k_{AT}, k_{BT}, k_{AB}\} \perp_{\mathbf{K}} P_* \\ \{k_{AT}, k_{BT}, k_{AB}\} \perp_{\mathbf{K}} Channels \end{array}\right\}$,
  - • $\beta_{A^2T^1B^0} \stackrel{\text{def}}{=} \left\{\begin{array}{l} u = k_{AB}, k_{AT}^{-1}(M_\circ) = \{k_{AB}\}, k_{BT}^{-1}(M_\circ) = \emptyset \\ k_{AB}^{-1}(M_\circ) = \{x\} \\ \{k_{AT}, k_{BT}, k_{AB}\} \perp_{\mathbf{K}} P_* \\ \{k_{AT}, k_{BT}, k_{AB}\} \perp_{\mathbf{K}} Channels \end{array}\right\}$,

- $\beta_{A^1T^2B^0} \stackrel{\text{def}}{=} \left\{\begin{array}{l} u = k_{AB}, k_{AT}^{-1}(M_\circ) = \{k_{AB}\}, k_{BT}^{-1}(M_\circ) = \{u\} \\ k_{AB}^{-1}(M_\circ) = \emptyset \\ \{k_{AT}, k_{BT}, k_{AB}\} \perp_{\mathbf{K}} P_* \\ \{k_{AT}, k_{BT}, k_{AB}\} \perp_{\mathbf{K}} Channels \end{array}\right\},$
- $\beta_{A^2T^2B^0} \stackrel{\text{def}}{=} \left\{\begin{array}{l} u = k_{AB}, k_{AT}^{-1}(M_\circ) = \{k_{AB}\}, k_{BT}^{-1}(M_\circ) = \{u\} \\ k_{AB}^{-1}(M_\circ) = \{x\} \\ \{k_{AT}, k_{BT}, k_{AB}\} \perp_{\mathbf{K}} P_* \\ \{k_{AT}, k_{BT}, k_{AB}\} \perp_{\mathbf{K}} Channels \end{array}\right\},$
- $\beta_{A^1T^2B^1} \stackrel{\text{def}}{=} \left\{\begin{array}{l} u = k_{AB}, v = u \\ k_{AT}^{-1}(M_\circ) = \{k_{AB}\}, k_{BT}^{-1}(M_\circ) = \{u\}, k_{AB}^{-1}(M_\circ) = \emptyset \\ \{k_{AT}, k_{BT}, k_{AB}\} \perp_{\mathbf{K}} P_* \\ \{k_{AT}, k_{BT}, k_{AB}\} \perp_{\mathbf{K}} Channels \end{array}\right\},$
- $\beta_{A^2T^2B^1} \stackrel{\text{def}}{=} \left\{\begin{array}{l} u = k_{AB}, v = u \\ k_{AT}^{-1}(M_\circ) = \{k_{AB}\}, k_{BT}^{-1}(M_\circ) = \{u\}, k_{AB}^{-1}(M_\circ) = \{x\} \\ \{k_{AT}, k_{BT}, k_{AB}\} \perp_{\mathbf{K}} P_* \\ \{k_{AT}, k_{BT}, k_{AB}\} \perp_{\mathbf{K}} Channels \end{array}\right\},$
- $\beta_{A^2T^2B^2} \stackrel{\text{def}}{=} \{y = x\}.$

A correctness of this marking can be justified by theorems from Sect. 2.7.

3 Application of Markings of Transition Graphs to the Problems of Verification of Distributed Processes

An **execution** of a DP P is a sequence $S_0, \ldots, S_n$ of states from Σ_P, such as $S_0 = \odot$, and either $n = 0$, or $\forall\, i = 0, \ldots, n-1 \quad S_i \to S_{i+1}$.

It is not difficult to see that each such sequence $S_0, \ldots, S_n$ corresponds to a path $G_P^0 = V^{S_0} \to \ldots \to V^{S_n}$ in TG G_P.

Some correctness properties of DPs have the following form: in each state S of an arbitrary execution of a DP P, the following implication holds:

$$S \models \beta \;\Rightarrow\; S \models \beta', \text{ where } \beta, \beta' \in Fm \text{ are given formulas.} \tag{9}$$

For example, for the DP P^*, where P is a DP presented in Sect. 2.5 by the scheme (4), one of the correctness properties has the following form: for arbitrary execution $S_0, \ldots, S_n$ of this DP, if $S_n \models (v_B^{S_n} = B^2)$, i.e. if B executed an action of receiving the message sent by A and wrote the received message in variable y, then $S_n \models (x = y)$, i.e. the received message is the same as the message x that A sent B.

Properties of the form (9) can be verified using a marking of TG G_P of an analyzed DP P as follows: a correct marking $(G, \{\beta_V \in Fm \mid V \in G\})$ of G_P is being built, and for each node $V \in G$, such that β_V implies β, the implication $\beta_V \to \beta'$ is being checked. To check the above statements, there is no need to fully build the TG G_P of the analyzed DP P. It is convenient to build the TG together with the construction of its marking as follows: if a formula β_V is built to mark the node V of G_P, and β_V implies an unrealizability of some edge outgoing from V, then this edge is discarded. For example, this can happen if a label of an edge outgoing from V is of the form $(c?\dot{y})_B$, and β_V contains the conjunctive term $M_c = \emptyset$.

As a result of such a construction with discarding unrealizable edges, a fragment of the TG G_P will be obtained. We shall call such fragment a **reduced TG**. It is not difficult to see that the reduced TG preserves all the properties of the TG G_P. In particular, for the solution of the verification problem described above for a property of the form (9), the corresponding reduced TG can be used instead of the TG G_P. An example of a reduction of a TG is presented below. The following edges in TG (5) (where actions $\alpha_i, \beta_i, \gamma_i$ $(i = 1, 2)$ are defined according to (3)) are unrealizable: $A^0T^0B^0 \stackrel{\beta_1}{\to} A^0T^0B^1$, $A^0T^0B^0 \stackrel{\gamma_1}{\to} A^0T^1B^0$, $A^1T^0B^0 \stackrel{\beta_1}{\to} A^1T^0B^1$, $A^1T^1B^0 \stackrel{\beta_1}{\to} A^1T^1B^1$, $A^1T^2B^1 \stackrel{\beta_2}{\to} A^1T^2B^2$, $A^2T^0B^0 \stackrel{\beta_1}{\to} A^2T^0B^1$, $A^2T^1B^0 \stackrel{\beta_1}{\to} A^2T^1B^1$. The reduced TG (5) has the form

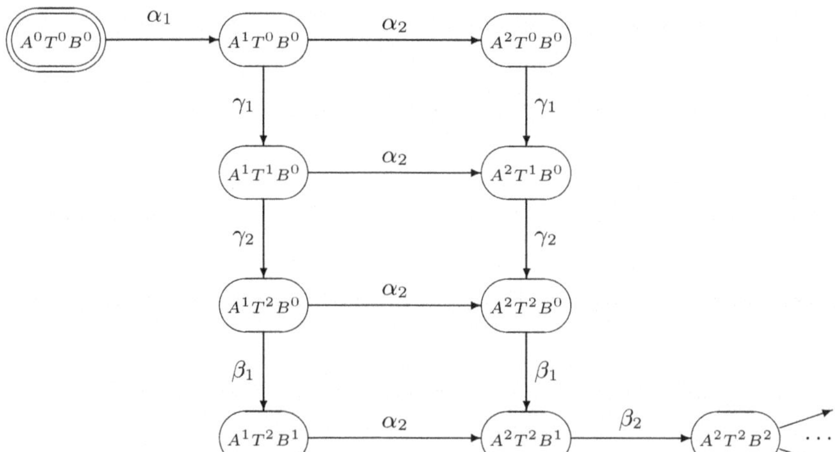

There is a correct markings of this reduced TG such that $\beta_{V^S} = (x = y)$. As stated above, this statement is a justification of the property $S \models (x = y)$. Thus, by building a suitable marking, we verified the following property of the considered DP: if B executed the action of receiving the message sent by A and wrote the received message in the variable y, then the received message is the same as the message that A sent B.

4 Conclusion

In the present work, a new model of cryptographic protocols (in which cryptographic protocols are represented by distributed processes) was built, and examples of its use for solving problems of verification of protocol integrity properties are shown. The objectives for further development of this model and verification methods based on it are the following:

1. an automation of synthesis of suitable markings in transition graphs of the analyzed protocols,
2. development of the language of specification of properties of cryptographic protocols, which allow to express e.g. properties of confidentiality (secrecy) of transmitted messages, i.e. the adversary's inability to extract any new

information about the content of messages intercepted by him, matching properties in authentication protocols, or zero knowledge properties, non-traceability properties in electronic payments, properties of correctness of the votes' counting in voting protocols,

3. construction of automated synthesis methods of cryptographic protocols by describing the properties which the cryptographic protocols must satisfy, etc.

References

1. Denning, D., Sacco, G.: Timestamps in Key distribution protocols. Commun. ACM **24**(8), 533–536 (1981)
2. Needham, R., Schroeder, M.: Using encryption for authentication in large networks of computers. Commun. ACM **21**(12), 993–999 (1978)
3. Needham, R., Schroeder, M.: Authentication revisited. Oper. Syst. Rev. **21**(1), 7 (1987)
4. Cervesato, I., Jaggard, A.D., Scedrov, A., Tsay, J.-K., Walstad, C.: Breaking and fixing public-key Kerberos. Inf. Comput. **206**(2–4), 402–424 (2008)
5. Lowe, G.: Breaking and fixing the needham-schroeder public-key protocol using FDR. In: Margaria, T., Steffen, B. (eds.) TACAS. LNCS, vol. 1055, pp. 147–166. Springer, Heidelberg (1996). https://doi.org/10.1007/3-540-61042-1_43
6. Kerberos: The Network Authentication Protocol. MIT Kerberos. 10 September 2015. http://web.mit.edu/kerberos/. Accessed 31 Oct 2015
7. Burrows, M., Abadi, M., Needham, R.: A logic of authentication. ACM Trans. Comput. Syst. **8**(1), 18–36 (1990)
8. Thayer, F.J., Herzog, J.C., Guttman, J.D.: Strand spaces: proving security protocols correct. J. Comput. Secur. **7**(2/3), 191–230 (1999)
9. Guttman, J.D., Thayer, F.J.: Authentication tests and the structure of bundles. Theor. Comput. Sci. **283**, 333–380 (2002)
10. Guttman, J.D.: State and progress in strand spaces: proving fair exchange. J. Autom. Reason. **48**(2), 159–195 (2012)
11. Cortier, V., Kremer, S.: Formal Models and Techniques for Analyzing Security Protocols. Now Publishers Inc., Hanover (2014)
12. Syverson, P., van Oorschot, P.C.: On unifying some cryptographic protocol logics. In: Proceedings of the 1994 IEEE Computer Security Foundations Workshop, vol. VII, pp. 14–29. IEEE Computer Society Press (1994)
13. Syverson, P., Meadows, C.: A logical language for specifying cryptographic protocol requirements. In: Proceedings of the 1993 IEEE Computer Security Symposium on Security and Privacy, pp. 165–177. IEEE Computer Society Press (1993)
14. Paulson L.: Proving properties of security protocols by induction. In: Proceedings of the IEEE Computer Security Foundations Workshop, vol. X, pp. 70–83. IEEE Computer Society Press (1997)
15. Brackin S.: A state-based HOL theory of protocol failure. ATR 98007, Arca Systems Inc. (1997). http://www.arca.com/paper.htm
16. Abadi, M., Gordon, A.: A calculus for cryptographic protocols: the Spi calculus. In: Proceedings of the Fourth ACM Conference on Computers and Communications Security, pp. 36–47. ACM Press (1997)
17. Abadi, M., Blanchet, B., Fournet, C.: The applied Pi calculus: mobile values, new names, and secure communication. [Research Report] ArXiv, p. 110. hal-01423924 (2016). https://arxiv.org/abs/1609.03003

A Method of Search Space Adaptation in the False Pseudonyms Detection Procedure

Anna Klimenko(✉) and Iakov Korovin

Scientific Research Institute of Multiprocessor Computer Systems of Southern Federal University, 2, Chekhov Street, 347928 Taganrog, Russian Federation

Abstract. The paper deals with the problem of the false pseudonyms detection in the Internet. This problem relates to the victim behavior of individuals, so, it is quite relevant under the conditions of the lack of physical representation and deanonymizing. The stylomentry approach has showed good results in the authorship analysis in the field of the doppelgangers finding, yet, in case of big data analysis the methods are needed to improve the capabilities of the false pseudonyms search procedure. In this paper fuzzy cognitive map and some elements of the distributed ledger technology are proposed as a facility to enhance the search space according to the assumption about the audiences intersection. The method proposed aims two goals: to improve the search results by means of cognitive analysis of the WEB-resources relations and to decrease the search time, eliminating the weakly-integrated resources.

Keywords: Cognitive analysis · False pseudonyms · Distributed ledger

1 Introduction

A problem of the false pseudonyms detection is quite relevant and is related to the issue of fake accounts. Yet, while fake accounts can be created by bots and is an integral part of the fake news phenomena, false pseudonyms are used by real individuals not only to implement the malicious intentions, but also to hide from bullies and to restore the reputation in the communities.

Yet, some malicious intentions connected with amoral and aggressive behaviour, as well as hiding from bullies relate to the so-called victim behavior, which affect the mental health and the real-life behavior of individuals.

Although some policies and methods do exist to detect the false pseudonyms, very little attention has been paid to the cognitive-assistant-like services which could detect the false pseudonyms having the initial search space and using the authorship attribution methods. The features of adaptiveness of the cognitive assistant can be implemented in the possibility to enhance or reduce the search space of the pseudonyms detection procedure, considering it as the procedure, which deals with the "big data". The key idea, architecture and basic methods of the cognitive assistant in-question are considered in our previous works [1–3]. Also the adaptiveness feature of the cognitive assistant is

© The Author(s), under exclusive license to Springer Nature Switzerland AG 2021
R. Silhavy (Ed.): CSOC 2021, LNNS 229, pp. 110–120, 2021.
https://doi.org/10.1007/978-3-030-77445-5_10

considered as a possibility to reduce the time needed for the big data processing and to improve the search quality.

In the current paper a method of search space adaptation is proposed. It is based on the distributed ledger elements and fuzzy cognitive map (FCM) usage. A basic approach to the particular type of the cognitive map construction is proposed and described, based on the initial search space given as an input data for the cognitive assistant. Also some simulations are made to provide the illustrations of the FCM usage.

2 The Approaches to the False Pseudonyms Detection Problem

The stylometry approach to the authorship attribution problem shows quite promising results in the false pseudonyms detection [4–9]. So, in our previous works a decision was made about the expediency of the cognitive assistant development based on the stylomentry approach and authorship analysis [1]. Also it was showed that the adaptiveness feature of the cognitive assistant allows to reduce time needed for the pseudonyms search procedure.

It must be mentioned that the stylometry approach is not the unique technique for false pseudontms detection.

There is a range of publications devoted to the fake accounts detection by means of other techniques.

For example, in [10] a list of features is presented which allows to detect false accounts. Features are divided into two classes: the attributes related to the account behavior and attributes and non-related to the account's behavior. The term of "network of trust" is considered – this is an overall activity network, which is created by the individuals by their communication in the Social Networks.

This network of trust has its statistical measures. They are, for example, the number of communities (pages, groups, events) a user belongs to, indegree (friend requests received), out degree (friend requests sent), betweeness centrality, closeness centrality etc. Based on these network attributes some models for the fakes detection are proposed.

In [11], the authors use the behavioural feature like 'pages liked by the user' and how similar the user likes pages (Rate of like activity) for the detection of analogous user behaviour in the Facebook network. If the frequency of page likes within a specific time interval is very high or all the pages liked by a user strictly belong to a single category, the user may be considered as suspicious.

In [12], a machine learning based approach has been used for the detection of social bots in twitter network. The author has used non behavioural features like number of friends, number of followers and followee ratio of a user profile.

Another area of the fake account detection is the techniques based on the content analysis.

In [13] authors are used content based features like the use of capital or lowercase, quotation count and punctuation mark. Another study [14] uses attributes like number of reposts, number of comments, number of likes, number of hash tags and urls in the post in order to detect the spammers on the network.

Although the range of false accounts detection techniques is quite considerable, sometimes the problem of false pseudonyms (doppelganger accounts) detection belongs to another field: for the given text example, it is needed to say, whether or not the author of the example is the author of the other texts. To the best of our knowledge, there is a considerable number of related works, too [15–17].

3 An Authorship Analysis Used by the Cognitive Assistant for the Victim Behavior Prevention

As has been mentioned earlier, the cognitive assistant proposed [1, 2] is considered as a technical facility for the victim behaviour prevention. The term "Cognitive assistant" presupposes the adaptive nature of the technical system which enhances the human's capabilities. Adaptiveness features of a cognitive assistant provide a high-level interconnection with the environment and enhancement of the results.

The cognitive assistant in question presupposes the following scenario of the authorship analysis usage:

1. A user has some text samples for cognitive assistant to create a text feature vector.
2. A user gives an initial search area where the cognitive assistant service has to detect the texts of the most similar authors.
3. Cognitive assistant service implements the search in the given search space of web-resources and returns the potentially doppelganger pseudonyms.

The generic architecture of the system functioning is presented in Fig. 1 and described in detail in our previous works [1–3].

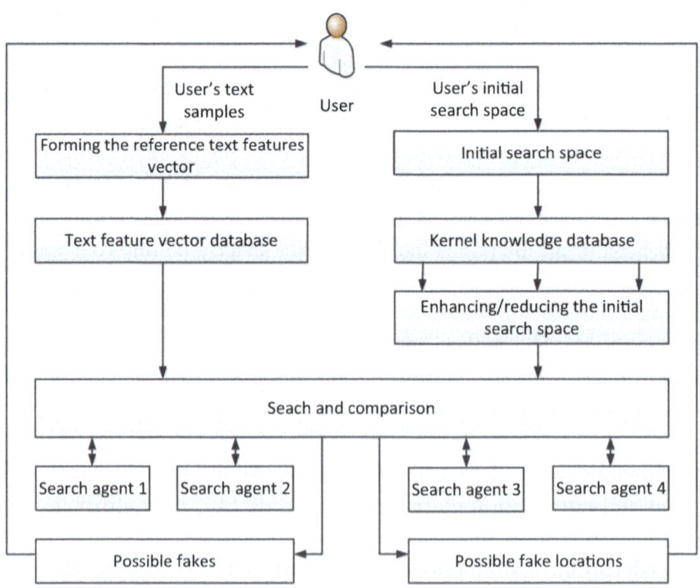

Fig. 1. The generic architecture of the cognitive assistant

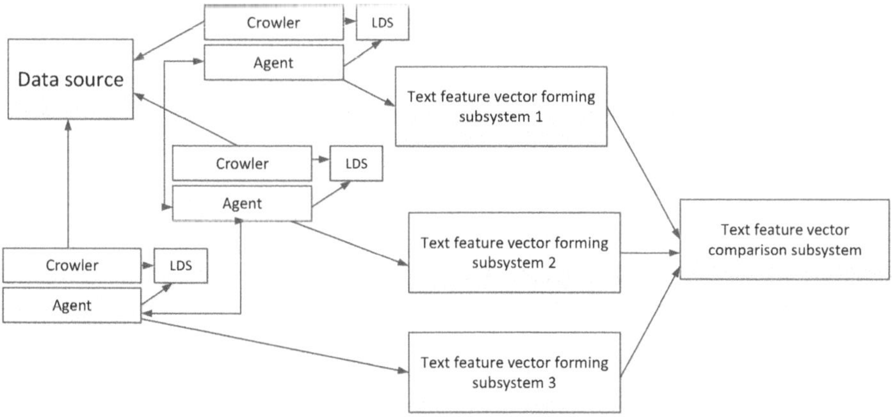

Fig. 2. The generic architecture of the search subsystem

It is also shown (Fig. 2) that the distributed search subsystem consists of some crowlers which are controlled by software agents. The results of the crowling can be collected in the Local Data Storages (LDS). Than the text feature vectors are formed concurrently. Those vectors are compared with the reference text feature vector. So, taking into account the values of the data to be processed, it is extremely important to choose an appropriate initial search space to get the results of a good quality. As the initial search space can be formed under uncertain conditions, so, there should be a possibility for the cognitive assistant to improve the situation.

We try to adapt the search space, summing up the input data about the initial search space and the collected data about the false pseudonyms, detected earlier. Also we assume that some web resources may have certain intersections of the audiences.

4 Search Space Adaptation Based on the Distributed Ledger Elements and Fuzzy Cognitive Map Usage

Consider a User, who has text samples of a known author and an initial search space described as a list of WEB-resources to analyze. Also consider that every user has his own copy of data, which describes some statistics of false pseudonyms found. This copy of data is organized as a distributed ledger due to its advantages. In our particular case they are: low network load and time of the "read" operations (data analysis) and the possibility and convenience to organize an inter-user data exchange. When a false pseudonym is detected, the information about this event is put into the ledger and so every user can make a new search with the usage of these data, updated by some other users.

The information about the pseudonym detected is put into the ledger in the following format:

<Acc1, t1, Resource1, Acc2, t2, Resource2>, where
Acc1 and Acc2 – pseudonyms of the same author;
t1 and t2 – time moments, when those pseudonyms were created (the need of this particular data is clarified below).

The scheme of the method proposed is presented in Fig. 3.

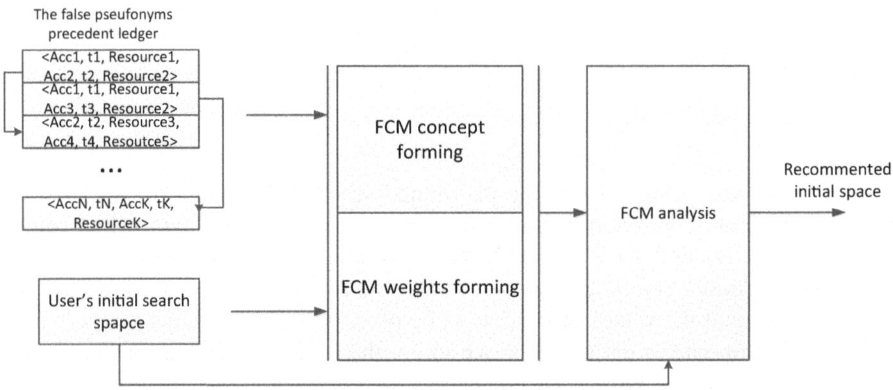

Fig. 3. A method of the search space adaptation based on the distributed ledger concept and the cognitive analysis

In the method proposed we also assume that audiences of the web-resources get intersected very often (and it is proved in [4]). So, having some statistic data on the audiences intersections through the web resources, which are contained in the ledger, we can improve the pseudonyms search in the following ways:

- to enhance the initial search space taking into account the resources, which are tightly coupled;
- having the results of the cognitive analysis, the resources can be arranged so as to improve the time of concurrent crawling/data analysis.

5 FCM Construction

A cognitive map can be described as a qualitative model of how a given system operates. The map is based on defined variables and the causal relationships between these variables (Fig. 4).

Many studies have used cognitive mapping to look at decision-making [18–20], Kosko [21] modified Axelrod's cognitive maps, which were binary, by applying fuzzy causal functions with real numbers in $[-1, 1]$ to the connections, hence the term fuzzy cognitive map (FCM) has emerged.

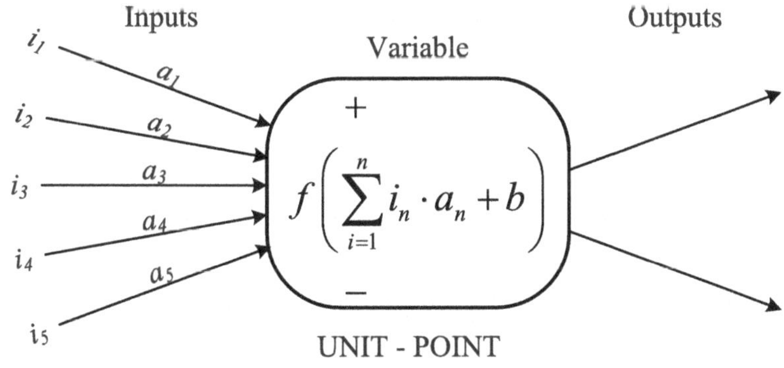

Fig. 4. FCM unit

In our case, a cognitive map construction has its peculiarities. Firstly, FCM is a directed graph, where the directions of the ridges are determined by the casual relation between the concepts. To construct FCM, the following steps were performed:

1. Determine the threshold numbers, which describe the "power" of the relations between the resources (0.8 – tightly coupled, 0.5 – average coupling; 0.2 and less – weakly coupled, for example).
2. To take the input data – the initial search space;
3. For each item on the list of the resources given do:

   – add the "$resourse_i$" to the FCM, if there is no such a concept;
   – search the ledger for the records with "$resourse_i$";
   – add all the concepts, which are connected with the "$resourse_i$".

4. For all the concepts in the FCM perform the following:

   – create a causal relation between "$resource_i$" and "$resourse_{i+1}$" taking into account, how much false pseudonyms of the same authors were detected there. For example, if there are 1000 accounts on the "$resourse_i$" and 800 accounts of the same authors on the "$resourse_{i+1}$", and the accounts on the "$resourse_i$" were created earlier, than the causal relation "$resourse_i$ - $resourse_{i+1}$" is "tightly coupled" and directed to the "$resource_{i+1}$". So, the time of the accounts creation determines the direction and the weight of the graph ridge.

Here a few worlds must be said about the principle of the FCM casual relations forming. If the audiences of two web resources intersect, there must be a cause and a consequence on the FCM. We consider the "cause"-resource as the one which has more accounts registered earlier, than the accounts of the same authors on the "consequence"-resource. So, to build the FCM, it is sufficient to analyze the time moments of the false pseudonyms creation on the resources.

6 Simulation Results: The Usage of the FCM

We created the FCM matrix (Fig. 5) (with the usage of FCMapper software) [22].

Resources	A	B	C	D	E	F	G	H	I	J	K	L	M
A	0,00	0,80	0,80	0,90	0,60	0,00	0,00	0,00	0,00	0,00	0,00	0,00	0,00
B	0,20	0,00	0,00	0,00	0,00	0,80	0,00	0,00	0,00	0,00	0,00	0,00	0,00
C	0,30	0,00	0,00	0,00	0,00	0,00	0,00	0,00	0,00	0,00	0,00	0,00	0,00
D	0,10	0,00	0,00	0,00	0,00	0,00	0,70	0,90	0,00	0,00	0,00	0,00	0,00
E	0,20	0,00	0,00	0,00	0,00	0,00	0,00	0,00	0,50	0,00	0,00	0,00	0,00
F	0,10	0,00	0,00	0,00	0,00	0,00	0,00	0,00	0,00	0,00	0,00	0,00	0,00
G	0,00	0,00	0,00	0,20	0,00	0,00	0,00	0,00	0,00	0,00	0,00	0,00	0,00
H	0,00	0,00	0,00	0,20	0,00	0,00	0,00	0,00	0,00	0,00	0,00	0,90	0,90
I	0,00	0,00	0,00	0,20	0,00	0,00	0,00	0,00	0,00	0,80	0,80	0,00	0,00
J	0,00	0,00	0,00	0,00	0,80	0,00	0,00	0,00	0,50	0,00	0,00	0,00	0,00
K	0,00	0,00	0,00	0,00	0,00	0,00	0,00	0,00	0,50	0,00	0,00	0,00	0,00
L	0,00	0,00	0,00	0,00	0,00	0,00	0,00	0,30	0,00	0,00	0,00	0,00	0,00
M	0,00	0,00	0,00	0,00	0,00	0,00	0,00	0,30	0,00	0,00	0,00	0,00	0,00

Fig. 5. FCM matrix

As is shown in Fig. 6, it contains the causal loops (resources A-B, A-C, A-F, e.g.)

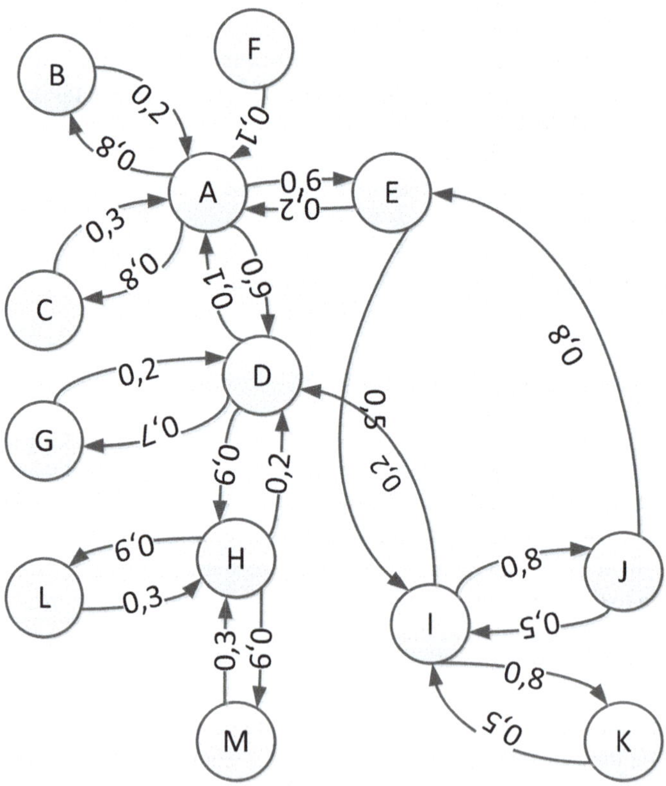

Fig. 6. WEB-resources FCM

The goal of the simulation is to determine if the resource FCM allows to enhance or arrange the web-resources.

To test the FCM we used the following scenarios, describing the initial search space (Figs. 7, 8 and 9):

1. (A, K) vertexes activation;
2. (D, G) vertexes activation;
3. (B) vertexes activation.

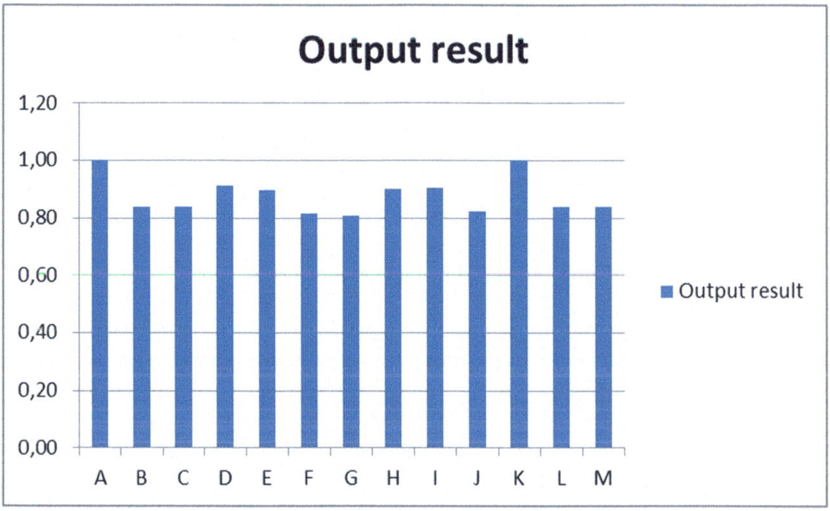

Fig. 7. (A, K) vertexes activation outputs

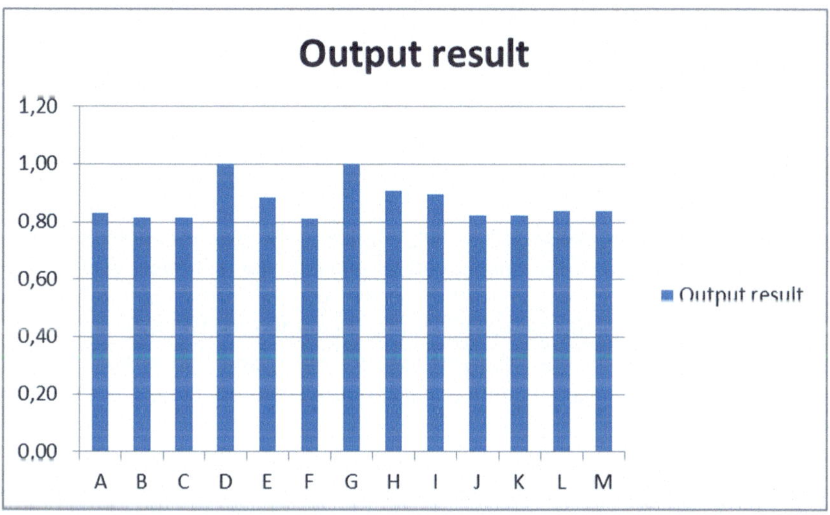

Fig. 8. (D, G) vertexes activation outputs

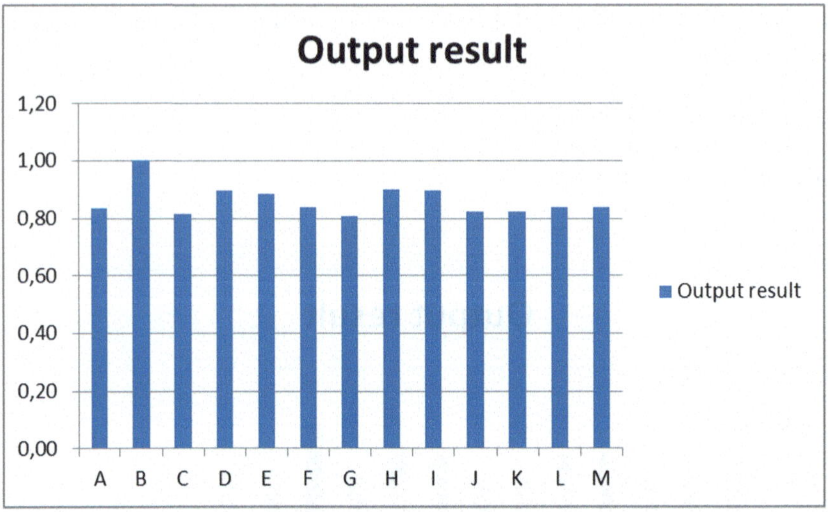

Fig. 9. (B) vertex activation outputs

The vertexes, which have the maximum output, describe the web resources, which are tightly connected with the initially given ones in terms of audience intersection. For example, with activation of vertex B the vertexes H and I must be explored too, and after them the vertex D should be analyzed. Then, arranging these resources by the strength of relations, we can get a range (B, H,I,D). Knowing the number of registered pseudonyms, the assistant search procedure can be initiated from the resource with the smallest number of pseudonyms, and so the following advantages are reached:

- the quality of the search is higher (potentially more pseudonyms can be found because of the related resources analysis);
- the time of the first results producing is reduced (due to the STF schedule type).

7 Conclusions

In this paper a method of search space adaptation for the false pseudonyms detection is proposed. The problem of the false pseudonyms detection is topical as it is and can be considered from the various points of view. We consider it from the victim behavior angle, because cases of victim behavior relate to the multiple false pseudonyms.

While the search of false pseudonyms requires much time and resources, we consider the method for the search results quality improvement by means of search space adaptation.

Adaptiveness of the search space is implemented by means of FCM with the usage of distributed ledger elements. Having the initial search space and the data of pseudonyms found in the ledger copy as an input data, a FCM is constructed. The relations between the concepts are formed with the information about the time of pseudonyms registering and their quantity on the WEB-resources. The method of Search space adaptation based on the distributed ledger elements and fuzzy cognitive map usage is the main contribution of this paper. The simulation conducted shows how the FCM implements the adaptation of the pseudonyms search space:

- it can enhance it, when there are tight relations between resources, built by means of ledger copy analysis;
- the possibility to arrange the WEB-resources according to the general number of the pseudonyms allows to reduce the search time and receive the first results faster.

Acknowledgements. The paper has been prepared within the RFBR projects 18-29-22093.

References

1. Melnik, E., Korovin, I., Klimenko, A.: A cognitive assistant functional model and architecture for the social media victim behavior prevention. In: Silhavy, R. (ed.) CSOC 2019. AISC, vol. 985, pp. 51–61. Springer, Cham (2019). https://doi.org/10.1007/978-3-030-19810-7_6
2. Melnik, E.V., Korovin, I.S., Klimenko, A.B.: The improvement of the stylometry-based cognitive assistant performance in conditions of big data analysis. In: Silhavy, R. (ed.) CSOC 2020. AISC, vol. 1226, pp. 85–99. Springer, Cham (2020). https://doi.org/10.1007/978-3-030-51974-2_8
3. Melnik, E.V., Klimenko, V.V., Klimenko, A.B., Korobkin, V.V.: The technique of data analysis tasks distribution in the fog-computing environment. In: Kovalev, S., Tarassov, V., Snasel, V., Sukhanov, A. (eds.) IITI 2019. AISC, vol. 1156, pp. 142–151. Springer, Cham (2020). https://doi.org/10.1007/978-3-030-50097-9_15
4. Afroz, S., Caliskan-Islam, A., Stolerman, A., Greenstadt, R., McCoy, D.: Doppelgänger finder: taking stylometry to the underground. In: Proceedings of IEEE Symposium on Security and Privacy (2014). https://doi.org/10.1109/SP.2014.21
5. Afroz, S., Brennan, M., Greenstadt, R.: Detecting hoaxes, frauds, and deception in writing style online. In: Proceedings of IEEE Symposium on Security and Privacy (2012). https://doi.org/10.1109/SP.2012.34
6. Brennan, M., Afroz, S., Greenstadt, R.: Adversarial stylometry: circumventing authorship recognition to preserve privacy and anonymity. ACM Trans. Inf. Syst. Secur. **15**, 1–23 (2012). https://doi.org/10.1145/2382448.2382450
7. McDonald, A.W.E., Afroz, S., Caliskan, A., Stolerman, A., Greenstadt, R.: Use fewer instances of the letter "i": toward writing style anonymization. In: Fischer-Hübner, S., Wright, M. (eds.) PETS 2012. LNCS, vol. 7384, pp. 299–318. Springer, Heidelberg (2012). https://doi.org/10.1007/978-3-642-31680-7_16
8. Almishari, M., Tsudik, G.: Exploring linkability of user reviews. In: Foresti, S., Yung, M., Martinelli, F. (eds.) ESORICS 2012. LNCS, vol. 7459, pp. 307–324. Springer, Heidelberg (2012). https://doi.org/10.1007/978-3-642-33167-1_18
9. Almishari, M., Kaafar, M.A., Oguz, E., Tsudik, G.: Stylometric linkability of tweets. In: Proceedings ACM Conference on Computer and Communications Security (2014). https://doi.org/10.1145/2665943.2665966
10. https://arxiv.org/ftp/arxiv/papers/1705/1705.09929.pdf
11. Viswanath, B., et al.: Towards detecting anomalous user behavior in online social networks. In: 23rd USENIX Security Symposium (USENIX Security 2014), pp. 223–238 (2014)

12. Wang, A.H.: Detecting spam bots in online social networking sites: a machine learning approach. In: Foresti, S., Jajodia, S. (eds.) DBSec 2010. LNCS, vol. 6166, pp. 335–342. Springer, Heidelberg (2010). https://doi.org/10.1007/978-3-642-13739-6_25
13. Solorio, T., Hasan, R., Mizan, M.: A case study of sockpuppet detection in wikipedia. In: Workshop on Language Analysis in Social Media (LASM) at NAACL HLT, pp. 59–68 (2013)
14. Zheng, X., Zeng, Z., Chen, Z., Yu, Y., Rong, C.: Detecting spammers on social networks. Neurocomputing **159**, 27–34 (2015)
15. Ochab, J., Essler, H.: Stylometry of literary papyri. In: DATeCH2019: Proceedings of the 3rd International Conference on Digital Access to Textual Cultural Heritage, pp. 139–142 (2019). https://doi.org/10.1145/3322905.3322930
16. Nirkhi, S., Dharaskar, R.V., Thakare, V.M.: Authorship verification of online messages for forensic investigation. Phys. Procedia **78**, 640–645 (2016). https://doi.org/10.1016/j.procs.2016.02.111
17. Canbay, P., et al.: Deep combination of stylometry features in forensic authorship analysis. Int. J. Inf. Secur. Sci. **9**(3), 154–163 (2020)
18. Axelrod, R.: Structure of Decision: The Cognitive Maps of Political Elites. Princeton University Press, Princeton (1976)
19. Bougon, M., Weick, K., Binkhorst, D.: Cognition in organizations: an analysis of the Utrecht Jazz Orchestra. Admin. Sci. Quart. **22**, 606–639 (1977)
20. Brown, S.M.: Cognitive mapping and repertory grids for qualitative survey research: some comparative observations. J. Manag. Stud. **29**, 287–307 (1992)
21. Kosko, B.: Fuzzy cognitive maps. Int. J. Man-Machine Stud. **1**, 65–75 (1986)
22. http://www.fcmappers.net/joomla/index.php?option=com_content&view=article&id=52&Itemid=53

Analytical Studies of Behavior of Users of the Moscow Electronic School Service

E. V. Lavrenova[1], Yu. V. Frolov[2], and T. M. Bosenko[2](✉)

[1] Institute of Digital Education, Moscow City Pedagogical University, 4 Vtoroy Selskohoziajstvenny proezd, Moscow 129226, Russia
lavrenovaev@mgpu.ru

[2] Department of Computer Science, Management and Technology, Moscow City Pedagogical University, Sheremetyevskaya Street, 28, Moscow 129594, Russia
bosenko@bmstu.ru

Abstract. The article presents the results of the study, which analyzed the database of the Moscow Electronic School (MES) in the amount of about 50 thousand observations. The relevance of the study is that with the emergence of viral threats and the transition of the education system to remote learning processes, MES educational content is becoming increasingly popular. In this regard, it is important to plan work to improve the quality of user characteristics of the accumulated content and its impact on the indicators of the remote educational process. The key subject on which the motivation of the students depends and the results of their education is the teacher. Teachers in the MES system act simultaneously in different roles - both as creators of new educational content and as its users and evaluators. The aim of the study was to identify "hidden" groups of users with different typical behaviors in the MES database. The behavior of users in this work has been investigated by decomposing it into six parameters. The study used the MES database, as well as statistical, neural network methods and application software packages implementing them, which solved data clustering and visualization tasks. Based on the results of the study, the relationships between the behavioral parameters of users were determined and 5 behavioral types of MES users were identified. The analysis of used methods of statistical analysis and clustering of MES data made it possible to assess their applicability in the prediction of user groups and to identify factors that significantly influence the achievement of the goal of improving the efficiency of the MES platform and design in the perspective of the recommendation system. #CSOC1120

Keywords: Moscow electronic school · Database · MES users · Indicators of behavior of users · Data clustering methods · Visualization of analytical research results

1 Introduction

The Moscow Electronic School is a cloud-based Internet platform containing educational materials, tools for their creation and editing, as well as educational scripts and programs

© The Author(s), under exclusive license to Springer Nature Switzerland AG 2021
R. Silhavy (Ed.): CSOC 2021, LNNS 229, pp. 121–132, 2021.
https://doi.org/10.1007/978-3-030-77445-5_11

constructors [1]. At present in the MES system: more than 520,000 educational materials are placed; approximately 55,000 teachers use the MES resources; there are more than 41,540 interactive lesson scripts, 348 electronic textbooks, 1,270 electronic educational programs for objects, created by teachers of Moscow schools in various subjects.

This study purpose was to identify the MES teachers (users) groups who exhibit different behaviours when interacting with educational content. Also, the work purpose was to propose and experimentally test (by examples) measurable indicators that would allow dividing users into clusters with homogeneous behaviour and which could be used in the future in the design of a recommender system.

The work scientific problem consisted in formalizing complex dependencies between the MES users' behavioural parameters and in identifying measurable indicators that would allow, firstly, to separate the MES users by their behaviour types in the system and, secondly, to play the role of filters - criteria in the design process recommender system. According to the problem terms, the users' behaviour in the MES system is written using six variable parameters, which values change when moving from cluster to cluster. This knowledge kind, according to the authors, will make it possible to more effectively manage the choosing educational content and technologies complex processes for working in the system by new users, in particular, young specialists - teachers by referring them to users' already existing groups with identical features in behaviour and professional interests.

This type research' relevance, especially in setting the task terms, the tools used are also due to their possible use as real-life cases in the educational process of training data analytics specialists, whose competencies should cover not only knowledge and skills in analysis techniques (machine learning), but also the ability to set research tasks and interpret their results [2]. Also, it is important to train University students studying in the pedagogy and business Informatics programs to use new analytics tools, such as applied statistical analysis packages and business analytics platforms [3].

2 Literature Review

The MES users (teachers) classification is based on their unequal participation in the generation and consumption of educational content.

In e-learning, user roles and functions are usually unevenly distributed. Users' relatively small percentage create content. A participants' slightly larger number is moderately active users who perform, for example, such actions as "like", evaluate, copy for use in their activities. Most of the users are passive observers [4].

In the literature, this observation is called the rule 1-9-90 [5–10]. According to this rule, approximately 1% of users create the content bulk, 9% are less active in generating the content, and 90% of users are content consumers created by others. In this regard, this work task consisted, in particular, in the MES platform users' behavioural models study, belonging to different groups according to their activity degree criterion in working with content.

To plan activities for a cloud-based e-learning system development, which an example is the Moscow E-School, it is important to identify patterns in such an ecosystem formation and self-regulation. In this context, one of the managing the network community priority tasks, which is formed around the educational objects' repository, is the users clustering by typical behavioural indicators, predicting the user groups behaviour, as well as supporting users to increase their activity in the system [4].

3 Methods

The study used cluster analysis methods. Cluster analysis is a multivariate statistical procedure that collects data containing information about an objects sample and then arranges the objects into relatively homogeneous groups (clusters). A cluster is an element group characterized by a common property, the cluster analysis main goal is to find similar objects groups in a sample [11–14].

There is no generally accepted clustering methods classification, but an approaches number can be distinguished. The most common clustering methods, which more detailed description is given in [15], are the following algorithms: 1) Method k-averages [16, 17]; 2) EM-algorithm [18]; 3) Neural network Kohonen, which is designed to solve vector classification problems [19, 20].

For the tasks categories accepted for study in this study, the following tools were used:

- applied software package Anaconda (Python 3.7): studying data structure; descriptive statistics exploration; using correlation methods between parameters; preparing data for clustering; cluster analysis methods implementation.
- IBM SPSS Statistics v.26 2019 [21], STATISTICA TIBCO [22]: input values correlation; studying the means of descriptive statistics; multiple statistical analysis means to study and results' evaluation; cluster analysis studying methods; deep data processing methods study, data clustering using the Kohonen map method, results' visualization [23].

4 The Problem Statement

The study problem statement included a cluster analysis based on data from the MES system with the filled cells number – 49,390 × 8. The MES platform users' behaviour was decomposed into parameters - options for user actions with objects in the system, presented in Table 1.

In the study course, the task was also set to consider the relationships between users based on records in the MES system about their actions with the repository educational objects based on data mining methods.

Table 1. Accepted for the study behavioural indicators (options for user actions) on the MES platform.

Category no	Name
1	Id_Autors
2	Views number
3	Number of additions to favourites
4	Marks awarded number
5	Number of scenarios used in the remote sensing (RS)
6	Copies created number
7	Scripts created number

5 The Study Results'

The study results' first stage, the MES user base data structure was examined.

As is known [9, 10], to conduct cluster analysis, the data must meet the following criteria:

- data distribution should be close to normal;
- indicators (measurements, data) should not correlate with each other;
- indicators (measurements, data) must meet the "stability" requirement, which is understood as the influence absence on their random factors values;
- the data sample should be homogeneous, not contain "outliers".

Using PPP Anaconda (Python 3.7), the MES platform users' behavioural indicators statistical indicators were determined. The results were provided in Table 2.

Table 2, the following designations are adopted: mean - the behavioural indicator average value in the sample 49,389 observations; std is the indicator value standard deviation, which characterizes the data spread in the sample; 25% -, 50% -, 75% - a specific behavioural indicator all observations' percentage associated with its value; min, max - the behavioural indicator (observation) minimum and maximum values.

Shown in Table 2 the results, in particular, show that each indicator (user action option) average value is determined by approximately three-quarters (75%) of this action option values from them in the MES user database total number.

To generate scatterplots and histograms on the MES user base, the Python 3.7 package was used.

The diagrams analysis obtained made it possible to draw the following conclusions. First, there is a linear relationship between the created copies number values and the created scenarios number. Secondly, there is a value "the scenarios uses number nonlinear correlation in the remote sensing" relative to all other parameters (user actions). Thirdly, you can notice additions to favourites action number weak correlation with three parameters – "the scripts created number", "the copies created number", "the marks given number".

Table 2. Behavioural indicators (options for user actions) statistical indicators on the 49,389 observations data general population.

	Views number	Number of Additions to favourites	Marks awarded number	Number of scenarios used in the remote sensing	Copies created number	Scenarios created number
Mean	236.138	43.204	12.846	2.193	23.426	25.861
Std	584.769	147.93	108.185	21.172	95.074	91.227
Min	0	0	0	0	0	0
25%	28	4	0	0	1	3
50%	84	11	1	0	4	7
75%	233	35	6	0	12	17
Max	36,968	9,679	10,777	1,301	6,197	4,762

In the cluster analysis context, it is important to understand how closely the studied input values (user actions) are interconnected. To test whether there is a statistically significant relationship between the behavioural indicators of the MES user base, correlation coefficients were calculated (Table 3).

From the presented results (Table 3), it can be seen that there is a strong correlation between the additions to favourites number, on the one hand, and the copies created number, the scripts created number, on the other hand. The results also show that the copies created number and the scripts created number are also strongly related to.

To check the unusual observations influence (to which include outliers in the studied behavioural indicators values) on the MES user base clustering quality, sample G was studied (with a 28340 observations volume, which was obtained by removing "unusual" observations from the MES user database general population. On this sample, was tested the six clusters presence hypothesis. The data clustering obtained results on sample G are illustrated in Table 4 and Table 5.

The results obtained (Table 4 and Table 5) show, firstly, that observations in the cleared outliers sample are distributed relatively evenly and, secondly, the only one parameter changes the value to a greater extent from cluster to cluster – "views". At the same time, other parameters remain practically unchanged.

It was concluded that the sample from the general population G does not allow dividing MES users into groups based on the behavioural indicators set. Therefore, further studies were carried out on a 49389 observations general population.

Table 3. Input quantities (Pearson's method) correlation.

Input values	Views number	Additions to favourites number	Marks number	Scenarios used number in remote sensing	Copies created number	Scripts created number
Views number	1	0.339	0.325	0.068	0.314	0.377
Additions to favorites number	0.339	1	0.483	0.041	0.721	0.725
Marks number	0.325	0.483	1	0.015	0.365	0.380
Scenarios used number in remote sensing	0.068	0.041	0.015	1	0.033	0.037
Copies created number	0.314	0.721	0.365	0.033	1	0.978
Scripts created number	0.377	0.725	0.380	0.037	0.978	1

Table 4. The sample G users' distribution by clusters from the MES user base general population.

Cluster	Observations in each cluster number
1	1,165
2	10,584
3	2,843
4	1,856
5	4,773
6	7,119
Total	28,340

In the work, numerical experiments were carried out to determine the average values of indicators (options for user actions) and the users' number in each cluster by sequentially increasing the clusters' number from four to ten.

Table 5. Sample G clustering results from the MES user base general population.

Average parameter values						
Cluster	1	2	3	4	5	6
Views number	213	13	113	157	73	40
Additions to favourites number	16	4	14	15	12	9
Marks number	3	1	2	2	2	1
Scenarios used number in remote sensing	0	0	0	0	0	0
Copies created number	6	2	5	6	5	4
Scripts created number	9	3	8	9	8	6

Table 6 shows an MES user base study results on the clusters ten number.

Table 6. Ten clusters (100 iterations, K-means method) Analysis.

The cluster №.	Observations in a cluster number	Indicator for filtering data	Views number	Additions to favourites number	Marks number	Scenarios used number in remote sensing	Copies created number	Scripts created number
1	1	M	305	11	9,718	0	11	17
		kpi	305	11	9,718	0	11	17
2	2	M	6,573	8,260	6,297	0	5,354	5,356
		kpi	3,286.5	4,129.75	3,148.5	0	2,676.75	2,677.75
3	68	M	9,424	200	272	14	167	274
		kpi	138.58	2.93	4	0.2	2.45	4.03
4	54	M	2,512	2740	719	6	1,617	1,548
		kpi	46.5	50.73	13.31	0.1	29.94	28.67
5	2	M	33,753	533	2,622	45	250	432
		kpi	16,876.3	266.5	1,310.75	22.5	125	216
6	1	M	15,579	5,137	10,777	0	4,633	4,636
		kpi	15,579	5,137	10,777	0	4,633	4,636
7	566	M	3,150	209	122	9	144	186
		kpi	5.5	0.3687	0.2153	0.0165	0.2542	0.3282
8	43507	M	114	26	6	1	12	14
		kpi	0.0026	0.0006	0.0001	0	0.0003	0.0003
9	3	M	7,635	4,461	7,090	0	103	100
		kpi	2,545	1,487	2,363.22	0	34.22	33.44

(*continued*)

Table 6. (*continued*)

The cluster №.	Observations in a cluster number	Indicator for filtering data	Views number	Additions to favourites number	Marks number	Scenarios used number in remote sensing	Copies created number	Scripts created number
10	5185	M	888	153	43	5	90	95
		kpi	0.1712	0.0294	0.0082	0.001	0.0174	0.0184

Notes: Clusters 1, 2, 5, 6, 9 were excluded from the analysis due to the observations low number in these clusters: cluster 1 - one observation; cluster 2 - 2 observations; cluster 5 - 2 observations; cluster 6 - one observation; cluster 9 - 3 observations.

Table 6, clusters are identified that have observations low number and were therefore excluded from the analysis.

Each behavioural indicator is presented in the table by two numerical (measurable) indicators - its average value in cluster "M" and this average value ratio to the users' number in a particular cluster – "kpi". These two indicators allow, firstly, to divide the MES users according to their behaviour types in the system and, secondly, they will serve as filters - criteria in the recommender system design.

The performed multivariate analysis made it possible to identify five clusters (Table 6). So, for example, the motivated users' behaviour (cluster 4, Table 6) is a vector that is influenced by such parameters as "The scripts created number", "The copies created number", "The marks given number", "The additions to favourites number" and "Views number". At the same time, the five indicated parameters four average values have the highest values in comparison with other behavioural types - highlighted in Table 6 clusters.

Table 7. Parameters (user actions) average values in the MES database (Kohonen map, 10 clusters, the sample of 49390 observations).

The cluster №.	1	2	3	4	8	9	10
Observations in a cluster number	4,068	20	155	774	44,166	61	134
Views number	746	2,715	617	1,286	161	378	1,939
Number of add. to favourites	165	3,213.5	117	261	19.5	63	1,199
Marks number	41	857	28	140	6.7	14.7	279.8
Scenarios used number in remote sensing	3	11.3	318	14.5	0.98	1.3	1.6
Copies created number	92	2,679	116	123	7.5	30	844
Scripts created number	97	2,558	58	119	10.1	34.8	816

Notes. Clusters №. 5, 6, 7 were excluded from the analysis due to the observations low number in these clusters: cluster 5 – 8 observations; cluster 6–2 observations; cluster 7 – one observation.

In general, the regularities in the MES users behaviour identified using the K-means clustering method (Table 6) confirm the MES database analytics results performed with the Kohonen neural network (Table 7).

6 Discussion

Let's analyze the results presented in the Table 6. Clusters 1, 2, 5, 6, and 9 with 1, 2, and 3 users are outliers, so they are excluded from the analysis. Thus, the analyst's research object can be the remaining 5 clusters, namely: clusters 3, 4, 7, 8 and 10.

The most numerous are cluster 8 with 43507 users. The users in this cluster can be characterized as *"passive MES users"*, since the most important in this cluster is the user's action option - the views number. Moreover, this behavioural indicator value is minimal compared to its values in all other clusters.

At the opposite extreme in activity terms are users from cluster 4 who actively create content, copy it, and add it to favourites. This cluster has a total of 54 users. They can be called *"motivated users" or "experts"*, because they not only generate content but also actively rate it. Thus, the behavioural indicator "the marks given number" value is the maximum for the studied sample, namely: the M average value is 719, and the kpi value is 13.31.

It should be noted that the previously revealed correlation between the indicators (options for user actions) "the created scenarios number" and "the copies created number" is confirmed. If we refer to the results presented in Table 6, then a pattern is observed, namely: these two parameters have the same order values in each cluster and change synchronously when passing from cluster to cluster.

Approximately the users (as in cluster 4) the same number, namely 68, are in the third cluster. However, these users' behaviour on the MES platform differs from the users' behaviour represented in cluster 4. Cluster 3 users are characterized, first, by high activity in viewing "foreign" content created by other users. This indicator value is the maximum value for the entire sample – 9,424 or per user 138.5. Also, these users show relatively high activity, evaluating the other users content (272 and 4), ranking second after experts.

The second place after experts is the users who entered cluster 3, also in created number parameters terms by the script and the copies created number. Therefore, they can reasonably be referred to as the MES system *active users* group.

The second-largest cluster (after cluster 8) is occupied by a cluster of 10 – 5,185 users. The differences in user behaviour from cluster 10 versus the behaviour shown by users from the largest cluster 8 is that they are more active in browsing content, adding content to favourites, and creating and copying scripts. Therefore, these users can be called *users with a motivation threshold level* to work in the MES system.

Finally, let's analyze the users' behaviour from cluster 7, which includes 566 users. These users are intermediate in activity on the MES platform between active users and users with a threshold motivation level. They show moderate activity in their actions in the system, performing operations on viewing content, and favourites adding content to, and setting ratings, and generating and copying scripts. On this basis, the users included in cluster 7 can be assigned to the *users' group with moderate activity in the system.*

Based on the analysis performed using the results example from Table 6 you can build the users five groups following row according to their activity degree in the system (in activity ascending order from minimum to maximum): passive *MES users; users with a motivation threshold level; users with moderate activity in the system; active users; motivated users (experts).*

From the analysis presented in Table 6 results obtained using the K-means method, it can be concluded that the actions (parameter) option "The scenarios uses a number in the remote sensing" is not a feature that could divide the system users according to their behaviour types.

The clustering results obtained using Kohonen maps (Table 7) show that in the sample dividing case into 10 clusters, 7 clusters with more than 20 users can be identified. However, there are some peculiarities. For example, the type *motivated users (experts)* can include those users who are located in clusters 2 and 10 by the neural network. In contrast to the results from Table 6, the neural network identified motivated users two subgroups (Table 7): 1) a smaller one (cluster 2) with the views number maximum values, the additions' number to favourites, the marks given number, as well as with the copies created and created scenarios largest observed values and 2) a subgroup (cluster 10), slightly inferior to users from the first subgroup in all the above indicators.

The second difference from the results according to the K-means method is that the neural network split the passive users largest *group* into two clusters - cluster 8 and cluster 9. Users, whom the neural network has identified in the smaller cluster 9, show slightly higher activity in the views number terms, the copies created number and the scripts created number.

The MES users who fell into cluster 1 (Table 7) can be referred to the group *users with the motivation threshold level.* These users main behaviour is that they are more active than passive users in activities such as viewing content, adding to favourites, making copies and creating scripts.

The neural network classified users included in cluster 4 to the active users' *group* (Table 7). They are slightly inferior to experts in the content views number terms, and also show relatively high activity in rating terms, copying and scripting actions.

And, finally, Kohonen's neural network ranked users who formed cluster 3 to the group *with moderate activity in the system* (Table 7). A distinctive feature in these users' behaviour, revealed by the neural network, is that these users are very active in using content for homework. This indicator "318" average value significantly exceeds the values characterizing user behaviour in other clusters.

Based on the study results using the K-means method (Table 6) it was noted that the action option (parameter) "script uses a number in the remote sensing" is not a feature that could divide system users by their behaviour types. Therefore, the result revealed by the neural network requires additional research.

7 Conclusion

In studying the data structure process characterizing the MES platform users' behaviour, scatter diagrams and histograms were built. It has been established that there is a linear relationship between the indicators "Copies created number" and "Scripts created number" values.

Options correlation analysis for user actions by Pearson's method is performed. From the presented results (Table 3), it can be seen that there is a strong correlation between the additions to favourites number, on the one hand, and the copies created number, the scripts created number, on the other hand. The results also show that the copies created number and the scripts created number are also strongly related to.

Research has been carried out on a sample from the general population G with a volume of 28340 observations, obtained by clearing the MES user base from "atypical" observations. It was found that the clusters probable number in the sample G is 6 possible clusters. Based on the studies performed results, it was found that the sample from the general population G does not allow dividing the MES users into groups according to the behavioural parameters set.

Numerical (determined by calculation) indicators are proposed: 1) the behavioural indicator (user action) average value in the cluster and 2) this average value ratio to the users' number in the cluster. These two indicators set in the numbers form allow, firstly, to divide the MES users into groups under their behaviour types in the system and, secondly, they will presumably serve as filters - criteria in the designing a recommender system process for MES users.

Based on the research performed, user behaviour five types in the MES system were identified (in increasing their activity in the system order): "passive MES users", "Users with a motivation threshold level"; "Users with moderate activity in the system"; "Active users"; "Motivated users (experts)".

Based on the study results using the K-means clustering method, it was determined that the action option (parameter) "scenario uses a number in the remote sensing" is not a feature that could divide system users by their behaviour types. However, the results obtained by the Kohonen neural network show that this behavioural indicator is included as an important factor in the users' cluster with moderate activity in the system.

The performed studies, according to the authors, not only create preconditions for the design of a recommendation system for MES users in the future but also allow making proactive decisions to improve the MES platform efficiency and service capabilities as a tool for distance (mixed) education.

References

1. Moscow Electronic School. https://mes.mosedu.ru
2. Frolova, E.V., Ryabova, T.M., Rogach, O.V.: Digital technologies in education: problems and prospects for Moscow electronic school project implementation. Eur. J. Contemp. Educ. **8**(4), 779–789 (2019)
3. Frolov, Y.V.: effective technologies of economic choice of MSPU (2001)
4. Frolov, Y.V., Bosenko, T.M.: Training of personnel for the development of innovative entrepreneurship. Acad. Entrep. J. **26**(327), 1–6 (2020)
5. Skrypnikova, N.N.: Technology of mixed training: relevance and problems. Prof. Educ. Labor Mark. **3**, 74–78 (2018)
6. Rezankova, H.: Cluster analysis of economic data. Statistica **94**(1), 73–86 (2014)
7. Mierlo, T.: The 1% rule in four digital health social networks: an observational study (2014)
8. Bandura, A.: Social foundations of thought and action: a social cognitive theory (1986).
9. Leshed, G.: Posters, lurkers, and in between: a multidimensional model of online community participation patterns. In: Proceedings of HIC (2005)

10. Saraçli, S., Doğan, N., Doğan, I.: Comparison of hierarchical cluster analysis methods by cophenetic correlation. J. Inequ. Appl. **203**, 1–8 (2013)
11. Rheingold, H.: The Virtual Community: Homesteading on the Electronic Frontier. MIT Press, Cambridge (2000)
12. Han, J., Kamber, M., Pei, J.: Data Mining: Concepts and Techniques, 3rd edn., p. 703. Elsevier Science Ltd. Publication (2011)
13. Ayvazyan, S.A., Buchstaber, V.M., Enyukov, I.S., Meshalkin, L.D.: Applied statistics: classification and dimension reduction, p. 607. Finance and Statistics, Moscow (1989)
14. Mandel, A.D.: Cluster analysis, p. 176. Finance and statistics, Moscow (1988)
15. Tyurin, A.G., Zuev, I.O.: Cluster analysis, clustering methods and algorithms. Bull. MGTU MSIRTEA **2**(3), 86–97 (2014)
16. Van Raisin, J: Classification and cluster, p 390. Mir, Moscow (1980)
17. Neal, R., Hinton, G.: A view of the EM algorithm that justifies incremental, sparse, and other variants. In: Jordan, M.I. (ed.) Learning in Graphical Models: Journal, pp. 355–368. MIT Press, Cambridge (1999)
18. Kohonen, T.: Self-organization and Associative Memory. Springer, Berlin (1987)
19. Burakov, M.V.: Nejronnye seti i nejrokontrollery: ucheb. Posobie. SPb.: GUAP, p. 284 (2013)
20. Software platform for statistical analysis IBM SPSS STATISTICS 26. https://www.ibm.com/ru-ru/products/spss-statistics
21. STATISTICA TIBCO Data Science software platform for statistical analysis and machine learning. https://www.tibco.com/products/data-science/downloads
22. Tableau Desktop software platform. https://www.tableau.com/products/desktop

Information and Control Systems with Distributed Ledger Usage: A Reliability Issue

Anna Klimenko[1(✉)] and Eduard Melnik[2]

[1] Scientific Research Institute of Multiprocessor Computer Systems of Southern Federal University, 2, Chekhov Street, 347928 Taganrog, Russian Federation
[2] Federal Research Centre the Southern Scientific Centre of the Russian Academy of Sciences, 41, Chekhov Street, 344006 Rostov-on-Don, Russian Federation

Abstract. The paper deals with the issue of reliability for the Information and Control systems with the distributed ledger usage. A considerable number of ICSs uses fog computing concept and, at the same time, the usage of the distributed ledger technologies is quite relevant. Yet, the question of the ICS reliability is not studied well, though the reliability function is quite important for technical complexes. The main contribution of this paper is the analytical models, which allow to estimate the nodes reliability with the usage of distributed ledger as an integral part of a system. Also some selected simulation results are presented and described.

Keywords: Information and control systems · Reliability · Distributed ledger

1 Introduction

Nowadays information and control systems are intensively used by a wide range of complex technical objects, including those ones, which are distributed geographically. The advert of Internet of Things and the fog computing made it possible to relocate the computational load and so to offload, for example, cloud nodes, or edge devises. In addition to the fog computing concept, various distributed ledger technologies have emerged to provide the consistent and up-to-date information to geographically distributed users. Besides, distributed ledger copies allow to conduct intensive read-operations with relatively low network load, if the ledger copy is in the fog or at the edge of the network. Being convenient, the usage of distributed ledger as a part of ICS, still provokes the question of the system reliability. Obviously, providing data consistency leads to extra data exchange through the network. So the reliability of the computational nodes can degrade dramatically. As the field of the ICS reliability with the distributed ledger usage has not been studied yet, in this paper some models for the reliability function estimation are presented. Also some selected simulation results are demonstrated.

© The Author(s), under exclusive license to Springer Nature Switzerland AG 2021
R. Silhavy (Ed.): CSOC 2021, LNNS 229, pp. 133–144, 2021.
https://doi.org/10.1007/978-3-030-77445-5_12

2 Information and Control Systems in the Fog: The Reliability Issue

Fog computing concept was proposed in 2012 [1, 2], its usage areas grow intensively last decade, and many applications use the fog to offload the edge devices, or to reduce the workload of the cloud. Some information and control systems use the fog computing concept, a small part of them is mentioned in [3–5].

However, very little attention has been paid to the reliability issue of the Information and Control Systems (ICSs), which use the fog computing. To the best of our knowledge, the particular reliability questions are considered in [6–10], and the potential of the workload shifting to the edge is considered partially in [11–14]. The issue of ICS reliability related to nodes has not been considered in the literature so far, though the computational nodes are an integral part of ICS.

Our previous works [15–18] propose the following approach to the system reliability: the reliability function of the computational node depends on its temperature, which, in its turn, depends on the computational workload and is represented by the following equation:

$$P(t) = e^{-\lambda_0 \cdot 2^{dL/10} t}, \tag{1}$$

where

λ_0 - is a failure rate;
L - is a node workload;
d - is a ratio between the workload and the temperature of the node.

In general, as every node has higher values of the reliability function, the system behaves in the same way. So, when the workload is distributed evenly, there is a higher possibility that every element of the system will work without failures for the longer time period.

Besides, in our previous works we have considered the issue of the so-called transit load: this is an additional workload, which the node has, when it is on the data transmission route. Simulation results show that in case of workload distribution according to the fog-computing concept, it is quite important to pay attention not only to the computational tasks positioning and the nodes performance, but also to that nodes, which transfer the data: in some cases the application of the fog computing concept to the system can decrease the reliability functions of the nodes dramatically due to the high volumes of the transit data. This phenomena has been described in our works [19, 20].

So, with the usage of distributed ledger the issue of the reliability of ICS is quite relevant, because the distributed ledger copies generate extra network workload, and, dealing with the transit workload, there is a range of questions, such as:

- which parameters affect the ledger copy positioning in the fog;
- which volumes of information for reading/writing to the ledger are preferable;
- which performance the transit nodes should have not to worsen the reliability of the system in general, etc.

The main contribution of this paper is analytical models, which allow to estimate the full node workload, including the transit one, and so to choose the optimal positioning of the ledger copy in terms of reliability.

In the following section we will give a short overview of the contemporary distributed ledger technologies.

3 Distributed Ledger Technologies and Information Exchange Model of the Blockchain-Based Ledger

The distributed ledger is a consensus of replicated, shared, and synchronized digital data geographically spread across multiple sites, countries, or institutions [21]. Nowadays the distributed ledger is widely used in cryptocurrencies and various applications [22–24].

There is a wide range of data storage types and consensus types. Data can be stored as:

- SQL or noSQL data storage;
- Blockchain structures;
- Blocklattice structure;
- Hashgraph;
- DAG, etc.

As to consensus type, the following types were developed:

- Proof-based algorithms;
- Vote-based algorithms.

The first group, proof-based, includes such well-known consensus types as: Proof of work, Proof of Stake, Delayed Proof of Work, Delegated Proof of Stake, Proof-of-Authority, Proof of weight (Proof of reputation, Proof of Space), Proof of elapsed time, Proof-of-Capacity, Proof-of-history, Proof of Stake Velocity, Proof of Importance, Proof of Burn, Proof of Identity, Proof of Activity, Proof of Time, Proof of Existence, Proof of believeability and others.

The second group includes Practical Byzantine Fault tolerance (PBFT), Paxos, Zab, Viewstamped Replication, Raft. An interesting example of consensus reaching is Hedera Hashgraph based on the Swirlds hashgraph consensus algorithm with its virtual voting.

Yet, in this work we consider the proof-based consensus as the most studied and widely applied. Also we assume some generalization and simplification of the ledger functioning. We consider the model of ledger software, which implements the following functions:

- to receive the transaction data;
- to form the transaction block;
- to implement the consensus;
- to implement the information exchange with other ledger replicas;
- to implement the data storage on the node;
- to deliver data to the user (e.g. information about the balance);
- to deliver data for the transaction verification;
- transaction verification.

The information interconnections are presented in Fig. 1.

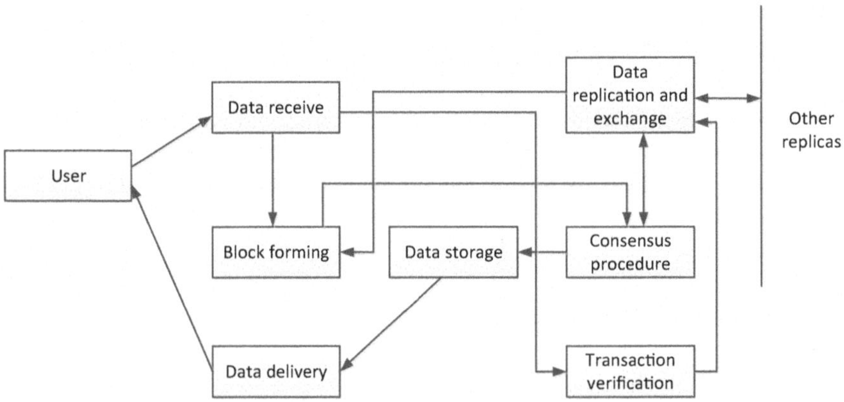

Fig. 1. The scheme of the information exchange within the ledger copy

The scheme can be generalized as follows in terms of external data exchange (Fig. 2).

Fig. 2. The general scheme of external data exchange of the ledger copy

In the next section a model of information exchange within the ledger is presented and described.

4 Workload Modeling for the System with Distributed Ledger Usage

Consider data volumes generated by the ledger replica:

k – is the number of users, which work with one ledger replica (assume that k can be >1, as, for example, it is implemented in sets of wallets);

V_w – is a data volume per one write operation;

H_w – is the number of data transfers from one user in the borders of one data block, which size is a constant;

V_r – is the data volume which can be read from the ledger replica per one user read operation;

H_r – is the the number of read operations accordingly.

So, the information interaction between the users on the network edge and the ledger copy can be described as:

$$V_u = k(v_r h_r + v_w h_w). \tag{2}$$

Consider the information exchange between ledger replicas. Through the ledger functioning there are two main types of inter-replicas data exchanges:

- transaction transmission through the all nodes;
- the "winner" block dissemination.

For the model simplification consider the case, when all transaction are successfully verified and there are no forks in the blockchain, so, only one block is accepted as a winner of competition.

Then the data volume, which is transfered by the ledger replica to other replicas can be estimated as:

$$V_l = kV_w h_w + V_{block}; \tag{3}$$

Then assuming that all transactions sent to the ledger are in the block, the data volume is estimated as follows:

$$V_l = 2kV_w h_w. \tag{4}$$

Consider the ledger replicas interaction on the simplified network topology and configuration, assessing the transit load generated by the ledger replica.

The network topology is presented in Fig. 3. We pay attention to the nodes 7 and 11 with performances *p1* and *p2* accordingly. The performance of the cloud nodes is *p3*.

Node №7 is situated between the ledger replica and the users, so, it transits data to write to the ledger replica and to read accordingly. The transit workload of the node 7 can be assessed as follows:

$$L_{tr}^u \sim k(v_w h_w + v_r h_r); \tag{5}$$

Consider ξ as a coefficient, which is a ratio between the transit data volume and the workload, generated by these data transmission.

$$L_{tr}^u \sim k\xi(v_w h_w + v_r h_r). \tag{6}$$

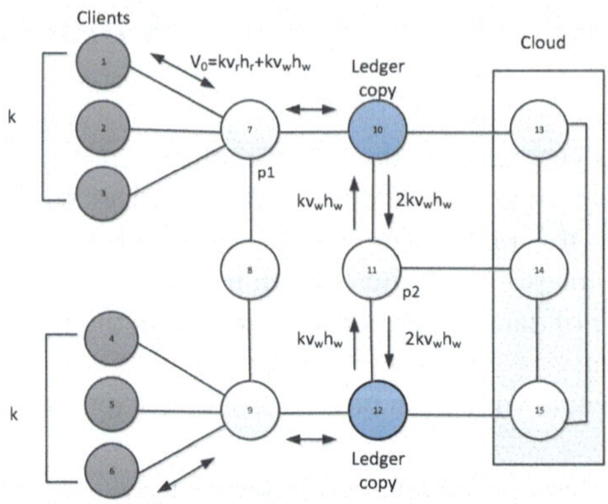

Fig. 3. The example network topology

With the p1 and T (the time of block forming) given, the estimation of the transit workload of the node 7 can be calculated as it is shown in Eq. 7.

$$L_{tr}^{u} = k\xi(v_w h_w + v_r h_r)/p_1 T. \tag{7}$$

Consider the estimation of the transit workload of the node 11.

For the network topology given, whith two ledger copies, the following data exchange takes place:

- Both of the nodes send transactions to each other;
- By the fact of competition win, one of the nodes sends a "winner" block to other replica.

So, node 11 has the following transit workload:

$$L_{tr}^{l} = \frac{3\xi k v_w h_w}{p_2 T}. \tag{8}$$

So, if the node is a transit one, and transits data between N ledger copies, iys workload will be:

$$L_{tr}^{l} = \frac{2\xi N k v_w h_w + k v_w h_w}{p_2 T}. \tag{9}$$

Summing up, the nodes, which are between the users pool and the ledger copy, have a transit workload, which can be estimated by the following equations:

For the bus topology, as in our example network topology, the transit load for the transit nodes is:

$$L_{tr}^{u} = k\xi(v_w h_w + v_r h_r)/p_1 T; \tag{10}$$

In case of multiple routes, and if the routing process splits the data to the parallel branches, accordingly:

$$L_{tr}^{u} = k\xi(v_w h_w + v_r h_r)/p_1 Tr, \tag{11}$$

where r is the number of parallel routes.

Nodes, which are on the routes interconnecting the ledger copies, have the following transit load:

$$L_{tr}^{l} = \frac{2\xi Nkv_w h_w + kv_w h_w}{p_2 T}, \tag{12}$$

and

$$L_{tr}^{l} = \frac{2\xi Nkv_w h_w + kv_w h_w}{p_2 Tr} \tag{13}$$

for the parallel routes with split data transmission volumes.

Yet, further we will use the Eqs. (10) and (12), because, in fact, these equations are estimations of the transit workload of the most pessimistic scenario of the data transfer and routing process.

Consider the workload for the other nodes of the ICS.

The workload of the nodes 10, 12 with ledger copies is described as follows:

$$L_{lo} = (3\xi kv_w h_w + \xi v_r h_r + W_l)/p_2 T. \tag{14}$$

If there are some elements with reliability functions $P_1(t)$, $P_2(t)$,...$P_n(t)$, then, with the enhancement of each $P_i(t)$ within the time period given, the overall $P(t)$ will be improved too, because of:

$$P(t) = P_1(t) * P_2(t) * \ldots * P_i(t). \tag{15}$$

So, for the system showed in Fig. 3, the overall reliability function can be estimated as:

$$P(t) = P_7(t) * P_8(t) \ldots P_{14}(t). \tag{16}$$

To illustrate the example (Fig. 3), consider the situation, when the ledger copies are positioned on the nodes 10, 12, and the routes and node performances are constant, while the write/read operations can be planned to adapt the transit workload of the system.

Considering Eq. (1), the following estimations are conducted for the nodes:

$$P_{7,9}(t) = e^{-\lambda_0 \cdot 2^{\frac{dk\xi(v_w h_w + v_r h_r)}{10 p_1 T}} t}; \tag{17}$$

$$P_{11}(t) = e^{-\lambda_0 \cdot 2^{\frac{d(2\xi Nkv_w h_w + kv_w h_w)}{10 p_2 T}} t}; \tag{18}$$

$$P_{10,12}(t) = e^{-\lambda_0 \cdot 2^{\frac{d(3\xi kv_w h_w + v_r h_r + W_l)}{10 p_1 T}} t} \tag{19}$$

So, the reliability function of the system presented in Fig. 3 is estimated by the following equations.

$$P_0(t) = P_7(t) \cdot P_{10}(t) \cdot P_{11}(t) \cdot P_{12}(t) \cdot P_9(t) \tag{20}$$

Consider the case when the ledger copies are positioned on the cloud nodes. Then, nodes 7, 10, 12, 9 are nodes, which transit data between users and the ledger copies. The main peculiarity of the ledger in the cloud is that there can be no transit nodes between ledger copies in the cloud, moreover, the copies can be set within one physical computational node. So, the overall reliability can be estimated as follows:

$$P_0(t) = P_7(t) \cdot P_{10}(t) \cdot P_{13}(t) \cdot P_{15}(t) \cdot P_{12}(t) \cdot P_9(t), \tag{21}$$

where

$$P_{13,15}(t) = e^{-\lambda_0 \cdot 2^{\frac{d(3\xi k v_w h_w + v_r h_r + W_l)}{10 p_3 T}} t}; \tag{22}$$

$$P_{7,9,10,12}(t) = e^{-\lambda_0 \cdot 2^{\frac{dk\xi(v_w h_w + v_r h_r)}{10 p_1 T}} t}. \tag{23}$$

The following section contains the simulation results of the reliability estimations for cases when the distributed ledger copy is in the fog and when the distributed ledger copy is in the cloud.

5 Simulation Results

We conducted experiments for the following conditions, which are presented in the Table 1.

Table 1. Experiment conditions

Operations intensiveness	Ledger positioning in the fog-layer	Ledger positioning in the cloud
Intensive read operations	Ledger nodes 10, 12	Ledger nodes 13, 15
Intensive write operations	Ledger nodes 10, 12	Ledger nodes 13, 15
Intensive read and write operations	Ledger nodes 10, 12	Ledger nodes 13, 15

With the usage of the variable values depicted, the following graphs were conducted (Figs. 4, 5 and 6).

One can see that with the conditions given and with the heterogeneous performance of the nodes in different layers the positioning of the ledger copies in the fog layer is more preferable in terms of system reliability.

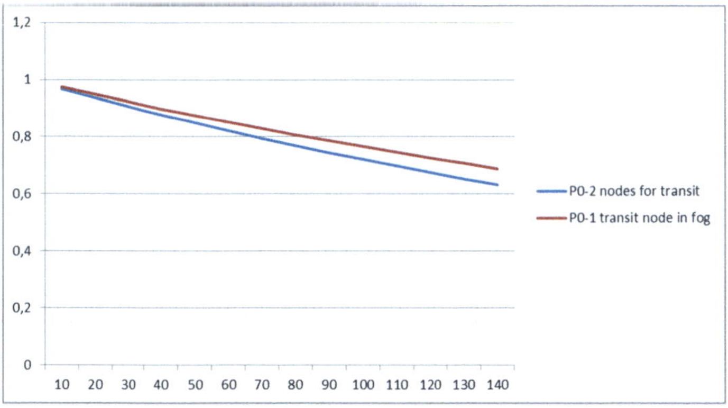

Fig. 4. Intensive read operations with ledger positioning on the nodes 10, 12 and 13, 15.

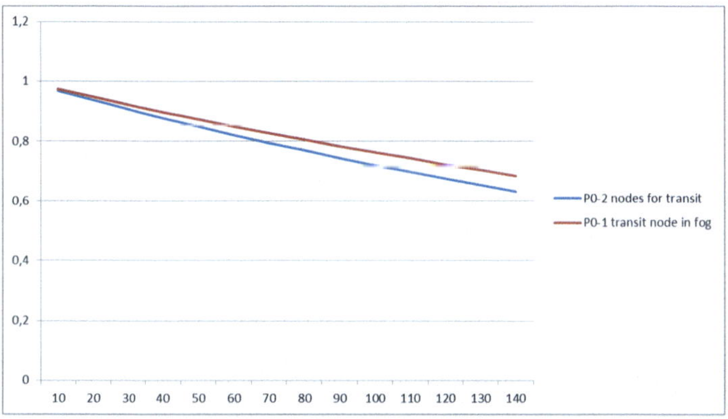

Fig. 5. Intensive write operations with ledger positioning on the nodes 10, 12 and 13, 15.

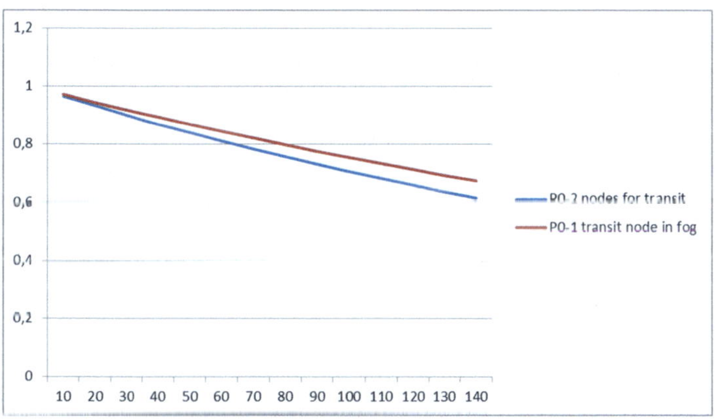

Fig. 6. Intensive write and read operations with ledger positioning on the nodes 10, 12 and 13, 15.

Yet, the performance of the fog nodes is less than the one of the cloud ones. So, with constant read-write operations intensiveness, the growth of the workload generated by the ledger itself (e.g., the Proof of Work procedure) can restrict the possibility to use the fog layer for the ledger copy positioning.

The following graph depicts the tendency of the reliability function value degradation with the ledger workload growth (Fig. 7).

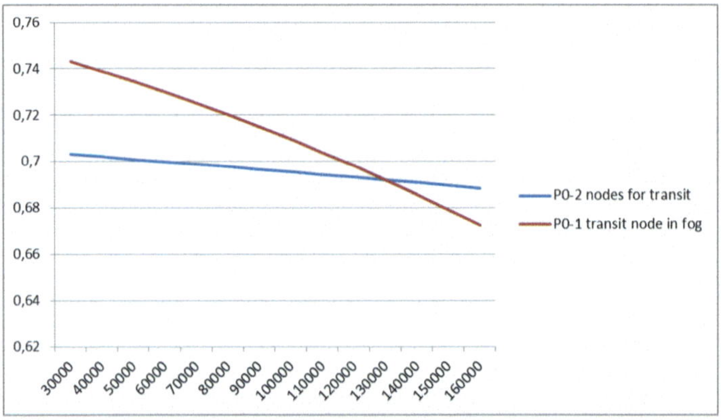

Fig. 7. The reliability function values depending on the ledger operational workload growth for the constant time = 100(h) and ledger positioning on the nodes 10, 12 and 13, 15.

So, one can see that there can be cases, when it is more preferable to use the cloud to position the ledger copies, though, the most part of the ledger positioning cases shows the advantages of the fog-layer positioning.

6 Conclusions

As many applications use the distributed ledger technology as well as fog computing concept, so a question of system reliability arises. Considering the ICSs as an integration of the functional tasks and computational nodes, this research focuses on that, how the transit data, generated by the distributed ledger functioning, affects the system reliability.

The main contribution of this paper is a complex of analytical estimations, which allow to position the ledger copies paying attention to the system reliability. The models proposed are of three types:

- For the transit nodes, which are between the clients (sensors) and the ledger copy;
- For the nodes with ledger copies;
- And for the nodes, which are the transit ones between the ledger copies.

Our estimations show that in the most cases the positioning of the ledger copies in the fog layer of the network is more preferable (at least under the conditions of blockchain-based ledger). Yet, with the increase of the workload generating by the consensus procedure and the database maintenance, the positioning of the ledger in the fog is hardly expedient because of the restricted performance of the fog nodes.

Acknowledgements. This study is supported by RFBR projects №. 18-29-22086 and 18-05-80092.

References

1. Alrawais, A., Alhothaily, A., Hu, C., Cheng, X.: Fog computing for the internet of things: security and privacy issues. IEEE Internet Comput. **21**, 34–42 (2017). https://doi.org/10.1109/MIC.2017.37
2. Moysiadis, V., Sarigiannidis, P., Moscholios, I.: Towards distributed data management in fog computing. Wirel. Commun. Mob. Comput. (2018). https://doi.org/10.1155/2018/7597686
3. Cao, Y. Chen, S., Hou, P., Brown, D.: FAST: a fog computing assisted distributed analytics system to monitor fall for stroke mitigation. In: Proceedings of 2015 IEEE International Conference on Networking, Archit. Storage, NAS 2015, pp. 2–11 (2015). https://doi.org/10.1109/NAS.2015.7255196
4. Aazam, M., Huh, E.N.: E-HAMC: leveraging Fog computing for emergency alert service. In: 2015 IEEE International Conference on Pervasive Computing Communication Workshops, PerCom Workshops 2015, pp. 518–523 (2015). https://doi.org/10.1109/PERCOMW.2015.7134091
5. Gia, T.N., Jiang, M., Rahmani, A.M., Westerlund, T., Liljeberg, P., Tenhunen, H.: Fog computing in healthcare internet of things: a case study on ECG feature extraction. In: Proceedings - 15th IEEE International Conference on Computer and Information Technology, CIT 2015, 14th IEEE International Conference on Ubiquitous Computing Communication, IUCC 2015, 13th IEEE International Conference on Dependable, Auton. Se, pp. 356–363 (2015). https://doi.org/10.1109/CIT/IUCC/DASC/PICOM.2015.51
6. Madsen, H., Albeanu, G., Burtschy, B., Popentiu-Vladicescu, F.: Reliability in the utility computing era: towards reliable fog computing. In: International Conference on Systems, Signals, Image Processing (2013). https://doi.org/10.1109/IWSSIP.2013.6623445
7. Pereira, J., Ricardo, L., Luís, M., Senna, C., Sargento, S.: Assessing the reliability of fog computing for smart mobility applications in VANETs. Future Gener. Comput. Syst. (2019). https://doi.org/10.1016/j.future.2018.11.043
8. Popentiu-Vladicescu, F., Albeanu, G.: Software reliability in the fog computing. In: 2017 Proceedings International Conference on Innovations in Electrical Engineering and Computational Technologies, ICIEECT 2017 (2017). https://doi.org/10.1109/ICIEECT.2017.7916578
9. Liu, Y., Fieldsend, J.E., Min, G.: A framework of fog computing: architecture, challenges, and optimization. IEEE Access (2017). https://doi.org/10.1109/ACCESS.2017.2766923
10. Sood, S.K.: SNA based QoS and reliability in fog and cloud framework. World Wide Web **21**(6), 1601–1616 (2018). https://doi.org/10.1007/s11280-018-0525-x
11. Inaltekin, H., Gorlatova, M., Mung, C.: Virtualized control over fog: interplay between reliability and latency. CoRR. abs/1712.0 (2017). https://arxiv.org/pdf/1712.00100.pdf
12. Mukherjee, M., Kumar, S., Shojafar, M., Zhang, Q., Mavromoustakis, C.X.: Joint task offloading and resource allocation for delay-sensitive fog networks. In: IEEE International Conference on Communications (2019). https://doi.org/10.1109/ICC.2019.8761239
13. Bittencourt, L.F., Lopes, M.M., Petri, I., Rana, O.F.: Towards virtual machine migration in fog computing. In: Proceedings of 2015 10th International Conference on P2P, Parallel, Grid, Cloud Internet Computing, 3PGCIC 2015 (2015). https://doi.org/10.1109/3PGCIC.2015.85

14. Alam, M.G.R., Tun, Y.K., Hong, C.S.: Multi-agent and reinforcement learning based code offloading in mobile fog. In: International Conference on Information Networking (2016). https://doi.org/10.1109/ICOIN.2016.7427078
15. Melnik, E., Klimenko, A., Klimenko, V.: A recovery technique for the fog-computing-based information and control systems. In: Silhavy, R., Silhavy, P., Prokopova, Z. (eds.) CoMeSySo 2018. AISC, vol. 860, pp. 216–227. Springer, Cham (2019). https://doi.org/10.1007/978-3-030-00184-1_20
16. Melnik, E., Klimenko, A., Korobkin, V.: Reconfigurable distributed information and control system multiagent management approach. In: Abraham, A., Kovalev, S., Tarassov, V., Snasel, V., Vasileva, M., Sukhanov, A. (eds.) IITI 2017. AISC, vol. 680, pp. 84–95. Springer, Cham (2018). https://doi.org/10.1007/978-3-319-68324-9_10
17. Melnik, E.V., Klimenko, A.B.: A workload distribution problem model and online constraint forming technique for the control systems in the fog-computing environment. J. Phys. Conf. Ser. (2019). https://doi.org/10.1088/1742-6596/1333/4/042014
18. Melnik, E.V., Klimenko, A.B.: A novel approach to the reconfigurable distributed information and control systems load-balancing improvement. In: Proceedings of 11th IEEE International Conference on Application of Information and Communication Technologies, AICT 2017 (2019). https://doi.org/10.1109/ICAICT.2017.8687069
19. Klimenko, A.B., Melnik, E.V.: An experimental study of the fog-computing-based systems reliability. In: Silhavy, R. (ed.) CSOC 2020. AISC, vol. 1225, pp. 438–449. Springer, Cham (2020). https://doi.org/10.1007/978-3-030-51971-1_36
20. Melnik, E.V., Klimenko, A.B.: A condition of reliability improvement of the system based on the fog-computing concept. J. Phys.: Conf. Ser. **1661**, 012007 (2020)
21. Distributed Ledger Technology: beyond block chain (PDF) (Report). UK Government, Office for Science. January 2016. Accessed 29 Aug 2016
22. Walport, M.: Distributed ledger technology: beyond block chain. Gov. Off. Sci. (2015)
23. Burkhardt, D., Werling, M., Lasi, H.: Distributed ledger. In: Proceedings of the 2018 IEEE International Conference on Engineering, Technology and Innovation, ICE/ITMC 2018 (2018)
24. Maull, R., Godsiff, P., Mulligan, C., Brown, A., Kewell, B.: Distributed ledger technology: applications and implications. Strateg. Chang. **26**, 481–489 (2017)

A Modified Ontology-Based Method of Workload Relocation Problem Solving for Monitoring and Forecasting Systems

E. V. Melnik[1], I. B. Safronenkova[1(✉)], and A. B. Klimenko[2]

[1] Federal Research Center the Southern Scientific Center of the Russian Academy of Sciences, 41, Chekhov Street, 344006 Rostov-on-Don, Russian Federation

[2] Scientific Research Institute of Multiprocessor Computer Systems of Southern Federal University, 2, Chekhov Street, 347922 Taganrog, Russian Federation

Abstract. The current paper deals with the problem of workload relocation in distributed monitoring and forecasting systems in terms of their reliability and system response. The research of publications presented allows us to conclude that existing methods of workload relocation problem-solving do not ensure getting of a satisfying result. Also, some specific characteristics of such systems are not incorporated in known methods. In this paper the modification of the existing workload relocation problem-solving method was proposed. In this context, a new domain ontological model and a new set of production rules were developed to adopt the method, which was proposed for distributed CAD systems. Some simulations for two different scenarios were conducted and theoretical conclusions were confirmed. A particular interest is the fact that the higher fog "deepness" value, the higher the effectiveness of developed method.

Keywords: Monitoring system · Forecasting system · Workload relocation · Ontology · Fog-computing · Distributed system

1 Introduction

The influences of human activities along with the cyclical movements of nature are the main reasons for emergency incidents escalation. It leads to expansion in the number of disasters such as floods, forest and peat fires, deforestation and desertification. Herewith, dangers and threats are complex and interconnected in the contemporary world, so one threat can involve a chain of threats. The complex and global character of threats demands new approaches to safety support for countries and regions. The monitoring and forecasting systems usage is the promising solution of the problem [1]. The efficiency of such systems depends on implemented methodological base, which has to provide the following requirements [2]:

- noninterruptable and effective data collecting;
- relevant and qualitative data analysis,

© The Author(s), under exclusive license to Springer Nature Switzerland AG 2021
R. Silhavy (Ed.): CSOC 2021, LNNS 229, pp. 145–152, 2021.
https://doi.org/10.1007/978-3-030-77445-5_13

- appropriate monitoring and forecasting data reporting;
- computing and technologic process (collecting, representation, forecasting) automation.

Today, the most advanced technology of distributed monitoring and forecasting systems implementation is a fog-computing concept [3–5]. In that respect, the question of effective operation, which involves reliability enhancement and reduces the system response time, emerges. As far as the fog-computing concept assumes the possibility of workload relocation to the "edge" of a network, the time reduce of workload relocation problem solving seems the one of the ways of system performance improvement.

In [6] the method of workload relocation problem-solving for information control systems, which operate in fog-computing environment, is proposed. This method involves "on-line" restrictions forming on computing nodes set using a "local device group". Nevertheless, iterative character of the proposed method does not ensure the satisfying result getting under the limited time conditions.

In [7] authors proposed a method, focused on time reducing of workload relocation problem solving in distributed CAD systems, based on fog-computing concept. The key idea of this method is the procedure of ontological analyses. The result of such a procedure is the reduction of a search space for workload relocation. However, population methodology of workload relocation in distributed CAD systems can not to be applied to monitoring and forecasting systems field, because it does not take into account the specific characteristics of such systems operation.

In that respect, the development of new methods and algorithms of workload relocation problem solving in fog-computing environment for monitoring and forecasting systems is an actual scientific problem.

2 A Modified Method of Workload Relocation Problem Solving

Here, a new method of workload relocation problem solving in monitoring and forecasting systems based on fog-computing concept is proposed. The basis of this method is a modified methodology, described in [7].

In this context, time reduction of workload relocation problem solving is achieved by means of opportunity to "eliminate" those nodes, which do not require to resource demands of a relocated problem. It was proposed to use an ontological analysis, which is a key component of this method.

Further, we use the following terms.

Candidate node – a computing node in the fog layer, which can potentially be used for relocated problem.

Leader node – a computing node, which is elected from the set of nodes, where the problem was initially performed, by any leader election algorithm.

An ontology-based analysis procedure involves the following steps:

- incoming input data;
- input data classification in accordance with domain ontology classes, which represent relocated problem and candidate nodes data;

- production rules appliance to ontology classes to limit nodes set, which is a result of leader node operation, that involves collection of information about existing resources;
- decision-making on limited set of nodes for workload relocation.

The flow-chart of ontology-based analysis algorithm is presented in Fig. 1.

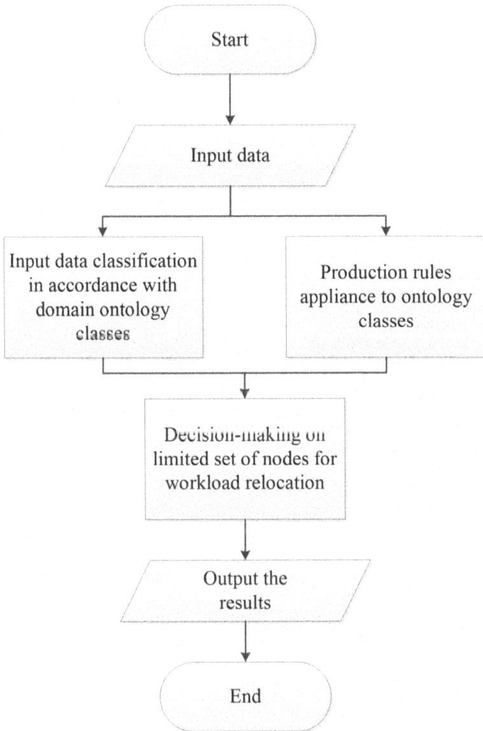

Fig. 1. An ontology-based analysis algorithm flowchart.

Consider each step of ontology-based analysis. In the context of monitoring and forecasting system we presuppose the following input data for ontology-based analysis:

1. Parameters the problem to be relocated, which are:

   - functional type: monitoring (subtype: collecting, preprocessing, storage), forecasting (subtype: cognitive modeling, forecast, computation, expert's opinion based on sensor data);
   - data type: sensory, service, modeling/decision-making/computation/evaluation, data for external systems.
   - initial problem localization on the Internet: cloud, fog, edge.
   - processing problem time, [second].

2. Candidate nodes parameters, which are:

- efficiency [number of tasks/ time unit].
- workload, [%];
- remoteness from a leader node [hop];
- computing node type: server, user computer, expert computer.

An ontological model of location of computational problems of monitoring and forecasting systems on candidate nodes was developed [8, 9]. Model development was made in Protégé 5.5.0. In Fig. 2 all can see an ontology class hierarchy.

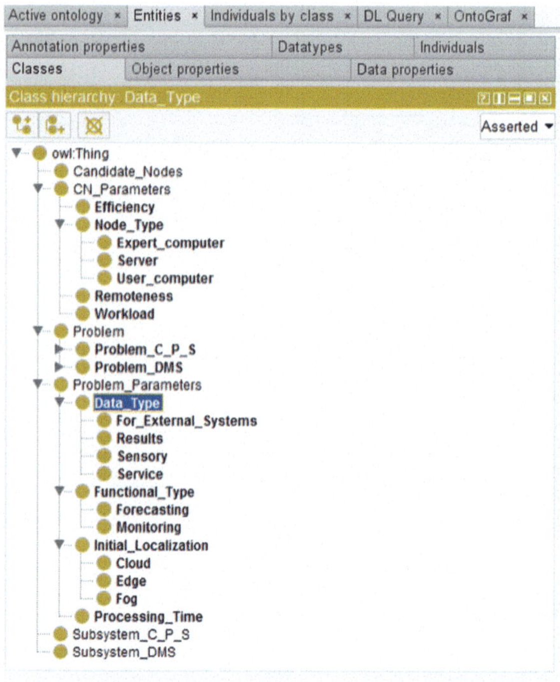

Fig. 2. An ontology class hierarchy.

The fragment of developed ontological model of location of computational problems of monitoring and forecasting systems on candidate nodes is presented in Fig. 3.

Ontology-based analysis provides to reduce a number of candidate computational nodes for workload relocation due to domain knowledge, which is represented in an ontological model.

Input data classification in accordance with domain ontology classes can be implemented using any known classification facilities, including the naive Bayesian classifier, the support vector method, the k-nearest neighbor method [10].

A logical conclusion which is a part of decision-making on a preferable set of computational nodes for workload relocation is implemented using a set of production rules, which connects an ontological class, containing knowledge about the relocated problem, and a potential set of nodes for workload relocation.

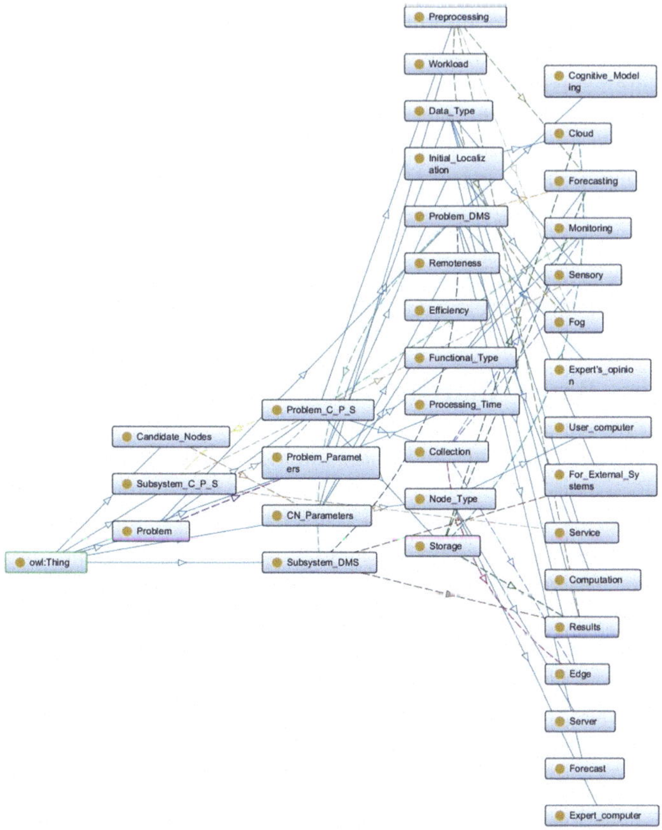

Fig. 3. The fragment of ontological model of computational problems of monitoring and forecasting systems location on candidate nodes.

Consider an example of a set of production rules for workload relocation in monitoring and forecasting systems. This set of production rules includes the main principles of effective workload distribution in heterogeneous computing systems [11].

IF functional_type = monitoring **AND** problem_subtype = collecting **AND** data_type = sensory **AND** problem_localization = edge **AND** execution_time = T, **AND** resource_requirements**=** threshold_value, **THEN** $\forall d_i \rightarrow$ min.

where d_i – a distance between computational node, where a problem is initially implemented, and a candidate node for its relocation.

IF functional_type = forecasting **AND** problem_subtype = cognitive_modeling **AND** data type = modeling results **AND** problem localization = cloud **AND** execution_time = T, **AND** resource_requirements = threshold_value, **THEN** $\forall d_i \rightarrow$ min.

where d_i – a distance between computational node, where a problem is initially implemented, and a candidate node for its relocation.

IF functional_type = monitoring **AND** problem_subtype = preprocessing **AND** data_type = sensory **AND** problem_localization = fog **AND** execution_time = T, **AND** resource_requirements**=** threshold_value, **THEN** $\forall d_i \rightarrow$ min.

This set of production rules has been developed according to the main principle of a fog-computing concept, which includes the minimization of informational exchanges between the computational nodes [12].

The final step of an ontology based analysis involves a decision-making on a limited set of candidate nodes for workload relocation in a monitoring and forecasting system. Then, a modeling procedure of problem location on a limited set of nodes takes place. At last, a decision is made according to the selected restrictions.

3 Ontology - Based Analysis Simulation Results

To test the theory a program ontological module was developed in Python 3.8.3 [13]. To present the effectiveness of the proposed method some simulations were carried out. We varied the fog "deepness" parameter, which means the maximum distance from a leader node to a computational node in the fog layer, where workload can be relocated, [hop].

1st scenario. We consider the fog "deepness" (Θ) equals 2…5 [hop], available computational nodes (F) in the fog layer is 100.

2nd scenario. We consider the fog "deepness" (Θ) equals 2…30 [hop], available computational nodes (F) in the fog layer is 100.

The effectiveness of proposed algorithm is calculated by the following formula:

$$\Delta = \frac{F - F_o}{F} \times 100\% \tag{1}$$

where F – a number of available nodes in the fog layer,

F_o- a number of nodes after an ontology-based analysis.

The comparison of effectiveness of ontology-based analysis is presented in Table 1 and Fig. 4.

Table 1. The comparison of effectiveness of ontology-based analysis.

Test number	A number of available nodes, (F)	A number of nodes after the ontology-based analysis for Θ [2…5], hop	A number of nodes after the ontology-based analysis for Θ [2…30], hop	Δ [2; 5], %	Δ [2;30], %
1	100	14	3	86	97
2	100	11	**1**	89	**99**
3	100	15	3	85	97

(*continued*)

Table 1. (*continued*)

Test number	A number of available nodes, (*F*)	A number of nodes after the ontology-based analysis for Θ [2…5], hop	A number of nodes after the ontology-based analysis for Θ [2…30], hop	Δ [2; 5], %	Δ [2;30], %
4	100	8	2	92	98
5	100	17	3	83	97
6	100	14	3	86	97
7	100	15	2	85	98
8	100	10	**1**	90	**99**
9	100	11	2	89	98
10	100	21	2	79	98

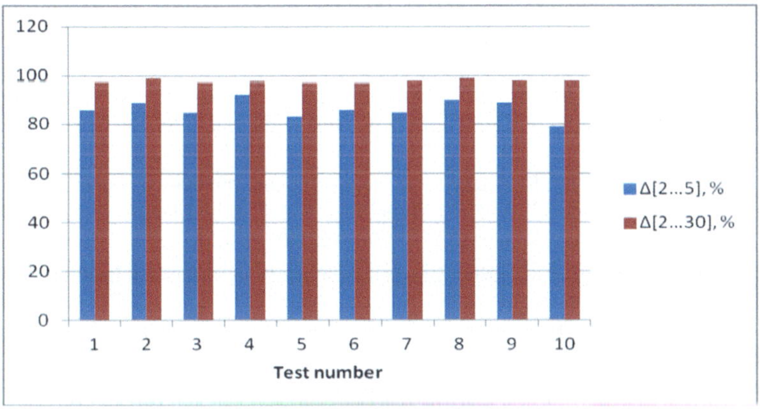

Fig. 4. The plots of effectiveness of ontology-based analysis for different Θ.

4 Discussion and Conclusions

Summarizing the simulation conducted, the following conclusions can be made:

- the ontological analysis method is effective in both scenarios for different values of *Θ*, [hop];
- the proposed method is more effective in the 2nd scenario;
- the highest effectiveness is achieved in the 2nd scenario and reaches 99% in 2nd and 8th test numbers.
- Basing on the theoretical assumptions, which were approved by the simulation results, we conclude, that:
- the proposed method can be used to reduce the time of workload relocation problem solving in the environmental threats monitoring systems domain;

- the higher fog "deepness" value, the higher the effectiveness of developed method;
- the program module, which implements ontology-based analysis, can be used as a functional part in a distributed monitoring and forecasting system to reliability enhancement and reduce the system response time.

Acknowledgements. This study is supported by the by the RFBR project 18-05-80092, 18-29-03229 and the GZ SSC RAS N GR project AAAA-A19–119011190173-6.

References

1. Puchkov V.A.: Modern systems for monitoring and forecasting emergency. FSI Center for Strategic Research of Civil Protection of the Ministry of Emergency Situations of the Russian Federation, Moscow (2013)
2. Vorobyev, Yu.L.: National security and strategic risk management in Russia. Risk Manage. (Special issue), 4–9 (2002)
3. Melnik, E.V., Orda-Zhigulina, M.V., Orda-Zhigulina, D.V., Rodina, A.A.: Reliability method by reconfiguration resources in monitoring and diagnostic systems of hazardous phenomena. Izvestiya Tula State Univ. **2**, 18–26 (2020)
4. Orda-Zhigulina, M.V., Melnik, E.V., Ivanov, D.Ya., Rodina, A.A.: Combined method of monitoring and predicting of hazardous phenomena. Adv. Intell. Syst. Comput. **984**, 55–61 (2019)
5. Qiao, H., Wang, T., Wang, P.: A tool wear monitoring and prediction system based on multiscale deep learning models and fog computing. Int. J. Adv. Manuf. Technol. **108**(7–8), 2367–2384 (2020). https://doi.org/10.1007/s00170-020-05548-8
6. Melnik, E.V., Klimenko, A.B., Ivanov, D.Ya.: A model of device group forming of informational control systems in the fog-computing environment. In: XIII All-Russian Meeting on Control Problems 2019 Proceedings, pp. 2979–2984. Institute of Control Sciences of Russian Academy of Sciences, Moscow (2019)
7. Kureychik, V., Safronenkova, I.: Efficiency enhancement method for distributed CAD systems in fog-computing environments. J. Phys.: Conf. Ser. **1333**, 082006 (2019)
8. Protégé. https://protege.stanford.edu. Accessed 03 Dec 2020
9. Ontology Development 101: A Guide to Creating Your First Ontology. https://protege.stanford.edu/publications/ontology_development/ontology101.pdf. Accessed 03 Dec 2020
10. Machine Learning Classifiers. https://towardsdatascience.com/machine-learning-classifiers-a5cc4e1b0623. Accessed 03 Dec 2020
11. Stojmenovic, I., Wen, S.: The fog computing paradigm: scenarios and security issues. In: Proceedings of 2014 Federated Conference on Computer Science and Information Systems, vol. 2, pp. 1–8 (2014)
12. Cisco Fog Computing Solutions: Unleash the Power of the Internet of Things. https://www.cisco.com/c/dam/en_us/solutions/trends/iot/docs/computing-solutions.pdf. Accessed 03 Dec 2020
13. Safronenkova, I.B., Melnik, E.V., Kureychik, V.M.: Certificate of state registration of a computer program 2020616523 Russian Federation. A module of ontological analysis for distributed CAD system (2020)

Intelligent Data Analysis Using a Classification Method for Data Mining Knowledge Discovery Process

Kamelia Shoilekova(✉)

Angel Kanchev University of Ruse, 8 Studentska Str., 7000 Ruse, Bulgaria
kshoylekova@uni-ruse.bg

Abstract. This paper presents problems, which are connected with data processing issues. An intelligent analysis is made over unstructured data. Based on the analysis, a model is designed that unambiguously juxtaposes the values of the attributes–predictors with the class where to a specified object belongs.

Keywords: Data mining · Data analysis · Classification model

1 Data Mining

Organizations nowadays refer to different information sources and normally generate big amounts of data for the purpose of their daily tasks. They have realized that the human capital and the disposable data become their most valuable resources. The effective use of available data and relevant information is an important factor for establishing both proper functioning and prosperity of organizations.

The effective utilization of large amounts of available data and its further transformation into information and knowledge, needed for the successful growth of organizations, requires application of appropriate aids for analysis and processing of big data volumes having variable data formats, i.e. the usage of innovated methods and tools for Data Mining [1, 2].

At present, Bulgarian universities undergo many changes and face a number of serious problems. Each subsequent year the number of potential applicants for higher universities decreases resulting in a strengthened competition among educational institutions. That circumstance leads to the necessity of providing up-to-date information to assist relevant bodies in making timely and important management decisions. Modern universities collect large data amounts, which in most cases need to be effectively analyzed and used. The problem can be solved with the help of effective data mining methods and techniques [6, 7].

2 Education Data Mining

Educational Data Mining is a field in the area of knowledge discovery from big data volumes intended to solve problems of targeted marketing related to attracting suitable candidates, students' retention and prevention from dropping out, higher quality of

© The Author(s), under exclusive license to Springer Nature Switzerland AG 2021
R. Silhavy (Ed.): CSOC 2021, LNNS 229, pp. 153–157, 2021.
https://doi.org/10.1007/978-3-030-77445-5_14

educational processes [4, 9, 10], management of candidate campaigns, improvement of organizational effectiveness, alumni management. The presented analysis of literature sources in the field of Educational Data Mining shows that the prediction of students' progress appears to be an essential current problem for universities. Students' progress is predicted usually by using data mining methods and tools, as in most cases the predicted parameter is a nominal variable, i.e. it is a problem of classification that ought to be solved. A widespread approach in Educational Data Mining is to find classification models for predicting students' potentials. The most popular classification methods involve decision trees, neural networks, cluster analysis, etc. [3, 8].

3 Classification Model

The purpose of a classification model is to identify the relationship patterns among input variables, which describe objects/records belonging to one and the same class. The detected relationships are further transformed into classification rules for predicting the class of objects/records for which the values of the predicting/input variables are known. Classification rules can be represented by various forms depending on the type of the applied model [5] (Fig. 1).

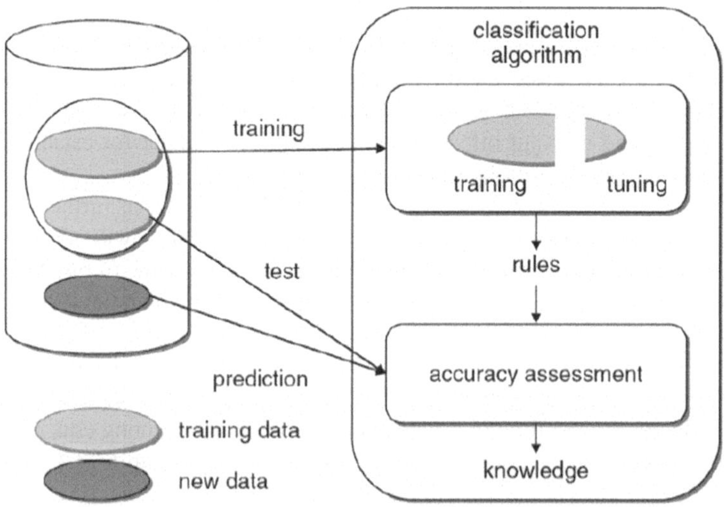

Fig. 1. Phases in the training process included in the classification algorithm [5]

Classification model designing includes three phases:

- Training – the classification algorithm is applied to a training sample aiming to retrieve the classification models and rules, which allow defining the targeted class for each observation.
- Testing – the rules generated in the process of training are used for classifying the examples from the test sample. To estimate the accuracy of the classification model, the

actual class of every example from the test sample is compared to the class predicted by the classifier. For obtaining a realistic assessment, the training and the test sample must be independent, i.e. examples from the training sample should not be contained in the test sample.
- Predicting – application of models and rules generated in the process of training with regard to the input variables describing the new examples.

A tuning test sample obtained from the training one can be added to these phases. It is mostly used in finding the optimal values of some parameters.

The analyzed training sample includes all the students who were enrolled for the master degree program of Informatics and Information Technologies in Education at the University of Ruse within the period of 2010–2020. Initially, the analysis was performed by successively investigating students' distribution according to specific characteristics:

- Distribution by gender (males and females);
- Distribution by places of residence (the town of Ruse and other locations);
- Distribution by the form of training (full-time and part-time);
- Students distribution in percentage rate with regard to the training completion (graduate students and non-graduates);
- Distribution by the number of graduates employed in accordance with the completed specialty (IT teachers) or similar, i.e. teachers of other subjects, and on other hand, graduates working in different subject areas.

The following conclusions can be drawn based on the initially analyzed data:

Regarding gender classification it is concluded that the number of female applicants enrolled in the program is significantly prevailing. According to the places of residence, it is observed that the number of students from Ruse is almost equal to that of students from other locations in Bulgaria.

An interesting moment appears with regard to the training form. The bigger portion of students prefers part-time studying. The results from the research show that students opt for the part-time training program, in view of the percentage of employees being considerably higher than the percentage of students who have not started their professional realization. Furthermore, the number of students employed in the educational area, i.e. teachers of subjects different from the IT program, is much higher than the number of students having different job positions.

The goal of the classification process is to construct a model that unambiguously juxtaposes the values of the attributes-predictors with the class where to the object belongs. The main stages of the classification process present the model designing and its implementation.

3.1 Model Designing (Training Phase)

1. Each object from the sample is connected with one of the predefined classes by the "class label" attribute;
2. A training set is chosen;

3. The model is presented as a system of classification rules, decision trees or some other formal set (Fig. 2).

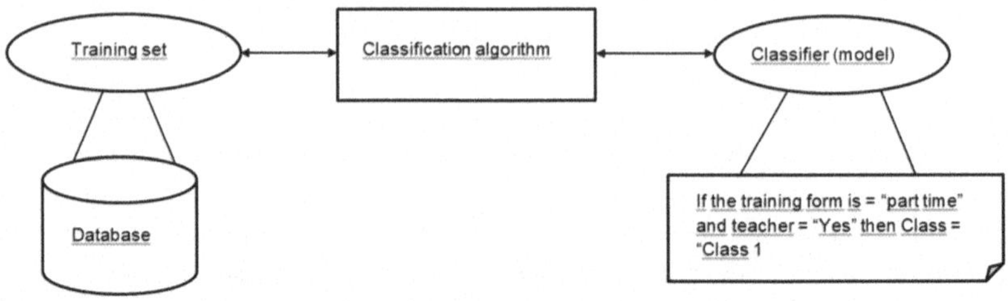

Fig. 2. Classification process. Model designing

3.2 Model Application – for Classifying New Objects

1. Evaluation of model accuracy with the help of a "test set";
2. The model accuracy is determined by the percentage of the properly classified objects in the test set;
3. The test set must not coincide with the training set;
4. If the accuracy is acceptable, the designed model is applied for the practical solving of the task (Fig. 3).

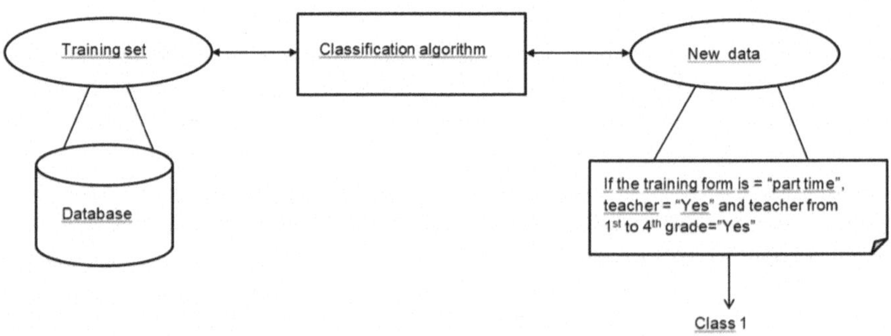

Fig. 3. Classification process. Model application

The assessment of the designed model effectiveness and accuracy is an essential stage for solving the classification task. The designed classification model is verified over a test data set using known targeted values of the classification attribute (known classes). The verification involves a comparison of the classification groups' values predicted by the model with their actual values taken from the test data set (Fig. 4).

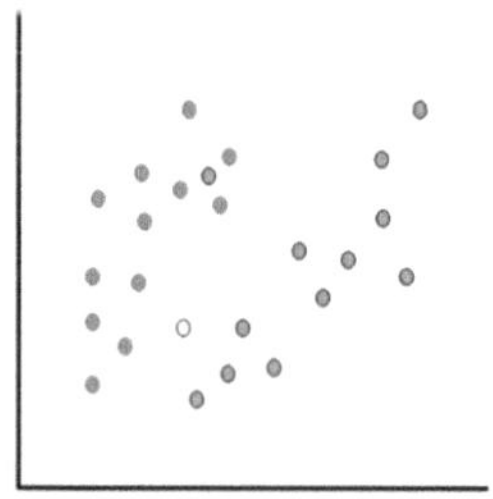

Fig. 4. A set of analyzed objects presented in 2D space

Conclusion: The knowledge obtained by the data mining process provides a possibility for detecting the hidden links and patterns contained in large data volumes. The models can be used for predicting the future behavior and/or the development of a given program (organization). For realizing the advantages of the proposed method for analysis, it is necessary to choose sufficiently effective techniques with regard to the proper selection of the input data.

References

1. Zaki, M.J., Meira, W., Jr.: Data Mining and Analysis: Fundamental Concepts and Algorithms. Cambridge University Press, Cambridge (2014)
2. Olson, D.L., Delen, D.: Advanced Data Mining Techniques. Springer, Heidelberg (2008)
3. Pena-Ayala, A.: Educational Data Mining, p. 37. Applications and Trends. Springer, Heidelberg (2014)
4. Ancheva, V., Voinohovska, V.: Integrating the stem methodology in robotics education in after school enrichment classes for students. In: 12th International Conference of Education, Research and Innovation, Seville, SPAIN, pp. 4610–4616 (2019)
5. Vercellis, C.: Business Intelligence: Data Mining and Optimization for Decision Making. John Wiley, Chichester, West Sussex, UK (2009)
6. Vialardi, C., Bravo, J., Shafti, L., Ortigosa, A.: Recommendation in higher education using data mining techniques. In: Conference Proceedings of the 2nd International Conference on Educational Data Mining (EDM 2009), Cordoba, Spain, 1–3 July 2009, pp. 190–199 (2009)
7. Wang, M.: Using data mining techniques to predict student development and retention. In: Presentation Included in the SAS Conference Proceedings: Midwest SAS User Group 2007, 28 Oct 2007/30 Oct 2007, Des Moines, Iowa, USA (2007)
8. Luan, L.: Data Mining and Knowledge management in Higher Education – Potential Applications. In: Presentation at AIR Forum, Toronto, Canada, 2002 (2002). www.cabrillo.edu/services/pro/oir_reports/DM_KM2002AIR.pdf
9. Voinohovska, V., Goranova, E.: Student peer teaching – an innovative instructional method to foster student problem-solving and in-depth learning. In: 10th annual International Conference of Education Research and Innovation Seville, Spain, pp. 741–748 (2017)
10. Velikova, E., Mierlus-Mazilu, I., Vasileva-Ivanova, R., Georgieva, D.: About the stem Education. In: Proceedings of University of Ruse - 2018, vol. 57, pp. 10–14, book 6.4 (2018). ISSN 1311-3321

Multimodal Transportation Overview and Optimization Ontology for a Greener Future

Oliviu Matei[1,2], Rudolf Erdei[2(✉)], and Daniela Delinschi[2]

[1] Technical University of Cluj Napoca, North University Centre of Baia Mare, Baia Mare, Romania
oliviu.matei@cunbm.utcluj.ro
[2] HOLISUN, Baia Mare, Romania
{rudolf.erdei,daniela.delinschi}@holisun.com

Abstract. Reducing the carbon footprint of our daily activities, as a race, is one of the most important priorities that we can have today. Optimizations play a key role in this process, as just swapping one fuel with another is not enough. An important part of managing and engineering a sustainable and economical transportation system, specifically in the field of multimodal transportation, is the complete overview of the entire system with it's parts. With the growing complexity of such systems, along with the many constraints that they hold, means that manual design becomes impossible. For such purposes, the engineers need a complete database of all the aspects of the transportation system, complete with the existing constraints. In this paper, we present the structure and development of an ontology can be moulded on the specific needs of a company, even large ones, in order to understand and improve the system as a whole.

Keywords: Ontology · Transportation · Multimodal

1 Introduction

As a direct consequence of continuous growth of global economics and globalisation in general, transportation has today become ubiquitous. From personal transport, to public transportation that suits our needs and freight transported from the other side of the Globe, humanity depends everyday more and more on transportation means. It is expected that by 2025 the Freight Transport Market will grow in value by 75% compared to 2019[1].

In order to sustain this very large growth, a number of large-scale changes are necessary at the organizations levels. The industry will have to rely less on fossil-fuels and more on sustainable source, but, more important, the overall energy consumption of the system will have to drop to manageable and sustainable amounts.

© The Author(s), under exclusive license to Springer Nature Switzerland AG 2021
R. Silhavy (Ed.): CSOC 2021, LNNS 229, pp. 158–172, 2021.
https://doi.org/10.1007/978-3-030-77445-5_15

Optimizing such a system implies, first of all, a complete overview and understanding of the transport system with all it's components and all of it's weak points. A mathematical formalisation and modelling will aid scientists to develop better algorithms that will optimize the transport system and reduce human intervention, enabling fast and cost effective transport, and, most importantly, reducing emissions. This is where the advantages of using ontologies are revealed.

Ontologies can be defined as ways to represent knowledge in a way that is search-able, query-able and understand-able by machine code. Ontologies can also be viewed as abstractions of pieces of knowledge, rather than just plain data. Ontologies can be used to infuse computer applications with knowledge and enable them to reason, in a sense making them smart, or even to formalize knowledge inside a company, in order to make organizing and finding easy.

This article presents an ontology for modelling multimodal transport. The requirements have been obtained from EKOL, a company with annual revenues of 710 M EUR. Ekol is an integrated (multimodal) logistics company, established in 1990. They provide international freight, warehousing, domestic distribution, foreign trade, customs and supply chain management service in 15 countries across the entire Europe.

Ekol is one of Europe's major logistics providers, boasting distribution centers with 1 000 000 m^2 of total indoor space in Turkey, Germany, Italy, Greece, France, Ukraine, Bosnia-Herzegovina, Romania, Hungary, Spain, Poland, Czech Republic, Bulgaria, Slovenia and Sweden, in addition to allowing for multimodal transportation with the utilization of its 2 Ro-Ro vessels, 52 block trains per week and 6,000 vehicles.

Overviewing and optimizing such a system requires a lot of work, planning and usage of the state-of-the-art tools available. An IT platform based on a ontology is a natural next step in this area.

The rest of the article is structured as follows: Sect. 2 shows the works attempting to develop an ontology suitable for expressing the semantics related to multimodal transport; Sect. 4 describes the methodology used by us in developing this ontology; Sect. 5 describes the structure we engineered for our implementation of the ontology, the main concepts; Sect. 6 describes all the phases of development.

2 Related Work

A comprehensive survey of transportation ontologies is presented in the article by Katsoumi and Fox [2].

Benaissa et al. [3] describe an ontological model for knowledge capitalization, a very important part that aids the formalization of the field, whereas Bendriss and Benabdelhafid [4] present two modelling approaches for the design of goods traceability solutions.

Heterogeneity is the main difference that Anand et al. [5] describe when comparing urban freight transport and traditional logistics operators. They classify urban logistics into eight general high-level classes and introduce another actor in this scene: the Municipality, that interacts directly with the Carrier, limiting or facilitating (it depends on the situation) it's working conditions.

Baazaoui and Besbes [6] discuss the importance of personalization and give the power of creating a preferred route to the desired destination to the user, a subject also addressed in [7,8]. Becker and Smith [9] present an ontology for Multimodal Transportation Planning and Scheduling that extends the OZONE ontology.

Other ontological systems for intelligent design of transportation systems are presented in [10–13] and also in the case study [14]. Different original approaches are presented in [15–18].

All the studied candidate ontologies lacked the actual framework the company uses, either by over or under-designing the ontology, or by modelling certain aspects (like the cargo) very differently, rendering the said ontologies useless. For example, an ontology suited for military use, will by no means be suitable for our use case.

3 Requirements for the Ontology

Optimization is the main goal of any piece of software written for transportation. This ontology is no different. So, in order to establish some valid and relevant optimisation scenarios, we need to define universal and unitary measurement units, used throughout the ontology and resulting project implementations, one example being the total duration of a shipment. Shipment companies often promise fast deliveries (e.g. 24 h inside a country). In order to keep this promise, all handling and transit times must be very well documented and tracked and all delays minimized.

3.1 Relevant Optimisation Requirements

From an engineering point of view, there are several optimisations that are relevant in the context of multimodal transportation [19], optimisations that were discussed and elicited together with the end-user:

- Optimizations relevant to each package
  - Total Transit Time
  - Total Transit Cost
  - Transport waiting time efficiency (percentage of time lost waiting)
- Optimizations relevant to the shipment company as a whole
  - Total amount of green-gas emission per year
  - Total amount of waiting time per year
  - Transportation System overall efficiency (percentage of time lost waiting)

Total Transit Time. We define the total duration of a transport as:

$$t = \sum_{i=1}^{n} (Tt_{i,i+1} + U_t + S_t + L_t + I_t) \quad (1)$$

where:

- $Tt_{i,i+1}$ - Transit time from i to i+1 checkpoint
- U_t - Unload time at checkpoint i;
- S_t - Sort time at checkpoint i;
- L_t - Load time at checkpoint i;
- I_t - Internal (checkpoint) transit time at checkpoint i;

Except the Transit time, all other parameters are optional (they can have zero value). Checkpoints can be warehouses, transfer zones or the final destination (the client).

As it can be noticed, all the internal operations that take place in any checkpoint can be considered delays. Also, from the companies experience, each checkpoint has a very low time efficiency, as packages mostly wait for handlers or automated systems to operate on them (unloading, sorting, loading). On the contrary, Transit operations have a very high time efficiency, close to 100%, only eventual defects to the mean of transport or traffic-related delays subtracting from the perfect score.

In conclusion, optimizing multimodal transport has little to do with traffic and routes, but more with dealing with transfer point (checkpoint) delays, minimizing waiting times and operator's efficiency.

Total Transit Cost. Similar to the Total Transit Time, the Total Transit Cost can be defined as a sum of all the costs involved:

$$c = \sum_{i=1}^{n} (Tc_{i,i+1} + U_c + S_c + L_c + I_c) \quad (2)$$

where:

- $Tc_{i,i+1}$ - Transit cost from i to i+1 checkpoint
- U_c - Unload cost
- S_c - Sort cost
- L_c - Load cost
- I_c - Warehouse costs

If dealing with unpredictable dispatching patterns, checkpoints must deploy a warehouse that acts as a buffer. The unpredictability of transportation modes/loading times result in the need for a large warehouse, that is expensive to maintain.

This means that costs optimizations include reducing as much as possible all the unpredictable factors as well as the delays (known or not) that may appear during a shipment.

Transport Waiting Time Efficiency

$$t_{\%} = \frac{M_t - W_t}{W_t} * 100 \tag{3}$$

where:

- $t_{\%}$ - Transit Time Efficiency (in percent)
- M_t - Total Moving Time (transit time minus waiting time)
- W_t - Total Waiting Time

Keeping this efficiency at or close to 100% can be almost impossible and may turn out very costly for the company, as very short transfer times mean a large number of handlers or high performance automated systems.

Optimisations Regarding the Company as a Whole. Company-wide optimizations have their importance, in the context of increasing demand and globalization. The optimization of company carbon footprint can be a matter of survival both for mankind, on the long run, as well as the company, in the current economic context. Lower service costs (due to lower wear-and-tear) and lower fuel consumption results in a more economic and efficient system that has less operational costs, and, of course, a lower carbon footprint. This in turn can mean lower costs for the clients as well.

Optimizations for the company as a whole are a simple sum for corresponding aspects for each transport. These factors are also taken into consideration when selecting routes and vehicles, for example, as the system aims for optimized choices. For example, at each step we select the most efficient mode of transport (and vehicle) that can deliver a container to it's next checkpoint, comparing it's **cost per distance unit** with other options. Efficiency values are normalized per mode of transport and vehicle (cost per container/ton and distance), so at each step it can choose the best available option.

4 Methodology for Ontology Development

As no existing ontology was perfectly suitable for our use case and adapting any of them would require a significant amount of work, we ended up developing an ontology from scratch. Our ontology will have to respond to the needs of our use case provider and end user, Ekol Logistics[1]. Several brainstorming sessions have been conducted with representatives of the end user, as well as other partners from the project, in order to best model all the entities present in the logistics process conducted at Ekol.

[1] https://www.ekol.com/en/.

The advantages of developing an ontology from scratch, rather than adapting one, are:

- a better adaptation to the end-user's needs;
- a better integration into the end-user's transportation process, rather than forcing a large company adapt its process to an existing ontology;
- development time can be shorter, if existing ontologies would need substantial adaptations;
- custom ontologies have been proven very powerful instruments for efficiently exploring the existing solution space [20,21].
- the existing ontologies are obsolete/incomplete, as the multimodal transport is very dynamic due to the infrastructure and communication related changes;
- the transport becomes more and more complex, the globalization is a reality and classical databases are unable to snapshot the semantics associated with the current realities.

With this in mind, the logistic system of Ekol was studied in detail, including its relations with third parties (like local railroad operators). We analyzed problematic concepts and areas to model and conceptual differences (in measuring units, for example) that we had to unify, so that the project in its entirety would speak a common language.

4.1 Conceptual Modelling

Describing a complex universe using a modelling language can be understood as **abstraction**. In his article, Oberle [22] describes **Conceptual Modelling**, and defines the high-level abstractions used in all ontologies, like **Classes**, **Properties**, **Instances** and so on. He suggests an **agile** approach for the development of an ontology, as there is no way that a complex universe can be understood perfectly from the beginning.

An **agile** approach enables the developers and engineers to work closely with the end-users at all times, including their feedback in every step of the development [23]. The approach means that after every iteration (called a **sprint**), a basic working MVP will be available, of course including only the known information at that point.

The **agile** approach also has the advantage of generating a result that better suits the end-user, even if its views of the product will change during the development, even in latter stages, an aspect often overlooked but very important, as no client has a perfect understanding of what an ontology is and how it can be best exploited.

An **Order**, in Ekol's conception, is comprised from one or more **Containers** that need to be transported from a **Sender** to a **Receiver**. Each Order can have only one Sender and only one Receiver, but can have more than one Container, depending on the need.

From the transportation point of view, each order has a **Route** that starts either with the Sender's address, or with the first hub (the dispatching hub).

The **Route** consists of one or, usually, more **Legs**. Each leg starts at a checkpoint, that can be either a hub or the originating/destination locations. Hubs are intermediary destinations that are used for routing the cargo. The succession of hubs is computed beforehand, along with the timetable for that specific order, in order to be able to assess problems and optimize the operations afterwards.

These basic high-level concepts will be modelled and queried by the system in order to generate all the necessary knowledge.

4.2 Methodology Application

When designing the ontology, we used the high-level DERA methodology [24,25] along with MOD [26] standards. This approach formalizes ontology development to enable reusability and flexibility. The selected domain for this ontology is of course **transportation**, more exactly **multimodal transportation**. The usage of DERA means that our ontology has a clear domain of application and a complete set of entities and relationships.

5 Structure of the Ontology

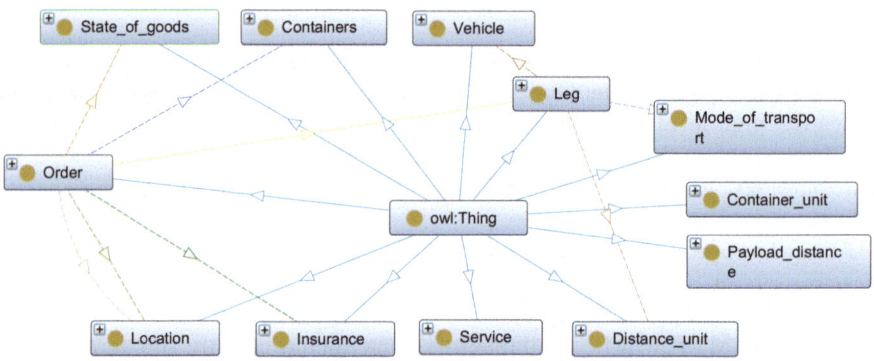

Fig. 1. Root classes and connections

The main concepts of the ontology (its root classes) are presented in Fig. 1. As it can be noticed, the top-level classes are formal definitions of standard high level abstractions. These abstractions can be objects (e.g. containers, vehicles), units of measurement, or other conceptual variations of items, specific to the (multimodal) transportation area of interest (e.g. legs).

The main classes, along with their connections and properties will be discussed in the following subsections.

5.1 Main Abstracted Classes

Defining any ontology needs to begin with the formalization of some high level concepts, like units for measurement, that will be used throughout the ontology. These concepts will be used as **properties** for other classes and usually they do not extend or implement any other high level classes.

- **Location** - this high level class refers to physical locations, and can be: the sender address (starting point), the destination address (end point) and other checkpoints/hubs that are on the route;
- **Various measurement units** like Distance units and Container units. These are definitions of measurements and scales to be used.
- **Vehicle** - This class includes vehicle types classified by mode of transport like Aircraft, Landcraft and Watercraft.
- **Operations** - are the actions that take place with the goods, namely loading, unloading, storage, transfer and transport.

5.2 Order

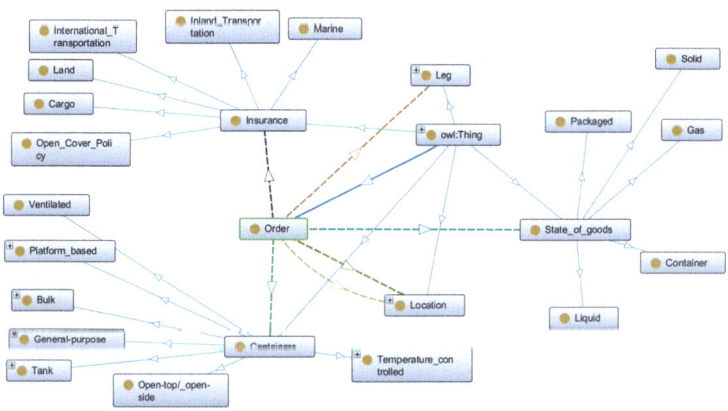

Fig. 2. Order class and connections

The order class, depicted in Fig. 2, is the most important class of this ontology and the main atomic unit that the company works with. Each order has some properties that make it unique and indivisible, like:

- **The Sender**, contains information about the sender of the package, complete with an originating location.
- **The Recipient**, contains information about the recipient of the package, complete with location information.
- **The Cargo** is a container of goods that is to be transported between two locations. Usually, the container is standardised both in dimensions and in weight, although the latter may, of course, vary.

- **Insurance** is a policy that offers coverage on the insured's property while it is in transit from one location to another on any necessary mode of transport.
- **The State of goods** includes the types of goods according to their physical properties, e.g. liquids, gases, solids, etc.
- **Legs**. A succession of **legs** form a **route**, complete with the hubs that will be transited. A more complete description of the concept is in Subsect. 5.4.

5.3 Hub

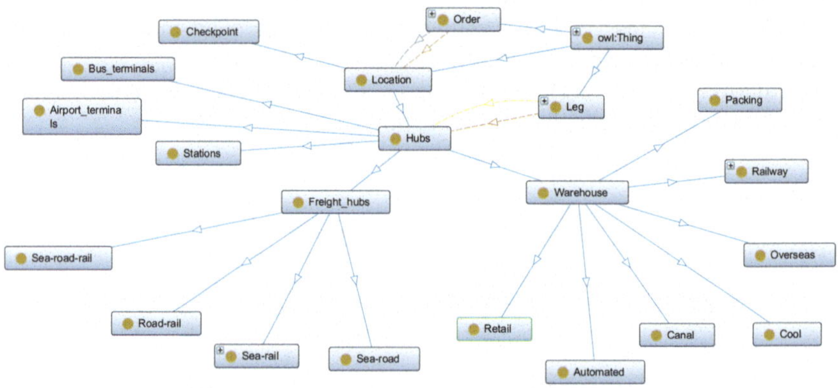

Fig. 3. Hub class and connections

- In the general sense, transportation hubs, or interchanges, are physical places where means (e.g. vehicles) or modes of transport are changed. This means that there must be at least one operator that does one or more of the next operations (Fig. 3):

- **Unloading** of the cargo from the transportation vehicle. This operation is the end of the previous **Leg**;
- **Storing** the cargo, if necessary;
- **Transferring** the cargo, also if necessary, if the physical places for unloading and loading are at a distance;
- **Loading** of the cargo onto the next transportation vehicle. This operation also starts the next **Leg**.

All these operations have some properties:

- **Operator** - the entity that is subcontracted (if it is the case) to do a particular operation or series of operations. These are carried out by qualified personnel with specialised machinery;
- **Start time** - a date and time when the operation is scheduled to begin. This property is also used when generating the timetable for a particular **Order**;

- **Duration** - the duration of the operation, in minutes. This property depicts the time that an operation should last;
- **Actual Start time** - for a particular order, this time depicts the date and time the operation actually started, and it's helpful in calculating delays, for optimization purposes.
- **Actual duration** - for a particular order, this field will record the effective duration of the said operation. Large differences between duration and actual duration means that an operation will introduce delays or other problems (e.g. the need to store the cargo) and will help pinpointing problems in the entire transportation process.

5.4 Leg

A leg can be defined as a part of a shipment order, that is between two hubs (locations) and during which the mode of transport does not change. The initial hub of a leg is the destination hub of the previous, so these hubs are in fact links between the legs.

A set of legs in a specific succession defines a route. Using this definition, the total distance can be calculated, complete with the total estimated time to deliver and also the complete timetable for a specific transport.

Modelling a leg involved establishing its properties and links to other classes. Here are some of the important properties of a leg:

- **Start hub** defines the originating point for a specific leg. This can be either a hub on the route, or the sender's address. Each leg can have only one starting hub;
- **Stop hub** defines the destination in a specific leg. This can be either a hub on the route, or the destination address of the cargo. Each leg can have only one ending hub;
- **Operation** contains the list and order of operations that are to be executed in order to complete a leg;
- **Mode of transport** contains the indirect connection to the vehicle that will transport the cargo between two hubs;
- **Distance unit** includes the types of units for measuring distance, for land being measured in km or miles, and on water in nautical miles.

A collection of legs define a complete transport. The starting location of the first leg is the sender's location, while the ending point of the last leg is the recipient's location (Fig. 4).

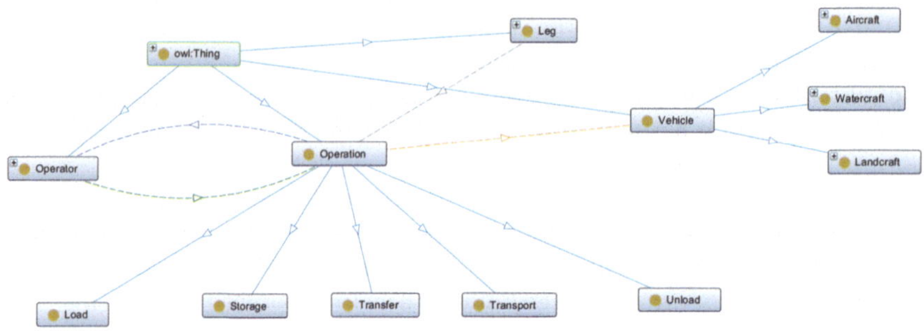

Fig. 4. Operation class and connections

5.5 Operations

The operations list of the **Leg** contains a custom ordered list with the operations to be carried out during that leg. Each operation has the following properties:

- an exact **position** that affects it's sequence. The position affects what neighbouring operations it has and also the exact moment (date and time) when it can start. Each operation has to wait for all the operations before it to end.
- an **estimated start date and time**, that can be calculated beforehand, by summing all the durations that must happen before.
- a **duration**, that defines the estimated amount of time needed for the current operation to end. The succeeding operation must wait for this duration to end, so its starting time must be at least the current estimated starting time plus the duration.

By analyzing the times and durations for all the operations of all the legs in a order, the system can detect if there are problems:

- **Are all the starting times set correctly? Do all the operations end before starting the next one?** Disregarding this constraint would result in inconsistencies, as it's physically impossible for two or more operations to take place at the same time or to overlap, and these inconsistencies lead to delays or incorrect total time values for the transport.
- **Are there any time gaps between 2 succeeding operations?** This would translate into freezing of the system, as machine queues could potentially mean deadlocking the system. Also, large gaps between unloading and loading, for example, can mean the necessity of storing the cargo (in a warehouse).

Detecting such situations is fairly easy and cost-effective resource-wise, the algorithm having $O(n)$ complexity. The order of operations is fixed (one cannot store something that was not unloaded previously), but some of these operations are optional. Operations that will not take place will have a duration of zero minutes. All these durations must take into account the possible time needed for transferring the cargo from one internal location to another (if it is the case) (Fig. 5).

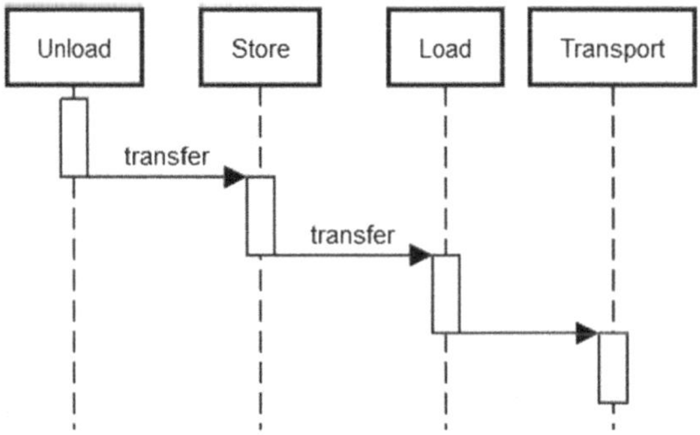

Fig. 5. Time diagram of operations in a hub

6 Results

For implementation, we used the latest version of **Protege**[2] (5.5.0 at the time of writing). Its advantages are the Open Source license and the fact that it's supported by academic, industry and government users.

The resulting ontology has the following statistics:

Property	Value
Classes	107
Data properties	21
Object properties	23
Ontology depth	6
Ontology width	39
Axiom	346
Logical axiom count	196
Declaration axiom count	150

7 Conclusions and Further Work

Ontologies have been proven to be a unique, powerful and very useful tool for conceptualizing and modelling domain knowledge, thus bridging the gap of transforming data into useful information and knowledge for machines. Actually, as the demand for knowledge formalization grows, they are used more and more in several fields, such as product design [20], healthcare [27], architecture and construction [28], business [29] and of course biology [30].

[2] https://protege.stanford.edu/.

The power of ontologies relies in their expressiveness of the semantics and the equal usability by humans as well as by machines (so much, that a new field, of evolutionary ontologies has emerged in the last few year [31]).

We have shown that the ontology developed can serve the internal needs of a multimodal transport company and has been validated on EKOL's specific case. Preliminary internal test results have proven the power and flexibility that an ontology based system provides into pinpointing potential problems and aiding existing or new optimization systems.

As the field of multimodal transportation is transforming everyday, so the used ontology will have to evolve along with it, adapting and changing to address challenges. Containers could be equipped with a large array of sensors that upload data into the cloud, for the use in the ontology. Such enhancements could provide an instant updated overview of the entire system, making managing and optimizing it easier, even on-the-fly.

A.I. self-learning ontology based systems, enhanced with evolutionary algorithms, could mean complex and complete cloud based solutions for extremely large transportation systems. The benefits of such a complex system could be a lower chance of human error, limiting the need of human overseers and technicians.

Acknowledgement. This work has received funding from the CHIST-ERA BDSI BIG-SMART-LOG and UEFISCDI COFUND-CHIST-ERA-BIG-SMART-LOG Agreement no. 100/01.06.2019.

References

1. Freight transport management market: growth, trends, and forecast (2020–2025)
2. Katsumi, M., Fox, M.: Ontologies for transportation research: a survey. Transp. Res. Part C: Emerg. Technol. **89**, 53–82 (2018)
3. Benaissa, E., Charfeddine, I., Bendriss, S., Benabdelhafid, A.: Ontological model for knowledge capitalization in multimodal transport domain. In: 2014 International Conference on Advanced Logistics and Transport (ICALT), pp. 35–40. IEEE (2014)
4. Bendriss, S., Benabdelhafid, A.: Data and ontological modelling approach to allowing traceability through freight transportation chain. Int. J. Collabor. Enterp. **2**(2–3), 184–202 (2011)
5. Anand, N., Yang, M., van Duin, J., Tavasszy, L.: GenCLOn: an ontology for city logistics. Expert Syst. Appl. **39**(15), 11944–11960 (2012)
6. Baazaoui Zghal, H., Besbes, G.: An ontology-based approach for personalized itinerary search. In: International Conference on Information Systems Development (ISD) (2019)
7. Houda, M., Khemaja, M., Oliveira, K., Abed, M.: A public transportation ontology to support user travel planning. In: 2010 Fourth International Conference on Research Challenges in Information Science (RCIS), pp. 127–136. IEEE (2010)
8. Jain, G.V., Jain, S., Parida, M.: Public transport ontology for passenger information retrieval. Int. J. Transp. Eng. **2**(2), 131–144 (2014)
9. Becker, M., Smith, S.F.: An ontology for multi-modal transportation planning and scheduling. Technical report CMU-R I-TR-98-15 (1997)

10. Larioui, J., EL Byed, A.: Towards a semantic layer design for an advanced intelligent multimodal transportation system. Int. J. **9**(2) (2020)
11. Nandini, D., Shahi, G.K.: An ontology for transportation system. Kalpa Publ. Comput. **10**, 32–37 (2019)
12. Romanenko, V., Mokhov, V., Yarkin, E.: System analysis and the task of ontological modeling of multimodal transportation. In: IOP Conference Series: Earth and Environmental Science, vol. 272, p. 022231. IOP Publishing (2019)
13. Sadeghi, M., et al.: Sprint: semantics for performant and scalable interoperability of multimodal transport. In: 8th Transport Research Arena TRA 2020, pp. 1–10 (2020)
14. Saad, S., Zgaya, H., Hammadi, S.: The flexible negotiation ontology-based knowledge management system: the transport ontology case study. In: 2008 3rd International Conference on Information and Communication Technologies: From Theory to Applications, pp. 1–7. IEEE (2008)
15. Tamani, N., Liétard, L., Rocacher, D.: Bipolarity in flexible querying of information systems dedicated to multimodal transport networks. In: 2011 10th International Symposium on Programming and Systems, pp. 108–115. IEEE (2011)
16. Warden, T., Porzel, R., Gehrke, J.D., Herzog, O., Langer, H., Malaka, R.: Towards ontology-based multiagent simulations: the plasma approach. In: ECMS, pp. 50–56 (2010)
17. Zgaya, H., Hammadi, S.: Multi-agent information system using mobile agent negotiation based on a flexible transport ontology. In: Proceedings of the 6th International Joint Conference on Autonomous Agents and Multiagent Systems, pp. 1–3 (2007)
18. Zhang, J., Liao, F., Arentze, T., Timmermans, H.: A multimodal transport network model for advanced traveler information systems. Procedia Soc. Behav. Sci. **20**, 313–322 (2011)
19. Prause, G., Schröder, M.: KPI building blocks for successful green transport corridor implementation. Transp. Telecommun. J. **16**(4), 277–287 (2015)
20. Matei, O., Contras, D.: Automated product design and development using evolutionary ontology. In: Silhavy, R., Senkerik, R., Oplatkova, Z., Silhavy, P., Prokopova, Z. (eds.) Artificial Intelligence Perspectives in Intelligent Systems, vol. 464, pp. 47–57. Springer, Cham (2016). https://doi.org/10.1007/978-3-319-33625-1_5
21. Oliviu, M., Contras, D., Pop, P.: Applying evolutionary computation for evolutionary ontology. In: Proceedings of IEEE Congress on Evolutionary Computation (CEC-2014), pp. 1520–1527 (2014)
22. Oberle, D.: How ontologies benefit enterprise applications. Semant. Web **5**(6), 473–491 (2014)
23. Peroni, S.: A simplified agile methodology for ontology development. In: Dragoni, M., Poveda-Villalón, M., Jimenez-Ruiz, E. (eds.) OWL: Experiences and Directions-Reasoner Evaluation. LNCS, vol. 10161, pp. 55–69. Springer, Cham (2016). https://doi.org/10.1007/978-3-319-54627-8_5
24. Giunchiglia, F., Dutta, B.: DERA: a faceted knowledge organization framework. Technical report, University of Trento (2011)
25. Giunchiglia, F., Dutta, B., Maltese, V.: From knowledge organization to knowledge representation. KO Knowl. Organ. **41**(1), 44–56 (2014)
26. Dutta, B., Nandini, D., Shahi, G.K.: MOD: metadata for ontology description and publication. In: International Conference on Dublin Core and Metadata Applications, pp. 1–9 (2015)

27. Matei, O.: Defining an ontology for the radiograph images segmentation. In: 9th International Conference on Development and Application Systems (2008)
28. Hartoonian, G., Frampton, K.: Ontology of Construction: On Nihilism of Technology in Theories of Modern Architecture. Cambridge University Press Cambridge, Cambridge (1994)
29. Kappelman, L.A., Zachman, J.A.: The enterprise and its architecture: ontology & challenges. J. Comput. Inf. Syst. **53**(4), 87–95 (2013)
30. Ashburner, M., et al.: Gene ontology: tool for the unification of biology. Nat. Genet. **25**(1), 25–29 (2000)
31. Matei, O., Contras, D., Pintescu, A.: Why evolutionary ontologies are a completely different field than genetic algorithms. Carpath. J. Electron. Comput. Eng. **7**(1), 19 (2014)

A Complex Cognitive-Based Technique for Social Tension Detection in the Internet

Iakov Korovin[1], Alexander Pavlenko[2], Anna Klimenko[1](✉), and Irina Safronenkova[3]

[1] Scientific Research Institute of Multiprocessor Computer Systems of Southern Federal, University, 2, Chekhov Street, 347922 Taganrog, Russian Federation

[2] A.P. Chekhov Institute of Taganrog (branch of) Rostov State University of Economics, 69, B. Sadovaya Street, 344022 Rostov-on-Don, Russian Federation

[3] Federal Research Center The Southern Scientific Center of the Russian Academy of Sciences, 41, Chekhov Street, 344006 Rostov-on-Don, Russian Federation

Abstract. With the enhancement of the usage of the Internet, various problems of social state shifted to the network. As a result, many techniques which were used by sociologists and psychologists lost their relevance. In this paper a problem of the automatic social tension detection is considered in terms of combined content, including video and static pictures with texts. We presuppose that social tension can be detected by means of particular language markers and individuals emotional states identification.

The main contribution of this paper is as follows:

- a set of language markers and emotional state markers, which allow to detect the social tension and unrest in video- and static Internet-content;
- forming of the cognitive maps which can be used as clear classifier in the field of the social tension detection.

Keywords: Monitoring system · Forecasting system · Workload relocation · Ontology · Fog-computing · Distributed system

1 Introduction

Nowadays, in the situation of pandemic, life has changed dramatically: more and more business processes are shifted to the information space of the Internet. It results in the increase of relevance of monitoring of social unrest manifestations, aggressive and victim behavior of individuals, various forms of cyberbullying, etc. Although there are works and systems dealing with the problems above, they are not capable of solving the problem of social tensions detection. As is known, the following tasks have already been solved quite successfully:

- Facial expression recognition [1, 2];
- Speech tone analysis [3, 4];
- Content sentiment analysis [5];

© The Author(s), under exclusive license to Springer Nature Switzerland AG 2021
R. Silhavy (Ed.): CSOC 2021, LNNS 229, pp. 173–184, 2021.
https://doi.org/10.1007/978-3-030-77445-5_16

- Emotional states recognition [6, 7].

Besides, the accompanying problem is that of classification, which is implemented by such techniques as artificial neural networks, fuzzy logic techniques, support vector machines, etc.

Each of classification techniques has its own pros and cons, yet, the neural networks seem to be the most efficient and widely used.

Detecting of social tensions and social unrest in the Internet is a particularly relevant problem. With the development of media-content (e.g., video), the text content analysis as such seems to be insufficient. So, a set of attributes should be formed, which allows to detect social tension in the Internet communities by means of complex text and video analysis.

In this paper a novel technique of the social tension detection in the Internet is presented. It combines both the detection of newly-formed set of attributes to detect social tension and the usage of cognitive maps to classify the attribute set as social unrest features.

The main contribution of this paper is as follows:

- language markers and the emotional states markers selection, which allow to detect the social tension and unrest in video- and static Internet-content;
- forming of the cognitive maps which can be used as clear classifier in the field of the social tension detection.

2 Language Data Markers

As is known, governmental and corporate structures have always had to deal with the tasks of identifying markers of social unrest and forecasting its development, the relevance of the latter increasing significantly in times of growing societal crises. Scientific and technological progress turns the control of communication channels used by opposition forces and a criminal element into a high-tech sphere and, while for centuries perlustration and military censorship of personal correspondence had been used, the 19th century saw these activities complemented by telegram checking and wiretapping. In the era of the information revolution, the complexity of the tasks facing law enforcement agencies and security forces, public and private, has increased immeasurably, and their range of responsibilities is replenished by tracking negative impacts in the digital environment, including services and technologies such as email, instant messengers, blogs and social media [8, 9].

The principal methodology of the studies in question is the analysis of time series of activity in social media on a specific subject matter [10], keywords reflecting aggression and protest topics being a crucial component of the filtered data [11]. Usually, such markers are extracted from the text as lexical means of expressing aggression and/or forming the general emotional coloring of the post, along with times, dates, places, etc. [12, 13].

Social movements differ in their goals, number of participants, methods and other indicators [14], but their success obviously depends on the efficient recruitment of new participants. The role of social media in this became evident already during the Arab Spring [15].

In this connection, among the markers of social tension a number of researchers highlight those known as "calls to action" (CTA), which are an effective tool of mobilization in the digital environment [16, 17]. It is these markers, as the most symptomatic object of censorship, that law enforcement agencies and corporate security services pay special attention to [18, 19]. Despite this, as far as we know, for the time being only one article has been devoted to the study of CTA as a type of data for automatic detection and prediction of social aggression and protest events [11].

Rogers et al. identify 14 main and borderline types of political calls to action and point out the fact that the activity in social media demonstrates a relatively positive correlation with actual attendance at rallies [11].

Typical CTAs are, predominantly, imperatives that induce the addressee to perform some action, e.g., "Stop everything to show – We cannot be stopped!" (Tikhanovskaya, Belarus). This example can be easily identified as CTA, but the whole range of such calls is not limited to this. There are many borderline cases, since CTAs can have not only direct but also indirect expression.

Rogers et al. 2019 identifies 5 main and 9 borderline cases of political calls to action [11]. Obviously, any other calls to action, such as those contained in commercial advertisements, should be excluded from consideration. Also, any messages related to protests, but not appeals by their nature, should be excluded as well.

Of the main and borderline cases of calls to action, we decided to consider 8 as actual calls to action.

According to [11], calls to action in protest movements can be classified as follows:

1. Calls to participating in protests;
2. Calls to broad changes/impact on society;
3. Instructions for specific protest events;
4. Calls to disseminating information;
5. Calls to activism and organizing protest groups;
6. Declarations of future actions with the intention to mobilize;
7. Suggestions of collective actions;
8. Indicating the necessity or desirability of some action;
9. Indication of possibility of an action;
10. Sharing organizational information for the protest group;
11. Opinions on the suggested actions;
12. Imperatives addressed to the opponents;
13. Questions (rhetorical or not);
14. Declarations of future achievements by the activists;
15. Calls to other kinds of actions.

This list is not universal as a whole but rather culture-dependent, as what may be a CTA in one culture is not necessarily in another.

On the basis of the CTA-categories above a list of topic-related keywords is formed as social tension detection attributes:

- … join us…(when) (where)!
- If you …, start the fight with….!
- Do not (do something)!
- Invite (somebody) – let them see (something)!
- (Somebody) (somewhere), please respond!
- …we will tell (somebody) what we think of (him, her, them)!
- I suggest we (do something)!
- On (date) we should come (quantity)!
- You can (do something) (where).
- We will start ….when we …..
- I (do) like…., but I think….. I'm against….
- Out with you, (name)! And take (names) with you…
- Is… really………? (rhetorical) Who wants to (help, participate, etc.)? (factual)
- Together we will get rid of (names, positions)!
- Everybody, come to …..! [according to Rogers et al. 2019, 39]

3 A Set of Emotional States Markers Relating to Social Tention

The problem of emotional state recognition is quite topical, so, a wide range of different solutions is proposed in this area. As far as we know, the major projects are based on the following concepts and technologies [20–22]:

- facial expression recognition, based on Ekman works;
- vibroimage technique;
- speech analysis, speech tone analysis, etc.;
- heart rate variability and brain activity analysis;
- decoding Spontaneous Emotional States in the Human Brain, and some others.

A concept of facial expression recognition is based on the fact that specific configurations of facial-muscle movements appear as if they broadcast a person's emotions. American psychologist Paul Ekman noticed that some facial expressions corresponding to certain emotions are common for all the people independently of their gender, race, education, ethnicity, etc. He proposed a discrete emotional model using six universal emotions: happiness, surprise, anger, disgust, sadness and fear. Nowadays It is proved that facial movements are reliable and can support the emotional state identification. To the best of our knowledge till nowadays the mostly used technique of emotion recognition is based on the well-known FACS and AFIA systems of facial expressions related to the emotional states. Besides, the following has been used: Cohn-Kanade expression database, CK+ expression database, and Rafd facial expression database established by CMU Robotics Research Institute, USA.

In general, the problem of the emotional state identification is that of classification.

As we consider the emotional states, potentially following the social unrest process, the following states were involved into the estimations:

- Fear;

- Excitement;
- Anger;
- Suffering.

All these emotional states can be detected on the content of vlogs with the usage of the following individual elements states, which are based on the Eckman's theory (Table 1).

Table 1. Individual elements states

Element	Element state	Emotional state	Detection method
Upper eyelid is lifted	Is up	Fear	The length of vertical line connecting the eyelids is more than 0.6 of eye width
Glance direction			
	Shifting glance	Excitement	The frequency of shifts is more than 0,4 Hz
	To the floor	Excitement	The glance direction is down and does not change for 60% of total time
Lips	Opened	Fear, anger	The ratio between lips height and width is more than 0.5
	Tightly compressed	Anger	1. Maximum lips height is calculated; 2. If lips height is less than 0.5 of maximum lips height than they are compressed
Eyebrows	Drawn together	Fear, anger	1. Normal eyebrow state is detected; 2. If mass centers are shifted to each other, then the eyebrows are drawn together
	Up	Fear	1. Normal eyebrow state is detected; 2. If mass centers are shifted up, then the eyebrows are up

(*continued*)

Table 1. (*continued*)

Element	Element state	Emotional state	Detection method
	Inner sides are lifted	Suffering	1. Normal eyebrow state is detected; 2. If 3 eyebrow points near the nose are up, the inner sides are lifted
Glabellar space	Wrinkled	Anger	Darkened area between eyebrows
Forehead center	Not wrinkled	Normal state, anger	Normal lightened area
	Wrinkled	Fear	Darkened area
Skin color	Pale	Fear, excitement	Color coefficient is less than normal
	Red	Anger, excitement	Color coefficient is higher than normal
	Red fragments	Anger, excitement	Non-uniform coloring
Hands	Tremor	Excitement, fear, anger	Hands coordinates change with high frequency within the limits of 20 mm
Fists	Clenched	Anger, excitement	The ratio between hand height and hand width is in 0.8–1.2
Shoulders	Lifted	Fear	Y coordinate of shoulders is more than Y coordinate of neck
Head	Lowered	Fear	The angle between the vertical axis of the body and the axis of the neck-head is less than 140 degrees
Pupil	Widened/narrowed	Anger, excitement	The ratio between the pupil diameter and the iris diameter is smaller than 60 or more than 80%

As one can see in the table, some element states relate to more than one emotional state, that is why it is useful to select a range of the most possible emotional states.

4 A Complex Cognitive-Based Technique for Social Tension Detection in the Global Network

To estimate the level of social tension we used a cascade of the connected fuzzy cognitive maps (FCM) as is shown in Fig. 1 [23]. The key idea of such a technique is that a social tension in the network can be detected by means of combined analysis of the selected language markers and the emotional states of the individuals. It is relevant for video content, as well as for static pictures with the text following.

Emotional states FCM is a bipartite graph, in which concepts describing the element state (e.g., eyes-opened) are connected to the concepts describing the individual emotional state (e.g., "anger" state). All outputs of the Topic-related keywords FCM and of the Emotional states FCM are connected to the output concepts, which are levels of "low danger", "medium danger" and "high danger".

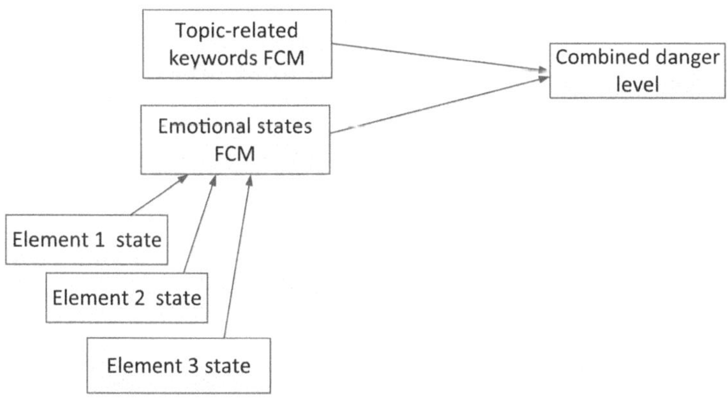

Fig. 1. A cascade of the connected FCM.

The topic-related keywords FCM is a fuzzy cognitive map, which estimates the danger level of topic-related keywords, listed in Sect. 2 of this paper. The relations between the concepts and their weights are generated by means of the expert activities.

The example of the topic-related keywords FCM is presented in the following Fig. 2.

A fragment of the Emotional state FCM is presented in Fig. 3.

The integration of these FCMs is as follows (Fig. 4). So we conduct the outputs, which are proportional to every danger level concept. As this presents a fuzzy solution, the operator should make a conclusion about the danger level.

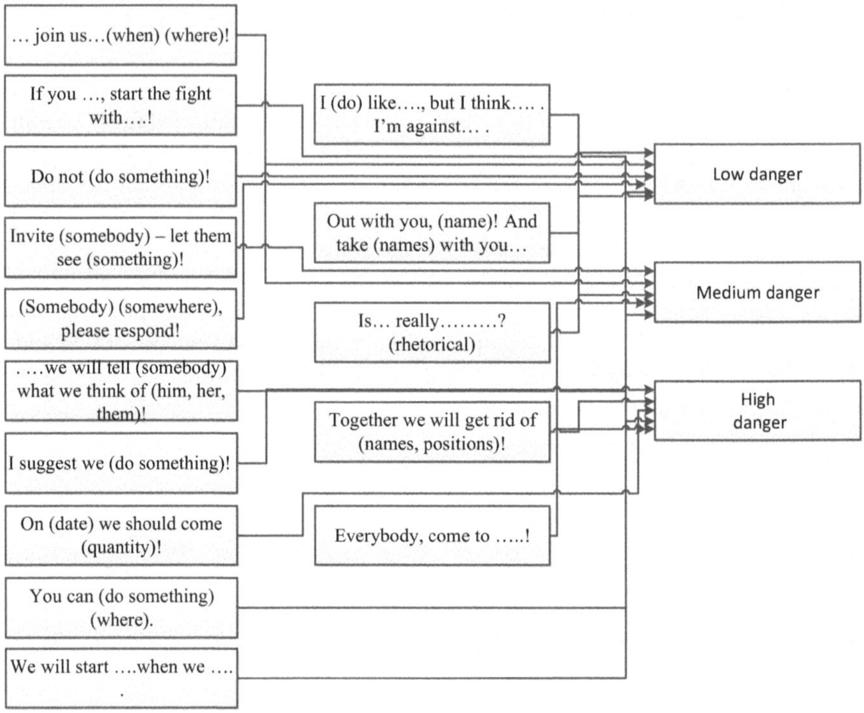

Fig. 2. Topic-related keywords FCM.

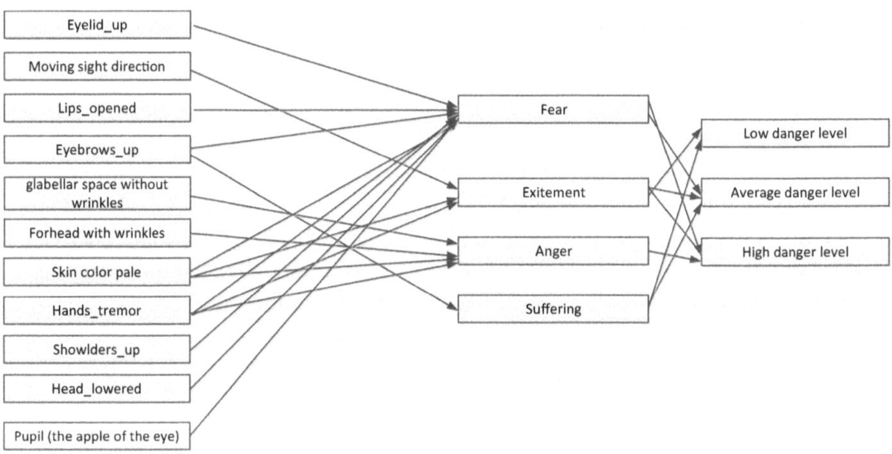

Fig. 3. A fragment of the Emotional state FCM.

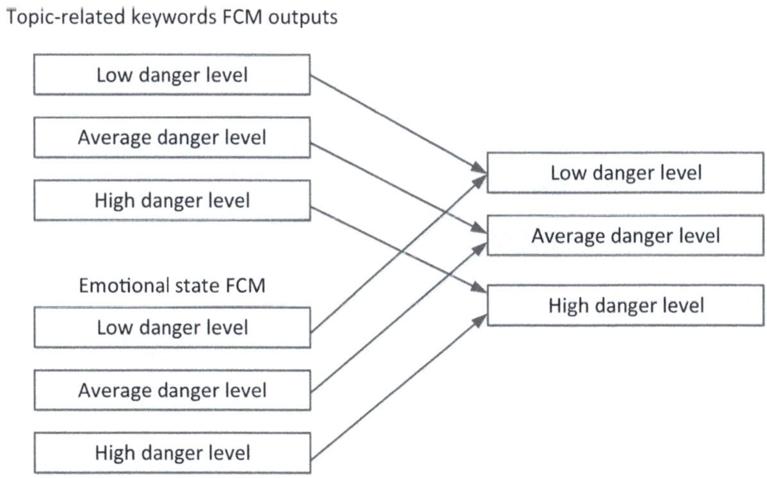

Fig. 4. The FCM integration scheme.

5 Simulation Results

To build the cascade of FCMs the software FCMapper has been used [24]. Based on the experts estimations, the basic relations were generated. Assuming that "fear" emotional state relates to the "medium" danger level, the following outputs were conducted (see "FCM_general" trend). One can see that the resulting state of the analyzed object is of average danger level (Fig. 5).

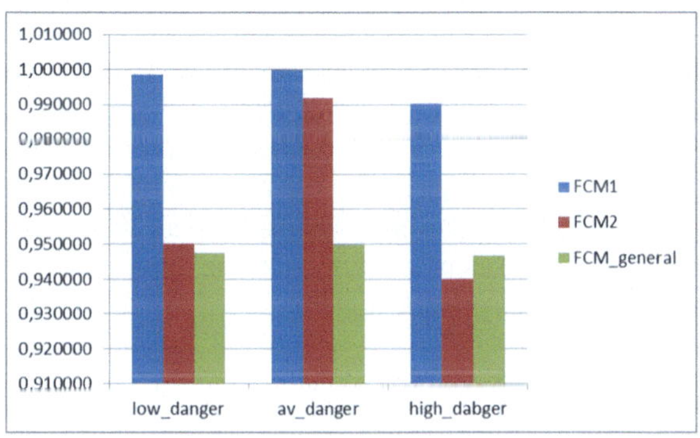

Fig. 5. Medium danger level

Increasing the danger level of the language markers, with the same inputs of the emotional state, the following graph has been conducted (Fig. 6).

Though the concept "av_danger" has the maximum value, in this figure the value of "high danger" is more than that of a "low_danger" concept. Finally, with the increase of the output FCM2_high_danger the following results are conducted.

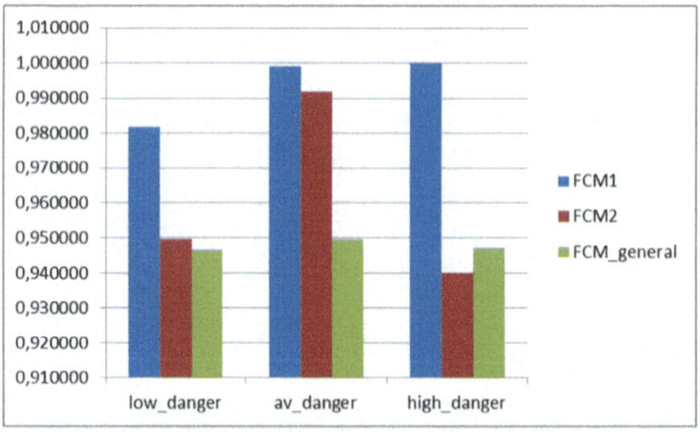

Fig. 6. The increase of danger level

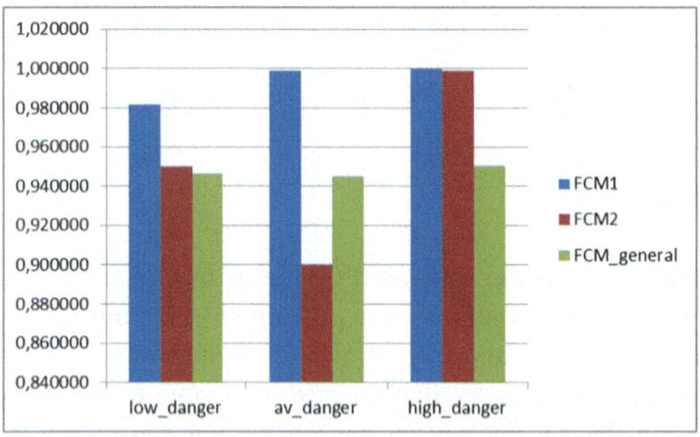

Fig. 7. The high danger level

One can see that in the Fig. 7 the FCM cascade allows to make a conclusion about the high level of danger in the field of the social tension.

So, we tested the cognitive maps connected with the layer of the concepts describing the overall danger level of the social unrest detected. The simulation results show that the cascade of the cognitive maps allows to choose those outputs, which are of the maximum values and to range them. It makes possible to make an appropriate decision about the state of the social tension in the network content.

6 Conclusions

As there more and more human activities shift to the Internet, the problem of social tension detection shifts as well. Yet, to the best of our knowledge, there are no particular techniques, which are oriented to the social tension detection in the Internet Media content. The main contribution of this paper is:

- A complex of language and emotional features for the social tension detection forming;
- The usage of cognitive maps as classifier;

- The incorporating of the features selected to the cognitive maps, which form a quantitative estimation of the social unrest danger level.
- The results conducted include:
- the complex of language and emotional markers for the social unrest detection;
- the structures of the cognitive maps, which are used as classifiers.

The usage of the FCMs is expedient in terms of their visual perspicuity and modification ability. Simulation conducted has shown the applicability of the cognitive maps for the social unrest detection in complex with markers formed, so, FCMs are prospective in terms of more complex systems in this field.

Acknowledgements. The reported study was funded by RFBR, grants №20–04- 60485 and 18-29-22086.

References

1. Facial Expression Recognition using Facial Landmark Detection and Feature Extraction via Neural Networks. https://arxiv.org/pdf/1812.04510.pdf. Accessed 14 Dec 2020
2. Tian, Y., Kanade, T., Cohn, J.F.: Facial expression recognition. In: Li, S., Jain, A. (eds.) Handbook of Face Recognition. Springer, London (2011)
3. Edoho, M., Ekpenyong, M., Inyang, U.: Speech features analysis for tone language speaker discrimination systems. In: Latifi, S. (ed.) Information Technology – New Generations. AISC, vol. 738, pp. 433–442. Springer, Cham (2018). https://doi.org/10.1007/978-3-319-77028-4_57
4. Schertz, J., Chow, C.T.Y., Kamal, N.S.N.: The influence of tone language experience and speech style on the use of intonation in language discrimination. J. Acoust. Soc. Am. **146**, EL58–EL64 (2019)
5. Paltoglou, G.: Sentiment analysis in social media. In: Agarwal, N., Lim, M., Wigand, R.T. (eds.) Online Collective Action. LNSN, pp. 3–17. Springer, Vienna (2014). https://doi.org/10.1007/978-3-7091-1340-0_1
6. Emotional State Recognition Based on Physiological Signals. https://ikuz.eu/materials/supervision/Artem-Bachynskyi-Emotional-State-Recognition-Based-on-Physiological-Signals.pdf. Accessed 14 Dec 2020
7. Khan, A.M., Lawo, M.: Recognizing emotional states using physiological devices. In: Stephanidis, C. (ed.) HCI 2016. CCIS, vol. 618, pp. 164–171. Springer, Cham (2016). https://doi.org/10.1007/978-3-319-40542-1_26
8. Ramakrishnan, N., et al.: Beating the news with EMBERS: forecasting civil unrest using open source indicators. In: Proceedings of 20th ACM SIGKDD International Conference on Knowledge Discovery and Data Mining, pp. 1799–1808. ACM, New York (2014)
9. Manrique, P., Qi, H., Morgenstern, A., Velásquez, N., Lu, T., Johnson, N.: Context matters: Improving the uses of big data for forecasting civil unrest: Emerging phenomena and big data. In: 2013 IEEE International Conference on Intelligence and Security Informatics, Seattle, WA, 2013, pp. 169–172. IEEE, NJ, USA (2013)
10. Hua, T., et al.: Analyzing civil unrest through social media. Computer **46**(12), 80–84 (2013)
11. Rogers, A., Kovaleva, O., Rumshisky, A.: Calls to action on social media: potential for censorship and social impact. In: Proceedings of the 2nd Workshop on NLP for Internet Freedom: Censorship, Disinformation, and Propaganda, Hong Kong, China, 4 November, pp. 36–44.Association for Computational Linguistics, USA (2019)

12. Benkhelifa, E., Rowe, E., Kinmond, R., Adedugbe, O.A., Welsh, T.: Exploiting social networks for the prediction of social and civil unrest: a cloud based framework. In: 2014 International Conference on Future Internet of Things and Cloud, pp. 565–572. IEEE, NJ, USA (2014)
13. Bahrami, M., Findik, Y., Bozkaya, B., Balcisoy, S.: Twitter Reveals: Using Twitter Analytics to Predict Public Protests, pp. 1–24 (2018). arXiv:1805.00358
14. Snow, D.A., Soule, S.A., Kriesi, H.: Mapping the terrain. In: The Blackwell Companion to Social Movements, pp. 3–16. Blackwell Pub. Ltd., Oxford (2004). https://doi.org/10.1007/978-3-319-64382-3_1
15. The Impact of Social Media on Social Unrest in the Arab Spring. https://publicpolicy.stanford.edu/publications/impact-social-media-social-unrest-arab-spring. Accessed 14 Dec 2020
16. Guidry, J., Waters, R., Saxton, G.D.: Moving social marketing beyond personal change to social change: strategically using twitter to mobilize supporters into vocal advocates (May 23, 2014). J. Soc. Mark. **4**, 240–260 (2014)
17. Savage, S., et al.: Botivist: Calling Volunteers to Action using Online Bots. ArXiv abs/1509.06026 (2016)
18. King, G., Pan, J., Roberts, M.E.: How censorship in china allows government criticism but silences collective expression. Am. Polit. Sci. Rev. **107**(2), 326–343 (2013)
19. King, G., Pan, J., Roberts, M.E.: Reverse-engineering censorship in China: randomized experimentation and participant observation. **345**(6199) 1251722 (2014)
20. Mather, M., Thayer, J.: How heart rate variability affects emotion regulation brain networks. Curr. Opin. Behav. Sci. **2018**, 98–104 (2018). PMID: 29333483; PMCID: PMC5761738
21. Screening of Passengers by Observation Techniques. https://how-emotions-are-made.com/notes/Screening_of_Passengers_by_Observation_Techniques. Accessed 14 Dec 2020
22. Hernandez-Matamoros, A., Bonarini, A., Escamilla-Hernandez, E., Nakano-Miyatake, M., Perez-Meana, H.: A facial expression recognition with automatic segmentation of face regions. In: Fujita, H., Guizzi, G. (eds.) Intelligent Software Methodologies, Tools and Techniques. SoMeT 2015. Communications in Computer and Information Science, vol. 53, pp. 529–540. Springer, Cham (2015)
23. Kobets, N., Kovaliuk, T., Mozoliova, D.: Cognitive modeling and cognitive map applying to the knowledge management in the higher education system. In: Wilimowska, Z., Borzemski, L., Świątek, J. (eds.) Information Systems Architecture and Technology: Proceedings of 40th Anniversary International Conference on Information Systems Architecture and Technology – ISAT 2019. ISAT 2019. Advances in Intelligent Systems and Computing, vol. 1052, pp. 63–73. Springer, Cham (2020)
24. Welcome to FCMappers! http://www.fcmappers.net/joomla/. Accessed14 Dec 2020

Method of Crystallization of Alternatives for Solving the Problem of Placing VLSI Elements

Boris K. Lebedev, Oleg B. Lebedev(✉), and Ekaterina O. Lebedeva

Southern Federal University, Rostov-on-Don, Russia

Abstract. This paper presents a new developed metaheuristics of intellectual optimization. This metaheuristics is based on the mechanisms of collective intelligence modeling. The peculiarity of this metaheuristics is the ability to take into account and apply alternative options from the list of best received solutions. The total amount of data on options and their estimates is a scattering of alternatives. The discriminant analysis of alternatives in the process of evolutionary collective adaptation is named by analogy with the processes of isolating objects (crystal formation) by crystallization. This metaheuristics was used to solve optimization problems of topology design, in particular, placement of VLSI elements. Experimental studies have shown that solving optimization problems using the proposed approach makes it possible to obtain a better solution in an acceptable amount of time. The time complexity of the algorithm, obtained experimentally, lies in the range of $O(n^2)$–$O(n^3)$.

Keywords: Collective intelligence modeling · Collective adaptation · Collective evolutionary memory · Multi-agent system · Algorithm · Placement of elements · VLSI · Crystallization alternative field

1 Introduction

The development of new search algorithms consists in the study and use of metaheuristics embedded in natural decision-making mechanisms [1]. In the work, by analogy with the metaheuristics on which the swarm algorithms are built, metaheuristics is used that takes into account the tendency to use alternatives (component options) from the best solutions found [1–3]. In the process of evolutionary collective adaptation, discriminatory analysis methods are used to assess the fitness of alternatives. The fitness of alternatives is considered as the probability of its use in the formed decision. The totality of data on alternatives and their estimates is a scattering of alternatives [4]. The discriminant analysis of alternatives in the process of evolutionary collective adaptation is called by analogy with the processes of isolating objects (crystal formation) by crystallization. In other words, in the process of evolutionary collective adaptation, the most adapted alternatives are extracted from the set of options. Hence the name of the optimization method – the method of crystallization of alternatives field [4, 5]

© The Author(s), under exclusive license to Springer Nature Switzerland AG 2021
R. Silhavy (Ed.): CSOC 2021, LNNS 229, pp. 185–193, 2021.
https://doi.org/10.1007/978-3-030-77445-5_17

The placement problem is one of the most important stages in the design of VLSI, since the quality of the resulting solution significantly affects the execution of subsequent trace connections, which in turn greatly affect the performance of the circuit [6–9]. The placement problem belongs to the class of *NP*-complete ones; therefore, there is no algorithm to obtain its solution in polynomial time. Studies of existing approaches to its solution show that existing algorithms produce solutions that are far from optimal [8, 9], which means that this task remains relevant.

2 Formulation of the Placement Problem

The placement problem can be formulated as follows. Given a set of elements (modules) with the terminals (pins) located on them. The set of circuits connecting the terminals of the modules is specified. A commutation field is specified with a description of the positions in which the elements can be placed. It is necessary to place elements on the control box with optimization of some quality criteria. Input information includes: a description of the modules, which indicates the shape, size, location of the terminals on the modules, a list of circuits indicating interconnections of the modules and a description of the control gear [10]. The output coordinates of the positions on the commutation field for the placement of all modules.

To solve the problem of placing elements in positions on the switching field, we use the following model. Let the set of elements $A = \{a_j | j = 1, 2, \cdots, n\}$ and the set of positions $P = \{p_i | i = 1, 2, \cdots, c\}$ on the commutation field be given. The hypergraph $H = (X, E)$ is used as a model of the scheme, where $X = \{x_i | i = 1, 2, \cdots, n\}$ is the set of vertices modeling the elements, $E = \{e_j | e_j \subset X, j = 1, 2, \cdots, m\}$ is the set of hyperedges that model the chains connecting the elements. To place all the elements, it is necessary to fulfill the condition $c \geq n$. Arbitrary placement of elements in positions is a permutation $S = n(1), n(2), \cdots n(i), \cdots n(c)$, where $n(i)$ sets the number of the element that is assigned to position p_i. Each element of the set A corresponds to one single element of the set S and vice versa. Depending on the selected criterion, the objective function $F(P)$ is introduced to evaluate the placement results. The placement problem is to find the optimal value of the function F on the set of permutations P [8–10].

3 The Placement Mechanisms on the Base Alternative Crystallization Field Method

In the alternative crystallization field method [5], each Q_k solution is formed (represented) by a set of agents $A = \{a_i | i = 1, 2, \cdots, n_k\}$, where n_a is the number of agents. Each agent a_i corresponds to many alternative states $S_i = \{s_{ij} | j = 1, 2, \cdots, n_p\}$, where n_i is the number of states of agent a_i. Each a_i agent may be in one of the alternative states. The solution Q_k is determined by the set of alternative states of the set of agents [4, 5].

The use of any metaheuristics for solving a combinatorial problem is to present the original formulation of the problem in the form of metaheuristics components. In the considered allocation problem, the agents are the elements, and the alternatives are the set of positions $P = \{p_j | j = 1, 2, \cdots, n_p\}$, where n_p is the number of positions in which the elements are placed.

In other words, the set of agents $A = \{a_i | i = 1, 2, \cdots, n_a\}$ corresponds to the set of all elements. The set of alternative states $S_i = \{s_{ij} | j = 1, 2, \cdots, n_p\}$ of agent a_i corresponds to the set of alternative placement options for agent a_i.

The solution Q_k of the placement problem is represented in the form of a bijective map $F_k = A \rightarrow P$, each element of the set A corresponds to one unique element of the set P and vice versa. The position number p_j in which the element ai is placed is determined by the relation $p_j = F_k(a_i)$.

Under the field of alternatives (FA) of a solution, the work refers to a data structure that carries information about the alternatives of agents in this solution and the usefulness of the solution used as a collective evolutionary memory (CEM) [5].

We represent the data structure for mapping the solution Q_k formed by the set of agents A in the form of a set of vectors $R_k = \{R_{ki} | i = 1, 2, \cdots, n_a\}$, which we will call a scattering of alternatives. Each vector $R_{ki} = \{R_{kij} | j = 1, 2, \cdots, n_p\}$ corresponds to agent a_i. The dimension of the vector R_{ki} is determined by the number of possible states of the agent a_i. In the vector R_{ki}, only one element r_{kij} corresponding to the state s_{ij} in which the agent a_i is located has a non-zero value and this value is equal to $\delta_k = f(\xi_k)$, where ξ_k is the estimate of the solution Q_k, δ_k is the estimate of the utility of this solution. The remaining elements of the vector R_{ki} have zero values. Thus, the vector R_{ki} stores information about the state implemented by agent a_i in the solution Q_k and about the utility estimate δ_k of this state.

Example. The Q_k solution is formed by six agents. Agents have 6 possible conditions. In the Q_k solution, the agents implemented the following states (alternatives): a_1 in state s_{12}, $a_2 - s_{25}$, $a_3 - s_{31}$, $a_4 - s_{43}$, $a_5 - s_{56}$, $a_6 - s_{64}$. The value of the objective function is $\delta_k = 6$. Then the field of alternatives for the solution R_k has the form shown in Fig. 1.

$R_k=$

P	p_1	p_2	p_3	p_4	p_5	p_6
R_{k1}	0	**6**	0	0	0	0
R_{k2}	0	0	0	0	**6**	0
R_{k3}	**6**	0	0	0	0	0
R_{k4}	0	0	**6**	0	0	0
R_{ki5}	0	0	0	0	0	**6**
R_{ki6}	0	0	0	**6**	0	0

Fig. 1. Decision search graph H

Let a set of solutions be constructed to solve the allocation problem, $Q = \{Q_k | k = 1, 2, \cdots, n_k\}$. For each solution Q_k, the estimates ξ_k and δ_k are calculated and an individual field of alternatives R_k is formed. The integrated field of alternatives R^* is formed by combining all the field of alternatives:

$$R^* = \{R_i^* | i = 1, 2, \cdots, n_a\};$$

$$R_i^* = \{r_{ij}^* | j = 1, 2, \cdots, n_p\}; \quad (1)$$

$$r_{ij}^{*} = \sum\nolimits_{k} (r_{kij}),\ k = \{1, n_k\}.$$

In fact, r_{ij}^{*} is the total value of the utility of the solutions in which agent a_i implemented the alternative s_{ij}. Let us integrate the solutions and construct an integral field of alternatives for the set of solutions $Q = \{Q_1, Q_2, Q_3, Q_4, Q_5, Q_6\}$. Let the agents in each solution implement the following alternatives:

$$(Q_1 : a_1 - 2, a_2 - 5, a_3 - 4, a_4 - 1, a_5 - 4,\ a_6 - 1),$$

$$(Q_2 : a_1 - 4, a_2 - 3, a_3 - 2, a_4 - 1, a_5 - 4,\ a_6 - 1),$$

$$(Q_3 : a_1 - 2, a_2 - 5, a_3 - 4, a_4 - 3, a_5 - 4,\ a_6 - 1),$$

$$(Q_4 : a_1 - 2, a_2 - 5, a_3 - 4, a_4 - 3, a_5 - 4,\ a_6 - 1),$$

$$(Q_5 : a_1 - 2, a_2 - 5, a_3 - 4, a_4 - 3, a_5 - 4,\ a_6 - 1),$$

$$(Q_6 : a_1 - 2, a_2 - 5, a_3 - 4, a_4 - 3, a_5 - 4,\ a_6 - 1),$$

$$(f_1 = 6,\ f_2 = 8, f_3 = 4, f_4 = 2, f_5 = 9,\ f_6 = 5).$$

The field of alternatives $\{R_1, R_2, R_3, R_4, R_5, R_6\}$, of the set of solutions are presented in Fig. 2.

$R_1=$

P	p_1	p_2	p_3	p_4	p_5	p_6
R_{k1}	0	6	0	0	0	0
R_{k2}	0	0	0	0	6	0
R_{k3}	6	0	0	0	0	0
R_{k4}	0	0	6	0	0	0
R_{ki5}	0	0	0	0	0	6
R_{ki6}	0	0	0	6	0	0

$R_2=$

P	p_1	p_2	p_3	p_4	p_5	p_6
R_{k1}	0	8	0	0	0	0
R_{k2}	0	0	0	8	0	0
R_{k3}	8	0	0	0	0	0
R_{k4}	0	0	0	0	0	8
R_{ki5}	0	0	8	0	0	0
R_{ki6}	0	0	0	0	8	0

$R_3=$

P	p_1	p_2	p_3	p_4	p_5	p_6
R_{k1}	0	4	0	0	0	0
R_{k2}	0	0	0	4	0	0
R_{k3}	4	0	0	0	0	0
R_{k4}	0	0	0	0	0	4
R_{ki5}	0	0	4	0	0	0
R_{ki6}	0	0	0	0	4	0

$R_4=$

P	p_1	p_2	p_3	p_4	p_5	p_6
R_{k1}	0	0	0	0	2	0
R_{k2}	0	2	0	0	0	0
R_{k3}	0	0	0	2	0	0
R_{k4}	0	0	3	0	0	0
R_{ki5}	0	0	0	0	0	2
R_{ki6}	2	0	0	0	0	0

$R_5=$

P	p_1	p_2	p_3	p_4	p_5	p_6
R_{k1}	9	0	0	0	0	0
R_{k2}	0	0	0	9	0	0
R_{k3}	0	9	0	0	0	0
R_{k4}	0	0	0	0	9	0
R_{ki5}	0	0	9	0	0	0
R_{ki6}	0	0	0	0	0	9

$R_6=$

P	p_1	p_2	p_3	p_4	p_5	p_6
R_{k1}	0	0	0	0	5	0
R_{k2}	0	0	0	0	0	5
R_{k3}	0	0	5	0	0	0
R_{k4}	0	0	0	5	0	0
R_{ki5}	0	5	0	0	0	0
R_{ki6}	5	0	0	0	0	0

Fig. 2. The field of alternative solutions $\{R_1, R_2, R_3, R_4, R_5, R_6\}$.

The best value of the solution estimate is $f^{\#} = 9$, the worst value is $f^{0} = 2$. The parameter π lying within the boundaries $f^{0} \leq \pi \leq f^{\#}$ is selected. Let $\pi = 2$. Let us calculate the values of the elements r^*_{ij} of the integral placer of alternatives according to formula 1. The integral field of alternatives is shown in Fig. 3.

$R^*=$

P	p_1	p_2	p_3	p_4	p_5	p_6
R^*_1	9	18	0	0	7	0
R^*_2	0	2	0	21	6	5
R^*_3	10	9	13	2	0	0
R^*_4	0	0	8	5	9	12
R^*_{k5}	0	5	13	0	8	8
R^*_{i6}	10	0	0	6	4	14

Fig. 3. The Integral field of alternatives R^*

The operation of the placement algorithm is based on the use of collective evolutionary memory (CEM), which means any kind of information that reflects the past history of development and is stored independently of individuals. The algorithm associated with evolutionary memory seeks to memorize and reuse ways to achieve better results. The CEM of the placement algorithm is a set of statistical indicators that reflect, for each fragment of the solution, the number θ of the occurrences of the fragment in the best solutions at previous iterations of the algorithm and the number δ indicating the usefulness of the fragment when constructing solutions at the previous iterations of the algorithm. The basis of a multi-agent system is self-organization, ensuring the achievement of the collective goals of the team on the basis of low-level interaction. The team does not have centralized management, and its features are the presence of indirect exchange of information. Indirect exchange – stigmergy is a time-distributed interaction in which one agent changes some areas of the CEM, and other agents later use the already changed information in these areas when they get into it [3].

The decision making process is iterative. Each iteration l includes three steps. At the first stage of each iteration, a constructive algorithm generates n_k solutions of Q_k. The formation of each solution Q_k is carried out by the set of agents A, by means of choosing by each agent a_i the position p_j. Each solution Q_k is a bijective mapping $F_k = A \rightarrow P$ and is formed by sequentially assigning elements to positions. The estimate ξ_k of the solution Q_k and the utility estimate δ_k of the set of positions P_k chosen by the agents in the solution Q_k are calculated. At the second stage, the agents increase the integral utility of the set of P_k positions by δ_k in the integral field of alternatives R^*. At the third stage, the utility estimates δ_k of the integral placer of alternatives are reduced by δ^*. The paper uses a cyclic method of forming decisions. In this case, the building up of estimates of the integral utility δ_k of the set of positions P is performed after the complete formation of the set of decisions Q. The process of assigning an element to a position involves two stages. At the first stage, an element is selected, and at the second stage, the position p_j. In this case, the restriction must be satisfied: each element of the set A corresponds to one single element of the set P and vice versa.

4 The Placement Algorithm on the Base Alternative Crystallization Field Method

1. The volume of the training solution population is set – N_k.
2. Generation of the initial set of solutions $Q = \{Q_k | k = 1, 2, \cdots, n_k\}$ by randomly choosing alternatives s_{ji}. Calculation of the estimates ξ_k and δ_k for all solutions Q_k of the set Q.
3. Narrowing the generated solution set Q to a given size $|Q| = n_k$ by selecting n_k best solutions (with a maximum utility rating value of δ_k). Determining the generated solution set $Q^{\#}$. With the best evaluation value $\xi^{\#}$ and Q^0 with the worst value ξ^0.
4. Formation of an individual field of alternatives R_k for each solution $Q_k \in Q$.
5. Formation of the initial integral field of alternatives $R^*(1)$ by combining all fields of alternatives R_k:

$$R^* = \{R_i^* | i = 1, 2, \cdots, n_a\};$$

$$R_i^* = \left\{r_{ij}^* | j = 1, 2, \cdots, n_p\right\};$$

$$r_{ij}^* = \sum\nolimits_k (r_{kij}),\ k = \{1, n_k).$$

6. $l = 1$ (l is the iteration number).
7. $k = 1$. (k is the solution number).
8. The memory is brought to the initial state:

$A(t) = A$, $A(t)$, is the Set of Unplaced Elements at Step t.
$P(t) = P$, $P(t)$, is the Set of Free Positions at Step t, $t = 1$.

9. At the first stage. For each $a_i \in A(t)$ in the corresponding vector R_i^*, the set $P_i^*(t)$ of positions, $|P_i^*(t)| = a$ with the maximum values of the utility estimate $\delta_{ij} = r_{ij}^*$ are determined with the restriction: $\cup|P_i^*(t)| = P(t)$, in other words, the set of subsets $P_i^*(t)$ covers the set of positions $P(t)$. If the restriction for the selected number a is not satisfied, then the parameter a increases until the restriction is observed.
10. For each $P_i^*(t)$, the sum σ_i^* of utility estimates $\delta_{ij} = r_{ij}^*$ at the positions of the set $P_i^*(t)$ is calculated.
11. Among the set of unplaced elements $A(t)$ with probability $P_i = \sigma_i / \sum_i(\sigma_i)$ proportional to σ_i, the element a_i^* is selected.
12. In the second stage. Among the set of positions $P_i^*(t)$ corresponding to the selected element a_i^* with probability $P_{iy} = \delta_{ij} / \sum_j(\delta_{ij})$ proportional to the value of the utility estimate $\delta_{ij} = r_{ij}^*$, the position p_j^* is chosen.
13. The element a_i^* is placed at the position p_j^* and is removed from $A(t)$.
14. The position p_j^* is removed from all the sets $P(t)$ and $P_i^*(t)$.
15. If $t < n_a$, then $t = t + 1$ and go to 9, otherwise go to 16.
16. The solution estimate ξ_k and the utility estimate δ are calculated.
17. A field of alternatives R_k for Q_k is formed. If $k < n_k$, then $k = k + 1$ and go to 8, otherwise go to 18.

18. Summation of utility estimates of the integral field of alternatives R^* with all placers of alternatives R_k.
19. Decrease in utility estimates $\delta_{ij} = r^*_{ij}$ of all positions, the integral utility of the set of positions by δ^*.
20. If $l < n_a$, then $l = l + 1$ and go to 7, otherwise go to 21.
21. Completion of work. Fixation and conclusion of the best solution.

5 Experimental Research

The study of the placement algorithm by the Crystallization of alternatives field (PCAF) method consists in the formation of test tasks for the placement problem with the available optimal result (PEKO). Optimal PEKO results are available online [11–14]. All circuits in PEKO are local, i.e. the length of the conductors of each circuit has the lowest possible value. PEKU kit circuits consist of PEKO-style local circuits. For experimental studies of the developed placement program, PEKU kit schemes with the well-known F_{opt} optimum were used. For comparison, modern placement algorithms were selected. Comparative analysis PCAF with other decision algorithms was carried out using standard test cases (benchmarks).

In the Table 1 shows the results of comparative experiments and their analysis. The proposed PCAF allocation algorithm is compared with PSO (Particle swarm optimization) [11] and IBA (Improved Bat Algorithm) [13] on benchmarks (FT06, FT10 and FT20) and (from LA01 to LA07). These tests are available on the OR-Library Website.

The contents of the table include the name of each test problem (Probl), the scale of the problem (Size), the value of the best known solution for each problem (Opt), the value of the best solution found using the compared algorithms (best), the average value of the solution (Aver), percentage deviations in relation to the most famous solution (RPE%). RPE = (best-Opt)\Opt) × 100, (RPE) RelativePercentageError.

The values of the Opt, best, Aver parameters reflect the total costs in arbitrary units.

Table 1. The results of comparative experiments.

No	Probl	Size $n \times m$	Opt	PSO			IBA			PCAF		
				best	Aver	RPE %	best	Aver	RPE %	best	Aver	RPE %
1	Ft06	6×6	55	55	58.9	0	55	56.93	0	55	56.3	0
2	Ft10	10×10	930	1007	1086	8.2795	991	1012.6	6.5591	975	976.57	4.8387
3	Ft20	20×5	1165	1242	1296	6.6094	1177	1188.5	1.0300	1181	1183.7	1.3733
4	La01	10×5	666	681	705	2.2522	666	666	0	666	675,9	0
5	La02	10×5	655	694	729.7	5.9541	668	677.5	1.9847	662	676.966	1.0526
6	La03	10×5	597	633	657.5	6.0301	613	622.6	2.6800	611	613.466	2.3450
7	La04	10×5	590	611	648.1	3.5593	611	611	3.5593	605	607.3	2.5423
8	La05	10×5	593	593	601.1	0	593	593	0	593	593.0	0
9	La06	15×5	926	926	940,2	0	926	926	0	926	927.5	0
10	La07	15×5	890	890	940.1	0	890	890	0	890	890.933	0

For tasks (La05, La06, and La07), all three algorithms give the best known values (best = Opt), the average value of Aver solutions for PCAF is better than the average value for PSO, but slightly worse than for IBA. Compared to PSO the FT20 – PCAF gives better results for all indicators; Compared to IBA, the slightly worse result in terms of best, but the best result in terms of Aver. For tasks (FT10, La02, La03, and La04), PCAF gives results for all indicators (best, Aver, RPE) better than PSO and IBA. The results show that the proposed approach improves the quality of the solution in relation to other algorithms and is an acceptable alternative for solving the problem of placing elements. Another important task of the ongoing test study is to determine the convergence of the developed algorithm. Testing determined that the developed algorithm for placing by method of crystallization of alternatives field (PCAF) on average has a convergence of 125 iterations. Testing showed that the probability that the optimal solution will be found is on average –0.9.

A tool that will improve the operation of the placement algorithm can be more balanced and carefully selected parameters that affect the entire operation of the algorithm.

6 Conclusion

Based on a comparative analysis of existing approaches and methods for solving the placement problem, multi-agent methods of intellectual optimization are used.

A new metaheuristic of intellectual optimization has been developed, taking into account the tendency to use alternatives from the best solutions found, based on the modeling of collective intelligence. The totality of data on alternatives and their assessments is a field of alternatives. The key points of the analysis of alternatives in the process of evolutionary collective adaptation, called by analogy with the processes of isolating objects (crystal formation) crystallization, are considered. The algorithm based on the crystallization of a placer of alternatives has been successfully applied to solve the placement problem. Such an approach is an effective way to search for rational solutions for optimization problems that can be interpreted as a scattering of alternatives. Experimental studies have shown that algorithms based on the proposed approach can give better results than when using the methods of bee and ant colonies separately.

Acknowledgements. This research is supported by grants of the Russian Foundation for Basic Research of the Russian Federation, the project № 20-07-00260.

References

1. Clerc, M.: Particle Swarm Optimization. ISTE, London, UK (2006)
2. Kennedy, J., Eberhart, R.: Particle swarm optimization. In: Proceedings of IEEE International Conference on Neural Networks, pp. 1942–1948 (1995)
3. Karpenko, A.P.: Modern Search Engine Optimization Algorithms. Algorithms Inspired by Nature: A Tutorial. Publishing House MSTU. N.E. Bauman, 448 p. (2014)
4. Gupta, J.N., Ho, J.: New Heuristic Algorithm for the One-dimensional Bin-packing Problem. Production Planning & Control, pp. 598–603 (1999)

5. Lebedev, B.K., Lebedev, V.B.: The method of crystallization of placers of alternatives. In: Collection of Scientific Papers of the VII International Scientific-Practical Conference "Integrated Models and Soft Computing in Artificial Intelligence", vol. 2, pp. 903–912. Publishing house Fizmatlit, Moscow (2013)
6. Wang, X.: Hybrid nature-inspired computational analysis/Doctoral dissertation. Helsinki University of Technology, TKK Dissertations, 161 p. (2009)
7. Raidl, G.R.: A unified view on hybrid metaheuristics. In: Lecture Notes In Computer Science, pp. 1–12. Springer, Verlag (2006)
8. Lebedev, O.B.: Models of Adaptive Behavior of Ants Colony in Design Problems, p. 199. Publishing house of SFEDU, Taganrog (2013)
9. Ross, P., Marin, B., Schulenburg, S., Hart, E.: Learning a procedure that can solve hard bin-packing problems: a new GA-Based approach to hyper-heurstics. In: Proceeding of the Genetic and Evolutionary Computation Conference, GECCO: – Chicargo Illinois USA, pp. 1295–1306 (2003)
10. Engelbrecht, A.P.: Fundamentals of Computational Swarm Intelligence. John Wiley & Sons Chichester, UK, 478 p. (2005)
11. Sha, D.Y., Cheng, Yu.: A hybrid particle swarm optimization for job shop scheduling problem. In: Computers & Industrial Engineering, pp. 791–808 (2006)
12. Alvim Adriana, C.F., Glover, F.S., Ribeiro, C.C., Aloise, D.J.: Local search for the bin-packing problem. J. Heuristics 205–229 (2004)
13. Zaher, H., Ragaa, N., Sayed, H.: A novel improved bat algorithm for job shop scheduling. Int. J. Comput. Appl. – BibTeX **164**(5), 24–30 (2017)
14. Job shop scheduling benchmarks. OR Library online. http://mscmga.ms.ic.ac.uk

Resilient Rerouting in IoT Systems with Evolutionary Computing

Ayoade Akeem Owoade[1] and Isaac Olusegun Osunmakinde[2](✉)

[1] School of Computing, College of Science, Engineering and Technology, University of South Africa, Pretoria 0001, South Africa
50781529@mylife.unisa.ac.za

[2] Computer Science Department, College of Science, Engineering and Technology, Norfolk State University, Norfolk, VA 23504, USA
ioosumakinde@nsu.edu

Abstract. Due to the vulnerability of Internet of Things (IoT) to random failures, this study investigates resilience of an evolutionary computing model based on capacity efficiency and on-demand restoration to swiftly address node-link failures. Researches have been conducted on IoT network single failures, but multiple node-link failures have not received enough attention. The model was experimented on node-link failures at various locations on big networks. The proposed model could generate excellent alternate routes and the bandwidth needed for rapid rerouting of multimedia traffic. As multiple node-link failures increase, bandwidth usage also increases which slows down multimedia traffic routing. The proposed model outperforms common proactive and reactive models, in terms of throughput and running time. The survival solution paths show that the proposed model works well in avoiding data loss in transit in spite of multiple node-link failures. This new model is essential in this COVID-19 pandemic, which often requires emergency communications.

Keywords: IoT · Resilient · Node-link failure · Rerouting · Multimedia · Evolutionary computing · Enhanced genetic algorithm

1 Introduction

The Internet can be explained as a communication network that connects people to information, while the Internet of Things (IoT) is an interconnected system that addresses physical objects with different degrees of processing, sensing and performance capabilities sharing the ability to interact and communicate via the Internet as their joint platform [1]. Therefore, the primary aim of the Internet of Things is to make it possible for objects to be linked at any time or anywhere to other objects, people, via any network, route or service. The IoT network must be secured to provide high performance in order to render nodes and links free of multiple communication failures. Therefore, shielding these networks from repeated communication failures is a significant concern that cannot be overlooked. Some of the new commercially-off-the-shelf IoT devices have

© The Author(s), under exclusive license to Springer Nature Switzerland AG 2021
R. Silhavy (Ed.): CSOC 2021, LNNS 229, pp. 194–211, 2021.
https://doi.org/10.1007/978-3-030-77445-5_18

software-level security solutions, but not enough to secure the whole IoT network. It is also noted that the software level security is inherently fundamentally flawed when an IoT network is considered because of the different use patterns [2]. Since most times these problems typically lead to node-link failure, therefore the EGA-based model is proposed in resolving such a failure in order to have a resilient IoT network system in the presence of failures.

This work is streamlined to resolving multiple communication failures of IoT network problem. The Internet was initially used to transfer data packets between users and data sources with a specific IP address. Due to advancements, the Internet is being used to share data among different small, resource constrained devices connected in billions to constitute the Internet of Things (IoT). A large amount of data from these devices impose overhead on the IoT network. Hence, this study is required to provide solutions to various network related problems in IoT including routing, energy conservation, congestion, heterogeneity, scalability, reliability, quality of service (QoS) and security to optimally make use of the available network [3].

IoT networks are used in diverse critical aspects of our society, including commerce, banking, healthcare and mission critical services. Multiple associated failures triggered by natural phenomena, malwares, software updates, latent failures, and deliberate attacks are challenges to the existing infrastructure of IoT networks. These events can cause IoT network services to degrade for a prolonged period of time. In the event of failures, a survival plan to preserve the integrity of a network is obviously vital [4]. The IoT network is widely used to send data, video and voice at a very high speed. IoT network integrates multiplexing and switching functions and allows communication between devices that operate at different speeds.

There have been many network failures in IoT network and interconnected systems, this provides strong motivation for this research. Such events often caused significant disruptions and have exacerbated their effect in terms of active service, economic loss and human sensitivities, such as trust in innovation, by their slow detection and treatment. All possible causes of IoT failures normally lead to node-link failures. Broadly, node-link failure is classified either as an individual node-link failure or a shared node-link failure. To distinguish between an individual node-link and a shared node-link failure depends on whether either one or several connections fail simultaneously. Shared failures demonstrate that the involved node-link share a common element of the network that fails [5].

1.1 Research Questions and Contributions

Various techniques have been suggested for failure recovery in literature and they vary in the network models used [6]. The gaps identified informed the following research questions investigated herein:

- How can a system model on evolutionary algorithm be effectively developed to survive multiple node-link failures disrupting traffic flow in IoT networks?
- To what extent can the traffic flow in IoT networks be resilient in the event of multiple node-link failures?

The main contributions of this paper are as follows:

- Assessing a newly proposed capacity efficiency and on demand recovery enhanced genetic algorithm (EGA) model which would assist IoT network users to survive multiple node-link failures; and
- Performance of comprehensive experiments to find optimal paths to reroute multimedia traffic on IoT big networks ranging from no failure to multiple node-link failures with the needed bandwidth.

2 Related Work

This study suggests an energy-efficient collection of recovery and backup nodes for IoT systems accompanied by energy-efficient preparation of messages [7]. The paper focuses on scheduling messages in an IoT environment where IoT subgroups cluster things/sensors, each subgroup has a message broker that delivers the messages from the group to the ultimate sensed data recipient. At the broker level, the message scheduler operates to determine which message to be first transmitted. The optimal handling of the disabled or failed nodes has become the key function of the proposed scheduling algorithm in order to keep the flow of services provided by these items/sensors continuous and non-disruptive. The authors consider energy as a scarce resource and the cost of deployment of the backup nodes as a result of the node failure occurring in the underlying environment in order to find the correct level of replacement nodes deployed for the sensor network. The results of the simulation demonstrate the efficacy and effectiveness of the proposed message scheduling considering the node failure with the recovery and replacement technique.

The authors [8] suggested the reliability assessment of an IoT infrastructure disaster recovery solution. The investigator recommended that the use of efficient fault-tolerant methods such as disaster recovery (DR) solutions must be planned for IoT systems. A Petri net-based methodology for modeling and evaluating DR solutions for IoT infrastructures is proposed in the work. The proposed models allow important DR measures to be evaluated, such as system availability, cost, and objective recovery time. To prove the approach viability, the authors introduced a case study in which a real-world IoT healthcare system is modeled and evaluated. Sensitivity analysis is often carried out to determine the impact of model parameters on the availability of the device.

Taking into account these three features of the Fog-IoT environment, [9] presents a fault tolerant approach: (i) dynamic infrastructure where entities enter and leave without synchronization, (ii) high heterogeneity in terms of functions, communication models, network, processing and storage capacities, and (iii) cyber-physical interactions that introduce non-deterministic and physical interactions. By saving the state of the application in an uncoordinated manner, fault tolerance is achieved. Notifications are propagated when a failure is observed to minimize the effect of errors and reconfigure the framework dynamically. For recovery, data processed during the state saving process is used, taking into account continuity in relation to the physical environment. The strategy was validated on a smart home platform through realistic experiments.

In view of the reviews, the research herein investigates resilience based on evolutionary computing model to multiple node-link failures in IoT networks.

3 Resilient Model to Node-Link Failures in IoT

3.1 Main Objective

Multimedia traffic needs to be rerouted immediately there is a failure occurrence in IoT network. Alternative paths must be provided with reduced delay coupled with transmission restrictions such as needed bandwidth (B_{req}) to transmit and minimum path cost (P_c) to achieve the shortest path in order for rerouting to be achieved. Thus, in the event of failure that poses risks to business operations around the world, it is the main thrust of this paper to find an optimal alternative path to reroute traffic. The causes of IoT network failure are attributed to hardware and software failure, configuration error, fire outbreak and power failure. We propose an EGA model that can be used to create an optimal alternate path to redirect multimedia traffic as soon as possible in the event of failure.

With the EGA-based model, optimization problem is solved. Either the minimum or maximum solution could be the solution. Both conditions are problems with optimization. There are three components to an optimization problem: decision variables, objective function and restrictions [10].

(i) *Decision variables:* These are the factors which can be regulated within a model. If there are n decision variables, they are represented as $z_1, z_2, \ldots, z_n$.
(ii) *Objective function*: This is the task that we want to optimize or mitigate. An objective function is written as $f(z_1, z_2, \ldots, z_n)$. If we want to maximize this function, we write $\max_{z_1, z_2, \ldots, z_n} f(z_1, z_2, \ldots, z_n)$ and, if the task should be mitigated, we express it as $\min_{z_1, z_2, \ldots, z_n} f(z_1, z_2, \ldots, z_n)$.
(iii) *Restrictions*: They are issues or restrictions of the task. The mathematical functions are written as follows:

$$k_1(z_1, z_2, \ldots) \leq 0, \quad k_2(z_1, z_2, \ldots) \leq 0, \quad k_3(z_1, z_2, \ldots) \leq 0$$

3.2 Mathematical Definition

In case there is a node-link failure within an IoT network when a multimedia traffic is being sent from a web application system to another called a recipient web application system, then in order to achieve quick rerouting, the optimal path (O_P) should be created to reroute the traffic to reduced queuing delay. This means that the rerouting route (O_P) to be maximized is based on the following restrictions: (i) There is a restricted amount of bandwidth (B_{req} Kbytes/s) due to the occurrence of connection loss. It is therefore necessary to decide the appropriate bandwidth (B_{req}) that the multimedia traffic will use to redirect the message. (ii) The cost of the route must be reduced in order to achieve the shortest rerouting path and reduce the queuing delay. (iii) It is important to determine the capacity efficiency (C_{ef}) that revolves around the sufficient bandwidth to route and redirect the traffic [10]. The minimal path cost obtained through the shortest path reduces the transmission delay D_L (s). This study seeks to optimize positive rewards by selecting optimal values. This is a linear programming (LP) problem, which is a problem of optimization where linear functions express the objective functions and all the restrictions. A linear function is expressed by $f(z_1, z_2, \ldots..) = a_1 z_1 + a_2 z_2 + \ldots + a_0$ where $a_1, a_2, \ldots.., a_0$ are constants. In general, an LP problem that maximises the objective function in this work is represented as follows:

Maximize $P_1 z_1 + P_2 z_2 + \cdots + P_n z_n$
(Objective function)
Subject to restrictions $z_1 + z_2 + \cdots z_n \leq B_w$
(Limited bandwidth)
$z_1 + z_2 + \cdots z_n \leq P_{cost}$ (Limited Path cost)
$z_1 + z_2 + \cdots z_n \leq C_{ef}$ (Limited network capacity efficiency)
Non-negativity restrictions: $z_1 \geq 0,\ z_2 \geq 0, \ldots, z_n \geq 0$ (The model cannot transmit non signals)

We use the proposed EGA-based capacity efficiency model to solve this optimization problem to produce the optimal alternative path along with the heuristics added to determine the necessary bandwidth needed to transmit the message.

3.3 Establishing a Framework of Resilience to Node-Link Failures

The IoT network must be well structured in such a way as to accommodate a restoration strategy. The survivable model is structured to generate an alternative path when there are multiple node-link failure occurrences in the IoT network environment by using an enhanced genetic algorithm. In this case, the network requires intelligence and knowledge of a failure occurrence to resolve it by consulting every node on the network for system information in order to achieve capacity efficiency and fast restoration. The proposed framework is described in Fig. 1 aimed at building a survivable IoT network.

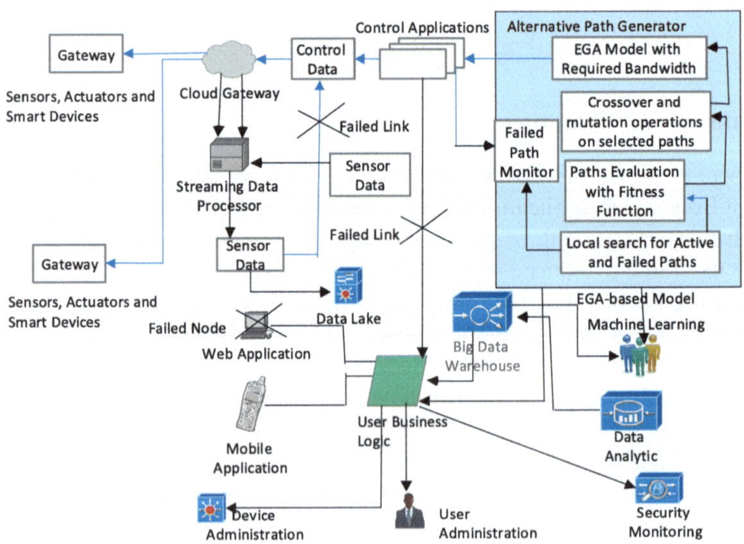

Fig. 1. Framework of resilience to multiple node-link failures in IoT networks (adapted from [11])

For an IoT framework to be reliable and efficient, some minimal set of measures should be met to achieve IoT integration and interoperability. As the IoT paradigm itself is still in an evolving state, we propose a set of minimum measures to be met by IoT integration frameworks. These are capacity efficiency and fast restoration. This EGA model is connected to the user applications system where the model gains access to fail and active nodes and links in IoT network. The user application allows users to connect to an IoT device and offers the options to track and manage their smart stuff. Device manager manages data routing and transmission over the Internet to various IoT hubs and computers. This is where the system experiences failures most times. Whenever there is node-link failure, the EGA model resolves this by getting failure information from device manager and resolves the failure generating optimal alternative path. The capacity efficiency and fast restoration capability of the model is achieved through EGA-based model by generating an alternative path and ensuring that the available bandwidth is capable of rerouting the size of the multimedia traffic (see Table 2).

3.4 Mathematical Analysis of EGA-Based Model

The proposed EGA is described below to model the resilient IoT network.

- **Genetic Crossover Operator**

The crossover operator blends two parent chromosome sub-parts and creates offspring bearing some parts of the genetic materials of both parents. There are primarily two forms of crossover: single-point crossover and multi-point crossover [12].

- **Genetic Mutation Operator**

The crossover process can produce degenerated populations. Mutation operations are performed to reverse this. A mutation operation can be inversion, insertion, reciprocal, omitting, exchange or others. In this work, omitting mutation operator [12] is used and done by randomly choosing one node in the new generation.

- Selection from the Collection of Network Paths

Ranking introduces a uniform scaling across the collection of network paths as shown in Eq. (1).

$$Let \;\; p_i = \frac{fitness_i}{\sum fitness_j}. \tag{1}$$

Where p_i is the probability that individual i will be selected, $fitness_i$ is the fitness of individual 'i', and $\sum fitness_j$ represents the sum of the fitness of all individuals in the network paths

- **Fitness Evaluation of Network Paths**

The quality of the path that involves computational effectiveness and accuracy is defined in Eq. (2) [13].

$$Fitness_i = \frac{1}{\sum_{j=1}^{l_i-1} C_{g_i(j), g_i(j+1)} + \sum_{j-1}^{l_i-1} B_{g_i}(j) g_i(j+1)} \quad (2)$$

Where F_i represents the fitness value of the i-th chromosome (path), l_i is the cost of the link in a path; $g_i(j)$ represents the gene (node) in the path, C is the path cost of a chromosome and B is the bandwidth.

- **Bandwidth Requirements for Alternative Paths Generated**

To determine the minimum bandwidth that would convey a multimedia traffic over two points [14].

$$\mathrm{B}_{g_i} = \frac{M_s}{D_L} \quad (3)$$

For calculating the required bandwidth between two connected network points, Eq. (3) can be used. Here, the size of a multimedia traffic is M_s, and the traffic delay is D_L. Traffic delay could be written as Eq. (4)

$$D_L = \frac{M_s}{B_{g_i}} \quad (4)$$

Then, the required bandwidth whose path consists of two or more nodes are shown in Eqs. (5) and (6),

$$p = f(g_i(j), C) = \alpha g_i(j) + \beta C \quad (5)$$

$$\mathrm{B}_{g_i} = \frac{M_s}{D_L - \mathrm{p}} \quad (6)$$

Where '*p*' is the unknown function depending on $g_i(j)$ and C, here, $g_i(j)$ is the number of nodes in the routing path and $\mathrm{C} = \sum_{\mathrm{n}} \mathrm{I_n}$, which is the sum of link cost of a routing path.

3.5 Towards Optimal Generation of the Transmission Route

Using an improved genetic algorithm, the module deals with finding the optimal path. The input to this module is the set of random paths generated. Each route is termed a chromosome and a node or link is termed a gene. A specific algorithmic scheme presented in algorithm 1 follows an optimal alternative route in an EGA model.

Algorithm 1: EGA model for rerouting on the failed IoT network

Input: network data (s, d, nodes, failed node(s), active nodes, links, failed link(s), active links)
EGA parameters (*pop*, *Gen*, *p*M, *p*C)
Where "pop" represents the population, *C* represents path cost, *g* represents the gene, which is the nodes in the chromosomes, M_s represents the size of video traffic, D_L represents packet delay, g_i = **number of nodes in a chromosome**, p=number of nodes in a chromosome + path cost of the chromosome, *ch* represents chromosome, 'S_n' represents the source node and 'd_n' represents the destination node, *pC* represents the crossover probability and *p*M represents the mutation probability, O_{AL} represents an optimal alternative link, M_{PC} represents minimal path cost, i, j=1,2,3….n and B_{g_i} represents the bandwidth required to move messages from one node to another respectively.

Output: O_{AL}, M_{PC}, B_{g_i}, D_L, T_t

```
1.  Begin
2.  pop = Generate Random Population as Candidate Solutions
3.  Generate Chromosomes, which are candidate solution
4.  Initialize C, S_n, d_n, B_{g_i}, D_L, Ch
5.  Calculate C=Sum (l_1 + l_2 + ··· +l_n); g_j=Sum (g_1 + g_2 + ··· +g_n)
6.  Calculate fitness = 1/(SUM(C_{g(j)} * g_i(j + 1)) + SUM(B_{g_i} * g_j(j + 1))
7.  Calculate D_L = M_s/ B_{g_i}
8.  Calculate p= C + g_j
9.  Calculate B_{g_i} = M_s/(D_L − p); Parent-1, Parent-2, Child-1, Child-2
10. Generate the New population [New pop]
11. Evaluate (ch) =fitness_i/SUM(fitness_i)      (\*Selection by Ranking)
12. For i = 1 to size (pop) Do
          Select [parent1, parent2] from pop
          Crossover (parent1, parent2) with probability 0.5= [child1, child2]
    End Do
13. If Parent Fitness > Child Fitness then
          Apply Mutation Operator
          Replace Old pop
    End if
14. While the termination condition is not met Do 12 & 13
15.       Generate new Pop Else
16.       Generate Output: O_AL, M_PC, B_{g_i}, D_L, T_t
    End While
```

4 Experimental Evaluations

4.1 Experimental Setup: Rerouting in Thirty (30) Nodes Network

Nodes are named A through Z and A1 through D1 and experiments are carried out at different locations with varying numbers of failed nodes and failed links, as seen in experiments 1–8. In the simulated experiments, the circles represent the nodes on the network, the line connecting a node to another represents a link and the failed nodes and failed links are indicated in the figure labels and the optimal alternative paths produced are also shown. Figures, tables and results are shown in experiments 1, 5 and 7 while results are only shown in experiments 2–4 and (6&8) in Table 2. Link costs are generated randomly. Multimedia traffics are sent from source node A to destination node X. The transition diagram describing an IoT network of thirty nodes in the experimental setup is shown in Fig. 2. Due to its suitability in network programming, the experiment is implemented with Java programming language.

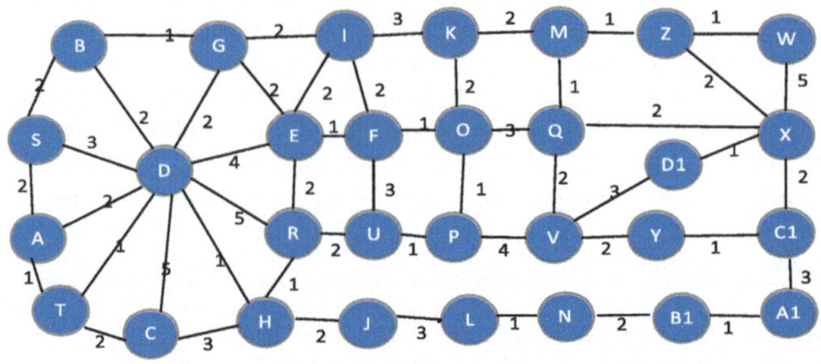

Fig. 2. Transition diagram representing a thirty nodes network

The parameter specification for the EGA-based model in experiments 1–8 is shown in Table 1.

Table 1. Parameters specification

EGA Properties	Number of Node	Crossover Probability	Multimedia Size (Bytes)	Initial Time(s)	IP Header	Node/Link Failure
Properties	varied	0.5	Varied	0.0050	20 Bytes	varied

The following experiments are carried out: Experiment 1: Considers an IoT network without node/link failure, Experiment 2: with one node and one link failures, Experiment 3: with one node and two links failure, Experiment 4: with two nodes and three links failure, Experiment5: with two nodes and four links failure, Experiment 6: with three nodes and five links failure, Experiment 7: with three nodes and six links failure and Experiment 8: with four nodes and seven links failure respectively. (See result in Table 2).

4.2 Performance Analysis of Thirty (30) Nodes Network with Different Number of Node-link Failures

Table 2 shows the performance analysis of thirty nodes network with different number of node-link failure.

Table 2. Performance analysis of experiments 1–8

Experiment number	Number of (node, link) failure	Transmissi-on Time (seconds)	Required Bandwidth (Kbytes/s)	Fitness	Delay (ms)	Path Cost	Multim-edia Size (Bytes)	Path Generated
1	(0,0)	0.737	33.33	0.0026	40	14	600	A – D – G – I– F– O – Q– X
2	(1,1)	0.782	50	0.0023	40	15	700	A – D – R–U – P– V– Q –X
3	(1,2)	0.828	46.43	0.0019	40	18	650	A –D –R –U –F – O – Q –X
4	(2,3)	0.89	18.75	0.0029	45	20	300	A–S–D –E –I –K –M Z X
5	(2,4)	1.156	30	0.0027	40	16	480	A–D –R –U– P – O – Q – X
6	(3,5)	1.17	22.22	0.0028	45	18	400	A– D– E– F– I– K–M–Z –X
7	(3,6)	1.185	23.81	0.0029	45	15	500	A–D–E– F– O– K–M–Q –X
8	(4,7)	1.203	28.95	0.0004	45	17	550	A– D– E–F– O –P–V–D1–X

Table 2 is a reflection of the resolution to the optimization problem. This is observed in experiments 1–2 and 7, for example, with the results of minimal path cost values recorded from the optimal alternative paths generated which in turn minimizes running time, thereby making rerouting of multimedia traffic faster.

The details of experiments 1, 5 and 7 are shown in subsequent sections.

4.3 Experiment 1: Consider an IoT Network Without Node-Link Failures

In Fig. 3, a multimedia traffic of size 600 Bytes is set for transmission from source node A to destination node X.

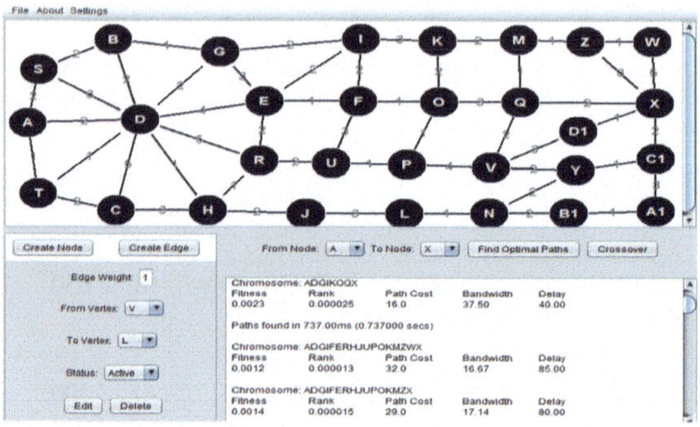

Fig. 3. IoT Network without Node-link Failure

Table 3 shows the paths generated from Fig. 3.

Table 3. Initial generated paths (first generation)

Serial No	Chromosomes	Fitness	Rank	Path Cost	Bandwidth (Kbytes/s)	Delay (s)
1	A – D – G – I – F – O – Q – X	0.0026	0.000028	14	33.33	40
2	A – D – G – E – I – F – O – K – M – Q - X	0.0021	0.000023	19	24.00	55
3	A – D – G – E – I – F – O – Q – V – D1 – X	0.0019	0.000021	21	26.09	55
4	A – D – G – E – I – F – O – Q – X	0.0023	0.000025	17	31.58	45
5	A – D – G – I – E – R – U – P – O – Q – X	0.0021	0.000023	19	24.00	55
6	A – D – G – I – E – R – U – P – V – D1 - X	0.0019	0.0022	21	26.09	55
7	A – D – G – I – E – F – O – K – M – Q – X	0.0023	0.000025	17	22.22	55
8	A – D – G – I – E – F – O – Q – X	0.0026	0.000027	15	28.57	45
9	A – D – G – I – K – M – Z – W – X	0.0022	0.000023	18	33.33	45
10	A – D – G – I – K – M – X	0.0026	0.000027	15	35.29	40

(continued)

Table 3. (*continued*)

Serial No	Chromosomes	Fitness	Rank	Path Cost	Bandwidth (Kbytes/s)	Delay (s)
11	A – D – C – H – R– E – F – O – Q – X	0.0020	0.000021	20	30.00	50
12	A – D – H – R– U – P – V – Q – X	0.0026	0.000027	15	28.57	45
13	A – D – H – R– U – F – I – K – M – Q – X	0.0021	0.000023	19	24.00	55
14	A – D – H – R– U – F – 0 – K – M – Z – X	0.0022	0.000024	18	23.08	55
15	A – D – H – R– U – F – O – K – M – Q – X	0.0023	0.000025	17	22.22	55
16	A – D – H – R– U – F – O – P – V – Q – X	0.0021	0.000023	19	24.00	55
17	A – D – H – R– U – F – O – Q – X	0.0026	0.000027	15	28.57	45
18	A – D – E – G– I – F – O – K – M – Q – X	0.0019	0.000021	21	26.09	55
19	A – D – E – F– U – P – O – Q – X	0.0023	0.000025	17	31.58	45
20	A – D – E – F– U– P – V – Y – C1 – X	0.0020	0.000021	20	30.00	50
21	A – D – R – U– P – O – K – M – Z– X	0.0021	0.000023	19	28.57	50
22	A– D – R – U– P – O – Q – X	0.0020	0.000025	15	37.50	40

New Population

Table 4 shows the different generations produced in the evolutionary cycle of experiment 1.

The routing path is taken from the last generated population (fourth generation), the path with higher ranking and lower path cost is considered, the path is A–D–G–I–F–O–Q–X and it uses a bandwidth of 33.33 Kbytes/s for transmission. The time it takes to generate the optimal alternative path is 0.7370 s.

Table 4. Evolutionary cycle in experiment 1

Second Generation	Fitness	Rank	Path Cost	Bandwidth (Kbytes/s)	Delay(s)
A – D – G – I – F – O – Q – X	0.0026	0.000028	14	33.33	40
A – D – H – R– U – P – V – Q – X	0.0026	0.000027	15	28.57	45
A – D – H – R– U – F – O – Q – X	0.0026	0.000027	15	28.57	45
A – D – G – I – K – M – X	0.0026	0.000027	15	35.29	40
A – D – G – E – I – F – O – Q – X	0.0023	0.000025	17	31.58	45
A – D – G – I – E – F – O – K – M – Q – X	0.0023	0.000025	17	22.22	55
A – D – H – R– U – F – O – K – M – Q – X	0.0023	0.000025	17	22.22	55
A – D – E – F – U – P – O – Q – X	0.0023	0.000025	17	31.58	45
A – D – H – R – U – F – O – K – M – Z – X	0.0022	0.000024	18	23.08	55
Third Generation	Fitness	Rank	Path Cost	Bandwidth (Kbytes/s)	Delay(s)
A – D – G – I – F – O – Q – X	0.0026	0.000028	14	33.33	40
A – D – H – R– U – P – V – Q – X	0.0026	0.000027	15	28.57	45
A – D – H – R– U – F – O – Q – X	0.0026	0.000027	15	28.57	45
A – D – G – I – K – M – X	0.0026	0.000027	15	35.29	40
Fourth Generation	Fitness	Rank	Path Cost	Bandwidth (Kbytes/s)	Delay(s)
A – D – G – I – F – O – Q – X	0.0026	0.000028	14	33.33	40
A – D – G – I – K – M – X	0.0026	0.000027	15	35.29	40

4.4 Experiment 5: Consider an IoT Network with Two Nodes and Four Links Failure (Failed Nodes: H and G, Failed Links: H – J, G – I, D – H and D – G)

In Fig. 4, a multimedia traffic of size 480 Bytes is set for rerouting from source node A to destination node X.

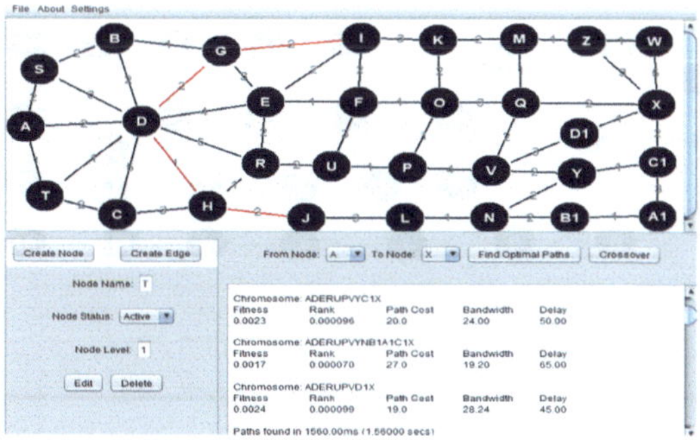

Fig. 4. Network with two nodes and four links failure

Initial Population: Five chromosomes (paths) are generated from Fig. 4 which form the initial population.

New Population

Table 5 shows the different generations produced in the evolutionary cycle of experiment 5.

Table 5. Evolutionary cycle in experiment 5

Second Generation	Fitness	Rank	Path Cost	Bandwidth (Kbytes/s)	Delay(s)
A – D – E – F – O – K – M – Z – X	0.0028	0.000117	16	24.00	45
A – D – R – E – F – O – Q – X	0.0027	0.000114	16	30.00	40
Third Generation	Fitness	Rank	Path Cost	Bandwidth (Kbytes/s)	Delay(s)
A – D – R – E – F – O – Q – X	0.0027	0.000114	16	30.00	40

The rerouting path is taken from the last generated population (third generation), the path with lower path cost and lower delay is considered, which is A–D–R–E–F–O–Q–X and it uses a bandwidth of 30.00 Kbytes/s for transmission. The time it takes to generate the optimal alternative path is 1.1560 s.

4.5 Experiment 7: Consider an IoT Network with Three Nodes and Six Links Failure (Failed Nodes: H, G and I, Failed Links: H – J, G – I, D – H, D – G, I – E and H – R)

In Fig. 5, a multimedia traffic of size 500 Bytes is set for rerouting from source node A to destination node X.

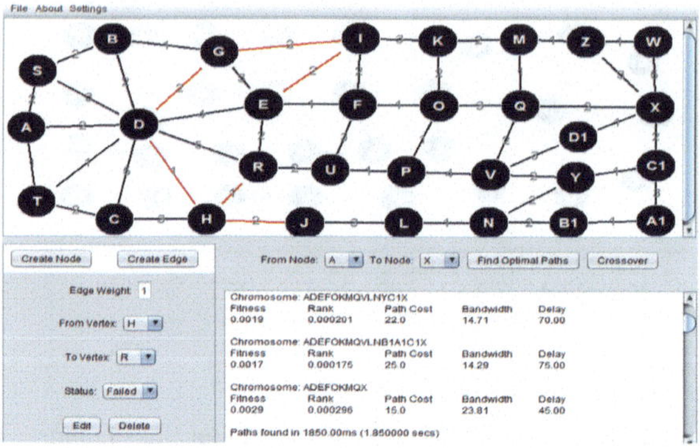

Fig. 5. Network with four nodes and six links failure

Initial Population: Six chromosomes (paths) are generated from Fig. 5 which form the initial population.

New Population

Table 6 shows the different generations produced in the evolutionary cycle of experiment 7.

Table 6. Evolutionary cycle in experiment 7

Second Generation	Fitness	Rank	Path Cost	Bandwidth (Kbytes/s)	Delay(s)
A – D – E – F – O – K – M – Q – X	0.0029	0.000296	15	23.81	45
A – D – E – F – O – K – M – Z – X	0.0027	0.00028	16	25	45
A – D – R – U – P – O – Q – X	0.0026	0.000274	16	31.25	40
Third Generation	Fitness	Rank	Path Cost	Bandwidth (Kbytes/s)	Delay(s)
A – D – E – F – O – K – M – Q – X	0.0029	0.000296	15	23.81	45
A – D – E – F – O – K – M – Z – X	0.0027	0.00028	16	25	45
Fourth Generation	Fitness	Rank	Path Cost	Bandwidth (Kbytes/s)	Delay(s)
A – D – E – F – O – K – M – Q – X	0.0029	0.000296	15	23.81	45

The routing path is A–D–E–F–O–K–M–Q–X and it uses the bandwidth of 23.81 Kbytes/s for transmission. The time taken to generate the optimal alternative path is 1.18500 s.

5 Benchmarking the Proposed Model with Popular Routing Algorithms

NS2 conducted a comparison and performance assessment for the three routing protocols: reactive (AODV), proactive (DSDV) [16] and the proposed model. Average throughput is the metric used to measure performance [15]. Simulation experiment result based on average throughput with the following parameters (see Table 7). The simulation results describe the average throughput of the (AODV), (DSDV) schemes and the proposed model as indicated in Table 8.

Table 7. Simulation environments

Environment	Value
Simulation area	550 m * 550 m
Simulation time	100–500 s
Transmission range	20 m
MAC protocol	IEEE 802.15.4
Movement speed	5 m/s
Packet size	512 Bytes
Number of nodes	100
Number of UDP connection	5–20
Data sending rate	4 kBps

Table 8. Average throughput result for 30 nodes network

Average Throughput (Kbps)	100	200	300	400	500
EGA	621.62	618.50	610.56	607.22	589.98
DSDV	555.94	542.95	540.70	532.35	515.57
AODV	501.00	480.20	475.30	468.60	460.32

Figure 6 represents the graphical interpretation of Table 8.

Based on the simulation results as shown in Fig. 6, the throughput value of EGA model consistently shows better performance with respect to throughput among two other routing protocols. At time 100ms, the highest throughput is attained because there is no node or link failure here and it is observed that as rerouting time is increasing, the node-link failures also increases while the throughput value is decreasing thus resulting in bad transmission. Rerouting is slower at 500 ms as a result of rising node-link failure, so this degrading average throughput.

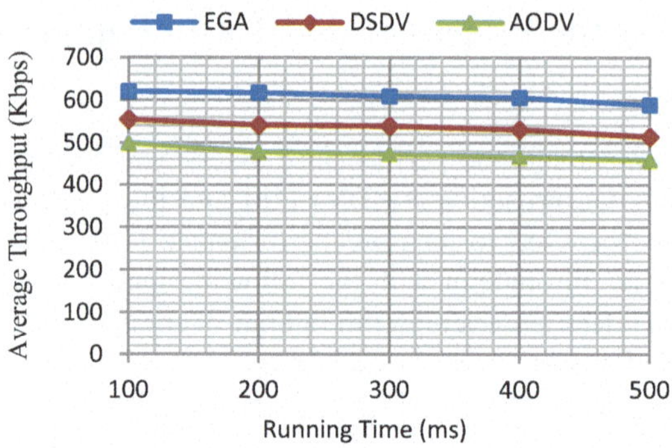

Fig. 6. Average throughputs of routing models against running time

6 Conclusion

The evolutionary computing model (EGA), which is resilient to multiple IoT node-link failures, was provided in this paper by generating an optimal alternative route with needed bandwidth to redirect multimedia traffic when a transmission path failed. It has been shown, among others, from simulation experiments 1–8, that connection failure increases the use of bandwidth and causes transmission delay that slows down rerouting of multimedia traffic. Further research has shown that the larger the size of the network, the higher the bandwidth required to effectively redirect a traffic, and the longer the rerouting time. The assessment of other similar routing models was the focus of the comparison in Sect. 5. The proposed model was found to outperform the proactive and reactive restoration models in terms of running time and average throughput. Unlike in other assessed restoration models, the proposed model reroute multimedia traffic in real-time as shown in experiments 1–8.

The prototype is proposed to be a pilot for providers of IoT services. In solving node-link failures within IoT networks, the proposed EGA-based model is recommended to make IoT networks resilient in this pandemic period where emergency communications are inevitable. The authors are also planning to test the proposed model for high-performance computing (HPC) in order to improve efficiency.

References

1. Hussein, A.H.: Internet of Things (IOT): research challenges and future applications. Int. J. Adv. Comput. Sci. Appl. **10**(6), 77–82 (2019)
2. Porras, J., Pankalainen, J., Knutas, A.: Security in the internet of things – a systematic mapping study. In: Proceedings of the 51st Hawaii International Conference on System Sciences (2018)
3. Srinidhi, N.N., Kumar, S.D., Venugopal, K.R.: Network optimizations in the Internet of Things: a review. Eng. Sci. Technol. Int. J. (Elsevier) **22**, 1–21 (2019)
4. Xie, L., Heegaard, P.E., Jiang, Y.: Modeling and quantifying the survivability of telecommunication network systems under fault propagation. In: Bauschert, T. (eds.) Advances in Communication Networking. EUNICE 2013. Lecture Notes in Computer Science, vol. 8115. Springer, Heidelberg (2013). https://doi.org/10.1007/978-3-642-40552-5_3

5. Mallika, N.M.: Link failure recovery in WDM networks. Int. J. Comput. Sci. Electron. Eng. **1**(5), 2320–4028 (2013)
6. Saeed, M.K., et al.: Connectivity restoration techniques for wireless sensor and actor network (WSAN): a review. Int. J. Adv. Comput. Sci. Appl. (IJACSA) **9**(9), 8 (2010)
7. Abdullah, S., Yang, K.: An energy efficient message scheduling algorithm considering node failure in IoT environment. Wirel. Pers. Commun. **79**(3), 1815–1835 (2014)
8. Andrade, E., Nogueira, B.: Dependability evaluation of a disaster recovery solution for IoT infrastructures. J. Supercomput. **76**(3), 1828–1849 (2018). https://doi.org/10.1007/s11227-018-2290-0
9. Ozeer, U., et al.: Resilience of stateful IOT applications in a dynamic fog environment. In: EAI International Conference on Mobile and Ubiquitous Systems: Networking and Services, New York, United States. (MobiQuitous 2018), pp. 1–10 (2018)
10. Owoade, A.A., Osunmakinde, I.O.: Surviving node-node failures within wireless networks for a near optimal ant colony system message rerouting. Int. J. Mob. Netw. Design Innov. **9**(3–4), 153–182 (2019)
11. Grizhnevich, A.: Process Automation and IoT, ScienceSoft (2018)
12. Gopalakrishnan Nair, T.R., Sooda, K., Yashoda, M.B.: Enhanced genetic algorithm approach for solving dynamic shortest path routing problems using immigrants and memory schemes. In: International Conference on Frontiers of Computer Science (2011)
13. Fadil, Y.A.: Routing using genetic algorithm for large networks'. J. Eng. Sci. **3**(2), 55–70 (2013)
14. Sultanov, T.G., Sukhov, A.M.: Simulation technique for available bandwidth estimation. IEEE Comput. Soc. **1**(2010), 490–495 (2010)
15. Taneja, S., Kush, A.: A survey of routing protocols in mobile ad hoc networks'. Int. J. Innov. Technol. Manag. **1**(3), 279 (2010)
16. Abdulleh, M.N., Yussof, S., Jassim, H.S.: Comparative study of proactive, reactive and geographical MANET routing protocols. Commun. Netw. **7**(2), 125–137 (2015)

Urban Monthly Water Consumption Forecasting Based on Signal Decomposition and Optimized Extreme Learning Machine

Jianjun Lei(✉) and Jiapeng Li

School of Computer Science and Technology, Chongqing University of Posts and Telecommunications, Chongqing 400065, China
leijj@cqupt.edu.cn

Abstract. With the development of information technologies, the smart water is getting to be popular. Water consumption forecasting plays a significant role in the smart water system. This paper proposes a PSO-ELM (Extreme Learning Machine Optimized by Particle Swarm Optimization) coupled with SDEMD (Secondary Decomposition of Empirical Mode Decomposition by Variational Mode Decomposition) for monthly water demand forecasting. Firstly, we use the EMD (Empirical Mode Decomposition) to decompose the water consumption data and the VMD (Variational Mode Decomposition) to process the first sub-time series. Then, PSO-ELM is employed to forecast the sub-time series separately, the forecasting results of every sub-time series are combined together to form the final forecasting result. The experimental results reveal that the SDEMD-PSO-ELM model can effectively forecast the monthly water consumption for strong nonlinearity and non-stationary data and obtains better performance than ELM (Extreme Learning Machine), RBFNN (Radial Basis Function Neural Network) and VMD-ELM models.

Keywords: Water consumption forecasting · Signal decomposition · Extreme learning machine · Particle swarm optimization

1 Introduction

Because of population growth, global warming and rapid economic development, the consumption of water in urban is quickly increased. Therefore, there is a shortage of water resources in recent years, many areas are facing the problem of running out of water resources, so the planning and management of water resources is very important. Urban water consumption forecasting plays an important role in urban construction planning and optimal operation of water distribution system [1]. According to the different forecast frequency, the water consumption forecast model can be divided into long-term forecast model and short-term forecast model. The long-term forecasting model (1 to 10 years' forecast horizon) focuses on the plan and design of water distribution systems. The short-term forecasting model (1 day to 1 month's forecast horizon) pays attention on the real-time water demands of the existing water distribution systems, which is generally

© The Author(s), under exclusive license to Springer Nature Switzerland AG 2021
R. Silhavy (Ed.): CSOC 2021, LNNS 229, pp. 212–223, 2021.
https://doi.org/10.1007/978-3-030-77445-5_19

used for the operation of water plants and pump stations[2]. In this study we target the short-term model which is based on signal decomposition and optimized ELM (Extreme Learning Machine) to forecast monthly water consumption. The contributions of this paper are as follows:

1) Based on EMD and VMD, we propose a signal decomposition method SDEMD that decompose the data and deal with the time-series data respectively. The main idea of SDEMD is using VMD to decompose the first sub-series of EMD to reduce signal frequency and obtain better prediction performance.
2) Present the framework of the hybrid forecasting model of SDEMD-PSO-ELM which is combined by SDEMD and PSO-ELM to build the monthly water consumption forecasting model and improves accuracy.
3) To simultaneously ensure forecasting accuracy and stability, a new method of feature selection is proposed in this paper. We extract the lag time of the water consumption data (1–12 months) and use R-square to extract the lag time suitable as the feature of the forecasting model, which can use effectively the feature of data itself and thus improve the performance of the forecasting model.

2 Related Work

Many researches have already proposed forecasting lots of models for water consumption forecasting. These models can generally be classified into traditional methods and machine learning algorithms [2]. Earlier period of researches used traditional statistical models to forecast water consumption, such as LR (Linear Regression), ES (Exponential Smoothing), and ARIMA (Auto-Regressive Integrated Moving Average) [3, 4]. These models are easy to understand and implement, so they are very popular in early research and have been widely used in practical projects. However, traditional models can only forecast linear changes in water consumption, and hardly can accurately predict its nonlinear changes.

To solve this problem, more complex models of machine learning algorithms based on artificial intelligence have been used in recent years. The model using machine learning algorithms is a data-driven nonlinear model, which is based on historical data to establish the relationship between water consumption and relevant variables to predict water consumption. Such as [5], the paper used SVR (Support Vector Regression) to predict water consumption and RF (Random Forest) was used in [6]. The other kind of water consumption prediction method in machine learning is neural networks [7–9]. In particular, due to the ease of modeling and good performance of the ELM model, ELM was chosen as the prediction model for decomposing water demand [12–14]. Compared with the traditional neural network model, ELM greatly improves the learning speed and has better generalization ability. However, the accuracy of the traditional single neural network prediction is still not enough to meet the practical needs, researchers began to focus on the combined model. Since the time series is a fluctuating line which looks like a signal, researchers found that using signal decomposition algorithms just like EMD (Empirical Mode Decomposition) [11] and VMD (Variational Mode Decomposition) [10] to process data can achieve higher prediction accuracy. Such as, a hybrid model was built in [12] combined VMD and ELM (Extreme learning machine) to forecast water consumption. This paper will focus on EMD, VMD and ELM to build a monthly water consumption forecasting model, and discuss the performance of the model in the prediction of monthly water consumption in Shapingba, Chongqing, China, and compare it with RBFNN (Radial Basis Function Neural Network), ELM, EMD-ELM, VMD-ELM and PSO-ELM.

3 Methodology

The overall framework of SDEMD-PSO-ELM monthly water consumption forecasting model proposed in this paper is shown in Fig. 1. Obviously, according to the framework, SDEMD-PSO-ELM algorithm is classified into two steps:

1) Use SDEMD to decompose original time series data. The original time series are decomposed into the EMD model, several Eimfs are output (e.g., Eimf1, Eimf2 in Fig. 1), and then secondary decomposed Eimf1 by VMD model to obtain the specified number of Vimfs.
2) PSO-ELM is used to model and forecast each sub-series (e.g., Eimf2, Vimf1 in Fig. 1) to obtain the prediction result of each IMF (Intrinsic Mode Functions). Finally, the results of all sub-series are integrated to obtain the final forecasting result.

SDEMD-PSO-ELM model proposed in this paper will be introduced from above two steps.

3.1 Secondary Decomposition of EMD by VMD (SDEMD)

EMD is a classical algorithm of signal decomposition which was proposed by Dr. Norden E. Huang [11], but it has some disadvantages for time series prediction. In order to solve this problem, Dragomiretskiy and Zosso proposed VMD [10] which is a non-recursive adaptive time-frequency analysis method based on EMD, and VMD is more suitable for time series prediction than EMD because of its non-recursive. However, the IMFs of VMD are different from the IMFs of EMD, we can get them by updating Eq. (1)–(3) until convergence:

$$\hat{\boldsymbol{h}}_k^{n+1}(\boldsymbol{x}) = \frac{\hat{\boldsymbol{f}}(\boldsymbol{x}) - \sum_{i<k} \hat{\boldsymbol{h}}_i^{n+1}(\boldsymbol{x}) - \sum_{i>k} \hat{\boldsymbol{h}}_i^{n}(\boldsymbol{x}) + \hat{\boldsymbol{\alpha}}^n(\boldsymbol{x})/2}{1 + 2\boldsymbol{\beta}(\boldsymbol{x} - \boldsymbol{x}_k^n)^2} \tag{1}$$

$$x_k^{n+1} = \frac{\int_0^\infty x \left| \hat{h}_k^{n+1}(x) \right|^2 dx}{\int_0^\infty \left| \hat{h}_k^{n+1}(x) \right|^2 dx} \tag{2}$$

$$\hat{\alpha}^{n+1}(x) = \hat{\alpha}^n(x) + \eta \left[\hat{f}(x) - \sum\nolimits_k \hat{h}_k^{n+1}(x) \right] \tag{3}$$

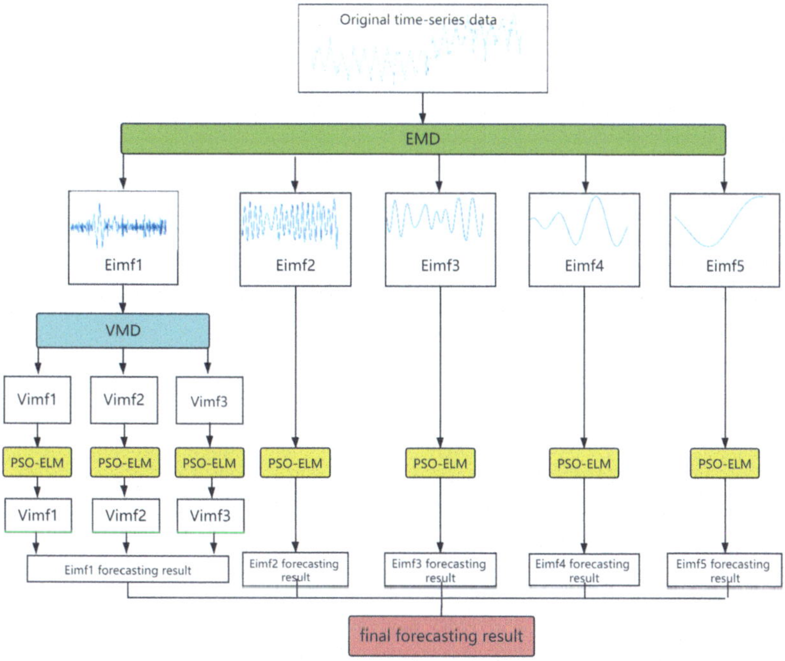

Fig. 1. Overall framework of monthly water consumption forecasting model

Where $\hat{h}_k^{n+1}(x)$ is the k mode in n round, $\hat{f}(x)$ is the Fourier transform of $f(x)$, x_k^n is the k center frequency in n round, β is the quadratic penalty factor, $\hat{\alpha}(x)$ is Lagrange multiplier in n round, η is the time step.

For time series prediction, the high-frequency series is often more difficult to forecast than the low-frequency series. The more complex, the more random the curve is, the less easy it is to predict. That is, in IMF1(the first signal sub-series) of EMD algorithm, because of the complexity of the original signal, the envelope is also complex, So the IMF1 of EMD is hard to forecast. According to previous studies, VMD is more suitable for prediction than EMD, but the IMFs (Intrinsic Mode Functions) except IMF1 in EMD decomposition are smoother than those in VMD, and the frequency is lower. It means they can make more accurate prediction than IMFs of VMD. Therefore, we need the low-frequency component of EMD and the characteristics which is VMD suitable for time series prediction. According to the above reasons, this paper proposes the Secondary Decomposition of EMD by VMD. Aiming at the complex problem of IMF1 decomposed by EMD, VMD is used to secondary decompose IMF1. Figure 2 shows the pseudocode of SDEMD.

```
Algorithm 1: SDEMD decompose IMFs
Require: This algorithm runs all time.
Input: x(t), k, beta
Output: IMFs[t] and Resi
Update: set IMFs[t] store the each IMF of signal x(t)
Temporary variables: x_up(t), x_low(t), m(t), c(t), resi(t),imf1(t)
Initialize x(t)
Identify the local extreme point of x(t)
calculate x_up(t) and x_low(t)
m(t) ← [x_up(t) + x_low(t)] / 2
while x(t) = resi(t) do
    c(t) ← resi(t) - m(t)
    if [x_up(t) + x_low(t)] < c(t) then
        resi(t)←x(t) - c(t)
        if [x_up(t) - x_low(t)] < 1 then
            IMFs[t]←c(t)
        end if
    else
        x(t)←c(t)
    end if
end while
imf1(t)←IMFs[1]
VMDs[t]←imf1(t) decompose by VMD
return VMDs[t] + IMFs[t] - IMFs[1]
```

Fig. 2. SDEMD Algorithm

3.2 Extreme Learning Machine Optimized by Particle Swarm Optimization (PSO-ELM)

ELM [15] is a new feedforward neural network algorithm which is used to solve the problems of the traditional BP neural network, such as long training time, easy to get into the local optimal solution, and harsh choice of parameters. The difference between ELM and BP neural network is that the connection weight and the hidden layer bias of ELM are randomly generated. Therefore, as long as the number of neurons in the hidden layer is determined, training and learning can be conducted. The operation of ELM is convenient and the generalization ability is strong.

Although neural networks such as ELM [15] have strong generalization ability, they all have a significant disadvantage which is the random value of weight and bias. When the least square method approaches the optimal weight and bias, it may inevitably fall into the local optimum, which affects the network performance, in order to solve these problems, PSO [16] is used to optimize ELM. PSO is a process in which each particle is used to search for the best location, and then the global optimal solution is obtained through continuous iteration. The location of each particle can be represented by a corresponding fitness function, and the direction and distance of each particle's motion can be represented by its retained velocity. Specifically, PSO can be used to find the optimal value, optimize the input weights and bias values of ELM, so that they are no longer randomly assigned, the optimal weights and bias values can be obtained, and then the optimal weights and bias values can be input into ELM to achieve the optimal network performance of ELM. Figure 3 shows the flow chart of PSO-ELM.

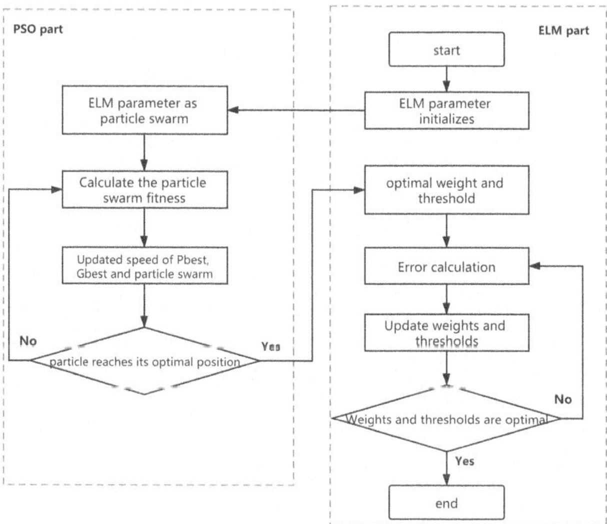

Fig. 3. The flowchart of PSO-ELM

4 Experimental Results and Analysis

4.1 Dataset and Data Processing

Dataset

The dataset of this paper is from the monthly water consumption data of Shapingba in Chongqing, China from 2000.1 to 2018.5. In order to train the model effectively, the water consumption data is scaled to the range from 0 to 1. Generally speaking, the ratio of training set to testing set is about 3:1, so in this paper, the training set is the monthly water consumption from 2000.1 to 2014.1, and the testing set is the monthly water consumption which is from 2014.2 to 2018.5.

Data Decomposition

In order to establish a water consumption model based on SDEMD, first we use EMD to decompose the water consumption time series, and then use VMD to secondary decompose the Eimf1 sub-series. The time series of water consumption are decomposed by EMD as shown in Fig. 4.

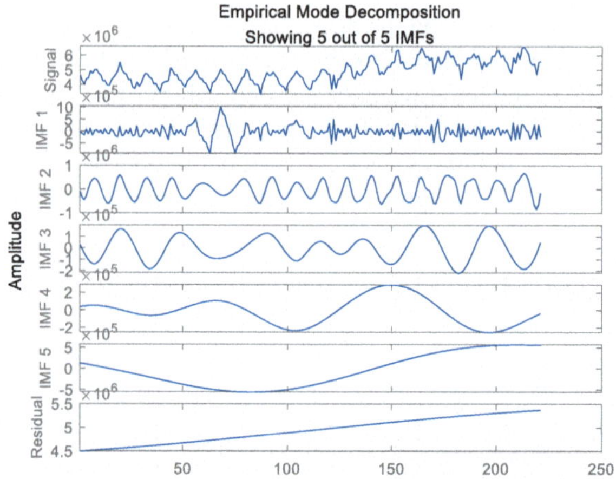

Fig. 4. Time series of water consumption decomposed by EMD

Unlike the EMD function, which only needs to input a time series, the number of VMD sub-series K and the quadratic penalty factor β are the input parameter of the VMD function, so in order to determine the value of K and β, this article uses different K and β values for VMD decomposition to compare optimal K and β values. According to Fig. 5, the optimal K and β values for this Eimf1 decompose by VMD is K = 3, β = 5. Figure 6 shows the Eimf1 sub-series after VMD decomposition.

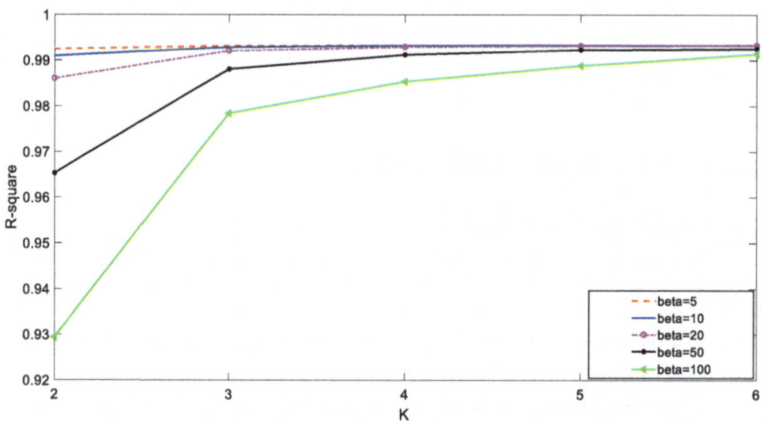

Fig. 5. Different K and β values of VMD decomposition

4.2 Model Performance Evaluation Indicators

The quality of a model is usually measured by performance evaluation indicators. In terms of accuracy evaluation of water consumption forecasting models, variety of measures are available to characterize the performance of the models [2]. Since there are

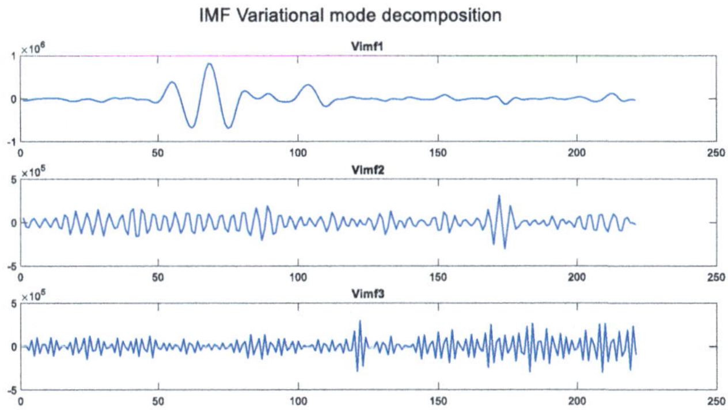

Fig. 6. Eimf1 after VMD decomposition (K = 3, β = 5)

no general performance indicators, it is best to choose the indicators suitable for a specific model. In addition, multiple performance indicators are usually used because each has advantages and disadvantages. In this paper, R-Square, MSE (Mean Square Error) and RMSE (Root Mean Square Error) are used as performance evaluation indicators. R-square is the determination coefficient of the equation, and its value ranges from 0 to 1. The closer it is to 1, the stronger accurate the model will be. Both MSE and RMSE are evaluation Indicators for calculating model errors, so the smaller they are, the better the model performance will be. The formula of R-Square is given in (4), the formulas of MSE and RMSE are as follows (5), (6):

$$R^2 = 1 - \frac{\sum_i (y_i - \hat{y}_i)^2}{\sum_i (y_i - \overline{y})} \tag{4}$$

Where y_i is observe value, $\hat{y}_2$ is the forecasting value, $\overline{y}$ is the mean of y_i.

$$MSE = \frac{1}{M} \sum\nolimits_{m=1}^{M} (y_m - \hat{y}_m)^2 \tag{5}$$

where M is the number of target data, y_m is the original data and $\hat{y}_m$ is the forecasted data.

$$RMSE = \sqrt{\frac{1}{M} \sum\nolimits_{m=1}^{M} (y_m - \hat{y}_m)^2} \tag{6}$$

4.3 Feature Selection

When neural networks such as ELM are used to build predictive models, potential input variables should be selected in advance. As many researchers have proved, a forecasting model that only considers historical water consumption data can obtain reliable forecasting results [17, 18]. In this paper, the method first collects historical water consumption data, extracts the feature of historical water consumption data and selects appropriate information as the input of the prediction model, then trains and tests the prediction model with the selected historical water consumption data. This paper considers the different lag times of the monthly water consumption time series. The optimal lag time is determined by the R-square. The structure of the input of the SDEMD-PSO-ELM model is as follows:

$$P_t = f(P_{t-1}, P_{t-2}, \ldots, P_{t-12}) \tag{7}$$

Where P_t is water consumption to be predicted for the current month, P_{t-i}, $1 \leq i \leq 12$ represents monthly water consumption in the previous i months of the current month. According to historical experience, R-Square greater than 0.5 should be a qualified feature. The paper selects the appropriate lag time as the feature for the monthly water consumption time series as follow Table 1.

Table 1. Feature of the monthly water consumption time-series

Sub-series		Feature
Eimf1	Vimf1	$f(Vimf1_{t-1}, Vimf1_{t-2}, Vimf1_{t-3}, Vimf1_{t-11}, Vimf1_{t-12})$
	Vimf2	$f(Vimf2_{t-4}, Vimf2_{t-5}, Vimf2_{t-8}, Vimf2_{t-9}, Vimf2_{t-12})$
	Vimf3	$f(Vimf3_{t-2}, Vimf3_{t-3}, Vimf3_{t-5}, Vimf3_{t-10}, Vimf3_{t-12})$
Eimf2		$f(Eimf2_{t-1}, Eimf2_{t-2}, Eimf2_{t-3}, Eimf2_{t-4})$
Eimf3		$f(Eimf3_{t-1}, Eimf3_{t-2}, \ldots, Eimf3_{t-5})$
Eimf4		$f(Eimf4_{t-1}, Eimf4_{t-2}, \ldots, Eimf4_{t-12})$
Eimf5		$f(Eimf5_{t-1}, Eimf5_{t-2}, \ldots, Eimf5_{t-12})$
Eresidual		$f(Eres_{t-1}, Eres_{t-2}, \ldots, Eres_{t-12})$

4.4 Experimental Results and Discussion

The lag time selected above is used as the feature to input into the SDEMD-PSO-ELM model. After integrating the prediction results of each sub-series, the final result is shown in Fig. 7. The red line represents the original water consumption data, and the green line represents the water consumption data forecasted by the SDEMD-PSO-ELM model. It is not difficult to notice that SDEMD-PSO-ELM fits well both in the overall trend and the specific month.

This paper compares SDEMD-PSO-ELM with ELM, PSO-ELM, EMD-ELM, RBFNN and VMD-ELM [12], and their R-square, MSE and RMSE are shown in Table 2. As can be seen from Fig. 8, compared with single ELM, PSO-ELM, EMD-ELM, RBFNN and VMD-ELM, the scatter point of SDEMD-PSO-ELM model is closer to the 1:1 slope line, so SDEMD-PSO-ELM has the most accurate prediction results in these models.

For a single ELM and RBFNN model, all the evaluation indicators are the lowest in the experimental models, which show that although ELM and RBFNN have been

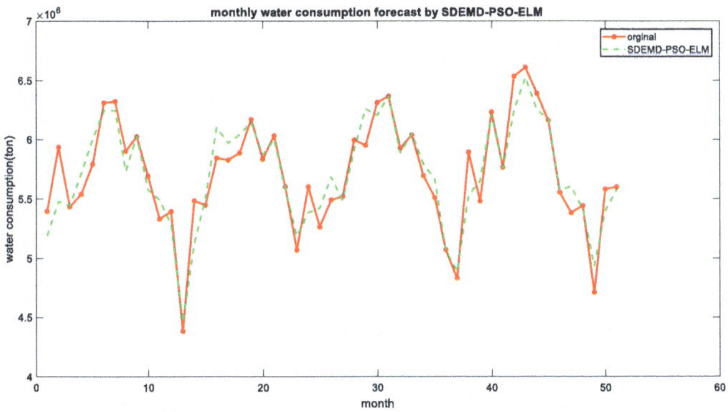

Fig. 7. Monthly water consumption forecasting by SDEMD-PSO-ELM

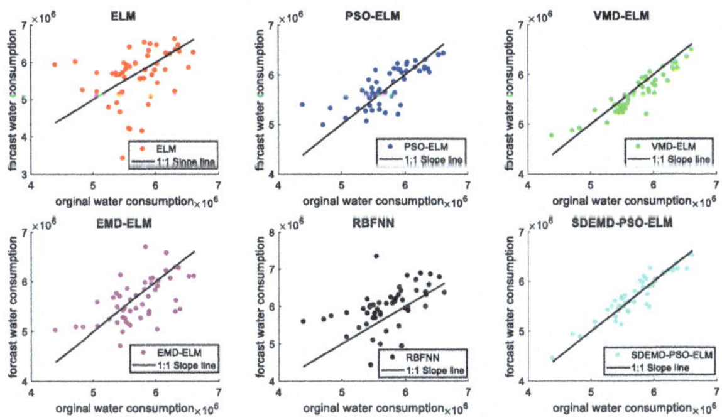

Fig. 8. Scatter diagram of ELM, PSO-ELM, VMD-ELM, EMD-ELM, RBFNN and SDEMD-PSO-ELM

improved based on the traditional neural network model, the generalization ability of the data set with less data volume is not strong. The evaluation indicators of PSO-ELM model seem to be much better than that of single ELM model. This is because under the optimization of PSO, ELM can find the optimal weight and bias. After a single ELM randomly assigns the weight and bias, there will be no back-propagation error, so the initial weight and bias are very important for ELM model. The prediction effect and evaluation indicators of EMD-ELM are not as good as that of PSO-ELM. The reason is very simple. The problem lies in the high frequency component Eimf1. According to Fig. 4, after the water consumption time series of the current month is decomposed by EMD, the frequency of Eimf1 component is even higher than that of the original data. As we all know, the higher the frequency, the more difficult the prediction of time series. Compared with Eimf1, the other Eimfs frequency is much lower, and the prediction effect is also very good, so this is the reason why the secondary decomposition of Eimf1 component is proposed in this paper. As for VMD-ELM, all the evaluation indexes are higher than those of the previous models, which also show that VMD is better than EMD for time series forecasting.

Therefore, in the secondary decomposition of Eimf1, this paper uses VMD to decompose the high-frequency time series in favor of prediction. SDEMD-PSO-ELM proposed in this paper is the best in all the experimental models. This model has the predictability of EMD low-frequency sub-series, VMD time series decomposition intermediate frequency sub-series, and PSO optimization of ELM, which further enhances the performance of the model.

Table 2. Performance evaluation indicators of each model

Model	R-square	MSE	RMSE
SDEMD-PSO-ELM	0.87418	0.00015	0.0125
EMD-ELM	0.46243	0.00098	0.0316
VMD-ELM	0.7945	0.00025	0.0160
ELM	0.02880	0.0043	0.0658
RBFNN	0.01523	0.0114	0.0874
PSO-ELM	0.61941	0.00047	0.0218

5 Conclusions

Monthly water consumption forecasting is great significant to operate water supply station efficiently. In order to obtain more accurate forecasting results, this paper proposes a forecasting framework combining forecasting method PSO-ELM and signal decomposition SDEMD, and identifies the performance of this forecasting model in the monthly water consumption prediction. The composite framework consists of two modules: data decomposition module and forecasting module. The data decomposition module is established by SDEMD algorithm, and in the forecasting module, PSO-ELM algorithm is used to respectively predict the sub-series obtained from the data decomposition module, and finally all results are integrated to obtain the final water consumption prediction result. In this paper, combined with the actual data of different feature in Chongqing, the combinatorial prediction framework is tested. The results show that the correct time series decomposition and neural network model optimization can reduce the prediction error and improve the performance of the model. Therefore, SDEMD-PSO-ELM can effectively predict the nonlinear and unstable monthly water consumption, which is conducive to the efficient operation of urban water supply.

References

1. Anele, A., Hamam, Y., Abu-Mahfouz, A., Todini, E.: Overview, comparative assessment and recommendations of forecasting models for short-term water demand prediction. Water **9**(11), 887–898 (2017)

2. Wu, S., Han, H., Hou, B.: Hybrid model for short-term water demand forecasting based on error correction using chaotic time series. Water **12**(6), 1683 (2020)
3. Odan, F.K., Reis, L.F.R.: Hybrid water demand forecasting model associating artificial neural network with Fourier series. Water Resour. Plan. Manag. **138**, 245–256 (2012)
4. Das, M., Ghosh, S.K., Chowdary, V.M., Saikrishnaveni, A., Sharma, R.K.: A probabilistic nonlinear model for forecasting daily water level in reservoir. Water Resour. Manag. **30**(9), 3107–3122 (2016). https://doi.org/10.1007/s11269-016-1334-6
5. Candelieri, A., Soldi, D., Archetti, F.: Short-term forecasting of hourly water consumption by using automatic metering readers data. Procedia Eng. **119**, 844–853 (2015)
6. Chen, G., Long, T., Xiong, J., Bai, Y.: Multiple random forests modelling for urban water consumption forecasting. Water Resour. Manag. **31**(15), 4715–4729 (2017). https://doi.org/10.1007/s11269-017-1774-7
7. Santos, C.C.D., Filho, A.J.P.: Water demand forecasting model for the metropolitan area of so Paulo Brazil. Water Resour. Manag. **28**(13), 4401–4414 (2014)
8. Romano, M., Kapelan, Z.: Adaptive water demand forecasting for near real-time management of smart water distribution systems. Environ. Model. Softw. **60**, 265–276 (2014)
9. Adamowski, J.F.: Peak daily water demand forecast modeling using artificial neural networks. J. Water Resour. Plan. Manag. **134**(2), 119–128 (2008)
10. Dragomiretskiy, K., Zosso, D.: Variational mode decomposition. IEEE Trans. Signal Process. **62**(3), 531–544 (2014)
11. Huang, N.E., Shen, Z., Long, S.R.: The empirical mode decomposition and the Hilbert spectrum for nonlinear and non-stationary time series analysis. Proc. Math. Phys. Eng. Sci. **454**(1971), 903–995 (1998)
12. Seo, Y., Kwon, S., Choi, Y.: Short-term water demand forecasting model combining variational mode decomposition and extreme learning machine. Hydrology **5**(4), 54 (2018). https://doi.org/10.3390/hydrology5040054
13. Deo, R.C., Şahin, M.: An extreme learning machine model for the simulation of monthly mean streamflow water level in eastern Queensland. Environ. Monit. Assess. **188**(2), 90 (2016). https://doi.org/10.1007/s10661-016-5094-9
14. Yaseen, Z.M., et al.: Stream-flow forecasting using extreme learning machines: a case study in a semi-arid region in Iraq. J. Hydrol. **542**, 603–614 (2016). https://doi.org/10.1016/j.jhydrol.2016.09.035
15. Huang, G.-B., Zhu, Q.-Y., Siew, C.-K.: Extreme learning machine: theory and applications. Neurocomputing **70**(1–3), 489–501 (2006). https://doi.org/10.1016/j.neucom.2005.12.126
16. Eberhart, R., Kennedy, J.: A new optimizer using particle swarm theory. In: Mhs95 Sixth International Symposium on Micro Machine & Human Science. IEEE (2002)
17. Guo, G., Liu, S., Wu, Y.: Short-term water demand forecast based on deep learning method. J. Water Resour. Plan. Manag. **144**(12), 1–11 (2018)
18. Bakker, M., Vreeburg, J.H.G., Schagen, K.M.V.: A fully adaptive forecasting model for short-term drinking water demand. Environ. Model. Softw. **48**, 141–151 (2013)

Comparison of Control Systems Based on PID Controllers and Based on Fuzzy Logic Using MatLab and Simulink

Sergey Adadurov[1], Anatoly Khomonenko[2,3](✉), and Alexander Krasnovidov[2](✉)

[1] JSC "VNIIZHT", Moscow, 3rd Mytischinskaya Street, 10, 129626 Moscow, Russia
adadurov.sergey@vniizht.ru

[2] Emperor Alexander I St. Petersburg State Transport University, Moskovsky prospekt, 9, Saint Petersburg, Russia

[3] VKA them. A.F. Mozhaisky, st. Zhdanovskaya, 13, 197198 St. Petersburg, Russia

Abstract. The persistent complication of various instruments, devices and assemblies leads towards the complexity of the algorithms for their automatic control. Many interrelationships existing inside automatic control systems (ACS), as well as external influences on these systems, are becoming increasingly difficult to describe with assistance of classical precise binary logic or differential equations. The matter is that a large number of real technical solutions are based on expert experience extremely problematic to formalize. The mathematical apparatus currently used for the synthesis and analysis of automatic control systems is based on differential equations. Full consideration of various inner connections and outdoor factors leads to the complication of these equations. Intelligent systems using would help to avoid these problems. There is no need to build the control process complex model in such kind of systems. An approach based on the use of a fuzzy logic controller in such systems considered in the article. A comparison of the control system with a fuzzy logic controller and a system using a traditional proportional-integral-differentiating (PID) controller is made. The systems were simulated using MatLab and Simulink.

Keywords: MatLab system · Automatic control systems · Fuzzy logic controller

1 Introduction

As the experience of the development of the control system shows, recently they have been developing in direction towards increasing their adaptability, that is, the ability to change their parameters depending on control actions and external conditions [1]. It's possible to mark three main stages in the development of such systems.

The first stage is characterized by using the expert's classical methods of the theory of automatic control and creating devices of two-digit (precise) logic, based on elements of hydraulic and pneumatic automation. However, such devices operating on the "prohibit-permit" principle have been deprived the possibility of the reconfiguration, even if have been forced to adapt to the changed conditions.

© The Author(s), under exclusive license to Springer Nature Switzerland AG 2021
R. Silhavy (Ed.): CSOC 2021, LNNS 229, pp. 224–237, 2021.
https://doi.org/10.1007/978-3-030-77445-5_20

At the second stage, microprocessor control with electronic programmable components appeared. Computer facilities, on-board computers and microcomputers made it possible to load into their storage various control programs, capable of adapting to changes in various outdoor conditions [2].

At the third, current, stage, intelligent control systems began to be created to simultaneously take into account the huge variety of information, various circumstances and situations, the characteristics of control actions, the outdoor environment and mechanisms in use. Their algorithms are based on fuzzy logic, resembling human thinking processes [3]. These are already adaptive systems in the classical sense.

Therefore, the issues of using fuzzy logic controllers in control systems are currently quite relevant. The advantages of using fuzzy logic methods and the implantation of intelligent technologies have been actively discussed over the past several years. This advantages are associated with the improvement of operation of various systems and equipment in comparison with those that using classical regulators [1, 4].

The model of the control object is creating either from the analysis of the physical processes taking place in, or on the basis of a combination of models of the processes that crop up in different parts of the control object. In general, the equations of a mathematical model of an object or a control system that set the relationship between input and output and variables are called equations of motion. These equations can be defined analytically on the basis of the laws of physics underlying at the base the operation of the control object and system elements, or experimentally using modeling methods. Thus, the synthesis of an automatic control system (ACS) on the basis of the traditional approach is reduced to the compilation of a differential equation or a system of such equations [2, 5].

The degree of the differential equations can be quite high, as both input and output quantities and its rate and acceleration can be related to a certain dependence. It should be in mind that only a small part of systems and processes are linear in nature. Most systems in practice are nonlinear, and therefore the equations of the dynamics of such systems turn out to nonlinear. In order to simplify the solutions, nonlinear equations are replaced by linear ones, which approximately (with a certain precision) describing the dynamic processes in the system. But even in fairly simple cases, such as in the system for maintaining a given speed of movement of some object, the solution of the equation of dynamics

$$m\frac{dV(t)}{dt} = k(V_0 - V(t)) + f_r,$$

where V_0 – is the specified speed, m – is the mass of the object; k is the coefficient of proportionality (gain), f_r – is the force of resistance to movement, is the function

$$V(t) = (1 - e^{-\lambda t})(V_0 + \frac{f_r}{k}).$$

It is clear that with increasing time t, the current speed

$$V(t) \to V_0 + \frac{f_r}{k}$$

that is, the set speed is kept. However, there will be no exact match. The speed will be kept with an fault, since during the movement the object is affected by a resistance force, the magnitude of which is influenced by various factors, such as aerodynamic resistance, rolling friction, etc. so, the set speed is kept. However, there will be no exact match. The speed will be kept with fault, since during the movement the object is affected by a resistance force, the magnitude of which is influenced by various factors, such as aerodynamic resistance, rolling friction, etc. The divergence can be made to very small if the gain k would increase. However, such increasing can lead to various technical difficulties: unrealizability of purely linear amplifiers, system swinging up to loss of stability due to increased random noise [5].

When synthesizing an ACS with a fuzzy controller, the bulk of the work falls on the construction of a base of fuzzy productive rules. A fuzzy controller implements state control if the base of fuzzy productive rules built using linguistic variables that characterize the state of the object. In this case, the goal of control in the form of the desired state of the control object is implicitly entered into the rule base by an expert at the stage of forming the base of fuzzy production rules [1, 4].

Input variables characterizing the real and desired state of the control object are made to fuzzified and used in the fuzzy inference system. The fuzzy inference system contains a base of fuzzy production rules of a generalized form of type IF "state of an object" THEN "impact on an object". In this base, the knowledge of experts on managing some process of changing the state of an object is presented in a fuzzy form. This state is changing from the current to the required one on the basis of knowledge about the values that characterize the actual state of the object. As a result of fuzzy inference and subsequent defuzzification, clear values of the output variables are obtained, which are used for subsequent control of the object.

To compare the two methods for constructing controllers, the article uses the Matlab system with the Simulink extension package. Environment (Package) Simulink is an application to the MatLab system and is for simulate linear and nonlinear, discrete and continuous systems that can operate both in computation mode and in real time. Programming and entering the characteristics of the simulated systems can be performed in the dialog mode, by assembling on the screen the connection diagram of elementary (standard or custom) links. A system model is obtained as a result of such an assembly, which in MatLab terminology is called an S - model. This model is saved in a file with the.mdl extension. Thus, Simulink implements the principle of visual programming. In accordance with this the entire model is built from standard library blocks and blocks modified or developed by the user. Each block implements its own mathematical function.

To implement the fuzzy modeling process in the MatLab environment, a special extension package Fuzzy Logic Toolbox is [7]. Inside of this package, the user can perform the necessary actions for the development and use of fuzzy models in one of the following modes:

- in interactive mode using graphical tools for editing and visualization of all components of fuzzy inference systems;
- in the command mode by entering the names of the corresponding functions with the necessary arguments directly into the command window of the MatLab system;
- in program mode by creating appropriate file - functions and script - files.

A fuzzy logic controller created in the Fuzzy Logic package is avaliable in the Simulink program. For this, a set of blocks from the Fuzzy Logic ToolBox section is used. So, the Fuzzy Logic Controller block – a fuzzy controller makes it possible to simulate systems controlled by a fuzzy controller.

A comparative analysis of automatic control systems using a PID regulator and a fuzzy controller has made in the article.

2 Model of an Automatic Control System Using a PID Controller

A proportional-integral-derivative controller or PID controller is a device with a feedback used in automatic control systems to maintain desired parameter value. Due to their versatility, they are broadly applied in various technological processes. A generalized diagram of an automatic control system using a PID controller shown in Fig. 1.

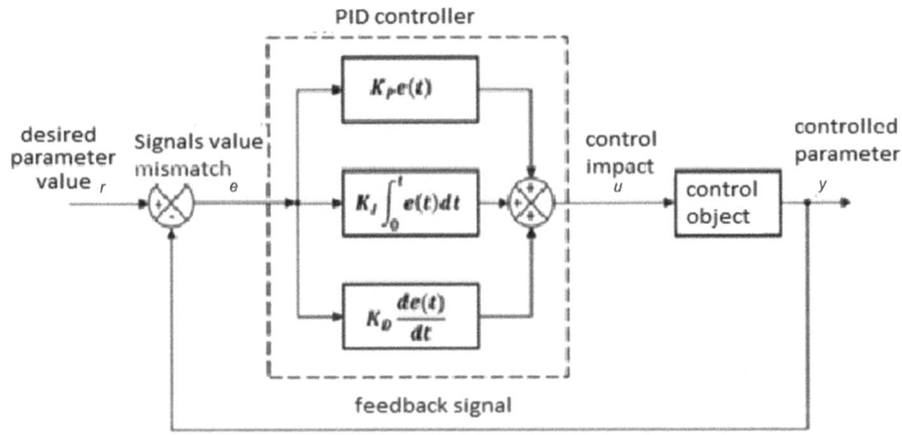

Fig. 1. Generalized PID controller circuit

During the PID controller operating the proportional component usually provides the main part of the control action. Integral and differential components are mainly designed to improve the accuracy and quality of the transient. PID controllers are often applied in industrial and household appliances. They are an integral part of devices such as frequency converters for motor controlling, temperature controllers (thermostats), air conditioning systems, etc.

As an example, the article discusses a system keeping the stable speed of an object. The actuator of such an object can be an engine, an electric motor rotor, etc. When modeling, such an object is usually represented by an aperiodic link of the first order [8], the transfer function of which is:

$$W(p) = \frac{1}{Tp + 1}$$

Taking this into account, the scheme of the considered automatic control system has the form shown in Fig. 2.

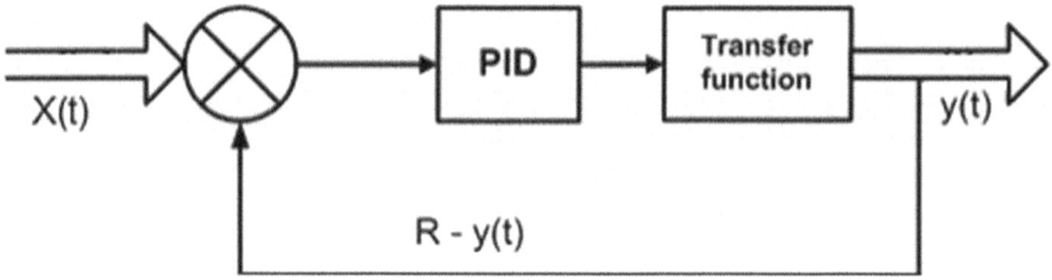

Fig. 2. Block diagram of the automatic control system

Here *x* (*t*) is the input action; *y* (*t*) – control action; *R* is the set value of the controlled variable, *R*-*y* (t) is the discrepancy (error) signal.

In accordance with the block diagram, a dynamic diagram of the ACS model had been built by means of visual programming Simulink. This diagram is shown in Fig. 3.

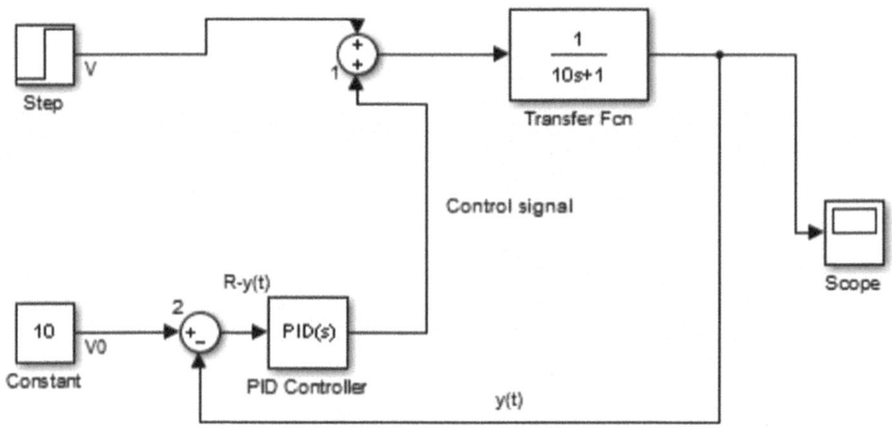

Fig. 3. Dynamic diagram of the automatic control system

The ***Step*** block forms the outdoor action in the form of a step function, the ***Constant*** block is used to set the value of the managed variable, the ***adders*** 1 and 2 form the mismatching signal and the compensated outdoor action, respectively, the ***Transfer Fun*** block forms (determines) the transfer function of the control object, the ***Scope*** block is used to visualize the simulation results.

The modeling process itself includes two successively performed stages:

1. ***Setting the parameters of the PID controller.*** At this stage, the selection of the coefficients of the integrating, differentiating and proportional components of the controller is carried out (see Fig. 1). To solve this problem, a step function is used as an input. As a result of solving this task, the coefficients KP, KI and KD of the controller and the system transient process are found. The character of the process completely depends on coefficients had detected and transfer function. Perhaps a number of

sequential steps would need to achieve fitting values of controller coefficients. The values of the coefficients obtained as a result of these actions are shown in Table 1.

Table 1. Values of the PID controller coefficients

Controller Parameters

	Tuned	Block
P	1.3185	1.3185
I	0.14871	0.22306
D	-4.1031	-2.7354
N	0.32135	0.48203

The transient process in the modeled system, that is, its response to the impact of a step function is shown on Fig. 4.

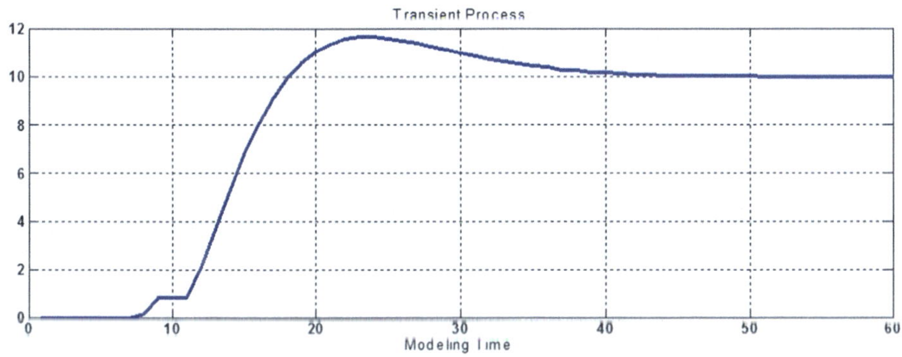

Fig. 4. Transient process in the simulated system

Analysis of Fig. 4 shows that the values of the transient process tend to the value of the set constant with increasing time. Obviously, the obtained coefficients fits for solving the assigned control problem.

2. ***Simulation of the system in dynamic mode.*** At this stage, instead of the Step block, which forms the outdoor action in the form of a step function, it's possibly use various generators of input actions. The rest of the dynamic schema stays not changed. The article uses a saw tooth input signal as shown in Fig. 5.

As a result of simulation in dynamic mode, the process y (t) at the output of the system had been obtained, shown in Fig. 6.

Analysis of Fig. 6 shows that after transient process is completed at about the 23rd cycle of the model time, the system enters into a steady state, in which the values of the

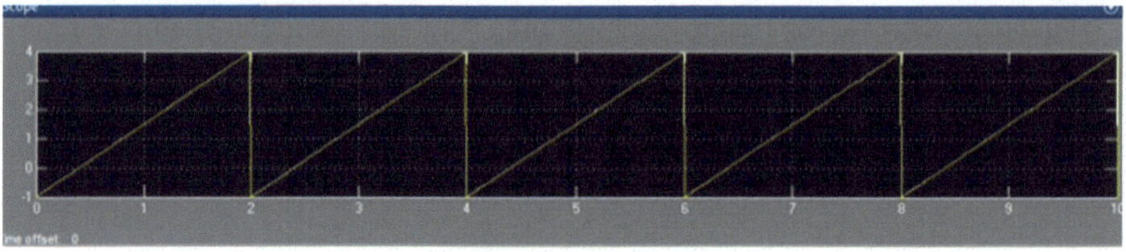

Fig. 5. The saw tooth input action

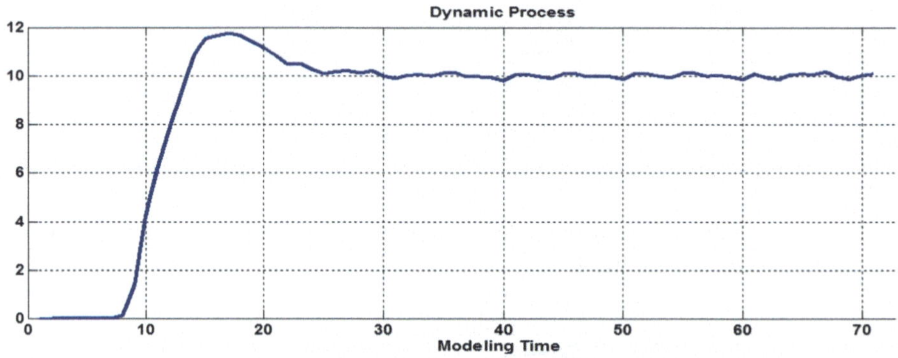

Fig. 6. Process at the exit of the system

output process y (t) deviate slightly from the set value equal to 10. The character and magnitude of these deviations are shown in Fig. 7.

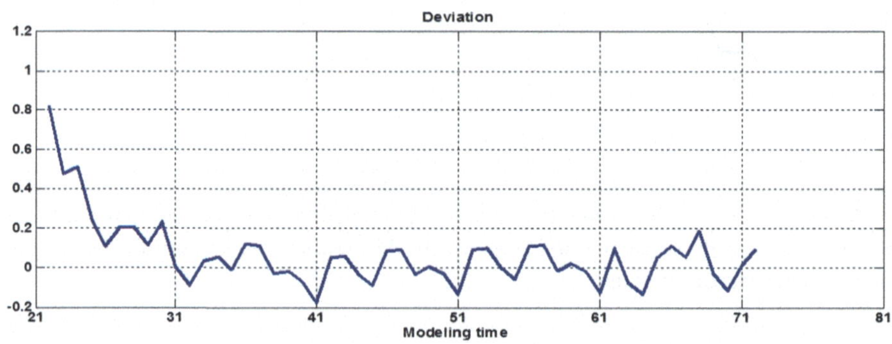

Fig. 7. Deviations of the values of the output process from the set value

The average deviation is 0.07; the variance of the deviation is 0.1716.

3 Model of an Automatic Control System with a Fuzzy Controller

All systems with fuzzy logic operate according to the same principle: the readings of measuring devices make to fuzzy, processed, defuzzified. All algorithms that use fuzzy inference consist of the following sequence of steps:

- Forming of rule base of fuzzy inference systems forming
- Fuzzification of input variables.
- Aggregation of sub conditions in fuzzy production rules.
- Activation of sub conclusions in fuzzy production rules.
- Accumulation of conclusions in fuzzy production rules
- Defuzzification of output variables.

The block diagram of the fuzzy logic controller in Fig. 8 is presented.

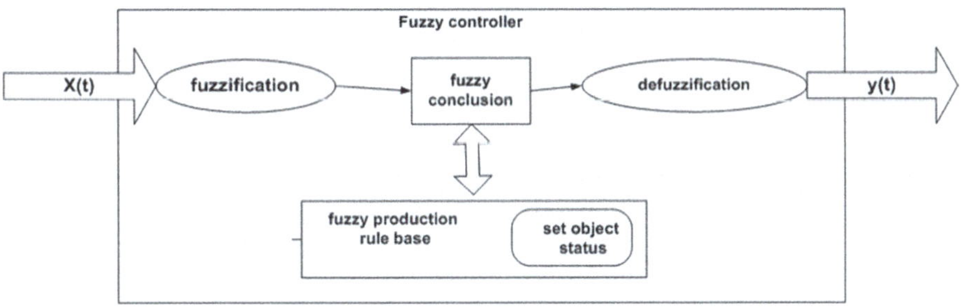

Fig. 8. The block diagram of the fuzzy logic controller

Fuzzification is the process of converting ordinary crisp variables into fuzzy ones. The range of variation of the variable is divided into sets (subsets), within each of which the function of belonging of the variable to each of the sets is constructed. Generally, membership functions are triangular. This form is the most common one, although in the general case they can be any based on the meaning of the problem being solved [1, 8]. To perform the control function on fuzzy variables, operations formulated in the form of fuzzy rules must be performed. The set of fuzzy rules and fuzzy variables is used to implement fuzzy logic inference, result of which is a control action on the control object. The block diagram of an automatic control system with a fuzzy logic controller in Fig. 9 is presented.

The incoming information of the fuzzy inference system is the input variables measured in some way. These variables correspond to real variables of the control process. The information forming at the output of the fuzzy inference system corresponds to the output variables, which are the control variables of the control process. The developing a fuzzy logic controller is to develop some expert system and its implementation. It had noted above that the most convenient means of building a model of such a system is MatLab and the Simulink and Fuzzy Logic Toolbox extension packages. They implement various programming principles, including visual one. Fuzzy Logic Toolbox is used in

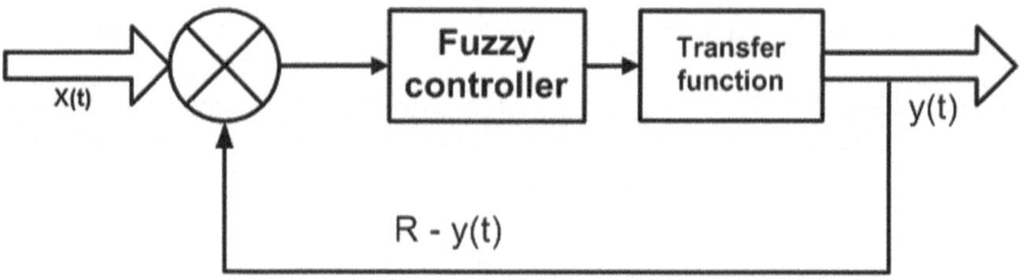

Fig. 9. Block diagram of an ACS with a fuzzy logic controller

MATLAB and is a package of applied programs related to the theory of fuzzy sets and allowing designing fuzzy expert and/or control systems.

The process of developing a fuzzy inference system in an interactive mode involves performing the following steps [6, 8]: Representation of empirical data in the form of the following heuristic production rules. In this case, the rules are as follows:

1. If the mismatch is large, then the control signal is large.
2. If the mismatch is average, then the control signal is average.
3. If the mismatch is small, then the control signal is small.

Since the fuzzy controller should be a functional analogue of the PID-controller discussed above, it should one input and one output as well. Therefore, the input variable fuzzy linguistic 'Deviation' - the value of the mismatch signal and the output linguistic variable 'Control' - the value of the control signal should be determined. The set T1 = {Big, Middle, Slight} will be used as the term - set of the linguistic variable Deviation. The set T2 = {Powerful, Medium, Small} will be used as the term - set of the linguistic variable Control. In this case, each of the terms of the input variable estimates in the range ($0 \div 150$), and each of the terms of the output variable estimates in the range ($0 \div 70$). Therefore, the input subjective information in the form of the following fuzzy productions can be presented:

Rule_1: if (Deviation is Big) then (Control is Powerful).
Rule_2: if (Deviation is Middle) then (Control is Medium).
Rule_3: if (Deviation is Slight) then (Control is Small).

The scheme was obtained using Fuzzy Logic Toolbox [8, 9] means. Using these, it's possible customize the developed fuzzy logic system. In particular, you can select the most appropriate membership functions. For this model, the Gaussian functions turned out to be the most fitting. They are presented in Fig. 10 as an example.

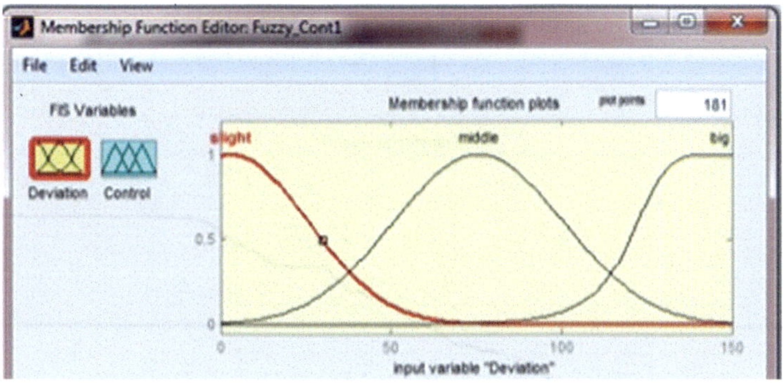

Fig. 10. Kind of membership functions for the input variable

After the configuration of the controller is completed it must be implemented into the dynamic circuit of the automatic control system (Fig. 11).

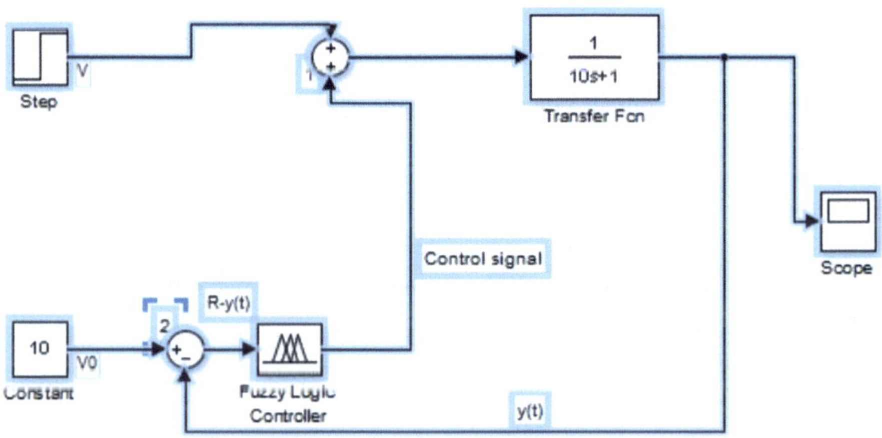

Fig. 11. Dynamic diagram of an ACS with a fuzzy controller

As a result of the simulation, the following results were obtained:

- Transient process in the simulated system (Fig. 12).

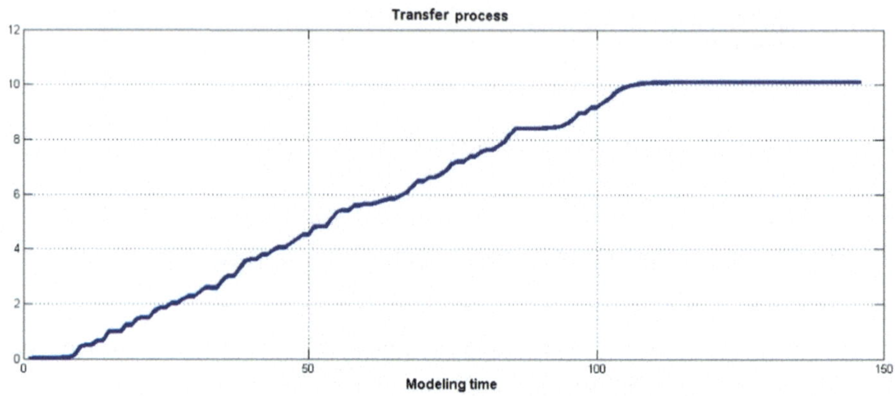

Fig. 12. Transient process in a simulated system with a fuzzy controller

- Process at the output of the system in a dynamic mode (Fig. 13).

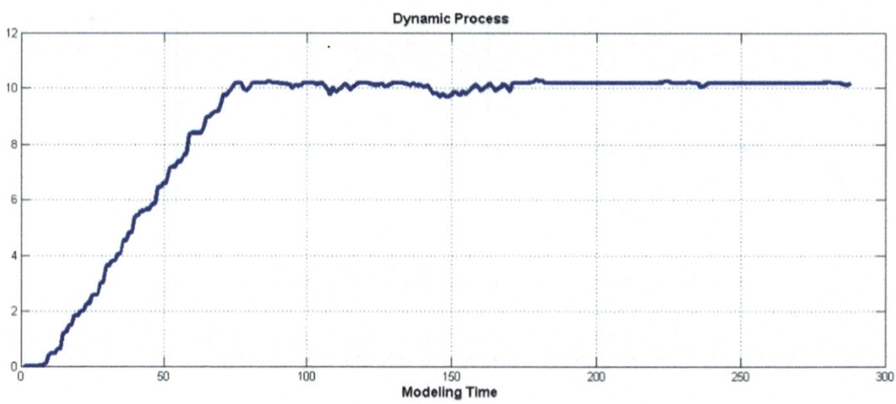

Fig. 13. Dynamic process y(t) at the system output

The input action is shown in Fig. 5.

- Deviations of the values of the output process from the set value (Fig. 14.).

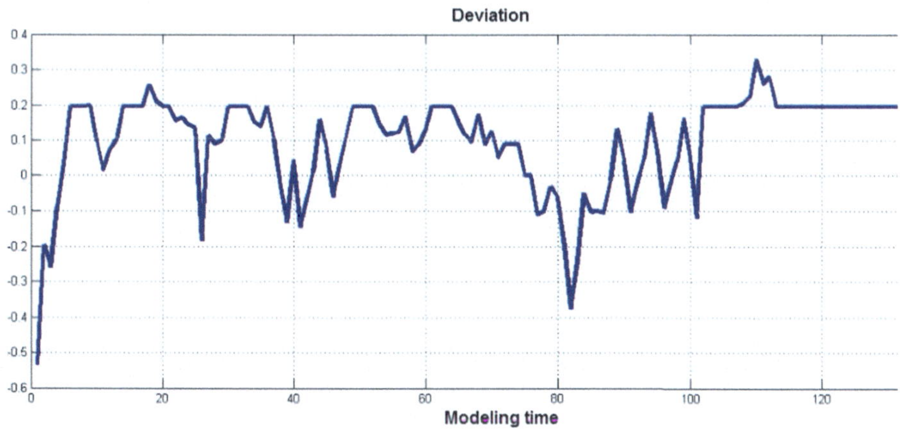

Fig. 14. Deviations of the output process values from the set value

The mean deviation is 0.09, the variance of the deviation is 0.14.

4 Conclusion

The two methods for constructing a controller of an automatic control system – a system for maintaining the speed of an object has compared. Analysis of the results obtained allows us to draw the following conclusions (Table 2).

Table 2. Comparative characteristics of the methods investigated

	PID controller	Fuzzy controller
Transient process	The duration is short, but there is an overshoot part (Fig. 4)	The duration is longer, but there is no overshoot. Transition to steady state is smooth
Dynamic process	The dynamic mode is characterized by small fluctuations relative to the set value (Fig. 6)	The dynamic mode is characterized by small fluctuations relative to the set value (Fig. 14)
Average deviation	The mean deviation is 0.07	The mean deviation is 0.09
Variance deviation	the variance of the deviation 0.17	the variance of the deviation 0.14

It is clear that both systems have their own advantages and disadvantages. So, a system with a PID regulator quickly goes into the steady state, but this transition is characterized by the presence of an overshoot section. A system with a fuzzy controller has less reactivity, but the transition to the steady state is smooth. The steady state is characterized by insignificant fluctuations relative to the set value in both cases. Fuzzy – the controller provides less variance of these fluctuations. Nevertheless, the use of fuzzy logic controllers has significant advantages in their implementation:

- PID – the regulator has only one input and one output, which makes it difficult to implement control systems that are affected by several input influences. Fuzzy – the controller can have several inputs and outputs, which simplifies the implementation of complex systems [10];
- Fuzzy control can be implemented using the mechanism of neural networks, which makes it possible to implement with their help various adaptive systems [11].

A historic fact: In 1973 at the University of London in England, Professor M. Mamdani and student S. Assilian tried to stabilize the speed of a small steam engine built by the student. They had the equipment to gain this — a PDP-8 mini-computer and the usual digital controls. But they were not able control the engine as they wanted. Overshoot had been observed, and the required speed reached only after a series of oscillations. By applying fuzzy logic, they got better results.

References

1. Introduction to C# and the .NET Framework. https://docs.microsoft.com/ru-ru/dotnet/csharp/getting-started/introduction-to-the-csharp-language-and-the-net-framework. Accessed 15 January 2020
2. Schildt, H.: C# 4.0: The Complete Reference. Osborne/McGraw Hill, New York (2010). ISBN 0-07-174117-8
3. MathWorks. https://www.mathworks.com/. Accessed 15 January 2020
4. Dyakonov, V.P.: MATLAB. Complete tutorial. M.: DMK Press, vol. 768, p. Ill (2012)
5. MATLAB documentation. https://docs.exponenta.ru/. Accessed 15 January 2020
6. Smolentsev, N.K.: MATLAB: Programming in Visual C #, Borland C #, JBuilder, VBA: Training Course (+ CD) - M: DMK Press; SPb .: Peter. - 464 p. Ill (2019)
7. Raboso, M., Jiménez, M.I., Val, L.D., Izquierdo, A., Villacorta, J.J., Codes, M.: MATLAB COM integration for engineering applications. In: Vasilios N. Katsikis (eds.) MATLAB - A Fundamental Tool for Scientific Computing and Engineering Applications, IntechOpen, vol. 2 (2012).https://doi.org/10.5772/46471. https://www.intechopen.com/books/matlab-a-fundamental-tool-for-scientific-computing-and-engineering-applications-volume-2/matlab-com-integration-for-engineering-applications
8. Krasnovidov, A.V., Khomonenko, A.D., Zabrodin, A.V., Smirnov, A.V.: On the peculiarities of the exchange of data between applications in high-level languages and MATLAB functions. In: 2018 CEUR Workshop Proceedings. (2018). St. Petersburg, Russia, 8–9 November 2018. vol. 2341, pp. 33–41 (2018)
9. Fomenko, Y.S., Abou Hasan, R., Khomonenko, A.D.: Logical data integration on the example of organizing queries to the database air transportation. Intellectual Technologies on Transport, no 4, pp. 39–47 (2019)

10. Kosykh, N.E., Khomonenko, A.D., Bochkov, A.P. and Kikot, A.V.: Integration of big data processing tools and neural networks for image classification. In: 2019 CEUR Workshop Proceedings. St. Petersburg, Russian Federation, 4–5 December, vol. 2556, pp. 52–58 (2019)
11. Khomonenko, A.D., Adadurov, S.E., Krasnovidow, A.V., Novikov, P.A.: Indoor mobile positioning using neural networks and fuzzy logic control. In: Ramakrishnan, S. (ed.) Modern Fuzzy Control Systems and Its Applications. InTech, Chapter 11, pp. 202–223 (2017)

The Impact of Promotion and Price Perception on Buying Decision and Repurchase Intention Through Online Shopping

Dam Tri Cuong(✉)

Industrial University of Ho Chi Minh City, Ho Chi Minh City, Vietnam
damtricuong@iuh.edu.vn

Abstract. In the past several years, online shopping in Vietnam has become more and more popular, and consumers have changed their buying habits from traditional shopping to more online shopping. Furthermore, consumer' repurchasing is crucial for online shopping stores' success and creates profitability. Thus, this research examines the influence of promotion and price perception on buying decisions and repurchasing intention through online shopping. We proposed the study model from prior researches. We investigated 316 consumers who active online buying in Ho Chi Minh City, Vietnam. The Partial Least Squares (PLS) technique was applied to measure the measurement and structural models. PLS method is a variance-based structural equation modeling (SEM) approach that is broadly used in marketing and social science. The finding of the research revealed that promotion has positively associated with purchasing decisions and repurchase intention. The finding also confirmed that price perception has positively related to buying decisions and repurchase intentions. Moreover, the results of the research approved that purchase decision has entirely connected to repurchase intention.

Keywords: Promotion · Price perception · Purchase decision · Repurchase intention · Online shopping

1 Introduction

In the past several years, online shopping in Vietnam has become more and more popular and received significant attention from consumers, especially after the outbreak of the covid-19 epidemic; consumers have changed their buying habits from traditional shopping to more online shopping. With nearly 45 million people participating in online shopping, the Vietnam Government aims to have 55% of the online shopping population by 2025 and sales about 35 billion USD. This growth is an attractive opportunity for businesses to choose to sell online, making this method a trend in the 4.0 era [1]. Another author also stated that the covid-19 epidemic happened, which has been changing consumer trends and was a very logical reason for the booming online shopping environment in Vietnam. Businesses will succeed if they change quickly and seizes the fastest opportunity. A recent Nielsen survey released at the Online Marketing Forum 2020 showed that the number of consumers shopping online increased by 25%, while

© The Author(s), under exclusive license to Springer Nature Switzerland AG 2021
R. Silhavy (Ed.): CSOC 2021, LNNS 229, pp. 238–250, 2021.
https://doi.org/10.1007/978-3-030-77445-5_21

traditional channels such as supermarkets, markets, and groceries only increased respectively 7%, 3%, and 6% [2]. Previous studies also proved that promotion was affected by customer buying decisions in online shopping [3–6]. Previous research also declared that promotion involved repurchasing intention in online buying [7]. Some scholars demonstrated that product price perception had an essential antecedent in purchasing decisions and impacted buying decisions [5, 6, 8]. Prior studies also said that price perception had a significate predictor of repurchasing intention in online purchasing [9, 10].

In the literature of consumers' purchase decisions for online shopping, some scholars stated that online shopping had joined a global range with increasing popularity, which pointed to more and more users become accustomed to it and utilize it for online purchasing [8]. Fachmi et al. (2019) disclosed factors (trust, promotion, and e-service quality) influencing consumers' buying decisions. Edwar et al. (2018) showed that factors (trust, price, and time) impacted online buying decisions [8]. Suhaily & Soelasih (2017) confirmed the influence of e-service quality on customer satisfaction and repurchase intention, while repurchase intention negatively impacts. Besides, the price had no impact on customer satisfaction but have powered to repurchase intention. Experiential marketing did not influence customer satisfaction and repurchase intention. Client satisfaction had a positive effect on repurchase intention. The impact of e-service quality and experiential marketing through customer satisfaction as a mediation variable did not influence repurchase intention, while price perception influenced repurchase intention [9]. While Wijaya and Astuti (2018) disclosed e-service quality positively impacted customer satisfaction, satisfaction positively affected trust and brand image; likewise, trust and brand image positively influenced rebuying intention [11]. Yet, little research has investigated the effect of promotion and price perception on buying decisions and repurchase intention through online shopping, especially in the Vietnam market. Therefore, this study's purpose to empirically analyze the impact of promotion and price perception on purchasing decisions and repurchase intention through online shopping.

2 Literature Review and Research Hypotheses

2.1 Promotion

In the theory of marketing, promotion relates to activities that communicate product advantage and convince target consumers to buy it [12]. Promoting has been a type of communication technology that companies have used to create awareness among consumers. Businesses have designed the promotional campaign for their product/brand through television, radio, billboard, web, print media, social media, etc., to attract clients and make up a linking and an image in the clients' mind [13]. Fachmi et al. (2019) stated that promotion activities include keeping discounts, providing vouchers, and several other promotional schemes [3]. The main promotional types cover advertising, sales promotions, personal selling, public relations and publicity, and direct marketing [14]. However, Sulaiman et al. (2017) disclosed that in online-based promotion, the promotion method was more restricted and particular to the support of technology and the Internet. Some of the e-promotion ways were online advertising, email marketing, search engine optimization, affiliate marketing, and social media marketing [15]. Businesses have applied sales promotion to support advertising and public relations plans that influence purchasers' images, beliefs, and attitudes to goods and brands, impacting their buying behaviors [16]. Previous empirical researches revealed that promotion was an antecedent of buying decisions and positively related to purchasing decisions [3–5]. Besides, Sulaiman et al. (2017) said that online sales promotions' continued appearance could affect the online purchaser buying intention and make them a compulsive purchaser. Sales promotion plans boost customers to go to the websites again and intend to repurchase it in a larger volume [15]. Some scholars demonstrated that promotion was a predictor of repurchasing intention and linked positively to buying intent in online shopping [7]. Based on literature and some prior studies, we suggest the hypotheses as follows:

H1: Promotion is related to positively buying decisions.
H2: Promotion is positively associated with rebuying intention.

2.2 Price Perception

From the consumers' view, price perception is something given up or sacrificed by consumers to get the product [17]. Price perception is customer evaluations, and related emotional types of whether the seller's price and the price compared to others are reasonable or acceptable [18]. Every product's price perception decided the purchaser's expense of money will require to get the products/services. In traditional business, price perception involves getting and adding the cost, identifying the purchaser's overall financial willingness to buy, and comparing rival prices. Internet growth has created price perception more complex and competitive as it generates more competitors that not only between traditional rivals' products but also on the retail and online shop. The Internet has also dropped some expenses such as store and staff costs, building more opportunities and benefits for online shop buyers. Still, at the same time, this also creates pressures for the traditional retail shop buyer [15].

Price perception is an essential factor in the buying decision, particularly for products that are often bought; thus, consumers use it to choose which shop, products, and brands [19]. Some prior empirical researches depicted that price perception was significantly associated with buying decisions [5, 6, 8]. Price perception has been considered to be an essential element in interpreting customer behaviors. Studying on price perception implies that consumers frequently utilize price as a signal when assessing consumers' experiences with a product/service and forming their attitude toward a supplier [20, 21]. Previous empirical studies disclosed that price perception was an antecedent of repurchase intention; it significantly affected repurchase intention [9, 10].

Therefore, we recommend hypotheses as follows:

H3: Price perception is significantly associated with purchase decisions.
H4: Price perception is significantly related to repurchase intention.

2.3 Purchase Decision and Repurchase Intention

Purchase behavior has recently got much recognition from marketers and scholars, as it has shown the role in forecasting operational results and achieving a sustainable competitive advantage [17, 22]. A purchase decision is a buyer's decision-making phase in which a consumer chooses to buy a real product being considered [14]. A purchase decision is also defined as an assessment phase, in which preferences' buyers towards a brand are established, and they buy the most popular brand through this stage [23]. Other scholars demonstrate that a purchase decision is a series of choices formed by the buyer before making to buy for matching their demand. The buyer will decide where to purchase, the desired brand, purchase quantity, time to purchase, the total money to be spent, and the payment way. Marketers could influence these decisions by providing information about their products/services for the buyer's evaluation stage [24].

Repurchase intention has been scrutinized widely by marketing scholars and is typically defined as the intention to purchase a specific product/service again over time [25]. Repurchase intention is understood as the customers' perception of rebuying from the same store, based on their previous experience [26]. Other scholars state that repurchase intention has considered the consumers' repeat purchasing behavior in a particular period because of their product's favorite [10]. In the online context, repurchase is the reuse of the online channel to buy from a specific retailer. It also can be viewed as continuance behavior (i.e., continue to buy from the same internet shop) [27].

The prior empirical study illustrated that purchase decision had an essential antecedent of repurchase intention and was significantly related to repurchase intent [28].

Thus, we propose the hypothesis as follows:

H5: Purchase decision is significantly related to repurchase intention.

Based on the purpose of the present research, literature reviews, and developing hypotheses, Fig. 1 illustrates the suggested research model.

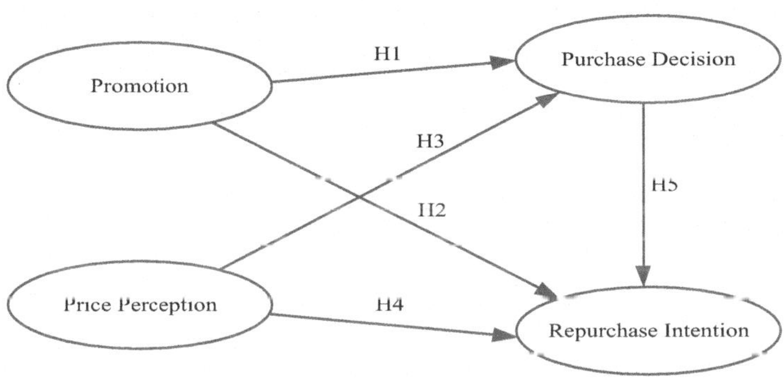

Fig. 1. The proposed research model

3 Research Methodology

3.1 Measures

Measurement items are selected and modified from previous researches to fit into the study circumstances. We use a five-point Likert scale from 1 (strongly disagree) to 5 (strongly agree) to estimate the items. In the present research, we adjust four indicators of promotion from Lau et al. (2006) [16], four items of price perception from Ali and Bhasin (2019) [29], Widyastutir and Said (2017) [30] five indicators of purchase decision from ALHuwaishel & AL-Meshal (2018) [31], and Hanaysha (2018) [24], and four items of repurchase intention from Ali and Bhasin (2019) [29].

3.2 Sample and Data Collection

The sample data were collected from consumers in Ho Chi Minh City, Vietnam, by convenience sampling method. The respondents were selected who active online shopping. We survey offline and online in Ho Chi Minh City. After removing the responses were not full information. There were 316 responses used for the ultimate analysis. The sample included 125 male consumers (39.6%) and 191 female consumers (60.4%). Consumers were in the age group 18 to 25 made 49.1% of the respondents; 29.7% were in the period group 26 to 35; 14.6% in the age group 36 to 45; and 6.6% were greater than 45.

3.3 Analytical Approach

Partial least squares (PLS) approach was performed in this research as this way had the most significant reasons for applying this approach was: (1) non-normal data, (2) small sample sizes, and (3) formatively estimated variables [32]. The partial least squares (PLS) path was a variance-based structural equation modeling (SEM) approach widely performed in the marketing and social sciences. It is a new statistical tool for the research method [33]. The PLS method was applied to measuring the proposed study model and hypotheses. Analyzing the recommended study model and hypotheses were conducted in two steps: (1) Estimation of the measurement model and (2) Estimation of the structural model [34].

4 Results and Discussion

4.1 Results

4.1.1 Estimation of the Measurement Model: Reliability and Validity

Table 1 exhibits the sub-variables measure of the constructs' findings. Cronbach's alpha coefficient and composite reliability (C.R.) values estimated the study variables' reliability. Cronbach's alpha (α) coefficient of the constructs should be greater than 0.70, and composite reliability (C.R.) criteria should be larger than 0.70, which means that there is suitable internal consistency among the constructs [34]. Table 1 depicts that Cronbach's alpha coefficients of the independent variables are over 0.70. Consequently, these concepts get internal consistency reliability.

Table 1. The analyzing results of constructs and sub-variables

Constructs and sub-variables	Factor loading	α	CR	AVE
Promotion		0.836	0.891	0.672
1. Advertisements from the website give various promotion programs (such as free-ship, flash sale, voucher, etc.)	0.833			
2. Window displays from online shops are attractive	0.845			
3. Advertisements for online goods are good-looking	0.839			
4. Advertisement for online goods excites me to buy	0.758			
Price perception		0.794	0.865	0.616
1. Price from the online shop is affordable	0.818			
2. Price from the online shops will be probably saving money	0.698			
3. Price from the online stores may be cheaper than other channels	0.801			
4. Product Price from online shops matches the product quality	0.818			
Purchase decision		0.824	0.877	0.587
1. I think my right decision when purchasing products from the online stores	0.737			
2. I often buy the product from online stores	0.798			
3. I want to buy products from online stores	0.756			
4. I would still purchase from online stores, even if other channels have a promotion	0.757			
5. General, I am satisfied with my purchasing products from online stores	0.781			
Repurchase intention		0.808	0.875	0.638
1. I will purchase products online from the site I use frequently	0.717			
2. If I would like to purchase something, I will consider buying it from an online shop	0.837			
3. I will use the priority online store site I frequently use for future buying	0.881			
4. I will suggest the online purchasing site to others that I use frequently	0.750			

We use the factor loading of all indicators values and the average variance extracted (AVE) values to evaluate convergent validity. The factor loading and AVE values should be larger than 0.50 [34, 35]. In this current research, the factor loading of all indicators and the AVE criteria is higher than 0.50. Therefore, the constructs' convergent validity is fit.

Cronbach alpha (α) 's formula as follows:

$$\alpha = \frac{k}{k-1}\left(1 - \frac{\sum \sigma_i^2}{\sigma_x^2}\right) \quad (1)$$

k: factor; σ_i^2: the observation's variance; σ_x^2: total scores' variance.

CR and AVE's formula as follows:

$$CR = \frac{\left(\sum \lambda_i\right)^2}{\left[\left(\sum \lambda_i\right)^2 + \sum 1 - \lambda_i^2\right]} \quad (2)$$

$$AVE = \frac{\sum \lambda_i^2}{\left[\sum \lambda_i^2 + \sum 1 - \lambda_i^2\right]} \quad (3)$$

λ_i: the outer (element) loading to an item.

$1 - \lambda_i$: in case of standardized items.

In addition, we estimate discriminant validity through the Fornell-Larcker criteria [36]. It describes the square root of the AVE coefficient with the latent variable. Correctly, the AVE's value square root should be higher than its highest correlation with any other notions [34]. Table 2 reveals that AVE's square root values reflect variables of promotion, price perception, buying decision, and repurchase intention is larger than the corresponding latent concepts correlation. Therefore, the discriminant validity of these concepts is verified.

Table 2. Discriminant validity

	Price perception	Promotion	Purchase decision	Repurchase intention
Price perception	0.785			
Promotion	0.570	0.820		
Purchase decision	0.570	0.668	0.766	
Repurchase intention	0.580	0.603	0.574	0.799

4.1.2 Assessing Structural Model Fit and Hypotheses Examining

(1) **Assessing Structural Model Fit**

Figure 2 and Table 3 demonstrate the suggested model fit outcomes. The outcomes in Table 3 describe that the Chi-square = 510.624 was significant at the 0.05 level (p = 0.00). SRMR (standardized root mean square residual) estimates the suggested study model fit. A model was fit when SRMR was less than 0.08 [37]. The summary outcomes in Table 3 show that this model has an SRMR coefficient of 0.073 < 0.08. Thus, the suggested research model fits the research data. Additionally, testing of a multicollinearity affair indicates that all VIF points were below the cut-off of 5. Hence, there was no multicollinearity affair in the structural model [34].

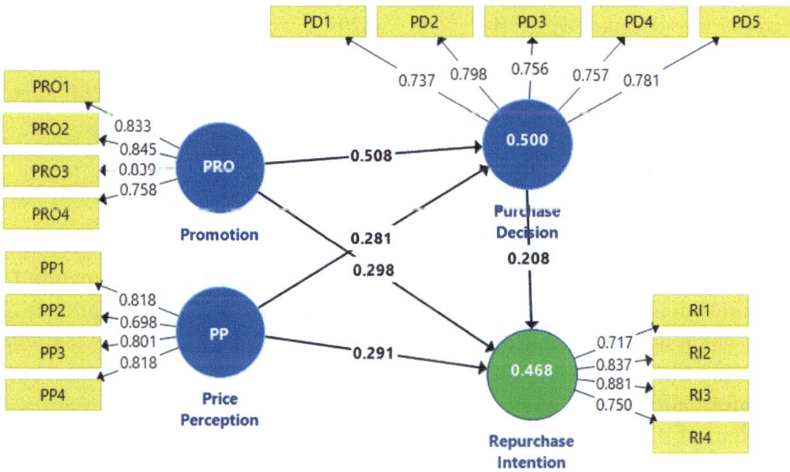

Fig. 2. The structural model's outcomes (PLS-SEM)

Table 3. The model fit's results.

	SRMR	d_ULS	d_G1	d_G2	Chi-square	NFI
Saturated model	0.073	0.820	0.327	0.272	510.624	0.806

(2) **Hypothesis Examining**

Table 4 presents the outcomes of the hypothesis examination. We implement 5000 resamplings bootstrapping outcomes with corresponding 95 percent confidence intervals for the proposed model exhibits that all parameter measures of the connections are in confidence intervals. Moreover, these confidence interval weights of these relationships do not include the zero value. In other words, all parameter measures of these connections in the proposed model are statistically significant at the 5% level. Therefore, these hypotheses H1, H2, H3, H4, and H5 are supported.

Table 4. The results of hypothesis testing

Links	Hypotheses	Path coefficients	95% CI	Results
Promotion → Purchase Decision	H1	0.508	[0.391–0.615]	supported
Promotion → Repurchase Intention	H2	0.298	[0.152–0.438]	supported
Price Perception → Purchase Decision	H3	0.281	[0.158–0.398]	supported
Price Perception → Repurchase Intention	H4	0.291	[0.167–0.413]	supported
Purchase Decision → Repurchase Intention	H5	0.208	[0.076–0.338]	supported

(3) **R^2(explained variance) and f^2 (effect size)**
The determination's coefficient (R^2) is the overall impact level estimation for the structural model [38]. The R^2 standard is between 0 to 1, with a greater coefficient, which means more accuracy in forecasts. The (f^2) influence extent explains estimating the independent factor contributing to the dependent variable.
The information for R^2 and f^2 are as follow:

- R^2 values of 0.67, 0.33, 0.19 are mentioned as significant, moderate, low [39].
- f^2 values of 0.35, 0.02, 0.15 are expressed as high, medium, low [40].
- The f^2's formula as follows

$$f^2 = \frac{R^2_{included} - R^2_{included}}{1 - R^2_{included}}. \tag{4}$$

In this study, the R^2 coefficient for the general model is 0.468 (see Fig. 2 and Table 5), below 0.67, which is considered a moderate impact. In other words, promotion, price perception, and buying decision explain 46.8% of the variance on repurchase intention; we notice that promotion has the most significant influence (0.298), following by price perception (0.291) and buying decision (0.208). Additionally, promotion and price perception describe 50% of the variance on buying decision; we also show that the promotion has a more critical influence (0.508) than price perception (0.281).

Table 5 illustrates the f^2 effect size. The high f^2 effect level occurs for the relationship of promotion → purchase decision (0.348). The average-low f^2 effect size appears for the linking of price perception → purchase decision (0.106) and price perception → repurchase intention (0.097). Likewise, the small f^2 effect sizes happen for the relationship of promotion → repurchase intention (0.084) and purchase decision → repurchase intention (0.041).

Table 5. R^2 and f^2

Links	Path coefficients	f^2	Construct	R^2
Promotion → Purchase Decision	0.508	0.348	Buying Decision	0.500
Promotion → Repurchase Intention	0.298	0.084		
Price Perception → Purchase Decision	0.281	0.106	Repurchase Intention	0.468
Price Perception → Repurchase Intention	0.291	0.097		
Purchase Decision → Repurchase Intention	0.208	0.041		

4.2 Discussion

This research has analyzed the effect of promotion and price perception on buying decisions and repurchasing intentions in Vietnam's context. The finding of this research demonstrated that the five hypotheses in the proposed research model were supported.

The research finding confirmed that promotion has significantly positively related to buying decisions. The promotion was an antecedent of purchasing decisions. The f^2 effect size of the relationship between promotion and buying decisions was large (0.348). The previous empirical study affirmed this research [3–5]. The findings also revealed that promotion has positively correlated with repurchase intention. The promotion was a predictor of repurchase intention; however, the f^2 impact size of linking promotion and repurchase intention was small (0.084). The previous study demonstrated this research [7].

Additionally, the study results also confirmed that price perception was positively associated with the purchase decision. Price perception was an antecedent of the purchase decision. The effect extent of the relationship between price perception and purchase decision was medium-small (0.106). The previous empirical investigations described this research outcome [5, 6, 8]. Likewise, the findings also reinforced that price perception was positively associated with repurchasing intention. Price perception was a predictor of repurchasing intention; though, the f^2 effect level of the connection of price perception and repurchasing intention was average weak (0.097). The previous empirical studies affirmed this research [9, 10].

Furthermore, the findings also verified that buying decisions positively affected repurchasing intention. Buying decisions was an antecedent of repurchasing intention; however, the effect extent of the connection of buying decisions and repurchasing intention was small (0.041). The f^2 previous empirical analysis confirmed this research [28].

5 Conclusions and Limitations

In line with the previous studies, this present study has explained the influence of promotion and price perception on buying decisions and repurchasing intentions through online shopping in Vietnam's context. Hence, practitioners should design marketing programs that improve purchasers' perception of promotion and price perception to improve buying decisions and intent to rebuy online shopping.

This study helps practitioners to understand the meaning of promotion on shopping decisions and repurchase intentions. Thus, practitioners should design promotional activities that communicate product/website advantage to persuade target customers to buy products or visit the site. Some of the features that managers should remark for the promotion plans include giving various promotion programs (such as free-ship, flash sale, voucher, etc.), displaying products and attractive window displays, and stimulating consumers to buy the product. If purchasers recognize promotion activities positively, they will tend to favor buying decisions and intent to rebuying in online shopping.

Furthermore, practitioners should give strategic plans of price perception to enhance buying decisions and repurchasing intention. For instance, providing an affordable price and showing that consumers save money or take costs cheaper than other online purchasing channels. Significantly, the price should match with product quality. When the clients evaluate the seller's price and the rival's price is accepted, they will incline to buy and intend to repurchase the product.

Lastly, the study will assist practitioners in realizing the vital of buying decisions on rebuying intent. So, practitioners should offer promotion activities and price strategy so that clients recognize that they were the online shopping right decision. When they think that they have good choices, they will intent to repurchase.

This research has some limitations. First, research data was conducted from the online shoppers living in Ho Chi Minh city, so this study may not represent all online shoppers in Vietnam. Future studies should be continued collecting the data from online shoppers in other cities. Second, this study only analyzed the impact of promotion and price perception on purchase decision and repurchase intention; thus, future examining should focus on other constructs such as perceived risk, product quality, loyalty, etc.

References

1. Dung, K.: Online shopping in the domestic country is becoming more and more popular, (Vietnamese). https://dangcongsan.vn/kinh-te/mua-sam-truc-tuyen-trong-nuoc-dang-ngay-cang-tro-nen-pho-bien-562026.html. Accessed 20 Aug 2020
2. Loan, H.: An explosion of online shopping during the Covid-19 era, (Vietnamese). https://vneconomy.vn/bung-no-mua-sam-online-thoi-covid-19-20201102124838357.htm. Accessed 02 Nov 2020
3. Fachmi, M., Setiawan, I.P., Hidayat, A.: Analysis of factors affecting consumer purchase decision at online shops. Int. J. Innovative Sci. Res. Technol. **4**, 118–121 (2019). https://doi.org/10.31219/osf.io/egysw
4. Kore, E.L.R., Lamalewa, F., Mulyaningsih, A.: The influence of promotion, trust, and convenience to online purchase decisions. Int. J. Mech. Eng. Technol. **9**, 77–83 (2018)

5. Indriani, I.: Analysis the factors influence consumer buying decision on online shopping clothing for consumer in manado. Jurnal EMBA: Jurnal Riset Ekonomi, Manajemen, Bisnis dan Akuntansi. **4**, 1166–1177 (2016). https://doi.org/10.35794/emba.v4i1.11874
6. Rakesh, S., Khare, A.: Impact of promotions and value consciousness in online shopping behaviour in India. J. Database Mark. Cust. Strategy Manag. **19**, 311–320 (2012). https://doi.org/10.1057/dbm.2012.30
7. Suryaningsih, I.B., Farida, L., Revanica, O., Kusuma, A.A.M.: The effect of coupon sales promotion, online customer review and perceived enjoyment on repurchase intention in e-commerce shopee. Int. J. Sci. Technol. Res. **8**, 435–440 (2019)
8. Edwar, M., Diansari, R.A.A., Winawati, N.F.: The factors that affecting the product purchasing decision through online shopping by students of surabaya state university. Int. J. Educ. Res. Rev. **3**, 54–64 (2018). https://doi.org/10.24331/ijere.432350
9. Suhaily, L., Soelasih, Y.: What effects repurchase intention of online shopping. Int. Bus. Res. **10**, 113–122 (2017). https://doi.org/10.5539/ibr.v10n12p113
10. Yulistyarini, D., Subagio, A., Paramu, H., Irawan, B.: Customer repurchase intention and satisfaction in online shopping. Int. Bus. Manag. **11**, 215–221 (2017)
11. Wijaya, H., Astuti, S.R.: The effect of trust and brand image to repurchase intention in online shopping. KnE Soc. Sci. **3**, 915 (2018). https://doi.org/10.18502/kss.v3i10.3182
12. Kotler, P., Keller, K.L.: Marketing Management. Pearson Education Inc, Upper Saddle River, New Jersey (2016)
13. Azad, A., Shamsher, R.: User perception toward brand loyalty of mobile phones : the Bangladesh context. J. Bus. Stud. **XXXV**, 123–139 (2014)
14. Kotler, P., Armstrong, G.: Principles of Marketing. Pearson Education Limited, Harlow, England (2016)
15. Sulaiman, Y., Yusr, M.M., Ismail, K.A.: The influence of marketing mix and perceived risk factors on online purchase intentions. Int. J. Res. Bus. Stud. Manage. **4**, 11 (2017)
16. Lau, M.M., Chang, M.T., Moon, K.L., Liu, W.S.: The brand loyalty of sportswear in Hong Kong. J. Text. Apparel, Technol. Manage. **5**, 1–13 (2006)
17. Zeithaml, V.A.: Consumer perceptions of price, quality, and value: a means-end model and synthesis of evidence. J. Mark. **52**, 2–22 (1988). https://doi.org/10.1177/002224298805200302
18. Lee, S., Illia, A., Lawson-Body, A.: Perceived price fairness of dynamic pricing. Ind. Manag. Data Syst. **111**, 531–550 (2011). https://doi.org/10.1108/02635571111133533
19. Faith, D., Edwin, A.: A review of the effect of pricing strategies on the purchase of consumer goods. Int. J. Res. Manage. Sci. Technol. **2**, 88–102 (2014)
20. Varki, S., Colgate, M.: The role of price perceptions in an integrated model of behavioral intentions. J. Serv. Res. **3**, 232–240 (2001). https://doi.org/10.1177/109467050133004
21. Han, H., Kim, W.: Outcomes of relational benefits: restaurant customers' perspective. J. Travel Tour. Mark. **26**, 820–835 (2009). https://doi.org/10.1080/10548400903356236
22. Shareef, M.A., Kumar, U., Kumar, V.: Role of different Electronic-Commerce (Ec) quality factors on purchase decision: a developing country perspective. J. Electron. Commer. Res. **9**, 92–113 (2008)
23. AL-Ghaswyneh, O.F.M.: Factors affecting the consumers decision behavior of buying green products. ESIC MARKET Econ. Bus. J. **50**, 389–418 (2019). https://doi.org/10.7200/esicm.163.0502.4
24. Hanaysha, J.R.: An examination of the factors affecting consumer's purchase decision in the Malaysian retail market. PSU Res. Rev. **2**, 7–23 (2018). https://doi.org/10.1108/prr-08-2017-0034
25. Liu, Y., Pu, B., Guan, Z., Yang, Q.: Online customer experience and its relationship to repurchase intention: an empirical case of online travel agencies in China. Asia Pac. J. Tour. Res. **21**, 1085–1099 (2016). https://doi.org/10.1080/10941665.2015.1094495

26. Hellier, P.K., Geursen, G.M., Carr, R.A., Rickard, J.A.: Customer repurchase intention: a general structural equation model. Eur. J. Mark. **37**, 1762–1800 (2003). https://doi.org/10.1108/03090560310495456
27. Khalifa, M., Liu, V.: Online consumer retention: contingent effects of online shopping habit and online shopping experience. Eur. J. Inf. Syst. **16**, 780–792 (2007). https://doi.org/10.1057/palgrave.ejis.3000711
28. Ali, H.: Purchase Decision and Repurchase Models: Product Quality and Process Analysis (Case Study of House Ownership Credit Financing in Permata Sharia Bank Jakarta). Scholars Bull. **05**, 526–535 (2019). https://doi.org/10.36348/sb.2019.v05i09.006
29. Ali, A., Bhasin, J.: Understanding customer repurchase intention in e-commerce: role of perceived price, delivery quality, and perceived value. Jindal J. Bus. Res. **8**, 142–157 (2019). https://doi.org/10.1177/2278682119850275
30. Widyastutir, S., Said, M.: Consumer consideration in purchase decision of SPECS sports shoes product through brand image, product design and price perception. Int. J. Supply Chain Manage. **6**, 199–207 (2017)
31. ALHuwaishel, N.S., AL-Meshal, S.A.: The impact of perceived value, quality, and loyalty on purchase decision in the accessories department: study on saudi females. Br. J. Mark. Stud. **6**, 21–31 (2018)
32. Hair, J.F., Sarstedt, M., Hopkins, L., Kuppelwieser, V.G.: Partial least squares structural equation modeling (PLS-SEM): an emerging tool in business research. Eur. Bus. Rev. **26**, 106–121 (2014). https://doi.org/10.1108/EBR-10-2013-0128
33. Henseler, J., Hubona, G., Ray, P.A.: Using PLS path modeling in new technology research: updated guidelines. Ind. Manag. Data Syst. **116**, 2–20 (2016). https://doi.org/10.1108/IMDS-09-2015-0382
34. Hair, J.F., Hult, G.T.M., Ringle, C.M., Sarstedt, M.: A Primer on Partial Least Squares Structural Equation Modeling (PLS-SEM). SAGE Publications Inc, Los Angeles, CA (2017)
35. Bagozzi, R.P., Yi, Y.: On the evaluation of structural equation models. Acad. Mark. Sci. **16**, 74–94 (1988)
36. Fornell, C., Larcker, D.F.: Evaluating structural equation models with unobservable variables and measurement error. J. Mark. Res. **18**, 39–50 (1981)
37. Hu, L., Bentler, P.: Fit indices in covariance structure modeling: Sensitivity to underparameterized model misspecification. Psychol. Methods **3**, 424–453 (1998)
38. Garson, G.D.: Partial Least Squares: Regression and Structural Equation Models. Statistical Associates Publishers, Asheboro, NC (2016)
39. Chin, W.W.: The partial least squares approach to structural equation modeling. In: Macoulides, G.A. (ed.) Modern Methods for Business Research, pp. 295–336. Lawrence Erlbaum Associates, Mahwah, NJ (1998)
40. Henseler, J., Ringle, C.M., Sinkovics, R.R.: The use of partial least squares path modeling in international marketing. Adv. Int. Mark. **20**, 277–319 (2009). https://doi.org/10.1108/S1474-7979(2009)0000020014

Review Rating Prediction Based on Text Polarity in the Czech and Slovak Languages

Ján Mojžiš(✉) and Marcel Kvassay

Institute of Informatics, Slovak Academy of Sciences, Bratislava, Slovakia
{jan.mojzis,marcel.kvassay}@savba.sk

Abstract. Online product reviews represent a valuable source of information about customer preferences. To process such reviews automatically, sentiment analysis is usually employed. Our primary goal in this paper is to determine the utility of an emotional lexicon capturing affective and attitudinal values of individual words for sentiment analysis of product reviews written in the Czech and Slovak languages. Both languages are highly inflectional and synthetic, and morphologically more complex than English. The specific challenge that we address is the prediction of the overall review rating from its text. (The rating is expressed on the scale of one to five stars.) We analyse the contribution of the "emotional" features derived from the emotional lexicon to the classification accuracy in the context of product reviews that are already pre-structured into positive and negative text segments by the reviewers. Contrary to our initial expectations, our investigation showed that the independent contribution of the "emotional" features towards classification accuracy was virtually nil in this context, and rather modest (about 7%) otherwise. These results imply that providing customers with an option to structure their product reviews into positive and negative segments might simplify their subsequent analysis significantly.

Keywords: Sentiment analysis · Review rating prediction · Emotional lexicon · Machine learning · Text polarity

1 Introduction

Online presence is becoming increasingly important for companies of all kinds. A vital part of preserving online relevance is the ability to react quickly to customer feedback. The feedback may come in many ways, formal as well as informal. While small businesses can hope to handle it manually, for larger ones an automatic processing and evaluation of such feedback is imperative. In order to facilitate automatic processing, many online marketplaces actively encourage their customers to provide structured feedback on the items they have bought. Their predefined feedback forms typically include some kind of overall rating, e.g. on the scale from zero to five stars. Sometimes, there is also an option to separate positive and negative parts (segments) of the feedback. Having customer reviews structured into positive and negative segments helps other customers more easily to decide which products to buy, and also greatly facilitates their automatic

© The Author(s), under exclusive license to Springer Nature Switzerland AG 2021
R. Silhavy (Ed.): CSOC 2021, LNNS 229, pp. 251–261, 2021.
https://doi.org/10.1007/978-3-030-77445-5_22

processing. Moreover, such collections of rated reviews can serve as valuable training sets for models meant to automatically process less structured forms of feedback, particularly those without explicit overall rating. An important challenge in this context is the ability to predict the overall review rating from its text. This is typically accomplished with the help of sentiment analysis, which exploits affective and attitudinal values of words, phrases or sentences in each feedback piece in order to predict its resulting overall rating.

Most published work in sentiment analysis concerns English and other major world languages. Smaller languages like Slovak and Czech are seriously underrepresented and suffer from a lack of standardized and readily available tools for their automated processing. Nevertheless, it is important that researchers of these smaller languages also make an effort to catch up with the latest developments in the field.

We first entered this challenging research area in the context of a Master's thesis [1] that one of us (J.M.) supervised. The thesis seemed to indicate that the use of an emotional lexicon capturing the affective and attitudinal values of individual words might systematically and significantly improve the accuracy of the overall review rating prediction in the Czech and Slovak languages. Hoping to validate those results on a larger dataset and to improve them yet further, we embarked on a mission that soon turned out to be much more demanding and complicated than we had initially expected. In this paper we describe our systematic (and ultimately unsuccessful) attempts to validate the findings of the student's thesis, analyse the reasons behind this unexpected result, and formulate a plan for further research.

In order to measure and compare the results and performance of various approaches in general, various metrics are used, such as accuracy [2–5], F-measure [6], mean square error [7] and others. In our field of research of sentiment analysis and review rating prediction, accuracy is the traditional measure, though sometimes F-measure also is used. In order to compare our results to other published work, we adopted the same metrics as in the cited works, namely Accuracy and, in one instance, macro F-measure.

2 Related Work

Regarding sentiment analysis in English, Turney [2] proposed a simple unsupervised learning algorithm for classifying reviews as recommended (thumbs up) or not recommended (thumbs down). The class of each review was predicted by the average semantic orientation of its phrases that contained adjectives or adverbs. The accuracy of his classification varied between 66% (movie reviews) and 84% (automobile reviews). However, his approach discriminated between two classes only. He actually used 410 reviews out of which 170 (41%) were of the type "not recommended" while 240 (59%) were "recommended". He duly noted that always guessing the major class would lead to the accuracy of 59%, so his improvement in accuracy above this baseline was 7% to 25%.

In a similar vein, Pang and Lee [3] considered the problem of classifying documents by their overall sentiment, that is, by distinguishing whether a review is positive or negative, yielding again a binary classification. They too used the movie review dataset like Turney, but included 2 053 reviews in two balanced classes. With classification accuracy varying between 72.8% and 82.9% (depending on the feature set and the classifier), they achieved the accuracy improvement of 22.8% to 32.9%.

Slovak language is not as intensively investigated as English and, consequently, there are fewer works dealing with sentiment analysis in Slovak. In this relatively restricted niche, Krchnavy and Simko [4] analyzed sentiment in social network posts in Slovak. They concluded that Slovak language required additional preprocessing in order to make the data usable for further analysis. In their view, lemmatization was a crucial step. But they also resorted to reconstructing accents in order to improve accuracy (both Slovak and Czech use accented characters with diacritical marks, which tend to be omitted in online contexts). In this manner, they were able to classify two classes with 82.16% accuracy, three classes with 66% accuracy, four classes with 50.23% accuracy and five classes with 40.93% accuracy. Their classes were not equally distributed, however, so always guessing the major class in their context would yield 30% accuracy for five classes, 37% for four classes, 45% for three classes and 55% for two classes.

In a subsequent work [5], Pecar, Simko and Bielikova tried to predict the overall sentiment of customer reviews split into 3 or 7 sentiment classes. For 7 classes their best accuracy was 72% and for 3 classes their best accuracy was 73%. (Both were obtained by combining a bidirectional LSTM, pre-trained embedding vectors and tokenized text with reconstructed diacritics.) Again, not having equally distributed classes, always guessing the major class in their case would yield 49% accuracy for seven classes and 55% accuracy for three classes. Their accuracy improvement above the baseline was thus 22% for seven and 17% for three classes.

Czech language is in many respects very close to Slovak. Given the scarcity of published research of this kind for Slovak language, we exploit this closeness and compare ourselves to (as well as draw inspiration from) the published research focusing on Czech. Thus, for example, Habernal, Ptáček and Steinberger [6] analyzed the sentiment of three different kinds of Czech social media and classified them into several sentiment classes. They classified social media posts, movie reviews and product reviews. For social media posts they achieved the highest macro F-measure of 90%. This was for two unequal classes, where the macro F-measure of "nearly always" guessing the majority class might be as high as 36%, so their improvement beyond this baseline was 54%.

3 The Data

Our original motivation for this work was to validate and further extend the results of [1]. We therefore collected our data from the same source and followed the same data preparation methodology for compatibility, but extended our datasets considerably.

Our data source was Alza (www.alza.sk), a major Czech and Slovak online marketplace for various types of products. Its customers can write reviews only about the products they have bought, and must specify an overall star rating from 1 to 5 stars. Next, they can (but need not) fill text segments (or blocks) of two polarities. They can fill several blocks of negative features of the product as well as several blocks of positive properties. Also, if they wish, they can fill one unmarked or neutral segment whose polarity is not expressly indicated.

In order to develop a model for review rating prediction based on its text, we distinguished reviews based on their language. We have used national specific (accented) letters to discriminate between Czech and Slovak. Using these specific national letters, we created two language-specific lists of words. We also created a third list of English only words. We then separated reviews based on their language. Together we have collected 263 155 reviews, out of which 172 455 were in Czech and 90 455 in Slovak. There were also some reviews in English, which we omitted from further analysis, see Table 1.

As each review may contain not only positive and/or negative segments of text, but also a neutral one, we selected only the reviews with at least one positive or one negative block (polarity block, PB). We expected that this would increase the probability of correctly detecting the overall review polarity (either negative or positive), and thus also the accuracy of our prediction models. We used two stop word filters (one for each language) and, analogously, two filters of "noise". The noise lists contained words (unigrams, bigrams, trigrams) which occurred frequently in negative blocks but did not carry any semantic value (e.g. the words and phrases like "nothing", or "no problem", etc. instead of actual product complaints). After the application of these filters we imposed another condition: there had to be at least one word from a positive or a negative block. Table 2 lists the subsets of our Czech and Slovak reviews after applying all these filters and conditions. The reviews in each language were then divided into five classes (1-star ones to 5-star ones) as listed in Table 3. For visual comparison, see Fig. 1.

Table 1. An overview of our raw datasets before filtering.

Review language	Review count	Review author count
Czech reviews	172,455	54,774
Slovak reviews	90,455	28,951
English reviews	245	158

Table 2. Datasets after filtering out stop words and "noise" as well as imposing a condition of having at least one polarity block per review.

Review language	Review count	Review author count
Czech reviews	169,565	51,873
Slovak reviews	88,607	27,880

Due to the highly inflectional and synthetic nature of both Czech and Slovak, we applied tokenization and lemmatization with language models for Morphodita[1,2]. We also employed the Emotional lexicon [8] in order to capture the affective and attitudinal values of individual words in the reviews. Each word in this lexicon is described by a ten-dimensional vector (with binary values in each dimension) capturing its association with (1) positive sentiment, (2) negative sentiment, (3) anger, (4) anticipation, (5) disgust, (6) fear, (7) joy, (8) sadness, (9) surprise and (10) trust.

[1] Czech models https://lindat.mff.cuni.cz/repository/xmlui/handle/11234/1-1836.
[2] Slovak models https://lindat.mff.cuni.cz/repository/xmlui/handle/11234/1-3278.

Table 3. Counts of reviews in each rating class of our datasets.

Class	Language	
	Czech reviews	Slovak reviews
"a" (5 stars)	107,497	57,476
"b" (4 stars)	31,371	15,590
"c" (3 stars)	10,369	5,423
"d" (2 stars)	5,881	2,729
"e" (1 star)	9,560	4,607

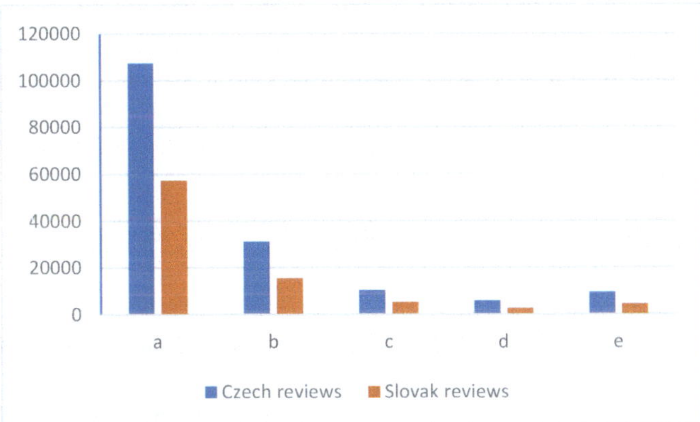

Fig. 1. Count of reviews in each rating class of our datasets. Y-axis represents the number of reviews, X-axis the review rating class.

This lexicon was compiled with the help of Amazon's Mechanical Turk service, which is a crowdsourcing platform. The lexicon's originators decided to have each word annotated by five different human annotators. Control questions were also supplied alongside each task to filter out untrustworthy annotators. However, the original language of this lexicon was English, and Google Translate then converted it to a number of other languages. This introduced additional sources of error due to Google Translate's limited ability to translate individual words presented out of context. For example, we discovered an error in the Slovak version of the lexicon for the English word 'plated', which was wrongly translated as 'á' into Slovak, but correctly as 'pokrytý' into Czech. Another example was the word 'ail', acceptably translated into Slovak as 'bolieť' but incorrectly into Czech as 'ail' (an acceptable translation would have been e.g. 'bolet'). Moreover, Google Translate considered some English words untranslatable: out of the total of 14,182 words in the full lexicon, there were no translations into Slovak for 271 words and into Czech for 270 words. In spite of these imperfections, we nevertheless expected the lexicon to help us increase the overall sentiment detection accuracy of our models.

After cleaning and preprocessing our raw data, we prepared two experimental setups, S1 and S2[3], which differed by the type and number of their attributes:

- S1 consisted of plain statistical information about the review structure, such as the number of positive and negative blocks as well as of words in them (word counts were both absolute and relativized with respect to the total number of words in each review). Table 4 lists all the nine "basic" features comprising S1.
- S2 was more elaborate and consisted of 36 features. It kept the nine "basic" S1 attributes and added to them new "emotional" ones based on the emotional lexicon. The numbers of words associated with each of the ten emotional dimensions of the lexicon were summed separately for positive and negative blocks, resulting in 20 new attributes for each review. To these, seven "hybrid" attributes were added: the key among them was the number of emotional dimensions associated with at least one word in the analysed text. This was again calculated separately for positive and negative blocks (count_positive_emotions and count_negative_emotions) and used in its absolute form as well as relativized with respect to word count and emotion count. (Emotion count = count_positive_emotions + count_negative_emotions.)

Table 4. Nine features comprising the "basic" setup S1.

Feature number	Feature name	Feature description
1	count_positive_words	word count for positive segments
2	count_positive_words_relative	word count for positive segments divided by the total word count
3	count_negative_words	word count for negative segments
4	count_negative_words_relative	word count for negative segments divided by the total word count
5	count_neg_w_divide_count_pos_w	ratio of negative and positive segment word counts
6	count_positive_blocks	positive segment count
7	count_negative_blocks	negative segment count
8	average_positive_words	count_positive_words divided by count_positive_blocks
9	average_negative_words	count_negative_words divided by count_negative_blocks

A class (or target) feature was also supplied for each setup. This feature represented the overall review rating. Since the users rated their reviews using five levels (stars), we defined five different final classes for prediction (1 - 5 stars). (We omitted the reviews

[3] These setups were identical to those used in [1], whose results we were trying to validate and extend.

that did not earn any star, because there were very few of them.) In all our experiments, we used the Random Forest classifier with default settings and 10-fold cross validation in Weka 3.4.

We also performed several additional experiments in order to compare our results to the recent publications cited above. As they were using different class distributions (i.e. their class numbers and size ratios were different from ours), we performed each comparison separately and each time adjusted our class distribution to make it compatible with the cited work.

4 Results and Discussion

As we have already mentioned, our initial motive for this work was the desire to validate the results of the Master's thesis [1], which analysed 4,214 Czech and 1,416 Slovak reviews from Alza.sk. Both Slovak and Czech reviews were split into two equal classes labelled "a" (5-star ones) and "b" (4-star ones). The remaining classes earning less than four stars were much smaller and were therefore excluded from detailed examination. This decision actually made the student's task harder because, essentially, he was not distinguishing positive reviews from the negative ones, but rather the very positive (5-star) ones from the positive (4-star) ones, so the expected differences in their language expressions were not very high. The classification accuracies that the student achieved for two different feature setups, S1 and S2, are listed in Table 5. In both cases, he used the Random Forest classifier.

Since the student's random accuracy baseline (i.e. the probability of correctly assigning reviews into one of two equal classes by pure chance) was 50%, his contribution to accuracy beyond this baseline ranged from 12.86% to 26.27%, which was quite in line with the results reported by other researchers. Moreover, the student's results seemed to indicate that the new features in S2 based on the emotional lexicon systematically improved the classification accuracy by about 6% regardless of the language. (The results appeared particularly encouraging for Slovak, where the best accuracy of 76.27% was achieved.)

Table 5. Accuracies (%) of experiments reported in [1], which consisted of binary classification of Czech and Slovak reviews in two different setups, S1 and S2. S1 consisted of 9 features listed in Table 4 and mainly captured plain statistical information. S2 included S1 along with additional features based on the emotional lexicon [8]. (Both setups were described in detail in the previous section.)

Language	Setup	
	S1	S2
Czech	62.86	69.34
Slovak	70.83	76.27

Unfortunately, this promising tendency for the "emotional" features completely disappeared when we repeated the experiments on larger datasets, as shown in Table 6.

In this case, the expanded datasets contained 31,180 Slovak reviews and 62,742 Czech reviews from Alza.sk, with reviews in each language again divided into two equal classes, "a" (5-star ones) and "b" (4-star ones). For compatibility, we again used the Random Forest classifier as in [1].

The results in Table 6 were a big disappointment for us. Not only had the positive influence of emotional features in S2 disappeared, but they actually seemed to harm the accuracy somewhat, a tendency characteristic of redundant features. Far from validating the student's thesis, it became apparent that the emotional values of individual words did not carry any additional useful information on top of that which was already present in the reviews pre-structured into explicitly marked positive and negative segments or blocks of text.

Our next experiment demonstrated that polarity blocks of text (represented by the setup S1 in Table 6) in fact carried considerably more information than the emotional lexicon: When the reviews were not pre-structured into textual blocks of positive and negative polarity, the features based on the emotional lexicon only enabled us to achieve the accuracy of 57.27% for Czech reviews (i.e. 7.27% above the random baseline) and 56.94% for Slovak reviews (6.94% above the random baseline). (We again used the Random Forest classifier for compatibility.) Both of these figures are considerably lower than 17.73% to 19.59% improvement above the baseline implied for the polarity blocks in the basic setup S1 by Table 6.

After a detailed investigation, we discovered errors in the student's data preparation procedure. (We used a slightly different one of our own making.) The structure of the student's reviews was damaged after lemmatization and, as a result, some positive and negative blocks of text were attributed to wrong reviews. This naturally had a negative impact on the accuracy achievable solely through polarity blocks in the setup S1. It seems that the use of emotional lexicon was able to compensate for this error, which created a false impression that there was independent information present in the "emotional" features that we included in the setup S2. Likewise, the higher precision achieved by the student for Slovak reviews seems to have been just a statistical fluke made possible by the limited number of Slovak reviews at his disposal.

At this point we decided to find out how our techniques compared to those used by the other authors. In these comparisons, we always used our larger (and therefore more representative) dataset of Czech reviews. Whenever the other authors did not use balanced classes, we too tweaked our class balance to the same size proportion (or ratio) between the largest and the smallest class so as to produce comparable results.

Table 6. Accuracies (%) of our repeated experiments with binary classification of Czech and Slovak reviews for expanded datasets using the same setups S1 and S2.

Language	Setup	
	S1	S2
Czech	69.59	69.08
Slovak	67.73	67.14

Thus, for example, Krchnavy and Simko in [4] reported the classification accuracy of 40.93% for social network posts divided into 5 classes, which was 10.93% above the trivial baseline of always guessing the majority class. As shown in Table 7, our approach achieved roughly the same accuracy when we tweaked our class distribution to make it compatible with theirs.

In a subsequent work [5], Pecar, Simko and Bielikova reported the accuracy of 73% for the prediction of overall sentiment of customer reviews split into 3 classes. In order to compare our work to theirs, we merged our classes "e" (1-star reviews) and "d" (2-star reviews) into one common "negative polarity" class, classes "a" (5-star reviews) and "b" (4-star reviews) into one common "positive polarity" class and used our original class "c" (3-star reviews) as the third, neutral class. We also adjusted the number of reviews in our most populous class ("positive polarity") so as to keep it in the same size proportion with respect to our smallest, neutral "c" class as in [5]. After training the Random Forest classifier on these three classes in the setups S1 and S2, we achieved the classification accuracies listed in Table 8. In this case, our methods appeared to outperform those of Pecar, Simko and Bielikova in [5] though, in all fairness, we should also take into account the fact that our reviews were already pre-structured into positive and negative polarity blocks by the authors of the reviews, which gave us a natural advantage.

Our last comparison is with Habernal, Ptáček and Steinberger [6], who reported the highest macro F-measure of 90% for classifying social media posts into two unequal classes. We approximated their class distribution by joining our 5- and 4-star reviews into one "positive" class, our 1- and 2-star reviews into one common "negative" class, and by adjusting their size ratio to that used in [6]. For this experiment, we also changed our quality metric to macro F-measure in order to be compatible with them. Our results are listed in Table 9.

Table 7. Comparison of classification accuracy of Krchnavy and Simko in [4] for five classes with the accuracies that we achieved with the Random Forest classifier in the setups S1 and S2 on Czech reviews with a compatible class distribution.

Setup	Accuracy (%)
Our setup S1	40
Our setup S2	41.74
Krchnavy and Simko in [4]	40.93

In this case, our methods were clearly underperforming. We believe this was mainly due to the more sophisticated methods of data preprocessing, feature construction and feature selection used in [6]. Our options here were more limited, because our main purpose has been the validation of the results of [1], but we certainly intend to incorporate the suggestions formulated in [6] in our future work.

Table 8. Comparison of classification accuracy of Pecar, Simko and Bielikova in [5] for three classes with the accuracies that we achieved with the Random Forest classifier in the setups S1 and S2 on Czech reviews with a compatible class distribution.

Setup	Accuracy (%)
Our setup S1	76.92
Our setup S2	77.49
Pecar, Simko and Bielikova in [5]	72.96

Table 9. Comparison of macro F-measure achieved by Habernal, Ptáček and Steinberger in [6] for binary classification of social media posts with our own results based on the feature setups S1 and S2 and the Random Forest classifier applied to Czech reviews with a compatible class distribution.

Setup	Macro F-measure
Our setup S1	83.5
Our setup S2	84.45
Habernal, Ptáček and Steinberger in [6]	90

5 Conclusion

In this paper we investigated the use of Emotional lexicon [8] and various structural features based on positive and negative segments of reviews from the online market Alza.sk. Our main goal was to validate the seemingly promising results of the Master's thesis [1] that indicated a significant contribution of the "emotional" features derived from the emotional lexicon to the classification accuracy of the overall polarity (star rating) of those reviews.

Contrary to our initial expectations, our investigation showed that the independent contribution of the "emotional" features towards classification accuracy was virtually nil for reviews that were already pre-structured into positive and negative segments, and rather modest (about 7%) otherwise.

Our subsequent verification experiments demonstrated that our methods achieved comparable accuracy to other published work for reviews in Slovak and Czech language, although there was undoubtedly room for further improvement, especially with respect to the results reported in [6].

For online sellers our results imply that providing customers with an option to structure their product reviews into positive and negative segments might pay back handsomely by simplifying their subsequent analysis. For us this work provides a valuable benchmark for developing more powerful analysis methods for reviews in the Slovak and Czech languages that are not pre-structured into such polarity blocks.

In the future, we intend to experiment with new ways of data preprocessing, feature construction and feature selection, drawing inspiration from other published research concerning online reviews, particularly that which focuses on the Czech and Slovak languages, such as [4–6].

One reason for the disappointing results regarding the use of the emotional lexicon [8] may have been its own flaws and imperfections. We therefore intend to try to improve it or create an entirely new one in order to determine the limits of classification accuracy for online reviews in Slovak and Czech achievable mainly or solely by adequate representation of affective and attitudinal values of their individual words (unigrams). In parallel, we also plan to exploit other more advanced features, such as bigrams, parts of speech tagging, emoticons and character n-grams suggested by [6] and other published research.

Acknowledgment. This research was supported by the projects VEGA 2/0125/20 and 2/0155/19.

References

1. Slaminka, E.: Predikcia hodnotiacich známok z textu hodnotenia v slovenskom a českom jazyku. Master's thesis, 59 (2020)
2. Turney, P.D.: Thumbs up or thumbs down? Semantic orientation applied to unsupervised classification of reviews. In: Proceedigs of the ACL (2002)
3. Pang, B., Lee, L., Vaithyanathan, S.: Thumbs up? Sentiment classification using machine learning techniques. In: Proceedings of the Conference on Empirical Methods in Natural Language Processing (EMNLP), pp. 79–86 (2002)
4. Krchnavy, R., Simko, M.: Sentiment analysis of social network posts in Slovak language. In: 2017 12th International Workshop on Semantic and Social Media Adaptation and Personalization (SMAP). IEEE, pp. 20–25, July 2017
5. Pecar, S., Simko, M., Bielikova, M.: Sentiment analysis of customer reviews: impact of text pre-processing. In: 2018 World Symposium on Digital Intelligence for Systems and Machines (DISA). IEEE, pp. 251–256, August 2018
6. Habernal, I., Ptáček, T., Steinberger, J.: Supervised sentiment analysis in Czech social media. Inf. Process. Manage. **50**(5), 693–707 (2014)
7. Kenyeres, M., Kenyeres, J.: Average consensus over mobile wireless sensor networks: weight matrix guaranteeing convergence without reconfiguration of edge weights. Sensors **20**(13), 3677 (2020)
8. Mohammad, S., Turney, P.: Emotions evoked by common words and phrases: using mechanical turk to create an emotion lexicon. In: Proceedings of the NAACL HLT 2010 workshop on computational approaches to analysis and generation of emotion in text, pp. 26–34, June 2010

Parameters of Recognition Algorithms for the Background Subtraction of Color Medical Images

I. G. Shelomentseva[1,2](✉), S. V. Chentsov[2], and N. V. Yakasova[2,3]

[1] Krasnoyarsk State Medical University Named After Professor V.F. Voino-Yasenetsky, 1, Partizan Zheleznyak Avenue, Krasnoyrsk 660022, Russia
[2] Siberian Federal University, 79, Svobodny Avenue, Krasnoyrsk 660041, Russia
[3] Khakassia State University named after N. F. Katanov, 90, Lenin Avenue, Abakan, Republic of Khakassia 655017, Russia

Abstract. Mycobacterium tuberculosis infection remains a major public health issue of global morbidity and mortality. One of the widely used methods for the finding of mycobacterium tuberculosis is the Ziehl-Neelsen method of microscopy. In this paper, a method for removing noise without producing image distortion for Ziehl-Neelsen stained images of sputum smear samples obtained using a light microscope is presented. The proposed approach is based on the convolution of the original image with the Laplacian of a Gaussian filter enhanced by high-frequency filtering. Used Laplacian of Gaussian filter was discretized as a 9x9 convolution kernel. If the original image is filtered with a simple Laplacian of Gaussian, the resulting output is rather noisy. Combining this result of filtration with the enhanced by high-frequency filtering will reduce the noise and will keep of mycobacterium tuberculosis for further analysis by automated medical diagnostic systems. To deal with the automatic determination of filtering quality the normalized color difference was proposed. The developed method of background subtraction can be used for images of microscopy and dermatology.

Keywords: Laplacian of Gaussian · Unsharp masking · Ziehl-Neelsen method · Color medical images

1 Introduction

The current study aims to find such reprocessing algorithms for the automated diagnostic system, which would provide a background subtraction without losing the object of analysis and did not require large computational resources. One of the tasks to achieve the goal is to choose the best parameters in the filtration of images of microscopy of tuberculosis patients without loss and analyze it to reduce noise.

Light microscopy images of the results of sputum analysis serve as research materials. The microscopy procedure used a ToupCam UCMOS01300KPA digital camera with a resolution of 0.3 MP at a magnification of 10x60.

© The Author(s), under exclusive license to Springer Nature Switzerland AG 2021
R. Silhavy (Ed.): CSOC 2021, LNNS 229, pp. 262–267, 2021.
https://doi.org/10.1007/978-3-030-77445-5_23

Figure 1 provides an example of an image of a sputum analysis painted by the Ziehl-Neelsen method, where you can observe the typical problem of color similarity between the object (mycobacterium tuberculosis) and the background, as well as the presence of a large number of high-intensity blue artifacts. The figure shows the three Mycobacterium tuberculosis in the image.

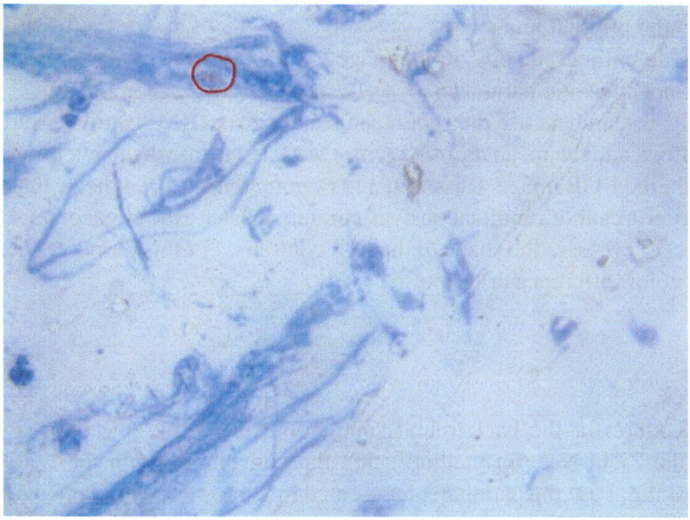

Fig. 1. ZN (Ziehl-Neelsen) stained image of sputum smear samples

2 Materials and Methods

Preliminary studies related to the task of segmentation without loss have shown that the laplacian convolution operation improves the sharpness of the image, and the use of Laplacian of Gaussian (LOG) filter with a sigma option equal to 3.05 identifies all the ROI of interest [1–4].

LOG filter (1) usually is used to find boundaries in a grayscale image, but some studies apply this method to color images, dividing its channels according to the chosen color model (RBG, CIELAB, etc.) [5–7]. If we look at this method from the point of view of the RGB color model, then applying the LOG filter to individual channels, and then summing them, we get a reinforcement of the standard line of colors and the background subtraction.

$$\nabla^2 f(x,y) = \frac{x^2 + y^2 - 2\sigma^2}{\sigma^4} \cdot e^{-\frac{x^2+y^2}{\sigma^2}} \tag{1}$$

x, y – coordinates,

σ – parameter of the Gaussian function.

This procedure means subtracting the defocused copy of the image from the originals image. Gonzales and Woods [8] point out that a generalization of unsharp masking is high-frequency filtering. The filtering procedure using high frequencies and using the LOG filter is given by formula (2).

$$f'(x,y) = \begin{cases} Af(x,y) - \nabla^2 f(x,y), \text{если } w(0,0) < 0 \\ Af(x,y) + \nabla^2 f(x,y), \text{если } w(0,0) > 0 \end{cases} \quad (2)$$

$f'(x, y)$ – the result of high-frequency filtering,
$f(x, y)$ – the original image,
$\overline{f(x, y)}$ – an unfocused copy of the image,
A – non-negative real parameter.

Typically, researchers use the root mean square error (RMSE) to measure the noise reduction effect, and the mean absolute error (MAE) to measure the retention of contour shape in a filtered RGB image. Russo [9] and Ponomarenko [10] indicate that the RMSE and MAE criteria cannot simultaneously accurately measure the effect of noise reduction and the effect of preserve the shape of the ROI contours. For this purpose, I use the NCD normalized color difference criterion.

3 Results

For the procedure for the background subtraction of microscopic images of sputum stained by the Ziehl-Neelsen method using the convolution with the LOG filter and unsharp masking, I use the parameters presented in Table 1. The fundamental parameter is σ, the dimension with the grid step are the parameters for discretization of the LOG filter, A - real parameter for filtering high frequencies. It is optimally selected based on the NCD filtering quality criterion.

Table 1. Parameters for the background subtraction of microscopic images of sputum stained by the Ziehl-Neelsen method

Parameter	Value
σ	3,05
dimension	9×9
grid step	2.4
A	$NCD < NCD_{min}$

Applying the high-frequency effect to the found LOG filter mask, one can make sure that the filtering quality improves - the amount of noise decreases, and mycobacterium tuberculosis remains in place (Fig. 2).

Figure 3a shows the filtering of the image with the parameter value A = 4 (formula 5). In this case, the shape of the contour practically did not change, and the number of segments is still 3. Figure 4b shows the filtration of the original image with the parameter A = 6.

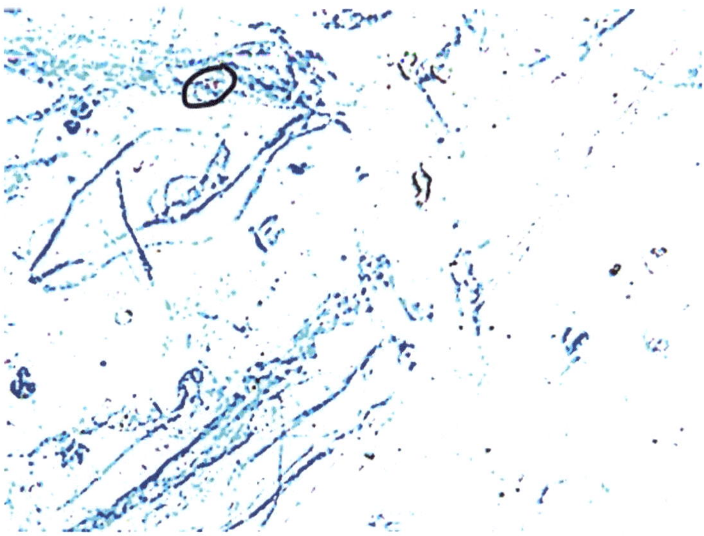

Fig. 2. ZN (Ziehl-Neelsen) stained image of sputum smear samples after applying the LOG filter at $\sigma = 3.05$

In this case, there are significantly fewer background objects, but the shape of the mycobacterium contour has changed crucially. At the same time, the filtration quality was evaluated both by the expert method and using the normalized color difference criterion NCD. It is necessary to select numerical filtering quality criteria to minimize the NCD criterion with some error allowed for changing the shape of the object to automate this process.

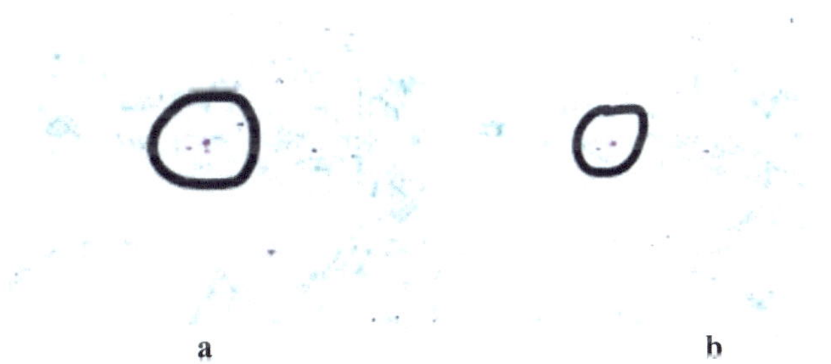

Fig. 3. ZN (Ziehl-Neelsen) stained image of sputum smear samples after applying the LOG filter and the high-frequency filtering: a - parameter A = 4, b - parameter A = 6

The developed method of background subtraction can be used not only for images of sputum analyzes of patients stained by the Ziehl-Neelsen method but also for images of microscopy and dermatology that have the same background (Fig. 4). You can also find the association between the parameters of method, the size of the desired objects, and the background color of the image.

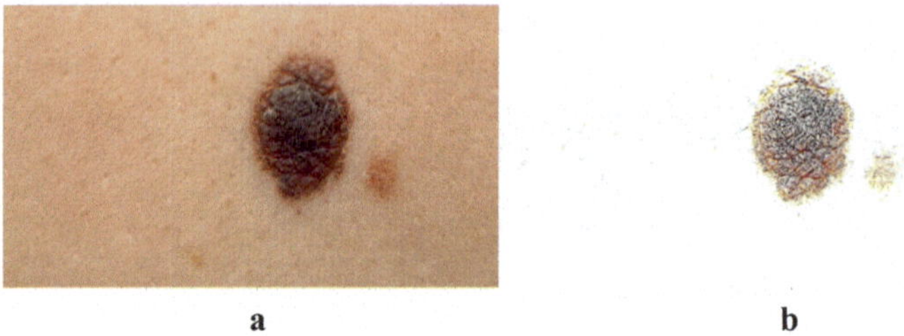

Fig. 4. Image of melanoma sample before (a) and after applying the LOG filter with the high-frequency filtering parameter A = 5 and $\sigma = 3.05$ (b)

4 Conclusions

The Laplacian of Gaussian operator can be applied to color images for background subtraction. At the same time, it is optimal to use the convolution of the original image with a discrete kernel of the LOG filter - such an operation is computationally simple, with good performance, and the popular computer vision libraries include it.

To discretize the Laplacian of Gaussian filter, you can use such parameters in formulas: the value of the sigma, the dimension of the filter, and the shape with or without diagonal elements. To improve the LOG filtering efficiency (increase sharpness, reduce the number of artifacts), you can use the effect of high frequencies.

In the case of using the Laplacian of Gaussian operator for filtering microscopy images of sputum analyzes stained by the Ziehl-Neelsen method, I recommend using a 9x9 convolution kernel with high frequency and diagonal components.

It is best to use the normalized color difference (NCD) criterion with a given shape change error as a filter quality criterion for simultaneously tracking the noise reduction effect and preserving the shape of contours.

References

1. Narkevich, A., Shelomentseva, I., Vinogradov, K., Sysoev, S.: Comparison of segmentation methods for digital microscopic images of sputum stained by the Ziehl-Neelsen method. Eng. J. Don **4**, 1–11 (2017)
2. Narkevich, A., Vinogradov, K., Koretskaya, N., Soboleva, V.: Segmentation of microscopic images of sputum stained by the Ziehl-Nielsen method using the Mexican Hut wavelet transform. Acta Biomed. Sci. **2**(5), 141–146 (2017)
3. Shelomentseva, I.: Results of filtration and segmentation of the images of analysis of sputum stained by Ziehl-Neelsen method. Int. J. Adv. Stud. **7**(4–2), 110–114 (2017)
4. Shelomentseva, I.: Parameters of filtering by LOG filter of microscopic images of sputum stained by Ziehl –Neelsen method. Econ. Inf. Technol. **47**(2), 362–371 (2020)
5. Millan, M., Valencia, E.: Laplacian filter based on color difference for image enhancement. In: Marcano, A., Paz, J. Proceedings SPIE. 5622, 5th Iberoamerican Meeting on Optics and 8th Latin American Meeting on Optics, Lasers, and Their Applications, pp. 1259–1264. SPIE, USA (2004)
6. Bhairannawar, S.S., Patil, A.N., Janmane, A.S., Huilgol, M.V.: Color image enhancement using Laplacian filter and contrast limited adaptive histogram equalization. Innovations in Power and Advanced Computing Technologies (i-PACT), pp. 1–5. IEEE, Vellore (2017)

7. Dey, N., Ashour, A., Shi, F., Balas, V.: Soft Computing Based Medical Image Analysis. Academic Press (2018)
8. Gonzales, R.C., Woods, R.E.: Digital Image Processing. 2nd edn. Pearson Education. (2002)
9. Russo, F.: Performance Evaluation of Noise Reduction Filters for Color Images through Normalized Color Difference (NCD) Decomposition. IRSN Machine Vision, 1–11 (2014)
10. Ponomarenko, N., Battisti, F., Egiazarian, K., Astola, J., Lukin, V.: Metrics performance comparison for color image data-base. In: Proceedings of the 4th International Workshop on Video Processing and Quality Metrics for Consumer Electronics, pp. 1–6. Scotts-dale, Arizona (2009)

Classification of Plant Diseases Using Convolutional Neural Networks

Ammar Alhaj Ali(✉), Bronislav Chramcov, Roman Jasek, Rasin Katta(✉), and Said Krayem

Faculty of Applied Informatics, Tomas Bata University in Zlin, Zlin, Czech Republic
{alhaj_ali,chramcov,jasek,Katta,krayem}@utb.cz

Abstract. 2020 was declared as the International Year of Plant Health, plant disease is a nightmare of any farmer, as it threatens their business and food security. The wide deployment and penetration of smartphones accompanied by computer vision models development created an economical and easy opportunity for using image classification in agriculture. Convolutional Neural Networks (CNNs) is the cutting edge in image recognition by providing a prompt and definite diagnosis.

In this paper, we will use some of the pre-trained models to detect selected common diseases of the cassava plant, as it is considered the major source of calories and carbs for people in developing countries.

A dataset containing 21397 images is used for model training and validation. Results show that the proposed method can achieve a high accuracy level. This demonstrates the technical feasibility of CNNs in identifying plant diseases and presents a perfect option for AI solutions for smallholder farmers.

Keywords: Deep learning · CNN · Classification · Transfer learning · Plant disease

1 Introduction

Plant health and food safety are two sides of a coin. The Food and Agriculture Organization of the United Nations (FAO) estimates that pests and diseases is responsible for the loss of 40% of global food production, creating a threat to food security [1]. In 2019, more than 20% of all global crop productions were lost due to pests and other pathogens [2].

In December 2018, The United Nations General Assembly adopted a Resolution declaring 2020 as the International Year of Plant Health (IYPH). It is considered as an opportunity to raise global awareness on how protecting plant health can help end hunger, reduce poverty, protect the environment, and boost economic development [3].

The cassava plant is the second-largest provider of carbohydrates in Africa and it is widely planted by poor smallholder farmers, this raised the importance of strengthening monitoring and early warning systems in order to protect plants and plant health using economical, fast and easy to use tool as the traditional method of human analysis by visual inspection is no longer feasible for crop diseases classification.

© The Author(s), under exclusive license to Springer Nature Switzerland AG 2021
R. Silhavy (Ed.): CSOC 2021, LNNS 229, pp. 268–275, 2021.
https://doi.org/10.1007/978-3-030-77445-5_24

The successful classification of plant diseases is crucial to improve agricultural production and reduce the undesirable use of chemical treatment such as fungicide/herbicide. Therefore, it is an emerging research topic to advance agricultural automation [4].

The development of computer vision models offers a quick, standardized, and accurate solution to this issue. A trained classifier can be easily utilized as an easy-to-use application. All what is required is a camera-equipped smartphone and an internet connection.

Therefore, such kind of research will improve agricultural automation and align to the sustainable development strategies of united nations goal to feed the growing global population by 2050 [5].

During the last decade Convolutional Neural Network noticeably proved its excellence in many pattern recognition topics, from image processing to voice recognition. This achievement has triggered both researchers and developers to move towards larger models aiming to solve complex tasks, as this was not possible with classic Artificial Neural Networks ANNs [6, 7].

Recently Several Convolutional neural network models have been proposed for plant disease classification only few reached admirable results, some of state-of-the-art architectures which results were promising are AlexNet, VGG, DenseNet, Inception-v4, and ResNet.

The rest of the paper is organized as follows: Sect. 2 we presented the concept of Transfer Learning, in Sect. 3 we presented the materials and methods used like datasets, software and hardware, the deep learning architectures and the training specifications, in Sect. 4 training results were demonstrated, finally we closed our research in the conclusion Sect. 5 and we listed the references in Sect. 6.

2 Transfer Learning

The status quo is encouraging the concept of do not reinvent the wheel, for this only few people would train an entire Convolutional Network from scratch (with random initialization), while having a dataset of sufficient size is seldom to happen. As an alternative, most common approach is to use pre-trained models that are already existing and are already trained on large benchmark datasets. One of the most common datasets is ImageNet, which contains 1.2 million images with 1000 categories. The concept of pre-trained models reuse is called transfer learning. Transfer learning helps to build accurate models in an efficient way (Rawat & Wang 2017) [8].

Having these open-source models was of great benefit for the Deep Learning community, the pre-trained models allowed to accelerate the computer vision research. After model selection the Convolutional Network is used either as an initialization or as a fixed feature extractor for the targeted task [9, 10].

3 Materials and Methods

3.1 Dataset

All the models were trained on a publicly available dataset which contains 21,367 labeled Cassava plant leaves images, the images size is 512 × 512 × 3. These images were collected during a regular survey in Uganda, most images were crowd sourced from farmers taking photos of their plants and marked by experts at the National Crops Resources Research Institute (NaCRRI) in collaboration with the AI lab at Makerere University, Kampala [11].

In Fig. 1 we can see images containing 5 different healthy/diseased leaves related to the cassava plant:

- Cassava Bacterial Blight
- Cassava Brown Streak Disease
- Cassava Green Mottle
- Cassava Mosaic Disease
- Healthy

Fig. 1. The plant diseases for cassava plant.

While Fig. 2 shows the images class distribution graph.

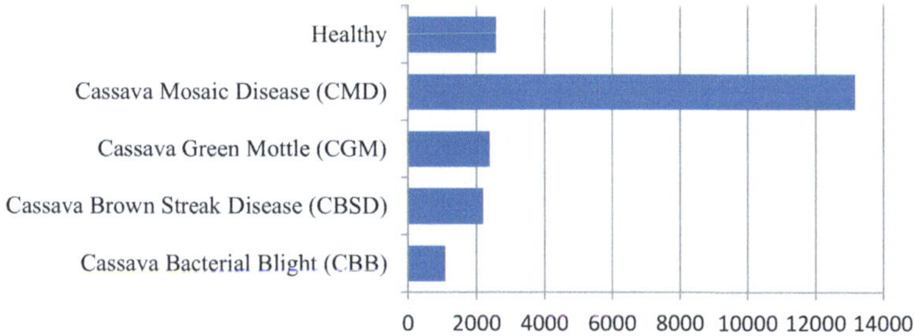

Fig. 2. The number of pictures in each class.

3.2 Software and Hardware Specifications

The deep learning architectures were programmed in Python language due to the fact that Python offers great flexibility and provides very useful libraries and deep learning frameworks. We used PyTorch to build the architectures as recently PyTorch is considered a strong player in the field of deep learning and artificial intelligence. In PyTorch, we use tensors to encode the inputs, outputs, and the model's parameters, after building tensors, the tensors can run on GPUs or other specialized hardware to accelerate the computing process [12].

All the experiments were executed on a Graphical Processing Unit GPU NVIDIA TESLA P100 having the Specifications: 16 GB memory, 3584 CUDA cores, 1190 MHz core clock, and with a memory bandwidth of 732.2 GB/sec [13].

3.3 Deep Learning Architectures

For this study we have taken twelve recognized convolutional neural networks (CNNs) architectures for the identification and classification of diseases in the cassava plant leaf.

We applied transfer learning concept by using existing CNN models in this study to speed up learning, the pre-trained models on ImageNet dataset helps to shorten the training time of the deep learning models [14, 15].

In this paper, we will cover the twelve pre-trained models for Image Classification which are:

- GoogleNet,
- Resnet18,
- Resnext50_32x4d,
- And wide_resnet50_2.
- Efficient Net (B0 – B7).

3.4 Training Specifications

All the models were trained on the cassava plant dataset, and the Adaptive moment estimation method (Adam) was used to train all the models. Adam optimizer has combined advantages of two extended versions of the Stochastic Gradient Descent (SGD) method that are Adagrad and RMSProp [16]. In Addition, we used image data augmentation that artificially expands the size of a training dataset through creating modified versions of images in the dataset.

Table 1 show models names and number of parameters for each model which we used in the training.

Table 1. Training specifications

#	Model name	Batch size	Parameters	Epochs	Learning rate
1	GoogleNet	16	5,605,029	25	0.0001
2	Resnet18	16	11,179,077	25	0.0001
3	Resnext50_32x4d	16	22,990,149	25	0.0001
4	wide_resnet50_2	16	66,844,485	25	0.0001
5	EfficientNet B0	16	4,013,953	25	0.0001
6	EfficientNet B1	16	6,519,589	25	0.0001
7	EfficientNet B2	16	7,708,039	25	0.0001
8	EfficientNet B3	16	10,703,917	25	0.0001
9	EfficientNet B4	16	17557581	25	0.0001
10	EfficientNet B5	16	28351029	25	0.0001
11	EfficientNet B6	8	40,747,229	15	0.0001
12	EfficientNet B7	4	63,799,765	10	0.0001

In both models EfficientNet B6 and EfficientNet B7 we were forced to use a smaller batch size due to limited RAM size in the used hardware.

4 Results and Discussion

All the results were evaluated in terms of training, validation accuracy/loss, and F1-score, Table 2 shows the performance of our models.

In general, we got acceptable results, yet more effort is required to improve the performance of the models by applying some techniques such as parameters tuning.

Furthermore, we will try to improve performance by modifying the architecture of models either by adding new layers or by combining some models together like (Resnext50_32×4d and EfficientNetB4).

Figures 3, 4, 5 and 6 show the Accuracy/loss curves for the models GoogleNet, Resnet18, EfficientNet B6, and EfficientNet B7 respectively.

Table 2. Models Performance Results

Model name	Training time (hours)	Training accuracy	Training loss	Val accuracy	Val loss	F1-score
GoogleNet	3.11	0.8823	0.3424	0.8848	0.3587	0.7911
Resnet18	2.96	0.8859	0.3276	0.8853	0.3685	0.7949
Resnext50_32x4d	5.19	0.91424	0.25212	0.889	0.38429	0.7951
wide_resnet50_2	8.2	0.90027	0.29066	0.885	0.37007	0.7944
EfficientNet B0	3.56	0.90933	0.25298	0.866	0.42043	0.7687
EfficientNet B1	3.93	0.92055	0.22688	0.878	0.43331	0.7748
EfficientNet B2	4.18	0.92703	0.20839	0.877	0.44765	0.7766
EfficientNet B3	4.72	0.93837	0.18177	0.879	0.46938	0.7811
EfficientNet B4	5.961	0.95186	0.14492	0.882	0.44889	0.7855
EfficientNet B5	7.86	0.95969	0.11677	0.877	0.53181	0.7773
EfficientNet B6	6.87	0.90512	0.27746	0.889	0.386	0.7974
EfficientNet B7	7.74	0.90109	0.28709	0.886	0.3408	0.7961

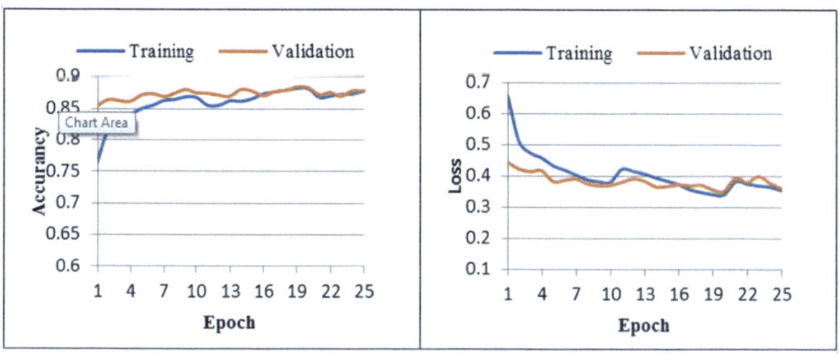

Fig. 3. Accuracy/loss curves of GoogleNet model

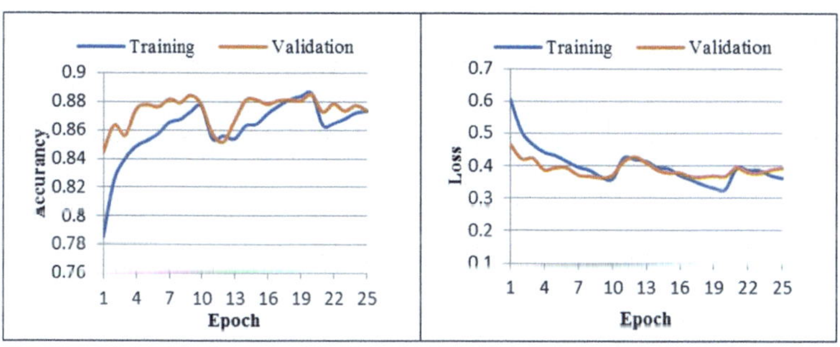

Fig. 4. Accuracy/loss curves of Resnet18 model

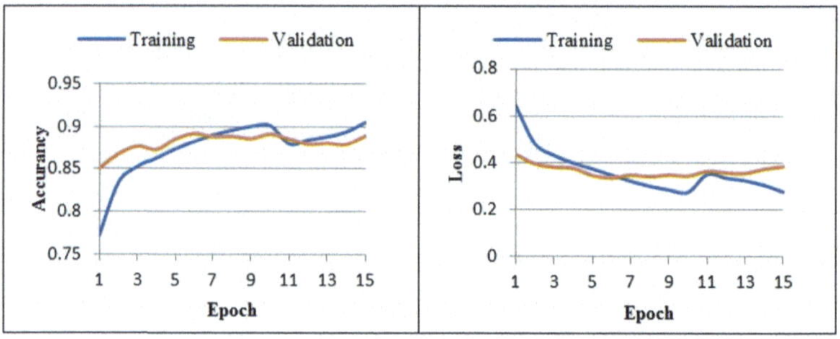

Fig. 5. Accuracy/loss curves of EfficientNet B6 model

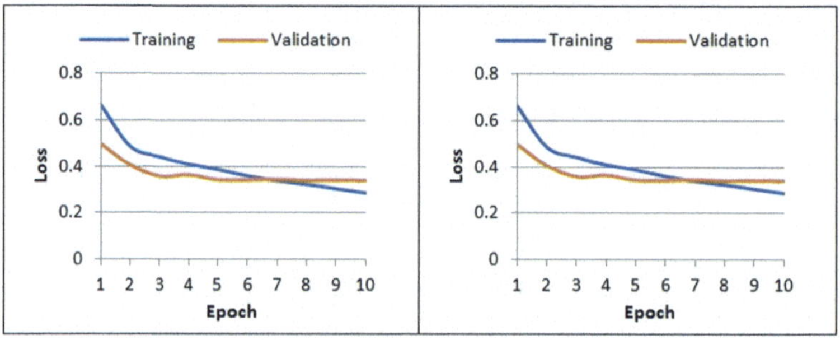

Fig. 6. Accuracy/loss curves of EfficientNet B7 model

5 Conclusions

Smallholder farmers need a fast, accurate and cheap solution in order to early discover Cassava plant leaf diseases. In this study, we used a pre-trained Convolutional Neural Network; our goal is to reach a legitimate level of accuracy in order to have the model ready for deployment as a free to use application.

In this paper, we applied various CNN-based image classification transfer learning models to the Plant Pathology Classification task. We discovered that the results vary in respect to used models, some models outperform other models significantly, however, in future research we will focus more on model tuning techniques taking in consideration best performing models from our current study.

In our study we relied on in-field photos taken using nonprofessional cameras, this has its impact on the quality and type of training data and the model's abilities as this makes the classifier usable for real case scenarios.

CNN's are still in its early development phase. Fantastic opportunities are still out there to be discovered by utilization of current algorithms/architectures and further optimization and tuning for current used methods to solve more complex problems.

The small-holder farmers desperately need such kind of research in their fight against plant disease, the CNN promises to open the black box of deep learning and offer legitimate solutions for plant researchers which in their turn need to select the best models that fit with their research and interpret the outcoming results to contribute to a better sustainable and secure food production.

References

1. Boulent, J., et al.: Convolutional neural networks for the automatic identification of plant diseases. Front. Plant Sci. **10**, 941 (2019)
2. Keh, S.S.: Semi-Supervised Noisy Student Pre-training on EfficientNet Architectures for Plant Pathology Classification.: arXiv preprint arXiv:2012.00332 (2020)
3. FAO International year of plant health 2020. http://www.fao.org/plant-health-2020/about/en/. Accessed 16 Jan 2021
4. Saleem, M.H., Potgieter, J., Arif, K.M.: Plant disease classification: a comparative evaluation of convolutional neural networks and deep learning optimizers. Plants **9**(10), 1319 (2020)
5. United Nation Resolution A/RES/73/252. https://www.un.org/en/ga/search/view_doc.asp?symbol=A/RES/73/252. Accessed 16 Jan 2021
6. Albawi, S., Mohammed, T.A., Al-Zawi, S.: Understanding of a convolutional neural network. In: 2017 International Conference on Engineering and Technology (ICET). IEEE, pp. 1–6 (2017)
7. Toda, Y., Okura, F.: How convolutional neural networks diagnose plant disease. Plant Phenomics **2019**, 9237136 (2019)
8. Rawat, W., Wang, Z.: Deep convolutional neural networks for image classification: a comprehensive review. Neural Comput. **29**(9), 2352–2449 (2017)
9. Shin, H.-C., et al.: Deep convolutional neural networks for computer-aided detection: CNN architectures, dataset characteristics and transfer learning. IEEE Trans. Med. Imaging **35**(5), 1285–1298 (2016)
10. Convolutional Neural Networks for visual recognition, Transfer Learning. https://cs231n.github.io/transfer-learning/. Accessed 16 Jan 2021
11. Kaggle, Cassava Leaf Disease Classification. https://www.kaggle.com/c/cassava-leaf-disease-classification/data. Accessed 16 Jan 2021
12. PyTorch, Tensors. https://pytorch.org/tutorials/beginner/blitz/tensor_tutorial.html. Accessed 16 Jan 2021
13. NVIDIA® TESLA® P100 GPU Accelerator. https://images.nvidia.com/content/tesla/pdf/nvidia-tesla-p100-PCIe-datasheet.pdf. Accessed 16 Jan 2021
14. Ahmad, I., Hamid, M., Yousaf, S., Shah, S.T., Ahmad, M.O.: Optimizing pretrained convolutional neural networks for tomato leaf disease detection. Complexity **2020** (2020)
15. Atila, Ü., Uçar, M., Akyol, K., Uçar, E.: Plant leaf disease classification using EfficientNet deep learning model. Ecol. Inf. **61**, 101182 (2021)
16. Kingma, D.P., Ba, J.: Adam: A method for stochastic optimization. arXiv 2014, arXiv:1412.6980

Pairwise Separability in the Problem of Identifying Objects in Photographs Using the Example of Cultural and Historical Architectural Buildings

Vladimir Klyachin(✉) and Aleksei Klyachin

Volgograd State University, University Avenue, 100, Volgograd 400062, Russia
{klyachin.va,aleksey.klyachin}@volsu.ru

Abstract. This article examines the problem of object recognition in photographs. Objects of historical or cultural significance were selected as objects. In the work, the research was limited to two objects. It is assumed that the pictures can be taken at different times of the day and from different angles. Based on this, for one approach to solving the problem, it is proposed to extract the contours on the image of the object and calculate its characteristics. The moments of the curves of the contours up to the third order inclusive, the central moments, the normalized moments, and also the invariant moments of Hu are used as numerical values. In addition to this, the integral of the total curvature of the contour curve is calculated, the geometric meaning of which is that this integral is equal to the total variation of the slope of the tangent line of curve. The second approach is based on splitting the original image into rectangular cells and calculating the same moments as for the contours, but for the brightness function. Finally, in the third approach, the original image is replaced by the image of the outlines themselves. The article obtained results for several machine learning models, including convolutional neural networks, the method of nearest neighbors, and also the gradient boosting method is used to improve the results. #CSOC1120.

Keywords: Identifying objects on image · Machine learning model · Contour · Total curvature · Moments

1 Preliminaries

1.1 Introduction

In connection with the widespread use of artificial intelligence systems in practice, the tasks of constructing and implementing computer vision systems are becoming relevant. One of the tasks of computer vision is the task of identifying objects in images. As an example of use, we can cite the development of mobile applications with augmented reality, in which it is necessary to identify an object in an image captured from a video stream of a device camera. Object identification in such applications results in additional information about the object or a three-dimensional model of the object. Applications

© The Author(s), under exclusive license to Springer Nature Switzerland AG 2021
R. Silhavy (Ed.): CSOC 2021, LNNS 229, pp. 276–287, 2021.
https://doi.org/10.1007/978-3-030-77445-5_25

that provide the user with information about historically significant objects or objects of cultural heritage are especially relevant. The tasks of identifying objects in the image are also of great importance in autonomous driving systems, where it is required to determine the type of this transport type in order to calculate the spatial orientation of vehicles on the roadway. It is also possible to present the tasks of 3D reconstruction of models of architectural structures from photographs or images of sketches. In these tasks, the actual is the definition and identification of architectural elements on the facades of buildings, for their subsequent inclusion in a three-dimensional model.

The problem of identifying an object by its image can be reduced in some cases to a classification problem. Of course, this can be done in cases where many classes are known a priori. In such a case, various machine learning methods can be applied. For example, in [1], a training scheme is proposed on a set of images, based on the functional representation of the image of an object and all its images as a conical set of Euclidean space. Note that a more general approach for morphological analysis of images using oblique projections in image spaces was developed by Yu.P. Pyt'yev [2–4]. Also, we note that the morphological methods based on k-means approach have been used for fish detection on image in [5]. In [6] an optimal architecture that accurately classifies low-quality images using DCNNs architectures was investigated. Three component Integrity Cognition Network was designed in [7] for solution the salient detection object problem. In papers [8]–[12], various methods of improving the identification of objects in images in various practical problems are proposed.

In this paper, we investigate the characteristics extracted from the image in order to apply these characteristics in the problem of identifying objects in the image. The extraction of these characteristics is based on the calculation of the contours of objects in the image, as well as simple and invariant moments of various parts of the images.

1.2 Mathematical Basics

The basic idea for extracting image characteristics is based on the following assumptions. First, the object must be identified regardless of the lighting and time of day in the photograph. Under the same conditions, the identification of an object by its image is considered in [1]. Secondly, the object is required to identify the object in the photograph taken from different angles - different viewing angle and distance to the object. Based on these conditions, when analyzing images, the following actions are performed:

- images are reduced to a single (sample-averaged) size,
- images are multiplied using both affine and perspective transformations of the original images,
- images are converted to grayscale.

After this preprocessing of the images, characteristics are calculated. Let's consider this issue in more detail.

Extraction of Contours. In most cases, an object can be visually easily identified by its outline. Moreover, the contour line of the object does not depend on its illumination.

Therefore, it was decided to use contours as one of the initial data for extracting characteristics. The contours are extracted on the image according to the standard scheme: first, the Canny filter is applied to the grayscale image, then binarization is applied with a suitable threshold value. Further, the contours are calculated using a scanning algorithm (Fig. 1).

a) b)

Fig. 1. The original image (a) and its numbered outlines (b).

The Canny filter uses the $G = (G_x, G_y)$ gradient of the image brightness, calculating the horizontal G_x and vertical G_y using Sobel filters. The set of selected contours is truncated in accordance with the inequalities

$$L(C) > L_0, \quad S(C) > S_0.$$

Here $L(C)$ and $S(C)$ denote the length of the contour C and the area of the region that it limits. The values of the thresholds L_0, S_0 are determined experimentally. After filtering the contours, their set is sorted in descending order of $L(C)$, and then in ascending order of the ordinate of the geometric center of the contour. The geometric center (x_0, y_0) of the contour C is calculated by the formulas

$$x_0 = \sum x/N, \; y_0 = \sum y/N,$$

where the sum is over all pixels of the contour, and N is the number of these pixels. The final contours are then used to extract image characteristics.

Moments. Let $f(x,y)$ denote the brightness of the image I pixel (x, y). The moments M_{ij} of order $p = i + j$ of the subset $A \cap I$ of the image are determined by the formulas

$$M_{ij} = \sum f(x, y) x^i y^j,$$

where the summation is over all pixels of the set A. Denoting by

$$x_0 = M_{10}/M_{00}, \; y_0 = M_{01}/M_{00}$$

the coordinates of the center of gravity, the central moments $M^c{}_{ij}$ are determined by the expressions

$$M_{ij}^c = \sum f(x, y)\,(x - x_0)^i\,(y - y_0)^j.$$

It is easy to see that the central moments are invariant with respect to the parallel translation of the set *A*. The normalized moments are determined in terms of the central moments by the formula

$$\mu_{ij} = M_{ij}^c/(M_{00})^{(i+j)/2\,+1}.$$

The normalized moments, as it is easy to see, are invariant under scale transformations. In our work, we additionally used seven Hue moments $I_1, I_2, \ldots I_7$ [13], which are expressed in terms of normalized moments and are invariant under transformations of displacement, scaling and rotation:

$$\begin{gathered}I_1 = \mu_{20} + \mu_{02},\ I_2 = (\mu_{20} - \mu_{02})^2 + 4\mu_{20}^2,\ I_3 = \alpha^2 + \beta^2,\ I_4 = \gamma^2 + \delta^2,\\ I_5 = \alpha\gamma\left(\gamma^2 - 3\delta^2\right) + \beta\delta\left(3\gamma^2 - \delta^2\right),\\ I_6 = (\mu_{20} - \mu_{02})\left(\gamma^2 - \delta^2\right) + 4\,\mu_{11}\,\gamma\delta,\\ I_7 = \beta\gamma\left(\gamma^2 - 3\delta^2\right) - \alpha\delta\left(3\gamma^2 - \delta^2\right),\end{gathered}$$

where $\alpha = \mu_{30} - 3\mu_{12}$, $\beta = 3\mu_{21} - \mu_{03}$, $\gamma = \mu_{30} + \mu_{12}$, $\delta = \mu_{21} + \mu_{03}$. In computer vision systems, Hue moments are used to identify geometric shapes of contours in images. Note, that the weighted moments are used for identification geometric shapes on images [14].

Total Curvature. The total curvature of the curve *C* in the plane is given by the integral.

$$k(C) = \int |k(s)|\,ds,$$

where the integration is along the curve *C*, and s is its natural parameter. It is well known that the value of $k(C)$ is equal to the total variation of the angle of inclination of the tangent to the curve. In particular, for convex closed curves $k(C) = 2\pi$. To approximately calculate this value for the contours in the images, we mean that each contour is an ordered set of points $p_1, p_2, \ldots, p_n$, so that their order corresponds to the location of points on the contour. Therefore

$$k(C) \approx \sum \alpha_i,$$

where α_i is the positive value of the angle between the segments p_{i-k}, p_i, and $p_i p_{i+k}$.

1.3 Machine Learning Models

The study of the possibility of identifying objects in images was carried out using the following machine learning models:

- deep convolutional neural networks,
- nearest neighbors method,

- naive Bayesian classifier,
- decision tree method,
- support vector machine.

When modeling convolutional neural networks, we used their parameters such as the number of feature filters in the convolutional layers, the number of convolutional layers, the size of the filters, the number of neural dense layer that is located in front of the output layer. The general architecture of the neural network is shown in Fig. 2.

In addition to the machine learning models listed above, this paper considers complex models in the form of a Gradient Boosting Tree Model, and attacks using the XGBoost gradient boosting implementation.

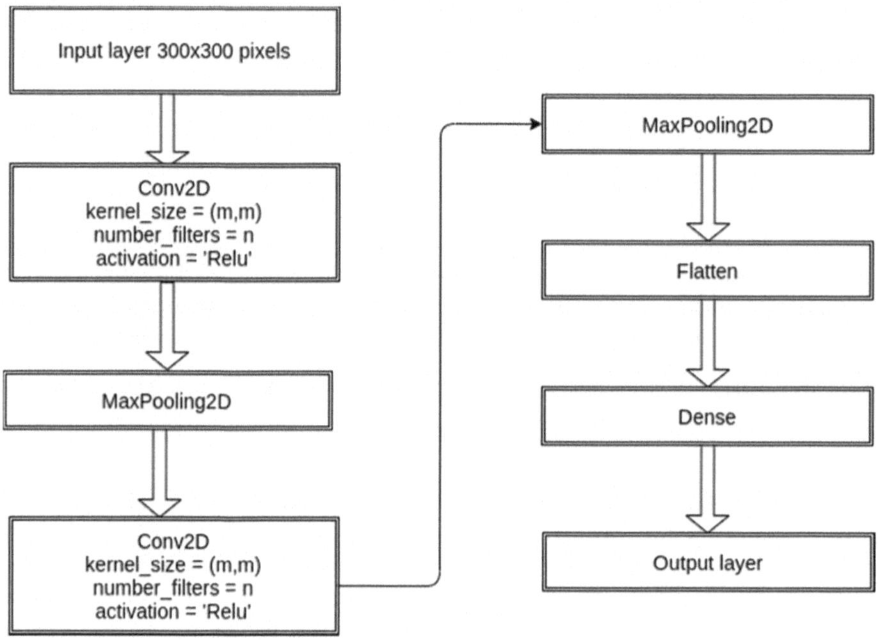

Fig. 2. Common architecture of a neural network.

1.4 Dataset

The tb2019 dataset containing photographs of 12 landmarks, approximately 200 photographs per object, was taken as objects for identification in the images. Examples of images are shown in Fig. 3.

To study the capabilities of machine learning models for solving the identification problem, two objects were selected, marked in the dataset with the labels 'castell_cartoixa' and 'estacio_nord'. These objects are shown in Fig. 4.

Information about the object shown in the image is recorded in a CSV file, in which each line contains two values: the name of the image file and the object's label.

Fig. 3. Examples of photo images from the tb2019 dataset

a) b)

Fig. 4. Objects from the tb2019 dataset: 'castell_cartoixa' (a) and 'estacio_nord' (b)

As mentioned above, images from the dataset are multiplied by performing rotation transformations around the center of the image, a perspective transformation, which is a fractional linear display, which is a composition of spatial rotation around the vertical axis and a central projection. In this case, for each image, 4 new ones are generated for each type of rotation with a random angle value in the range from -π/8 to π/8. As a result, a set of about 3200 images was obtained to identify the two selected objects. In addition, the tb2019 dataset has a tagged set of images for testing.

The resulting images are the source for data directly used for machine learning models. We used the following schemes for vectorizing these sets.

Matrix Data A matrix is assigned to each image, the dimensions of which coincide with the size of the image, and the values correspond to brightness levels in the range from 0 to 1. In this case, the image consists only of selected and filtered outlines. Such arrays are packed into one three-dimensional array, which is saved into one file in the npz format - the numpy library format for storing multidimensional arrays.

Vector of Contour Moments and Total Curvature. For each selected contour in the image, a vector of 31 moments is calculated. This vector includes 10 moments of the 0th, 1st, 2nd, 3rd orders, 7 central moments of the 2nd and 3rd orders, 7 normalized moments and 7 Hue moments. In addition we calculate total curvature for every contour on image. The complete vector of moments is packed into a two-dimensional numpy array and saved to an npz file.

Vector of Moments of Rectangular Cells. Each image is divided into rectangular cells, and the same moments are calculated for each cell as for the contours. The two-dimensional array saved to the nzp file is constructed in a similar way. In our experiments, a 300×300 image is split into 144 rectangular cells of 25×25 pixels.

2 Results

To present the research results, we introduce the following abbreviations. Multiplication (augmentation) of images will be performed as indicated above in two versions: using perspective transformations (Augmentation with Perspective Transformation APT) and using only rotation around the center of the image (Augmentation with Image Rotation – ARI). We will also use abbreviations for the used types of vectorization of a set of images, which are described above. Namely, for matrix data – MD (Matrix Data), for the vector of contour moments ant total curvature – CM (Contour Moments) and for the vector of moments of rectangular image cells – RCM (Rectangle Cell Moments).

2.1 Results for Convolution Neural Network

Neural networks were trained using Python libraries Tensorflow and Keras (Table 1).

Table 1. Experimental results using a convolutional neural network

Structure of convolutional layers	Number of convolutional layers	The number of neurons in the dense layer	Accuracy
50, kernel (3,3) 50, kernel (3,3) 32, kernel (3,3)	3	1024	78%
100, kernel (3,3) 50, kernel (3,3) 32, kernel (3,3)	3	1024	83%
100, kernel (7,7) 50, kernel (3,3) 32, kernel (3,3)	3	1024	72%
50, kernel (7,7) 100, kernel (3,3) 32, kernel (3,3)	3	1024	78%

(*continued*)

Table 1. (*continued*)

Structure of convolutional layers	Number of convolutional layers	The number of neurons in the dense layer	Accuracy
50, kernel (7,7) 50, kernel (3,3) 32, kernel (3,3) 32, kernel (3,3)	4	1024	75%
50, kernel (7,7) 50, kernel (3,3) 32, kernel (3,3) 12, kernel (3,3)	4	1024	85%
50, kernel (7,7) 100, kernel (3,3) 32, kernel (3,3) 12, kernel (3,3)	4	1024	93%

2.2 Results for Support Vector Machine

The experimental studies were built on the basis of the Python Scikit-learn library (Table 2).

Table 2. Support Vector Machine Experimental Results

Kernel	Vector data representation type	Augmentation type	Accuracy
rbf	CM	ARI	77%
linear	CM	ARI	75%
poly	CM	ARI	78%
rbf	CM	APT	77%
linear	CM	APT	72%
poly	CM	APT	75%
rbf	RCM	ARI	71%
linear	RCM	ARI	63%
poly	RCM	ARI	68%
rbf	RCM	APT	72%
linear	RCM	APT	63%
poly	RCM	APT	72%

2.3 Results for Naive Bayes Classifier

The naive Bayesian classifier is implemented in Scikit-learn in three flavors: MultinomialNB, BernoulliNB, GaussianNB. We tested our data for BrenoulliNB and GussaianNB (Table 3).

Table 3. Experimental results for a naive Bayesian classifier

Classifier type	Vector data representation type	Augmentation type	Accuracy
BernoulliNB	CM	ARI	60%
BernoulliNB	CM	APT	60%
BernoulliNB	RCM	ARI	83%
BernoulliNB	RCM	APT	81%
GaussianNB	CM	ARI	65%
GaussianNB	CM	APT	71%
GaussianNB	RCM	ARI	62%
GaussianNB	RCM	APT	62%

2.4 Results for k Nearest Neighbors

When using the k nearest neighbors method, we use controlling parameter which is the number of k neighbors (Table 4).

Table 4. Experimental results for the nearest neighbors method

Neighbors number	Vector data representation type	Augmentation type	Accuracy
3	CM	ARI	62%
7	CM	ARI	68%
15	CM	ARI	72%
3	CM	APT	57%
7	CM	APT	68%
15	CM	APT	71%
3	RCM	ARI	80%
7	RCM	ARI	80%
15	RCM	ARI	80%
3	RCM	APT	82%
7	RCM	APT	82%
15	RCM	APT	80%

2.5 Results for Decision Tree Classifier

The parameter max_depth is the maximum depth of the constructed decision tree (Table 5).

Table 5. Experimental results for the decision tree method

max_depth	Vector data representation type	Augmentation type	Accuracy
2	CM	ARI	60%
4	CM	ARI	77%
8	CM	ARI	77%
12	CM	ARI	78%
16	CM	ARI	72%
2	CM	APT	81%
4	CM	APT	74%
8	CM	APT	68%
12	CM	APT	78%
16	CM	APT	74%
2	RCM	ARI	72%
4	RCM	ARI	69%
8	RCM	ARI	63%
12	RCM	ARI	71%
16	RCM	ARI	71%
2	RCM	APT	74%
4	RCM	APT	71%
8	RCM	APT	65%
12	RCM	APT	62%
16	RCM	APT	60%

3 Using Gradient Boosting

Gradient boosting is a machine learning technique based on building a sequential ensemble of models to improve prediction. The basic idea of gradient boosting is to take into account the training errors of the previous model when training the next one. We selected two boosting implementations in the form of the Gradient Boosting Tree Model of the scikit-learn library and a separate implementation in the form of the XGBoost library (Table 6 and 7).

Table 6. Experimental results for the Gradient Boosting Tree Model method

Maximal depth	Vector data representation type	Augmentation type	Accuracy
2	CM	ARI	81%
4	CM	ARI	87%
8	CM	ARI	90%
12	CM	ARI	92%
16	CM	ARI	89%
2	CM	APT	84%
4	CM	APT	92%
8	CM	APT	91%
12	CM	APT	92%
16	CM	APT	84%
2	RCM	ARI	80%
4	RCM	ARI	78%
8	RCM	ARI	71%
12	RCM	ARI	69%
16	RCM	ARI	69%
2	RCM	APT	81%
4	RCM	APT	84%
8	RCM	APT	78%
12	RCM	APT	65%
16	RCM	APT	63%

Table 7. Experimental results for the XGBoost method

Vector data representation type	Augmentation type	Accuracy
CM	ARI	89%
CM	APT	92%
RCM	ARI	71%
RCM	APT	71%

4 Conclusion

The results obtained allow us to draw the following conclusions:

- Some models improve their results if we used perspective transforms instead of rotations during reproduction. It must be said that these transformations emulate spatial rotations around the vertical axis.
- For some models, the advantage is given to calculating the characteristics of individual contours (CM values in the tables), for other models, on the contrary - the best results are obtained for calculating moments on rectangular cells.

Acknowledgments. This work was supported by the Ministry of Education and Science of Russia (the project "Development of Virtual 3D Reconstruction of Historical Objects Technique", scientific theme code 2019–0920, project number in the research management system FZUU-0633–2020-0004).

References

1. Zubyuk, A.V., Fedotov, A.B.: An algorithm for object identification based on hyperplane separation but tolerant to illumination and viewing angle variations. Moscow Univ. Phys. Bull. **70**(6), 479–483 (2015)
2. Pyt'ev, Y.P., Chulichkov, A.I.: Estimating the parameters of images and signals by morphological analysis. Measur. Tech. **59**(6), 584–588 (2016)
3. Pyt'ev, Y.P.: Oblique projectors and realtive forms in image morphology. Comput. Math. Math. Phys. **53**(12), 1916–1937 (2013)
4. Vizilter, Y.V., Pyt'Ev, Y.P., Chulichkov, A.I., Mestetskiy, L.M.: Morphological image analysis for computer vision applications. Int. Syst. Ref. Libr. **73**, 9–58 (2015)
5. Saifullah, S., Suryotomo, A.P., Yuwono, B.: Fish Detection Using Morphological Approach Based-on K-Means Segmentation. *arXiv preprint* arXiv:2101.06352 (2021)
6. Kabir, M.M., Ohi, A.Q., Rahman, M.S., Mridha, M.F.: An Evolution of CNN Object Classifiers on Low-Resolution Images. In: 2020 IEEE 17th International Conference on Smart Communities: Improving Quality of Life Using ICT, IoT and AI (HONET), pp. 209–213 (2020)
7. Zhuge, M., Fan, D.P., Liu, N., Zhang, D., Xu, D., Shao, L.: Salient Object Detection via Integrity Learning. *arXiv preprint* arXiv:2101.07663 (2021)
8. Goupilleau, A., Ceillier, T., Corbineau, M.C.: Active learning for object detection in high-resolution satellite images. *arXiv preprint* arXiv:2101.02480 (2021)
9. Chen, Y., Carlinet, E., Chazalon, J., Mallet, C., Duménieu, B., Perret, J.: Combining Deep Learning and Mathematical Morphology for Historical Map Segmentation. *arXiv preprint* arXiv:2101.02144 (2021)
10. Zhang, T., Ma, W., Wang, G.: Six channel Image Representation for Cross-domain Object Detection. *arXiv preprint* arXiv:2101.00561 (2021)
11. Gale, E.M., Martin, N., Blything, R., Nguyen, A., Bowers, J.S.: Are there any 'object detectors' in the hidden layers of CNNs trained to identify objects or scenes? Vision. Res. **176**, 60–71 (2020)
12. Auliasari, K.: Mira Orisa algorithm for identifying objects in the relief image using watershed segmentation. Sci. J. Inf. **4**(2), 104–114 (2017)
13. Hu, M.K.: Visual pattern recognition by moment invariants. IRE Trans. Info. Theor. **IT-8**, 179–187 (1962)
14. Lazarou, M., Li, B., Stathaki, T.: A novel shape matching descriptor for real-time hand gesture recognition. *arXiv preprint* arXiv:2101.03923 (2021)

Cognitive Logic's Principles

V. P. Savnykh[1,2] and V. Ya Tsvetkov[1,3](✉)

[1] Russian Technological University (RTU MIREA), Vernadsky Prospekt, 78, 119454, Moscow, Russia

[2] Moscow State University of Geodesy and Cartography (MIIGAiK), 4, Gorokhovsky lane, Moscow 105064, Russia

[3] Research and Design Institute of Design Information, Automation and Communication on Railway Transport, 27, bldg 1 Nizhegorodskaya Street, 119454 Moscow, Russia

Abstract. This work purpose is to study and describe the cognitive logic's principles. The article leitmotif is to show the mathematical and cognitive logic generality. The relationship between mathematical logic and cognitive logic is described. Shown, that mathematical logic uses oppositional analysis and oppositional variables. Cognitive logic primarily uses dichotomous analysis and dichotomous variables. The content of cognitive dichotomous logic is revealed. Comparison of oppositional and dichotomous division is given. The basic laws of cognitive logic are described, which are derived from the mathematical logic laws. It is shown that cognitive logic is trinitarian, while mathematical logic is dyadic. The cognitive logic application examples based on precedents and stereotypes are described. #CSOC1120.

Keywords: Artificial intelligence · Logic · Cognitive logic · Expert analysis · Mathematical logic · Oppositional analysis · Dichotomous analysis · Cognitive analysis

1 Introduction

The motive for writing this article, on the one hand, was the need to generalize the cognitive logic concept [1–4]. On the other hand, this article authors have a categorical disagreement with some authors viewpoint [5], opposing cognitive logic and mathematical logic. Cognitive logic development is due to the objective need for science evolution. For example, complex problems have appeared that do not have simple algorithms for solving and require the heuristic solutions use such as the second kind algorithms. Non-deterministic algorithms and multi-agent algorithms appeared, excluding description using mathematical logic. The idea of an algorithm as a strict sequence has changed. The artificial intelligence methods development also requires cognitive logic use [6], which goes beyond the mathematical logic framework. The artificial neural networks (ANNs) use revealed the fact that ANNs can use different algorithms to correctly solve the problem. It also precludes the mathematical logic simple unambiguous methods use to describe such situations. The ternary logic and constructive logic emergence [7, 8] is also dictated by the mathematical imperfection logic, based on the rigid "the third

© The Author(s), under exclusive license to Springer Nature Switzerland AG 2021
R. Silhavy (Ed.): CSOC 2021, LNNS 229, pp. 288–296, 2021.
https://doi.org/10.1007/978-3-030-77445-5_26

exclusion law". Cognitive logic is an even greater generalization compared to ternary logic and constructive logic. Therefore, its research is an urgent task. It is difficult to describe all the cognitive logic features in one article. Therefore, this work describes the cognitive logic principles, which show the general and difference with mathematical logic, but deny their opposition possibility.

2 The Research Methodology

The research is based on system's analysis, logical analysis, dichotomous analysis and oppositional analysis. The main research method is the dichotomous analysis method. The materials used were publications in the field of mathematical logic, constructive logic, dichotomous analysis, cognitive logic and artificial intelligence.

3 Research Results

3.1 The Relationship Between Mathematical Logic and Cognitive Logic Is Described

Cognitive logic is close to intuitionistic logic [9] and modal logic [10]. Cognitive logic allows for uncertainty and describes uncertainty. It is based mainly on dichotomous analysis [11], but admits the oppositional analysis possibility. For example, a parse tree is built on the cognitive logic basis. Mathematical logic formally excludes the uncertainty presence. Informally, mathematical logic includes the uncertainty presence, for example, disjunction ($\vee$) contains it, but "exclusive or" ($\oplus$) does not.

Mathematical logic is based on only oppositional analysis, for example, a binary tree is based on oppositional signs. A parse tree can transform into a binary tree under certain conditions, but not vice versa.

If we consider the logical square [12] of Aristotle, then the cognitive logic statements can be attributed to general and generally affirmative judgments. In comparison with cognitive logic, mathematical logic statements refer to particularly affirmative statements. This does not give grounds to oppose them but gives grounds to consider them as general and particular.

Cognitive logic uses dichotomous division, that is, it uses the difference and relationship between the parts of the whole, as well as the relationship between the whole and its parts. Variables of cognitive logic are dichotomous variables that, under additional conditions, can transform into oppositional variables. From these positions, cognitive logic is based on trinitarian relations.

Mathematical logic uses oppositional division [13, 14], that is, it uses the difference and relationship between opposites. Mathematical logic variables are oppositional variables. Mathematical logic relies on dyadic relationships.

The cognitive logic basis is the difference. The mathematical logic basis is difference and opposition. This does not give grounds to oppose them but gives grounds to consider them as general and particular.

Cognitive logic relies on the precedents' method, psychology and the tacit knowledge extraction [13]. In this aspect, it resembles algebra, while mathematical logic is closer to arithmetic, especially in the propositional calculus field. Mathematical logic excludes the implicit knowledge concept, cognitive logic allows it.

Common to mathematical and cognitive logic is the tendency in an objective situation construction that describes reality, if possible, excluding the subject opinion. In ordinary logic, such objectification is achieved through decomposition and detailing. In cognitive logic, such objectification is achieved through decomposition, generalization, and a stereotype or precedent use. Conventional logic uses the step by step paradigm based on formal language and certainty. Cognitive logic uses the "step by example" or "by example" paradigm and allows for ambiguity as a variable kind. Cognitive logic is used to build cognitive maps.

Cognitive logic is used to solve problems that can be reduced to objects and relationships between them, tasks that can be expressed in terms of objects and relationships between them. Cognitive modelling is based on taking into account the judgments' ideas or the substantive truth content that are transformed during modelling. Let's make a comparison.

Formal logic operates with formal statements, uses some quantifiers and logical connectives, and uses universal inference rules. Cognitive logic: operates with semantics, explicates a concept structure, studies the inference rules dependence on a concept structure.

Formal logic sets as its goal the sequences' construction taking into account syntactic and formal language in situations that are well formalized by the logic methods. Cognitive logic aims to construct sequences taking into account syntactic, semantics, associations of argumentation, which are poorly formalized by the logic methods.

A cognitive analysis important question is the question: how does cognitive logic affect logical consequence and its structure formation? The cognitive approach consists of the contents or statements syntactic and semantic connections' analysis. This makes the cognitive logic apparatus more voluminous in comparison with formal logic.

Cognitive logic allows building sequences or logical consequence models, taking into account the relationships in texts or speech peculiarities. However, the logical consequences generated by the cognitive approach are always contextually and interpretively dependent. Because of this, they can lead to different results for different contexts. Cognitive logic is similar to how an artificial neural network works. For one task, using a neural network, different implementations of the solution are possible, but the result is correct for any implementation. Cognitive logic allows logical consequences in different types, but the result is the same.

Cognitive logic allows building sequences or logical consequence models, taking into account the relationships in texts or speech peculiarities. However, the logical consequences generated by the cognitive approach are always contextually and interpretively dependent. Because of this, they can lead to different results for different contexts. Because of this, the cognitive logical reasoning result depends on expert and expert knowledge.

Cognitive logic is similar to how an artificial neural network works. For one task, using a neural network, different implementations of the solution are possible, but the result is correct for any implementation. Cognitive logic allows logical consequences in different types, but the result is the same. The premises' role in cognitive logic is played by statements that contain semantics and can be formalized with varying degrees of detail, depending on the chosen language.

Recently, the logical approach has been brought closer to a reasoning scheme using the non-monotonic and paraconsistent logics apparatus. In particular, non-monotonic logics include the natural reasoning property, which consists of changing deducibility as additional information is included in the reasoning. This property has been used for a long time in mathematics.

What does the cognitive approach bring new to reasoning schemes? Cognitive modelling is based on the statements' semantics or the subject's representations associative structure. This means that the subject's knowledge base content and its interpretation methods are important for generating a system of sequences within the cognitive approach framework. Within the cognitive approach framework, the key point is a logical information units system generation as an interpretation tool. Eliminate detailed evidence by referring to a precedent or black box.

In a sense, cognitive logic stands in alternative to the formally logical or syntactic. If the formal logical approach is based on the reasoning correctness' formal syntactic, then the cognitive logical approach seeks to reproduce the argumentation using the analysis of semantic models and semantics, context and associations.

Cognitive modelling of argumentation is based on taking into account the content of representations or the meaningful truth of judgments that are transformed in the argumentation course. Cognitive logic is used in living systems, in multi-agent systems, in heterogeneous systems. This new direction requires further research and development.

3.2 Cognitive Dichotomous Logic

The cognitive logic basis is a dichotomous analysis and dichotomous divisions. Cognitive logic can often include oppositional logic. The dichotomous analysis allows us to identify systemic features of the object of research, but also to assess its complexity [15]. According to the dichotomy, objects can be divided into groups "simple - complex".

When constructing algorithmic schemes and decomposing tasks, dichotomous and oppositional divisions are used, as well as the corresponding variables. An oppositional analysis is based on the opposition variables use (*x1, x2*) [14]. Such variables are based on the scheme "yes or no", "object - not object", "relation - relation of the opposite", "connection - opposite connection", "vector - opposite vector". An oppositional variables special case is the mathematical logic "true-false" values. Therefore, oppositional analysis is carried out by mathematical logic methods Oppositional analysis uses rigid division and opposition.

A dichotomous analysis is based on a dichotomous qualitative scale and dichotomous variables use (*y1, y2*). A dichotomous analysis is not as rigid as an oppositional analysis. The main thing in it is the difference and inequality axioms. Dichotomous variables are used in the nominal qualitative scale "object - another object", "color - another color", "object - part of an object", "one part of an object - another part of an object", "relation - complementary relation", "connection - additional connection","covariance vector - contravariant vector","function - co-function".

Relationships serve as the basis for the situations logical description in dichotomous analysis and obtaining the analysis results. The algorithm must be logically correct, therefore it must be created according to the laws of logic and using the logic language. Consider the logical rules for dichotomous variables.

Dichotomous algorithmization and dichotomous division is softer and allows many conditions. The main condition is distinguishability, but not the antagonism that follows from the "law of negation for oppositional variables." For dichotomous variables, there are logical rules that allow some logical situations. If *y1, y2* are dichotomous variables, then there are logical rules for them that follow from their basic tautologies. For dichotomous variables, there is a third concept, which we will call a certain integer (*Int*).

Trinitarian dichotomy law

$$y1 + y2 = Int$$

It means there is a triad: two parts (*y1, y2*) that belong to the whole (*Int*).

The Law of difference for dichotomous variables

$$y1 \neq y2; \ \neg y1 \neq \neg y2;$$

It means that the parts that belong to the whole and their negations may not be equal to each other.

The dichotomous law of identity

$$y2 \oplus y1 \subset Int \equiv 1$$

It means that the parts belong to the whole.

The dichotomous law of double negation

$$\neg(\neg y1) \equiv y1; \ \neg(\neg y2) \equiv y2; \ \neg(\neg y1) \neq y2$$

It means that double negation of each part gives that part but does not give the other part.

Dichotomous law of logical contradiction

$$y1 \wedge (\neg\, y1) \equiv 0; \ y2 \wedge (\neg y2) \equiv 0; \ y1 \wedge (\, y2\,) - \mathit{acceptable};$$

It means, that the law of logical contradiction applies to each part, but may not apply between the parts.

The dichotomous law of the excluded third

$$y1 \vee (\neg y1) \equiv 1; \ y2 \vee (\neg y2) \equiv 1; \ y1 \vee (y2) \equiv 1;$$

It means, that the law of the excluded third applies to each part, but may not apply between parts.

For dichotomous variables, some laws of mathematical logic are valid.

Idempotency of a conjunction

$$y1\, y1; \ y2\, y2; \ y1\, y2;$$

Idempotency of disjunction

$$y1 \vee y1 \equiv y1; \ y2 \vee y2 \equiv y2;$$

Commutativity of a conjunction

$$y1 \wedge b \equiv b \wedge y1; \ y2 \wedge a \equiv a \wedge y2$$

Disjunction commutativity

$$y1 \vee B \equiv B \vee y1; \ A \vee y2 \equiv y2 \vee A \ ;$$

These rules follow from 28 basic tautologies of mathematical logic. The slight difference between dichotomous and oppositional variables is due to their qualitative difference.

3.3 Oppositional and Dichotomous Division

The oppositional division is used in mathematical logic, and dichotomous division is used in cognitive logic. If we designate the procedure of oppositional division of the object *O* as *OD (O)*, then we can compose expressions

$$OD(O) \rightarrow x \wedge (\neg x) \equiv 0; x \vee (\neg x) \equiv 1; \neg\neg x \equiv x$$

Sometimes the notation $x = x1$ is introduced; $\neg x = x2$. Fundamentally, the oppositional division gives only two opposite parts $x1 = \neg x2$. $OD(O) \rightarrow x1; x2$. Fundamentally, the oppositional division extends to a simple two-part object.

Dichotomous division can refer to an object and a set. Accordingly, cognitive logic can be extended to objects and sets. If we consider a certain countable set *M* with elements x, then dichotomous division leads to the selection of the element x_i and the formation of a new set M', which x_i does not contain. We denote the procedure for the dichotomous division of the set *M* as *DD (M)*, then we can compose some expressions.

$$DD(M) \rightarrow M', \ xi$$

$$DD(M') \rightarrow M'', \ xk$$

$$DD(M'') \rightarrow M''', \ xj$$

Moreover, $xi \neq xk \neq xj$, $M > M' > M'' > M'''$.

The dichotomous division' consistent application reduces the original set and selects from it elements that may be in different relationships with each other. Fundamentally, the dichotomous division gives many complementary parts. Fundamentally, the dichotomous division extends to a complex object with many parts or many.

3.4 Case and Codified Reasoning

An example of the application of cognitive logic is a law, which is built on precedents. In Russia and some other countries, the law has been codified. There is a code or other regulation, which applies to "all occasions" and prescribes punishment for a particular offence. Case law is based on the search for existing precedents, comparing the current precedent with those found. Comparison of the measure of punishment of the current precedent with the measure of the found precedent. The precedent method develops logic and rhetoric and relies on public opinion represented by the jury. The codified method does not develop logic and creates the arbitrariness of officials, who subjectively interpret the act at their discretion. A stereotype is a closer concept to cognitive logic since it requires less analysis than a precedent and uses the "by example" rule. Currently, in the field of information technology and programming, a stereotype or pattern is a kind of model that serves to describe a situation and for standardized actions. One of the functions of a stereotype is to speed up the process of reasoning or decision-making by using a pattern. A stereotyped approach in logical reasoning and decision-making is an approach in which there is an opportunity to use a certain investigated and verified by experience situation or regularity as a model for which a decision is known in such a situation. This sample is used for: excluding some (their number may be different) processes of analysis and making a ready-made decision from a typical situation for the current situation. Decision making and cognitive logic are based on this approach. The logical sequence for applying the stereotype is as follows:

1. Stereotypical situational perception of reality,
2. Building a chain of stereotypes of analogues to simulate a real situation
3. Transferring solutions to a chain of analogues of stereotypes into a chain of reasoning for a real situation and building based on this reasoning.
4. Using a chain of stereotyped decisions as a logical follow-up for solving in a real situation without its detailed analysis.

This approach excludes the decomposition of the real situation to the level of logical information units. Decomposition is carried out to the level of stereotypes. When analyzing a text for this approach, such a comparison is appropriate: the text is analyzed not to the level of words and relationships between them, but the level of well-known phrases containing sentences and relationships between phrases. This approach significantly reduces the time and volume of analytical processes. Cognitive logic with a stereotyped approach is built according to the rule "If *A*, then *B*". This means that if there is a stereotyped information situation described by the parameter system $A(a_1, a_2, a_3, a_n)$, then the action described by the parameter system $B(b_1, b_2, b_3, b_m)$. Such a logical sequence is described by one sequence link

$$A \to B. \tag{1}$$

Expression (1) formally contains an implication. But in logic, the implication is carried out between boolean variables or parameters. Expression (1) implies one system of parameters into another. From the point of view of formal logic, expression (1) should be preceded by the procedures' proof

$$A = a_1 \wedge a_2 \wedge a_3, \wedge a_n; \tag{2}$$

$$B = b_1 \wedge b_2 \wedge b_3, \wedge b_m; \tag{3}$$

$$a_1 \to b_1; \tag{4}$$

$$a_2 \rightarrow b_2; \tag{5}$$

$$a_3 \rightarrow b_3; \tag{6}$$

$$a_{n-2} \wedge a_{n-1} \wedge a_n \rightarrow b_m; \tag{7}$$

Expressions (2), (3) correspond to the decomposition of situation *A* and solution *B* into logical variables. Expressions (4–6) correspond to the direct transformation of parameters. Expression (7) describes the procedure for preliminary complexation (integration) of parameters and their subsequent transformation. All these procedures are excluded by cognitive logic. It should be especially noted that the parameters number in *A* and *B* is often different and an additional problem arises which parameters should be transformed directly and which should be combined. However, the cognitive logic peculiarity is that it can lead to different conclusions depending on the intelligence and argumentation of the subject applying it. One subject will choose situation *A* as a stereotype, another situation *A'*, the third situation *A**. The situations are similar but different. This will lead to a difference in the results *B*, *B'*, *B**.

4 Conclusion

Cognitive logic does not contradict mathematical logic but is its generalization. Cognitive logic introduces the concept of a situation, which can have different descriptions. Cognitive logic allows for different initial descriptions that lead to the same conclusions. Cognitive logic allows for a plurality of situations and analyzes such a set. Cognitive logic is effective in solving problems of the second kind when the initial situation decomposition is impossible to the elements' level, but positive experience has been accumulated in solving problems in such situations on the generalized reasoning basis. Cognitive logic is an important component of a situational analysis system. Cognitive logic allows to solve many problems, of problems which two groups be distinguished. The first group of tasks is the analysis of situations for consistency, based on existing experience and knowledge. The second task is the task of analyzing sequences within a logical chain for complementarity and efficiency. It can be stated that the division of the situation into sequences is not the only one. You can make up related sequences' different types that are related by the experience presence and the subject cognitive level. Cognitive logic includes the subject and his abilities in the inference system. Semantic modelling, which is also part of cognitive logic, remains outside the scope of this article.

References

1. Novaes, C.D.: Formal Languages in Logic: A Philosophical and Cognitive Analysis. Cambridge University Press, Cambridge (2012)
2. Tsvetkov, V.Y.: Cognitive aspects of building virtual educational models. Prospects Sci. Educ. **3**, 38–46 (2013)

3. Tsvetkov, V.Y.: Cognitive logic. ITNOU. Inf. Technol. Sci. Educ. Manage. **1**(11), 106–10 (2019)
4. Hoang, P.N.: Interesting cognitive logic of vietnamese and English perception verbs. Linguistics **5**(4), 95–98 (2017)
5. Wang, P.: Cognitive logic versus mathematical logic. In: Proceedings of the Third International Seminar on Logic and Cognition (2004)
6. Lieto, A., Pozzato, G.L.: A description logic framework for commonsense conceptual combination integrating typicality, probabilities and cognitive heuristics. J. Exp. Theor. Artif. Intell. **32**(5), 769–804 (2020)
7. Fandinno, J., del Cerro, L.F.: Constructive logic covers argumentation and logic programming. In: Sixteenth International Conference on Principles of Knowledge Representation and Reasoning (2018)
8. Spinks, M., Veroff, R.: Paraconsistent constructive logic with strong negation as a contraction-free relevant logic. Don Pigozzi on Abstract Algebraic Logic, Universal Algebra, and Computer Science, pp. 323–379. Springer, Cham (2018). https://doi.org/10.1007/978-3-319-74772-9_13
9. Minghui, M., et al.: A graphical deep inference system for intuitionistic logic. Logique et Anal. (N.S.) **245**, 73–114 (2019)
10. Odintsov, S.P., Wansing, H.: Disentangling FDE-based paraconsistent modal logics. Stud. Logica. **105**(6), 1221–1254 (2017)
11. Tsvetkov, V.Y.: Dichotomous systemic analysis. Life Sci. J. **11**(6), 586–90 (2014)
12. Vernant, D.: Existence presupposition and the logical Square. Al-Mukhatabat **345**(4130), 1–11 (2016)
13. Sigov, A.S., Tsvetkov, V.Y.: Tacit knowledge: oppositional logical analysis and typologization. Herald Russ. Acad. Sci. **85**(5), 429–33 (2015)
14. Tsvetkov, V.Y.: Opposition information analysis. Eur. J. Technol. Design **4**(6), 189–96 (2014)
15. Tsvetkov, V.Y.: Dichotomous analysis of the complexity of the system. Prospects Sci. Educ. **2**(8), 15–20 (2014)

A Machine Learning Firefly Algorithm Applied to the Resource Allocation Problems

Hernan Pinto[1], Alvaro Peña[1], Carlos Maureira[2(✉)], Matías Valenzuela[1], and Gabriel Villavicencio[1]

[1] Escuela de Ingeniería en Construcción, Pontificia Universidad Católica de Valparaíso, Valparaíso, Chile
{hernan.pinto,alvaro.pena,matias.valenzuela, gabriel.villavicencio}@pucv.cl

[2] Escuela de Tecnologías de la Información, Facultad de Ingeniería y Negocios, Universidad de las Américas, Santiago, Chile
carlos.maureira.pasten@edu.udla.cl

Abstract. In the fields of engineering and science, there are many combinatorial optimization problems. Many of these problems are NP-hard problems, which are difficult to solve with complete techniques. Therefore, the design of binary algorithms based on swarm intelligence continuous metaheuristics is an area of interest in operations research. In this article, we use a general binarization mechanism based on the k-means technique. We apply k-means technique to the firefly algorithm to solve the multidimensional knapsack problem (MKP). Design experiments to prove the practicality of k-means technique in binarization.

Keywords: Combinatorial optimization · Knapsack · Metaheuristics · k-means

1 Introduction

The combinatorial problems have great relevance at industrial level, we find them in various areas such as Civil Engineering [1–5], Operational Research [6–8], resource allocation [9–11], scheduling problems [12–15], routing problems [16–19] among others. On the other hand, in recent years, algorithms inspired by nature phenomena to solve optimization problems have been generated. As examples of these algorithms we have Cuckoo Search, Black Hole, Bat Algorithm, and firefly algorithms among others. Many of these algorithms work naturally in continuous spaces and therefore must be adapted to solve combinatorial problems. In the process of adaptation, the mechanisms of exploration and exploitation of the algorithm can be altered, having consequences in the efficiency of the algorithm.

Several binarization techniques have been developed to address this situation. In a literature search, the main binarization methods used correspond to transfer functions, angle modulation and quantum approach, for more detail refer to [20].

© The Author(s), under exclusive license to Springer Nature Switzerland AG 2021
R. Silhavy (Ed.): CSOC 2021, LNNS 229, pp. 297–305, 2021.
https://doi.org/10.1007/978-3-030-77445-5_27

In this article we present a new binarization method that uses the clustering k-means technique to group the solutions and then perform the binarization process. To verify the efficiency of our method, we used the firefly algorithm.

To check our binary firefly algorithm (BKFA), we use the well-known knapsack problem. Experiments were developed using a random operator to validate the contribution of the k-means technique in the binarization process of the firefly algorithm. In addition, local search operator is used to strengthen the results.

2 KnapSack Problem

The multidimensional knapsack problem (MKP), models resource allocation situations. The goal is to find the subset of objects that produce the greatest benefit by satisfying a set of constraints. MKP is one of the most studied NP-hard class combinatorial problems, however it remains as a challenge due to the difficulties encountered in solving medium and large size instances. A search for the last few years in the literature, we find that MKP has been solved for example in [21] using a quantum binarization technique, in [22] applied the k-means clustering technique to perform binarization, in [23] used a differential algorithm with transfer functions, and in [24] a modification of the PSO equations was used. MKP can be set as:

$$\text{maximize} \sum_{j=1}^{n} p_j x_j \tag{1}$$

$$\text{subjected to} \sum_{j=1}^{n} c_{ij} x_j \leq b_i \ , \ i \in \{1, ..., m\}. \tag{2}$$

With $x_j \in \{0,1\}$, $j \in \{1,...,n\}$. p_j corresponds to the profit of element j. c_{ij} represents a cost associated with dimension i and element j. The constraints in each dimension i are represented by b_i. The solution can be modelled using a binary representation, in this representation a 0 means the element is not included in the knapsack.

3 Binary Firefly Algorithm

The application of statistics and data mining, can be found in many areas such as transports, smart cities, agriculture and computational intelligence [22,25–29]. In this section, we explore the application of the k-means technique to the binarization of continuous metaheuristics.

The binary firefly algorithm(BKFA) uses 4 operators to achieve its goal. An initialization operator, A local search operator, the binary operator using the k-means technique, and a repair operator for the case of solutions that do not meet any of constraints. The general scheme of the algorithm is shown in Fig. 1.

In the first step, BKFA initializes the solutions, this is detailed in Sect. 3.1. Once we have all the solutions initialized, we check the compliance of the stopping criterion of the algorithm associated to the maximum number of iterations. The firefly algorithm is then run in conjunction with the k-means operator, this is described in Sect. 3.2. Subsequently once the binarization was performed, the results are compared with the best solution obtained so far. In the case of finding a superior solution, it replaces the previous solution and is submitted to a local search operator. In the following subsection we will explain in detail the initialization method, the binary operator and the repair operator.

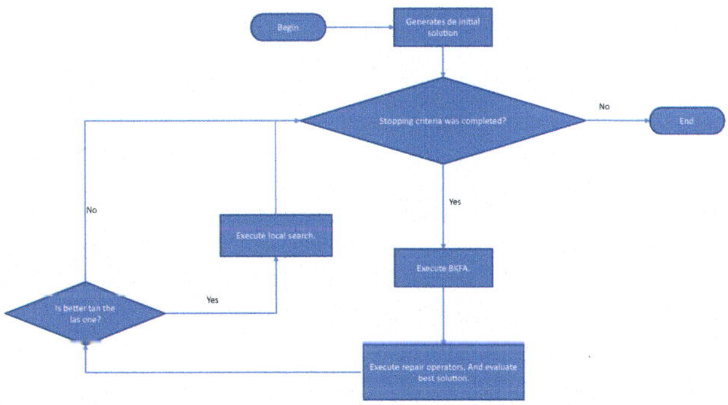

Fig. 1. Flowchart of the binary firefly algorithm.

3.1 Initialization and Element Weighting

As BKFA is a swarm algorithm, to begin the exploration and exploitation of the search space, the list of solutions must be initialized. For the generation of each solution, an element is randomly chosen first. The next step is to check if other elements can be incorporated, for this we must evaluate the constraints of our problem. To select the new element, we generate a list of possible elements that comply with the constraints. Each element is calculated a weight and the best weight element is selected. This procedure is repeated until no additional element can be incorporated.

To calculate the weight of each element, several methods have been developed. In [30] a pseudo-utility in the surrogate duality approach was proposed. The way to calculate it is shown in Eq. 3. In this equation the variable w_j corresponds to the surrogate multiplier whose value is between 0 and 1. This multiplier can be interpreted as a shadow prices of the j-th constraint.

$$\delta_i = \frac{p_i}{\sum_{j=1}^{m} w_j c_{ij}} \tag{3}$$

A more intuitive measure focused on the average resource occupancy was proposed by [31]. It is shown in Eq. 4.

$$\delta_i = \frac{\sum_{j=1}^{m} \frac{c_{ij}}{mb_j}}{p_i} \tag{4}$$

In this article we used a variation of this last measure focused on the average occupation and proposed in [22]. this variation considers the elements that exist in knapsacks to calculate the average occupancy. In each iteration depending on the selected items in the solution the measure is calculated again. The expression of this new measure is shown in Eq. 5.

$$\delta_i = \frac{\sum_{j=1}^{m} \frac{c_{ij}}{m(b_j - \sum_{i \in S} c_{ij})}}{p_i} \tag{5}$$

3.2 K-Means Transition Ranking Operator

Consider our metaheuristic is a continuous and population based. Due to its iterative nature, it needs to update the position of particles at each iteration. When the metaheuristic is continuous, this update is performed in $\mathbb{R}^n$ space. In Eq. 6, the update position is presented in a general manner. The x_{t+1} variable represents the x position of the particle at time t+1. This position is obtained from the position x at time t plus a Δ function calculated at time t+1. The function Δ is proper to each metaheuristic and produces values in $\mathbb{R}^n$. For example in Cuckoo Search $\Delta(x) = \alpha \oplus Levy(\lambda)(x)$, in Black Hole $\Delta(x) = \text{rand} \times (x_{bh}(t) - x(t))$ and in the Firefly, Bat and PSO algorithms Δ can be written in simplified form as $\Delta(x) = v(x)$.

$$x_{t+1} = x_t + \Delta_{t+1}(x(t)) \tag{6}$$

In the k-means transition ranking operator, a model for transitions in a discrete space is proposed. This model is based on considering the movements generated by the metaheuristic in each dimension for all particles. $\Delta^i(x)$ corresponds to the magnitude of the displacement $\Delta(x)$ in the i-th position. Subsequently these displacement are grouped using the magnitude of the displacement $\Delta^i(x)$. This grouping is done using the k-means technique where k represents the number of clusters used. Finally, a generic P_{tr} function given by the Eq. 7 is proposed to assign a transition probability.

$$P_{tr} : \mathbb{Z}/k\mathbb{Z} \rightarrow [0, 1] \tag{7}$$

A transition probability through the function P_{tr} is assigned to each group. Naturally, this P_{tr} function is modelled as a cumulative probability function. For the case of this study, we particularly use the linear function given in Eq. 8. In this equation, $\text{N}(x^i)$ indicates the location of the group to which $\Delta^i(x)$ belongs. The α coefficient, corresponds to the transition probability and β to the transition separation coefficient. Both parameters are estimated in each metaheuristic. For our particular case, $N(x^i) = 0$ corresponds to elements belonging to the group that has the lowest Δ^i values. $N(x^i) = 7$ corresponds to the group of elements that have the greatest Δ^i values.

$$P_{tr}(x^i) = P_{tr}(N(x^i)) = \alpha + \beta N(x^i)\alpha \tag{8}$$

The k-means transition ranking operator starts calculating Δ^i for each of the particles. This step is specific in each metaheuristic. Subsequently the particles are grouped using k-means clustering technique and the magnitude of Δ^i as distance. With the group assigned to each particle we obtain the probability of transition using Eq. 8. Afterwards the transition of each particle is performed. In the case of firefly algorithm the rule 9 is used to perform the transition, where $\hat{x}^i$ is the complement of x^i. Finally, each solution is repaired using the repair operator.

$$x^i(t+1) := \begin{cases} \hat{x}^i(t), \text{ if } rand < P_{tg}(x^i) \\ x^i(t), \quad \text{otherwise} \end{cases} \tag{9}$$

3.3 Repair Operator

Local search and binary k-means technique can generate infeasible solutions. There are different mechanisms to address these infeasible solutions. In this article, the repair of the solutions was considered. To perform the repair, the measure described in the Eq. 5 was used. If the solution requires repair, the element with the maximum measure is chosen and it is removed from the solution, this process is iterated until a valid solution is obtained. The solution is then improved by exploring the possibility of incorporating new elements. This stage of improvement is iterated until there are no elements that can be incorporated without violating the constraints

4 Results

4.1 Insight of BKFA Algorithm

This section aims to find out the contribution of the binary k-means operator to the performance of the algorithm. To carry out the comparison problems cb.5.250 from the OR-library were selected. Violin charts and the Wilcoxon non-parametric signed-rank test were used to perform the statistical analyzes. In the charts, the X axis identifies the studied instances and the Y axis the % -Gap described in the Eq. 10. The Wilcoxon test is run to determine if the results obtained by BKFA have significant difference with respect to other algorithms. The parameter settings and browser ranges are shown in Table 1.

$$\% - Gap = 100\frac{BestKnown - SolutionValue}{BestKnown} \tag{10}$$

Table 1. Setting of parameters for Binary firefly Algorithm.

Parameters	Description	Value	Range
N	Number of solutions	30	[20, 25, 30]
G	Number of clusters	5	[4, 5, 6]
Iteration Number	Maximum iterations	1000	[1000]

Evaluation of k-Means Binary Operator. A random operator was designed to evaluate the contribution of the k-means binary operator. This random operator has the peculiarity of performing the transitions with a fixed probability of 0.5 without considering in which cluster the variable is located. Two configurations were studied: The first one where the local search operator is included and the second where the local operator is not considered. BKFA corresponds to our standard algorithm. *random.ls* is the random variant that includes the local search operator. *BKFA.wls* corresponds to the version with k-means binary operator without local search operator. Finally *random.wls* describes the random algorithm without local search operator.

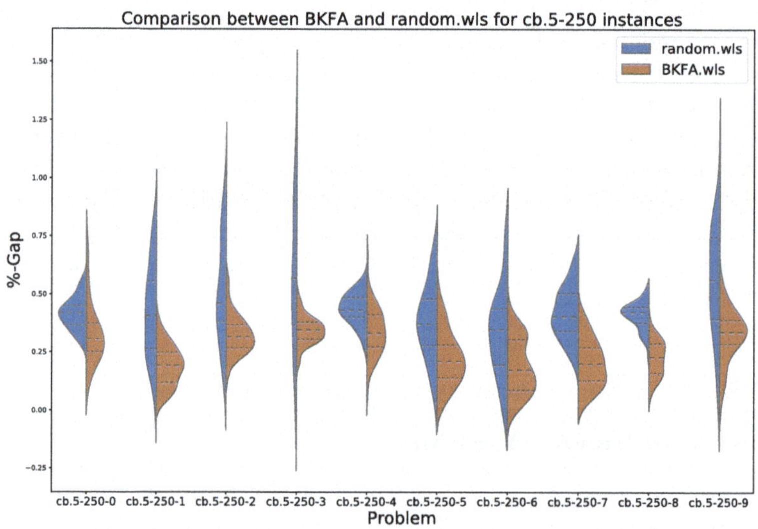

Fig. 2. Evaluation of k-means binary operator with Local Search operator

When we compared the Best Values between BKFA and *random.ls* which are shown in Table 2. BKFA outperforms to *random.ls*. However the Best Values between both algorithms are very close. In the Average comparison, BKFA outperforms *random.ls* almost in all problems. The comparison of distributions is shown in Fig. 2. We see the dispersion of the *random.ls* distributions are bigger than the dispersion of BKFA. In particular this can be appreciated in the problems 1, 5, 6 and 7. Therefore, the k-means binary operator together with local search operators, contribute to the precision of the results. Finally, the BKFA distributions are closer to zero than *random.ls* distributions, indicating that BKFA has consistently better results than *random.ls*. When we evaluate the behaviour of the algorithms through the Wilcoxon test, this indicates that there is a significant difference between the two algorithms.

Table 2. Evaluation of k-means binary operator

Set	Best Known	Best *random.ls*	Best BKFA	Best *random.wls*	Best *wls*	Avg *random.ls*	Avg BKFA	Avg *random.wls*	Avg *wls*
cb.5.250-0	59312	59211	59225	59158	59175	59132.1	59149.1	59071.8	59132.1
cb.5.250-1	61472	61435	61435	61409	61409	61324.6	61385.6	61288.3	61370.7
cb.5.250-2	62130	62036	62074	61969	61969	61894.4	61974.2	61801.6	61924.6
cb.5.250-3	59463	59367	59446	59365	59349	59257.8	59323.8	59136.1	59269.3
cb.5.250-4	58951	58914	58951	58883	58930	58725.6	58820.1	58693.6	58758.5
cb.5.250-5	60077	60015	60056	59990	60015	59904.6	59955.6	59837.8	59945.9
cb.5.250-6	60414	60355	60355	60348	60349	60208.2	60334.7	60230.6	60311.1
cb.5.250-7	61472	61436	61436	61407	61407	61290.8	61330.5	61233.9	61342.3
cb.5.250-8	61885	61829	61885	61790	61782	61737.1	61785.8	61644.9	61738.1
cb.5.250-9	58959	58832	58866	58822	58787	58769.1	58781.4	58653.7	58770.4
Average	60413.5	60343	60372.9	60314.1	60317.2	60221.4	60283.4	60159.1	60256.3
p-value								2.4 e−04	5.01 e−05

5 Conclusions

In this article, the k-means technique is used to perform the binarization of the GOA algorithm. It should be noted that the k-means technique can be applied in the binarization of any continuous swarm-intelligence algorithm. We used the knapsack problem to evaluate our binary algorithm. The contribution of the k-means binary operator was studied, observing that this operator contributes to the precision and quality of the solutions obtained.

As a future line of research, it is interesting to compare the performance of the k-means operator with the transfer functions, all applied in the binarization of the firefly algorithm. Also, it is interesting to use the k-means operator to binarize other algorithms in addition to solving other NP-hard problems. From the theoretical point of view, it would also be interesting to understand how the exploration and exploitation of space are altered when we introduce the k-means operator. Another interesting research line is to explore adaptive techniques to automate the selection of parameters.

References

1. Yepes, V., Martí, J.V., García, J.: Black hole algorithm for sustainable design of counterfort retaining walls. Sustainability **12**(7), 2767 (2020)

2. García, J., Martí, J.V., Yepes, V.: The buttressed walls problem: an application of a hybrid clustering particle swarm optimization algorithm. Mathematics **8**(6), 862 (2020)
3. Crawford, B., et al.: Investigating the efficiency of swarm algorithms for bridge strengthening by conversion to tied-arch: a numerical case study on san luis bridge. Iranian J. Sci. Technol. Trans. Civil Eng. 1–13 (2020)
4. García, J., Yepes, V., Martí, J.V.: A hybrid k-means cuckoo search algorithm applied to the counterfort retaining walls problem. Mathematics **8**(4), 555 (2020)
5. Martínez-Muñoz, D., Martí, J.V., García, J., Yepes, V.: Embodied energy optimization of buttressed earth-retaining walls with hybrid simulated annealing. Appl. Sci. **11**(4), 1800 (2021)
6. Crawford, B., Soto, R., Monfroy,E., Astorga, G., García, J., Cortes, E.: A meta-optimization approach for covering problems in facility location. In: Workshop on Engineering Applications, pp. 565–578. Springer (2017)
7. García, J., Crawford, B., Soto, R., Astorga, G.: A percentile transition ranking algorithm applied to binarization of continuous swarm intelligence metaheuristics. In: International Conference on Soft Computing and Data Mining, pp. 3–13. Springer (2018)
8. García, J., Maureira, C.: A KNN quantum cuckoo search algorithm applied to the multidimensional knapsack problem. Appl. Soft Comput. **102**, 107077 (2021)
9. Garcia, J., Crawford, B., Soto, R., Astorga, G.: A percentile transition ranking algorithm applied to knapsack problem. In: Proceedings of the Computational Methods in Systems and Software, pp. 126–138. Springer (2017)
10. García, J., Altimiras, F., Peña, A., Astorga, G., Peredo, O.: A binary cuckoo search big data algorithm applied to large-scale crew scheduling problems. Complexity **2018** (2018)
11. García, J., Lalla-Ruiz, E., Voß, S., Droguett, E.L.: Enhancing a machine learning binarization framework by perturbation operators: analysis on the multidimensional knapsack problem. Int. J. Mach. Learn. Cybern. **11**(9), 1951–1970 (2020). https://doi.org/10.1007/s13042-020-01085-8
12. García, J., Crawford, B., Soto, R., García, P.: A multi dynamic binary black hole algorithm applied to set covering problem. In: International Conference on Harmony Search Algorithm, pp. 42–51. Springer (2017)
13. García, J., Crawford, B., Soto, R., Astorga, G.: A clustering algorithm applied to the binarization of swarm intelligence continuous metaheuristics. Swarm Evol. Comput. **44**, 646–664 (2019)
14. Crawford, B., Soto, R., Astorga, G., García, J.: Constructive metaheuristics for the set covering problem. In: International Conference on Bioinspired Methods and Their Applications, pp. 88–99. Springer (2018)
15. García, J., Astorga, G., Yepes, V.: An analysis of a KNN perturbation operator: an application to the binarization of continuous metaheuristics. Mathematics **9**(3), 225 (2021)
16. García, J., Peña, A.: Robust optimization: concepts and applications. In: Nature-Inspired Methods for Stochastic, Robust and Dynamic Optimization, p. 7 (2018)
17. Astorga, G., Crawford, B., Soto, R., Monfroy, E., García, J., Cortes, E.: A meta-optimization approach to solve the set covering problem. Ingeniería **23(3)** (2018)
18. García, J., et al.: A db-scan binarization algorithm applied to matrix covering problems. Comput. Intell. Neurosci. **2019** (2019)
19. Tapia, D., et al.: A q-learning hyperheuristic binarization framework to balance exploration and exploitation. In: International Conference on Applied Informatics, pp. 14–28. Springer (2020)

20. Crawford, B., Soto, R., Astorga, G., García, J., Castro, C., Paredes, F.: Putting continuous metaheuristics to work in binary search spaces. Complexity **2017** (2017)
21. Haddar, B., Khemakhem, M., Hanafi, S., Wilbaut, C.: A hybrid quantum particle swarm optimization for the multidimensional knapsack problem. Eng. Appl. Artif. Intell. **55**, 1–13 (2016)
22. García, J., Crawford, B., Soto, R., Castro, C., Paredes, F.: A k-means binarization framework applied to multidimensional knapsack problem. Appl. Intell. **48**(2), 357–380 (2017). https://doi.org/10.1007/s10489-017-0972-6
23. Liu, J., Wu, C., Cao, J., Wang, X., Teo, K.L.: A binary differential search algorithm for the 0–1 multidimensional knapsack problem. Appl. Math. Model. (2016)
24. Bansal, J.C., Deep, K.: A modified binary particle swarm optimization for knapsack problems. Appl. Math. Comput. **218**(22), 11042–11061 (2012)
25. García, J., Pope, C., Altimiras, F.: A distributed k-means segmentation algorithm applied to lobesia botrana recognition. Complexity **2017** (2017)
26. Graells-Garrido, E., García, J.: Visual exploration of urban dynamics using mobile data. In: International Conference on Ubiquitous Computing and Ambient Intelligence. pp. 480–491. Springer (2015)
27. Peredo, O.F., García, J.A., Stuven, R., Ortiz, J.M.: Urban dynamic estimation using mobile phone logs and locally varying anisotropy. In: Geostatistics Valencia 2016, pp. 949–964. Springer (2017)
28. Graells-Garrido, E., Peredo, O., García, J.: Sensing urban patterns with antenna mappings: the case of Santiago, Chile. Sensors **16**(7), 1098 (2016)
29. García, J., Moraga, P., Valenzuela, M., Pinto, H.: A db-scan hybrid algorithm: an application to the multidimensional knapsack problem. Mathematics **8**(4), 507 (2020)
30. Pirkul, H.: A heuristic solution procedure for the multiconstraint zero? One knapsack problem. Naval Res. Logist. **34**(2), 161–172 (1987)
31. Kong, X., Gao, L., Ouyang, H., Li, S.: Solving large-scale multidimensional knapsack problems with a new binary harmony search algorithm. Comput. Oper. Res. **63**, 7–22 (2015)

Market Acceptance of the Blockchain Technology on the Example of Bitcoin

Vladislav Rutskiy[1], Sarfaraz Javed[2], Viacheslav Rayk[3], Nadezhda Kulakova[4], Omer Faruk Derindag[5], Alexey Chikizov[1], Igor Kukhar[6], Alexey Knyazkov[1], and Roman Tsarev[1,7(✉)]

[1] Siberian Federal University, Krasnoyarsk, Russia
[2] Mazoon College, Muscat, Oman
[3] National Research University Higher School of Economics, Moscow, Russia
[4] Financial University under the Government of the Russian Federation, Kaluga, Russia
[5] Inonu University, Malatya, Turkey
[6] Reshetnev Siberian State University of Science and Technology, Krasnoyarsk, Russia
[7] V.A. Trapeznikov Institute of Control Sciences of Russian Academy of Sciences, Moscow, Russia

Abstract. The paper is aimed at revealing market acceptance of blockchain technology in the digital economy on the example of bitcoin. The authors study features if cryptocurrencies and the pricing factors of bitcoin in the period of the rising interest of global media in 2016–2017. With the help of regression tools, the authors demonstrate the interconnection of the bitcoin price with the factors of global financial markets and internal expectation factors of the crypto industry dynamics in the course of early adoption of bitcoin by various industries of the economy. #CSOC1120

Keywords: Blockchain · Cryptocurrency · Bitcoin · Market acceptance · Bitcoin price determinants

1 Introduction

The phenomenon of cryptocurrency originated at the intersection of economics, cryptography and philosophy. It appears that the mass adoption of bitcoin, which is the first and most valuable cryptocurrency (digital currency type based on blockchain technology using cryptography), happens due to people's distrust of fiat currencies [1, 2]. The cryptocurrency market developed especially actively in 2016–2018 when bitcoin was just beginning to spread outside the exchanges penetrating the other industries and there were places where cryptocurrency could be spent on shopping, on another currency, on travelling, or it could be sent to charity or even help to acquire citizenship [3, 4].

There is still a lot of controversy over whether cryptocurrencies are a new – digital – revolutionary version of currencies in the digital economy segment or something like pyramid schemes [4–6]. Nevertheless, many economists became interested in this issue. As a result, active studies of cryptocurrencies began and continue to this day.

© The Author(s), under exclusive license to Springer Nature Switzerland AG 2021
R. Silhavy (Ed.): CSOC 2021, LNNS 229, pp. 306–315, 2021.
https://doi.org/10.1007/978-3-030-77445-5_28

Researchers elaborate on their properties, features, the development path and forecasting their exchange rates (prices) [7, 8].

Today, the cryptocurrency market is still in growing demand and continues to develop. More than 50 million active investors carry out transactions with cryptocurrencies on more than 100 exchanges around the world [9].

This paper aims to find relationships between the bitcoin exchange rate as the most popular and valuable cryptocurrency and some selected indicators. We are to determine the degree of their influence on the bitcoin exchange rate using econometric data analysis. Modelling the behaviour of such currencies makes it possible to bring greater clarity about what factors underlie the formation of their exchange rate, which is an important task for market participants and a difficult task for researchers, given that the main value is formed by consumers and their demand for cryptocurrency [10].

To achieve the aim, the following tasks were resolved in the paper: to study the theoretical aspects and features of the cryptocurrencies; to highlight potential factors of the regression model that can presumably influence the bitcoin rate; to identify methods for evaluating data and to build a regression model for the selected variables.

2 Features of Bitcoin as a Technology and Asset

The emergence of the term "cryptocurrency" is closely related to the emergence of another term - "bitcoin". This concept appeared since the release of the paper published in Forbes in 2011, which revealed the main aspects and features of "something new" called "bitcoin" [11].

The term "cryptocurrency" still does not have a universal formulation that everyone would agree with. Until now, this phenomenon is treated differently: someone calls cryptocurrency private money or a unit of account [12], while someone officially recognizes it as a monetary surrogate, referring to the features similar to those of financial pyramids [13].

The main reason for the development of such digital currencies seemed to be the distrust of the people to the so-called fiat (fiduciary) money [1, 2]. These include various forms of national currencies, for example, banknotes secured by the authority of the issuer (national monetary bodies of the state), regardless of the cost of materials spent on such money production. In other words, these are money with a fictitious value indicated by the law, which obliges its user to accept them as a unit of account. These include the familiar national currencies such as rubles, dollars, euros and so on.

So the cryptocurrency was created considering the disadvantages of fiat money. Its creator, Satoshi Nakamoto, claimed he had invented Bitcoin to eliminate the need for a "third party" in transactions [11].

In the case of cryptocurrencies, a financial transaction is carried out by only two parties, which is possible due to the existence of blockchain technology. For example, if bitcoin is a piece of encrypted information, a digital file that stores records of transactions made from the moment of the emergence of crypto coins to the last transaction, then the blockchain is a chain of these records, grouped into blocks, each containing information about recent transactions not yet recorded in the register, as well as additional service information. This chain is public. In other words, it is always possible to check the transactions carried out earlier (from the very beginning) and any computer, being connected to the system, can identify the movements of each cryptocurrency coin, which excludes the possibility of counterfeiting for bitcoin and other digital currencies. Thus, cryptocurrency provides anonymity and inspires trust among many users.

When using ordinary money, there is a risk that your account may be blocked at any time, if a third party (intermediary) considers this action to be correct and eligible [14].

Also, bitcoin, as an example of cryptocurrencies, has some features (including advantages over ordinary money) that characterize cryptocurrency as a revolutionary phenomenon in the financial sector of economy.

1) Absence of inflation due to the limitation of the volume of issuance. There is simply no possibility of making changes to a certain computer algorithm (while in the case of fiat money, the state can freely regulate everything down to cost). In other words, this is ensured by the fact that usually one of the reasons for the growth in the rate of inflation is the growth of the money supply. In the case of bitcoin, it works the same way. Only since the number of bitcoins is limited, and the rate of their production (and, accordingly, the supply growth) decreases, then the bitcoin inflation not only does not grow but gradually tends to zero and possibly to negative values [15].
2) Decentralization [16].
3) The inability to cancel the completed transaction [17].
4) The exchange rate largely depends only on the supply and demand for a specific cryptocurrency (or on the network effect) [18].
5) Cryptocurrency is an international currency, which, accordingly, allows you to use it regardless of the rates of other currencies and regardless of the country in which, for example, a particular bitcoin was mined [19].

The disadvantages of digital currencies (specifically, bitcoin) include the following points that are ones of the most discussed [20]:

1) Possibility of growth of the illegal sector of the economy due to the anonymity of transactions (terrorism, drugs, and so on) [21, 22];
2) The possibility of hiding part of the income and non-payment of taxes in full;
3) Speculativeness. In other words, the very essence of bitcoin is such that its rate will invariably grow (which is associated with the constantly growing complexity of algorithms and, accordingly, mining). However, jumps in the growth of the crypto-currency rate at times later are accompanied by the collapse of this rate. Due to such instability, digital currencies appear to be a dubious "pillar" of a healthy economy [18];
4) The likelihood of a decrease in demand for the national currency. Some believe that, as a result, the banking system will weaken, which will lead the economy to an unstable state [23].

3 Bitcoin Price Factors

As mentioned earlier, the bitcoin rate is generally volatile, despite the trend towards continuous growth. For example, in 2016, which was called the year of bitcoin among miners and experts in the field of cryptocurrencies, the rate of this currency increased from $ 370 at the beginning of the year to $ 980 at the end of December. By January 5, this value exceeded the three-digit number and amounted to $ 1180, which is an increase of over 300%. Later in 2018, the growth was more than 40 times compared to the beginning of 2017 (calculated based on data from Investing.com).

In the papers on bitcoin and other cryptocurrencies, which study their prices (exchange rates), there are three groups of pricing factors investigated: economic, technical and media factors [24].

The first group includes, for example, inflation, which was already mentioned earlier. Researchers call Bitcoin transactions another economic factor. In short, this is the volume of bitcoins exchanged over a given period: the rate changes depending on the number of buyers who want to buy and sellers who want to sell their cryptocurrency.

Technical factors include the peculiarities of the system of the bitcoin network, which complicate the processes of mining and conducting operations with this cryptocurrency. So, for example, the rate is influenced by the fact that cryptocurrencies are not widespread to the same extent as fiat money at the moment. And one of the most important functions of digital money is money transfers. However, if bitcoins are transferred to a country with a low level of cryptocurrency distribution, then they are immediately sold, which leads to a decrease in the price.

As for media factors, it is obvious that the demand for currency (which appears to be the main factor determining the exchange rate) depends on the level of people's trust in this currency greatly, which predetermines the desire of people to acquire or mine bitcoins accordingly. So, according to some research studies, the demand for bitcoin was largely influenced by the information emerging in the media about the forecasts for the rate of this cryptocurrency.

In this paper, we have studied only economic factors, while others were taken as external variables (in the residuals of the model):

1) The exchange rate of the dollar against the euro;
2) The exchange rate of gold in dollars;
3) Price of the Hang Seng stock index in dollars;
4) The exchange rate of Ethereum (ETH) in dollars.

Before proceeding with the econometric analysis of the data, let us establish the meaning of some concepts.

The Hang Seng Index (HSI) is an index of the Hong Kong Stock Exchange [25], which is the weighted average of the prices of shares of the 34 largest joint-stock companies in Hong Kong. This index includes companies in four sectors: finance, utilities, land ownership and manufacturing (Bloomberg website).

Ethereum (Ethereum) is another cryptocurrency, named by the Canadian creator only as "Cryptocurrency 2.0" or "Bitcoin 2.0", which is still justified since at the moment this cryptocurrency has the second-largest capitalization after the oldest cryptocurrency bitcoin (as of 05/01/2020 this value was $ 23 billion against $ 162 billion of bitcoin capitalization). This cryptocurrency is much younger – its first release took place on July 30, 2015. However, according to its creator, this cryptocurrency appeared considering the errors of digital predecessors, despite the general similarity (for example, both systems – Bitcoin and Ethereum – are open: that is, they have security, anonymity and transparency of operations).

The main advantages that this cryptocurrency has gained over Bitcoin are as follows [26–28].

1) This system is designed for mining "at home". That is, the fact that this cryptocurrency can be mined by ordinary users using their personal computers (while, mining of bitcoin has become virtually impossible in such conditions due to the significantly increased complexity of bitcoin algorithms, so even if the computer will work the whole day, the miner will receive only a penny and will not even cover the costs of wearing out of equipment and payment for electricity);
2) Optimization of the Ethereum block made it possible to increase the speed of transactions by approximately 8–10 times;
3) The reward for the mined cryptocurrency does not decrease over time, as happens with bitcoin (its reward is reduced by 2 times every four years), since the volume of issuance of Ethereum in the system is not limited to the same 21 million units as bitcoin.

Overall, Ethereum has tremendous potential, perhaps even more than Bitcoin. It is Ethereum that is the ancestor of "smart contracts" which has already been noticed and started to be adopted by such large companies as Microsoft and IBM [26, 27].

4 Methods, Modelling, and Assessment of the Study Results

For our quantitative study, we selected data with an interval of one week, starting from March 16, 2016, to September 20, 2017 (80 observations for testing the model under conditions of progressive growth of the exchange rate of Bitcoin). Presumably, this period is associated not only with the rising interest in Bitcoin in the global media but also with the strengthening of the status of Ethereum as a reliable and convenient cryptocurrency for mining. All calculations were performed using the econometric software application package Gretl 1.9.92.

Before constructing the model, a number of the variables studied were checked for stationarity. Evaluation of a non-stationary series can lead to distortion of the results or the appearance of the so-called "fictitious regression", the estimates of which do not allow making adequate conclusions.

Based on the results obtained during the Dickey-Fuller test and the ACF and PACF graphs, it was found that the series are not stationary and it is necessary to take successive differences until stationarity is obtained as a result of verification.

The series was analyzed by evaluating the ARIMA (p, d, q) model, which is often used to predict prices in the stock markets [29]. Following the methodology, it is necessary to select different models with different values of the parameters p and q, and then choose the one that is most suitable for the information criteria.

After designing models with different specifications, it was found that most of the selected indicators are insignificant (with linear, square logarithmic and semi-logarithmic dependencies). As a result of the selection, ARIMA(2,2,2) turned out to be the model with the highest quality:

$$\begin{aligned} BTCUSD_t = & -1,04BTCUSD_{t-1} - 0,5BTCUSD_{t-2} + 2,8ETHUSD_{t-1} \\ & 7,8ETHUSD_{t-2} + 4,97ETHUSD_{t-3} + \varepsilon_t - 1,99\varepsilon_{t-1} - 0,99_2\varepsilon_{t-2}. \end{aligned} \quad (1)$$

Model (1) is a linear multiple regression model, in which the explanatory variables are the past values of the most dependent variable, and the regression residual is the moving average of white noise elements. The order of the moving average is determined by the number of previous values of random deviations taken into account in the model. All coefficients for this model are significant. A larger (or smaller) order of p or q leads to a decrease in the statistics of tests and decreases the level of significance of other parameters. Of all the models studied and modelled, we chose the model that has the best statistics and the minimum values of parameters p and q.

To calculate a retrospective forecast of the Bitcoin rate dynamics, we could transform the model by opening the differences or performing the opposite actions. The graphical results of forecasting are shown in Fig. 1.

This model gives us more obvious results in many respects. The fluctuations are not so noticeable, although they repeat the dynamics of the initial series of observations in general. The model can predict the Bitcoin exchange rate fluctuations in many ways, but it reacts to these fluctuations too strongly (perhaps, since the model with a certain coefficient should include the shocks of previous periods, which we are not able to calculate within the framework of this study).

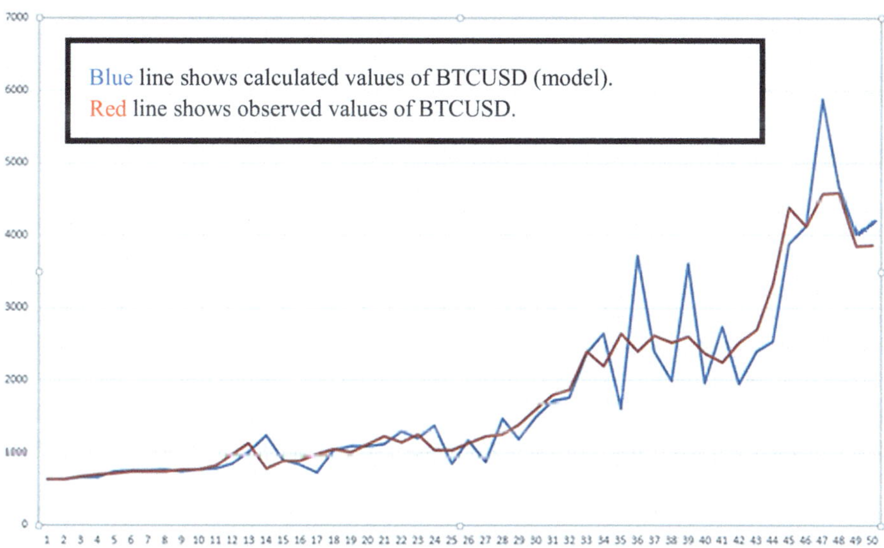

Fig. 1. Calculated and observed values of the dependent variable BTCUSD.

Despite the positive results of all tests and the significance of the model, we could assume that, given the calculated values, the model can only be the basis for a preliminary analysis of the Bitcoin rate factors. The dependencies are explained quite logically, however, the estimates may be somewhat distorted. This problem is most likely related to the results of checking for the ARCH processes in observed values.

The presence of ARCH processes implies the dependence of the conditional variance on the past values of the variance. Accordingly, the assumption about the absence of such relationships, as in the process called white noise, are violated, which leads us to somewhat distortion of the results and forecasts.

Usually, such outliers are explained by special periods of high volatility (usually short), which are features of ARCH processes. That is why, for a more accurate result and analysis in subsequent studies, it is necessary to design an improved model that would consider the presence of autoregressive conditional heteroscedasticity.

This study was aimed at identifying the main pricing factors of Bitcoin. Understanding such a process would make it possible to build a forecast for the future values of the Bitcoin exchange rate. This could help us conclude what could be the situation in various cryptocurrency markets, as far as the very nature of financial markets allows.

In the course of our research study, many tests were carried out and more than 600 models were built in various ways and with the help of various methods to determine the one that describes the behaviour of the bitcoin rate the best.

As a result, a model was built that included the lag values for the dependent variable – the Bitcoin rate of exchange to the US dollar, the Ethereum rate lags and the previous values of random deviations. All other variables turned out to be insignificant or significant at a significance level of about 10%, which was inappropriate for our study.

Thus, we can conclude that, according to the model, the bitcoin rate largely depends on its previous values (the previous two weeks). Presumably, exactly two weeks are the period of the full adaptation of the crypto market to the changing exchange rate. The third and fourth lags of the variable have a negative coefficient. This tells us that approximately every month (3–4 weeks) the bitcoin rate falls as if stabilizing and then goes up again. However, this phenomenon is not entirely regular and, according to the analysis, has no obvious factors for seasonality.

The Ethereum price also had a strong impact on the Bitcoin price as a rival cryptocurrency. These two currencies are developing in parallel, but many see more prospects in Ethereum. For this reason, Etherium has an unstable effect on the Bitcoin rate (both significantly positive and significantly negative in different periods). All this is possible due to the fluctuating demand of "rushing" buyers and the constant exchange and trading of these currencies on crypto exchanges.

The Bitcoin rate in the past period (a week ago) had the greatest impact on the current value of the Bitcoin rate to the dollar, which was quite predictable. During the week, people manage to follow the trend and, if the rate grew in the previous period, they strive to acquire as much of this cryptocurrency as possible, which increases demand and, accordingly, this leads to an inevitable increase in the rate in the future. This is how we get a consistent increase in the rate until it reaches too high values for a specific period. Then demand falls and the rate abruptly breaks off, trying to come to a more stable level, in which the number of buyers and sellers wishing to buy or sell bitcoin for a specific price will be the greatest.

The chosen model also includes the previous values (the previous two weeks) of random deviations. This, again, is extremely logical for the normal situation in the financial market and the situation in the cryptocurrency market, which is also influenced by the media. For example, the news about the ban of cryptocurrencies in China in September 2017 sharply cut off the bitcoin rate, as it expressed the fall of trust in this phenomenon in one of the largest economies due to government policy.

Thus, in the course of econometric analysis, a model was designed that described the variables that were the most influencing on the bitcoin rate. This model makes it possible to carry out a preliminary forecast of data, which promises (and confirms) the continuation of the tendencies of the Bitcoin exchange rate to growth in all subsequent periods (taking into account the periodic collapses of the exchange rate).

However, the model obtained in this study is not suitable for predicting the exact values of the bitcoin rate in future. The limitation of this study is the method. The analysis revealed the existence of ARCH processes typical for currency and financial markets. This may indeed be the case for bitcoin, since, as we noted, it is highly dependent on expectations and media factors and is highly susceptible to supply and demand shocks, especially. For further analysis, it is necessary to design a refined model that takes into account these phenomena, for example, an ARCH/GARCH model.

Earlier, the authors of this paper revealed the dependence of the bitcoin exchange rate on the main factors of demand for world reserve currencies in the period 2014–2019 [30]. During the simulation, it was found that the factor of the renminbi exchange rate as the national currency of China (the country with the largest share of miners - Bitcoin producers), is of particular importance. In other words, the supply factors of the cryptocurrency are important in addition to demand factors. The results of this study on the factors of the bitcoin exchange rate during the period of bitcoin penetration into various sectors of the economy in 2016–2017 complement the earlier research findings. So we conclude that during the shorter-term period the factors of the price of bitcoin were more related to the factors of demand for it when there was a significant increase in interest in cryptocurrency in the media.

The cryptocurrency market is outside the control of any regulator, which makes its analysis more difficult. In reality, not so many factors affect the rates of cryptocurrencies, while, according to the results of the analysis, the value of the bitcoin rate in previous periods has a greater influence on the current value of the bitcoin rate. However, those factors that still affect the rate have characteristics that are difficult to be analyzed and predicted. Nevertheless, the study allows using the resulting model to predict the overall dynamics of the bitcoin rate. However, the results to be obtained will not take the especially significant shocks inherent in this market into account. In many ways, the modelling results could be "more optimistic" than in reality. In other words, the bitcoin rate could be significantly overestimated during periods of growth or underestimated during periods of drops. Presumably, economic factors are sufficient to build a high-quality forecast, however, we assume to build a refined model for a better adaptation of the model to significant shocks in future studies.

5 Conclusions

The purpose of this study was to determine the degree of influence of the selected indicators on the bitcoin rate. In the course of the study, many indicators were found to be insignificant, surprisingly including the Hong Kong Stock Exchange Index, which was supposed to be of paramount importance in the entire model. All these insignificance can be attributed to the fact that bitcoin, like any other cryptocurrency, is a special financial asset not under the control of any "third party". When selecting variables that potentially have an impact on the bitcoin rate, it was assumed that the gold rate or the stock index rate could have an indirect effect on the bitcoin price as an alternative investment option or as an indicator of trends in financial markets. However, the results were different. The study of the variables influencing the bitcoin exchange rate to dollar confirmed the speculative nature of the demand for bitcoin. In many ways, the bitcoin rate in the current period depends on the previous values of the rate and various kinds of previous periods shocks. The dependence of the Bitcoin rate on the cost of Ethereum, the second-largest cryptocurrency, was also revealed, which proved their "rivalry".

A regression model was compiled, which allowed us to carry out a retrospective forecast of the bitcoin rate dynamics to detect the periods in which, there is a high probability of the rate to rise and the periods in which it could collapse. In general, the value of Bitcoin fluctuates around some constantly increasing trend to a "peak" point in the period. Then a sharp drop in demand happens, leading to a decrease in the rate, and then the cycle repeats until the next peak is reached, and so on. However, the cycles have different lengths, so predictive models can be useful in assessing course dynamics.

References

1. Nian, L.P., Chuen, D.L.K.: Introduction to bitcoin. Handbook of digital currency. Academic Press, 5–30 (2015)
2. DeVries, P.D.: An analysis of cryptocurrency, bitcoin, and the future. Int. J. Bus. Manage. Commer. **1**(2), 1–9 (2016)
3. Yermack, D.: Is Bitcoin a real currency? an economic appraisal. Handbook of digital currency. Academic Press, 31–43 (2015)
4. Leung, D., Dickinger, A.: Use of bitcoin in online travel product shopping: The European perspective. Information and communication technologies in tourism 2017. Springer, Cham, 741–754 (2017)
5. Cheah, E.T., Fry, J.: Speculative bubbles in Bitcoin markets? an empirical investigation into the fundamental value of Bitcoin. Econ. Lett. **130**, 32–36 (2015)
6. Demir, E., Gozgor, G., Lau, C.K.M., Vigne, S.A.: Does economic policy uncertainty predict the Bitcoin returns? Empirical Invest. Finance Res. Lett. **26**, 145–149 (2018)
7. Ciaian, P., Rajcaniova, M., Kancs, D.: The digital agenda of virtual currencies: can bitcoin become a global currency? IseB **14**(4), 883–919 (2016). https://doi.org/10.1007/s10257-016-0304-0
8. Atsalakis, G.S., Atsalaki, I.G., Pasiouras, F., Zopounidis, C.: Bitcoin price forecasting with neuro-fuzzy techniques. Eur. J. Oper. Res. **276**(2), 770–780 (2019)
9. Makarov, I., Schoar, A.: Trading and arbitrage in cryptocurrency markets. J. Financ. Econ. **135**(2), 293–319 (2020)

10. Hayes, A.S.: Cryptocurrency value formation: an empirical study leading to a cost of production model for valuing bitcoin. Telematics Inform. **34**(7), 1308–1321 (2017)
11. Nakamoto, S.: Bitcoin: a peer-to-peer electronic cash system. Bitcoin. https://bitcoin.org/bitcoin.pdf. Accessed 25 Nov 2020
12. Birch, D.G.: What does cryptocurrency mean for the new economy? handbook of digital currency. Academic Press, 505–517 (2015)
13. Vasek, M., Thornton, M., Moore, T.: Empirical Analysis of Denial-of-Service Attacks in the Bitcoin Ecosystem, pp. 57–71. International conference on financial cryptography and data security. Springer, Berlin, Heidelberg (2014)
14. Bouoiyour, J., Selmi, R., Tiwari, A.K., Olayeni, O.R.: What drives Bitcoin price. Econ. Bull. **36**(2), 843–850 (2016)
15. Schilling, L., Uhlig, H.: Some simple bitcoin economics. J. Monet. Econ. **106**, 16–26 (2019)
16. Ruffing, T., Moreno-Sanchez, P., Kate, A.: Coinshuffle: Practical Decentralized Coin Mixing for Bitcoin, pp. 345–364. European Symposium on Research in Computer Security. Springer, Cham (2014)
17. Gürcan, Ö., Ranchal Pedrosa, A., Tucci-Piergiovanni, S.: On cancellation of transactions in bitcoin-like blockchains. In: Panetto, H., Debruyne, C., Proper, H.A., Ardagna, C.A., Roman, D., Meersman, R. (eds.) OTM 2018. LNCS, vol. 11229, pp. 516–533. Springer, Cham (2018). https://doi.org/10.1007/978-3-030-02610-3_29
18. Sapuric, S., Kokkinaki, A.: Bitcoin is volatile! Isn't that right? International Conference on Business Information Systems, Springer, Cham, pp. 255–265 (2014).https://doi.org/10.1007/978-3-319-11460-6_22
19. Seetharaman, A., Saravanan, A.S., Patwa, N., Mehta, J.: Impact of bitcoin as a world currency. Account. Finance Res. **6**(2), 230–246 (2017)
20. Kethineni, S., Cao, Y., Dodge, C.: Use of bitcoin in darknet markets: examining facilitative factors on bitcoin-related crimes. Am. J. Crim. Justice **43**(2), 141–157 (2018)
21. Saito, T.: A microeconomic analysis of Bitcoin and illegal activities. Handbook of Digital Currency. Academic Press, pp. 231–248 (2015)
22. Yelowitz, A., Wilson, M.: Characteristics of Bitcoin users: an analysis of Google search data. Appl. Econ. Lett. **22**(13), 1030–1036 (2015)
23. Ciaian, P., Rajcaniova, M., Kancs, D.A.: The economics of BitCoin price formation. Appl. Econ. **48**(19), 1799–1815 (2016)
24. Ovsyannikova, P.: Bitcoin: factors affecting cryptocurrency volatility. Electr. Bull. Rostov Soc. Econ. Inst. **2**, 254–260 (2016)
25. Heng, S.: Index. Financial financial, software, data, and media company Bloomberg. https://www.bloomberg.com/quote/HSI:IND. Accessed 09 Dec 2020
26. Mohanty, D.: Deploying Smart Contracts, pp. 105–138. Ethereum for Architects and Developers. Apress, Berkeley (2018)
27. Mohanty, D.: Ethereum use cases. Ethereum for Architects and Developers. Apress, Berkeley, CA, pp. 203–243 (2018).https://doi.org/10.1007/978-1-4842-4075-5_9
28. Mensi, W., Al-Yahyaee, K.H., Kang, S.H.: Structural breaks and double long memory of cryptocurrency prices: a comparative analysis from Bitcoin and Ethereum. Financ. Res. Lett. **29**, 222–230 (2019)
29. Ariyo, A.A., Adewumi, A.O., Ayo, C.K.: Stock price prediction using the ARIMA model. In: 2014 UKSim-AMSS 16th International Conference on Computer Modelling and Simulation. IEEE, pp. 106–112 (2014)
30. Rutskiy, V., et al.: The price determinants of bitcoin as a new digital form of money. In: Silhavy, R., Silhavy, P., Prokopova, Z. (eds.) CoMeSySo 2020. AISC, vol. 1294, pp. 784–792. Springer, Cham (2020). https://doi.org/10.1007/978-3-030-63322-6_66

A Machine Learning Firefly Algorithm Applied to the Matrix Covering Problem

Gabriel Villavicencio[1(✉)], Matias Valenzuela[1], Leonardo Causa[2], Paola Moraga[1], and Hernan Pinto[1]

[1] Escuela de ingeniería en construcción, Pontificia Universidad Católica de Valparaíso, Valparaíso, Chile
{gabriel.villavicencio,matias.valenzuela,paola.moraga, hernan.pinto}@pucv.cl

[2] Facultad de Ingeniería y Negocios, Universidad de las Américas, Santiago, Chile
lcausamo@edu.udla.cl

Abstract. At the business level, there are a large number of combinatorial problems. A subset of these is NP-hard type. The study of algorithms that address this type of problem is of great interest. On the other hand, there are a large number of metaheuristic algorithms that naturally work in continuous spaces. Adapting the latter to solve combinatorial problems is of great interest at an industrial level. In this article, we explore a general binarization mechanism of continuous metaheuristics based on cauterization techniques. The experiments are designed to demonstrate the utility of the clustering technique in binarization. Besides, we verify the effectiveness of our algorithm through reference instances. The results indicate that the binary firefly optimization algorithm (MLFA) obtains adequate results when evaluated with a combinatorial problem such as the SCP.

1 Introduction

The combinatorial problems have great relevance at industrial level, we find them in various areas such as Operational Research [1–3], resource allocation [4–7], scheduling problems [8–11], routing problems [11–15], civil engineering [16–20], among others. On the other hand, in recent years, algorithms inspired by nature phenomena to solve optimization problems have been generated. As examples of these algorithms we have Cuckoo Search [21], Black Hole [22], Bat Algorithm [23], and Firefly Algorithm [24] among others. Many of these algorithms work naturally in continuous spaces and therefore must be adapted to solve combinatorial problems. In the process of adaptation, the mechanisms of exploration and exploitation of the algorithm can be altered, having consequences in the efficiency of the algorithm.

Several binarization techniques have been developed to address this situation. In a literature search, the main binarization methods used correspond to transfer functions, angle modulation and quantum approach. In this article we present a binarization method that uses the k-means technique to group the solutions and

© The Author(s), under exclusive license to Springer Nature Switzerland AG 2021
R. Silhavy (Ed.): CSOC 2021, LNNS 229, pp. 316–325, 2021.
https://doi.org/10.1007/978-3-030-77445-5_29

then perform the binarization process. To verify the efficiency of our method, we used the MLFA. This algorithm was proposed in [24].

To check our MLFA, we use the well-known set covering problem. Experiments were developed using a random operator to validate the contribution of the k-means technique in the binarization process of the Grasshopper algorithm. In addition, to verify our results, the JPSO algorithm developed by [25] and MDBBH developed in [8] were chosen. The results show that MLFA algorithm obtain competitive results.

2 Set Covering Problem

The set covering problem corresponds to a classical problem in combinatorics, and complexity theory. The problem aims to find subsets of a set. Given a set and its elements, we want to find subsets that completely cover the set at a minimum cost. The SCP is one of the oldest and most studied optimization problems. It is well-known to be NP-hard [26].

SCP is an active problem because medium and large instances often become intractable and cannot be solved any more using exact algorithms. Additionally, SCP due to its large number of instances, is used to verify the behavior of proposed new algorithms. In recent years, SCP has been approached by various continuous Swarm intelligence metaheuristics and using different methods to binarize metaheuristics. In [27], they used the teaching-learning-based optimisation with a specific binarization scheme to solve medium and large size instances. An SCP in fault diagnosis application was developed by [28]. In this article, the Gravitational Search algorithm was used and transfer functions were applied to perform the binarization. A rail scheduling problem application using SCP was developed in [8]. In [25] a Jumping PSO was used to solve SCP and in [29] CS metaheuristics were binarized using a k-means algorithm.

SCP has many practical applications in engineer, e.g., vehicle routing, facility location, railway, and airline crew scheduling [1,30–32] problems.

The SCP can be formally defined as follows. Let $A = (a_{ij})$, be a $n \times m$ zero-one matrix, where a column j cover a row i if $a_{ij} = 1$, besides a column j is associated with a non-negative real cost c_j. Let $I = \{1, ..., n\}$ and $J = \{1, ...m\}$, be the row and column set of A, respectively. The SCP consists in searching a minimum cost subset $S \subset J$ for which every row $i \in I$ is covered by at least one column $j \in J$, i.e.,:

$$\text{Minimize } f(x) = \sum_{j=1}^{m} c_j x_j \tag{1}$$

$$\text{Subject to } \sum_{j=1}^{m} a_{ij} x_j \geq 1, \forall i \in I, \text{ and } x_j \in \{0, 1\}, \ \forall j \in J \tag{2}$$

where $x_j = 1$ if $j \in S$, $x_j = 0$ otherwise.

3 Binary Firefly Algorithm

The application of statistics and machine learning techniques, are used in many areas such as transports, smart cities, agriculture and computational intelligence, examples of applications are found in [33–36, 36, 37]. In this article, we explore the k-means technique applied to binarization of continuous metaheuristics. As an algorithm to binarize, Firefly is used and as problem to solve, SCP. The Proposed MLFA has two modules. The first module corresponds to the initialization of the feasible solutions, and is detailed in the Sect. 3.1. Once the initialization of the particles is performed, it is consulted if the maximum if iterations is satisfied. Subsequently if the criterion is not satisfied, The MLFA is then run in conjunction with the k-means technique, this is detailed in Sect. 3.2). Once the transitions of the different solutions are made, we compare the resulting solutions with the best solution previously obtained. In the event that a superior solution is found, this replaces the previous one. The general algorithm scheme is detailed in Fig. 1.

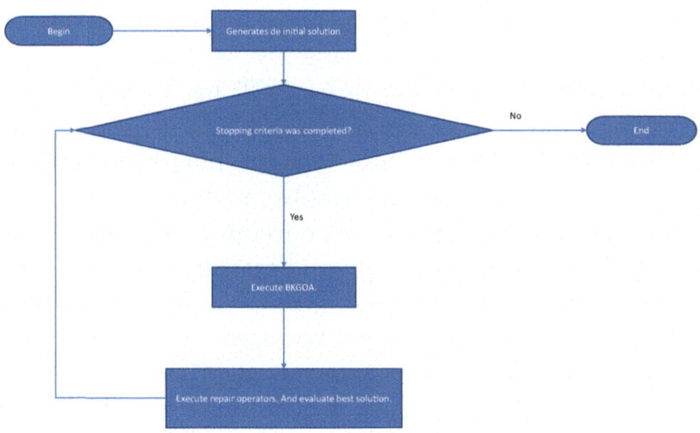

Fig. 1. Flowchart of the MLFA algorithm.

3.1 Initialization

For initialization of a new solution, a column is randomly chosen. It is then queried whether the current solution covers all rows. In the case of the solution does not meet the coverage condition, the heuristic operator is called (Sect. 3.3). This operator aims to select a new column. This heuristic operation is iterated until all rows are covered. Once the coverage condition is met, the solution is optimized. The optimization consists of eliminating columns where all their rows are covered by more than one column. The detail of the procedure is shown in Algorithm 1.

Algorithm 1. Initialization Operator

Function Initialization()
Input
Output Initialized solution S_{out}
$S \leftarrow$ SRandomColumn()
while All row are not covered **do**
    S.append(Heuristic(S))
end while
$S \leftarrow$ dRepeatedItem(S)
$S_{out} \leftarrow S$
return S_{out}

3.2 Machine Learning Transition Ranking Operator

Consider our metaheuristic is a continuous and population based. Due to its iterative nature, it needs to update the position of particles at each iteration. When the metaheuristic is continuous, this update is performed in $\mathbb{R}^n$ space. In Eq. 3, the update position is presented in a general manner. The x_{t+1} variable represents the x position of the particle at time $t+1$. This position is obtained from the position x at time t plus a Δ function calculated at time $t+1$. The function Δ is proper to each metaheuristic and produces values in $\mathbb{R}^n$. For example in Cuckoo Search $\Delta(x) = \alpha \oplus Levy(\lambda)(x)$, in Black Hole $\Delta(x) = \text{rand} \times (x_{bh}(t) - x(t))$ and in the Firefly, Bat and PSO algorithms Δ can be written in simplified form as $\Delta(x) = v(x)$.

$$x_{t+1} = x_t + \Delta_{t+1}(x(t)) \quad (3)$$

In the k-means transition ranking operator, a model for transitions in a discrete space is proposed. This model is based on considering the movements generated by the metaheuristic in each dimension for all particles. $\Delta^i(x)$ corresponds to the magnitude of the displacement $\Delta(x)$ in the i-th position. Subsequently these displacement are grouped using the magnitude of the displacement $\Delta^i(x)$. This grouping is done using the k-means technique where k represents the number of clusters used. Finally, a generic P_{tr} function given by the Eq. 4 is proposed to assign a transition probability.

$$P_{tr} : \mathbb{Z}/k\mathbb{Z} \rightarrow [0,1] \quad (4)$$

A transition probability through the function P_{tr} is assigned to each group. Naturally, this P_{tr} function is modelled as a cumulative probability function. For the case of this study, we particularly use the linear function given in Eq. 5. In this equation, $N(x^i)$ indicates the location of the group to which $\Delta^i(x)$ belongs. The α coefficient, corresponds to the transition probability and β to the transition separation coefficient. Both parameters are estimated in each metaheuristic. For our particular case, $N(x^i) = 0$ corresponds to elements belonging to the group that has the lowest Δ^i values. N(x^i) = 7 corresponds to the group of elements that have the greatest Δ^i values.

$$P_{tr}(x^i) = P_{tr}(N(x^i)) = \alpha + \beta N(x^i)\alpha \tag{5}$$

The k-means transition ranking operator starts calculating Δ^i for each of the particles. This step is specific in each metaheuristic. Subsequently the particles are grouped using k-means clustering technique and the magnitude of Δ^i as distance. With the group assigned to each particle we obtain the probability of transition using Eq. 5. Afterwards the transition of each particle is performed. In the case of grasshopper algorithm the rule 6 is used to perform the transition, where $\hat{x}^i$ is the complement of x^i. Finally, each solution is repaired using the repair operator.

$$x^i(t+1) := \begin{cases} \hat{x}^i(t), \text{ if } rand < P_{tg}(x^i) \\ x^i(t), \qquad \text{otherwise} \end{cases} \tag{6}$$

3.3 Heuristic Operator

The goal of the Heuristic operator is to select a new column for cases where a solution needs to be built or repaired. As input variables, the heuristic operator considers the incomplete solution S_{in} which must be completed. The operator obtains the columns that belong to S_in, then obtains the rows R that are not covered by the solution to S_in. With the set of rows not covered and using Eq. 7 we obtain in line 4 the best 5 rows to be covered. With this list of rows ($lRows$) on line 5 we obtain the list of the best columns according to the heuristic indicated in Eq. 8. Finally randomly in line 6 we obtain the column to incorporate.

$$WeightRow(i) = \frac{1}{L_i}. \tag{7}$$

where L_i is the sum of all ones in row i

$$WeightColumn(j) = \frac{c_j}{|R \cap M_j|}. \tag{8}$$

where M_j is the set of rows covered by Col j

Algorithm 2. Heuristic operator

1: **Function** Heuristic(S_{in})
2: **Input** Input solution S_{in}
3: **Output** The new column *colOut*
4: $lRows \leftarrow$ getBestRows(S_{in}, N=10)
5: $listcolumsnOut \leftarrow$ getBestCols($LRows$, M=5)
6: $colOut \leftarrow$ getCol($lcolsOut$)
7: **return** *colOut*

4 Results

In this section we detail the behavior of MLFA when it is applied to SCP. The contribution of the binary k-means operator was studied when solving the different SCP instances. Additionally, MLFA was compared with other algorithms that have recently resolved SCP. To solve the different SCP instances, a PC with Windows 10, core i5 processor and 16GB in RAM was used. The program was coded in Python 3.5. For the statistical analysis, the non-parametric Wilcoxon signed-rank test was used in addition to violin charts. The analysis of violin charts is performed by comparing the dispersion, median and the interquartile range of the distributions.

4.1 Insight of MLFA Algorithm

In this module we developed experiments that allowed us to study the contribution of the binary k-means operator with respect to the quality of the solutions obtained and the iterations used when solving SCP instances. To perform this evaluation, we used the instances of Balas and Carrera. In the comparison, a random operator was designed which replaces our binary k-means operator. This random operator instead of executing transitions assigning transition probabilities per group, uses a fixed value of 0.5. To compare the distributions of the results of the different experiments we use violin Chart. The horizontal axis X corresponds to the problems, while Y axis uses the measure % - Gap defined in Eq. 9

$$\% - Gap = 100\frac{BestKnown - SolutionValue}{BestKnown} \tag{9}$$

Furthermore, a non-parametric test, Wilcoxon signed-rank test is carried out to determine if the results of MLFA with respect to the random algorithm have significant difference or not. The parameter settings and browser ranges are shown in Table 1.

Table 1. Setting of parameters for Binary percentile firefly Algorithm.

Parameters	Description	Value	Range
N	Number of solutions	25	[20, 25, 30]
K	Number of clusters	5	[4,5,6]
Iteration Number	Maximum iterations	800	[700,800,900]

MLFA corresponds to our standard algorithm. *random.0.5* is the random variant. The results are shown in Table 2 and Fig. 2.

When we compared the Best Values between MLFA and *rnd.0.5* which are shown in Table 2. In the Average comparison, MLFA outperforms *rnd.0.5* in

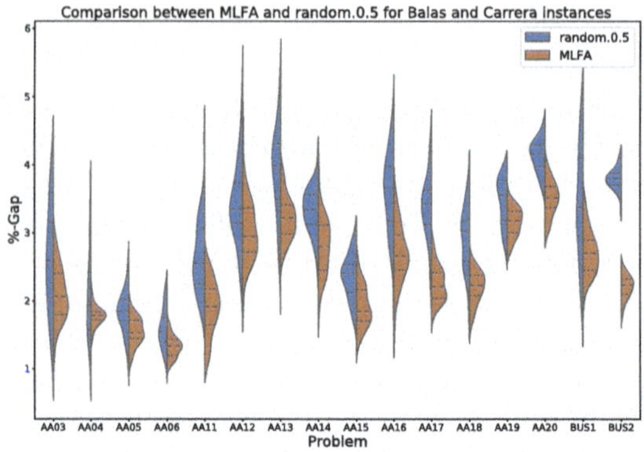

Fig. 2. Evaluation of k-means binary operator.

Table 2. Balas and Carrera instances.

Instance	Row	Col	Density	Best known	MLFA (avg)	Std	(Secs) (Secs)	*rnd.0.5* (avg)	Std Std
AA03	106	8661	4.05%	33155	33355.1	128.1	212	33693.5	247.6
AA04	106	8002	4.05%	34573	34989.7	57.4	201	34931	233.1
AA05	105	7435	4.05%	31623	31726.3	65.2	213	31889.2	113.8
AA06	105	6951	4.11%	37464	37675.9	56.1	216	37651.3	99.4
AA11	271	4413	2.53%	35384	35802.1	130.1	199	35901.2	196.3
AA12	272	4208	2.52%	30809	31430.8	152.7	217	31631.3	218.6
AA13	265	4025	2.60%	33211	34031.8	136.6	198	34201.4	236.8
AA14	266	3868	2.50%	33219	33765.3	127.2	157	33967.3	127.3
AA15	267	3701	2.58%	34409	35017.2	115.1	151	34846.8	166.1
AA16	265	3558	2.63%	32752	33204.8	136.4	185	33601.3	188.9
AA17	264	3425	2.61%	31612	32001.4	88.3	172	32387.7	152.1
AA18	271	3314	2.55%	36782	37176.8	77.3	147	37419.3	149.5
AA19	263	3202	2.63%	32317	33208.2	79.2	149	33108.3	147.8
AA20	269	3095	2.58%	34912	35853.6	99.4	141	35962.6	102.1
BUS1	454	2241	1.88%	27947	28406.1	82.1	197	28758.2	187.2
BUS2	681	9524	0.51%	67760	68823.6	84.2	174	68680.6	121.1
Average				35495.6	36029.3	101.1	183.1	36164.4	168.0
p-value								2.4e7−03	

all problems. The comparison of distributions is shown in Fig. 2. We see the dispersion of the *rnd.0.5* distributions are bigger than the dispersions of MLFA. Therefore, the k-means binary operator, contribute to the precision of the results. Finally, the MLFA distributions are closer to zero than *rnd.0.5* distributions, indicating that MLFA has consistently better results than *random.0.5*. When we evaluate the behaviour of the algorithms through the Wilcoxon test, this test indicates that there is a significant difference between the two algorithms.

5 Conclusions

In this work, the MLFA algorithm based on the k-means technique was proposed to perform binarization of FA metaheuristics. We must emphasize that this k-means technique can be applied in the binarization of any continuous metaheuristics of the swarm intelligence type. To evaluate the performance of Our k-means technique the set covering problem was used together with the Grasshopper algorithm. Experiments were designed to evaluate the contribution of the binary operator k-means of the algorithm. It was found that the operator contributes significantly to improve the accuracy and quality of the solutions, all these visualized through violin graphics. Finally, compared to the best metaheuristic algorithm that has solved SCP, our algorithm had a lower yield of 1.26 %, which is not a big difference considering that JPSO uses a particular adaptation mechanism for PSO and the k-means technique It can be easily adapted to any technique. As future works, we believe that the autonomous search for efficient configuration is a little-explored area and it could help to improve the performance of the binarizations both in the quality of the solutions and in the times of convergence. Additionally, it is interesting to see how this k-means technique works with other metaheuristics and other NP-hard problems. Finally, we believe that using machine learning we can generate indicators or obtain correlations that allow us to understand how the k-means algorithm translates the exploration and exploitation properties from the continuous space to the binary space.

References

1. Crawford, B., Soto, R., Monfroy, E., Astorga, G., García, J., Cortes, E.: A meta-optimization approach for covering problems in facility location. In: Figueroa-García, J.C., López-Santana, E.R., Villa-Ramírez, J.L., Ferro-Escobar, R. (eds.) WEA 2017. CCIS, vol. 742, pp. 565–578. Springer, Cham (2017). https://doi.org/10.1007/978-3-319-66963-2_50
2. García, J., Crawford, B., Soto, R., Astorga, G.: A percentile transition ranking algorithm applied to binarization of continuous swarm intelligence metaheuristics. In: Ghazali, R., Deris, M.M., Nawi, N.M., Abawajy, J.H. (eds.) SCDM 2018. AISC, vol. 700, pp. 3–13. Springer, Cham (2018). https://doi.org/10.1007/978-3-319-72550-5_1
3. Tapia, D., et al.: A q-learning hyperheuristic binarization framework to balance exploration and exploitation. In: Florez, H., Misra, S. (eds.) ICAI 2020. CCIS, vol. 1277, pp. 14–28. Springer, Cham (2020). https://doi.org/10.1007/978-3-030-61702-8_2
4. García, J., Moraga, P., Valenzuela, M., Pinto, H.: A db-scan hybrid algorithm: an application to the multidimensional knapsack problem. Mathematics **8**(4), 507 (2020)
5. García, J., Altimiras, F., Peña, A., Astorga, G., Peredo, O.: A binary cuckoo search big data algorithm applied to large-scale crew scheduling problems. Complexity **2018** (2018)

6. García, J., Lalla-Ruiz, E., Voß, S., Droguett, E.L.: Enhancing a machine learning binarization framework by perturbation operators: analysis on the multidimensional knapsack problem. Int. J. Mach. Learn. Cybern. **11**(9), 1951–1970 (2020). https://doi.org/10.1007/s13042-020-01085-8
7. García, J., Maureira, C.: A kNN quantum cuckoo search algorithm applied to the multidimensional knapsack problem. Appl. Soft Comput. **102**, 107077 (2021)
8. García, J., Crawford, B., Soto, R., García, P.: A multi dynamic binary black hole algorithm applied to set covering problem. In: Del Ser, J. (ed.) ICHSA 2017. AISC, vol. 514, pp. 42–51. Springer, Singapore (2017). https://doi.org/10.1007/978-981-10-3728-3_6
9. García, J., Crawford, B., Soto, R., Astorga, G.: A clustering algorithm applied to the binarization of swarm intelligence continuous metaheuristics. Swarm Evol. Comput. **44**, 646–664 (2019)
10. Crawford, B., Soto, R., Astorga, G., García, J.: Constructive metaheuristics for the set covering problem. In: Korošec, P., Melab, N., Talbi, E.-G. (eds.) BIOMA 2018. LNCS, vol. 10835, pp. 88–99. Springer, Cham (2018). https://doi.org/10.1007/978-3-319-91641-5_8
11. García, J., Astorga, G., Yepes, V.: An analysis of a KNN perturbation operator: An application to the binarization of continuous metaheuristics. Mathematics **9**(3), 225 (2021)
12. García, J., Peña, A.: Robust optimization: concepts and applications. In: Nature-inspired Methods for Stochastic, Robust and Dynamic Optimization, p. 7 (2018)
13. Astorga, G., Crawford, B., Soto, R., Monfroy, E., García, J., Cortes, E.: A meta-optimization approach to solve the set covering problem. Ingeniería **23**(3), 274–288 (2018)
14. García, J., et al.: A db-scan binarization algorithm applied to matrix covering problems. Comput. Intell. Neurosci. **2019** (2019)
15. García, J., Măntoiu, M.: Localization results for zero order pseudodifferential operators. J. Pseudo-Differ. Oper. Appl. **5**(2), 255–276 (2013). https://doi.org/10.1007/s11868-013-0084-y
16. Crawford, B., et al.: Investigating the efficiency of swarm algorithms for bridge strengthening by conversion to tied-arch: A numerical case study on san luis bridge. Iranian J. Sci. Technol. Trans. Civil Eng. 1–13 (2020)
17. García, J., Martí, J.V., Yepes, V.: The buttressed walls problem: an application of a hybrid clustering particle swarm optimization algorithm. Mathematics **8**(6), 862 (2020)
18. García, J., Yepes, V., Martí, J.V.: A hybrid k-means cuckoo search algorithm applied to the counterfort retaining walls problem. Mathematics **8**(4), 555 (2020)
19. Yepes, V., Martí, J.V., García, J.: Black hole algorithm for sustainable design of counterfort retaining walls. Sustainability **12**(7), 2767 (2020)
20. Martínez-Muñoz, D., Martí, J.V., García, J., Yepes, V.: Embodied energy optimization of buttressed earth-retaining walls with hybrid simulated annealing. Appl. Sci. **11**(4), 1800 (2021)
21. Yang, X.-S., Deb, S.: Cuckoo search via lévy flights. In: World Congress on Nature & Biologically Inspired Computing, 2009. NaBIC 2009, pp. 210–214. IEEE (2009)
22. Hatamlou, A.: Black hole: a new heuristic optimization approach for data clustering. Inf. Sci. **222**, 175–184 (2013)
23. Yang, X.-S.: A new metaheuristic bat-inspired algorithm. In: Nature Inspired Cooperative Strategies for Optimization (NICSO 2010), pp. 65–74 (2010)
24. Yang, X.-S.: Firefly Algorithm (2008)

25. Balaji, S., Revathi, N.: A new approach for solving set covering problem using jumping particle swarm optimization method. Natural Comput. **15**(3), 503–517 (2015). https://doi.org/10.1007/s11047-015-9509-2
26. Gary, M.R., Johnson, D.S.: Computers and intractability. In: A Guide to the Theory of NP-Completeness (1979)
27. Lu, Y., Vasko, F.J.: An or practitioner's solution approach for the set covering problem. Int. J. Appl. Metaheuristic Comput. (IJAMC) **6**(4), 1–13 (2015)
28. Li, Y., Cai, Z.: Gravity-based heuristic for set covering problems and its application in fault diagnosis. J. Syst. Eng. Electron. **23**(3), 391–398 (2012)
29. García, J., Crawford, B., Soto, R., Astorga, G.: A percentile transition ranking algorithm applied to knapsack problem. In: Silhavy, R., Silhavy, P., Prokopova, Z. (eds.) CoMeSySo 2017. AISC, vol. 662, pp. 126–138. Springer, Cham (2018). https://doi.org/10.1007/978-3-319-67621-0_11
30. Kasirzadeh, A., Saddoune, M., Soumis, F.: Airline crew scheduling: models, algorithms, and data sets. EURO J. Transp. Logist. **6**(2), 111–137 (2015). https://doi.org/10.1007/s13676-015-0080-x
31. Horváth, M., Kis, T.: Computing strong lower and upper bounds for the integrated multiple-depot vehicle and crew scheduling problem with branch-and-price. Central Eur. J. Oper. Res. **27**(1), 39–67 (2017). https://doi.org/10.1007/s10100-017-0489-4
32. Stojković, M.: The operational flight and multi-crew scheduling problem. Yugoslav J. Oper. Res. **15**(1) (2016)
33. García, J., Crawford, B., Soto, R., Castro, C., Paredes, F.: A k-means binarization framework applied to multidimensional knapsack problem. Appl. Intell. **48**(2), 357–380 (2017). https://doi.org/10.1007/s10489-017-0972-6
34. García, J., Pope, C., Altimiras, F.: A distributed k-means segmentation algorithm applied to Lobesia Botrana recognition. Complexity **2017** (2017)
35. Graells-Garrido, E., García, J.: Visual exploration of urban dynamics using mobile data. In: García-Chamizo, J.M., Fortino, G., Ochoa, S.F. (eds.) UCAmI 2015. LNCS, vol. 9454, pp. 480–491. Springer, Cham (2015). https://doi.org/10.1007/978-3-319-26401-1_45
36. Graells-Garrido, E., Peredo, O., García, J.: Sensing urban patterns with antenna mappings: the case of Santiago, Chile. Sensors **16**(7), 1098 (2016)
37. Peredo, O.F., García, J.A., Stuven, R., Ortiz, J.M.: Urban dynamic estimation using mobile phone logs and locally varying anisotropy. In: Gómez-Hernández, J.J., Rodrigo-Ilarri, J., Rodrigo-Clavero, M.E., Cassiraga, E., Vargas-Guzmán, J.A. (eds.) Geostatistics Valencia 2016. QGG, vol. 19, pp. 949–964. Springer, Cham (2017). https://doi.org/10.1007/978-3-319-46819-8_66

Introduction into Autonomous Mobile Robot Research and Multi Cooperation

Stanislav Alexovič(✉), Milan Lacko, Ján Bačík, and Daniela Perduková

Department of Electrical Engineering and Mechatronics, Technical University of Košice, Košice, Slovakia
stanislav.alexovic@tuke.sk

Abstract. This paper is intended as an introduction into the research area of Autonomous Mobile Robot (AMR). It describes in first part basic design and usage of AMR with related problems and already used solutions and strategies to solve these problems. Second part is dedicated to cooperation among multiple AMRs. Architecture design, communication among AMRs and algorithms which could be used to solve path planning problems.

Keywords: AMR (Autonomous Mobile Robot) · AGV (Autonomous Guided Vehicle) · Fleet

1 Introduction

With increased automatization in industry, Autonomous Mobile Robots are more and more used to perform various tasks in the logistics segment of factories. Usually, they are assigned to a task which contains moving from point A to point B and some basic manipulation. Usually, they are loaded with an operator or other arm robot then they are moving to destination stations where they are unloaded or if they contain some automatic load/unload device they are capable to do this task fully automatically [1].

Another area of usage are warehouses. This is a logical and straightforward application for AMRs because of their nature to carry stuff but the main problem here is that AMR must be capable of automatically loading and unloading items from and to shelfs. This could sound like an easy task but in reality it is a difficult operation. AMR must perfectly know its environment and items which are going to operate. But this could be partially solved with adjusting the environment for AMR. It means that AMR will operate only with items which are suitable for it.

And there is one new area of usage for these robots. It is in a civil environment. There are a lot of opportunities to carry some stuff but cooperation in this environment is hard because it is very variable, always changing, and interacting with people.

But again, we could find some subareas where AMR would fit. E.g., for cleaning surfaces in hospitals [2, 3] or offices during night to avoid interaction with people. Or a new concept of disinfection of surfaces with germicidal UV lamps (see Fig. 1).

© The Author(s), under exclusive license to Springer Nature Switzerland AG 2021
R. Silhavy (Ed.): CSOC 2021, LNNS 229, pp. 326–336, 2021.
https://doi.org/10.1007/978-3-030-77445-5_30

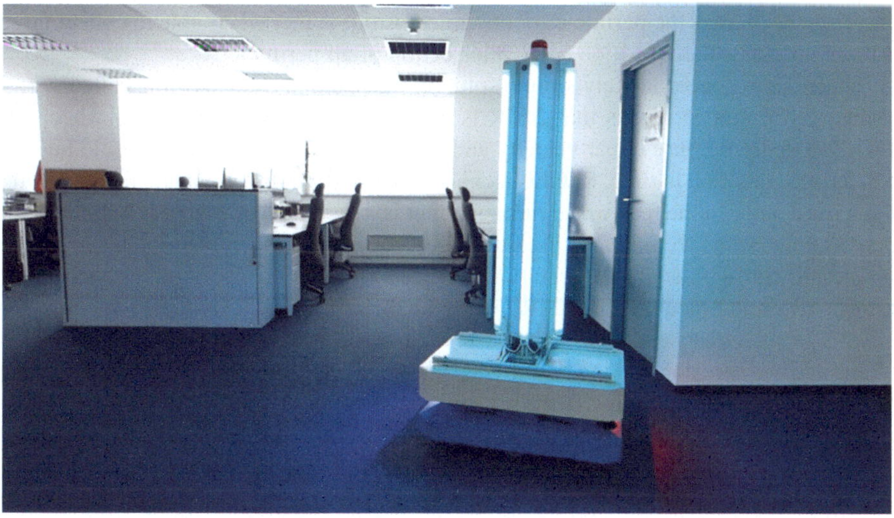

Fig. 1. AMR with germicide addon.

2 AMR (Autonomous Mobile Robot)

AMR or Autonomous Mobile Robot is the successor of AGV (Autonomous Guided Vehicle). AMR are considered as more intelligent robots which are capable of better navigation in its environment. AGVs are navigated with magnetic tape on the floor. It is closer to the railway system. While AMR is capable of navigation without this floor tape. Basically, you define a goal station and AMR should be able to find a path there and even avoid collisions with people or other objects which it meets during movement [4].

Eyes for AGV are only sensors for magnetic tape. And navigation is to follow tape until it reaches goal station which must be somehow marked on the floor. But AMR needs more sophisticated eyes and more intelligent navigation. There are two main components which can provide this functionality. 3D cameras and lidars. Both measure distance to the surrounding environment [2].

2.1 Navigation

Navigation is the process of finding a path from point A to point B while tracking the actual position. To know the actual position is an essential requirement of navigation. This could be achieved in various ways. Maybe the simplest solution is to use GPS but for many AMRs this is not an option. Because AMR usually operates under a roof where there is no GPS signal.

Another option is to use a known position of some signal beacon. E.g., Wi-Fi access points. If AMR is able to receive signals from at least three beacons then relative position to these beacons could be determined. This process is called trilateration. For these solutions AMR must be able to determine beacon straight distance. This could be achieved with measurement of time which is required for signals to travel from beacon to AMR.

One distance defines the perimeter of possible positions. If AMR is able to collect three of these distances then position is the intersection of all perimeters.

There is also one relatively simple solution for position tracking. It is called wheel odometry [6]. And it means that AMR is precisely tracking movement of wheels. If AMR is able to measure linear velocity and also angular velocity then it should be able to determine position in time. But this technique has one big disadvantage. It is precision of measurement. In the real-world measurement is never ideal and error is increasing in time. So longer this technique is used then a bigger inaccuracy in position is received.

But this could be a complementary technique to something else. And it is measuring distance to static objects in an AMR environment. Lidar and 3D cameras are used for this. But these sensors are weak if the observable environment is not changing. Imagine a long narrow corridor and AMR moving through. Even though AMR is moving all observable elements have always the same distance. So, for calculating position relative to static objects some inequality is required. And exactly in this case wheel odometry could be helpful to determine position.

Algorithm which can combine these two approaches is called AMCL (Adaptive Monte Carlo Localization). Is probabilistic localization system which is based on particle filter to track position against known map [5].

2.2 SLAM

All previous navigation approaches used some known entities. Position of beacons, starting position or map. But what to do if none of this is known? This area of research is called SLAM (Simultaneous Localization and Mapping).

SLAM is "is the computational problem of constructing or updating a map of an unknown environment while simultaneously keeping track of an agent's location within it." Input for SLAM are sensors. Mainly 3D cameras or lidars. If AMR can "see" its environment, then SLAM algorithms can determine position relative to seen objects. If the robot starts to move, then seen objects are observed from different angles. Therefore, the SLAM algorithm can determine the new position of AMR. It means that movement can be tracked.

But what to do if objects in the environment are also moving? Then it is hard to determine what is moving. If it is an AMR or observed object. For this reason, SLAM must use also other sensors for detecting its own movement. It is the accelerometers, gyroscopes and magnetometer. All together form Inertial Measurement Unit (IMU) (Fig. 2).

Which can provide orientation, position and velocity. So, if AMR with SLAM can detect its own pose change and surrounding changes due to movement then SLAM can provide exact pose inside the observed environment.

Fig. 2. Visualization of lidar lasers

To successfully operate AMR in some environment both of attitudes must be used. First stage is to use SLAM to map new unknown environment and create map from it (see Fig. 3). This could be done manually with controlled movement of AMR or automatically where AMR is roaming around to discover a new environment. And when a map is created the operator can perform a second stage where he/she specifies some points in the known map where AMR should go to perform some action. It is also possible to update map during the second stage if the observed environment is changing [7].

3 Fleet

This was a brief introduction into AMR navigation. But if somebody wants to use more AMRs simultaneously a new area of problems arises. How to do navigation, operations and management of more than one AMR? It is important to note that more AMRs does

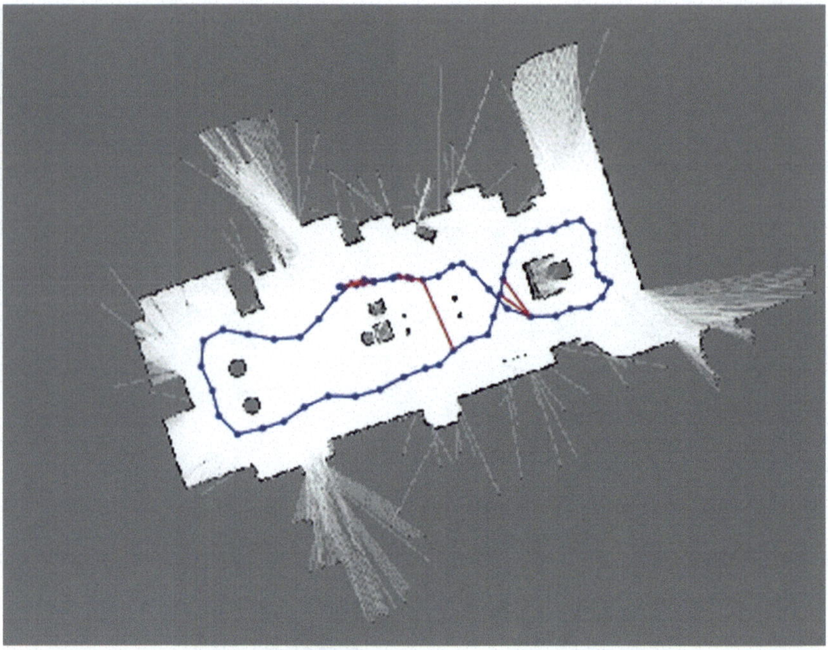

Fig. 3. Occupancy grid created by lidar SLAM.

not mean that the fleet system will be N times more effective than one AMR while N is count of AMRs. E.g., fleet can contain ten AMRs and if nine of them are blocked by one then the effectiveness of the whole fleet equals to only one. Therefore, management

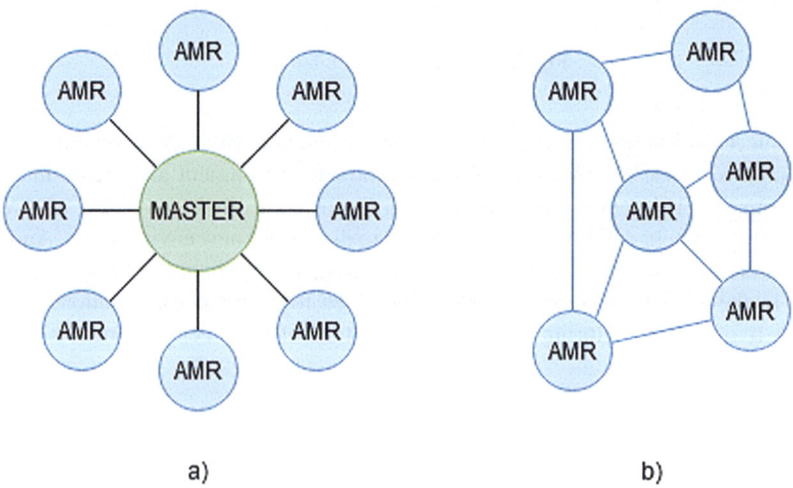

Fig. 4. a) centralized, b) decentralized architecture of fleet management system.

of the fleet needs to be designed in a robust way to use the whole potential of multiple AMRs together [8, 9].

The first question which needs to be answered is whether a centralized, decentralized (See Fig. 4) or distributed system for fleet management will be used. Both have their pros and cons.

3.1 Centralized

This approach consists of building one powerful master unit which can control each AMR. But it would be impractical if the master would have to control each step of AMR. But usually, the master sends the whole path which is followed by AMR. From this point of view the master needs to solve optimization problems of generating proper paths to achieve its goal. Because this approach knows the whole system it can find optimal solutions but even though this approach has some drawbacks [18].

- High network traffic - All AMRs communicate with master therefore it needs to be able to handle traffic from all AMRs simultaneously.
- Failure in master means failure of whole system.
- Lack of scalability - If the master system is scaled for a certain number of AMRs then adding some more could result in overload of the master system.

3.2 Decentralized

Another approach is to leave organization of AMRs to themself. This will lead to faster responses of AMRs because they must solve their goals be themselves. To use this approach could be also beneficial to avoid problems mentioned in the centralized approach. But a decentralized system is weaker to find optimal paths because of deadlocks [5].

3.3 Combined

It is usually a good approach to combine various approaches to get best results. In this case a centralized system could be used to plan global paths and goals. But reacting to environment events could be left for specific AMR. Collision avoidance is a good example. AMR could change the planned path to avoid obstacles and only notify the master system about this event [11].

4 Communication

Communication for AMR is essential functionality. It means that one AMR must be able to connect to another or to master system depending on fleet architecture.

Because of the nature of AMR communication must be wireless. For these purposes, a lot of wireless protocols could be used. E.g., Wi-Fi, Bluetooth, ZigBee. For main communication with master systems Wi-Fi is often used because of its widespread usage. But for communication between two AMR or direct communication with some other devices as charges, Bluetooth is more suitable [12].

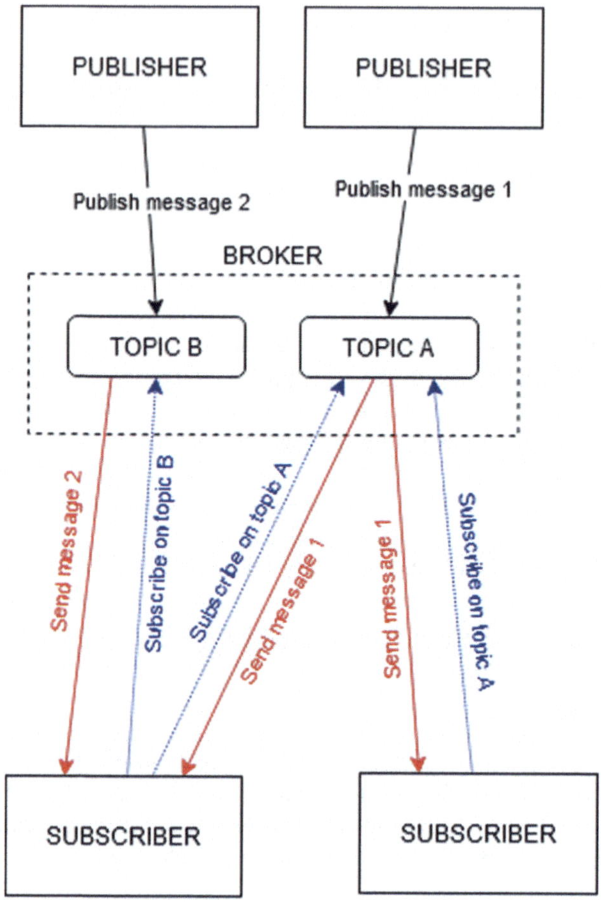

Fig. 5. Publish/Subscribe design pattern.

Other layer of communication transport. Transport layer is responsible for sending and receiving data from and to AMR. Wi-Fi physical layer is often used with TCP/IP protocol. This protocol is heavily used for computers local and wide network. Therefore, AMR could be easily connected, and they behave in network as another computer. This is an advantage because an AMR logic program could be used as tools which are also available for computers [13, 14].

Last layer of communication is the application layer. This layer describes the logic of communication. And it depends on the use case and requirements of the fleet. One popular pattern is publish/subscribe. It consists of publishers, subscribers and one broker (See Fig. 5). The major advantage of this pattern is that entities are decoupled. It means they do not know each other. Which support scalability and robustness of whole implementation. This pattern works in this way. Publishers only publish data. Usually the state of some component. (velocity, position). Subscriber subscribes for some specific data. It means that it receives data from the publisher. Communication entity between publisher and subscriber is called topic. Topic has its own name and on this Topic name publisher can publish something and subscribers are able to receive data from Topic. But they do not have to know who the publisher is and who is the subscriber. All this is managed by broker which is joining publisher to subscriber over topic [15].

5 Path Planning Algorithms

It is an intelligent process of finding a path from point A to point B while avoiding obstacles which are placed in the environment. The Resulting path could be defined as a curve or set of small segments connected [16]. This quite popular research subject which has been studied since 1970. And resulting algorithms are used in robotics but also in the game industry [17, 18]. All these algorithms are solving optimization tasks of finding the best path. The best path could not be considered as only the shortest but also time and cost factor needs to be taken in account. Sometimes the shortest path could lead through areas which will result in other costs for the robot or whole fleet of robots.

According [16, 17] path planning methods could be categorized according to environmental conditions as static and dynamic. Static is defined as an environment which is not changing in time. And dynamic is an environment which could be changed in time. It means that it could contain elements which are moving and are considered as obstacles. For that reason, path planning algorithms are divided into two categories. Global and local. Global algorithms are meant for static environments or at least for static parts of the environment. And their job is to find the optimal path through this environment. Whilst local planning algorithms are used for finding paths to avoid obstacles which were not considered during global planning. Local planners use robot sensors to detect

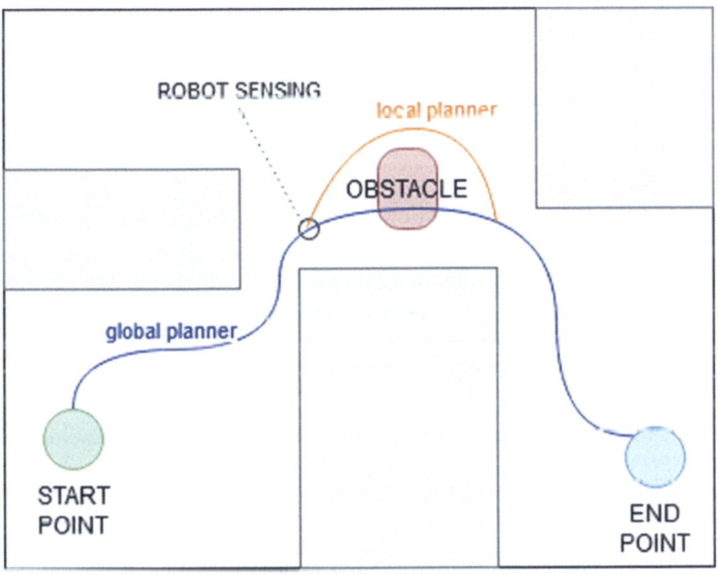

Fig. 6. Global planner with local planer for obstacle avoidance.

obstacles and avoid them. While a global planner is not able to see moving objects in the environment (see Fig. 6).

Path planning algorithms are in nature optimization problem algorithms. There is a wide range of heuristics methods which could be used to solve this problem. Such as Particle Swarm Optimization (PSO), Artificial Neural Networks (ANN), Fuzzy logic (FL) and Genetic Algorithm (GA) [20–22].

6 Path Planning in Fleet

Planning paths simultaneously for multiple AMRs is a complex problem. Because one AMR to another means obstacle. The main requirement of the fleet is to generate paths which will not collide, and each AMR time is used as effectively as possible. Another problem besides collision avoidance is deadlock. Meaning that group of AMRs is waiting for each other and none of them is moving. The problem is known as multiple robot path coordination problem [23] and leads researchers to finding advanced approaches to how to solve it. In [22] the author leaves bigger autonomy for specific robots. But they share the same network of paths. Each robot must acquire free paths to be able to use them. Fleet server is responsible for keeping track of acquired path segments. This approach should prevent collisions. But deadlocks and overall effectiveness is questionable.

Another approach was to plan paths in a static environment and later adjust velocities of robots to prevent collisions [24]. This approach was later tuned with prioritization in a way that the robot got priority level and therefore it was preferred during conflicting movement [25].

Prioritization is a useful strategy also in decentralized fleet architecture [26]. For decentralized system Leader-Follower algorithm could be used [27]. In this scenario one robot is the leader and knows the global goal and another robot or robots are only followers. Their responsibility is only to keep track of the leader and follow him. Leader supplements work on the master system for a group of robots.

7 Conclusion

This paper was written as an introduction into the Autonomous Mobile Robot research area. Intention was to collect information about usage of AMRs, related problems and methods used to solve these problems. Another part was about cooperation among multiple AMRs and identifying problems and possible solutions for these problems.

Intention of the author is to continue in research of multiple robot path coordination problem and try to design algorithm for effective usage of each AMR.

Acknowledgement. This work was supported by Slovak Research and Development Agency under project APVV-18-0436.

This work was supported by the Slovak Research and Development Agency under the Contract no. APVV-19-0210.

References

1. Vancea, A., Orha, I.: A survey in the design and control of automated guided vehicle systems. Carpathian J. Electron. Comput Eng. **12**, 41–49 (2019). https://doi.org/10.2478/cjece-2019-0016
2. Bačík, J., Ďurovský, F., Biroš, M., Kyslan, K., Perduková, D., Padmanaban, S.: Pathfinder–development of automated guided vehicle for hospital logistics. IEEE Access **5**, 26892–26900 (2017). https://doi.org/10.1109/ACCESS.2017.2767899
3. Antony, M., Parameswaran, M., Mathew, N., Sajithkumar, V.S., Joseph, J., Jacob, C.M.: Design and implementation of automatic guided vehicle for hospital application. In: 2020 5th International Conference on Communication and Electronics Systems (ICCES), COIMBATORE, India, pp. 1031–1036 (2020). https://doi.org/10.1109/ICCES48766.2020.9137867
4. Jia, J., Chen, W., Xi, Y.: Design and implementation of an open autonomous mobile robot system. In: Proceedings of IEEE International Conference on Robotics and Automation, ICRA 2004, New Orleans, LA, USA, vol. 2, pp. 1726–1731(2004). https://doi.org/10.1109/ROBOT.2004.1308073
5. Zhou, S., Cheng, G., Meng, Q., Lin, H., Du, Z., Wang, F.: Development of multi-sensor information fusion and AGV navigation system. In: 2020 IEEE 4th Information Technology, Networking, Electronic and Automation Control Conference (ITNEC), Chongqing, China, pp. 2043–2046 (2020). https://doi.org/10.1109/ITNEC48623.2020.9084687
6. Quan, S., Chen, J.: AGV localization based on Odometry and LiDAR. In: 2019 2nd World Conference on Mechanical Engineering and Intelligent Manufacturing (WCMEIM), Shanghai, China, pp. 483–486 (2019). https://doi.org/10.1109/WCMEIM48965.2019.00102
7. Chen, Y., Wu, Y., Xing, H.: A complete solution for AGV SLAM integrated with navigation in modern warehouse environment. In: 2017 Chinese Automation Congress (CAC), Jinan, pp. 6418–6423 (2017). https://doi.org/10.1109/CAC.2017.8243934
8. Zhang, Y., Mehrjerdi, H.: A survey on multiple unmanned vehicles formation control and coordination: normal and fault situations. In: International Conference on Unmanned Aircraft Systems (ICUAS) (2013)
9. Hoshino, S., Seki, H.: Multi-robot coordination for jams in congested systems. Robot. Auton. Syst. **61**(8), 808–820 (2013). https://doi.org/10.1016/j.robot.2013.04.011
10. LaValle, S., Hutchinson, S.: Optimal motion planning for multiple robots having independent goals. IEEE Trans. Robot. Autom. **14**(6), 912–925 (1998)
11. Digani, V., Sabattini, L., Secchi, C., Fantuzzi, C.: Hierarchical traffic control for partially decentralized coordination of multi AGV systems in industrial environments. In: 2014 IEEE International Conference on Robotics and Automation (ICRA), May 2014, pp. 6144–6149 (2014)
12. Robinson, S.H.: Living with the challenges to functional safety in the industrial Internet of Things. In: Living in the Internet of Things (IoT 2019), London, UK, pp. 1–6 (2019). https://doi.org/10.1049/cp.2019.0160
13. Chao, H., Yufeng, Z.: The wireless communication method between industrial robot and AGV based on SQLServer. In: 2019 3rd International Conference on Electronic Information Technology and Computer Engineering (EITCE), Xiamen, China, pp. 481–484 (2019). https://doi.org/10.1109/EITCE47263.2019.9095115
14. Vince, T., et al.: IoT implementation in remote measuring laboratory VMLab analyses. J. Univ. Comput. Sci. **26**(11), 1402–1421 (2020)
15. Deogirikar, J., Vidhate, A.: An improved publish-subscribe method in application layer protocol for IoT. In: 2017 International Conference on Smart Technologies for Smart Nation (SmartTechCon), Bangalore, pp. 1070–1075 (2017). https://doi.org/10.1109/SmartTechCon.2017.8358534

16. Kala, R., Shukla, A., Tiwari, R., Rungta, S., Janghel, R.R.: Mobile robot navigation control in moving obstacle environment using genetic algorithm, artificial neural networks and a* algorithm. In: WRI World Congress on Computer Science and Information Engineering, vol. 4, pp. 705–713 (2009)
17. Bhushan Mahajan, P.M.: Literature review on path planning in dynamic environment. Int. J. Comput. Sci. Netw. **2**(1), 115–118 (2013)
18. Meyer, J.-A., Filliat, D.: Map-based navigation in mobile robots: II. A review of map-learning and path-planning strategies. Cogn. Syst. Res. **4**(4), 283–317 (2003)
19. Ramirez-Serrano, H.L.A., Pettinaro, G.C.: Mobile robot localization in quasi-dynamic environments. Ind. Robot Int. J. **35**, 246–258 (2008)
20. Mohanty, P.K., Parhi, D.R.: Controlling the motion of an autonomous mobile robot using various techniques: a review. J. Adv. Mech. Eng. **1**, 24–39 (2013)
21. Roberge, V., Tarbouchi, M., Labonté, G.: Comparison of parallel genetic algorithm and particle swarm optimization for real-time uav path planning. IEEE Trans. Ind. Inform. **9**(1), 132–141 (2012)
22. Towards a robot fleet for intra-logistic tasks: combining free robot navigation with multi-robot coordination at bottlenecks
23. Robot Motion Planning. Kluwer, Boston (1991)
24. Kant, K., Zucker, S.W.: Toward efficient trajectory planning: the path-velocity decomposition. Int. J. Robot. Res. **5**(3), 72–89 (1986)
25. Erdmann, M., Lozano-Pérez, T.: On multiple moving objects. In: Proceedings of the IEEE International Conference on Robotics and Automation, San-Francisco, CA, pp. 1419–1424 (1986)
26. Alami, R., Robert, F., Ingrand, F.F., Suzuki, S.: Multi-robot cooperation through incremental plan-merging. In: IEEE International Conference on Robotics and Automation, Nagoya, Japan, pp. 2573–2578 (1995)
27. Desai, J.P., Ostrowski, J., Kumar, V.: Controlling formations of multiple mobile robots. In: Proceedings of IEEE International Conference on Robotics and Automation, pp. 2864–2869 (1998)

Equilibrium Commodity Flow Assignment in Simple Networks

Alexander Krylatov(✉) and Yulia Lonyagina

Saint-Petersburg State University, Universitetskaya emb. 7-9,
199034 Saint-Petersburg, Russia
a.krylatov@spbu.ru

Abstract. Today efficient commodity management seems impossible without the support of intelligent systems and computing based on mathematical models of commodity flow assignment since a modern market network is a large-scale system with many elements. The present paper is devoted to the equilibrium commodity flow assignment problem, which solution is valuable from decision-making perspectives. The paper attempts to fill the gap in the analytical relations between the equilibrium commodity assignment pattern and demand-supply parameters. We study commodity assignment in the case of linear demand, supply, and transportation functions. The problem is formulated in a form of nonlinear optimization program with dual variables reflecting supply and demand prices. The equilibrium commodity assignment pattern is obtained explicitly via equilibrium prices. For simple networks of three and four nodes we are lucky to obtain explicit conditions for identification of active commodity flows. In other words, once the demand-supply parameters are given, then active commodity flows can be identified explicitly.

Keywords. Nonlinear optimization · Equilibrium commodity assignment · Multi-commodity network

1 Introduction

The modern market networks was formed during several centuries. Actually, the market already lost its national and territorial boundaries and turned into a global market for commodities from spatial perspectives. The sale and purchase of commodities can occur at completely different prices, bounded above by the price of demand, and below – by the price of supply. Actual prices depend primarily on the structure of market and transaction costs, which incorporate the transportation costs. There are several structures of the market.

Y. Lonyagina—The reported study was funded by RFBR, project number 19-31-90109.

© The Author(s), under exclusive license to Springer Nature Switzerland AG 2021
R. Silhavy (Ed.): CSOC 2021, LNNS 229, pp. 337–346, 2021.
https://doi.org/10.1007/978-3-030-77445-5_31

- Perfect competition markets: many small firms with homogeneous products.
- Monopoly: there is only one company on the market that produces unique products.
- Monopolistic competition: there are many small firms on the market whose products are heterogeneous.
- Oligopoly: there are a small number of large firms with homogeneous or heterogeneous products on the market.

For each of the above structures, it is possible to determine such a situation (point) in the market, when neither the buyer nor the seller is interested in changing the current situation. The price at which the product offered on the market corresponds to the demand is called the equilibrium. The market mechanism begins to work, exerting pressure on prices from the lower and upper sides to achieve an equilibrium price. The study of the market, as well as the principles of its functioning and regulatory mechanisms, today seems to be relevant and necessary for understanding the essence of the socio-economic processes that are currently taking place throughout the world.

The first consideration of the spatial price equilibrium problem was made in [15]. The foundations for the study of spatial production, consumption, and trade of commodities was given in [17]. Up-to-date there exists a wide range of computational techniques for coping with such kind of problems [1,5,10,12,16]. Comprehensive mathematical models concerning spatial equilibrium are studied in [6,7,13]. Spatial equilibrium models are commonly exploited to solve the traffic assignment problem [4,8,9,18]. Moreover, its applications can be found in energy markets [2,14] and telecommunication markets [3].

In this paper, we study the spatial market equilibrium in the case of linear transportation costs. The problem is formulated in a form of nonlinear optimization program in Sect. 2. The supply and demand price functions are assumed to be given as well as the unit transaction cost functions, which are assumed to incorporate the unit transportation costs. The conditions on active equilibrium commodity flows are obtained explicitly in Sect. 3 for the commodity network of three nodes. Moreover, in Sect. 4 the conditions on active equilibrium commodity flows are obtained explicitly in for the commodity network of four nodes. Discussions and conclusions are given in Sects. 5 and 6 respectively.

2 Equilibrium Commodity Flow Assignment

Consider the set of suppliers M and the set of consumers N, which are associated with commodity production, distribution, and consumption. We denote by s_i the supply of $i \in M$, and by λ_i – the price of a unit of the i-th supply. By d_j we denote the demand of $j \in N$, and by μ_j – the price of a unit of the j-th demand. We believe that price λ_i smoothly depends on supply, i.e., $\lambda_i = r_i(s_i)$, $i \in M$, while price μ_j smoothly depends on demand, $\mu_j = p_j(d_j)$, $j \in N$. Finally, let $x_{ij} \geq 0$ be the commodity volume between a pair (i, j), while $c_{ij}(x_{ij})$ is the cost of the transaction of a unit of x_{ij}. Let us also introduce the indicators of market relations:

$$\delta_{ij}^t = \begin{cases} 1, \text{ for } x_{ij} > 0, \\ 0, \text{ for } x_{ij} = 0, \end{cases} \quad \forall (i,j) \in M \times N,$$

$$\delta_i^r = \begin{cases} 1, \text{ for } s_i > 0, \\ 0, \text{ for } s_i = 0, \end{cases} \quad \forall i \in M, \qquad \delta_j^p = \begin{cases} 1, \text{ for } d_j > 0, \\ 0, \text{ for } d_j = 0, \end{cases} \quad \forall j \in N.$$

Definition. The demand, supply, and commodity assignment pattern (s, x, d) is the *equilibrium* if

$$\begin{array}{ll} \lambda_i + c_{ij}(x_{ij}) = \mu_j \text{ for } & x_{ij} > 0, \\ \lambda_i + c_{ij}(x_{ij}) \geq \mu_j \text{ for } & x_{ij} = 0, \end{array} \quad \forall (i,j) \in M \times N,$$

where $\lambda_i = r_i(s_i)$, $i \in M$, and $\mu_j = p_j(d_j)$, $j \in N$.

Thus, if the sum of the ith supplier's price and the transaction costs between i and j exceeds the demand price of jth consumer, then the pair (i, j) will not have any market relations. The demand, supply, and commodity assignment pattern, which satisfies the following optimization problem

$$\min_{s,x,d} \sum_{i \in M} \int_0^{s_i} r_i(z)dz + \sum_{i \in M} \sum_{j \in N} \int_0^{x_{ij}} t_{ij}(z)dz - \sum_{j \in N} \int_0^{d_j} p_j(z)dz \tag{1}$$

subject to

$$\sum_{j \in N} x_{ij} = s_i, \quad i \in M, \tag{2}$$

$$\sum_{i \in M} x_{ij} = d_j, \quad j \in N, \tag{3}$$

$$x_{ij} \geq 0, \quad i \in M, \quad j \in N, \tag{4}$$

$$s_i \geq 0, \; d_j \geq 0, \quad i \in M, j \in N, \tag{5}$$

under

$$\sum_{i \in M} s_i = \sum_{j \in N} d_j, \tag{6}$$

is proved to be spatial market equilibrium [11,13].

Within the present paper, we examine equilibrium commodity assignment in case of linear demand, supply, and transportation functions. In other words, we assume that

$$r_i(z) = r_i^0 + k_i^r z, \quad r_i^0 \geq 0, \; k_i^r > 0, \; \forall i \in M,$$

i.e., if demand grows, so does the sale price,

$$p_j(z) = p_j^0 - k_j^p z, \quad p_j^0 \geq 0, \; k_j^p > 0, \; \forall j \in N,$$

i.e., if the supply increases, then the purchase price decreases,

$$t_{ij}(z) = t_{ij}^0 + k_{ij}^t z, \quad t_{ij}^0 \geq 0, \; k_{ij}^t > 0, \; \forall (i,j) \in M \times N,$$

i.e., transportation cost increases when the volume of transportation increases.

Proposition 1. *The equilibrium commodity assignment under linear demand, supply, and transportation functions is obtained by the following pattern:*

$$x_{ij} = \begin{cases} \frac{\mu_j - \lambda_i - t^0_{ij}}{k^t_{ij}}, & \mu_j - \lambda_i \geq t^0_{ij}, \\ 0, & \mu_j - \lambda_i < t^0_{ij}, \end{cases} \quad \forall i \in M, j \in N, \tag{7}$$

while equilibrium demand and supply are

$$d_j = \begin{cases} \frac{p^0_j - \mu_j}{k^p_j} & \mu_j \leq p^0_j, \\ 0 & \mu_j < p^0_j. \end{cases}, \quad s_i = \begin{cases} \frac{\lambda_i - r^0_i}{k^r_i}, & \lambda_i \geq r^0_i, \\ 0, & \lambda_i < r^0_i, \end{cases} \quad \forall i \in M, j \in N, \tag{8}$$

where (λ, μ) *are such that*

$$\begin{pmatrix} -B_\lambda & B \\ -B^T & B_\mu \end{pmatrix} \begin{pmatrix} \lambda \\ \mu \end{pmatrix} = q \tag{9}$$

where

$$B_\lambda = \text{diag}\left\{ \sum_{j \in N} \frac{\delta^t_{1j}}{k^t_{1j}} + \frac{\delta^r_1}{k^r_1}, \ldots, \sum_{j \in N} \frac{\delta^t_{mj}}{k^t_{mj}} + \frac{\delta^r_m}{k^r_m} \right\}$$

$$B_\mu = \text{diag}\left\{ \sum_{i \in M} \frac{\delta^t_{i1}}{k^t_{i1}} + \frac{\delta^p_1}{k^p_1}, \ldots, \sum_{i \in M} \frac{\delta^t_{in}}{k^t_{in}} + \frac{\delta^p_n}{k^p_n} \right\}$$

$$B = \begin{pmatrix} \frac{\delta^t_{11}}{k^t_{11}} & \frac{\delta^t_{12}}{k^t_{12}} & \cdots & \frac{\delta^t_{1n}}{k^t_{1n}} \\ \frac{\delta^t_{21}}{k^t_{21}} & \frac{\delta^t_{22}}{k^t_{22}} & \cdots & \frac{\delta^t_{2n}}{k^t_{2n}} \\ \vdots & \vdots & \ddots & \vdots \\ \frac{\delta^t_{m1}}{k^t_{m1}} & \frac{\delta^t_{m2}}{k^t_{m2}} & \cdots & \frac{\delta^t_{mn}}{k^t_{mn}} \end{pmatrix}, \quad q = \begin{pmatrix} \sum_{j \in N} \frac{t^0_{1j}\delta^t_{1j}}{k^t_{1j}} - \frac{r^0_1 \delta^r_1}{k^r_1} \\ \vdots \\ \sum_{j \in N} \frac{t^0_{mj}\delta^t_{mj}}{k^t_{mj}} - \frac{r^0_m \delta^r_m}{k^r_m} \\ \sum_{i \in M} \frac{t^0_{i1}\delta^t_{i1}}{k^t_{i1}} + \frac{p^0_1 \delta^p_1}{k^p_1} \\ \vdots \\ \sum_{i \in M} \frac{t^0_{in}\delta^t_{in}}{k^t_{in}} + \frac{p^0_n \delta^p_n}{k^p_n} \end{pmatrix}.$$

Proof. Let us consider Lagrangian of the problem

$$L = \sum_{i \in M} \int_0^{s_i} (r^0_i + k^r_i z) dz + \sum_{i \in M} \sum_{j \in N} \int_0^{x_{ij}} (t^0_{ij} + k^t_{ij} z) dz - \sum_{j \in N} \int_0^{d_j} (p^0_j - k^p_j z) dz$$

$$+ \sum_{i \in M} \lambda_i \left(\sum_{j \in N} x_{ij} - s_i \right) + \sum_{j \in N} \mu_j \left(- \sum_{i \in M} x_{ij} + d_j \right)$$

$$+ \sum_{i \in M} \sum_{j \in N} (-x_{ij}) \xi_{ij} + \sum_{i \in M} (-s_i) \eta_i + \sum_{j \in N} (-d_j) \nu_j,$$

where λ_i, μ_j, and $\eta_i \geq 0$, $\nu_j \geq 0$, $\xi_{ij} \geq 0$ for $i \in M$, $j \in N$, are multipliers of Lagrange. According to Kuhn–Tucker conditions

$$\frac{\partial L}{\partial x_{ij}} = t_{ij}^0 + k_{ij}^t x_{ij} + \lambda_i - \mu_j - \xi_{ij} = 0, \quad \forall i \in M, j \in N,$$

$$\frac{\partial L}{\partial s_i} = r_i^0 + k_i^r s_i - \lambda_i - \eta_j = 0, \quad \forall i \in M, j \in N,$$

$$\frac{\partial L}{\partial d_j} = -(p_j^0 - k_j^p d_j) + \mu_j - \nu_j = 0, \quad \forall i \in M, j \in N,$$

in particular, complementary slackness:

$$\begin{cases} \xi_{ij} x_{ij} = 0, \\ \xi_{ij} \geq 0, \end{cases} \quad \begin{cases} \eta_i s_i = 0, \\ \eta_i \geq 0, \end{cases} \quad \begin{cases} \nu_j d_j = 0, \\ \nu_j \geq 0, \end{cases} \quad \forall i \in M, j \in N,$$

we obtain the following expressions:

$$t_{ij}^0 + k_{ij}^t x_{ij} \begin{cases} = \mu_j - \lambda_i, \text{ if } x_{ij} > 0, \\ \geq \mu_j - \lambda_i, \text{ if } x_{ij} = 0, \end{cases} \quad \forall i \in M, j \in N,$$

$$r_i^0 + k_i^r s_i \begin{cases} = \lambda_i, \ s_i > 0, \\ \geq \lambda_i, \ s_i = 0, \end{cases} \quad p_j^0 - k_j^p d_j \begin{cases} = \mu_j \ d_j > 0, \\ \leq \mu_j \ d_j = 0. \end{cases} \quad \forall i \in M, j \in N$$

Therefore, we can express equilibrium demand, supply, and commodity assignment as (7)–(8). Moreover, requirements (2) and (3) lead us to

$$\sum_{j \in N} \frac{\mu_j - \lambda_i - t_{ij}^0}{k_{ij}^t} \delta_{ij}^t = \frac{\lambda_i - r^0 s_i}{k_i^r} \delta_i^r, \quad \forall i \in M$$

$$\sum_{i \in M} \frac{\mu_j - \lambda_i - t_{ij}^0}{k_{ij}^t} \delta_{ij}^t = \frac{p_j^0 - \mu_j}{k_j^p} \delta_j^p, \quad \forall j \in N,$$

that in matrix form can be expressed as (9). □

Therefore, due to proved proposition, one can find equilibrium demand, supply, and commodity assignment pattern via the system of linear equations.

3 Commodity Network of Three Nodes

Let us solve the system (9) to construct conditions identifying active flows in a commodity network of three nodes. First of all, we consider two suppliers and one consumer.

Proposition 2. *Assume $M = \{1, 2\}$ and $N = \{1\}$. The equilibrium commodity assignment x under linear demand, supply, and transportation functions satisfies the following conditions: $x_{11} > 0$ and $x_{21} > 0$ if and only if*

$$p_1^0 - r_1^0 - t_{11}^0 > \frac{k_1^p}{k_{21}^t + k_2^r}(r_1^0 + t_{11}^0 - r_2^0 - t_{21}^0), \tag{10}$$

$$p_1^0 - r_2^0 - t_{21}^0 > \frac{k_1^p}{k_{11}^t + k_1^r}(r_2^0 + t_{21}^0 - r_1^0 - t_{11}^0). \tag{11}$$

Proof. If $M = \{1, 2\}$ and $N = \{1\}$, then for $x_{11} > 0$ and $x_{21} > 0$ matrix Eq. (9) has the following form:

$$\begin{pmatrix} -\left(\frac{1}{k^t_{11}} + \frac{1}{k^r_1}\right) & 0 & \frac{1}{k^t_{11}} \\ 0 & -\left(\frac{1}{k^t_{21}} + \frac{1}{k^r_2}\right) & \frac{1}{k^t_{21}} \\ -\frac{1}{k^t_{11}} & -\frac{1}{k^t_{21}} & \left(\frac{1}{k^t_{11}} + \frac{1}{k^t_{21}} + \frac{1}{k^p_1}\right) \end{pmatrix} \begin{pmatrix} \lambda_1 \\ \lambda_2 \\ \mu_1 \end{pmatrix} = \begin{pmatrix} \frac{t^0_{11}}{k^t_{11}} - \frac{r^0_1}{k^r_1} \\ \frac{t^0_{21}}{k^t_{21}} - \frac{r^0_2}{k^r_2} \\ \frac{t^0_{11}}{k^t_{11}} + \frac{t^0_{21}}{k^t_{21}} + \frac{p^0_1}{k^p_1} \end{pmatrix}.$$

Let us express λ and μ from this system and substitute obtained expressions into (7):

$$\begin{aligned} x_{11} &= \frac{-\left(t^0_{11}k^t_{21}+k^t_{21}r^0_1+t^0_{11}k^r_2-k^t_{21}p^0_1+r^0_1k^r_2+t^0_{11}k^p_1-t^0_{21}k^p_1-k^r_2p^0_1+r^0_1k^p_1-r^0_2k^p_1\right)}{Q} > 0; \\ x_{21} &= \frac{-\left(t^0_{21}k^t_{11}+k^t_{11}r^0_2+t^0_{21}k^r_1-k^t_{11}p^0_1+r^0_2k^r_1-t^0_{11}k^p_1-k^r_1p^0_1+t^0_{21}k^p_1-r^0_1k^p_1+r^0_2k^p_1\right)}{Q} > 0, \end{aligned} \tag{12}$$

where $Q = k^t_{11}k^t_{21} + k^t_{11}k^r_2 + k^t_{21}k^r_1 + k^r_1k^r_2 + k^t_{11}k^p_1 + k^t_{21}k^p_1 + k^r_1k^p_1 + k^r_2k^p_1$. Since $Q > 0$, then (12) leads to

$$t^0_{11}k^t_{21} + k^t_{21}r^0_1 + t^0_{11}k^r_2 - k^t_{21}p^0_1 + r^0_1k^r_2 + t^0_{11}k^p_1 - t^0_{21}k^p_1 - k^r_2p^0_1 + r^0_1k^p_1 - r^0_2k^p_1 < 0,$$

$$t^0_{21}k^t_{11} + k^t_{11}r^0_2 + t^0_{21}k^r_1 - k^t_{11}p^0_1 + r^0_2k^r_1 - t^0_{11}k^p_1 - k^r_1p^0_1 + t^0_{21}k^p_1 - r^0_1k^p_1 + r^0_2k^p_1 < 0,$$

that can be re-written as (10)–(11). Note that goal function is convex, then Kuhn–Tucker conditions are necessary and sufficient. □

Secondly, we consider the case of one supplier and two consumers.

Proposition 3. *Assume $M = \{1\}$ and $N = \{1, 2\}$. The equilibrium commodity assignment x under linear demand, supply, and transportation functions satisfies the following conditions: $x_{11} > 0$ and $x_{12} > 0$ if and only if*

$$\left(p^0_1 - r^0_1 - t^0_{11}\right) > \frac{k^r_1}{k^t_{12} + k^p_2}\left(t^0_{11} - t^0_{12} + p^0_2 - p^0_1\right), \tag{13}$$

$$\left(p^0_2 - r^0_1 - t^0_{12}\right) > \frac{k^r_1}{k^t_{11} + k^p_1}\left(t^0_{12} - t^0_{11} + p^0_1 - p^0_2\right). \tag{14}$$

Proof. If $M = \{1, 2\}$ and $N = \{1\}$, then for $x_{11} > 0$ and $x_{21} > 0$ matrix Eq. (9) has the following form:

$$\begin{pmatrix} -\left(\frac{1}{k^t_{11}} + \frac{1}{k^t_{12}} + \frac{1}{k^r_1}\right) & \frac{1}{k^t_{11}} & \frac{1}{k^t_{12}} \\ -\frac{1}{k^t_{11}} & \frac{1}{k^t_{11}} + \frac{1}{k^p_1} & 0 \\ -\frac{1}{k^t_{12}} & 0 & \frac{1}{k^t_{12}} + \frac{1}{k^p_2} \end{pmatrix} \begin{pmatrix} \lambda_1 \\ \mu_1 \\ \mu_2 \end{pmatrix} = \begin{pmatrix} \frac{t^0_{11}}{k^t_{11}} + \frac{t^0_{12}}{k^t_{12}} - \frac{r^0_1}{k^r_1} \\ \frac{t^0_{11}}{k^t_{11}} + \frac{p^0_1}{k^p_1} \\ \frac{t^0_{12}}{k^t_{12}} + \frac{p^0_2}{k^p_2} \end{pmatrix}.$$

Let us express λ and μ from this system and substitute obtained expressions into (7):

$$\begin{aligned} x_{11} &= \frac{-\left(t^0_{11}k^t_{12}+k^t_{12}r^0_1+t^0_{11}k^r_1-t^0_{12}k^r_1-k^t_{12}p^0_1-k^r_1p^0_1+t^0_{11}k^p_2+k^r_1p^0_2+r^0_1k^p_2-p^0_1k^p_2\right)}{Z} > 0, \\ x_{12} &= \frac{-\left(t^0_{12}k^t_{11}+k^t_{11}r^0_1-t^0_{11}k^r_1+t^0_{12}k^r_1-k^t_{11}p^0_2+k^r_1p^0_1+t^0_{12}k^p_1-k^r_1p^0_2+r^0_1k^p_1-p^0_2k^p_1\right)}{Z} > 0, \end{aligned} \tag{15}$$

where $Z = k^t_{11}k^t_{12} + k^t_{11}k^r_1 + k^t_{12}k^r_1 + k^t_{11}k^p_2 + k^t_{12}k^p_1 + k^r_1k^p_1 + k^r_1k^p_2 + k^p_1k^p_2$. Since $Z > 0$, then (15) leads to

$$
\begin{aligned}
&t^0_{11}k^t_{12} + k^t_{12}r^0_1 + t^0_{11}k^r_1 - t^0_{12}k^r_1 - k^t_{12}p^0_1 - k^r_1p^0_1 + t^0_{11}k^p_2 + k^r_1p^0_2 + r^0_1k^p_2 - p^0_1k^p_2 < 0,\\
&t^0_{12}k^t_{11} + k^t_{11}r^0_1 - t^0_{11}k^r_1 + t^0_{12}k^r_1 - k^t_{11}p^0_2 + k^r_1p^0_1 + t^0_{12}k^p_1 - k^r_1p^0_2 + r^0_1k^p_1 - p^0_2k^p_1 < 0,
\end{aligned}
$$

that can be re-written as (13)–(14). Note that goal function is convex, then Kuhn–Tucker conditions are necessary and sufficient. □

Therefore, explicit conditions (10)–(11) allow one to identify active commodity flows in the case of two suppliers and one consumer, while explicit conditions (13)–(14) allow one to identify active commodity flows in the case of one supplier and two consumers.

4 Commodity Network of Four Nodes

Let us solve the system (9) to construct conditions identifying active flows in a commodity network of four nodes. First of all, we consider three suppliers and one consumer.

Proposition 4. *Assume $M = \{1, 2, 3\}$ and $N = \{1\}$. The equilibrium commodity assignment x under linear demand, supply, and transportation functions satisfies the following conditions: $x_{ij} > 0$, $(i, j) \in M \times N$, if and only if*

$$
p^0_1 - r^0_i - t^0_{i1} > \sum_{s \in M \setminus i} \frac{k^p_1}{(k^r_s + k^t_{s1})}(r^0_i - r^0_s + t^0_i - t^0_s) \quad \forall i \in M. \tag{16}
$$

Proof. If $M = \{1, 2, 3\}$ and $N = \{1\}$, then (9) has the following form:

$$
\begin{pmatrix}
-\left(\frac{\delta^t_{11}}{k^t_{11}} + \frac{\delta^r_1}{k^r_1}\right) & 0 & 0 & \frac{\delta^t_{11}}{k^t_{11}}\\
0 & -\left(\frac{\delta^t_{21}}{k^t_{21}} + \frac{\delta^r_2}{k^r_2}\right) & 0 & \frac{\delta^t_{21}}{k^t_{21}}\\
0 & 0 & -\left(\frac{\delta^t_{31}}{k^t_{31}} + \frac{\delta^r_3}{k^r_3}\right) & \frac{\delta^t_{31}}{k^t_{31}}\\
-\frac{\delta^t_{11}}{k^t_{11}} & -\frac{\delta^t_{21}}{k^t_{21}} & -\frac{\delta^t_{31}}{k^t_{31}} & \frac{\delta^t_{11}}{k^t_{11}} + \frac{\delta^t_{12}}{k^t_{12}} + \frac{\delta^t_{13}}{k^t_{13}} + \frac{\delta^p_1}{k^p_1}
\end{pmatrix}
\begin{pmatrix}
\lambda_1\\ \lambda_2\\ \lambda_3\\ \mu_1
\end{pmatrix}
$$

$$
= \begin{pmatrix}
\frac{t^0_{11}\delta^t_{11}}{k^t_{11}} - \frac{r^0_1\delta^r_1}{k^r_1}\\
\frac{t^0_{21}\delta^t_{21}}{k^t_{21}} - \frac{r^0_2\delta^r_2}{k^r_2}\\
\frac{t^0_{31}\delta^t_{31}}{k^t_{31}} - \frac{r^0_3\delta^r_3}{k^r_3}\\
\frac{t^0_{11}\delta^t_{11}}{k^t_{11}} + \frac{t^0_{21}\delta^t_{21}}{k^t_{21}} + \frac{t^0_{31}\delta^t_{31}}{k^t_{31}} + \frac{p^0_1\delta^p_1}{k^p_1}
\end{pmatrix}.
$$

If $x_{11} > 0$, then

$$
x_{11} = (k^r_2k^r_3p^0_1 + k^r_2k^t_{31}p^0_1 + k^r_3k^t_{21}p^0_1 + k^t_{21}k^t_{31}p^0_1 - k^p_1k^r_2r^0_1 - k^p_1k^r_3r^0_1 + k^p_1k^r_2r^0_3
$$

$$
+k^p_1k^r_3r^0_2 - k^p_1k^t_{21}r^0_1 - k^p_1k^t_{31}r^0_1 + k^p_1k^t_{21}r^0_3 + k^p_1k^t_{31}r^0_2 - k^r_2k^r_3r^0_1 - k^r_2k^t_{31}r^0_1 - k^r_3k^t_{21}r^0_1
$$

$$-k_{21}^t k_{31}^t r_1^0 - k_1^p k_2^r t_{11}^0 - k_1^p k_3^r t_{11}^0 + k_1^p k_2^r t_{31}^0 + k_1^p k_3^r t_{21}^0 - k_1^p k_{21}^t t_{11}^0 - k_1^p k_{31}^t t_{11}^0 + k_1^p k_{21}^t t_{31}^0$$
$$+ k_1^p k_{31}^t t_{21}^0 - k_2^r k_3^r t_{11}^0 - k_2^r k_{31}^t t_{11}^0 - k_3^r k_{21}^t t_{11}^0 - k_{21}^t k_{31}^t t_{11}^0)/W > 0,$$

where

$$W = k_1^p k_1^r k_2^r + k_1^p k_1^r k_3^r + k_1^p k_2^r k_3^r + k_1^p k_1^r k_{21}^t + k_1^p k_2^r k_{11}^t + k_1^p k_1^r k_{31}^t + k_1^p k_3^r k_{11}^t + k_1^p k_2^r k_{31}^t$$
$$+ k_1^p k_3^r k_{21}^t + k_1^r k_2^r k_3^r + k_1^p k_{11}^t k_{21}^t + k_1^p k_{11}^t k_{31}^t + k_1^p k_{21}^t k_{31}^t + k_1^r k_2^r k_{31}^t + k_1^r k_3^r k_{21}^t + k_2^r k_3^r k_{11}^t$$
$$k_1^r k_{21}^t k_{31}^t + k_2^r k_{11}^t k_{31}^t + k_3^r k_{11}^t k_{21}^t + k_{11}^t k_{21}^t k_{31}^t.$$

Since $W > 0$, then we come to

$$(p_1^0 - r_1^0 - t_{11}^0) > \frac{k_1^p}{(k_3^r + k_{31}^t)}(r_1^0 - r_3^0 + t_{11}^0 - t_{31}^0) + \frac{k_1^p}{(k_2^r + k_{21}^t)}(r_1^0 - r_2^0 + t_{11}^0 - t_{21}^0).$$

Analogically, for $x_{21} > 0$ we can obtain:

$$(p_1^0 - r_2^0 - t_{21}^0) \geq \frac{k_1^p}{(k_3^r + k_{31}^t)}(r_2^0 - r_3^0 + t_{21}^0 - t_{31}^0) + \frac{k_1^p}{(k_1^r + k_{11}^t)}(r_2^0 - r_1^0 + t_{21}^0 - t_{11}^0)$$

and for $x_{31} > 0$:

$$(p_1^0 - r_3^0 - t_{31}^0) \geq \frac{k_1^p}{(k_2^r + k_{21}^t)}(r_3^0 - r_2^0 + t_{31}^0 - t_{21}^0) + \frac{k_1^p}{(k_1^r + k_{11}^t)}(r_3^0 - r_1^0 + t_{31}^0 - t_{11}^0).$$

Therefore, we come to expression (16). Note that goal function is convex, then Kuhn–Tucker conditions are necessary and sufficient. □

Secondly, we consider the case of one supplier and three consumers.

Proposition 5. *Assume $M = \{1\}$ and $N = \{1, 2, 3\}$. The equilibrium commodity assignment x under linear demand, supply, and transportation functions satisfies the following conditions: $x_{ij} > 0$, $(i, j) \in M \times N$, if and only if*

$$p_j^0 - r_1^0 - t_{1j}^0 > \sum_{s \in N \setminus j} \frac{k_1^r}{(k_s^p + k_{1s}^t)}(p_s^0 - p_j^0 + t_{1j}^0 - t_{1s}^0) \quad \forall j \in N. \tag{17}$$

Proof. If $M = \{1\}$ and $N = \{1, 2, 3\}$, then (9) has the following form:

$$\begin{pmatrix} -\left(\frac{1}{k_{11}^t} + \frac{1}{k_{12}^t} + \frac{1}{k_{13}^t} + \frac{1}{k_1^r}\right) & \frac{1}{k_{11}^t} & \frac{1}{k_{12}^t} & \frac{1}{k_{13}^t} \\ -\frac{1}{k_{11}^t} & \frac{1}{k_{11}^t} + \frac{1}{k_1^p} & 0 & 0 \\ -\frac{1}{k_{12}^t} & 0 & \frac{1}{k_{12}^t} + \frac{1}{k_2^p} & 0 \\ -\frac{1}{k_{13}^t} & 0 & 0 & \frac{1}{k_{13}^t} + \frac{1}{k_3^p} \end{pmatrix} \begin{pmatrix} \lambda_1 \\ \mu_1 \\ \mu_2 \\ \mu_3 \end{pmatrix}$$
$$= \begin{pmatrix} \frac{t_{11}^0}{k_{11}^t} + \frac{t_{12}^0}{k_{12}^t} + \frac{t_{13}^0}{k_{13}^t} - \frac{r_1^0}{k_1^r} \\ \frac{t_{11}^0}{k_{11}^t} + \frac{p_1^0}{k_1^p} \\ \frac{t_{12}^0}{k_{12}^t} + \frac{p_2^0}{k_2^p} \\ \frac{t_{12}^0}{k_{13}^t} + \frac{p_3^0}{k_3^p} \end{pmatrix}.$$

If $x_{11} > 0$, then

$$x_{11}E = k_2^p k_3^p p_1^0 + k_2^p k_1^r p_1^0 + k_3^p k_1^r p_1^0 - k_2^p k_1^r p_3^0 - k_3^p k_1^r p_2^0 + k_2^p k_{13}^t p_1^0 + k_3^p k_{12}^t p_1^0$$
$$+k_1^r k_{12}^t p_1^0 + k_1^r k_{13}^t p_1^0 - k_1^r k_{12}^t p_3^0 - k_1^r k_{13}^t p_2^0 + k_{12}^t k_{13}^t p_1^0 - k_2^p k_3^p r_1^0 - k_2^p k_{13}^t r_1^0$$
$$-k_3^p k_{12}^t r_1^0 - k_{12}^t k_{13}^t r_1^0 - k_2^p k_3^p t_{11}^0 - k_2^p k_1^r t_{11}^0 - k_3^p k_1^r t_{11}^0 + k_2^p k_1^r t_{13}^0 + k_3^p k_1^r t_{12}^0 - k_2^p k_{13}^t t_{11}^0$$
$$-k_3^p k_{12}^t t_{11}^0 - k_1^r k_{12}^t t_{11}^0 - k_1^r k_{13}^t t_{11}^0 + k_1^r k_{12}^t t_{13}^0 + k_1^r k_{13}^t t_{12}^0 - k_{12}^t k_{13}^t t_{11}^0 > 0$$

where $E = k_1^p k_2^p k_3^p + k_1^p k_2^p k_1^r + k_1^p k_3^p k_1^r + k_2^p k_3^p k_1^r + k_1^p k_2^p k_{13}^t + k_1^p k_3^p k_{12}^t + k_2^p k_3^p k_{11}^t + k_1^p k_1^r k_{12}^t + k_2^p k_1^r k_{11}^t + k_1^p k_1^r k_{13}^t + k_3^p k_1^r k_{11}^t + k_2^p k_1^r k_{13}^t + k_3^p k_1^r k_{12}^t + k_1^p k_{12}^t k_{13}^t + k_2^p k_{11}^t k_{13}^t + k_3^p k_{11}^t k_{12}^t + k_1^r k_{11}^t k_{12}^t + k_1^r k_{11}^t k_{13}^t + k_1^r k_{12}^t k_{13}^t + k_{11}^t k_{12}^t k_{13}^t$. Since $E > 0$, then we come to

$$p_1^0 - r_1^0 - t_{11}^0 > \frac{k_1^r}{k_3^p + k_{13}^t}\left(p_3^0 - t_{13}^0 - p_1^0 + t_{11}^0\right) + \frac{k_1^r}{k_2^p + k_{12}^t}\left(p_2^0 - t_{12}^0 - p_1^0 + t_{11}^0\right).$$

Analogically, for $x_{12} > 0$ we can obtain:

$$p_2^0 - r_1^0 - t_{12}^0 > \frac{k_1^r}{k_3^p + k_{13}^t}\left(p_3^0 - t_{13}^0 - p_2^0 + t_{12}^0\right) + \frac{k_1^r}{k_1^p + k_{11}^t}\left(p_1^0 - t_{11}^0 - p_2^0 + t_{12}^0\right)$$

and for $x_{31} > 0$:

$$p_3^0 - r_1^0 - t_{13}^0 > \frac{k_1^r}{k_2^p + k_{12}^t}\left(p_2^0 - t_{12}^0 - p_3^0 - t_{13}^0\right) + \frac{k_1^r}{k_1^p + k_{11}^t}\left(p_1^0 - t_{11}^0 - p_3^0 + t_{13}^0\right).$$

Therefore, we come to expression (17). Note that goal function is convex, then Kuhn–Tucker conditions are necessary and sufficient. □

Therefore, explicit conditions (16) allow one to identify active commodity flows in the case of three suppliers and one consumer, while explicit conditions (17) allow one to identify active commodity flows in the case of one supplier and three consumers.

5 Discussions

Obtained results are important, since it can be used in heuristics procedures to estimate active commodity flows in large-scale market networks. Indeed, for any M one can examine if

$$p_1^0 - r_i^0 - t_{i1}^0 > \sum_{s \in M \setminus i} \frac{k_1^p}{(k_s^r + k_{s1}^t)}(r_i^0 - r_s^0 + t_i^0 - t_s^0) \quad \forall i \in M \tag{18}$$

or not. If (18) holds, then one faces the candidate flow to be active commodity flow. On the other hand, for any N one can examine if

$$p_j^0 - r_1^0 - t_{1j}^0 > \sum_{s \in N \setminus j} \frac{k_1^r}{(k_s^p + k_{1s}^t)}(p_s^0 - p_j^0 + t_{1j}^0 - t_{1s}^0) \quad \forall j \in N \tag{19}$$

or not. If (19) holds, then one faces the candidate flow to be active commodity flow.

6 Conclusion

In this paper, we study commodity assignment in the case of linear demand, supply, and transportation functions. The problem is formulated in a form of nonlinear optimization program with dual variables reflecting supply and demand prices. The equilibrium commodity assignment pattern is obtained explicitly via equilibrium prices. For simple networks of three and four nodes we are lucky to obtain explicit conditions for identification of active commodity flows. In other words, once the demand-supply parameters are given, then active commodity flows can be identified explicitly. Further work should be concentrated on a general commodity networks.

References

1. Allevi, E., Gnudi, A., Konnov, I.V.: Combined methods for dynamic spatial auction market models. Optim. Eng. **13**, 401–416 (2012)
2. Anderson, E.J., Philpott, A.B.: Optimal offer construction in electricity markets. Math. Oper. Res. **27**, 82–100 (2002)
3. Courcoubetis, C., Weber, R.: Pricing Communications Networks: Economics, Technology and Modelling. Wiley (2003)
4. Dafermos, S.: Traffic equilibrium and variational inequalities. Transp. Sci. **14**, 42–54 (1980)
5. Guder, F., Morris, J.G., Yoon, S.H.: Parallel and serial successive overrelaxation for multicommodity spatial price equilibrium problem. Transp. Sci. **26**, 48–58 (1992)
6. Konnov, I.: Equilibrium Models and Variational Inequalities. Elsevier (2007)
7. Krylatov, A., Zakharov, V., Tuovinen, T.: Optimization Models and Methods for Equilibrium Traffic Assignment. Springer International Publishing (2020)
8. Krylatov, A.Y., Zakharov, V.V.: Competitive traffic assignment in a green transit network. Int. Game Theory Rev. **18**(2), 1640003 (2016)
9. Krylatov, A.Y., Zakharov, V.V., Malygin, I.G.: Competitive traffic assignment in road networks. Transp. Telecommun. **17**(3), 212–221 (2016)
10. Migdalas, A., Pardalos, P., Storoy, S.: Parallel Computing in Optimization. Kluwer Academic Publishers, The Netherlands (1997)
11. Nagurney, A.: Network Economics: A Variational Inequality Approach. Kluwer Academic Publishers, The Netherlands (1993)
12. Nagurney, A., Nicholson, C.F., Bishop, P.M.: Massively parallel computation of large-scale spatial price equilibrium models with discriminatory ad valorem tariffs. Ann. Oper. Res. **68**, 281–300 (1996)
13. Patriksson, M.: The Traffic Assignment Problem: Models and Methods. Dover Publications (1994)
14. Popov, I., Krylatov, A., Zakharov, V., Ivanov, D.: Competitive energy consumption under transmission constraints in a multi-supplier power grid system. Int. J. Syst. Sci. **48**(5), 994–1001 (2017)
15. Samuelson, P.: Spatial price equilibrium and linear programming. Am. Econ. Rev. **42**, 283–303 (1952)
16. Sharpe, W.F.: Capital asset prices: a theory of market equilibrium under conditions of risk. J. Financ. **19**, 425–443 (1964)
17. Takayama, T., Judge, G.G.: Spatial and Tempoml Price and Allocation Models. North-Holland Pub. Co., Amsterdam (1971)
18. Zakharov, V., Krylatov, A.: Transit network design for green vehicles routing. Adv. Intell. Syst. Comput. **360**, 449–458 (2015)

Co-evolutionary Approach to Solving the Problem of VLSI Partitioning Using the Ant Colony Method

B. Lebedev Oleg[1(✉)], A. Zhiglatiy Artemy[1], and A. Purchina Olga[2]

[1] Southern Federal University, Rostov-on-Don, Russia
[2] Don State Technical University, Rostov-on-Don, Russia

Abstract. A coalition approach to the construction of a partitioning algorithm by the ant colony method is considered. The interpretation of the solution is a bipartite graph that defines the distribution of the set of vertices over the nodes. The evolving population is divided into subpopulations, each of which solves the initial optimization problem and has its own optimization strategy. Several subpopulations evolve simultaneously in the search space. Periodically, agents move from one subpopulation to another, passing on their experience. The co-evolutionary approach by adapting the strategy during the search provides a broader view of the solution space and a higher probability of localizing the global extremum of the problem.

Keywords: VLSI · Partitioning · Swarm intelligence · Adaptive behavior · Co-evolution · Subpopulation · Ant algorithm · Optimization

1 Introduction

When solving design problems arising in the design, one often has to face the problem of splitting integrated circuits into separate blocks. When solving such problems, it is widely used the division of large circuits into composite blocks, which include certain sets of components. This is due to the fact that a modern VLSI can contain tens of millions of transistors. Such a scheme cannot be designed in general due to limited computational capabilities (memory, speed). As a result of splitting, many blocks and interconnections between blocks are formed. In very large schemes, a hierarchical partitioning structure can also be used [1, 2]. Various heuristic methods are used to reduce the time required to solve the partitioning problem. As a circuit model, either the hypergraph $H(X, U)$ is used, where X is the set of vertices, $|X| = n$, U is the set of edges, or the bipartite graph $D = (X \cup W, Y)$, where X is the set of elements (first share), $|X| = n$, $W = \{w_i | i = 1, 2, \ldots, n\}$ is the set of nodes (second share) formed when splitting the set of elements X. Y is the set of edges connecting the vertices of the set X with vertices of the set W. When designing VLSI, two approaches are used to place the formed nodes on the chip. In the first approach, the circuit is divided into nodes without taking into account the possible placement of nodes on the chip. Formation of configurations of nodes and

© The Author(s), under exclusive license to Springer Nature Switzerland AG 2021
R. Silhavy (Ed.): CSOC 2021, LNNS 229, pp. 347–357, 2021.
https://doi.org/10.1007/978-3-030-77445-5_32

areas of their placement on the chip is carried out after solving the partition problem. In the second approach, the areas of placement of areas and their sizes are determined by the results of VLSI planning. Partitioning actually boils down to the distribution of the set of vertices over the areas located on the VLSI chip. Therefore, in the second approach, besides the traditional accounting for the number of links, the length of links is predicted. The existing partitioning algorithms are divided into two main classes: constructive and iterative [1–3]. The first class is characterized by relative speed, but at the same time, low quality of the solution. The second class is more laborious, but it allows you to get better quality solutions [3]. Among the iterative algorithms, one can distinguish deterministic and probabilistic. In deterministic algorithms, the change in the partition (solution) is implemented on the basis of a clear, deterministic dependence on the changeable solution. The disadvantage is that it often falls into the local optimum. In probabilistic algorithms, the transition to a new solution is carried out randomly. The disadvantage of algorithms that implement purely random search is significant labor intensity [4]. In recent years, a scientific direction called "Natural Computing" has been intensively developed, combining mathematical methods that incorporate the principles of natural decision-making mechanisms [4]. These methods include, first of all, methods of simulating annealing, the method of evolutionary modeling, genetic algorithms, evolutionary adaptation, algorithms of swarm intelligence and ant algorithms [5–8]. The paper proposes a hybrid approach, methods and methods for representing the problem of splitting VLSI, based on modeling the adaptive behavior of an ant colony. In hybrid algorithms, the advantages of one algorithm can compensate for the disadvantages of another. Integration of metaheuristics of population algorithms provides a broader view of the search space and a higher probability of localizing the global extremum of the problem. This approach allows to partially solve the problem of premature convergence, provides a way out of local optima and increases the speed of obtaining the result. To solve the partitioning problem, a co-evolutionary algorithm based on the ant colony method has been developed. The essence of the co-evolutionary approach is that the evolving population is divided into subpopulations that evolve in parallel [9, 10]. Periodically, agents move from one subpopulation to another, passing on their experience. The co-evolutionary approach provides a broader view of the solution space and a higher probability of localizing the global extremum of the problem.

2 Formulation of the Partition Problem

Let a graph $G(X, U)$ be given, where X is the set of vertices, $|X| = n$, U is the set of edges. It is necessary to split the set X into k nonempty and disjoint subsets X_i, $\cup X_i = X$, $X_i \cap X_j = \emptyset$, $X_i \neq \emptyset$. Constraints are imposed on the generated nodes (blocks, components): $|X_i| = n_i$, $\sum n_i = n$. The optimization criterion is the total number of links F between X_i. The goal of optimization is to minimize criterion F. Figure 1 shows a plan divided into chip regions and a model of connections between vertices x_i by vertices of the graph.

The coordinates of the region's center point are considered as the coordinates of the location of the vertex located in the region. When calculating the vertex score, the connections that connect the vertices placed in different areas are considered.

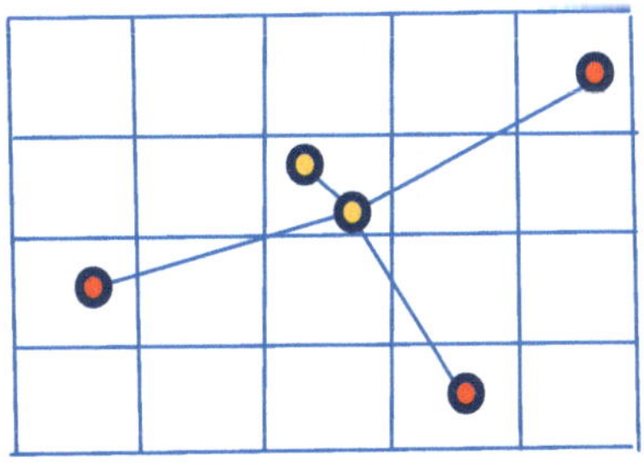

Fig. 1. Model of connections of vertex x_i

For a given location of regions on the chip and distribution of vertices over regions, the cost ξi of each vertex x_i is determined.

$$\xi_i = k_1 \cdot d_i + k_2 \cdot s_i,$$

where d_i is the total number of connections incident to the vertex x_i, s_i – total length of connections incident to vertex x_i, k_1, k_2 are coefficients.

For the second approach to partitioning, the estimate of the solution to the problem of distributing vertices over regions is defined as: $\xi = \sum_i(\xi_i)$. The metaheuristic of the ant algorithm is based on a combination of two techniques: the general scheme is built on the base method, which includes an inline procedure. The basic method is to implement an iterative procedure for finding the best solution, based on the mechanisms of the adaptive behavior of the ant colony. A built-in procedure is a constructive algorithm for the agent to construct a specific interpretation of the solution [7]. The ant algorithm is used to solve the problem of finding the shortest route in a complete graph. Each ant in the colony forms its own route, which is an interpretation of the solution of a certain problem on the complete graph of the search for solutions. As the ant moves, it marks a path with a pheromone, and this information is used by other ants to choose a path. The authors have developed modifications of the canonical paradigm of the ant algorithm: the structure of the graph of the search for solutions and the construction of the representation (interpretation) of the solution on it; methods of constructing an interpretation of the solution by an agent on the basis of the graph of the search for solutions (development of a constructive algorithm). To increase the efficiency, enhance the convergence of the algorithm and the ability to exit local optima, an approach to co-hybridization [9, 10] of the partitioning algorithm based on the model of adaptive behavior of an ant colony is proposed. At the same time, several subpopulations evolve in the search space for solving the problem, each of which solves the original optimization problem and has its own optimization strategy. The partitioning problem is solved by two groups A_1, A_2 of agents. Agents of the A_1 group build routes in accordance with the C_1 strategy. Accordingly: agents of the A_2 group – with the C_2 strategy. Strategies C_1 and C_2 differ in constructive algorithms for splitting the scheme by agents.

3 Development of a Co-evolutionary Partitioning Algorithm Based on Models of Adaptive Behavior of an Ant Colony

The complete bipartite graph $D = (X \cup W, E)$ is used as the solution search graph, where $X = \{x_i | i = 1,2, \ldots, n\}$ is the set of vertices (first share) corresponding to the set of vertices of the graph $H(X,U)$, and $W = \{w_v | v = 1,2, \ldots, s\}$ is the set of vertices (second part) corresponding to the set of nodes $|W| = s$. E is the set of edges (x_i, w_v) connecting the vertices $x_i \in X$ with the vertices of the set W. $|E| = n_s$.

The problem of a constructive algorithm is reduced to finding on a complete bipartite graph $D = (X \cup W, E)$ a bipartite subgraph $D^k = (X^k \cup W^k, E^k)$ for which the estimate F has a minimum value. The bipartite subgraph D^k is used as a representation of some k-th solution of the partition problem – it defines the distribution of the set of vertices X^k over the set of nodes W^k. The main restrictions of the presented bipartite subgraph $D^k = (X^k \cup W^k, E^k)$ are as follows:

- $E^k = \cup E^k{}_v$;
- the number of edges of the set $E^k{}_v$ incident to the vertex $w^k{}_v$ is equal to the number of vertices of the set $X^k{}_v$, $|X^k{}_v| = n_v$;
- each edge $(x_i, w^k{}_v) = e^k{}_{iv} \in E^k{}_v$, on the one hand, is incident to the vertex $w^k{}_v \in W^k$, and on the other hand, is incident to one and only one vertex $x_i \in X^k{}_v$; $\cup X^k{}_v = X^k = X$;
- the local degree of any vertex $x_i \in X^k$ is equal to one, $\rho(x_i) = 1$;
- the local degree of the vertex $w^k{}_v$ is equal to the cardinality of the set $X^k{}_v$, $\rho(w^k{}_v) = |X^k{}_v|$.

The paper considers two constructive partitioning algorithms included in the search partitioning algorithm. Agents of subpopulation A_1 build a solution using the first constructive algorithm on the set of vertices X^k. Agents of subpopulation A_2 using the second constructive algorithm on the set of vertices W^k. The process of finding solutions is iterative. Each iteration l includes three stages. At the first stage of each iteration, each agent z_k forms its own solution, at the second stage, the agents deposit a pheromone on the edges of the solution search graph D, and at the third stage, the pheromone is evaporated. The work uses the cyclic method of ant systems. In this case, the pheromone is deposited by the agents on the edges after the complete formation of the solution.

The subgraph D^k is formed by sequentially forming the set E^k of edges of the bipartite subgraph $D^k = (X^k \cup W^k, E^k)$, based on the set E of edges of the complete bipartite graph $D = (X \cup W, E)$ (step by step). The set of edges E^k on the basis of the original set of vertices $X \cup W$ form a bipartite partition graph $D^k = (X^k \cup W^k, E^k)$, which is a solution to the partition problem.

Consider the first constructive partitioning algorithm. Agents have memory. At each step t in the memory of agent z^k there is:

- the amount of pheromone $f_{iv}(t)$ deferred on each edge $e_{iv} = (x_i, w_v)$ of the graph D;
- the list of vertices $X_{1k}(t) \in X$ already included in the bipartite subgraph $D^k(t)$ being formed, and the list of the remaining (free) vertices $X_{2k}(t) \in X$, $X_{1k}(t) \cup X_{2k}(t) = X$;

– $\rho(w^k{}_v(t))$ is the local degree of the vertex $w^k{}_v(t)$ of the graph $D^k(t)$ at step t, $v = 1,2, \ldots, s$.

At the initial stage, the same (small) amount of pheromone $\Theta = \xi/\varepsilon$, where $\varepsilon = |E|$, is deposited on all edges of the graph $D = (X \cup W, E)$. $X_{1k}(0) = \emptyset$. $X_{2k}(t) = X$. $E^k = \emptyset$.

At the first stage of each iteration, each agent z_k forms its own bipartite subgraph D^k. The process of constructing the subgraph D^k by the agent is carried out by sequentially selecting and including n edges $e^k{}_{iv} \in E^k$ into the bipartite subgraph D^k being formed, such that $e^k{}_{iv} = (x_i, w_v)$, $x_i \in X$, $w_v \in W$, $\rho(w_v) = n_v$, $\rho(x_i) = 1$. Each of the agents z_k sequentially forms the set of edges E^k, where k is the agent's number.

Consider an algorithm for constructing a bipartite graph by an agent z_k – a solution $D^k = (X^k \cup W^k, E^k)$ based on the complete bipartite decision graph $D = (X \cup W, E)$, on the vertex set X^k. At the first stage of each iteration, the task of each agent is to sequentially form a set of edges E^k. The process of forming E^k includes two stages, performed at each step t. At the first stage of step t, for each vertex $x_i \in X_{2k}(t)$, a set of incident edges $U_i(t)$ is determined that connect x_i with all vertices w_v of the set W. $|U_i(t)| = |W| = s$.

For each edge $e_{iv} = (x_i, w_v) \in U_i(t)$ incident to the vertex $x_i \in X_{2k}(t)$, the parameter $f_{iv}(t)$ is determined – the total pheromone level on this edge e_{iv}. For each vertex $x_i \in X_{2k}(t)$, the average amount of pheromone per edge of the set $U_i(t)$ is determined:

$$f^*(U_i(t)) = \sum\nolimits_v (f_{iv}(t))/s.$$

The value $\mu_{ik}(t) = f^*(U_i(t))$ is declared the value of the vertex x_i. After that, agent z_k with probability $P_{ik}(t) = \mu_{ik}(t)/\sum_i(\mu_{ik}(t))$, proportional to $\mu_{ik}(t)$, chooses the vertex $x^*{}_i(t)$. At the second stage of step t, among the edges $U_i(t)$ incident to the selected vertex $x^*{}_i(t)$, with the probability $P_{vk}(t) = f_{iv}(t)/\sum_v(f_{iv}(t))$, the edge $e^*{}_i = (x^*{}_i(t), w_v)$, which is included in the set of edges E^k formed by the agent z_k.

Next, post procedures are performed. The edge $e^*{}_{iv} = (x^*{}_i(t), w^k{}_v)$ is included in the set $E^k(t)$. The vertex $x^*{}_i(t)$ is included in the set $X_{1k}(t)$ and excluded from the set $X_{2k}(t)$. The local degree of the vertex $w^k{}_v \in e*_{iv}$ in the graph D^k increases by 1, $\rho(w_v)(t) = \rho(w_v)(t) + 1$. Go to the next step. The process of forming the set E^k by the agent z^k ends when $X_{2k}(t) = \emptyset$.

Consider the second constructive partitioning algorithm on the vertex set W^k. Let us introduce the notation. Let $W_{1k}(t)$ be the list of vertices w_v considered at step t. $Q_v(t)$ is the set of edges $e_{iv} = (x_i, w_v)$ incident to the vertex $w_v \in W_{1k}(t)$ at step t. $\rho(w_v)(t)$ is the local degree of the vertex $w_v \in W_{1k}(t)$ at step t. At the initial stage, the same (small) amount of pheromone ξ/ε, where $\varepsilon = |E|$, is deposited on all edges of the graph $D = (X \cup W, E)$. $W_{1k}(0) = W$. For all $w_v \in W_{1k}(t)$, the initial local degree of the vertex is $\rho(w_v)(0) = n$. $E^k = \emptyset$. For each vertex $w_v \in W_{1k}(0)$, an initial set $Q_v(0) \in E$ of incident edges of the complete bipartite graph D is formed. At the first stage of step t, in each of the sets $Q_v(t)$ formed at the previous steps, for each $e_{iv} = (x_i, w_v) \in Q_v(t)$, the parameter f_{iv} is determined the total pheromone level on this edge e_{iv}

For each $Q_v(t)$, the average amount of pheromone per edge is determined $e_{iv} \in Q_v(t)$:

$$f(Q_v(t)) = \sum\nolimits_v f_{iv}/|Q_v(t)|.$$

The value $\sigma_{vk}(t) = f(Q_v(t))$ is declared the value of the vertex $w_v \in W_{1k}(t)$. After that, agent z_k with probability $P_{vk}(t) = \sigma_{vk}(t)/\sum_v(\sigma_{vk}(t))$, proportional to $\sigma_{vk}(t)$, selects the vertex $w^*_v \in W_{1k}(t)$ and the corresponding set of edges $Q^*_v(t)$. At the second stage of step t, among the edges $Q^*_v(t)$ incident to the selected vertex w^*_v, with the probability $P_{vk}(t) = f_{iv}(t)/\sum_i(\mu_{ik}(t))$, the edge $e^*_{iv} = (x^*_i, w^*_v)$, which is included in the set of edges E^k formed by the agent z_k.

Next, post procedures are performed. The vertex x^*_i incident to the selected edge $e^*_{iv} = (x^*_i, w^*_v)$ is determined. A subset $U^*_i(t)$ of edges incident to the selected vertex x^*_i is determined. The edges of the subset $U^*_i(t)$ are removed from all subsets of $Q_v(t)$. If after removing $U^*_i(t)$, for some $Q_v(t)$ the local degree $\rho(w_v)(t)$ of the vertex w_v at step t becomes equal to n_v, then the subset $Q_v(t)$ is excluded from consideration at further steps. The process of forming the set E^k by the agent z^k ends if for all $Q_v(t)$ the local degree $\rho(w_v)(t)$ of the vertex w_v at step t becomes equal to n_v and all $Q_v(t)$ are excluded from consideration.

After the agents form bipartite subgraphs D^k (each agent has its own bipartite subgraph D^k), at the second stage of the iteration, each agent deposits a pheromone on those edges of the complete bipartite graph D that correspond to the edges of the constructed bipartite subgraph D^k. After each agent has formed a solution and postponed the pheromone, at the third stage there is a general evaporation of the pheromone on the edges of the complete bipartite graph R in accordance with the formula:

$$h_{iv}(t) = h_{iv}(t) \cdot (1 - \rho),$$

where $h_{ij}(t)$ is the level of pheromone deposited by agents on the edge $e_{iv} \in E^k$, ρ is the update coefficient.

4 Co-evolutionary Partitioning Algorithm

1. In accordance with the initial data, a complete bipartite solution search graph $D = (X \cup W, Y)$ is formed.
2. The number of agents N_a in each of the subpopulations Z_1 and Z_2 is specified.
3. The values of the control parameters are set.
4. The number of iterations is set – N_l.
5. On all edges of the graph $D = (X \cup W, Y)$ the initial amount of pheromone is deposited. Coefficient values are set.
6. $l = 1$. (l is the iteration number)
   (*1 constructive algorithm*).
7. $k = 1$. (k – agent number).
8. The values of memory parameters for the agent z_k of the population Z_1 are formed. $X_{1k}(0) = \emptyset$. $X_{2k}(t) = X$. $t = 0$. $E^k(0) = \emptyset$.
9. $t = t + 1$. (t – step number).
10. (*Stage 1*). For each vertex $x_i \in X_{k2}(t)$, the set $U_i(t)$ of incident edges $e_{iv} = (x_i, w_v) \in U_i(t)$ is defined.
11. For each edge $e_{iv} = (x_i, w_v) \in U_i(t)$, the parameter $f_{iv}(t)$ is determined – the amount of pheromone placed on e_{iv}.

12. The average amount of pheromone per one edge of the set $U_i(t)$ is determined:

$$f^*(U_i(t)) = \sum\nolimits_v (f_{iv}(t))/s.$$

The value $\mu_{ik}(t) = f^*(U_i(t))$ is declared the value of the vertex $x_i \in X_{2k}(t)$.
13. Among the set of vertices $X_{2k}(t)$ with probability $P_{ik}(t) = \mu_{ik}(t)/\sum_i(\mu_{ik}(t))$, a vertex $x^*_i(t) \in X_{2k}(t)$ is chosen.
14. (*Stage* 2). Among the edges $U^*_i(t)$ incident to the selected vertex $x^*_i(t)$, with the probability $P_{vk}(t) = f_{iv}(t)/\sum_v(f_{iv}(t))$, the edge $e^*_i = (x^*_i(t), w_v)$, which is included in the set of edges $E^k(t)$ formed by the agent z_k.
15. Execution of post-procedures after step t. The vertex $x^*_i(t)$ is included in the set $X_{1k}(t)$ and excluded from the set $X_{2k}(t)$. $E^k(t) = E^k(t\text{-}1) \cup e^*_i$.
16. If $X_{2k}(t) = \emptyset$, then go to 17, otherwise go to 9.
17. Formation of the solution graph $D^k = (X^k \cup W^k, E^k)$ based on the set of edges $E^k(t)$ and the calculation of the estimate $F^k(l)$ of the partition given by the graph $D^k = (X^k \cup W^k, E^k)$.
18. On the edges E of the graph D corresponding to the found set of edges E^k of the graph D^k, a pheromone is deposited in an amount proportional to the estimate $F^k(l)$:

$$\Delta\tau_l(t) = Q/F_k(l).$$

19. The procedure of pheromone evaporation on the edges of graph G is performed.
20. If $k < N_a$, then $k = k + 1$ and go to 8, otherwise go to 21.
(*2 constructive algorithm*)
21. $k = 1$. (k – agent number).
22. The values of memory parameters for the agent z_k of the population Z_2 are formed.
$W_{1k}(t)$ is the list of vertices w_v considered at step t.
$Q_v(t)$ is the set of edges $e_{iv} = (x_i, w_v)$ incident to the vertex $w_v \in W_{1k}(t)$ at step t.
$\rho(w_v)(t)$ is the local degree of the vertex $w_v \in W_{1k}(t)$ at step t.
$W_{1k}(0) = W$. For all $w_v \in W_{1k}(t)$, the initial local degree of the vertex is $\rho(w_v)(0) = n$. $t = 0$. $E^k(0) = \emptyset$.
For each vertex $w_v \in W_{1k}(0)$, an initial set $Q_v(1) \in E$ of incident edges of the complete bipartite graph D is formed.
23. $t = t + 1$. (t is the step number).
24. *Stage 1*. In each of the sets $Q_v(t)$ formed at the previous steps for each edge $e_{iv} = (x_i, w_v) \in Q_v(t)$, the total level f_{iv} of the pheromone deposited on it is determined.
25. For each $Q_v(t)$, the average amount of pheromone per edge $e_{iv} \in Q_v(t)$ is determined:

$$f(Q_v(t)) = \sum\nolimits_v (f_{iv})/|Q_v(t)|.$$

The value $\sigma_{vk}(t) = f(Q_v(t))$ is declared the value of the vertex $w_v \in W_{1k}(t)$.
26. Agent z_k with probability $P_{vk}(t) = \sigma_{vk}(t)/\sum_v(\sigma_{vk}(t))$, proportional to $\sigma_{vk}(t)$, selects the vertex $w^*_v \in W_{1k}(t)$ and the corresponding set of edges $Q^*_v(t)$.
27. *Stage 2*. Among the edges $Q^*_v(t)$ incident to the chosen vertex w^*_v, with probability $P_{vk}(t) = f_{iv}(t)/\sum_i(\mu_{ik}(t))$, an edge $e^*_{iv} = (x^*_i, w^*_v)$, which is included in the set of edges $E^k(t)$ formed by the agent z_k.

28. Execution of post-procedures after step t. The vertex x^*_i incident to the selected edge $e^*_{iv}= (x^*_i, w^*_v)$ is determined.
    A subset $U^*_i(t)$ of edges incident to the selected vertex x^*_i is determined. The edges of the subset $U^*_i(t)$ are removed from all subsets of $Q_v(t)$.
    If after removing $U^*_i(t)$, for some $Q_v(t)$ the local degree $\rho(w_v)(t)$ of the vertex w_v at step t becomes equal to n_v, then the subset $Q_v(t)$ is excluded from consideration at further steps.
29. If all $Q_v(t)$ are excluded from consideration, then go to 30, otherwise go to 23.
30. Formation of the solution graph $D^k= (X^k \cup W^k, E^k)$ on the basis of the set of edges $E^k(t)$ and the calculation of the estimate $F^k(l)$ of the partition given by the graph $D^k= (X^k \cup W^k, E^k)$.
31. On the edges E of the graph D corresponding to the found set of edges E^k of the graph D^k, a pheromone is deposited in an amount proportional to the estimate $F^k(l)$: $\Delta\tau_l(t) = Q / F_k(l)$.
32. The procedure of pheromone evaporation on the edges of graph D is performed.
33. If $k < N_a$, then $k = k + 1$ and go to 22, otherwise go to 34.
34. If $l < N_l$, then $l = l + 1$ and go to 7, otherwise go to 35.
35. The end of the algorithm.

The time complexity of this algorithm at one iteration is $O(n^2)$. In the general case, the running time of this algorithm depends on the colony lifetime l (the number of iterations), the number of graph vertices n, and the number of agents m.

5 Experimental Research

When creating the program, we used the Visual Studio 2005 package for 32-bit operating systems of the Windows®9x family, and debugging, testing and experiments were carried out on an IBM® compatible computer with an AMD Athlon-64®3000+ processor and 512 MB DIMM PC-3200 RAM. The co-evolution partitioning (CP) program was created on the basis of the developed partitioning algorithms.

To carry out the experiments, we used the procedure for synthesizing test examples with a known optimum F_{opt} by analogy with the well-known BEKU method (Partitioning Examples with Tight Upper Bound of Optimal Solution) [11]. Examples with up to 1000 vertices were studied. The weight of the vertices was taken to be zero, and the weight of all edges was taken to be one. In this case, the graphs were "split" into two subgraphs with an equal number of vertices in each subgraph. Based on the processing of experimental studies, the average dependence of the quality of solutions on the number of iterations (Fig. 2) and the size of the population (Fig. 3) was constructed. The quality is assessed by the value F_{opt}/F, where F is the estimate of the obtained solution. It was established that the initial amount of pheromone Q should be 14 times greater than the average amount of the pheromone value $\tau_k(l)$ deposited by the agents at each iteration. Renewal coefficient $\rho = 0.95$.

As a result of experiments, it was found that with a population size of $M = 100$, the algorithm converges on average at 120 iterations. Moreover, in a number of experiments,

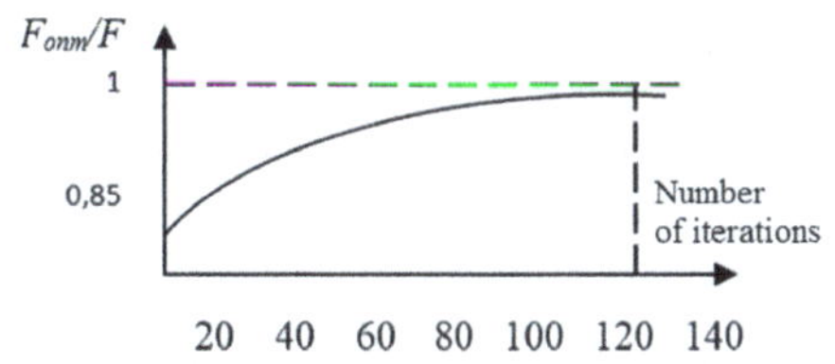

Fig. 2. Dependence of the quality of solutions of the CP algorithm on the number of iterations

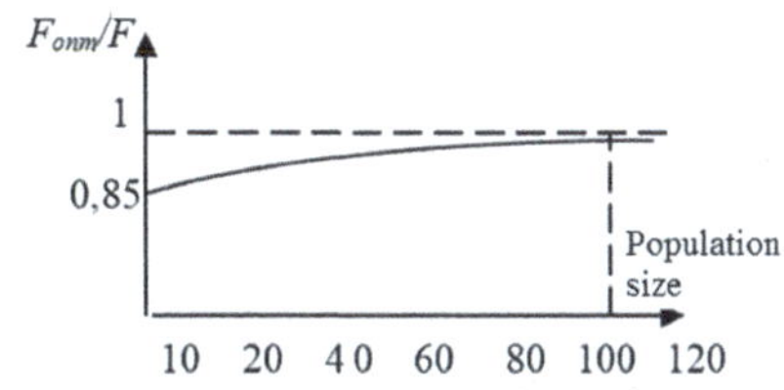

Fig. 3. Dependence of the quality of solutions of the CP algorithm on the population size

the number of iterations was less than the average up to 10%, and more, from the average to 35%. The time complexity of the procedure at one iteration is $O(n)$.

Comparative analysis with other partitioning algorithms was carried out using standard test examples and schemes (benchmarks) posted at: http://www.cad.polito.it/tools/9.html, http://www.cbl.ncsu.edu. These examples are standard industrial circuits (blocks, circuits, integrated circuits, large integrated circuits, VLSI). The comparison was made according to the following parameters:

- Cut – the number of chains caught in the cut;
- CPU – time spent looking for a solution.

A comparative assessment of the work of the developed algorithms was carried out on the benchmarks cordic(881), misex3(1349), X3(1369), c6288(2435), s15850(4321), frisc(4400), elliptic(4711) (the number of gates is given in brackets) with the results of the following algorithms (on the same diagrams): Pure hMetis [12]; Net&Path Based [13]; PPFF [13]; VPR [12]. Comparative estimates are in Table 1.

Table 1. The results of comparative experiments

Algorithm	F	Benchmarks						
		Cordic	misex3	X3	c6288	s15850	Frisc	Elliptic
Pure hMetis	Cut	327	543	211	182	617	838	508
	CPU	2	5	4	31	28	34	16
Net& Path	Cut	278	614	261	216	671	869	619
	CPU	5	13	6	80	37	61	22
PPFF	Cut	313	540	228	201	615	851	510
	CPU	6	14	8	79	43	66	26
VPR	Cut	319	543	216	204	622	843	513
	CPU	10	18	13	85	47	71	27
CP	Cut	280	610	253	209	651	842	567
	CPU	4	12	5	71	301	46	15

Analyzing the results of experiments, we can conclude that the developed algorithms find solutions that are not inferior in quality, and sometimes surpass their counterparts by an average of 3–4%. Perceptible improvement in co-evolution partitioning results compared to hMetis, which is a very strong algorithm. For example, on the Cordic scheme, co-evolution results.

On average, starting the program provides finding a solution that differs from the optimal one by less than 0.5%.

6 Conclusion

To increase efficiency, enhance the convergence of the algorithm and the ability to exit from local optima, a coalition approach to the construction of a partitioning algorithm is proposed. To solve the partitioning problem, the authors have developed a modified metaheuristic by analogy with the models of the adaptive behavior of ant colony. In this paper, in contrast to the canonical ant colony paradigm, the interpretation of the solution is the subgraph $D^k = (X^k \cup W^k, E^k)$, which specifies the distribution of the set of vertices X^k over the nodes of the set W^k. The work of the algorithm actually consists in forming a subset of E^k edges. At the same time, two subpopulations evolve in the search space of the optimization problem, each of which solves the same initial optimization problem, but with different search areas and optimization strategies. The co-evolutionary approach provides a broader view of the solution space and a higher probability of localizing the global extremum of the problem. Distinctive features of the interaction between co-evolving subpopulations are that they are based on the use of a common graph for finding solutions, common evolutionary memory, and forming a single interpretation of a solution in the form of a bipartite graph.

Acknowledgements. This research is supported by grants of the Russian Foundation for Basic Research of the Russian Federation, the project № 20-07-00260.

References

1. Nemudrov, V., Martin, G.: Systems-on-a-Chip. Design and Development. Tekhnosfera, p. 157 (2004)
2. Kazennov, G.G.: Fundamentals of Design of Integrated Circuits and Systems, p. 295. Knowledge Laboratory, Binom (2005)
3. Alpert, C.J., Mehta, D.P., Sapatnekar, S.S.: Handbook of Algorithms for Physical Design Automation, p. 245. Auerbach, Boston (2009)
4. Karpenko, A.P.: Modern Search Engine Optimization Algorithms. Algorithms Inspired by Nature: A Tutorial, p. 448. Publishing House MSTU. N.E. Bauman (2014)
5. Kureichik, V.M., Lebedev, B.K., Lebedev, O.B.: Hybrid partition algorithm based on natural decision-making mechanisms. Artif. Intell. Decis. Making 3–15 (2012)
6. Lebedev, B.K., Lebedev, O.B.: Ant partitioning algorithms using a task representation other than the canonical. Bull. Rostov State Transp. Univ. **2**(62), 71–77 (2016)
7. Lebedev, O.B.: Models of Adaptive Behavior of Ants Colony in Design Problems. Taganrog, p. 199. Publishing house of SFEDU (2013)
8. Lebedev, B.K., Lebedev, O.B.: The memetic algorithm for splitting. Bull. Rostov State Transp. Univ. **3**, 136–145 (2017)
9. Zhou, Y., Pei, Sh.: A hybrid co-evolutionary particle swarm optimization algorithm for solving constrained engineering design problems. J. Comput. (JCP) **5**(6), 965–972 (2010)
10. Vorobyeva, E.Yu., Karpenko, A.P., Seliverstov, E.Yu.: Co-hybridization of particle swarm algorithms. Sci. Life **4** (2012)

11. Cong, J., Romesis, M., Xie, M.: Optimality, scalability and stability study of partitioning and placement algorithms. In: Proceedings of the International Symposium on Physical Design, Monterey, CA, pp. 88–94 (2003)
12. Selvakkumaran, N., Karypis, G.: Multi-objective hypergraph partitioning algorithms for cut and maximum subdomain degree minimization. In: ICCAD, pp. 234–249 (2003)
13. Ababei, C., Selvakkumaran, N., Bazargan, K., Karypis, G.: Multi-objective circuit partitioning for cutsize and path-based delay minimization. In: ICCAD, pp. 118–137 (2002)

Evaluation of the Dependence of the Air Temperature Influence on the Reproductive Function of Cows Based on ASC-Analysis

Ekaterina N. Rud[1], Eugeny V. Lutsenko[2], Ksenia A. Semenenko[1](✉), Elena V. Kuzminova[1], and Marina P. Semenenko[2]

[1] Krasnodar Research Center for Animal Husbandry and Veterinary Medicine, Krasnodar, Russia
[2] Kuban State Agrarian University named after I.T. Trubilin, Krasnodar, Russia

Abstract. The aim of the work is the study of the dependence of the cattle reproductive function on various parameters associated with the ambient temperature. Traditionally, such tasks are solved using multivariate analysis. However, in this case, the application of this method is problematic for a number of reasons: the initial data are high dimensional and are measured in different units of measurement, the number of observations is much less than the number of factors, the factors depend on each other, the number of factors is too large. It is proposed to overcome these limitations by using the automated system-cognitive analysis (ASC-analysis) and its software tool – Eidos intelligent system. For this, the following tasks are being solved: to formulate the idea and concept of solving the problem; to justify the choice of method and tool for solving the problem; to apply the chosen method and tool to solve the problem; to evaluate the effectiveness of the proposed solution of the problem; to consider the limitations and disadvantages of the proposed solution to the problem and the prospects for its development by overcoming these limitations and disadvantages. # CSOC2021

Keywords: Automated system-cognitive analysis (ASC-analysis) · Eidos intellectual system · Reproductive function of cattle · Ambient temperature

1 Introduction

The weather and ambient temperature are the most important factors in the life of farm animals. These questions belong to the field of the science of zoometeorology, one of the important branches of agricultural meteorology, which studies the influence of meteorological conditions on farm animals. This is especially important at the present time, when, on the general background of climate warming, there is an increase in the frequency, intensity and duration of extreme weather and climatic events [1, 2].

The problem solved in the work is the study of the dependence of the cattle reproductive function on the ambient temperature.

The reproductive function of cattle is measured by a number of indicators, and first of all, the number of aborted cows and heifers, as well as the number of stillborn

© The Author(s), under exclusive license to Springer Nature Switzerland AG 2021
R. Silhavy (Ed.): CSOC 2021, LNNS 229, pp. 358–366, 2021.
https://doi.org/10.1007/978-3-030-77445-5_33

calves from cows and heifers. The parameters characterizing the ambient air temperature include the temperature determined by the months of the year, the average temperature for the quarter, as well as the total temperature for the quarter.

Thus, in total, the object of modeling describes five dimensional factors that determine 7 resulting indicators, which can be estimated using the automated system-cognitive analysis (ASC-analysis) and its software tool – Eidos intelligent system, which do not impose special requirements on the initial data and work with the data that is actually at the disposal of the researcher.

2 Materials and Methods

To fulfill the aim of the work, it is necessary to solve the following tasks:

Task 1: to formulate the idea of using a modern innovative method of artificial intelligence and the concept of solving the problem using the method of quantitative identification in a comparable form of the strength and direction of causal relationships in incomplete noisy data of a very large dimension of a numerical and non-numerical nature, measured in various types scales and units of measurement.

Task 2: to justify the choice of method and tool for solving the problem. The Eidos system is a universal system that can be applied in any subject area. It is in the public domain and we have all the corresponding source code. In addition, this system is one of the first domestic personal-level artificial intelligence systems, supports an online knowledge accumulation environment; it is widely used all over the world and supports a multilingual interface in 51 languages.

The advantage of applying this system is the use of an approach, the effectiveness of which does not depend on the subject area. It generates models directly from empirical data, not from our understanding of how patterns are implemented in that data.

All this determines the choice of ASC-analysis and the Eidos system as a tool for solving the problem. This method has been successfully applied to solve a large number of problems in various subject areas [3–5].

Task 3: to apply the chosen method and tool to solve the problem with the help of cognitive structuring of the subject area, its formalization, synthesis of statistical and system-cognitive models, their verification and selection of the most reliable model. Finally, to recognize and predict the most reliable diagnostic model.

Task 4: to evaluate the effectiveness of the proposed solution to the problem;

Task 5: to consider the limitations and disadvantages of the proposed solution to the problem and the prospects for its development by overcoming these limitations and disadvantages.

All these stages of the ASC-analysis, with the exception of the first one, are automated in the Eidos system. We consider them in the order of execution.

At the stage of cognitive-target structuring of the subject area, on the basis of knowledge of the ambient temperature, we evaluate the reproductive functions of cattle using classification scales (the results of the influence of factors) – aborted cows (animals), aborted heifers (animals), aborted cows and heifers in total (animals), dead calves from cows (animals), dead calves from heifers (animals), dead calves from cows and heifers in total (animals), milk yield per forage cow (kg) and descriptive scales (factors) – air temperature (°C) for the first month of the quarter, air temperature (°C) for the second month of the quarter, air temperature (°C) for the third month of the quarter, total air temperature for the quarter (°C), average air temperature for the quarter (°C).

Then on the developed classification and descriptive scales the source data is encoded, as a result of which a training sample is formed, which is essentially a source data base normalized using the classic ionic and descriptive scales and gradations (Table 1).

Table 1. Initial data (fragment)

Time period	Aborted cows (animals)	Aborted heifers (animals)	Aborted cows and heifers in total (animals)	Dead calves from cows (animals)	Dead calves from heifers (animals)	Dead calves from cows and heifers in total (animals)	Milk yield per forage cow (kg)	Air temperature (°C) for the first month of the quarter	Air temperature (°C) for the second month of the quarter	Air temperature (°C) for the third month of the quarter	Total air temperature for the quarter (°C)	Average air temperature for the quarter (°C)
2010, 1st quarter	1 389	351	1 740	690	1 486	2 176	1 410	0.1	3.4	5.8	9.3	3.1
2010, 2nd quarter	1 302	372	1 674	731	1 461	2 192	1 382	12.2	19.2	24.6	56	18.67
2010, 3rd quarter	1 405	401	1 806	837	1 530	2 367	1 301	26.8	27.7	21.7	76.2	25.4
2010, 4th quarter	1 513	313	1 826	790	1 231	2 021	1 358	11.5	12	7.2	30.7	10.23
Total in 2010	5 609	1 437	7 046	3 048	5 708	8 756	5 451	0	0	0	0	14.35
2011, 1st quarter	1 304	398	1 702	702	1 284	1 986	1 410	-0.1	-1.3	4.6	3.2	1.07
2011, 2nd quarter	1 198	415	1 613	763	1 407	2 170	1 399	10	17.1	22.6	49.7	16.57
2011, 3rd quarter	1 543	441	1 984	749	1 486	2 235	1 307	27.1	23.7	19.4	70.2	23.4
2011, 4th quarter	1 394	314	1 708	706	1 114	1 820	1 403	11.7	1.4	5.7	18.8	6.27
Total in 2011	5 439	1 568	7 007	2 920	5 291	8 211	5 519	0	0	0	0	11.83
2012, 1st quarter	1 183	288	1 471	682	1 302	1 984	1 490	-0.2	-5.1	3.1	-2.2	-0.73
2012, 2nd quarter	1 346	306	1 652	664	1 399	2 063	1 464	16.5	21.4	24.7	62.6	20.87
2012, 3rd quarter	1 412	357	1 769	736	1 487	2 223	1 426	25.8	25.2	21.3	72.3	24.1
2012, 4th quarter	1 351	297	1 648	723	1 267	1 990	1 509	16.8	8.3	2.3	27.4	9.13
Total in 2012	5 292	1 248	6 540	2 805	5 455	8 260	5 889	0	0	0	0	13.34
2013, 1st quarter	1 181	268	1 449	639	1 087	1 726	1 551	4.5	5.7	7.6	17.8	5.93
2013, 2nd quarter	1 273	317	1 590	703	1 192	1 895	1 500	14	21.7	23.5	59.2	19.73
2013, 3rd quarter	1 358	321	1 679	697	1 277	1 974	1 474	24.9	25.3	16.9	67.1	22.37
2013, 4th quarter	1 084	285	1 369	551	1 142	1 693	1 492	11.3	9	0.8	21.1	7.03
Total in 2013	4 896	1 191	6 087	2 590	4 698	7 288	6 017	0	0	0	0	13.77
2014, 1st quarter	1 017	242	1 259	490	1 006	1 496	1 651	0.9	2.6	8.5	12	4
2014, 2nd quarter	1 294	279	1 573	554	1 176	1 730	1 623	13.1	20.1	22	55.2	18.4

The Eidos system has many programming interfaces with external data sources (API) that automate the stage of formalizing the subject area. These interfaces can be grouped into three categories: interfaces with textual data; interfaces with tabular data; interfaces with graphic data. In this case, the universal programming interface for data entry from external data sources API-2.3.2.2 was used.

As a result of the operation of this software interface, classification and descriptive scales and gradations and a training sample were created (Tables 2, 3, 4).

Thus, the formalization of the subject area has been carried out and all conditions have been created for the implementation of the following stages of the ASC-analysis: synthesis and verification (reliability assessment) of models and solutions based on the most reliable of them various tasks.

Table 2. Classification scales and gradations (fragment)

CODE_CLS	NAME_CLS
1	ABORTED COWS (ANIMALS)-1/10-{927.0, 968.0}
2	ABORTED COWS (ANIMALS)-2/10-{968.0, 1078.0}
3	ABORTED COWS (ANIMALS)-3/10-{1078.0, 1181.0}
4	ABORTED COWS (ANIMALS)-4/10-{1181.0, 1229.0}
5	ABORTED COWS (ANIMALS)-5/10-{1229.0, 1294.0}
6	ABORTED COWS (ANIMALS)-6/10-{1294.0, 1318.0}
7	ABORTED COWS (ANIMALS)-7/10-{1318.0, 1358.0}
8	ABORTED COWS (ANIMALS)-8/10-{1358.0, 1412.0}
9	ABORTED COWS (ANIMALS)-9/10-{1412.0, 4591.0}
10	ABORTED COWS (ANIMALS)-10/10-{4591.0, 5609.0}
11	ABORTED HEIFERS (ANIMALS)-1/10-{122.0, 255.0}
12	ABORTED HEIFERS (ANIMALS)-2/10-{255.0, 270.0}
13	ABORTED HEIFERS (ANIMALS)-3/10-{270.0, 280.0}
14	ABORTED HEIFERS (ANIMALS)-4/10-{280.0, 301.0}
15	ABORTED HEIFERS (ANIMALS)-5/10-{301.0, 314.0}

Table 3. Descriptive scales and gradations (fragment)

CODE_ATR	NAME_ATR
1	AIR TEMPERATURE (°C) FOR THE FIRST MONTH OF THE QUARTER-1/10-{-0.2000000, 0.2000000}
2	AIR TEMPERATURE (°C) FOR THE FIRST MONTH OF THE QUARTER-2/10-{0.2000000, 2.1000000}
3	AIR TEMPERATURE (°C) FOR THE FIRST MONTH OF THE QUARTER-3/10-{2.1000000, 10.0000000}
4	AIR TEMPERATURE (°C) FOR THE FIRST MONTH OF THE QUARTER-4/10-{10.0000000, 11.1000000}
5	AIR TEMPERATURE (°C) FOR THE FIRST MONTH OF THE QUARTER-5/10-{11.1000000, 11.7000000}
6	AIR TEMPERATURE (°C) FOR THE FIRST MONTH OF THE QUARTER-6/10-{11.7000000, 12.3000000}
7	AIR TEMPERATURE (°C) FOR THE FIRST MONTH OF THE QUARTER-7/10-{12.3000000, 14.4000000}
8	AIR TEMPERATURE (°C) FOR THE FIRST MONTH OF THE QUARTER-8/10-{14.4000000, 23.0000000}
9	AIR TEMPERATURE (°C) FOR THE FIRST MONTH OF THE QUARTER-9/10-{23.0000000, 25.5000000}
10	AIR TEMPERATURE (°C) FOR THE FIRST MONTH OF THE QUARTER-10/10-{25.5000000, 27.1000000}
11	AIR TEMPERATURE (°C) FOR THE SECOND MONTH OF THE QUARTER-1/10-{-5.1000000, 1.4000000}
12	AIR TEMPERATURE (°C) FOR THE SECOND MONTH OF THE QUARTER-2/10-{1.4000000, 3.4000000}
13	AIR TEMPERATURE (°C) FOR THE SECOND MONTH OF THE QUARTER-3/10-{3.4000000, 4.8000000}
14	AIR TEMPERATURE (°C) FOR THE SECOND MONTH OF THE QUARTER-4/10-{4.8000000, 6.5000000}
15	AIR TEMPERATURE (°C) FOR THE SECOND MONTH OF THE QUARTER-5/10-{6.5000000, 9.0000000}

The next stage of the ASC analysis is the synthesis of statistical and system-cognitive models, carried out in the Eidos system, as a result of which three statistical models (Abs, Prc1, Prc2) and seven system-cognitive models (SC-models: Inf1-Inf7) are created. However, it is correct to apply these models for solving problems only if they are sufficiently reliable for this. Therefore, verification of the created models is the necessary stage of the ASC-analysis.

The assessment of the reliability of the models is carried out by solving the problem of classifying the objects of the training sample according to generalized images of classes and counting the number of true and false, positive and negative decisions according to the classical F-measure of van Rijsbergen as well as L1- L2-measures of prof. E.V.

Table 4. Training sample (full)

NAME_OBJ	N2	N3	N4	N5	N6	N7	N8	N9	N10	N11	N12	N13
2010, 1st quarter	8	17	28	36	48	58	62	1	12	23	31	41
2010, 2nd quarter	6	18	27	38	48	58	61	6	17	30	37	48
2010, 3rd quarter	8	18	28	39	49	59	61	10	20	28	40	50
2010, 4th quarter	9	15	29	39	47	57	61	5	16	24	36	45
Total in 2010	10	20	30	40	50	60	69	1	11	21	31	47
2011, 1st quarter	6	18	28	37	47	57	62	1	11	23	31	41
2011, 2nd quarter	4	18	25	38	48	58	61	3	16	29	36	47
2011, 3rd quarter	9	19	29	38	48	59	61	10	18	27	39	49
2011, 4th quarter	8	15	28	37	46	56	62	5	11	23	34	43
Total in 2011	10	20	30	40	50	60	69	1	11	21	31	45
2012, 1st quarter	4	14	24	36	48	57	63	1	11	22	31	41
2012, 2nd quarter	7	15	26	36	48	58	62	8	18	30	38	49
2012, 3rd quarter	8	17	28	38	49	58	62	10	19	27	40	50
2012, 4th quarter	7	14	26	37	47	57	64	8	15	21	35	44
Total in 2012	10	20	30	40	50	60	69	1	11	21	31	45
2013, 1st quarter	3	12	23	35	45	56	64	3	14	25	33	43
2013, 2nd quarter	5	16	25	37	46	56	63	7	18	30	38	48
2013, 3rd quarter	7	16	27	36	47	57	63	9	19	26	39	49
2013, 4th quarter	3	14	22	32	46	55	63	5	15	21	34	43
Total in 2013	10	19	30	40	50	60	69	1	11	21	31	46
2014, 1st quarter	2	11	22	32	45	53	65	2	12	25	32	42
2014, 2nd quarter	5	13	24	33	46	56	64	7	18	28	37	48
2014, 3rd quarter	7	15	26	34	46	56	63	9	20	27	40	50
2014, 4th quarter	1	12	21	31	45	54	64	4	13	22	34	43
Total in 2014	9	19	29	39	50	60	69	1	11	21	31	45

Lutsenko [3], which are fuzzy multiclass generalizations of the van Rijsbergen measure, invariant according to the size of the training sample (Fig. 1).

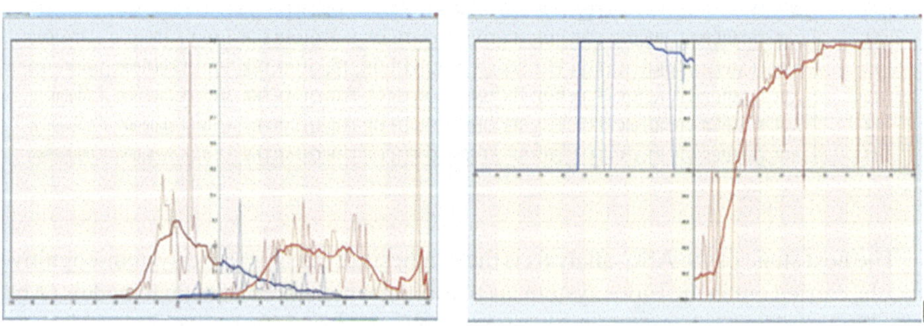

Fig. 1. Screen forms of the Eidos system model verification mode

All this provides us with the tools and criteria that allow solving the tasks posed in the work in the most reliable SC-model, the results of solving some of which we will briefly consider below.

3 Results and Discussion

As a result of the research and development, models have been created that correctly reflect the causal relationships between the factors and the results of their influence in the simulated subject area, and, therefore, these models can be correctly used to solve the tasks set in the work.

In accordance with the scheme for processing data, information and knowledge in the Eidos system, INF1 model was chosen as the current model. The model shows that the results of predicting the reproductive function of cattle based on information about the ambient temperature coincide very well with the fact.

Solving the problem of studying a simulated subject area by examining its model, in particular, has the following research opportunities: non-local neurons and neural network; cognitive diagrams of classes and factor values; automated cognitive SWOT analysis; cognitive functions, cognitive agglomerative clustering of classes and values of factors and a number of others.

All indicators of the reproductive function of cattle can be reasonably divided into two superclusters (i.e., the largest clusters) according to the influence of the ambient temperature on them. These superclusters form the poles of the class construct, opposite in the system of their determination by the ambient temperature.

Low moderate fertility rates (i.e., many abortions and dead calves) and low milk yield are highlighted in blue. Medium and high rates of reproductive function (i.e. few abortions and births of dead calves) and high milk yield are highlighted in red. Similarly, the ambient temperature indicators can be divided into clusters. The blue super cluster mainly includes very low temperatures for the 1st, 2nd and 3rd months of the quarter and very low total temperature for the quarter, as well as average values of the average temperature for the quarter; the red super cluster includes the average and high temperature indicators by month for the quarter and in total and in average for the quarter. Moreover, this super cluster is clearly divided into two subclusters with medium and very high ambient temperatures. This gives reasons to speak about the obvious presence of some specificity of the influence of very high ambient temperatures on the reproductive function of cattle and milk yield. This specificity is what we call "temperature stress".

Wherein, the dependence of the values of the output indicators on the values of the acting factors can be seen in cognitive functions (Fig. 2).

Cognitive functions reflect in color the amount of information that we receive from a certain value of a factor that the system under its influence will go into a state corresponding to different classes.

One of the significant negative manifestations in dairy cattle breeding is heat stress that occurs in the summer period and brings tangible economic losses. Heat stress in cows can be caused by high air temperature in combination with increased or, conversely, very low humidity and certain features of the technological process of milk production [6].

Prolonged exposure of the heat stress on an animal's body causes a deterioration in its health, a decrease in overall resistance, as well as reproductive qualities. The main reason for the economic losses in animal husbandry under conditions of high temperatures is a decrease in the adaptive abilities of the organism in maintaining the temperature balance of the body and tolerance to heat stress, which is facilitated by the selection of animals

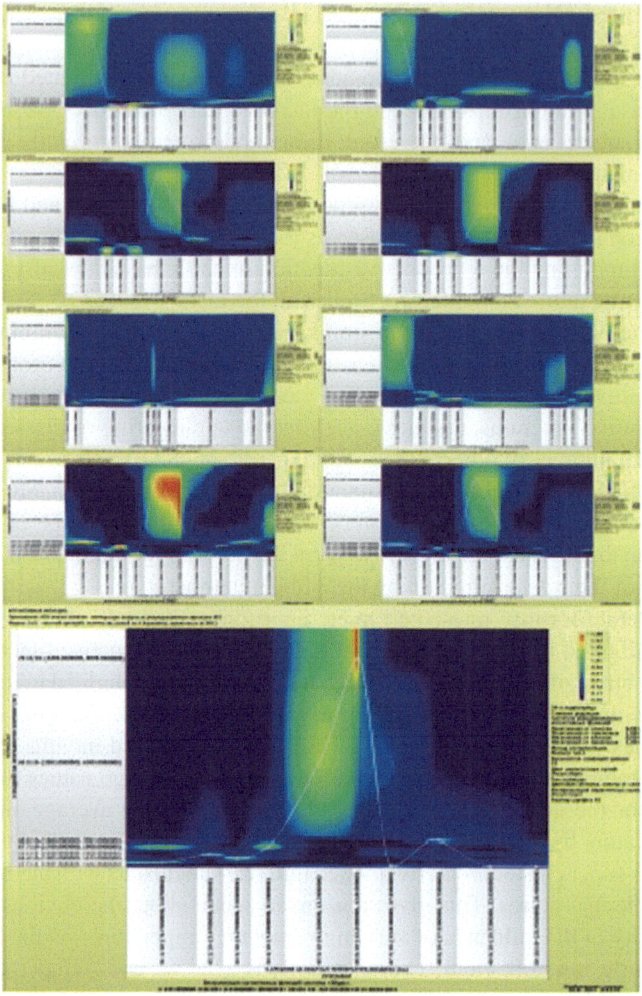

Fig. 2. Cognitive functions, reflecting the influence of ambient temperature on reproductive functions and milk yield of cattle (fragment)

for high productivity. At the same time, highly productive livestock has a super-intensive metabolism, during which the body forms its own metabolic heat, as well as imperfect mechanisms of water-electrolyte metabolism. This is especially true for dairy cows (milk productivity decreases to 35%), in which seasonal depression in the fertility rate also increases [7].

In dairy farming, milk production is reduced because of the heat stress, and impaired nutritional cows' supply is generally considered to be the main cause of decreased milk synthesis. However, some researchers note that the decrease in feed dry matter intake accounts for only 40–50% of the decrease in milk production during the heat stress. The remaining 50–60% is accounted for other changes triggered by overheating. In hot weather, not only productivity reduces, the content of fat and protein in milk decreases and the properties of the product deteriorate [8, 9].

In response to the heat stress, the body reacts by increasing sweat and salivation to lower body temperature. The respiratory rate also increases. As a result, part of the energy previously used for milk production or body weight gain is redirected to thermoregulation. In the long term, heat stress leads to a weakening of immunity and/or a decrease in reproduction and safety of livestock [10, 11].

The economic losses associated with the heat stress are caused by decreased productivity, increased mortality and reduced reproduction. At extremely high temperatures, milk yield decreases to 40%, animal mortality can reach 5% of the total number of cattle, and only 10–20% of inseminations of cows lead to pregnancy, and the possibility of abortion increases [12].

In the Krasnodar region, taking into account the temperate continental climate, hot summers and absolute temperature maximums from +38 to +42 °C, as well as high humidity, animals constantly find themselves outside this comfort zone; the heat stress in highly productive cows is a significant problem in the summer season in the Kuban region.

4 Conclusion

Based on the given output forms, the following conclusions can be reasonably made:

1. Very low or very high temperatures adversely affect the reproductive function of cattle, leading to the maximum number of abortions and stillborn calves. At the same time, although not maximum, but rather high milk yields are observed.
2. High ambient temperatures cause a significant decrease in cow productivity.
3. The ambient temperature from 9 to 14 °C is the most favorable for obtaining high milk yield per forage cow (in kg). The maximum milk yield is observed at an ambient temperature of 13–14 °C.

As the analysis of the results of system-cognitive modelling shows, the proposed and implemented in the Eidos system solution of the set tasks is quite effective; this allows reasonably asserting that the goal of the work has been achieved.

Acknowledgements. Special gratitude to Steffen Pirsig, member of the board of directors, chief architect, co-founder of Alaska Software Inc (https://www.xing.Com/profile/Steffen_Pirsig) for providing the author and the developer of the ASC-analysis and Eidos intelligent system prof. E.V. Lutsenko programming language Xbase ++ 2.0, Professional Edition, Ser. No. 127,980 free of charge.

Grant support. The reported study was funded by RFBR according to the research project № 20-316-90009.

References

1. Gringoff, I.G., Babushkin, O.L.: Climate, weather and animal breeding. In: Pasechnyuk, A.D. (ed.) Roshydromet. Obninsk, p. 352 (2010)

2. Malinin, I., Sadovnikova, N.: Heat stress: the rules of the game. Livestock Russia **3**, 32–35 (2016)
3. Lutsenko, E.V., Troshin, L.P., Zviagin, A.S., Milovanov, A.V.: Application of the automated system-cognitive analysis for solving problems of genetics. J. Mech. Eng. Res. Dev. **41**(2), 1–8 (2018)
4. Lutsenko, E.V., Semenenko, K.A., Snimschikova, I.V., Loiko, V.I., Semenenko, M.P.: Mathematical model of the influence of transnationalization on the russian agricultural machinery market. In: Silhavy, R. (ed.) CSOC 2020. AISC, vol. 1225, pp. 210–222. Springer, Cham (2020). https://doi.org/10.1007/978-3-030-51971-1_17
5. Lutsenko, E.V.: Conceptual principles of the system (emergent) information theory and its application for the cognitive modelling of the active objects (entities). In: Proceedings – 2002 IEEE International Conference on Artificial Intelligence Systems, ICAIS 2002, pp. 268–269, p. 1048109 (2002)
6. Kuzminova, E.V., Semenenko, M.P., Abramov, A.A., Rud, N.A., Rud, E.N.: The problem of heat stress in dairy farming. Vet. Med. Kuban **3**, 10–11 (2020). https://doi.org/10.33861/2071-8020-2020-3-10-11
7. Tuzov, I.N., Serous, K.G.: Influence of microclimate on milk productivity of cows. Collect. Sci. Works North Caucasian Res. Inst. Livestock **2**(3), 115–119 (2014)
8. Buryakov, N.P., Buryakova, M.A., Aleshin, D.E.: Heat stress and feeding characteristics of dairy cattle. Russ. Vet. J. **3**, 5–13 (2016)
9. Fomichev, Y., Sulima, N., Abilova, T., Bardin, O.: Heat stress in lactating dairy cows and methods of its prevention. Dairy Beef Cattle Breed. **3**, 24–26 (2013)
10. Brown-Brandl, T.M.: Understanding heat stress in beef cattle. Revista Brasileira de Zootecnia. 47 (2018). https://doi.org/10.1590/rbz4720160414
11. Fernandez-Novo, A., Pérez-Garnelo, S.S., Villagrá, A., et al.: The effect of stress on reproduction and reproductive technologies in beef cattle-a review. Animals (Basel) **10**(11), 2096 (2020). https://doi.org/10.3390/ani10112096
12. Bagath, M., Krishnan, G., Devaraj, C., et al.: The impact of heat stress on the immune system in dairy cattle: a review. Res. Vet. Sci. **126**, 94–102 (2019). https://doi.org/10.1016/j.rvsc.2019.08.011

Adaptive Control of the Movement of a Team of Mobile Robots in a Column

Gennady E. Veselov, Boris K. Lebedev, and Oleg B. Lebedev(✉)

Southern Federal University, Rostov-on-Don, Russia
gev@sfedu.ru

Abstract. The work solves the problem of managing a team of robots, which ensures the movement of a team of robots in a column. The vector $\mathbf{V_i} = (x_i, y_i, \varphi_i, V_i)$ of the state of each of the robots R_i in the absolute coordinate system is known. The $(x_i, y_i, \varphi_i, V_i)$ position and orientation of the robot in the absolute coordinate system, and the linear velocity of the robot, respectively. Management is carried out using an alternative collective adaptation algorithm based on the ideas of the collective behavior of adaptation objects. To implement the adaptation mechanism, the parameters of the vector are compared to the adaptation automata $AA - a_i^{\omega}, a_i^{xy}, a_i^{V}$, which simulate the behavior of adaptation objects in the environment. The structure of the process of alternative collective adaptation has been developed, under the control of which the movement of a group of robots in a formation is carried out.

Keywords: Swarm of mobile robots · Movement in a column · Group control · Alternative collective adaptation · Automatic adaptation

1 Introduction

This work is devoted to the topical problem of mathematical modeling and control theory: the problem of decentralized control of a multi-agent system consisting of agents modeling autonomous robots in order to ensure the movement of a group of agents in a formation having a given geometric shape [1–5].

A formation (formation) is the required arrangement of robots on a plane or in space. Depending on how the target formation is set, the combat mission may have a different setting.

For uniform distribution of agents in the mission area, maintaining stable communication within the group and avoiding collisions, robots must observe a certain geometric structure when moving: a certain location relative to each other within the formation or relative to the center of mass of the group, forming a certain geometric figure. There is a need to work out a mathematical model describing the movement of a group of robots, and to develop a decentralized control rule and an algorithm that would allow efficient control of the movement of agents while maintaining the geometric shape of the formation.

The purpose of this work is to develop a completely decentralized rule for controlling a group of agents modeling mobile robots, which would ensure the movement of a group

© The Author(s), under exclusive license to Springer Nature Switzerland AG 2021
R. Silhavy (Ed.): CSOC 2021, LNNS 229, pp. 367–375, 2021.
https://doi.org/10.1007/978-3-030-77445-5_34

of agents in compliance with a certain geometric structure of the formation (certain mutual distances relative to each other) under the conditions of complete autonomy of the agent and the possibility of obtaining information only from its nearest neighbors. In this case, the desired control rule must be efficient under the conditions of occurrence of the emergency situations described above.

The work solves the problem of managing a team of robots, which ensures the movement of a team of robots in a column. Structurally, the robot includes propellers, manipulators, various sensors, communication devices, etc. The type of robot propeller is determined by the way it moves in space, which, in turn, is determined by the environment in which the robot should function. The work assumes the use of propellers to move along the ground. When moving together, two or more robot technical complexes (RTC), it is important to know their relative position with high accuracy. Each of the robots needs a set of sensors to get information about the environment. Therefore, for the full-fledged operation of the mobile RTC, the robots are equipped with radio frequency sensors for determining the relative position. Each intelligent agent (mobile robot) includes a solver that receives data from other devices wirelessly. By processing them, the device generates commands for the autopilot. By processing the information read by the sensors, the motion control system solves the problem of constructing a path of movement [6–8].

2 Control of the Movement of a Group of Robots in a "Escort" Formation

Control of the movement of a group of robots in an "escort" formation is considered. The problem of localizing robots using a relative position has been solved. Special attention is paid to obtaining the law of control of the movement of robots in a column.

In the first setting, a column is considered – a system in which robots are located sequentially one after another at a given distance. The robots are numbered from R_0 – leading robot to R_n – closing the column. This type of movement requires that each robot R_i(except for the first – the leader), move along the trajectory along which the robot R_{i-1} moves, with some time delay T. The lead robot R_0 can be equipped a sensory system that allows him to plan a trajectory in an unpredictable environment, while the rest of the robots must simply follow the trajectory of their nearest neighbor. In Fig. 1 shows a diagram of the movement of a team of mobile robots in a column [9, 10].

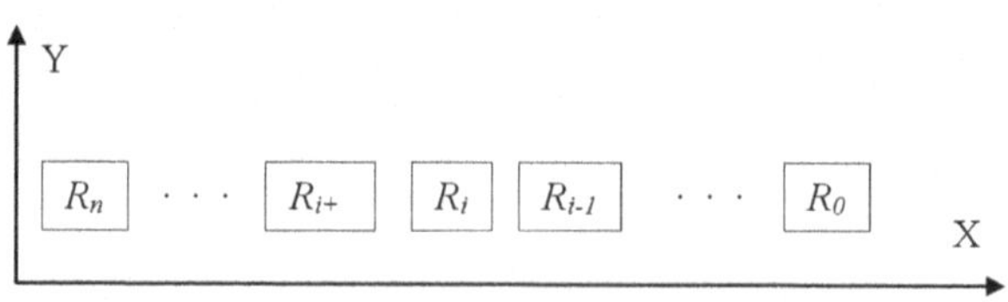

Fig. 1. Scheme of movement of a swarm of mobile robots in a column

The paper considers systems in which agents move in ordinary Euclidean space with discrete time. Let us now give a kinematic model of a mobile robot, as well as a developed method for controlling its movement along an arbitrary trajectory [11].

Figure 2 shows the location of the mobile robot R_i in the absolute coordinate system $0XY$ and the coordinate system $0_iX_iY_i$ associated with R_i.

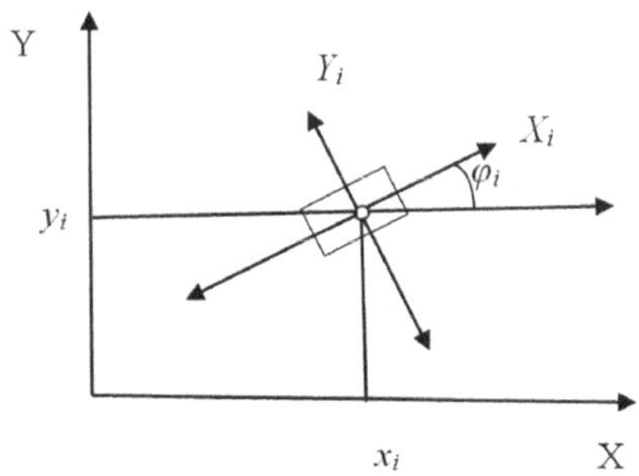

Fig. 2. Mobile robot coordinates

The initial tuning parameters are set. The vector $\boldsymbol{V_i}= (x_i, y_i, \varphi_i, V_i)$ of the state of each of the robots R_i in the absolute coordinate system is known. The $(x_i, y_i, \varphi_i, V_i)$ position and orientation of the robot in the absolute coordinate system, and the linear velocity of the robot, respectively. S_0 is a fixed distance between any pair of robots R_i and R_{i+1} adjacent in the column in absolute coordinate systems.

Let $S_i(t) = \sqrt{((x_i(t) - x_{i-1}(t))^2 + (y_i(t) - y_{i-1}(t))^2}$ is the distance between two neighboring robots R_i and R_{i+1} in the column.

$\omega_{i,i-1} = \varphi_i - \varphi_{i-1}$ is the angle between the axes 0_iX_i and $0_{i-1}X_{i-1}$ of the connected coordinate systems of neighboring robots R_i and R_{i+1}.

$V_i(t)$ is the linear speed of movement of the robot R_i at time t.

The control law is limited to considering the kinematic model. The inertial and design parameters of robots are not taken into account here.

If the vector $\boldsymbol{V_i} = (x_i, y_i, \varphi_i, V_i)$ of the state of each of the robots in the absolute coordinate system is known, then to calculate the robot speed R_i, it is sufficient to know the length of the projection of the robot vector R_i in the absolute coordinate system onto the axis of the associated coordinate system X_i. This circumstance greatly simplifies the implementation of the control law.

Let us now consider the problem of controlling slave robots. The main idea is that the goal of controlling slave robots is to fulfill equalities:

$$\omega_{i,i-1}(t+1) = 0;$$
$$V_i(t+1)cos\omega_{i,i-1}(t-1) = V_{i-1}(t);$$
$$S_i = S_0.$$

The control method consists in the fact that in the process of movement at each moment of time t, the parameters of movement of the robots following the robot R_{i+1} are rescheduled. The robot's error R_i is: $\delta_{Si} = S_i(t) - S_{0,}\omega_{i,i-1}(t) > 0, \delta x_i, \delta y_i$.

The challenge is for the R_i robot to follow the R_{i-1} robot precisely.

Strategy 1 for each robot $R_i(i > 0)$ is to correct the movement according to the current deviation from the given trajectory of the pseudo-target (robot R_{i-1}). In this case, all points must be passed with a minimum deviation.

Strategy 2 (approaching the pseudo-target) implements a “parallel approach” of the robot R_i with the trajectory of the pseudo-target, observing the specified distance between the pair of robots R_{i+1} and R_i adjacent in the column, and involves calculating the yaw angles to predict their meeting place. Exact passage of the robot through the reference points is not required. In Fig. 3 shows how the programmed trajectory of each slave robot is formed [12, 13].

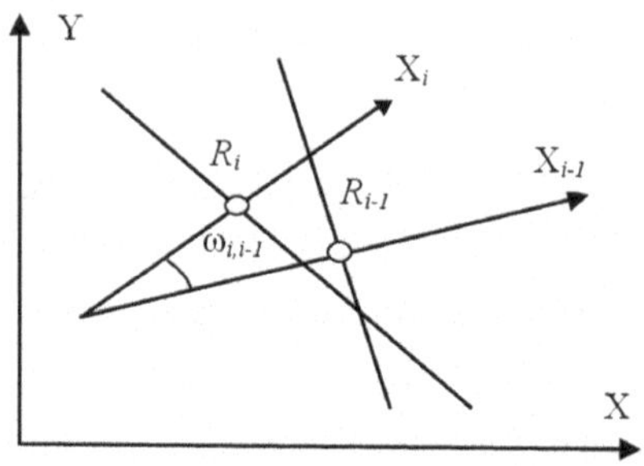

Fig. 3. The layout of two neighboring robots in the absolute coordinate system *OXY*

Let us now consider the corresponding models. The kinematic model of each of the robots is as follows:

$$\omega_{i,i-1}(t+1) = \omega_{i,i-1}(t) + \delta_\omega \omega_{i,i-1}(t); \tag{1}$$

$$V_i(t+1)cos\omega_{i,i-1}(t+1) = V_{i-1}(t+1) + \delta_V V_{i-1}(t+1); \tag{2}$$

$$x_i(t+1) = x_i(t) + \delta_x x_i(t); \tag{3}$$

$$y_i(t+1) = y_i(t) + \delta_y y_i(t); \tag{4}$$

The programmed trajectory of each slave robot is formed as follows. The control law is limited to considering the kinematic model. The inertial and structural parameters of the robot are not taken into account here.

The maximum values of the speed $V_i(t+1)$ and the yaw angle $\omega_{i,i-1}(t+1)$ are set by the control system within the permissible limits.

Let us now consider the corresponding models. Here are the applied production rules in order of priority of their execution:

Rule 1 (elimination of the divergence of the course of the robot R_{i+1}):

If $\omega_{i,i-1}(t+1) > 0$, then the angle is corrected between the coordinate axes 0_iX_i and $0_{i-1}X_{i-1}$ in order to decrease it, in accordance with the expression (1).

Rule 2: if the deviation and deviation processing time does not exceed the predetermined threshold values, then the speed $V_i(t+1)$ of movement is corrected according to the current deviation from the given trajectory of the pseudo-target, in accordance with expressions (2, 3, 4).

The first rule is intended to solve the control problem under the condition of exerting a control effect on the direction of movement of the agent, and the second rule is intended to provide a control effect on the agent's speed. For each control rule, its discrete analogue is presented.

Linear velocity $V_{i-1}(t)$, angle $\omega_{i,i-1}(t)$ and coordinates x_i, y_i the location of the pursued robot R_{i-1} are disturbances of an indefinite nature, and for the catching robot R_i – as parameters for controlling the movement of the robot.

To control the movement of a group of robots in an "escort" formation, an adaptive feedback algorithm has been developed, based on the compensation method and the classical concept of astatism of control systems. The approach to solving this problem is to formalize the concept of deviation of the current position of the robot from the required one based on visual information and search such control with feedback, which ensures the reduction of the norm of this deviation to zero. To control the movement of robots along a fragment of the trajectory of the previous robot, we used kinematic equations in state variables characterizing the fulfillment of the control goal. Solved the problem of localization of robots using relative position.

3 Algorithm of Movement of a Group of Robots in a "Escort" Type Formation Based on Collective Adaptation

Control of the movement of a group of robots in a "escort" type formation is carried out using an alternative collective adaptation algorithm based on the ideas of the collective behavior of adaptation objects [14–19].

The objects of adaptation of the catching robot R_i are the parameters of the vector $\mathbf{V_i} = (x_i, y_i, \varphi_i, V_i)$ considered as controls – the position and orientation of the robot in the absolute coordinate system, and the linear velocity of the robot, respectively. The collective of adaptation objects (their totality) of a group of robots in a "escort" type formation corresponds to an optimization object (OO).

To implement the adaptation mechanism, the parameters of the vector $\mathbf{V_i} = (x_i, y_i, \varphi_i, V_i)$ are compared to the adaptation automata AA – $a_i^{\omega}, a_i^{xy}, a_i^{V}$, which simulate the behavior of adaptation objects in the environment. The adaptation automaton (Fig. 4) has two groups of states: $C_1 = \{c_{1i} | i = 1, 2, ..., g\}$ and $C_2 = \{c_{2i} | i = 1,2, ..., g\}$, corresponding to two alternatives A_1 and A_2 behavior of the adaptation object in the environment: A_1 – change the size of the parameter, A_2 – leave it unchanged. Thus, the output alphabet of the adaptation automaton is $A = \{A_1, A_2\}$. The number of states in a group is specified by a g parameter called memory depth. The input alphabet $Q = \{+, -\}$ includes possible responses of the environment: "reward" – (+) and "punishment" – (–). The graph-diagram of AA transitions is shown in Fig. 4.

A signal "encouragement" or "punishment" is sent to the input of the adaptation machine, depending on the state of the adaptation object (parameter of the vector $\mathbf{V_i}$) in the environment. The sign (+) marks the transitions in AA under the influence of the "encouragement" signal, the sign (–) marks the transitions under the influence of the "punishment" signal. Without loss of generality, let us consider the principles of functioning of one AA. Initially, AA is in one of the initial states (in the figure, these states are in bold).

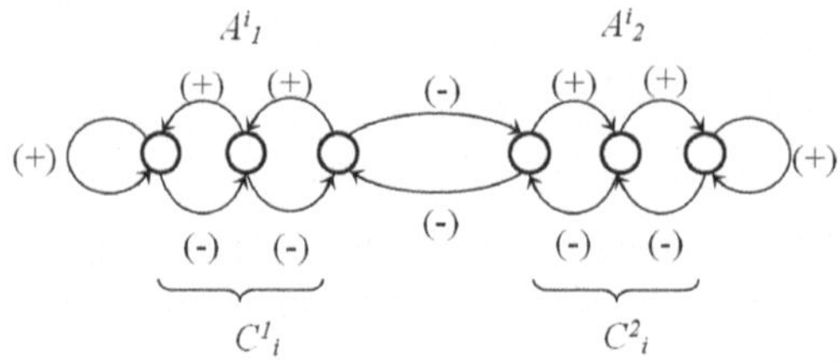

Fig. 4. Adaptation machine structure

Let us introduce the notation for the estimates (values) of the robot parameters R_i:

$$\rho_i^{\alpha}(t) \text{ for } \omega_{i,i-1}(t),\ \rho_i^{V}(t) \text{ for } V_{i-1}(t),\ \rho_i^{x}(t) \text{ for } x_i(t),\ \rho_i^{y}(t) \text{ for } y_i(t).$$

The local goal of the adaptation object is to achieve a state in which its estimate is $\rho_i^{-} = 0$. The global goal of the collective of adaptation objects is to achieve such a state *S* (distribution of vertices over nodes) such that $F(S) \rightarrow min$.

The process of alternative collective adaptation, which controls the movement of a group of robots in an "escort" type formation, at each step *t* is carried out in four steps (Fig. 5).

At the first step of collective adaptation for each R_i, the values of the parameters of the vector $\mathbf{V_i}$ are calculated.

On the second clock cycle, for each adaptation automaton a_i^{-}, the environment responds: "reward" or "punishment".

1. For each AA governing changes in the angle $\omega_{i.\ i-1}(t)$:
   If $sgn(\omega_{i,i-1}(t)) = 0$, then for AA $a^{\omega}{}_i(t+1)$, a signal "encouragement" (+) is generated.
   If $sgn(\omega_{i,i-1}(t)) \neq 0$, then for AA $a^{\omega}{}_i(t+1)$ the signal "punishment" (−) is generated.
2. For each AA managing changes x_i, y_i:
   If $sgn(S_i(t+1) - S_0) = 0$, then for AA $a^{xy}{}_i(t+1)$, a signal "encouragement" (+) is generated.
   If $sgn(S_i(t+1) - S_0) \neq 0$, then for AA $a^{xy}{}_i(t+1)$ the signal "punishment" (−) is generated.
3. For each AA controlling V_i changes.
   If $sgn(V_i(t) - V_{i-1}(t)) = 0$, then for AA $a^{xy}{}_i(t+1)$, a signal "encouragement" (+) is generated.
   If $sgn(V_i(t) - V_{i-1}(t)) \neq 0$, then for AA $a^{xy}{}_i(t+1)$ the signal "punishment" (−) is generated.

At the third step, in each adaptation automaton a_i, under the action of the response q_i fed to its input, a transition to a new state is carried out.

On the fourth cycle, for each adaptation object, an alternative is implemented in accordance with the AA outputs.

1. If AA $a^{\omega}{}_i(t+1)$ is in one of the states of the group $C^1{}_{ij}$, then $\omega_{i,i-1}(t)$ does not change.

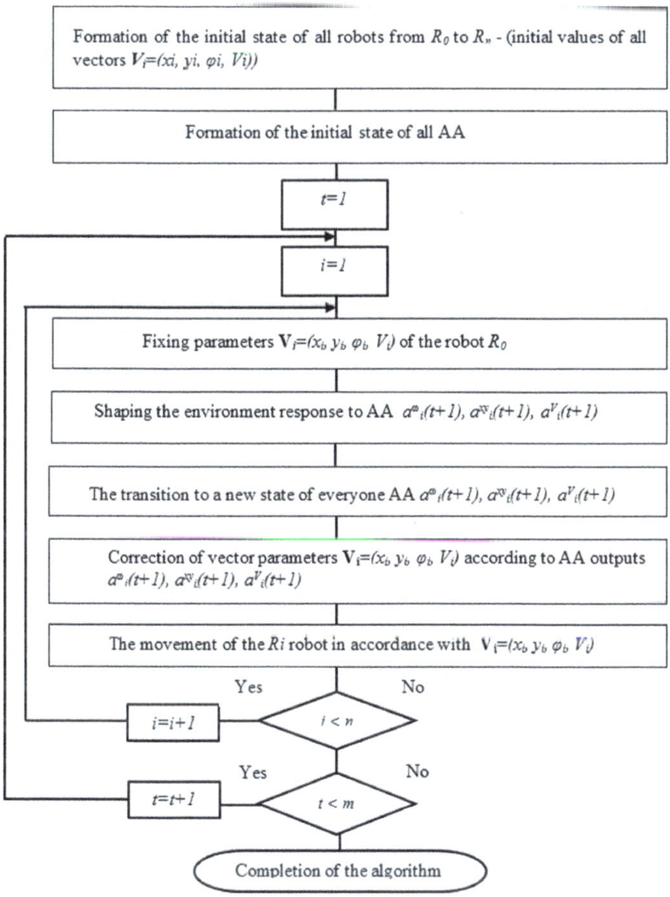

Fig. 5. Algorithm for controlling the movement of a team of robots in a column

If AA $a^{\omega}{}_i(t + 1)$ is in one of the states of the group $C^2{}_{ij}$, then $|\omega_{i,i-1}(t + 1))| = |\omega_{i,i-1}(t)| - \delta_{\omega}|\omega_{i,i-1}(t)|$.

2. If AA $a^{xy}{}_i(t + 1)$ is in one of the states of the group $C^1{}_{ij}$, then $x_i(t + 1) = x_i(t)$, $y_i(t + 1) = y_i(t)$.
   If AA $a^{xy}{}_i(t + 1)$ is in one of the states of the group $C^2{}_{ij}$, then:
   If $S_i(t) > S_0$, then $x_i(t + 1) = x_i(t) - \delta_x \cdot cos\varphi_i(t + 1)$, $y_i(t + 1) = y_i(t) - \delta_y \cdot sin\, \varphi_i(t + 1)$.
   If $S_i(t) < S_0$, then $x_i(t + 1) = x_i(t) + \delta_x \cdot cos\varphi_i(t + 1)$, $y_i(t + 1) = y_i(t) + \delta_y \cdot sin\, \varphi_i(t + 1)$.
3. If AA $a^V{}_i(t + 1)$ is in one of the states of the group $C^1{}_{ij}$, then $V_i(t + 1) = V_i(t)$.
   If $V_i(t) > V_{i-1}(t)$, then $V_i(t + 1) = V_i(t) - \delta_V \cdot cos\varphi_i(t + 1)$, $y_i(t + 1) = y_i(t) - \delta_x \cdot sin\varphi_i(t + 1)$.
   If $S_i(t) < S_0$, then $x_i(t + 1) = x_i(t) + \delta_x \cdot cos\varphi_i(t + 1)$, $y_i(t + 1) = y_i(t) + \delta_x \cdot sin\varphi_i(t + 1)$.

4 Conclusion

In this work, new mathematical methods have been developed for modeling the motion of the robots formation. To solve the problems of pitch control, original control rules for speed and acceleration have been developed, which have a number of advantages over other methods: complete decentralization of control in combination with dynamic correction of the parameters of robots that specify position and orientation of the robot in the absolute coordinate system, and the linear velocity of the robot, respectively. Management is carried out using an alternative collective adaptation algorithm based on the ideas of the collective behavior of adaptation objects which allows you to efficiently handle emergency situations, such as the failure of agents changes in the number of agents due to failure or sudden acquisition of connection with the next agent, as well as in the presence of measurement errors and noise that satisfy certain restrictions.

Acknowledgements. This research is supported by grants of the Russian Foundation for Basic Research of the Russian Federation, the project № 19-07-00645.

References

1. Karpov, V.E.: Collective behavior of robots. Desired and real. In: Modern Mechatronics. Proceedings of the All-Russian Scientific School, pp. 35–51 (2011)
2. Karpenko, A.P.: Modern Search Engine Optimization Algorithms. Algorithms Inspired by Nature: A Tutorial, p. 448. Publishing House MSTU. N.E. Bauman (2014)
3. Karpenko, A.P.: Robotics and Computer-Aided Design Systems: Textbook, p. 71. Publishing House MSTU. N.E. Bauman (2014)
4. Kalyaev, I.A., Gaiduk, A.R., Kapustyan, S.G.: Methods and Models of Collective Management in Groups of Robots, p. 280. Fizmatlit, Moscow (2009)
5. Kuznetsov, A.V.: A brief overview of multi-agent models. In: Management of Large Systems: Proceedings, pp. 6–44 (2018)
6. Morozova, N.S.: Decentralized control of the movement of robots in a dynamically changing environment. In: Artificial Intelligence and Decision Making, pp. 65–74 (2015)
7. Sotnikova, M.V.: Algorithm for automatic holding of a wheeled robot on a visually given line. Bulletin of St. Petersburg State University (2016)
8. Burdakov, S.F., Sizov, P.A.: Algorithms for controlling the motion of a mobile robot in the pursuit problem. Informatics, Telecommunications and Management, pp. 49–58 (2014)
9. Zenkevich, S.L., Nazarova, A.V.: Mobile robot control system. Vestnik MGTU. Ser. Instrumentation. Publishing House MSTU. N.E. Bauman (2006)
10. Zenkevich, S.L., Zhu, H.: Movement control of a group of robots in an "escort" type formation. Mechatronics, automation, control, pp. 30–34 (2017)
11. Zenkevich, S.L., Zhu, H., Zhai, M.: Controlling the movement of robots in a group based on trajectory smoothing. In: Extreme robotics: collection of abstracts of the International Scientific and Technical Conference, pp. 143–145. SPb. Publishing and Printing Complex "Gangut" (2017)
12. Guillet, A., Lenain, R., Martinet, P.: Adaptive and predictive formation control of autonomous vehicles. IEEE Robot. Autom. Mag. (2014)
13. Nazarova, A.B., Ryzhova, T.P.: Methods and algorithms for multi-agent control of a robotic system. Vestnik MGTU. Ser. Instrumentation, pp. 93–105. Publishing House MSTU. N.E. Bauman (2012)

14. Kuznetsov, A.V.: Model of motion and interaction of the system of intelligent agents. Bulletin of VSU. Series: System Analysis and Information Technology, pp. 130–138 (2018)
15. Rastrigin, L.A.: Adaptation of Complex Systems, p. 375. Zinatne, Riga (1981)
16. Kureichik, V.M., Lebedev, B.K., Lebedev, O.B.: Search Engine Adaptation: Theory and Practice, p. 272. Fizmatlit, Moscow (2006)
17. Clerc, M.: Particle Swarm Optimization, p. 187. ISTE, London, UK (2006)
18. Kennedy, J., Eberhart, R.C.: Particle swarm optimization. In: Proceedings of IEEE International Conference on Neural Networks, pp. 1942–1948 (1995)
19. Wang, X.: Hybrid nature-inspired computation method for optimization. Doctoral dissertation. Helsinki University of Technology, TKK Dissertations, Espoo, p. 161 (2009)

A Binary Firefly Algorithm Applied to Knapsack Problem

Hernan Pinto[1], Matias Valenzuela[1], Carlos Maureira[2(✉)], Luis Lopez[1], and Andrés Fernández[1]

[1] Escuela de ingeniería en construcción, Pontificia Universidad Católica de Valparaíso, Valparaíso, Chile
{hernan.pinto,matias.valenzuela,luis.lopez,andres.fernandez}@pucv.cl
[2] Escuela de Tecnologías de la Información, Facultad de Ingeniería y Negocios, Universidad de las Américas, Santiago, Chile
carlos.maureira.pasten@edu.udla.cl

Abstract. The NP-hard problems are of interest in operations research and particularly the combinatorial problems. Therefore, the design of efficient algorithms that address these combinatorial problems are active lines of research. Inspired by the above, this article designs a binarization method so that continuous metaheuristics can solve combinatorial problems. The binarization method uses the concept of percentile. This method of percentile is applied to the firefly algorithm. The multidimensional knapsack problem (MKP) was used to verify our algorithm.

Keywords: Combinatorial optimization · Multidimensional Knapsack problem · Metaheuristics

1 Introduction

Combinatorial problems frequently appear in the industry. For example we found combinatorial problems in civil engineering [1–5], Operations Research [6–9], allocation problems [10–13], vehicle routing problems [14,15], robust optimization, and problems of scheduling [16–18], among others.

In recent years we have witnessed the generation of a large number of nature-inspired algorithms that are capable of solving optimization problems efficiently. Examples of these algorithms are a black hole, gravitational search, and multi-verse algorithm, among others. Due to the nature of the phenomenon, we find that many of these algorithms work in continuous search spaces and must be adapted to solve the MKP since the latter is a combinatorial problem. The adaptation process must ensure that the exploitation and exploration mechanisms are preserved so that the efficiency of the algorithm is maintained.

When performing a state of the art, it is found that there are several binarization techniques that allow adapting algorithms that work in continuous spaces based on binary spaces. Among the main binarization methods used are the functions of transfer, angular modulation, and quantum focus; The reader can

© The Author(s), under exclusive license to Springer Nature Switzerland AG 2021
R. Silhavy (Ed.): CSOC 2021, LNNS 229, pp. 376–385, 2021.
https://doi.org/10.1007/978-3-030-77445-5_35

consult [19,20] for more details. In this article, we developed a binarization method based on the percentile technique to perform the binarization process.

Firefly metaheuristic algorithm was selected to verify the effectiveness of the percentile technique in the binarization process. Firefly algorithm has been successfully applied to various engineering and operations research problems. The binary version of the firefly algorithm will be denoted by BPFA and this version will be applied to MKP. Experiments were developed that account for the percentile operator contribution in the binarization process. For this, a random operator was built that serves as a baseline to measure the contribution.

2 KnapSack Problem

The allocation of resources which is modeled by MKP is a frequent problem at the industrial level. The objective of MKP is to identify a subset of objects, which produce the greatest benefit by satisfying a set of restrictions specific to each problem. MKP corresponds to one of the most studied problems that fall in the NP-hard category. When we look at the literature, we see that MKP has been solved using different techniques. For example, in [21] an unsupervised learning technique was applied. A quantum binarization was designed in [22] to address MKP. In [23] a fruit fly algorithm was adapted, and in [24] a differential algorithm was used. Formally MKP is defined as:

$$\text{maximize} \sum_{j=1}^{n} p_j x_j \tag{1}$$

$$\text{subjected to} \sum_{j=1}^{n} c_{ij} x_j \leq b_i \ , \ i \in \{1, ..., m\}. \tag{2}$$

With $x_j \in \{0,1\}$, $j \in \{1,...,n\}$. p_j represents the profit of the item j. c_{ij} corresponds a cost related to dimension i and element j. In each dimension, the constraints i are noted by b_i. The solution can be conceptualized using a binary representation, in this way a 0 means absence of the element inside the knapsack.

3 Binary Percentile Firefly Algorithm

The areas of data mining and statistics are cross-cutting disciplines and we can find applications of them in different situations, such as transports, agriculture, cities, and mathematics [14,21,25–29]. The percentile technique applied to the binarization of metaheuristics that work in continuous spaces is explored in this section. In Fig. 1, a general diagram of the solution is shown. As a first stage, the solutions must be initialized, then it is determined if the criterion of the maximum number of iterations has been met. In the case that it has not been fulfilled, the firefly algorithm is executed and using the solutions obtained, the percentile algorithm is applied to perform its binarization and subsequently a repair algorithm for the case that there is any restriction that is not complied with. Once the binarization has been carried out through the percentile technique, the solutions obtained are compared with the best. In the case of having a superior, a local search operator is executed. Otherwise, the previous steps are repeated.

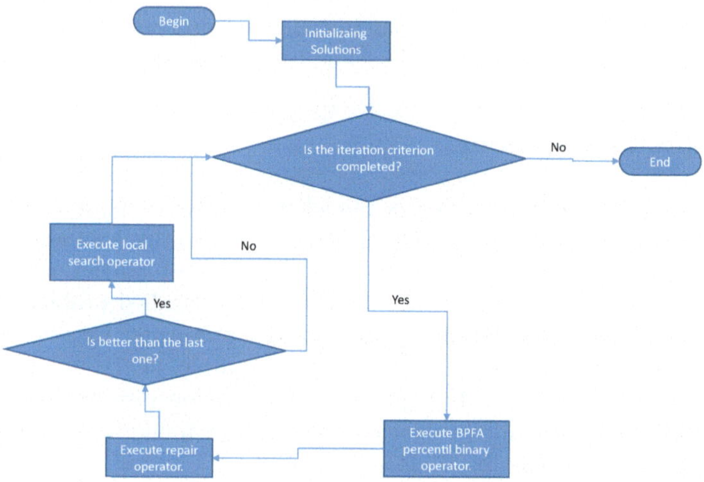

Fig. 1. Flowchart of the binary percentile firefly algorithm.

3.1 Initialization and Element Weighting

Since the bat algorithm is a swarm algorithm, to begin the search for the best solution, it is necessary to initialize the list of solutions. The mechanism used in the generation of each solution is as follows: first, a random element is chosen. Then, it should be verified if it is feasible to incorporate new elements, therefore, it is necessary to evaluate the restrictions of the problem. In the selection of the new element, if it is feasible, a list of possible candidates is generated that is consistent with the restrictions. Each item is assigned a previously calculated weight and the one with the best weight is selected. This procedure is repeated until no additional element can be incorporated.

In the calculation of the weight of an element, several heuristics have been developed. For example, in [30] a pseudo-utility was proposed. The equation that allows the calculation is written in Eq. 3. In this equation, the variable w_j corresponds to the surrogate multiplier whose value is between 0 and 1. This multiplier can be interpreted intuitively as a shadow price of the constraint j.

$$\delta_i = \frac{p_i}{\sum_{j=1}^{m} w_j c_{ij}} \tag{3}$$

A more intuitive measure focused on the average resource occupancy was proposed by [31]. It is shown in Eq. 4.

$$\delta_i = \frac{\sum_{j=1}^{m} \frac{c_{ij}}{mb_j}}{p_i} \tag{4}$$

In this article, a variation based on the average occupation will be used. This variation was proposed in [21]. In this proposal, the existing elements in the backpacks are also considered to calculate the average occupancy. Therefore, in each iteration, according to the elements selected for each solution, the measurement must be calculated again. The mathematical expression that defines this measure is shown in Eq. 5.

$$\delta_i = \frac{\sum_{j=1}^{m} \frac{c_{ij}}{m(b_j - \sum_{i \in S} c_{ij})}}{p_i} \tag{5}$$

3.2 Percentile Operator

Considering that the bat algorithm works in a continuous space and also it is iterative. Then, it results in the speed and position of each solution being updated in $\mathbb{R}^n$. In Eq. 6 the above is written in general. In this equation x_{t+1} corresponds to the position of the particle x at the time $t + 1$. To calculate the position, the *Delta* function, which is specific to each algorithm, must be considered additionally.

$$x_{t+1} = x_t + \Delta_{t+1}(x(t)) \tag{6}$$

In the case of the bat algorithm, binarization is done through a binary percentile algorithm. This algorithm considers that, given a solution x, we calculate the magnitude of its displacement $\Delta^i(x)$ in the i-th component and then group all the displacements in order to obtain the values for the percentiles {20, 40, 60, 80, 100}. In each percentile, a transition probability is associated that corresponds to the values shown in Eq. 7. Using these transition probabilities together with Eq. 8 the binarization of the solutions is performed. The algorithm is detailed in 1.

$$P_{tr}(x^i) = \begin{cases} 0.1, & \text{if } x^i \in \text{group } \{0, 1\} \\ 0.5, & \text{if } x^i \in \text{group } \{2, 3, 4\} \end{cases} \tag{7}$$

$$x^i(t+1) := \begin{cases} \hat{x}^i(t), & \text{if } rand < P_{tg}(x^i) \\ x^i(t), & \text{otherwise} \end{cases} \tag{8}$$

Algorithm 1. Percentile operator

Function pbinary(vLst, pLst)
Input vList, pList
Output pGroupValue
pValue = getPValue(vLst, pLst)
for each value in vList **do**
    pGroupValue = getPGroupValue(pValue,vLst)
end for
return pGroupValue

3.3 Repair Operator

Because the percentile operator and the local search operator can generate solutions that do not satisfy any of the constraints, it is necessary to apply a repair operator. The measure described in Eq. 5 was used to select the elements that perform the repair. In the case that repair is required, the element with the maximum measure is chosen and removed from the solution. This process is iterated until a solution that satisfies all constraints is obtained. The pseudocode is shown in Sect. 3.3.

Algorithm 2. Repair Algorithm

Function Rep(Sl_{in})
Input Input solution Sl_{in}
Output The Rep solution Sl_{out}
$S \leftarrow S_{in}$
while needRepair(S) == True **do**
    $s_{max} \leftarrow$ MaxWeight(S)
    $S \leftarrow$ removeElement(S, s_{max})
end while
state ← False
while state == False **do**
    $s_{min} \leftarrow$ MinWeight(S)
    if $s_{min} == Null$ **then**
        state ← True
    else
        $S \leftarrow$ addElement(S, s_{min})
    end if
end while
$S_{out} \leftarrow S$
return S_{out}

4 Results

4.1 Understanding BPFA Contribution

In this section, the experiments are developed to determine the contribution of the percentile binary operator in the binarization process. To determine this contribution, problems cb.5.250 of the OR library were selected. As comparison artifacts, violin plots, tables, and statistical comparisons were used. Particularly, the non-parametric Wilcoxon test was used to perform analyses. In the plots, the X-axis identifies the analyzed instances and the Y-axis the % -Gap defined in Eq. 9. The Wilcoxon test aims to verify whether the difference between the results obtained by BPFA is significant compared to the other algorithms. The exploration of the parameters is detailed in Table 1.

$$\% - Gap = 100\frac{BestKnown - SolutionValue}{BestKnown} \tag{9}$$

Table 1. Setting of parameters for Binary Percentile bat Search Algorithm.

Parameters	Description	Value	Range
N	Solutions	30	[20, 25, 30]
G	Percentiles	5	[4, 5, 6]
Iteration Number	Maximum number of iterations	1000	[1000]

Percentile Binary Operator Contribution. To understand the contribution of the percentile operator to the binarization process a random operator was constructed to be considered as a baseline. This random operator was configured to perform transitions with a fixed probability of 0.5 without considering in which percentile the speed variable is located. Two configurations were analyzed: the first configuration includes the local search operator and the second configuration does not consider this operator. The above with the objective of decoupling the contribution of the local search operator and the percentile. BPFA corresponds to our standard algorithm. *rd.ls* is the random variant that includes the local search operator. *BPFA.wls* corresponds to the version with a binary percentile operator without a local search operator. Finally, *rd.wls* describes the random algorithm without a local search operator.

The first comparison corresponds to comparing the best values between BPFA and *rd.ls*. These results are shown in Table 2. BPFA exceeds *rd.ls*. Despite being higher, the values are close. Our second comparison corresponds to considering the averages. In this case, BPFA exceeds *rd.ls* in almost all problems bluntly. The comparison of distributions through the violin plots shown in Fig. 2, indicates that the dispersion of the distributions *rd.ls* is greater than the dispersions of BPFA. In particular, this is notorious in problems 1, 2, 4, and 9. Therefore, the percentile and local search operators as a whole contribute to improving and making the results more robust. Finally, BPFA distributions are closer to zero than the *rd.ls* distributions, indicating that BPFA has consistently better results than *random.ls*. When we evaluate the behavior of the algorithms through the Wilcoxon test, this indicates that the difference is significant.

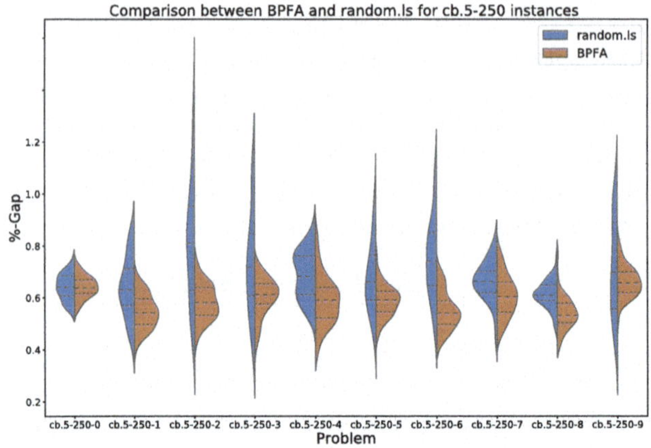

Fig. 2. Evaluation of percentile binary operator with Local Search operator

The following analysis aims to separate the contribution of the local search operator from the binary percentile operator, in order to be able to size the contribution of the percentile operator to the binarization process. For this, the comparison between the *wls* and *rd.wls* algorithms is made.

Table 2. Evaluation of percentile binary operator

Set	Best Known	Best *rd.ls*	Best BPFA	Best *rd.wls*	Best *wls*	Avg. *rd.ls*	Avg. BPFA	Avg. *rd.wls*	Avg. *wls*
cb.5.250-0	59312	59211	59225	59158	59175	59124	59129.7	59038.3	59101.7
cb.5.250-1	61472	61435	61472	61409	61409	61276.1	61334.1	61284.9	61344.1
cb.5.250-2	62130	62036	62074	61969	61969	61862.4	61968.4	61771.4	61900.7
cb.5.250-3	59463	59367	59446	59365	59349	59205.5	59258.1	59113.4	59214.5
cb.5.250-4	58951	58914	58951	58883	58930	58671.3	58804.5	58664.8	58747.5
cb.5.250-5	60077	60015	60056	59990	60015	59874.9	59903.8	59823.6	59924.1
cb.5.250-6	60414	60355	60355	60348	60349	60196.1	60284.2	60210.6	60266.2
cb.5.250-7	61472	61436	61472	61407	61407	61253.6	61281.7	61209.3	61341
cb.5.250-8	61885	61829	61885	61790	61782	61686.3	61735.3	61615	61683
cb.5.250-9	58959	58832	58866	58822	58787	58762.7	58670.2	58631.5	58735.5
Average	60413.5	60343.0	60380.2	60314.1	60317.2	60191.3	60233.2	60136.3	60225.8
p-value								1.23 e−04	4.25 e−04

The results are shown in Table 2: when analyzing the best results, we observe that *wpe* gets better results than *05.wpe* in all problems except 1, 6 and 7. However, the results are again very close. When we analyze the average, *wpe* exceeds *05.wpe* in all problems. The Wilcoxon test indicates that the difference is significant. This indicates that *wpe* is consistently better than *05.wpe*. Note that in this case we are not using the local search operator. The violin plots are shown in Fig. 3. This plots show that the dispersion of the solutions for the case of *05.wpe* is much greater than in the case of *wpe*. This indicates that the percentile binary operator applied to the firefly algorithm plays an important role in the accuracy of the results.

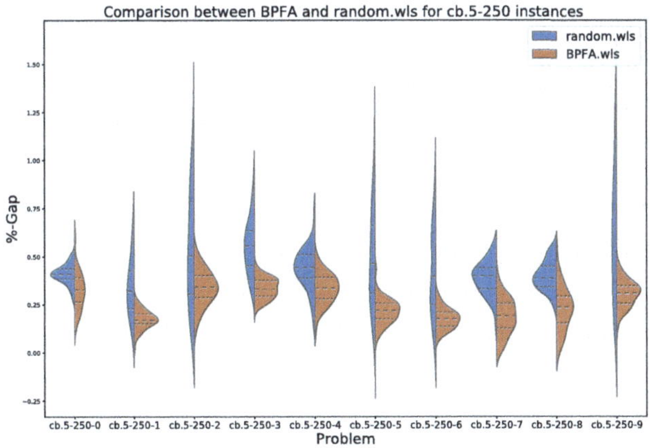

Fig. 3. Evaluation of percentile binary operator without Local Search operator

5 Conclusions

In this article, the percentile technique was explored as a method to perform the binarization of the bat algorithm. We must emphasize that the percentile technique is a general binarization technique and therefore can be applied in the binarization of any continuous swarm intelligence algorithm. MKP was used to evaluate the performance of the binary algorithm. The contribution of the percentile binary operator was analyzed, and it was concluded that this operator contributes to the accuracy and quality of the solutions obtained.

As a future line of research, it is interesting to use the bat algorithm in the three binarization mechanisms: Percentile, k-means, and transfer functions. In addition, it is interesting to binarize other metaheuristics with the percentile technique along with solving other NP-hard problems. Another line of longer-term research corresponds to exploring adaptive techniques to dynamically change the parameterization of metaheuristics, in this sense it is interesting to explore reinforced learning techniques.

References

1. García, J., Yepes, V., Martí, J.V.: A hybrid k-means cuckoo search algorithm applied to the counterfort retaining walls problem. Mathematics **8**(4), 555 (2020)
2. García, J., Moraga, P., Valenzuela, M., Pinto, H.: A db-scan hybrid algorithm: an application to the multidimensional knapsack problem. Mathematics **8**(4), 507 (2020)

3. García, J., Martí, J.V., Yepes, V.: The buttressed walls problem: an application of a hybrid clustering particle swarm optimization algorithm. Mathematics **8**(6), 862 (2020)
4. B. Crawford, et al.: Investigating the efficiency of swarm algorithms for bridge strengthening by conversion to tied-arch: A numerical case study on San Luis bridge. Iranian J. Sci. Technol. Trans. Civ. Eng. 1–13 (2020). https://doi.org/10.1007/s40996-020-00531-5
5. Martínez-Muñoz, D., Martí, J.V., García, J., Yepes, V.: Embodied energy optimization of buttressed earth-retaining walls with hybrid simulated annealing. Appl. Sci. **11**(4) (2021). https://www.mdpi.com/2076-3417/11/4/1800
6. Crawford, B., Soto, R., Astorga, G., García, J.: Constructive metaheuristics for the set covering problem. In: International Conference on Bioinspired Methods and Their Applications, pp. 88–99. Springer (2018)
7. García, J., Crawford, B., Soto, R., Astorga, G.: A percentile transition ranking algorithm applied to binarization of continuous swarm intelligence metaheuristics. In: International Conference on Soft Computing and Data Mining, pp. 3–13. Springer (2018)
8. Crawford, B., Soto, R., Monfroy, E., Astorga, G., García, J., Cortes, E.: A meta-optimization approach for covering problems in facility location. In: Workshop on Engineering Applications, pp. 565–578. Springer (2017)
9. García, J., Crawford, B., Soto, R., Astorga, G.: A clustering algorithm applied to the binarization of swarm intelligence continuous metaheuristics. Swarm Evol. Comput. **44**, 646–664 (2019). http://www.sciencedirect.com/science/article/pii/S221065021730528X
10. Garcia, J., Crawford, B., Soto, R., Astorga, G.: A percentile transition ranking algorithm applied to knapsack problem. In: Proceedings of the Computational Methods in Systems and Software, pp. 126–138. Springer (2017)
11. Astorga, G., Crawford, B., Soto, R., Monfroy, E., García, J., Cortes, E.: A meta-optimization approach to solve the set covering problem. Ingeniería **23**(3), 274–288 (2018)
12. García, J., Lalla-Ruiz, E., Voß, S., Droguett, E.L.: Enhancing a machine learning binarization framework by perturbation operators: analysis on the multidimensional knapsack problem. Int. J. Mach. Learn. Cybern. **11**(9), 1951–1970 (2020). https://doi.org/10.1007/s13042-020-01085-8
13. García, J., Maureira, C.: A KNN quantum cuckoo search algorithm applied to the multidimensional knapsack problem. Appl. Soft Comput. **102**, 107077 (2021)
14. García, J., Peña, A.: Robust optimization: concepts and applications. Nat.-Inspired Methods Stochastic Robust Dyn. Optim. 7 (2018)
15. García, J., Astorga, G., Yepes, V.: An analysis of a KNN perturbation operator: an application to the binarization of continuous metaheuristics. Mathematics **9**(3), 225 (2021)
16. García, J., Crawford, B., Soto, R., García, P.: A multi dynamic binary black hole algorithm applied to set covering problem. In: International Conference on Harmony Search Algorithm, pp. 42–51. Springer (2017)
17. García, J., Altimiras, F., Peña, A., Astorga, G., Peredo, O.: A binary cuckoo search big data algorithm applied to large-scale crew scheduling problems. Complexity **2018** (2018)
18. Crawford, B., Soto, R., Monfroy, E., Astorga, G., García, J., Cortes, E.: A meta-optimization approach to solve the set covering problem. Ingeniería **23**(3), 274–288 (2018)

19. Crawford, B., Soto, R., Astorga, G., García, J., Castro, C., Paredes, F.: Putting continuous metaheuristics to work in binary search spaces. Complexity **2017** (2017)
20. García, J., et al.: A db-scan binarization algorithm applied to matrix covering problems. Comput. Intell. Neurosci. **2019** (2019)
21. García, J., Crawford, B., Soto, R., Castro, C., Paredes, F.: A k-means binarization framework applied to multidimensional knapsack problem. Appl. Intell. **48**(2), 357–380 (2017). https://doi.org/10.1007/s10489-017-0972-6
22. Haddar, B., Khemakhem, M., Hanafi, S., Wilbaut, C.: A hybrid quantum particle swarm optimization for the multidimensional knapsack problem. Eng. Appl. Artif. Intell. **55**, 1–13 (2016)
23. Meng, T., Pan, Q.-K.: An improved fruit fly optimization algorithm for solving the multidimensional knapsack problem. Appl. Soft Comput. **50**, 79–93 (2017)
24. Liu, J., Wu, C., Cao, J., Wang, X., Teo, K.L.: A binary differential search algorithm for the 0–1 multidimensional knapsack problem. Appl. Math. Modell. **40**, 9788–9805 (2016)
25. García, J., Pope, C., Altimiras, F.: A distributed k-means segmentation algorithm applied to Lobesia Botrana recognition. Complexity **2017** (2017)
26. Graells-Garrido, E., García, J.: Visual exploration of urban dynamics using mobile data. In: International Conference on Ubiquitous Computing and Ambient Intelligence, pp. 480–491. Springer (2015)
27. García, J., Măntoiu, M.: Localization results for zero order pseudodifferential operators. J. Pseudo-Differ. Oper. Appl. **5**(2), 255–276 (2013). https://doi.org/10.1007/s11868-013-0084-y
28. Graells-Garrido, E., Peredo, O., García, J.: Sensing urban patterns with antenna mappings: the case of Santiago, Chile. Sensors **16**(7), 1098 (2016)
29. Peredo, O.F., García, J.A., Stuven, R., Ortiz, J.M.: Urban dynamic estimation using mobile phone logs and locally varying anisotropy. In: Geostatistics Valencia, pp. 949–964. Springer (2016)
30. Pirkul, H.: A heuristic solution procedure for the multiconstraint zero? One knapsack problem. Nav. Res. Logist. **34**(2), 161–172 (1987)
31. Kong, X., Gao, L., Ouyang, H., Li, S.: Solving large-scale multidimensional knapsack problems with a new binary harmony search algorithm. Comput. Oper. Res. **63**, 7–22 (2015)

Adapted Model Neural-Like Hopfield Network and the Algorithm of Its Training for Finding the Roots Systems of Linear Algebraic Equations

Alexander Gluhov[1], Anatoly Baranovskiy[1,2], Yulia Fomenko[1,2(✉)], and Alexander Bochkov[1]

[1] Emperor Alexander I St. Petersburg State Transport University, Moskovsky prospekt, 9, Saint Petersburg, Russia

[2] VKA Team. A.F. Mozhaisky, st. Zhdanovskaya, 13, 197198 St. Petersburg, Russia

Abstract. An approach for automatic formation of the structure of the Hopfield neural network model and its training (parameter settings) for solving systems of linear algebraic equations (SLAE) of arbitrary order is proposed. Adapted for solving of SLAE model of network is configured automatically in the environment for modeling Simulink, interacting with the Matlab computing system, which allows the user to vary the input data (order, constants vector and coefficients matrix system of linear algebraic equations). The results of research on the quality of the finding for solutions to SLAEs are presented.

Keywords: Computing system · Artificial Neural Networks · Hopfield network · Iterative methods · Systems of linear algebraic equations · Jacobi method · Seidel method

1 Introduction

One of the main ways for creating artificial intelligence (AI) based on building computer systems from neural-like elements (neural networks). This form of AI can be called (named) neuralintelligence. Artificial Neural Networks (ANN) are tools for computing and can be components of computers. The use of neural networks (NN) is reasonable for solving weakly specified tasks, that human intelligence is ineffective, also for time-consuming tasks [1]. The superiority of ANN over statistical methods consist in the high degree of parallelization of information processing and self-learning ability. Neural-like network are extremely effective means for solving tasks that are very difficult for traditional computing systems. The proof of this is comparative characteristic [2] computers based on the von Neumann principle with neural-like computing systems with the architecture of a biological neural system. This characteristic shows that the calculations of the first systems are centralized and sequential, the calculations of the second systems are distributed and parallel. These two criteria show that the use of neural networks can increase in productivity of computers.

The purpose of the work is justification of choice of the type of neural network for solving systems of linear algebraic equations (SLAE) and its automatic configuration in

© The Author(s), under exclusive license to Springer Nature Switzerland AG 2021
R. Silhavy (Ed.): CSOC 2021, LNNS 229, pp. 386–395, 2021.
https://doi.org/10.1007/978-3-030-77445-5_36

dependency on the variation of the input data. Environment for modeling Simulink was used to create the NN model, which interacts with the Matlab computing system and user interface created using the Matlab GUIDE toolbox.

2 Approaches to Solving System of Linear Algebraic Equations

Applying of system of linear equations plays an important role in different areas technical and science. Detailing work of physical systems can be expressed by a formal description in the form of mathematical models that can be presented by systems of linear algebraic equations. To applied tasks, using mathematical models in the form of SLAE include:

- data approximation problems;
- research of automatic control system;
- tasks of calculating and analyzing the operation of electrical devices in stationary modes, etc.

Numerical methods for solving SLAE are subdivided into two groups: direct and iterative (Fig. 1). Direct methods compute the solution to a problem in a finite number of steps. These methods applied for solving simple systems with a small number of equations. Iterative methods is a mathematical procedure that uses an initial value to generate a sequence of improving approximate solutions. When the prescribed accuracy is achieved, the iteration process stop.

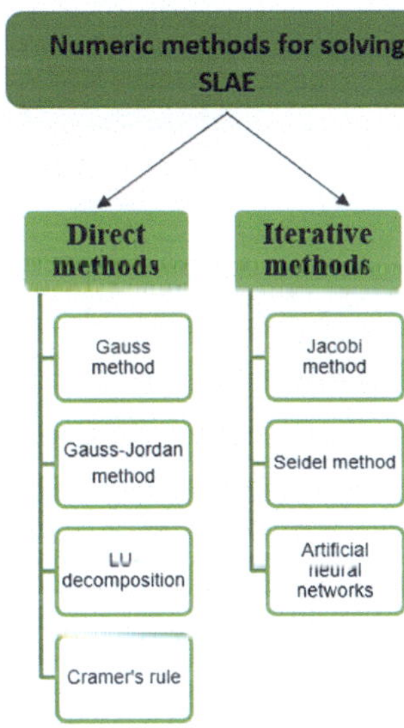

Fig. 1. Classification of numerical methods for solving SLAE

ANN are capable of performing tasks of different classes. Each network has unique properties that determinate its superiority in specific class. Typical tasks solved by neural networks include:

- classification;
- clustering (categorization);
- associative memory;
- forecasting;
- optimization.

Research shown that NN are capable of solving numerous tasks in matrix algebra. In [3, 4] NN for solving system of equations of a certain order, without data integration between applications is proposed. In [5, 6] NN with feedforward for solving matrix algebra tasks is proposed. In [7–9] NN with feedforward and recurrent was proposed for solving SLAE.

The choice of the type NN for neural-like computing system was carried out according to a number of criteria (Fig. 2).

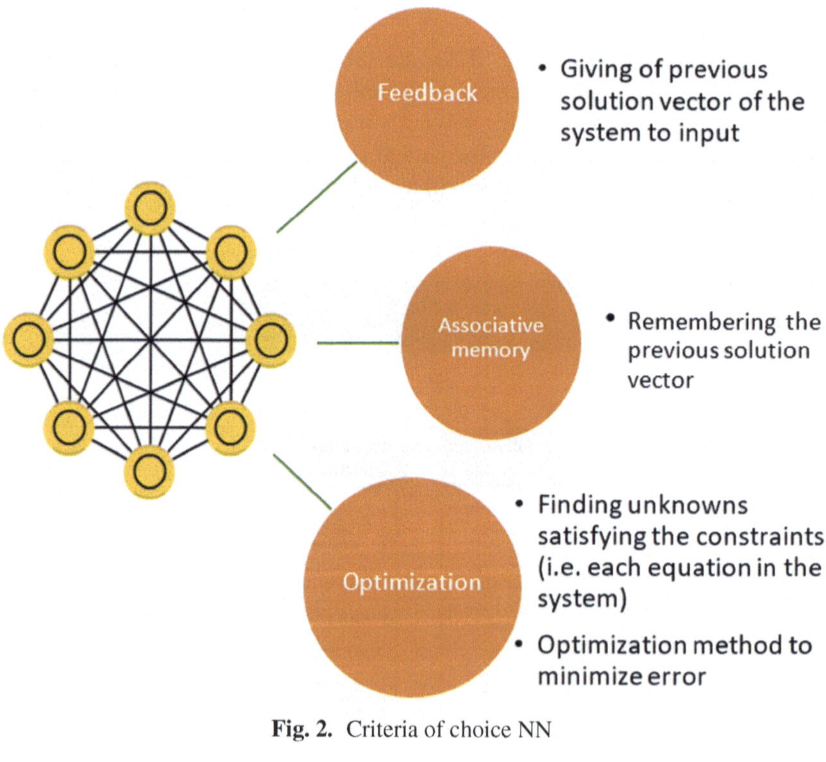

Fig. 2. Criteria of choice NN

Considering the main criteria for the development NN, the use of the Hopfield recurrent network seems to be a more preferable variant for the implementation of the tasks under study.

3 Task Formulation

Let user define the following input data:

- order of the SLAE;
- coefficient matrix A:

$$A = \begin{bmatrix} a_{11} \; a_{12} \; \ldots \; a_{1n} \\ a_{21} \; a_{22} \; \ldots \; a_{2n} \\ \ldots \; \ldots \; \ldots \ldots \\ a_{n1} \; a_{n2} \; \ldots \; a_{nn} \end{bmatrix} \tag{1}$$

- constants vector y:

$$y = \begin{bmatrix} y_1 \\ y_2 \\ \ldots \\ y_n \end{bmatrix}$$

- initial (approximate) values of the vector x:

$$X_0 = \begin{bmatrix} x_1 \\ x_2 \\ \ldots \\ x_n \end{bmatrix}$$

- number of clock cycles of the neural network.

Requires the determination modified Hopfield network architecture, implement automatic addition of Simulink blocks and setting parameters using Matlab.

4 Modified Hopfield Network and Parameter Definition

Matrix notation:

$$A \cdot x^* = y, \tag{2}$$

where A – is the matrix of coefficients for n equations in n unknowns; x^*– is the vector of unknowns; y – is the vector of constants.

It is proposed to description the Hopfield neural network for solving SLAE by the following equation:

$$\dot{x} = -D * x + W * f(x) + b, \tag{3}$$

where $\dot{x}$ – is the vector of the current state of the network; D – is the diagonal matrix (a square matrix where the diagonal elements (the elements where the two subscripts are equal) are nonzero); $f(x)$ – is the neural elements output function, b – is the displacement vector; W – is the matrix of connections of neural elements.

Modified Hopfield network architecture is presented on the Fig. 3.

In the proposed neural network model, the elements of the equation elements of the Eqs. (3) are calculated according formulas [3, 4]:

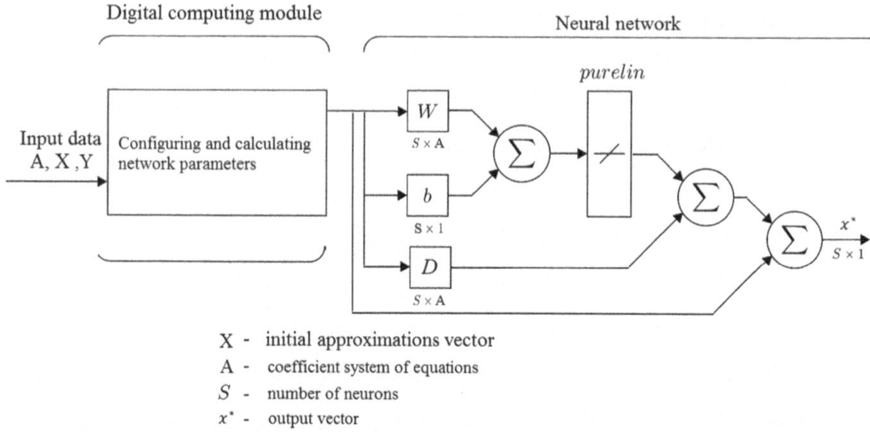

Fig. 3. Computational structure model with adapted Hopfield network

– connection matrix with a zero diagonal W

$$W_{ij} = \sum\nolimits_{k=1}^{n} a_{ki}a_{kj}, \tag{4}$$

where $w_{ii} = 0;\ w_{ij} = w_{ji}$.

– elements of the diagonal matrix D

$$d_i = \sum\nolimits_{k=1}^{n} {a_{ki}}^2 \tag{5}$$

– elements of the vector coefficients of external input

$$b_i = \sum\nolimits_{k=1}^{n} a_{ki}y_k \tag{6}$$

The Hopfield network fulfills its calculations asynchronously, i.e. after each iteration neurons simultaneously change their state.

In the classical Hopfield network for when implementing training procedures iterative process of updating matrix component values of connections of neuron elements repeated on a positively semidefinite matrix C, can become divergent, and the values of weights will tend to infinity [5]. In this case, the Hopfield network will not be able to reach its stationary state, and the solution of SLAE will not be obtained.

This situation arises when the network searches for solutions to a system of equations starting from the 5th order. To stabilize the process of finding solutions, it is proposed to decrease the matrix of coefficients of the system of Eqs. (1) by an order, which will lead to a decrease in the increment (7) of the network state vector at each step of the search for the roots of the SLAE.

5 Configuration of Network Model and Capacity Research

The model is configured automatically in the environment for modeling Simulink interacting with Matlab computing system through Matlab GUIDE user interface. This method of NN modeling relieves the user from the need to pre-generate the network for a certain order of the equation system, adding additional blocks and manual input parameters.

Computational model functionality:

- setting the clock cycles of the neural network;
- determining the order of SLAE and determining its parameters;
- dynamics of change of the output signal.

As an example, the search for solution of the system equations of the 10th order is considered. Window for entering system parameters presented on the Fig. 4.

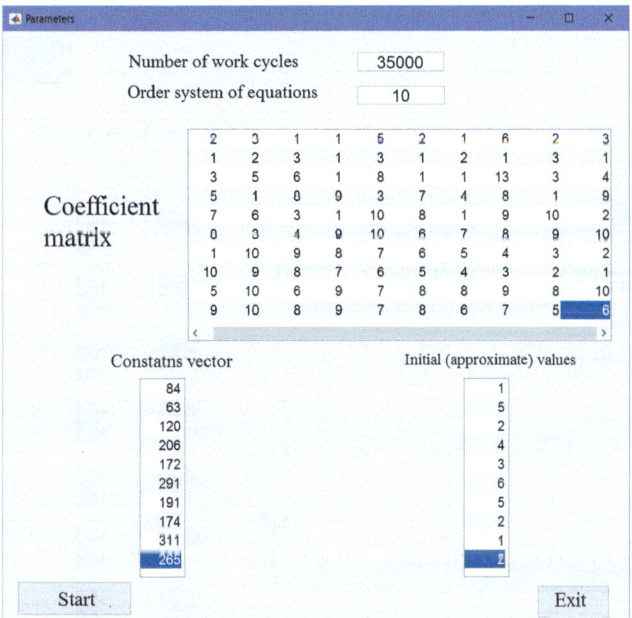

Fig. 4. The window setting parameters system of equations

The generated architecture of the modified Hopfield network in Simulink presented on the Fig. 5. The calculated values of the matrix W, D and vector b are transferred from Matlab workspace line-by-line to the conforming blocks (Fig. 6). Dynamic of changes output values of the neural network – finding solution system of SLAE – presented on the Fig. 7.

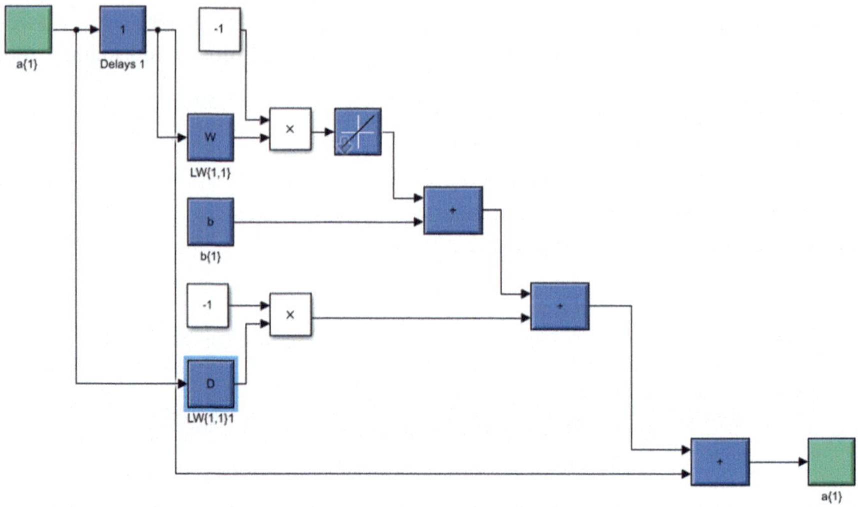

Fig. 5. Modified Hopfield network architecture in Simulink

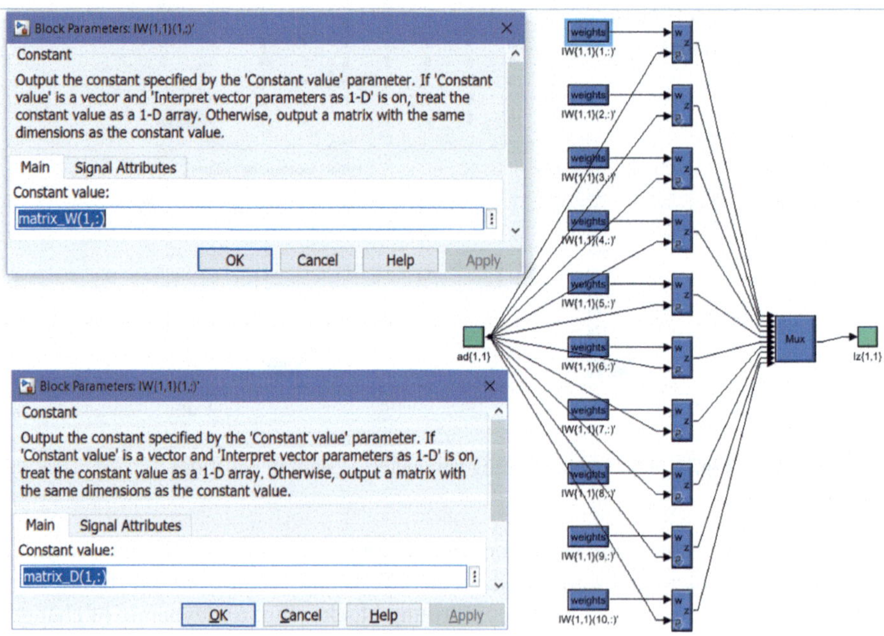

Fig. 6. Setting parameters

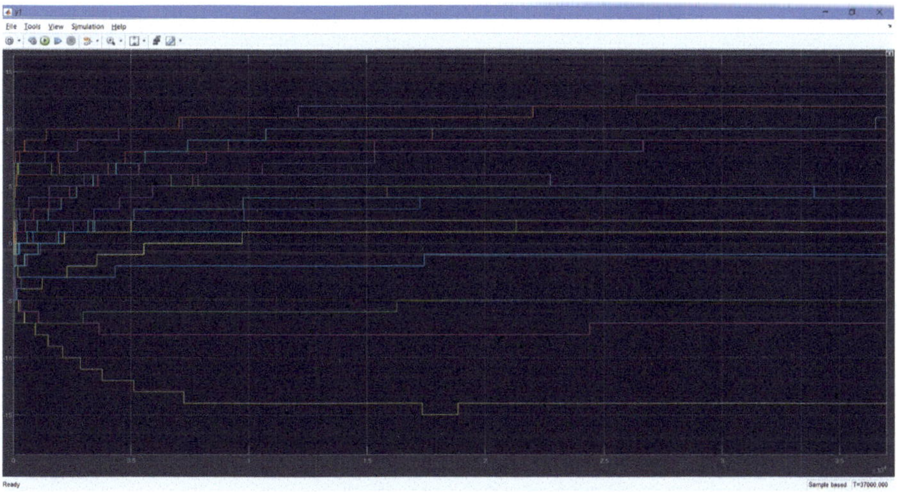

Fig. 7. The result of the neural network

Comparison of iterative methods for solving SLAE is presented in Table 1.

9 experiments were carried out in 3 experiments for each order the system of equations. SLAE of a certain order was randomly generated (4th, 6th, 8th), which had a unique solution. Using online service [6, 7] solutions system of equations were calculated by the Jacobi method and the Seidel method. An increase in the calculation error was observed exponentially. The calculation error is presented at 38 iteration. In contrast to the classical methods, the proposed Hopfield network model allows accurate solution result of SLAE at the end of the transient process in a limited number of iterations (horizontal lines on the graph - Fig. 7).

According to experiments the following conclusions can be drawn:

1. Adapted Hopfield neural-like network reached its equilibrium position in all 9 experiments. The influence of the order the system of equations on its operation was not revealed, which cannot be said about the Jacobi method and the Seidel method.
2. Convergence condition of the iterative process for the Jacobi method and the Seidel method for the 6th and 8th order equations was not achieved, as aresult, the iterative process became divergent which resulted in a huge error, no solution to the system of equations was obtained.
3. The results obtained for systems of equations of the 4th order by the Jacobi method and the Seidel method are not accurate. Answer with a large error. The proposed network structure provides an exact solution to an arbitrary order SLAE.
4. The most computationally complex iterative method for solving SLAE is the Jacobi method. When using the proposed computing architecture with a specialized neuromodule, only initial calculations are required to configure (set) the network parameters.

Table 1. Comparison of iterative methods for solving SLAE and solving the Hopfield network

Order of SLAE						
No	Hopfield network		Jacobi method		Gauss-Seidel method	
	Number of iterations	Convergence of processes	Number of iterations	Convergence of processes	Number of iterations	Convergence of processes
Fourth-order						
1	24	✓	24	−0.00288	15	−0.00444
2	8	✓	∞	−3.05E+14	15	−0.00136
3	5	✓	∞	−12788.3001	38	−2.87
Sixth-order						
1	3359	✓	∞	−3.8E+37	∞	−55601.936
2	2822	✓	∞	−2428.2	∞	−9829.33
3	1090	✓	∞	−7.2E+28	∞	−8.8E+17
Eighth-order						
1	2045	✓	∞	−1.6E+29	∞	−5.6E+59
2	3830	✓	∞	−7E+35	∞	−5.6E+33
3	3406	✓	∞	−4.6E+33	∞	−433.07

6 Conclusion

The proposed model and technique for the automatic formation of the structure of the Hopfield neural network model and one-step training procedure (parameter settings) for solving systems of linear algebraic equations of arbitrary order allows us to build a model of a hybrid computing system that includes a classical digital computer and the neural-like Hopfield network. Calculator function in terms of finding a solution of the SLAE will be reduced to setting up (one-step learning) the network. Finding solution to the SLAE will be autonomously carried out the trained network after setting its parameters and initial values (approximate) of roots, which by default can be considered equal to zero. The advantage of solving SLAE by the iterative method using a computing system based on the Hopfield neural network is obvious (Table 1). The network reaches a stationary state on a wide variety of input data. Dynamics of finding solutions allows use of intermediate values in the computing system to solve macro tasks.

Practical implementation is possible on the basis of native developments in the field of neurocomputing. In 2020, the Scientific and Technical Center (STC) "Module" released a computer in the form factor USB Flash drive – Module «NM Stick» [8]. The proposed modified Hopfield network can be implemented on this platform, since the main areas of application of this module are neural networks.

Relevant in future research, this neural network model can be adapted for solving systems of nonlinear algebraic equations or for solving transcendental equations that are of practical importance in a number of areas of science and technology. One of these areas is the problems of identifying complex objects and processes [9–11], the solution of which allows to optimally organize the management of objects and processes in various field of activity.

References

1. Ph. D. Wasserman Neural Computing: Theory and Practice, p. 230. Coriolis Group (1989)
2. Jain, A.K., Mao, J., Mohiuddin, K.M.: Artificial neural networks: a tutorial. IEEE Comput. **29**(3), 31–44 (1996)
3. Baranovskiy, A.M., Dorozhko, I.V., Krasilo, D.P., Kobzarev, S.A.: Hopfield neural network model for solving systems of linear algebraic equations. Certificate of registration of a computer program. Request no. 2015611642 (2015)
4. Baranovskiy, A.M., Silantyev, S.B., Smolitsky, K., Yafrakov, M.F., Bubnov, V.P.: Synthesis of a neural-like Hopfield network for solving systems of linear algebraic equations. Proc. High. Educ. Inst. Instrum. **37**(3–4), 47–51 (1994)
5. Akilova, I.M.: Neuroinformatics. Amur State University Press. Blagoveshchensk, p. 158 (2007)
6. Seidel method online. https://math.semestr.ru/optim/zeidel.php. Accessed 10 Nov 2020
7. Simple iteration method online. https://math.semestr.ru/optim/iter.php. Accessed 10 Nov 2020
8. Module «NM Stick». https://www.module.ru/products/2-moduli/nm-stick. Accessed 15 Nov 2020
9. Dmitriev, A.K., Yusupov, R.M.: Identification and technical diagnostics, p. 512. VIKKI (1987)
10. Baranovskiy, A.M.: Method of active identification of dynamic systems. Methods and algorithms for research and development of automatic control systems, pp. 50–54. VIKKI (1989)
11. Baranovskiy, A.M.: Active identification of stabilization systems. Proc. High. Educ. Inst. Instrum. **40**(8), 31–34 (1997)

Data Classification in Medicine and Healthcare Service

Olga Alekseevna Mitina[1(✉)] and Ivan Alexandrovich Yurchenkov[2]

[1] Department of Applied Mathematics, MIREA – Russian Technological University, st. Kantemirovskaya, 14, 115522 Moscow, Russia

[2] Department of Applied Mathematics, MIREA – Russian Technological University, st. Soviet, 116, apt. 7, 142620, region, Orekhovo-Zuevsky District, Kurovskoe, Moscow, Russia

Abstract. The field of medicine and healthcare service is developing rapidly. Thanks to advances in instruments that allow quantitative and qualitative measurements of physiological parameters, large volumes of medical data are accumulating. Data analysis is essential for improving workflow management, patient care, and predicting the development of various diseases, prescribing treatment and making diagnoses. Leading medical organizations strive to create and use systems that can help predict the duration and outcome of diseases, make diagnoses, and provide better medical care. #CSOC1120.

Keywords: Neural networks · Neural network training · Loginom analytical platform · Medicine and healthcare service · Correlation analysis

1 Introduction

The relevance of "classification" concept is due to its comprehensive application. The problem of medical diagnostics is one of the most urgent in the field of medicine and healthcare service. Having various information about a patient's health, one can classify the type of his disease, determine the most appropriate treatment method, or assess the risk of complications [1].

The purpose of the data classification problem is to assign different categories to a set of some objects or observations which are called class labels [2]. A class label is a discrete, unordered value that can be understood as belonging to a group of instances. Separation occurs using a classifier, an analytical model that solves the classification problem.

The classification problem is a supervised training type and is one of the basic tasks of artificial intelligence, machine learning and applied statistics.

The main task of supervised training is to build a model that will allow making predictions about previously unseen data using certain, pre-prepared and classified data.

© The Author(s), under exclusive license to Springer Nature Switzerland AG 2021
R. Silhavy (Ed.): CSOC 2021, LNNS 229, pp. 396–403, 2021.
https://doi.org/10.1007/978-3-030-77445-5_37

2 Materials and Methods

Classification can be found in many areas and tasks: natural language processing, pattern recognition, medical diagnostics, customer analysis in banking, geological exploration, speech recognition, document classification, etc.

Let's consider the basic concepts and definitions in more details.

There is a set of descriptions of objects $X = \{x_1, x_2, \ldots, x_n\}$, divided in some way into a set of class labels $Y = \{y_1, y_2, \ldots, y_n\}$.

A finite set of objects is given, for which it is known to which classes they belong $X \times Y$. Such a set will be called a training set. The class of the remaining objects is unknown.

The set of object descriptions is presented as a vector $x \in X$, each element of which is a feature. The vector $x = (f_1(x), \ldots, f_n(x))$ is called the feature description of the object x.

It is required to create a function f: $X \rightarrow Y$ that can correctly map each object description to its class label. This function is called a classifier [2, 3].

A neural network can be used as a classifier.

Artificial neural network (ANN) is a wide class of mathematical models that were originally created to simulate how the brain works. Their purpose was that they had to "learn" on the available data and then solve various problems of data analysis [4, 5]. At the moment, neural networks solve a huge range of problems associated with the identification of these dependencies in a multi-parameter system, allowing you to control some processes, such as risk assessment at the stages of the product life cycle [6], or to provide answers to incoming new requests or cases.

There are many different types of neural networks. They are divided according to the type of problem being solved – regression / forecasting, clustering, classification. Different types of model architectures and network learning algorithms are also distinguished.

The neural network training process involves weights selection for the connections in such a way as to obtain a value as close as possible to the value of the class label in the training set.

The general principle of network learning consists of several steps. The architecture of the neural network is chosen and weights are arbitrarily assigned to all connections – this step is called neural network initialization. Then, using the selected algorithm, training begins. The weights are changed iteratively in such a way as to minimize the total error E. One complete pass through the training sample is called the training epoch [2, 7].

When further changes in the weights do not give a result, errors percentage does not change, or quality criterion value satisfies task requirements, the model training process ends.

Thus, we can say that neural network training essence is to minimize some error function. The effectiveness of a trained model is determined by the percentage of erroneous predictions on the test data.

The parallelization of information processing and the ability to self-learn allow neural networks to solve very complex problems that were previously considered insoluble or intractable. The advantages of neural systems are that they can solve problems by detecting unknown patterns.

Also, the neural network is sufficiently resistant to noise in the input data, does not feel a special need for strict data filtering and has a higher data analysis speed than standard systems.

To solve the classification problem using a neural network, it is necessary to take into account that the features that will be fed to the network input must be weakly correlated with each other and strongly correlated with the class label.

According to definition, correlation is the dependence of two or more quantities. With the help of correlation analysis, it is necessary to exclude strongly correlated features of objects [3].

To do this, we will use the Pearson correlation coefficient, as well as the general system for classifying correlations by their strength. In our case, it is necessary to exclude all correlated variables that are not included in the interval $(-0.5; 0.5)$ [8, 9]. The variables included in this interval have a moderate level of correlation and are suitable for our task.

Thus, to solve the classification problem, we will create a classifier – a neural network. To improve the results of the constructed mathematical model, it is necessary to remove highly correlated variables. The neural network is trained using the prepared data.

The set studied in this article [2] contains records of cardiotocograms – the results of studies of fetus state, which are classified by three expert obstetricians. The data includes 21 features.

Three classes of fetal condition have been identified: normal – 1, suspicious – 2, pathological – 3.

Due to the fact that the data contains gaps, they must undergo preprocessing. For these purposes, the Loginom analytical platform is used.

The platform includes tools that allow you to shorten the development time as much as possible, quickly create and launch new applied solutions to the market. The creation of a complete solution takes a minimum of time: it is enough to get the data, define the processing scenario and set the place for exporting the results.

The next step is to select quality features. To select the features suitable for these features' analysis, we will use the feature selection method based on the correlation method [9].

Method feasibility demonstration: features are considered relevant if their values change systematically depending on their belonging to a particular category.

Thus, the data required for a neural network should contain features that are highly correlated with the target variable, and at the same time do not correlate with each other.

Let's carry out a correlation analysis and evaluate its results.

We calculate Pearson correlation coefficient between all the presented features and the class label – the target value of the classification (Fig. 1).

Three attributes strongly correlate with the target — "long-term slowdowns" or the number of long-term slowdowns per second, "abnormal short-term volatility" or percentage of time with abnormal short-term volatility, and "percentage of time with abnormal long-term volatility".

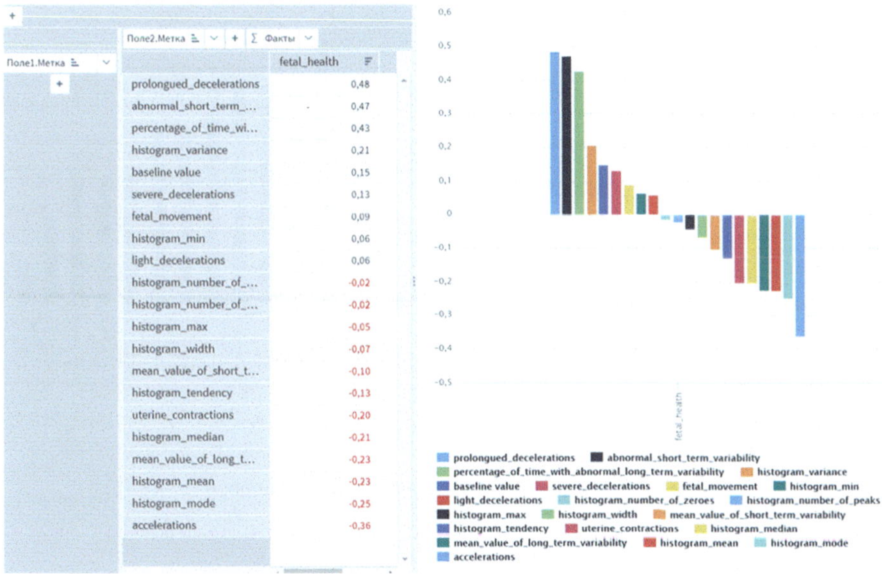

Fig. 1. Results visualization of the "Correlation analysis" node for the target variable.

The result of this analysis shows that the most important features for the classifier are the selected parameters of the sample objects, since they clearly have a great relationship with the indicator of interest to us, which is classification label.

Find features that strongly correlate with each other, and exclude them before starting training. Strongly correlated features interfere with the learning of the neural network.

The exclusion will be carried out taking into account the specified correlation boundaries. Let us group the variables according to the number of attributes correlating with them with a correlation value in modulus greater than 0.5. Figure 2 shows a visualization of the number of correlated variables above the threshold for each feature.

After filtering the features by the number of pairwise correlated features greater than 3, some characteristics of the histogram obtained as a result of grouping the pickup signal, as well as the "long deceleration" feature from the past analysis, will remain as input features (Fig. 1).

In order to more accurately select the parameters of the neural network, we will take into account the balance of the classes in the original sample (Fig. 3).

Three classes of fetal condition have been identified: normal – 1, suspicious – 2, pathological – 3.

We will train the neural network for classifying objects on a subsample using the Loginom platform due to the possibility of automatically building a training program for given hyperparameters – global training parameters (Fig. 4). These hyperparameters are responsible for neural network structure, the choice of a random distribution of the initial weights, the potential for weights changing, the number of layers in the network, the number of neurons in each of the layers, etc.

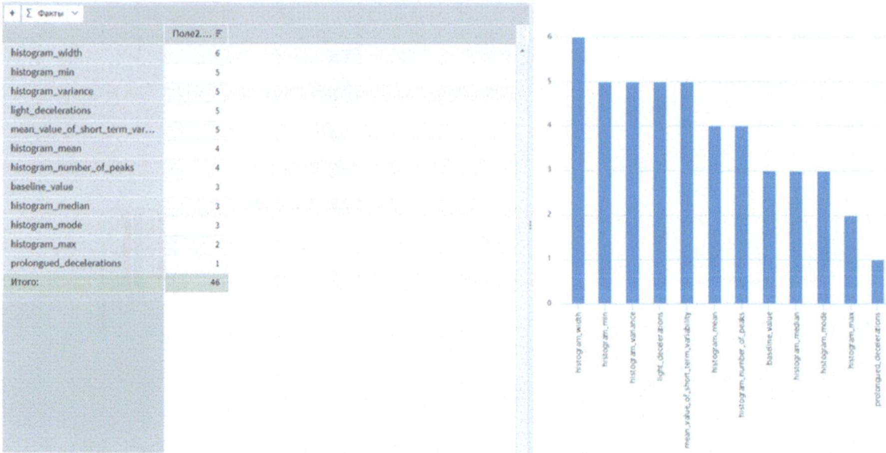

Fig. 2. Visualization of correlated variables number.

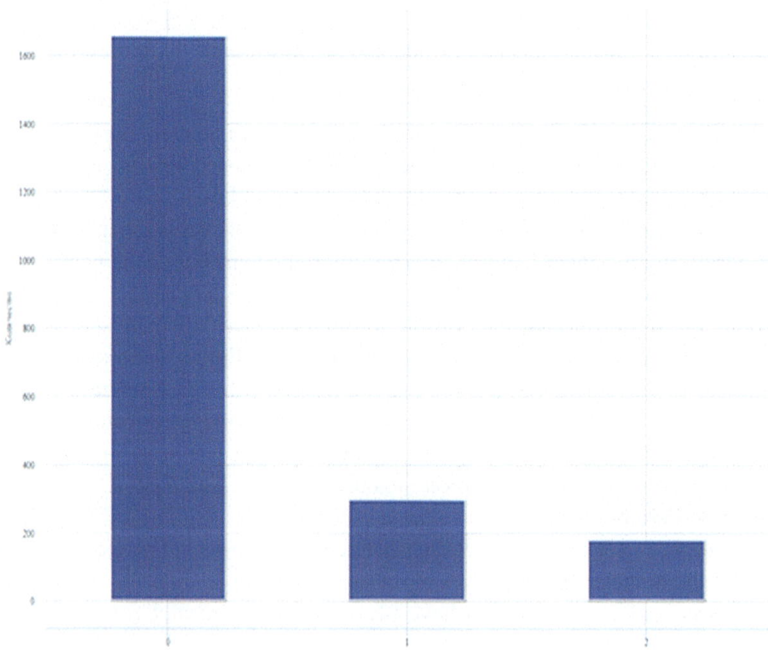

Fig. 3. The balance of classes in the original sample.

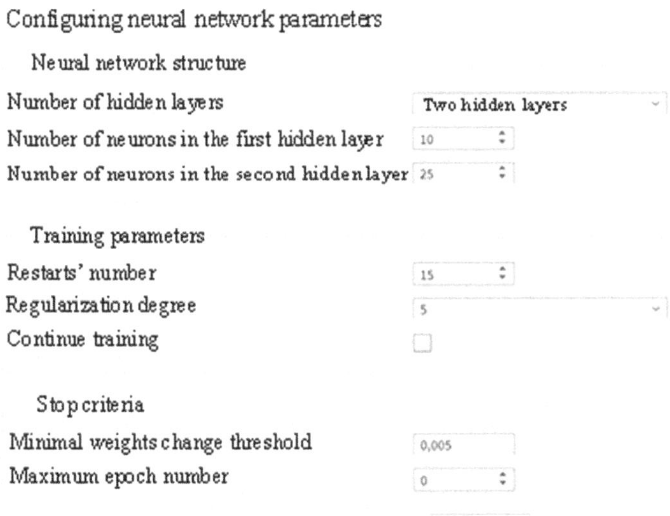

Fig. 4. Configuring neural network parameters in the Loginom platform.

For all the parameters of the objects remaining after the correlation analysis, we will standardize according to the rule:

$$\hat{x} = \frac{x - \bar{x}}{\sigma(x)}, \quad (1)$$

where x is the initial parameter (feature, description), $\bar{x}$ is the average value of the parameter over the entire training subsample, $\sigma\ (x)$ is the standard deviation of the parameter x in the training subsample.

The input parameters of the future neural network are standardized parameters on which the classification result will depend. Standardization helps to reduce the training time of the neural network and reduce the likelihood of training a "careless" algorithm.

We divide the original subsample with the data into disjoint subsets of 80% and 20% of the initial data for the training and test subsamples, respectively [10].

On the training subsample, the parameters of our neural network will be tuned using the backpropagation algorithm. On the test subsample, without resorting to changing the parameters, the performance of the neural network will be evaluated on the "new" data of our problem.

During training, the neural network hyperparameters are also tuned based on the cross-validation algorithm [11].

Having included the neural network component in the Loginom script, we will train the neural network for our task of classifying labels. The accuracy of solving the classification problem using a neural network in the classical version is estimated using the errors percentage in the classification of the trained model on the test set (Fig. 5). However, in the case of a sample in which there is clearly an imbalance in classes, it is worth checking the degree of model accuracy in several ways.

Name	Tag	Value
12 TotalSamples	Total samples	1701
12 TotalSelectedSamples	Total selected samples	1701
12 TrainSamples	Samples in training set	1701
9.0 TrainRMSError	Standard error on training set	0.20
9.0 TrainClsErrorPercentage	Classification error percentage on training set	7.94
9.0 TrainAvgCE	Average cross entropy on training set	0.21
9.0 CVRMSError	Cross validation Standard error	0.24
9.0 CVClsErrorPercentage	Cross validation classification error percentage	12.93

Fig. 5. Neural network training result.

In the case of a binary classification (two types of labels), an assessment of the form ROC-AUC is used. In the case of multiclass classification, along with the estimate of errors percentage on the test set (Fig. 6), the matrix of neural network predictions (Fig. 7) is considered, in which the true labels of the test set are in the columns, and the neural network prediction labels in the rows.

Name	Tag	Value
9.0 Examples	Total examples	420.00
9.0 Correct_predicts	Correct predicts	379.00
9.0 Percent	Error percentage on the set	9.76

Fig. 6. Percentage of classification errors on the test set.

12 fetal_health [forecast labels]	12 1	12 2	12 3
1	320	17	6
2	8	35	7
3	2	1	24

Fig. 7. Cross-table of true test set labels (columns) and forecast labels (rows).

The result of the neural network operation is successful and is 90.24% on the test set and 92.06% on the training set. It should be noted that the accuracy of the classifier depends on the size of the initial sample and the quality of parameters selection. The accuracy of the classifier can be increased by applying special algorithms for reducing the dimension of the input data.

3 Results

Thus, a multiclass model with a classifier as a neural network has been created. The algorithm divided the data into classes with high accuracy, thereby confirming the conclusion that neural networks can be used to solve the classification problem in medicine and healthcare service.

Data collection in the field of medicine and healthcare service and their subsequent processing allows you to create a holistic view of the patient's health, personalize treatment and investigate various diseases.

References

1. Silkina, U.I., Balandin, V.A.: Discriminant analysis of the parameters of variational heart rate monitoring. Russ. Technol. J. **8**(3), 81–91 (2020)
2. Goodfellow, Y., Benjio, I., Kurville, A.G.: Deep Learning. Trnsl. from English, Slinkina, A.A. (ed.) 2nd ed., rev, p. 652. DMK Press, Moscow (2018)
3. Barsegyan, A.A.: Analysis of data and processes: teaching guide. In: Barsegyan, A.A., Kupriyanov, M.S., Kholod, I.I., Tess, M.D., Elizarov, S.I. (ed.) 3rd ed., rev. and add, p. 512. BHV-Petersburg, St. Petersburg (2018)
4. Tarik, R.: We Create a Neural Network: trnsl. from English, p. 272. LLC "Alpha-kniga", St. Petersburg (2017)
5. Nikolenko, S., Kadurin, A., Arkhangelskaya, E.: Deep Learning, p. 480. Peter, St. Petersburg, Russia (2018)
6. Chesalin, A.N., Grodzensky, S.Ya., Van, Ty.F., Nilov, M.Yu., Agafonov, A.N.: Technology for risk assessment at the stages of the product life cycle using fuzzy logic. Russ. Technol. J. **8**(6), 167–83 (2020)
7. Kashirina, I.L., Demchenko, M.V.: Research and comparative analysis of optimization methods used in training neural networks, p. 10 VSU (2018)
8. Ivanter, E. V., Korosov, A. V.: Basics of Biometrics. Introduction to the Statistical Analysis of Biological Phenomena and Processes: Textbook for Universities, p. 163. PSU, Petrozavodsk (1992)
9. Mark, A.: Correlation-Based Feature Selection for Machine Learning, p. 200. Hall - Hamilton (1999)
10. Rashka, S.: Python and machine learning. Translator: Logunov, A.V., Movchan, D.A. (eds.), p. 408. DMK-Press, Moscow (2017)
11. Cross-validation. https://wiki.loginom.ru/articles/cross-validation.html. Accessed 02 Dec 2020
12. Machine Learning. http://www.machinelearning.ru/wiki/index.php?title=Classification. Accessed 02 Dec 2020
13. Kaggle. https://www.kaggle.com/andrewmvd/fetal-health-classification. Accessed 02 Dec 2020

Development of a Method for Automatic Generation and Optimization of Fuzzy Controller Parameters Based on an Adaptive Genetic Algorithm

Vladimir Vladimirovich Ignatyev(✉), Denis Aleksandrovich Beloglazov, Viktor Vladimirovich Soloviev, and Andrey Vladimirovich Kovalev

Southern Federal University, Bolshaya Sadovaya Str., 105/42, 344006 Rostov-on-Don, Russia
vvignatev@sfedu.ru

Abstract. The article proposes a method for automatic synthesis of a fuzzy controller and optimization of its parameters based on an adaptive genetic algorithm with combined operators of random changes. A distinctive feature of the proposed method is the method of forming a knowledge base of a fuzzy controller based on statistical information about the operation of a real industrial facility.

The use of an adaptive genetic algorithm makes it possible to quickly and efficiently optimize the parameters of a fuzzy controller in such a way as to provide the best indicators of its practical use: minimum duration, absence of oscillation and steady-state error of the transient process.

The proposed method is automated due to the development of a special software application in the MATLAB modeling environment, and requires minimal human participation in its work. Simulation modeling is performed and the results confirming the correctness of the proposed method and the possibility of its practical use are presented. The work of the method can be simplified as a sequence of the following stages: the formation of the initial parameters of the fuzzy controller; the search for the optimal lengths of the term-sets of input-output linguistic variables; search for optimal parameters of term-sets of input-output linguistic variables. #CSOC1120.

Keywords: Fuzzy controller · Genetic algorithm · Optimization · Parameters · Adaptivity · Controller

1 Introduction

Automatic control systems (ACS) of technical objects are used to withdraw and then hold them in a certain mode of operation, providing the best performance indicators. Efficiency can be understood as any criterion reflecting the requirements for the behavior of the control object (CO) at the moment. For example, the speed of reaction to disturbing influences, the amount of expended resources, the accuracy of keeping the controlled parameters in a given range, etc.

© The Author(s), under exclusive license to Springer Nature Switzerland AG 2021
R. Silhavy (Ed.): CSOC 2021, LNNS 229, pp. 404–416, 2021.
https://doi.org/10.1007/978-3-030-77445-5_38

The development of ACS requires certain knowledge and skills, the ability to analyze the available information about the automated CO. In the vast majority of cases, the basis of ACS is linear controllers that implement PID, PI, PD control laws. Their significant advantage is the simplicity of design and logical transparency, the disadvantage is the need to make significant assumptions in determining the numerical values of the tuning coefficients when used for nonlinear, multi-connected op-amps, which negatively affects the effectiveness of the application.

Alternative methods, such as the fuzzy logic apparatus, can be used to solve the problem of synthesizing a CO ACS with non-linearities and multi-connectivity. However, here, too, there is a problem of setting parameters, the number of which is much larger in comparison with PID, PI, PD controllers. The specific number of parameters to be configured for fuzzy controllers varies depending on the selected model, the number of linguistic variables and control rules.

The process of adjusting the ACS controllers can be largely automated using bioinspired algorithms, for example, genetic ones, which is confirmed by the presence of a large number of publications on this topic, for example [1–10].

The work [1] presents the results of using a genetic algorithm (GA) for training an artificial neuro-fuzzy network (NFN) that controls the manipulator. The NFN of Takagi-Sugeno architecture used by the authors, also referred to as ANFIS, is widely known and is often used as an alternative to classical fuzzy controllers (FC). This is due to its better adaptability for sharing with various optimization algorithms. The genetic algorithm is used in this paper to adjust the shape of the membership functions of fuzzy sets of NFN and the weight coefficients of its output layer.

The work [2] presents the results of using a genetic algorithm to adjust the parameters of a fuzzy controller, including the location of the vertices of triangular membership functions, and the combination of input and output fuzzy variables in the control rules. The purpose of the fuzzy controller is to hold a small-sized underwater robot on a given trajectory. The results obtained in [2] confirm the correctness of the ideas presented by the authors, but this does not negate the high similarity of the stated GA elements and the way it is used to find the optimal set of FC parameters with the works of other authors on this topic.

In work [3], a method for using the ANFIS NFN to solve the problem of constructing an adaptive PID controller for the anti-lock system of a car is proposed. The NFN parameters are determined before the start of system operation using a genetic algorithm and subsequently do not change.

In work [4], an example of the use of a genetic algorithm for searching and optimizing the parameters of a fractional PID controller that controls a manipulator is given. The structure and parameters of the used GA, including the type of random change operators, the method of forming a new population of solutions are close to the classical GA.

In work [5], an example of the use of GA for finding the optimal parameters of a fuzzy controller that controls the indoor climate is given. Control rules act as adjustable parameters, namely, a set of combinations of input and output fuzzy variables for each of them.

In work [6], a method for constructing a neuro-fuzzy PID controller of a voltage control system is proposed. The main idea of the method is to accumulate statistical data on the operation of an adaptive PID controller, the coefficients of which are adjusted by an artificial neural RBF network in real time, and use them to train a neural fuzzy network ANFIS. The RBF network training process is based on the application of a genetic algorithm. After training, ANFIS is used as the main controller of the control system.

In work [7], a method for constructing an adaptive PID controller that controls the spatial position of the telescope is proposed. The parameters of the PID controller are adjusted in real time by the fuzzy controller. The process of determining the parameters of a fuzzy controller, including the base of control rules, the term-sets of linguistic variables, is carried out on the basis of GA.

In work [8], a method for controlling the position of wind turbine blades based on the fuzzy controller of the Takagi-Sugeno model is proposed. To find the optimal parameters of the fuzzy controller, the GA is used.

In work [9], a method for controlling a robot manipulator based on a fuzzy controller is proposed. The initial parameters of the fuzzy controller are set based on expert experience and further optimized by GA.

The work [10] presents a method for constructing optimal fuzzy models using genetic algorithms. In the process of work, the GA adjusts the shape of the membership functions and the weight coefficients of the Takagi-Sugeno model in accordance with the original evaluation criteria proposed by the authors.

The works discussed above [1–10] are characterized by a high proportion of human designer participation in the search for optimal parameters of ACS controllers. This automatically means that the amount of time and effort spent on getting a result can be reduced if the share of human labor is reduced.

In this work, a method is proposed that allows one to almost completely automate the process of synthesis and tuning of fuzzy controllers. The method is based on the idea of replacing classical PID, PI, PD with fuzzy controllers, provided that there is statistical information in the form of input-output about the operation of the latter [11, 12]. To optimize the parameters of fuzzy controllers, an adaptive genetic algorithm with combined random change operators is used.

2 Method for Automatic Generation and Optimization of Fuzzy Controller Parameters

Fuzzy controllers are successfully used in various areas of human activity and have proven their effectiveness. One of the features of their design is the complexity of forming a knowledge base that includes control rules and term-sets of linguistic variables.

The solution to this problem can be significantly facilitated by automation provided the following conditions are met:

- availability of statistical data on the functioning of the automatic control system of the object;
- possibility of accumulating statistical data on the functioning of the automatic control system of the object.

Each of these conditions implies the possibility of obtaining a certain amount of data about the control object in the form of a value set:

Controlled parameters – The output signal of the control system.

The accumulation of the required amount of information is quite a long process, but in most modern enterprises that operate SCADA and similar systems, the accumulation of information occurs automatically and a large amount of data is already available today.

The sequence of actions performed during the automatic generation and optimization of the parameters of the fuzzy controller based on the proposed method has the following form [11, 12]:

- *Step 1.* Obtaining statistical information about the operation of a technical object as a set of value pairs of controlled parameters and output data of the control system;
- *Step 2.* User definition of the number of linguistic variables and their parameters (length of term-sets, range of variation);
- *Step 3.* User definition of the permissible degree of fuzzy variables overlap;
- *Step 4.* Automatic generation of term-sets of linguistic variables of the fuzzy controller;
- *Step 5.* Reading the statistical data of the automated control system application of the automated object, and then converting them into an array;
- *Step 6.* Obtaining the i th row of the array of statistical data of the automatic control system application, followed by determining the fuzzy variables that its values belong to, as shown in Fig. 1 (for an example of using an automated object of the PI controller in the control system);
- *Step 7.* Generating a control rule based on the numbers and quantities of fuzzy variables (if the numerical value of one of the parameters falls into two or more fuzzy variables, several rules are created);
- *Step 8.* Return to step 6 until the end of the statistics array is reached;
- *Step 9.* Saving the parameters of the fuzzy controller (a term-sets of linguistic variables, a bases of control rules) to a fuzzy model file in *.fis (the results of visualization of the generated fuzzy controller in the Fuzzy Logic Toolbox MATLAB editor are shown in Fig. 2.);
- *Step 10.* Optimization of the length of term-sets of input and output linguistic variables of a fuzzy controller based on a genetic algorithm;
- *Step 11.* Optimization of term-sets parameters of input and output linguistic variables of a fuzzy controller based on a genetic algorithm.

At *step 10*, the optimal number of fuzzy variable components of the term-sets of input and output linguistic variables of the fuzzy controller is determined. The main parameters of the applied GA are as follows: a combined adaptive crossing-over operator with a triggering probability of 0.7; a combined adaptive mutation operator with a triggering probability not exceeding 0.35 (the triggering probability is corrected during the GA operation); selection of solutions to a new population based on the roulette method.

The following analytical expression is used as a function for assessing the fitness of the solutions generated by the GA at step 10:

$$Fit_1 = \int_{t_0}^{t_i} \left((g - y)^2 + |y|\right)dt, \tag{1}$$

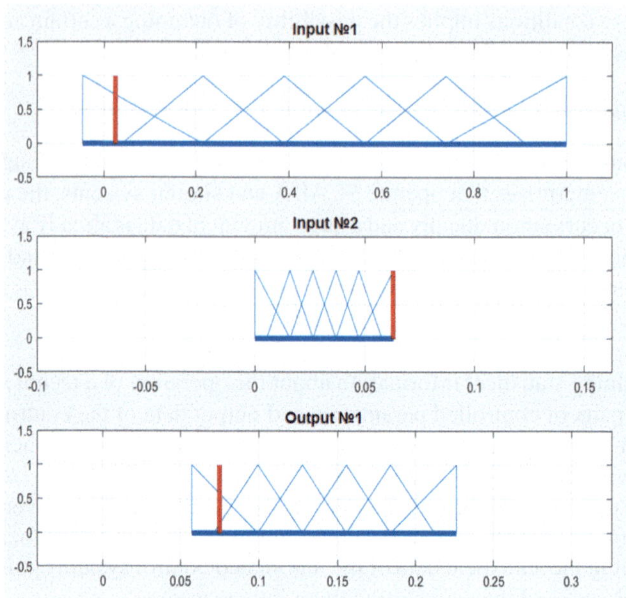

Fig. 1. Definition of fuzzy variables that belong to the numerical values θ, ∫ θdt, U *i*-th row of the data array of the PI controller application

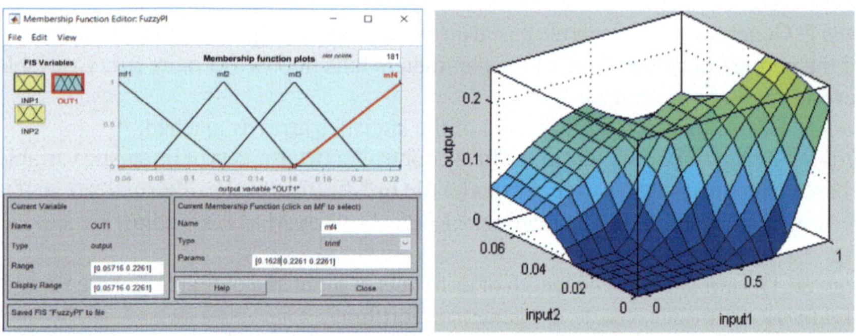

Fig. 2. Results of visualization of the generated fuzzy controller in the Fuzzy Logic Toolbox MATLAB editor

where *g* – the desired output signal of the fuzzy controller; *y* – the current output signal of the fuzzy controller.

The process of optimizing the lengths of term sets of input and output linguistic variables of a fuzzy controller based on GA is shown in Fig. 3.

At *step* 11, the GA searches for the optimal parameters of the fuzzy variables of term-sets of input and output linguistic variables of a fuzzy controller. The applied parameters are similar to those given earlier for *step* 10. The main difference lies in the way the chromosomes are evaluated. The structure describing the method of encoding the parameters of fuzzy variables in the GA chromosome is shown in Fig. 4.

The analytical expression for calculating the fitness of GA chromosomes at *step* 11 is as follows:

$$Fit_2 = \int_{t_o}^{t_i} \left((g - y)^2 + |y| \right) dt + w_{inp} \cdot LPInpError + w_{out} \cdot LoutError, \qquad (2)$$

where w_{inp}, w_{out} – weight coefficients; *LPInpError* – the value of the total positioning error of the vertices of fuzzy variables term-sets of input linguistic variables; *LPOutError* – value of the total positioning error of the vertices of fuzzy variables of the term-set of the output linguistic variable.

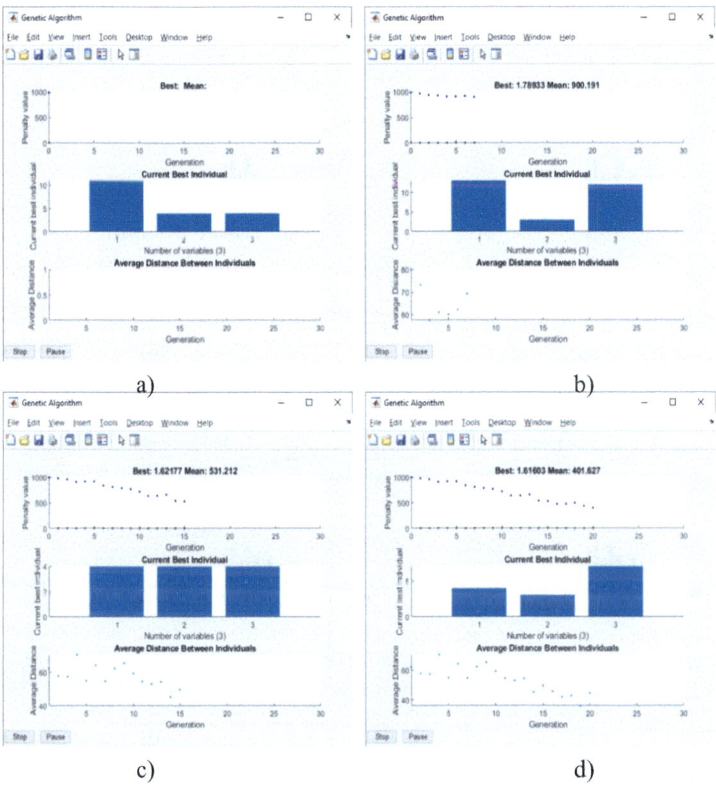

Fig. 3. An example of using GA to optimize the lengths of term-sets of input and output linguistic variables of a fuzzy controller

The process of optimizing the parameters of terms-sets of input and output linguistic variables of a fuzzy controller based on GA is shown in Fig. 5

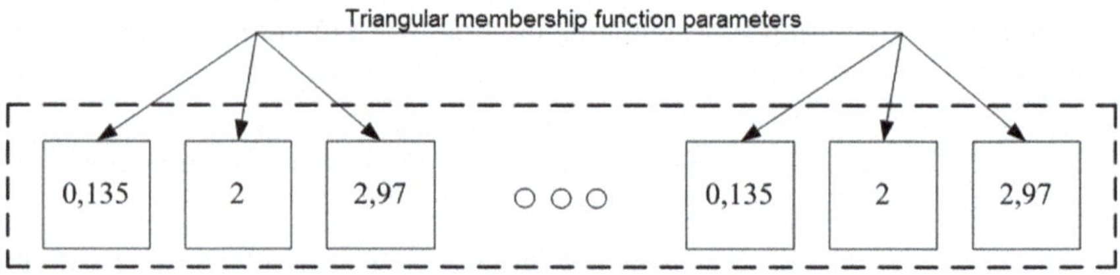

Fig. 4. The structure of the GA chromosome used to optimize the parameters of term-sets of input and output linguistic variables of a fuzzy controller

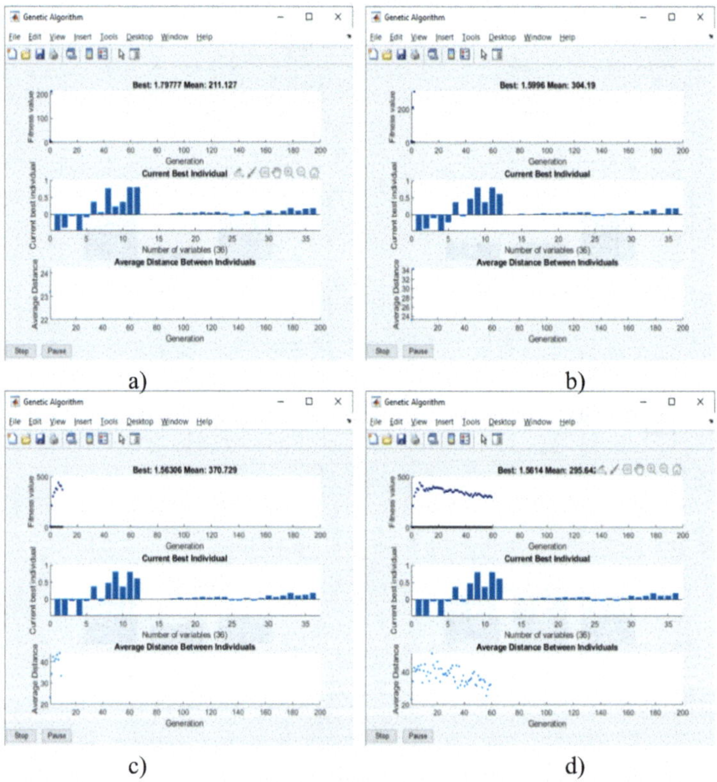

Fig. 5. An example of using GA to optimize the parameters of terms-sets of input and output linguistic variables of a fuzzy controller

3 Adaptive Genetic Algorithm

To optimize the parameters of the fuzzy controller at *step 10, 11*, item 2, an adaptive genetic algorithm is used, which has the following distinctive features: combined mutation operator; combined crossing-over operator; a fuzzy controller that changes the parameters of the combined random change operators(the probability of using the mutation operator and its forms, the probability of using the forms of the crossing-over operator) based on the fitness values of the solutions generated on their basis.

The combined mutation operator includes the following options for changing the parameters of triangular-shaped fuzzy variables:

- changing the position of the left border of the fuzzy variable;
- changing the position of the right border of the fuzzy variable;
- changing the position of the center of a fuzzy variable;
- changing the position of all parameters of a fuzzy variable.

Application of the combined mutation operator can be represented as the following sequence of steps:

Step 1. Formation of an array of chromosomes subjected to mutation with an indication of its type and the degree of fitness of the obtained solution to the conditions of the current problem;
- *Step 2.* Formation of an array of chromosomes that will be mutated using the probability of using the operator;
- *Step 3.* If the array of chromosomes selected at *step 2* has a nonzero length, then correct the probability of applying the forms of the mutation operator, as shown in Fig. 6 (initially, the probabilities of using each of the combined operator forms are set to be the same);
- *Step 4.* Selecting a specific form of the mutation operator, which will be used to randomly change the *i*-th chromosome (for each chromosome, its own mutation operator is selected);
- *Step 5.* Performing a random change of *i*-th chromosome based on the selected mutation operator.

Figure 7 illustrates the use of various forms of the combined mutation operator.

The combined crossing-over operator includes the following operations for changing the parameters of fuzzy variables of triangular shape:

- single-point crossing-over;
- two-point crossing over;
- two-point crossing-over with three parents.

Application of the combined crossing-over operator can be represented as the following sequence of steps:

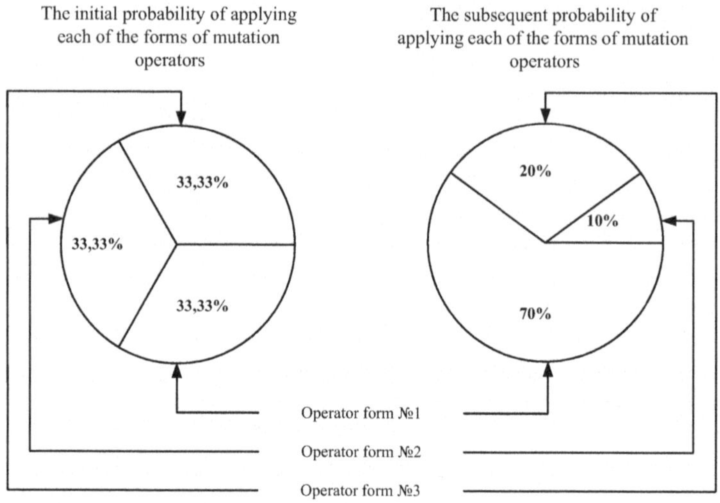

Fig. 6. Correction of the probability of applying each of the forms of mutation operators

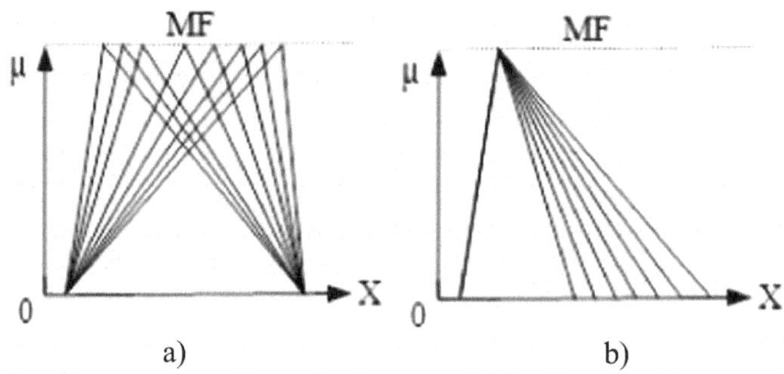

a) – changing the center position of a fuzzy variable; b) - changing the right border position of a fuzzy variable;

Fig. 7. Illustration of the use of different forms of the combined mutation operator

- *Step 1.* Formation of an array of chromosomes subjected to crossing-over with an indication of its type and the degree of fitness of the obtained solution to the conditions of the current problem;
- *Step 2.* Formation of an array of chromosomes that will be subjected to crossing-over (crossing) using the probability of using the operator;

- *Step 3.* If the array of chromosomes selected at *step 2* has a nonzero length, then correct the probability of applying the forms of the crossing-over operator, as shown in Fig. 6 (initially, the probabilities of using each of the combined crossing-over operator forms are set to be the same);
- *Step 4.* Selecting a specific form of the crossing-over operator, which will be used to randomly change the *i*-th chromosome (for each chromosome, its own crossing-over operator is selected);
- *Step 5.* Performing a random change of *i*-th chromosome based on the selected crossing-over operator.

Figure 8 illustrates the use of various forms of the combined crossing-over operator.

As input variables of the fuzzy controller that changes the parameters of applying the combined random change operators, the following parameters are used:

- the degree of genetic diversity of the chromosome population;
- the degree of success of the application *i*-th form of the crossing operator;
- the degree of success of the application *i*-th form of the mutation operator.

The output parameters of the FC will be:

- the probability of applying the mutation operator;
- the probabilities of application each of the forms of crossing operators, mutations.

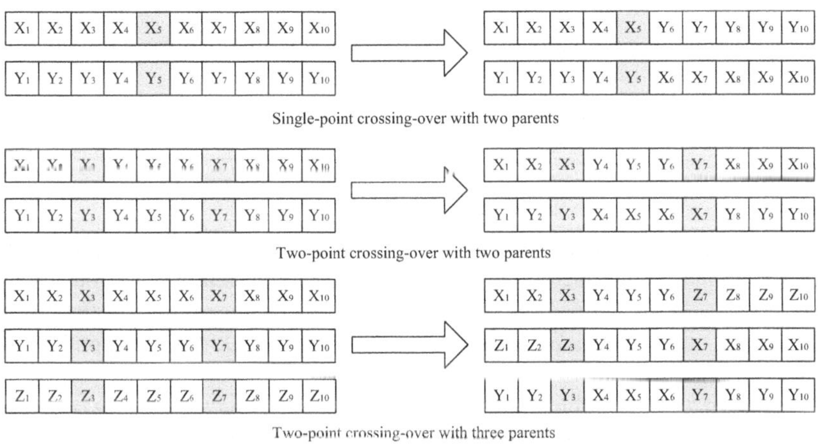

Fig. 8. Illustration of the use of different forms of the combined crossing-over operator

4 Result Analysis

The duration of the software operation on the computer with the characteristics (Intel Core i7 2.8 GHz, 16 GB of RAM, SSD hard disk) is at least 30 min if the classical genetic algorithm is used. In the case of using an adaptive genetic algorithm with combined random change operators, the duration of the software operation does not exceed 19 min. An example of the transition process, term-sets of linguistic variables and the solution surface of the resulting fuzzy controller are shown in Fig. 9, 10.

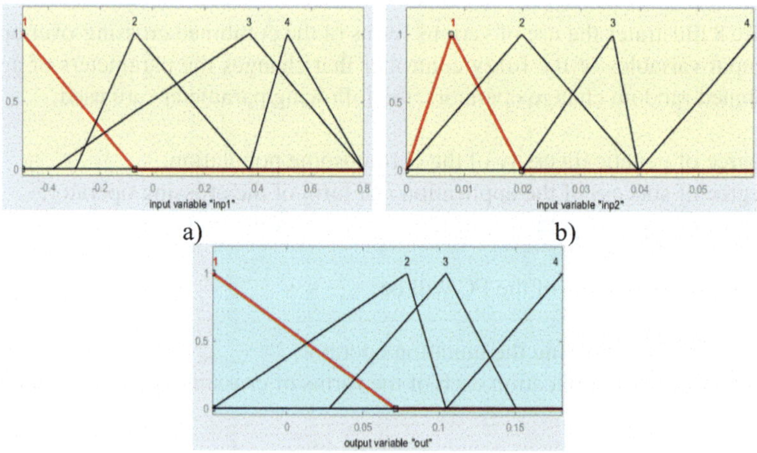

c)

Fig. 9. Term-sets of linguistic variables of the obtained fuzzy controller

5 Discussion

The paper presents a method for automating the generation of fuzzy controllers based on statistical data on the operation of automatic control systems for various technical objects. The results confirming the possibility of successful generation of fuzzy control systems based on information about the operation of PI, PD, and PID controllers are presented.

Automation of the searching process for the optimal parameters of a fuzzy controller is achieved through the use of an adaptive genetic algorithm with combined operators of random changes and special software developed in the MATLAB environment.

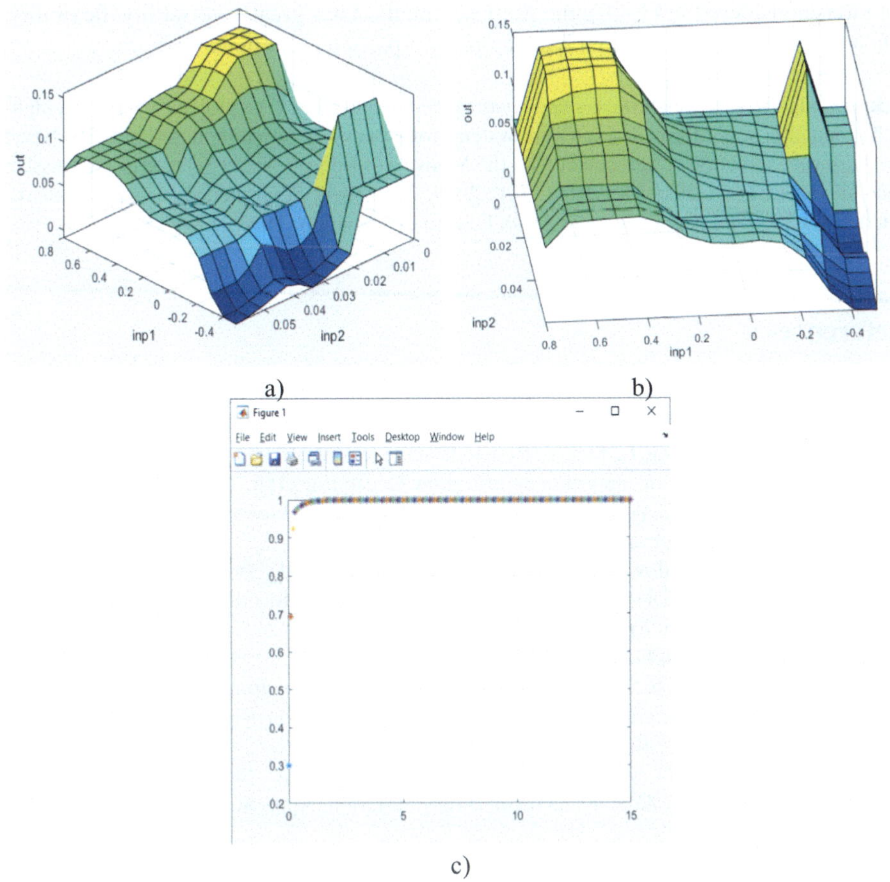

Fig. 10. An example of a transient and the solution surface of the resulting fuzzy controller

6 Conclusion

Analysis of recent works in the field of design and application of hybrid control systems using artificial intelligence reveals that methods of automatic adjustment of systems combining classical and fuzzy control technologies in conditions of uncertainty are not sufficiently developed.

The process of developing such systems is characterized by:

- the complexity of parameter setting;
- a high degree of human participation in the design process;
- low automation of the synthesis process of intelligent controllers;
- applied, highly specialized designation of the results obtained.

This article presents a method for the synthesis of fuzzy control systems based on statistical data on the operation of PI, PD, PID and other controllers. Compared with the solutions considered in [1–10], the proposed method has greater versatility, flexibility, efficiency and requires minimal human operator participation in its work.

Acknowledgements. Scientific research was carried out as part of the project "Creating a high-tech production of hardware and software systems for processing agricultural raw materials based on microwave radiation" (Agreement with the Ministry of Education and Science of the Russian Federation № 075-11-2019-083 dated 20.12.2019, Agreement South Federal University № 18 dated 20.09.2019, number of work in South Federal University № HD/19-25-RT).

References

1. Refoufi, S., Benmahammed, K.: Control of a manipulator robot by neuro-fuzzy subsets form approach control optimized by the genetic algorithms. ISA Trans. **77**, 133–145 (2018)
2. Chen, J., Zhu, H., Zhang, L., Sun, Y.: Research on fuzzy control of path tracking for underwater vehicle based on genetic algorithm optimization. Ocean Eng. **156**, 217–223 (2018)
3. BadieSharkawy, A.: Genetic fuzzy self-tuning PID controllers for antilock braking systems. Eng. Appl. Artif. Intell. **23**, 1041–1052 (2010)
4. Lazarevic, M.P., Batalov, S.A., Latinovic, T.S.: Fractional PID controller tuned by genetic algorithms for a three DOF's robot system driven by dc motors. In: 6th Workshop on Fractional Differentiation and Its Applications Part of 2013 IFAC Joint Conference SSSC, FDA, TDS Grenoble, France, 4–6 February 2013
5. Navale, R.L., Nelson, R.M.: Use of genetic algorithms and evolutionary strategies to develop an adaptive fuzzy logic controller for a cooling coil – Comparison of the AFLC with a standard PID controller. Energy Build. **45**, 169–180 (2012)
6. Al Gizi, A.J., Mustafa, M.W., Al-geelani, N.A., Alsaedi, M.A.: Sugeno fuzzy PID tuning, by genetic-neutral for AVR in electricalpower generation. Appl. Soft Comput. **28**, 226–236 (2015)
7. Demidova, G.L., Lukichev, D.V., Kuzin, A.Yu.: A genetic approach for auto - tuning of adaptive fuzzy PID control of a telescope's tracking system. Procedia Comput. Sci. **150**, 495–502 (2019)
8. Civelek, Z.: Optimization of fuzzy logic (Takagi-Sugeno) blade pitch angle controller in wind turbines by genetic algorithm. Eng. Sci. Technol. Int. J. **23**(1), 1–9 (2020)
9. Pizarro-Lerma, A.O., García-Hernández, R., Santibáñez, V.A.: Fine-tuning of a fuzzy computed-torque control for a 2-DOF robot via genetic algorithms. IFAC-PapersOnLine **51**(13), 326–331 (2018)
10. Amjad Hossain, M., Shill, P.C., Sarker, B., Murase, K.: Optimal fuzzy model construction with statistical information using genetic algorithm. Int. J. Comput. Sci. Inf. Technol. IJCSIT **3**(6), 241–257 (2011)
11. Vladimir Vladimirovich, I., Tudevdagva, U., Kovalev, A.V., Borisovich, S.O., Maksimov, A.V., Sergeevna, I.A.: Model of adaptive system of neuro-fuzzy inference based on PID- and PID-Fuzzy-controllers. In: Silhavy, R. (ed.) CSOC 2020. AISC, vol. 1225, pp. 519–533. Springer, Cham (2020). https://doi.org/10.1007/978-3-030-51971-1_43
12. Ignatyev, V.V., Soloviev, V.V., Beloglazov, D.A., Boldyreff, A.S.: Development of a method for automatic generation and optimization of fuzzy controller parameters using genetic algorithm. In: Proceedings of SPIE 11543, Artificial Intelligence and Machine Learning in Defense Applications II, 115430A, 20 September 2020. Event: SPIE Security + Defence, 2020 Online Only. Vol. 11543 115430A-1 (2020). https://doi.org/10.1117/12.2581481

Application of Machine Learning for Post Process Mining Analysis and Problem Detection in Bank

Andrey A. Bugaenko(✉)

Sberbank PJSC, St. Vavilova, 19, 105275 Moscow, Russia

Abstract. The research offers the method of definition of the "problem" sections in the business process after building of the network graph of the process based on the process mining methodology. Article discusses the features of selecting machine learning methods under "significance" and "redundancy" criteria, configuring hyperparameters, and selecting a clusterer that designates the 'problem' status on a grid graph. It also describes one of the ways to get the problem causes data by reading logs for additional descriptions, if any #CSOC1120.

Keywords: Machine learning · Model selection · Feature selection · Vectorization · Clustering · Process mining · Business process in bank

1 Introduction

Many large companies have already introduced the process mining technology at the moment, i.e. drawing of network graphs of business processes based on logs of actions from primary IT systems, rather than based on "handwritten" regulations, many of which very often are no longer consistent with the reality. However, the classic process mining is limited to the provision of an analyst with the analysis tool, with which he/she searches for some opportunities for optimization of a business process manually, using additional embedded tools. Even though this approach is much more effective than the analysis of the processes, which are built based on the regulations or according to employees' words, however it has a number of drawbacks as well. First, the analysis of the process network graph requires presence of a analyst, who has a high expertise level in the process scope. Second, the network graph analysis as such often consumes enormous time resource. Third, the software used to draw business processes demands plenty of resources, since the obtained network graph should "look nice" and be examination-friendly for the analyst. These drawbacks are supposedly minor if the process has just several nodes; however if the number of process nodes is great, and, accordingly the number of transitions between the nodes is even greater, they acquire marginal impact. Out of the processes analyzed in the bank divisions, the process which was maximal as to the number of nodes and transitions had 122 nodes and 909 edges on the network graph drawn based on logs.

© The Author(s), under exclusive license to Springer Nature Switzerland AG 2021
R. Silhavy (Ed.): CSOC 2021, LNNS 229, pp. 417–427, 2021.
https://doi.org/10.1007/978-3-030-77445-5_39

The "robotization" of the analyst's functions upon obtaining the "actual" network graph of the business process based on IT logs may serve as a logical solution to the abovementioned problem.

The most frequent tasks during the business processes analysis are:

- comparison of as-is and to-be business processes, i.e. the most probable path in the network graph with the optimal path in the network graph of the business process. This task may be solved using reinforcement learning (q-learning, cross-entropy end/or genetic algorithms); however, the consideration of this task in detail is beyond the scope of this article and is considered by me in detail in another publication.
- search for opportunities for optimization of the business process, i.e. search for specific "problem" nodes or transitions between the nodes of the process, and analysis of the reasons for such problems. The article will consider this particular task.

2 Application of Machine Learning

Let us call as "problem" the node of the process or transition between the nodes, which within the framework of some process instance has the number of repetitions that is higher than the average value for this node ("redundant" number of repetitions) and/or is "significant" in terms of the time of the process execution.

The "problem" status may be assigned to both the process node and transition between the nodes. Within the framework of this study, it was necessary to combine the task of the search for "problem" sections with the task of the search for the optimal path in the network graph using reinforcement learning(RL). This posed some requirements to the selection of the entity, the status of the "problem nature" of which we needed to deal with. The "environment" was set in the network graph in order to use reinforcement learning with the aim of defining the optimal path. In terms of RL:

- environment - neting graph of business process;
- state - node (stage of process);
- action - edge (transition between nodes);
- reward – bank's profit from one transaction (one business process example) (or fictive profit);
- loss – bank's operation expanses for every stage of process. [1]

Within the framework of such a "environment", a «agent» can influence the transition between the process nodes and cannot influence the process nodes themselves, since if he/she could influence the process nodes, we would be able to decompose the process node into the nodes and transitions between the nodes as well, and so on ad infinitum. The process node is for us a minimal entity. For this reason, it is senseless to mark the process node as the "problem" one, and the "problem" area was defined based on the transitions between the nodes only. However, if an additional manual analysis of problems is planned in the business process schedule (though this does not make sense in case of an automated analysis, which we perform), the process nodes may be marked additionally as "problem" ones as well.

The level of "redundant" for the transitions between the process nodes is defined as follows:

$$G_{ij} = \sum\nolimits_{k=1}^{m} MAX\left(\left(n_{ijk} - \left(\frac{1}{m}\sum\nolimits_{k=1}^{m} n_{ijk}\right)\right); 0\right), \tag{1}$$

where G_{ij} – level of "redundant" from the *i*-th node of the process to the *j*-th node of the process, $i,j\epsilon[1,h]$, where *h*- maximum possible number of nodes in the process, m-number of instances of the process in the sampling, $k \epsilon[1,m]$, n_{kij} – number of repetitions of the transition from the *i*-th node of the process to the *j*-th one in the *k*-th instance of the process. Actually, summarized the excess in the number of repetitions of the transition between the process nodes over the average value on all process instances for each transition. In the majority of cases, the average number of transitions between the process nodes equals one, however there are situations when repetition of some nodes is normal for a process. For example, speaking about the process of accepting payments, the node of bar code scanning can be performed several times for several payment documents in one and the same typical instance of the process.

It is proposed to define "significant" transitions between the process nodes using the model of dependence of time of execution of the process instance on the number of transitions between the process nodes. We will use the time of the process execution as the target variable of the model. The number of transitions between the process nodes was taken as the model factors (features):

$$\begin{aligned} t_1 &= f(n_{11}, n_{12}, \ldots n_{1ij}) \\ &\ldots \\ t_k &= f(n_{k1}, n_{k2}, \ldots n_{kij}) \\ &\ldots \\ t_m &= f(n_{m1}, n_{m2}, \ldots n_{mij}) \end{aligned} \tag{2}$$

where t_k – time of the execution of the k-th instance of the process, $i,j\epsilon[1,h]$, where *h* maximum possible number of nodes in the process, m-number of instances of the process in the sampling, $k \epsilon[1,m]$, n_{kij} – number of repetitions of the transition from the *i*-th node of the process to the *j*-th one in the *k*-th instance of the process. In order to build the model, the sampling was divided into three parts: teaching, validation, and test ones.

In our case, our task was solved for the following business process made in the bank divisions: acceptance of payments from individuals (Is); lodging cash into/withdrawal cash from an account; issue, change and closing of bank cards, sale of coins and bullions; effectuation of Is' bank transfers; regular payments towards the repayment of credits; opening/closing/replenishment of deposits; Is' currency exchange; lease of safe deposit boxes; sale of securities and certificates; cash change and change of spoiled banknotes; expert examination of token money, compensations and payments towards depositors of external banks. These processes were chosen for analysis, since they actually cover less than 99% of all transactions in the divisions, and since their analysis was required for business purposes of the bank. There were from 50ths. to 4mln. instances on each process (before filtration), i.e. statistical sampling contained from 50ths. to 4mln. elements. The model building procedure was carried out in three stages:

- selection of the machine learning method with optimal configurations of hyperparameters on full initial set of features;
- choosing of the method of factors (features) selection using the method selected in the previous clause;
- repeated selection of the machine learning method and reconfiguration of hyperparameters on the set of the features selected in the previous clause.

The methods from libraries «sklearn», «MLXtend», «xgboost» were taken as tested methods:

- GradientBoostingRegressor,
- DecisionTreeRegressor,
- KNeighborsRegressor,
- GaussianProcessRegressor,
- LinearRegression,
- Ridge,
- Lasso,
- MultiTaskLasso,
- LassoLars,
- BayesianRidge,
- OrthogonalMatchingPursuit,
- SGDRegressor,
- PassiveAggressiveRegressor
- TheilSenRegressor,
- RANSACRegressor,
- ElasticNet,
- MultiTaskElasticNet,
- Adaboost,
- Adaboost(estimator=ElasticNet),
- ARDRegression,
- HuberRegressor,
- SVR,
- IsotonicRegression,
- XGBRegressor.

For each of the methods, the set of the hyperparameters typical of it was tested. The peculiarities of the impact of hyperparameters of this or that method on the accuracy of definition of "significance" will be considered in a separate article. After selection of hyperparameters, the methods were tested against each other under the MAPE criterion (all values of time of the process execution are positive, so the criterion is applicable).

Further, in order to find the best method under the MAPE criterion, the methods of factors selection from libraries «BorutaPy», «BoostARoota», «sklearn», «mlxtend», «skfeature» are tested:

- Boruta,
- Boostaroota,

- RFECV,
- SequentialFeatureSelector,
- RFE,
- MRMR,
- CIFE,
- SelectFromModel,
- SelectKBest.

In a similar way, the best method of factors selection under the MAPE criterion is taken for this business process in order to assess "significance." It should be noted that fewer than all methods of factors selection are applicable for some methods of building the model. If we look at the allocation of the methods, which turned out to be the best ones for some business process, we will see that SequentialFeatureSelector was the best method for 3 tested business processes (out of 13 ones), RFECV was the best one in 3 cases, and each of the remaining ones was the best for 1 case (Fig. 1). This means that none of the tested methods can be used as a universal method for definition of the "significance" of transition between the nodes of the process, and each new business process will demand repeated tests in order to find out which of the methods of factors selection exactly turns out to be the best one for it.

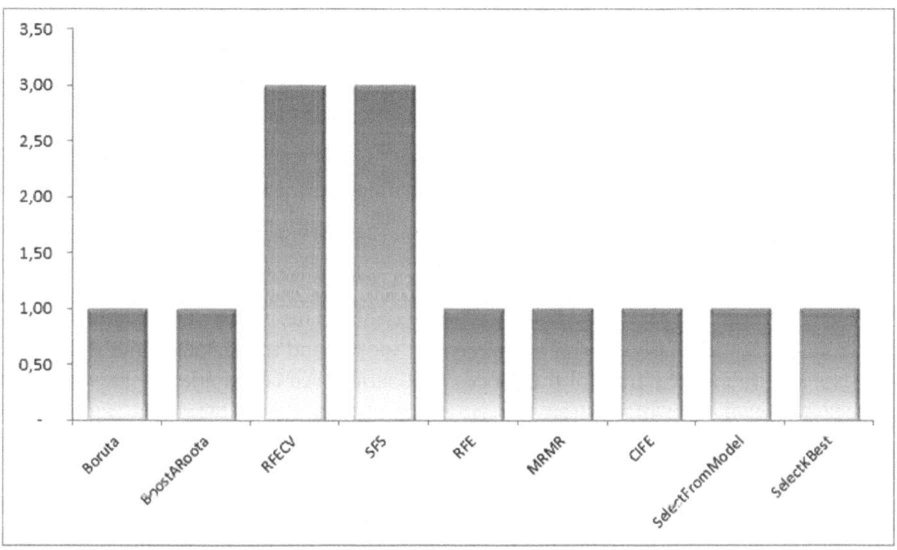

Fig. 1. Frequency of loss of a given method of factors selection as the best one for definition of the "significance" in the business process.

Further, the set of the factors chosen through the best method of factors selection serves as a basis for the repeated selection of the best method of building the model. Allocation of the best method of building the model to define the index "significance" is given on Fig. 2.

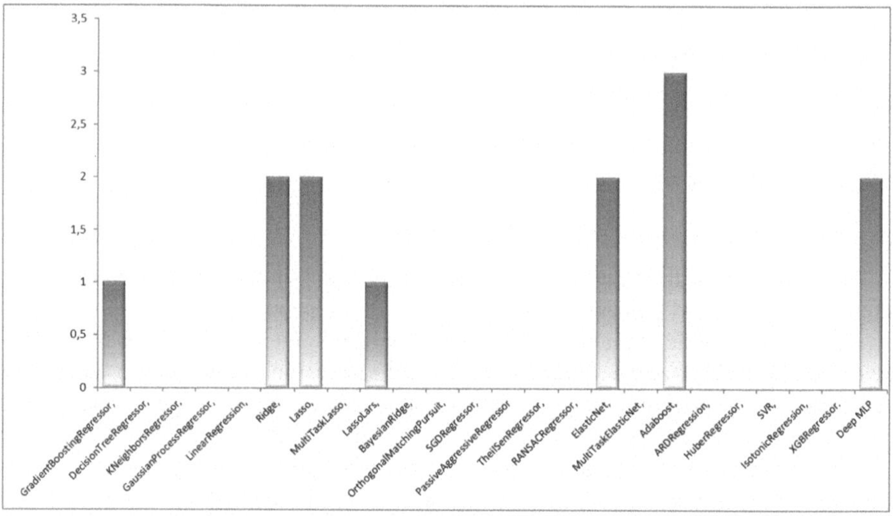

Fig. 2. Frequency of loss of a given method of building the model as the best one for definition of the "significance" in the business process.

As we see from the Fig. 2, the method Adaboost turned out to be the best one for 3 cases, Ridge, Lasso, ElasticNet, MLP was the best one in 2 cases, LassoLars and GradientBoostingRegressor – in 1 case. However, yet none of the methods turned out to be the best one to the extent allowing to call it universal for building the model of dependence of time of the process execution on the number of repetitions of transitions between the process nodes. Like in case of the method of factors selection, it is necessary to define the method of building the model individually for each new business process.

Actually, all transitions between the nodes under the grouping "significant" – level of "redundant" may be divided into 4 types:

- "Not significant" with a low level of "redundant." This means that this transition between the process nodes is not engaged in "loops" and at the same time it actually does not influence the time of the process execution. An automatic message about a successful execution of the transaction, which is fixed in the logs, may serve as an example here.
- "Significant" with a low level of "redundant." Such transitions between the nodes render material influence on the time of the process execution, but are not engaged in "loops" as well. This type of the transition is not the most interesting one from the point of view of the optimization potential, but it can be a "bottleneck." Here, both a long-lasting, but necessary node of filling in different forms in the primary IT system, and a regular delay (without repetitions and returns) on some node may serve as examples.
- "Not significant", but with a high level of "redundant." As a matter of fact, such transitions between the nodes are parts of "loops" in the process, but they actually do not influence its duration. Such types of transitions, maybe, need to be analyzed,

but not on a first-priority basis. A repetitive turning of the primary IT system to other different systems with a returning error may serve as an example here.
- "Significant" with a high level of "redundant." This is the most important type of transitions for the analysis purposes, which has the maximum optimization potential. It includes, for example, repeated long execution of some documents, in which mistakes are found, and which are sent back for re-execution.

The "Significant" transitions with a high level of "redundant" are definitely the most "problem" ones. And "not significant" transitions with a low level of "redundant" are definitely not "problem" ones. However, the situation is not that simple with the rest of the two categories; moreover, the quantity divider index between the "low" and "high" redundancy is not always obvious, which is an important matter as well.

In order to find out exactly whether the transition between the process nodes is "problem" (rejections) or not, 6 of clustering methods were tested:

- Robust covariance,
- One-class SVM,
- Isolation Forest,
- Local outlier Factor,
- DBSCAN,
- Kmeans.

Besides, a simple division of transitions between the nodes, which are ranged based on "redundancy" at the ratio of 20:80, was taken as a 7th method.

Recall, precision were calculated for each business process (out of 13) on the data mapped out manually and were averaged for all business processes; after that, f-score was calculated. The results are given in the Table 1.

Table 1. Results of clustering filters testing.

Method	Recall	Precision	f-score
Robust covariance	0.64	0.91	0.75
One-class SVM	0.77	0.79	0.78
Isolation Forest	0.51	0.78	0.62
Local outlier Factor	0.43	0.63	0.51
DBSCAN	0.94	0.7	0.80
Kmeans	0.05	0.99	0.10
20/80	0.31	0.99	0.47

If we assess the method based on the f-score parameter, which represents a combined criterion containing the evaluations of both recall and precision – DBSCAN is the best one, which is by 2 p.p. higher than the second-large One-class SVM and by 70 p.p. higher than the worst method KMeans. However, it is the high value of Recall which is much more important for our situation. If we assign the status "not problem" to the "problem" transition between nodes by mistake, we will miss an important place in the business process, which needs optimization. Alternatively, if we have a low value of Precision, i.e. if we mark transitions in the business process as "problem" ones, when they are not problem ones, we will just exclude them during a more detailed study. However, again, the best value of recall is that of DBSCAN 0,94, which is by 17 p.p. higher than in One-class SVM (which is the second one in terms of this index) and by 81 p.p. higher than in Kmeans (which is the worst one in terms of this index). This means that the DBSCAN method is the most efficient one for clustering filtration in order to define "problem" transitions between nodes based on the principle "rejection" – "non-rejection".

We have already obtained the list of "problem" transitions between the process nodes on this stage, which are automatically placed on the network graph (or BPMN/EPC diagram) of the business process, so that an analysts could see immediately the place of the problem in the process. However, in the majority of banking systems, when a manager's actions are logged, the system, besides a minimum set of the data for drawing the process: transaction time, manager's id, action id (id of the process instance was not present in the examples tested by us), writes down a field with more detailed information. If such a field is not present in the logs, further steps are assigned to an analyst, who starts to perform a detailed analysis of the reasons, knowing exactly the places of problems in the business process (we have marked them on the network graph as a "problem" transition). The majority of banking systems has such a field, and further steps of an analyst may be made automated as well.

The field called "description" was such a field in the banking systems, which were tested by us. This field contained the information, which was partially generated by the system automatically, and partially filled in by a manager with the transaction details, reasons for errors or some additional data, which are important for saving in the manager's opinion. Obviously, this field was extremely unstructured and contained information in a free format. If an analyst decided to read by himself/herself and study each of these descriptions, he/she would have to analyze more than 4mln. database entry in some of the researched business processes; and since all of them have different format, such a number cannot be reduced by simple groupings at all. Owing to the fact that we know what transitions between the process nodes are "problem", we can take for analysis purposes "detailed descriptions" in logs only, which correspond to "problem" transitions, which will reduce the analysis sampling by 80%–90%. It is logical to assume that it is senseless to study "detailed descriptions" relating to the areas in the process, where we already know that everything is all right. However, even if the amount of "detailed descriptions" is reduced 10 time, their amount remains enormous in any case, which is extremely labor-consuming for manual analysis.

Allocation of typical problem "detailed descriptions" is a solution here. We know what groups of "problem descriptions" refer to "problem" transitions between the process nodes. One "problem" transition between the process nodes may have several reasons for its "problem" nature. This means that the entire group of "problem descriptions" (obviously, the number of "problem descriptions" in one transition between the process nodes is equal to the number of instances of the process with this node) should be divided into some number of clusters. Definition of the optimal number of these very clusters is not a simple task and requires a deeper study. In the banking business processes considered in this article, the number of clusters empirically varied from 4 to 17. Further, it is necessary to perform vectorization of words for all problem descriptions in each cluster (in total, the number of clusters is equal to the number of "problem" transitions between the nodes, multiplied by the set value of clusters in the "problem" transition). 7 vectorizers from libraries «gensim» и «sklearn» were tested:

- GloVe, trained by Wikipedia,
- GloVe, trained by discription of data-set,
- Word2vec, trained by Wikipedia,
- Word2vec, trained by discription of data-set,
- Fastext, trained by Wikipedia,
- Fastext, trained by discription of data-set,
- TF-IDF.

Testing was performed on the mapped-out data set on each of 13 business processes. "Description", which is the description of the entire cluster, was marked manually. After identification of the "description", which is the "description" of the entire cluster, using each of the methods, it was studied, in how many cases a given vectorization method led to a correct choice; further, this share was averaged for all business processes (Table 2).

Table 2. The share of successfully determining the "description" of the cluster for each vectorizer

	A measure of classifier confidence (0–1)
GloVe, Wiki	0.41
GloVe, data-set	0.90
Word2vec, wiki	0.58
Word2vec, data-set	0.89
Fastext, wiki	0.56
Fastext, data-set	0.90
TF-IDF	0.55

Fastext and GloVe, which were taught based on the data set of the transactions logs, were the best ones, however the time of operation of GloVe turned out to be almost 30 times longer than the time of operation of all the other methods.

After vectorization of "descriptions", the average value of the vector all over the cluster was found as a simple average of the vectors of all "descriptions", which were present in the cluster.

$$\overrightarrow{Wm_{uv}} = \frac{1}{p_{uv}} \sum\nolimits_{f-1}^{p} \overrightarrow{W_{uvf}} \tag{3}$$

where $\overrightarrow{Wm_{uv}}$ - average vector of words in the v-th cluster of the u-th "problem" transition between the process nodes, p – number of instances of the process in the v-th cluster of the u-th "problem" transition between the nodes of the process, $\overrightarrow{W_{uvf}}$ - vector of words in the "description" in the f-th instance of the process in the v-th cluster of the u-th "problem" transition between the nodes of the process.

Further, the "description" of the entire cluster $\overrightarrow{W_{uv}}$ was defined. The "description" of $\overrightarrow{W_{uvf}}$ of the f-th instance of the process in the v-th cluster of the u-th "problem" transition between the nodes of the process, the vector of the words of which had the least Euclid distance with $\overrightarrow{Wm_{uv}}$ - average vector of words in the v-th cluster of the u-th "problem" transition between the nodes of the process was taken as the description.

As a result, we obtained the vector of the "descriptions" of the clusters $\overrightarrow{W_{uv}}$, to which these very "descriptions" corresponded, which were presented to an analyst as a list of the problems in the business process.

3 Conclusion

The research offers the method of definition of the "problem" sections in the business process after building of the network graph of the process based on the process mining methodology. It is proposed to assume as "problem" the transition between the nodes of the process, which is simultaneously "significant", i.e. it turns out to be a factor influencing the time of the execution of the process instance, and "redundant", i.e. it has an extremely high value of the number of excesses of repeats of the transition between the nodes of the process over the average value of such transitions.

In order to build the model of dependence of time of execution of the process on the number of transitions between the process nodes, 24 machine learning methods and 9 methods of selection of significant factors were tested. The study results showed that there is no any universal method, and a method should be chosen individually based on the validation sampling for each new business process.

7 methods of clustering filtration (6 ML methods and 1 basic split 20/80) were tested in order to define the feature "problem" for the transition between the nodes of the process under the factors "significant" and "redundant." The results showed that the method DBSCAN has the best values for all types of business processes by a wide margin as to the indices of recall and precision.

In case of presence of detailed text description to the instance of the process, it was proposed to use the method of automated compilation of the list of the "descriptions", which would show the problems of the business process. For this purpose, among other things, the methods of text vectorization were compared. fasttext, which was taught by the data set of the transaction logs descriptions themselves, turned out to be the best one. GloVe and word2ver showed approximately the same results, but GloVe works much slower.

Following the study results, the methodology and the results of the studies may be applied to similar tasks of optimization of other business processes. An analyst will not have to study by himself/herself the result of the application of process-mining – network graph of the business process (or bpmn/epc diagram); instead, he/she can obtain automatically the places of problem transitions in the process nodes and list of particular problems.

However, within the framework of the current task, there is a potential for further studies. When using the process mining methodology, there often appears a task of definition of an optimal order of a manager's steps in the business process. The application of reinforcement learning (RL) seems to be extremely efficient for this task. More specifically, definition of the RL method, which is the best one for this task, description of the principles of creation of "environment", "Loss", "Reward", consideration of the peculiarities of configuring hyperparameters are of interest. Besides, detailed study of the impact of configuration of the methods hyperparameters, which are used for definition of "significance" of the transition (building of the model of dependence of time of the process execution on the number of transitions), on the accuracy of definition of "significant" transitions is of interest.

References

1. Panfilov, M., Goncharenko, I., Bugaenko, A.: Application of DS methods for solving applied task in finance. In: AI Journey Conference, Kaliningrad (2019)
2. van der Aalst, W.M.P.: Process Mining: Discovery Conformance and Enhancement of Business Processes. Springer, Heidelberg (2011). https://doi.org/10.1007/978-3-642-19345-3ISBN 978 3 642-19344-6
3. van der Aalst, W.: Using process mining to deal with "events" rather than "numbers"? Spreadsheets for business process management (2017)
4. van der Aalst, W.: Process Mining, Data Science in Action, 2nd edn. Springer, Heidelberg (2016). https://doi.org/10.1007/978-3-662-49851-4_1
5. Born, M., Brelage, C., Markovic, I., Pfeiffer, D., Weber, I.: Auto-completion for executable business process models. In: Ardagna, D., Mecella, M., Yang, J. (eds.) BPM 2008. LNBIP, vol. 17, pp. 510–515. Springer, Heidelberg (2009). https://doi.org/10.1007/978-3-642-00328-8_51
6. Sarno, R., Sinaga, F., Sungkono, K.R.: Anomaly detection in business processes using process mining and fuzzy association rule learning. J. Big Data **7**(1), 1–19 (2020). https://doi.org/10.1186/s40537-019-0277-1
7. Andrews, R., van Dun, C.G.J., Wynn, M.T., Röglinger, M.K.F., ter Hofstede, A.H.M.: Quality-informed semi-automated event log generation for process mining. Decis. Supp. Syst. **132**, 113265 (2020)
8. Dijkman, R., Gao, J., Syamsiyah, A., van Dongen, B., Grefen, P., ter Hofstede, A.: Enabling efficient process mining on large data sets: realizing an in-database process mining operator. Distrib. Parallel Databases **38**(1), 227–253 (2019). https://doi.org/10.1007/s10619-019-07270-1
9. Leemans, S.J.J., Fahland, D.: Information-preserving abstractions of event data in process mining. Knowl. Inf. Syst. **62**(3), 1143–1197 (2019). https://doi.org/10.1007/s10115-019-01376-9
10. Kalenkova, A.A., van der Aalst, W., Lomazova, I.A., Rubin, V.: Process mining using BPMN: relating event logs and process models. Softw. Syst. Model. **16**(4), 1019–1048 (2017)
11. Tang, P., Qiu, W., Huang, Z., Lian, H., Li, Z.: Anomaly detection in electronic invoice systems based on machine learning. Inf. Sci. **535**, 172–186 (2020)
12. Angiulli, F., Basta, S., Lodi, S., Sartori, C.: Reducing distance computations for distance-based outliers. Expert Syst. Appl. **147**, 113215 (2020)
13. Liu, C., Yang, J., Wu, J.: Web intrusion detection system combined with feature analysis and SVM optimization. EURASIP J. Wirel. Commun. Netw. **2020**(1), 1–9 (2020). https://doi.org/10.1186/s13638-019-1591-1

Comprehensive Analysis of Cryptographic Protocol Implementations Using Formal Verification and Dynamic Analysis Using the PDA Language

Liudmila Babenko(✉) and Ilya Pisarev

Information Security Department, Southern Federal University, Taganrog, Russian Federation
lkbabenko@sfedu.ru

Abstract. Cryptographic protocols form the basis of the security of any system that uses data transfer between components. High-quality and in-depth analysis of cryptographic protocols at all stages of its development and implementation is important. The paper describes an integrated approach to the analysis of the security of the source codes of cryptographic protocols, which is distinguished by the highest efficiency and quality of analysis at all stages of the development and implementation of the protocol. A verification scheme is presented based on an integrated approach using formal verification and dynamic analysis of cryptographic protocol implementations to create protocols that are safe in practice. An approach to formal verification using the Avispa tool is described. A method of extracting the structure of a cryptographic protocol from the source code to obtain its description in the Alice-Bob language is shown, based on the use of the Roslyn parser for the C# programming language and building a tree from message elements. The method of dynamic analysis of protocol implementations and the PDA language are presented. The method is based on the principle of false termination and is based on simulating real attacks by an intruder when the system is operating in practice. An example of a protocol is described, and its comprehensive safety analysis is carried out using the proposed integrated approach. The comparison of the work of the approach proposed by the authors with the existing classical approaches to verification using the Scyther and Avispa tools is given #CSOC1120.

Keywords: Formal verification · Dynamic analysis · PDA · Avispa · Scyther · Analysis · Cryptographic protocols · Cryptography · Modeling · Attacks

1 Introduction

Verification of cryptographic protocols is an urgent and difficult task. In connection with the development of various e-commerce services, as well as mobile robots, unmanned aerial vehicles, the Internet of Things, computing resources in which in most cases are very limited and it is not always possible to use well-known protocols. Any protocol developed must go through a verification procedure. Mainly formal verification methods

© The Author(s), under exclusive license to Springer Nature Switzerland AG 2021
R. Silhavy (Ed.): CSOC 2021, LNNS 229, pp. 428–443, 2021.
https://doi.org/10.1007/978-3-030-77445-5_40

are used to assess the security of a cryptographic protocol. There are verifiers that allow you to check the protocol for vulnerabilities and issue a list of attacks (Scyther [1, 2], Avispa [3, 4], ProVerif [5, 6], CryptoVerif [7, 8], Tamarin Prover [9]). However, verification by these verifiers does not guarantee the security of the protocol implementation in the programming language. There are works in which the implementation of known protocols is verified [10–12]. However, they do not offer a general approach to applying these approaches to any protocol. There are also works where such an approach is proposed [13, 14]. The approaches described there allow you to extract the structure of cryptographic protocols from the source code. Basically, in these works, the protocol structure is extracted, most often on the basis of annotations left in the source code, after which the protocol description is translated to use the ProVerif verifier. Other verifiers can also be used. This approach, although it is more efficient compared to the verification of the security of the protocol before its implementation, but still has a number of disadvantages. The main disadvantage is that the healed structure is abstracted from some implementation details, which entails the likelihood of not detecting attacks on the protocol. In addition, the very limitation of the functionality of the verification tools does not allow checking all possible effects on the protocol, which entail the appearance of attacks. There are several works [15, 16], which offer a special IDE for creating protocols. Protocols are verified in real time when they are written, and then you can generate code sections in the selected programming language for use in the system. This approach also has drawbacks in the form, as mentioned earlier, of imperfections in formal verification tools, as well as the fact that the generated code sections must be embedded in the system being developed, which can also lead to errors and vulnerabilities in the protocol. Thus, verification of the source codes of cryptographic protocols is relevant and a method is needed that would allow directly analyzing the implementations of cryptographic protocols. In this paper, we propose an integrated approach based on the use of formal verification both at the beginning of the development of the protocol and after implementation in the programming language, as well as using dynamic analysis using the PDA language.

2 Proposed Verification Scheme

The verification scheme is shown in the Fig. 1. The proposed verification scheme requires:

1. Develop the protocol directly
2. Describe its message and their content, transmitted between the parties in the simplest language Alice-Bob in the form:

   A- > B: Ek (M1)
   B- > A. Ek (M2)
   ...

3. Translate the protocol from the Alice-Bob language to the CAS+ language for use in the Avispa verifier.

4. Conduct formal verification using the Avispa tool.
5. After formal verification, if attacks were found, we return to step 3 until the attacks are found.
6. If no attacks were found, then the protocol is implemented in the selected programming language in the context of the system being developed.
7. Next we work with the source code of the cryptographic protocols.
8. Further verification is split into two branches. Branch 1 - formal verification and branch 2 - dynamic analysis.

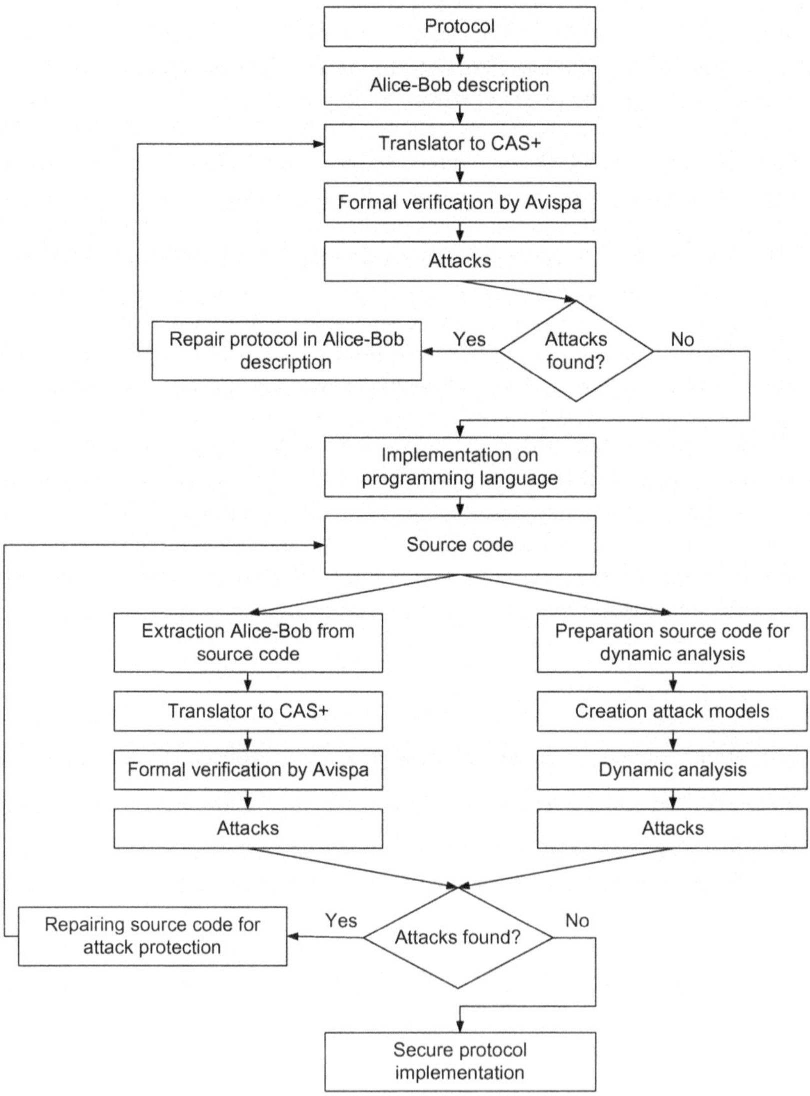

Fig. 1. Formal verification scheme

9. Description of branches.

    9.1. Branch 1:

        9.1.1. Extracting the structure of cryptographic protocols from source code into the Alice-Bob language
        9.1.2. CAS+ translation
        9.1.3. Formal verification with the Avispa verifier
        9.1.4. Protocol attacks

    9.2 Branch 2:

        9.2.1. Preparing the system source code for dynamic analysis
        9.2.2. Creating attack models using an active attacker
        9.2.3. Dynamic analysis
        9.2.4. Protocol attacks

10. If attacks were found, then the source code of the cryptographic protocols is changed and the return to point 7. This is repeated until no attacks are detected. As a result of this algorithm, the output will be a secure implementation of a cryptographic protocol in a programming language.

3 Formal Verification

There are many verifiers Scyther, ProVerif, Crypto Verif, Tamarin Prover, Avispa. The Avispa verifier was chosen for the current work due to its efficiency, speed of work and simplicity of the specification language. Avispa is a tool for automated security analysis of cryptographic protocols. Implemented for linux OS. The architecture includes: HLPSL2IF translator, which translates protocol description from HLPSL into a special IF (Intermediate format), 4 OFMC (on-the-fly Model-Checker) verification modules, AtSe (CL-based Attack Searcher), SATMC (SAT-based Model-Checker), TA4SP (Tree Automata-based Protocol Analyzer). The principle of operation of the OFMC module used in this work is based on the use of symbolic verification of models with the verification objectives previously described in the specification and knowledge of the participating parties. The modular formal language HLPSL or the higher-level CAS+ specification language is used to describe cryptographic protocols and their security properties. Avispa includes a Security Protocol ANImator (SPAN) subsystem, which has a translator for converting code from CAS+ to HLPSL and allows you to visualize the interaction patterns of the parties.

The CAS+ specification language is simpler in syntax and allows you to quickly describe the protocol. An example of a description of a part of the protocol in CAS+ is given below:

```
  protocol NeedhamSchroederPublicKey;
identifiers
A,B                 : user;
Na,Nb               : number;
KPa,KPb             : public_key;

messages
1. A -> B           : {Na, A}KPb
2. B -> A           : {Na, Nb}KPa
3. A -> B           : {Nb}KPb

knowledge
A          : A,B,KPa,KPb;
B          : A,B,KPa,KPb;
  ...
```

This fragment describes the roles of users A, B, the data that each user knows (random numbers Na, Nb, keys KPa, KPb), the order of transmission of messages and knowledge (knowledge) of the interacting parties A, B.

As a result of the check, the corresponding result will be given. If attacks are detected, the type of attack and its progress will be shown in the form of corresponding changes in messages by the attacker. If no attacks are found, then the program output will contain a corresponding message that everything is in order. Using the "Protocol simulation" button, you can see the scheme of interaction between the parties in the protocol. Using

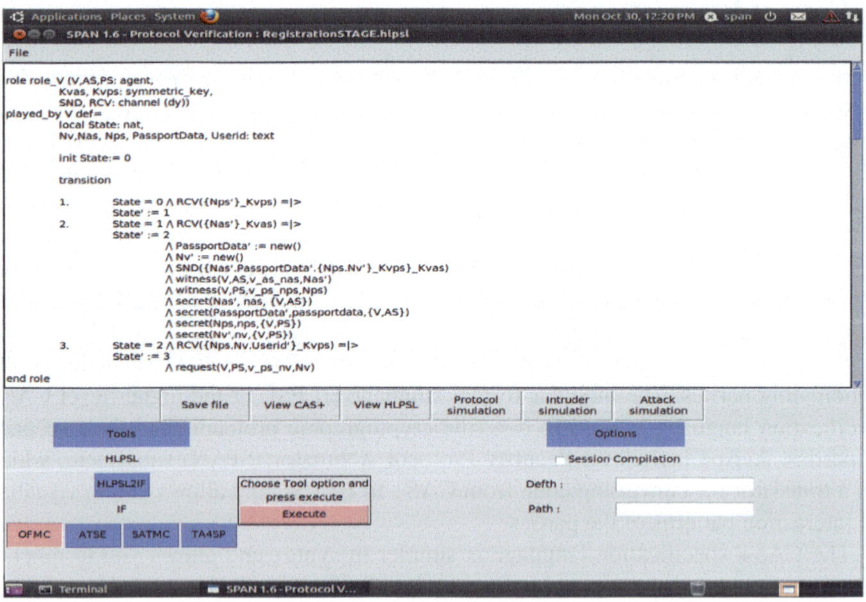

Fig. 2. Interface

the "Protocol simulation" button, you can see the scheme of interaction between the parties in the protocol. Interface is shown on Fig. 2.

Using the "Intruder simulation" button, you can see a similar scheme, only with the participation of the attacker, which will display the data intercepted by him. Using the "Attack simulation" button, you can see the attack pattern with the participation of an attacker, provided that any attack was detected in your protocol.

4 Formal Verification of Source Codes

When implementing the protocol in the source code, mistakes can be made and the message passing process can be redesigned so that the protocol can be used in practice. For example, messages can be added containing the length of the transmitted sequence, or any necessary information for the program to work. Such overhauls are not taken into account during the initial verification and can cause protocol vulnerabilities, which can lead to various errors in practice. As a result, despite the security of the protocol, which is confirmed by formal verification, it may be unsafe during program execution. In order to fix this problem, formal verification of the source codes of cryptographic protocols is required. To be able to do this, it is proposed to use the source code analyzer developed by the authors earlier [17]. It is based on using the Roslyn parser and building a tree from message elements. Thus, a cryptographic protocol in the Alice-Bob format is extracted from the source codes.

Table 1. Extraction Alice-Bob from C# source code

Source code (C# language)	Protocol on Alice-Bob language

...
byte[] M1enc;
using (RSACryptoServicePro-
vider RSA =
new RSACryptoServicePro-
vider())
{
RSA.ImportParameters(
rsaPB.ExportParame-
ters(false));
M1enc = RSA.Encrypt(M1,
true);
}
socA.Send(M1enc);
...
	1. A -> B: E_kb(Na, A)
	2. B -> A: E_ka(Na, Nb)
	3. A -> B: E_kb(Nb, k)
	4. B -> A: E_k(M1)

Further, a parser from the Alice-Bob language to the CAS+ language is used, also developed by the authors of this work earlier [18] (Table 2).

Table 2. Alice-Bob to CAS+ translation

Source code (C# language)	Protocol on Alice-Bob language
	protocol TEST; identifiers A,B : user; Na,Nb,M1 : number; Pka,Pkb : public_key;
1. A -> B: E_kb(Na, A) 2. B -> A: E_ka(Na, Nb) 3. A -> B: E_kb(Nb, k) 4. B -> A: E_k(M1)	K : symmetric_key; messages 1. A -> B: {Na, A}Pkb 2. B -> A: {Na, Nb}Pka 3. A -> B: {Nb, K}Pkb 4. B -> A: {M1}K knowledge A : A,B,Pka,Pkb; B : A,B,Pka,Pkb; ...

After that, formal verification is performed using the Avispa tool. To be able to automate the process, you can use a script for sequential call of programs, which will look like this:

1. Calling the translator from Alice-Bob to CAS+
2. Calling the translator from CAS+ to HLPSL.
3. Calling the translator from HLPSL to IF.
4. Calling the backend of the OFMC module for IF (intermediate format) in order to verify security.

5 Dynamic Analysis

According to the Dolev-Yao model [19], all messages transmitted between legal parties pass through the attacker. Dynamic analysis is based on the false termination principle. When implementing protocols, in the event of a malfunction, for example, data substitution, either a repeated request for the transmitted data or a break in the current session can be used. In most cases, it is the breaking of the current session and restarting the protocol from the beginning. The principle of false termination is that the attacker will modify the data, and if the protocol is correctly executed on all legal parties, then this will indicate an attack on this protocol. The correct execution of the protocol will be the successful completion marks, which were described in the previous paragraph. A diagram of an active attack by an attacker is shown in Fig. 3.

In this case, the protocol consists of three messages and is intended for interaction between two parties. The attacker replaces message 2 with a similar message received from the previous session. But despite the substitution, the protocol is correctly executed to the end on both sides and the success flags are set to True. This message can be, as

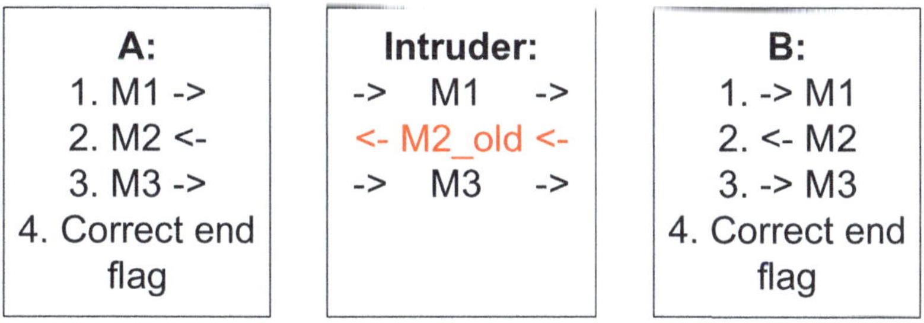

Fig. 3. Replay-attack example.

well as a part of the authentication of the parties, which, for example, will allow the attacker to authenticate instead of the legal party, or contain some semantic data that will be used in the system and subsequently may lead to its incorrect behavior.

Starting the dynamic analysis process using the example of communication between two parties looks like this:

1. The attacker's process I starts the processes of client A and server B. On client A, a round-robin attempt is made to connect to a server with a local IP address and some port1. Server B opens a connection with a local IP address and some port port2 and waits for a client to connect.
2. Next, a man's attack in the middle is modeled. The attacker connects as a client to the local address and port2 and starts communicating with server B, and also opens a connection to the local address and port1 and waits for client A to connect. By default, on the attacker's side, all messages sent through him are forwarded in the same form.
3. The consistency of the attack model is carried out by replacing message elements, order of forwarding, blocking channels, etc.
4. At the end of the protocol execution, the successful completion flags are checked. If both are present, it will indicate the success of the current simulated attack. The absence of one of the flags or both flags of successful completion will indicate the protection against attack. Depending on the implementation of a particular system, active attacks by an attacker can throw unhandled exceptions. In this case, in the absence of one or both flags of successful completion, it will be considered that the simulated attack was not successful, but the corresponding information that this attack causes unstable behavior of the system will be remembered.

6 PDA Language

PDA language (protocols dynamic analysis) is a specification language for carrying out dynamic analysis of the source codes of cryptographic protocols. It can be used to simulate any possible attack and test it in practice in the system under investigation. The language is represented by sections that describe the information required for modeling.

```
  Projects
  {
    Info1:    “project_client”,    “ProtocolCall(bool
mode)”, “123”, “project_client/protocol.cs”
    Info2:    “project_server”,    “ProtocolCall(bool
mode)” , “223”, “project_server/protocol.cs”
  }

  Parameters
  {
    A: Info1
    B: Info2
    I: INTRUDER
  }

  MainProtocol1 //Please use this as base for models
  {
    A=>B
    1 [A = “sertA.dat”, Na = RAND, kb = “keyB.dat”]
A->B: E_kB(Na, A)
    2 [B = “sertB.dat”, Nb = RAND, ka = “keyA.dat”,
Na = Get_1_Na] B->A: E_kA(Na, Nb, B)
    3 [kb = “keyB.dat”, Nb = Get_2_Nb] A->B: E_kB(Nb)
  }

  Model1
  {
    A=>I
    I=>B
    1 [A = “sertA.dat”, Na = RAND, kI = “keyI.dat”]
A->I: E_kI(Na, A)
    2 [A = Get_1_A, Na = Get_1_Na, kb = “keyB.dat”]
I->B: E_kB(Na,A)
    3 [B = “sertB.dat”, Nb = RAND, kA = “keyA.dat”,
Na = Get_2_Na] B->I: E_kA(Na, Nb, B)
    4 [Mess4 = Get_3_Message] I->A: Mess4
    5 [kI = “keyI.dat”, Nb = Get_3_Nb] A->I: E_kI(Nb)
    6 [kb = “keyB.dat”, Nb = Get_4_Nb] I->B: E_kB(Nb)
  }

  Run
  {
    Model1
  }
```

6.1 Section «Projects»

This section (lines 1–7) describes protocol implementation information. Initially, our dynamic analyzer only needs information about the paths to the parties' projects. Then this section is automatically filled in. The view of this section is as follows:

```
    [Info name]: [project path], [function with protocol],
[string number], [file path with protocol]
```

6.2 Section «Parameters»

This section describes the parties. The parameters of the party indicate information about the project. Here (lines 9–14) 2 draft legal parties A and B and information about the attacker's party I are indicated.

6.3 Section «MainProtocol1»

This section (lines 16–24) describes the basic protocol for the system. These sections can be multiple in case there are several protocols in the system. This section is helpful in designing an attack model, since the model must include at least one underlying protocol followed by the attack. The protocols on the legal sides are unchanged. The only thing that can change is the transmitted data. They can be changed by loading from files. The section initially describes the order of joining the parties (line 18), after which the order of message transmission and their content is described (lines 19–23). The message is specified in the following format:

```
Message index [message element = load path, …]
Side1->Side2: Message
```

6.4 Section «Model1»

Section describing the simulated attack. As mentioned earlier, a simulated attack must have the entire underlying protocol with the attack itself. This section (lines 26–39) contains Lowe's classic attack on the Needham-Schroeder protocol [20]. It is worth noting that during each simulation, copies of projects of legal parties are created with replaced data loads into message elements according to the model specification and successful completion marks, as well as a project with an attacker, in which messages are generated and elements are loaded from files also according to the specification specified in the current section… An attack is considered successful if, after completing its simulation, the protocol has been correctly and completely executed on all legal parties. There can be several models. All of them are checked in turn. It is also possible to use parallel sessions as well as post-session sessions. All this can be described in the body of the model.

6.5 Section «Run»

Starts modeling according to the body of the Model area. As a result of the simulation, the result is displayed in the form of whether the attack was successful or not:

Run model1 - Attack is working.
Run model2 - Secure.

The intruder role and its source code are generated. Forwarding messages according to the body of the region Model. The stripped-down system no longer exists. In places that participate in the protocol and require dynamic data input (loading from a file, keyboard input), these places are replaced by loading data from files.

7 Results

For testing, the following protocol was taken, based on the Needham-Schroeder protocol:

$A \to B : E_{pkB}(A, Na)$
$B \to A : E_{pkA}(Na, Nb, B)$
$A \to B : E_{pkB}(Nb, k)$
$B \to A : E_k(Mess1, Na)$
$A \to B : E_k(Mess2, Nb)$

In this protocol, two parties are mutually authenticated, after which party A sends the session key k for further use of symmetric encryption in messages (4) and (5). This protocol was implemented in the C # programming language in the form of communication between two components, a server and a client. The server side acted as side B, the client side - A. The following attack model was drawn up for the protocol:

1. $A \to B : E_{pkB}(A, Na)$
2. $B \to A : E_{pkA}(Na, Nb, B)$
3. $A \to B : E_{pkB}(Nb, k)$
4. $B \to A : E_k(Mess1, Na)$
5. $A \to I : E_k(Mess2, Nb)$
6. $I \to B : E_k(Mess1, Na)$

Section Model:

```
    Model1
    {
      A=>I
      I=>B
      1 [A = "sertA.dat", Na = RAND, kb = "keyB.dat"]
  A->I: E_kB(Na, A)
      2 [Mess2 = Get_1_Message] I->B: Mess2
      3 [B = "sertB.dat", Nb = RAND, ka = "keyA.dat",
  Na = Get_2_Na] B->I: E_kA(Na, Nb, B)
      4 [Mess4 = Get_3_Message] I->A: Mess4
      5  [kb  =  "keyB.dat",  Nb  =  Get_4_Nb,  k  -
  "key.dat"] A->I: E_kB(Nb,k)
      6 [Mess6 = Get_5_Message] I->B:
      7 [k = Get_6_k, M1 = "text1.txt", Na = Get_6_Na]
  B->I: E_k(M1,Na)
      8 [Mess8 = Get_7_Message] I->A: Mess8
      9  [k  =  "key.dat",  M2  =  "text2.txt",  Nb  =
  Get_8_Nb] A->I: E_k(M2,Nb)
      10 [Mess10 = Get_7_Message] I->B: Mess8
    }
```

The dynamic analysis revealed that the simulated attack on the protocol was successful. When reviewing the source code, it was determined that there was no action in the code when checking the returned random number Na and Nb in messages (4) and (5). As a result, the protocol was executed to the end, but instead of the required data in the M2 message, the party received data M1.

8 Discussion

To compare the proposed approach, we took the most well-known existing tools for conducting security verification of the Scyther and Avispa protocols. The Scyther tool described the protocol under test and verified the security of all properties. The protocol description is shown in Fig. 4.

As a result of verification, it was found that the protocol is secure. The test protocol was also described in the CAS+ language and translated into the HLPSL language for use in the Avispa verifier. Avispa is an automated security protocol analyzer. Supports the specification of protocols and security properties using a modular specification language. Integrates various back-ends implementing various modern automatic analysis methods for protocol verification (by looking for an attack on the input protocol). Abstraction based validation methods for both finite and infinite sessions. During the analysis, the OFMC mode was used. The protocol description in CAS+ language is shown in Fig. 5.

As a result of verification, it was found that the protocol is secure. As can be seen from the verification results, the existing tools did not detect the vulnerability of the protocol to a replay attack, since they do not analyze the source code. The protocol itself is inherently secure, however, the absence of key checks in the source code makes the protocol vulnerable in practice, which was discovered thanks to the approach proposed by the authors.

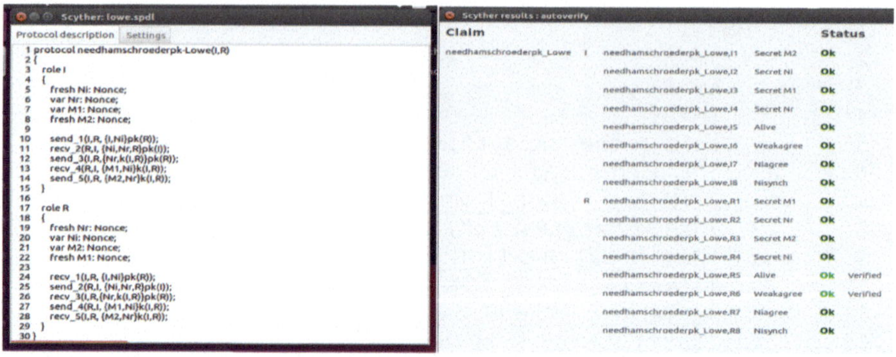

Fig. 4. Protocol description and verification results.

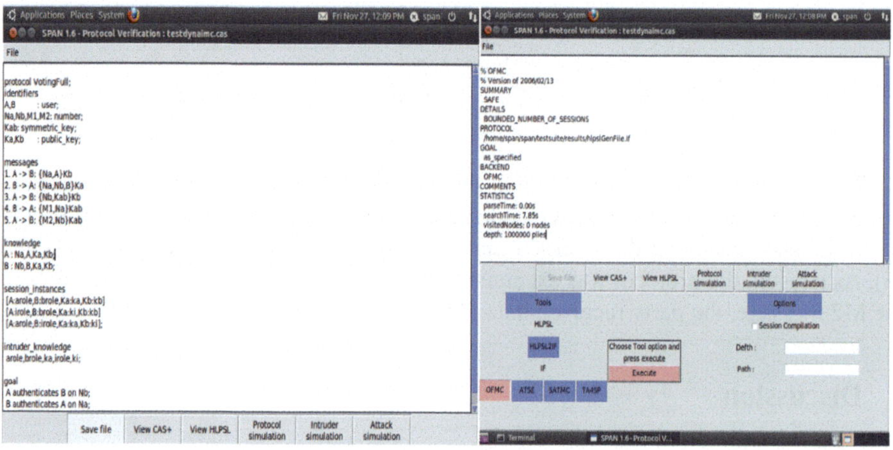

Fig. 5. Protocol description and verification results.

A comparison was made of the main indicators of the Scyther, Avispa verifiers and the dynamic PDA analyzer developed by the authors. Table 1 shows a comparison of the main indicators. The main advantages of the dynamic analysis proposed by the authors using the PDA language are the ability to use their own models and analysis of the direct source code of the cryptographic protocol. It is also an advantage to support all possible cryptographic primitives and functions, since the protocol implementation is analyzed directly. The disadvantages are the low speed of operation of even one model, not to mention the fact that a large number of models are required for a sufficient level of protocol analysis, as well as the impossibility of automated detection of attacks, since the models must be described independently. The speed of work in the future can be improved by using technologies for parallelizing processing on the processor, as well as by optimizing existing algorithms. Regarding automation of detection, it is planned to create an automated generator of model templates for testing various attacks (Table 3).

Table 3. Comparison of main characteristics

Indicator name	Scyther	Avispa	PDA
Using your own models	No	No	Yes
Source code analysis	No	No	Yes
Work speed (in the example given)	1.2 sec	7.8 sec	15 sec (one model)
Automated attack detection	Yes	Yes	No
Support for cryptographic primitives and functions	Timestamps, symmetric and asymmetric keys, signature, hashing.	Symmetric and asymmetric keys, non-atomic keys, key tables, key Diffie-Hellman convention, hash functions, algebraic functions typed and untyped data, signature.	Any

9 Conclusion

This paper presents an integrated approach to verification of the security of cryptographic protocols based on formal verification of source codes and dynamic analysis of source codes of protocols using the PDA language. A verification scheme is described, including formal verification of the protocol using the Avispa tool, implementation in a programming language, extraction of the structure of a cryptographic protocol from the source code to obtain a description in the Alice-Bob language, formal verification of the extracted protocol description, an approach to dynamic analysis based on the principle of false termination. The elements of the PDA language are described, the test protocol is presented and analyzed. A vulnerability was discovered in the implementation of the test protocol, which consisted in the lack of action when checking a random number of recent messages, which led to a replay attack. The comparison of existing verifiers with the dynamic analyzer presented by the authors is carried out. Existing verifiers have shown the safety of the test protocol. The comparison showed that the approach proposed by the authors is the most effective from the point of view of detecting all possible potential attacks, since the analysis of the source codes of cryptographic protocols is carried out according to a given model. Also, thanks to the analysis of source codes, it is possible to support all possible cryptographic functions and primitives, which makes it possible to detect vulnerabilities in protocols using them. However, the operating time is longer relative to other verifiers. In addition, at this stage of research, automated verification of source codes is not possible, since you need to describe your own verification models. In the future, the authors plan to reduce the speed of the analyzer, as well as support the possibility of automated application of pre-generated model templates to check the security of cryptographic protocols.

The reported study was funded by RFBR, project number 19-37-90151.

References

1. El Madhoun, N., Guenane, F., Pujolle, G.: An online security protocol for NFC payment: Formally analyzed by the scyther tool. In: 2016 Second International Conference on Mobile and Secure Services (MobiSecServ), pp. 1–7. IEEE (2016). https://doi.org/10.1109/MOBISECSERV.2016.7440225
2. Yang, H., Oleshchuk, V.A., Prinz, A.: Verifying Group Authentication Protocols by Scyther. J. Wirel. Mob. Netw. Ubiquitous Comput. Dependable Appl. **7**(2), 3–19 (2016)
3. Mir, O., van der Weide, T., Lee, C.-C.: a secure user anonymity and authentication scheme using AVISPA for telecare medical information systems. J. Med. Syst. **39**(9), 1–16 (2015). https://doi.org/10.1007/s10916-015-0265-8
4. Amin, R., et al.: Design of an enhanced authentication protocol and its verification using AVISPA. In: 2016 3rd International Conference on Recent Advances in Information Technology (RAIT), pp. 404–409. IEEE (2016). https://doi.org/10.1109/RAIT.2016.7507899
5. Cheval, V., Cortier, V., Turuani, M.: A little more conversation, a little less action, a lot more satisfaction: Global states in ProVerif. In: 2018 IEEE 31st Computer Security Foundations Symposium (CSF), pp. 344–358. IEEE (2018)
6. Chothia, T., Smyth, B., Staite, C.: Automatically checking commitment protocols in ProVerif without false attacks. In: Focardi, R., Myers, A. (eds.) POST 2015. LNCS, vol. 9036, pp. 137–155. Springer, Heidelberg (2015). https://doi.org/10.1007/978-3-662-46666-7_8
7. Blanchet, B.: Composition theorems for cryptoverif and application to TLS 1.3. In: 2018 IEEE 31st Computer Security Foundations Symposium (CSF), pp. 16–30. IEEE (2018). https://doi.org/10.1109/CSF.2018.00009
8. Eswaraiah, G., Vishwanathan, R., Nedza, D.: Automated proofs of signatures using bilinear pairings. In: 2018 16th Annual Conference on Privacy, Security and Trust (PST), pp. 1–10. IEEE (2018). https://doi.org/10.1109/PST.2018.8514201
9. Cremers, C.: Symbolic security analysis using the tamarin prover. In: 2017 Formal Methods in Computer Aided Design (FMCAD), p. 5. IEEE (2017)
10. De Ruiter, J., Poll, E.: Protocol state fuzzing of {TLS} implementations. In: 24th {USENIX} Security Symposium ({USENIX} Security 15), pp. 193–206 (2015)
11. Cohn-Gordon, K., et al.: A formal security analysis of the signal messaging protocol. In: 2017 IEEE European Symposium on Security and Privacy (EuroS&P), pp. 451–466. IEEE (2017)
12. Beringer, L., et al.: Verified correctness and security of OpenSSL {HMAC}. In: 24th {USENIX} Security Symposium ({USENIX} Security 15), pp. 207–221 (2015)
13. Kobeissi, N., Bhargavan, K., Blanchet, B.: Automated verification for secure messaging protocols and their implementations: a symbolic and computational approach. In: 2017 IEEE European Symposium on Security and Privacy (EuroS&P), pp. 435–450. IEEE (2017). https://doi.org/10.1109/EuroSP.2017.38
14. Dowling, B., et al.: A cryptographic analysis of the TLS 1.3 handshake protocol candidates. In: Proceedings of the 22nd ACM SIGSAC Conference on Computer and Communications Security, pp. 1197–1210 (2015)
15. Garcia, R., Modesti, P.: An IDE for the design, verification and implementation of security protocols. In: 2017 IEEE International Symposium on Software Reliability Engineering Workshops (ISSREW), pp. 157–163. IEEE (2017). https://doi.org/10.1109/ISSREW.2017.69
16. Modesti, P.: AnBx: automatic generation and verification of security protocols implementations. In: Garcia-Alfaro, J., Kranakis, E., Bonfante, G. (eds.) FPS 2015. LNCS, vol. 9482, pp. 156–173. Springer, Cham (2016). https://doi.org/10.1007/978-3-319-30303-1_10
17. Pisarev, I.A., Babenko, L.K.: C# parser for extracting cryptographic protocols structure from source code. Trudi instituta sistemnogo programmrovania RAN **31**(3), 191–202 (2019)

18. Babenko, L., Pisarev, I.: Translation of cryptographic protocols description from Alice-Bob format to CAS+ specification language. In: Kovalev, S., Tarassov, V., Snasel, V., Sukhanov, A. (eds.) Proceedings of the Fourth International Scientific Conference "Intelligent Information Technologies for Industry" (IITI'19), pp. 309–318. Springer , Cham (2020). https://doi.org/10.1007/978-3-030-50097-9_32
19. Baskar, A., Ramanujam, R., Suresh, S.P.: Dolev-Yao theory with associative blindpair operators. In: Hospodár, M., Jirásková, G. (eds.) CIAA 2019. LNCS, vol. 11601, pp. 58–69. Springer, Cham (2019). https://doi.org/10.1007/978-3-030-23679-3_5
20. Szymoniak, S., Siedlecka-Lamch, O., Kurkowski, M.: SAT-based verification of NSPK protocol including delays in the network. In: 2017 IEEE 14th International Scientific Conference on Informatics, pp. 388–393. IEEE (2017)

Stationary Distribution in the One Server Queuing System $M|M|1|\infty$ with the Stopping Intensity of the Input Flow

Gurami Tsitsiashvili(✉)

Institute for Applied Mathematics, Far Eastern Branch of Russian Academy Sciences, Vladivostok, Russian Federation
guram@iam.dvo.ru

Abstract. There is a lot of papers dealing with both theoretical and applied aspects and devoted to queuing models in a random environment. In these papers, the main focus is on calculation and analysis of stationary distributions that do not depend on the initial state. However, there are other models in which a random process, that determines the state of the medium, develops on a segment with reflecting and absorbing ends. The stationary distribution of such processes already depends on the initial state. In this paper, we consider the simplest version of such a model of a random environment with a queuing system placed in it. The stationary distribution of the queuing system in such a random environment is calculated and its dependence on the initial state is investigated. It is shown that such a random environment model is identical to the player's ruin model, and its study is reduced to solving a discrete analog of the Dirichlet problem. #CSOC1120.

Keywords: Random environment · Queuing system · Dirichlet problem

1 Introduction

A large number of papers dealing with both theoretical and applied aspects are devoted to queuing models in a random environment (see, for example, [1–7]). In these works, the main focus is on calculation and analysis of stationary distributions that do not depend on the initial state. However, there are other models in which a random process, that determines the state of the medium, develops on a segment with reflecting and absorbing ends (see, for example, [8]).

Due to the fact that the intensity of the input flow obeys the Markov chain with absorbing states, the process, describing the queuing system, converges to a process with a fixed intensity of the input flow. Moreover, there is an obvious dependence of the input flow intensity on the initial state of this Markov chain. The stationary distribution of such processes already depends on the initial state.

In this paper, we consider the simplest version of such a model of a random environment with a queuing system placed in it. The stationary distribution of

© The Author(s), under exclusive license to Springer Nature Switzerland AG 2021
R. Silhavy (Ed.): CSOC 2021, LNNS 229, pp. 444–449, 2021.
https://doi.org/10.1007/978-3-030-77445-5_41

the queuing system in such a random environment is calculated and its dependence on the initial state is investigated. It is shown that such a random environment model is identical to the player's ruin model, and its study is reduced to solving a discrete analog of the Dirichlet problem [9, ch. II, §§ 2-7].

2 Methods

Consider a single-server queuing system $M|M|1|\infty$ with an infinite number of waiting places and a service intensity μ. Random process $\lambda(t)$, which characterizes the intensity of the input flow, is given by a Markov chain $z_k,\ k = 0, 1, \ldots,$ with a set of states $\{0, \ldots, N\}$. Markov chain $z_k,\ k = 0, 1, \ldots,$ describes a game to ruin the player [10, ch. I, § 9] and defines a piecewise constant random process $z(t) = z_k,\ k \leq t < k+1,\ k = 0, 1, \ldots$ Random process $\lambda(t) = \Lambda(z(t)) > 0,\ t \geq 0,$ with a set of states $\{\Lambda(0), \ldots, \Lambda(N)\}$ is Markov. A queuing system $M|M|1|\infty$ with a so-defined input flow intensity $\lambda(t)$ will be denoted by $\mathbf{M}$.

Consider the Markov process $(z(t), k(t)),\ t \geq 0,$ that characterizes the random intensity of the input flow and the random number of customers in the system $\mathbf{M}$ at time t. Next, we assume that the inequalities are satisfied

$$\Lambda(0) < \mu, \ldots, \Lambda(N) < \mu. \tag{1}$$

Our task is to calculate the stationary distribution of the second component $k(t)$ of this process.

It is obvious that the stationary distribution of the Markov chain $z_k,\ k = 0, 1, \ldots,$ depends on the initial state z_0. At first glance, this dependence makes it difficult to solve the problem. However, the analogy of the Markov chain $z_k,\ k = 0, 1, \ldots,$ with the process describing the game for the player's ruin allows, on the contrary, to significantly simplify the calculation of the stationary distribution of the process in the system $\mathbf{M}$. Solution of this problem may be reduced to the solution of a discrete analogue of the Dirichlet equation and to the use of known formulas for the stationary distribution of the service process in a system with a constant input flow intensity.

Consider a Markov chain $z_k,\ k = 0, 1, \ldots,$ with the set of states $\{0, 1, \ldots, N\}$, with non-zero elements of the matrix $\Theta = ||\theta_{i,j}||_{i,j=0}^{N}$ transition probabilities

$$\theta_{i,i+1} = p,\ \theta_{i,i-1} = q,\ i = 1, \ldots, N-1,\ \theta_{0,0} = \theta_{N,N} = 1,\ 0 < p < 1,\ q = 1 - p. \tag{2}$$

Everywhere else, the probability $P_{z_0}(A)$ denotes the probability of the event A, provided that the Markov chain $z_k,\ k = 0, 1, \ldots,$ takes the initial value z_0. Denotes

$$\pi_{z_0}(k) = \sum_{z=1}^{N-1} P_{z_0}(z_k = z),\ \psi_{z_0}(k) = P_{z_0}(z_k = 0) + P_{z_0}(z_k = N),$$

it is obvious that the relation is valid

$$\pi_{z_0}(k) + \psi_{z_0}(k) = 1. \tag{3}$$

The formulas (2), (3) are executed for any $z_0 = 0, \dots, N$. The following inequalities follow from the relations (2)

$$P_{z_0}(z_0 = 0) \leq P_{z_0}(z_1 = 0) \leq \dots,$$

$$P_{z_0}(z_0 = N) \leq P_{z_0}(z_1 = N) \leq \dots$$

and so there are limits

$$\lim_{k\to\infty} P_{z_0}(z_k = 0) = \Psi_{z_0}(0), \lim_{k\to\infty} P_{z_0}(z_k = N) = \Psi_{z_0}(N),$$

satisfying the relation

$$\lim_{k\to\infty} \psi_{z_0}(k) = \Psi_{z_0}(0) + \Psi_{z_0}(N) = \Psi_{z_0} \leq 1.$$

Obviously, for $\gamma = p^N + q^N < 1$, the relation holds

$$\pi_{z_0}(kN) - \pi_{z_0}((k+1)N) \geq \pi_{z_0}(kN)\gamma$$

and means

$$\pi_{z_0}((k+1)N) \leq \pi_{z_0}(kN)\gamma, \ k = 1, 2, \dots$$

Hence we get the limiting ratio

$$\lim_{k\to\infty} \pi_{z_0}(k) = 0. \tag{4}$$

From the formulas (3), (4) we have the limiting ratio

$$\lim_{k\to\infty} \psi_{z_0}(k) = \Psi_{z_0}(0) + \Psi_{z_0}(N) = \Psi_{z_0} = 1. \tag{5}$$

From the definition of a Markov chain z_k, $k = 0, 1, \dots$, it follows that $\Psi_{z_0}(0)$ is the probability that leaving the status of z_0, Markov chain z_k, $k = 0, 1, \dots$, will ever be in state 0. Therefore, $\Psi_{z_0}(0)$ may be interpreted as the probability of ruin of the player with an initial amount of z_0, with the win probability p and the probability of losing q and the sum of the winnings in player's ruin model.

Then the function $\Psi_{z_0}(0)$ satisfies a system of linear algebraic equations

$$\Psi_{z_0}(0) = p\Psi_{z_0-1}(0) + q\Psi_{z_0+1}(0), \ z_0 = 2, \dots, N-2, \ \Psi_0(0) = 1, \ \Psi_N(0) = 0. \tag{6}$$

Using the well-known formulas for the probability of the player's ruin [10, ch. I, § 9], we can write out the following relations

$$\Psi_{z_0}(0) = \frac{(p/q)^{z_0} - (p/q)^N}{1 - (p/q)^N}, \ p \neq q; \ \Psi_{z_0}(0) = 1 - \frac{z_0}{N}, \ p = q; \ z_0 = 0, \dots, N, \tag{7}$$

In turn, the formula (5) implies that the probability of winning

$$\Psi_{z_0}(N) = 1 - \Psi_{z_0}(0). \tag{8}$$

The distribution $\Psi_{z_0}(0), \dots, \Psi_{z_0}(N)$ is a stationary distribution of the Markov chain z_k, $k = 0, 1, \dots$, with the initial state z_0. This distribution is concentrated at the points $0, N$, that are absorbing for this chain. It follows that the chain z_k, $k = 0, 1, \dots$, under any initial condition $z_0 = 0, \dots, N$, has a stationary distribution. However, this chain is not ergodic, since its stationary distribution depends on the initial condition. The same properties are possessed by the piecewise constant process $z(t)$, defined by the Markov chain z_k, $k = 0, 1, \dots$, and specifying the intensity of the input flow $\lambda(t) = \Lambda(z(t))$ of queuing system **M**.

3 Results

Under the initial condition $z(0) = z_0$ the stationary distribution of the process $\lambda(t) = \Lambda(z(t))$ satisfies the relations

$$\lim_{t\to\infty} P_{z_0}(\lambda(z(t)) = \Lambda(0)) = \Psi_{z_0}(0), \ \lim_{t\to\infty} P_{z_0}(\lambda(z(t)) = \Lambda(N)) = \Psi_{z_0}(N).$$

In other words, a random process $\lambda(t)$ has on the set of states $\{0, \dots, N\}$ a two-point stationary distribution $P(\lambda(t) \equiv \Lambda(0)) = \Psi_{z_0}(0)$, $P(\lambda(t) \equiv \Lambda(N)) = \Psi_{z_0}(N)$, depending on the initial state z_0. And the process $\lambda(t)$, falling into the state $\Lambda(0)$ (when entering the state $\Lambda(N)$) stops there and does not change further.

Let us now proceed to the calculation of the limit distribution of the number of customers in the system **M** under the condition (1). It is known that the stationary distribution $p(k)$ of the number of customers in the system $M|M|1|\infty$ with a deterministic intensity of the input flow λ and the service intensity μ satisfies the relations

$$p(k) = (1-\rho)\rho^k, \ k = 0, 1, \dots, \ \rho = \frac{\lambda}{\mu} < 1. \tag{9}$$

Denote $\rho(0) = \dfrac{\Lambda(0)}{\mu} < 1$, $\rho(N) = \dfrac{\Lambda(N)}{\mu} < 1$, then the stationary distribution $P_{z_0}(k)$, $k = 0, 1, \dots$, of the number of customers in the queuing system **M** satisfies the equality

$$P_{z_0}(k) = \Psi_{z_0}(0)(1-\rho(0))\rho^k(0) + \Psi_{z_0}(N)(1-\rho(N))\rho^k(N), \ k = 0, 1, \dots \tag{10}$$

If we assume that $A(i)$, $i = 0, \dots, N$, is the initial distribution of the random variable z_0, then, due to the formula (10), the stationary distribution $\Pi(k)$, $k = 0.1, \dots$, of the number of customers in the queuing system **M** has the form

$$\Pi(k) = \sum_{i=0}^{N} A(i)\left[\Psi_i(0)(1-\rho(0))\rho^k(0) + \Psi_i(N)(1-\rho(N))\rho^k(N)\right], \ k = 0, 1, \dots \tag{11}$$

Note that if $A(1) = \dots = A(N-1) = 0$, then for the existence of a stationary distribution, it is sufficient to require $\Lambda(0) < \mu$, $\Lambda(N) < \mu$.

4 Discussions

Let us now list the possible generalizations of the results obtained. From the simplest single-server system $M|M|1|\infty$, the underlying model $\mathbf{M}$, we can move to multi-server systems, systems with failures, to open queuing networks. The Markov chain z_k, $k = 0, 1, \ldots,$, which determines the intensity of the input flow, may be replaced by a random process with continuous time, for example, a diffusion process with absorption or partial absorption and reflection at the ends of a certain segment. This Markov chain can determine not only the intensity of the input flow, but also the intensity of the service. Then the probability of hitting the Markov chain z_k, $k = 0, 1, \ldots$, satisfies the discrete analog of the Dirichlet equation and may be solved by known methods of probability theory and mathematical physics. We can also use the stationary distribution of the process $k(t)$, $t \geq 0$, of the number of customers in the queuing system $\mathbf{M}$ (estimated, for example, from observations) to determine the initial state z_0 of the process $z(t)$, $t \geq 0$, that characterizes the intensity of the input flow at the initial time.

5 Conclusion

The random process model considered in this paper, which describes the intensity of the input flow, with absorbing states, makes it possible to analytically calculate the limit distribution of the number of customers in the queuing system. An important element of such a model is the inclusion of the Dirichlet problem in it. This makes it possible to combine the methods of queuing theory with the analytical and computational methods of mathematical physics in solving this problem.

Appendix: Springer-Author Discount

LNCS authors are entitled to a 33.3% discount off all Springer publications. Before placing an order, the author should send an email, giving full details of his or her Springer publication, to orders-HD-individuals@springer.com to obtain a so-called token. This token is a number, which must be entered when placing an order via the Internet, in order to obtain the discount.

6 Checklist of Items to be Sent to Volume Editors

Here is a checklist of everything the volume editor requires from you:

- [ ] The final LaTeX source files
- [ ] A final PDF file
- [ ] A copyright form, signed by one author on behalf of all of the authors of the paper.
- [ ] A readme giving the name and email address of the corresponding author.

References

1. Vyshnevskiy, V.M., et al.: Stochastic systems with correlated flows. Theory and Application in Telecommunications Networks, Moscow: Technosphere (2018). (In Russian)
2. Vyshnevskiy, V.M., Semenova, O.V.: Polling Systems: Theory and Application in Broadband Wireless Networks. Technosphere, Moscow (2007). (In Russian)
3. Vyshnevskiy, V.M., et al.: Principles of construction of mobile and stationary tethered high-altitude unmanned telecommunication platforms of long-term operation. Commun. Comput. Inf. Sci. **919**, 561–569 (2018)
4. Klimenok, V.I., et al.: A retrial queueing system with alternating inter-retrial time distribution. Commun. Comput. Inf. Sci. **919**, 302–315 (2018)
5. Vishnevskiy, V.M., et al.: Multiserver queieing system with map arrivals for modelling LTE cell with H2H and M2M communications and M2M aggregation. Commun. Comput. Inf. Sci. **918**, 63–74 (2017)
6. Korotaev, I.A., Spivak, L.R.: Queuing systems in a semi-Markov random environment. Autom. Telemech. **7**, 86–92 (1992). (In Russian)
7. Jernovoy, Y.V.: Queuibng system $M|M|n|r$, functioning in a synchronous random environment. Inf. Process. **9**(4), 352–363 (2009)
8. Bondrova, O.V., et al.: Derivation of Kolmogorov-Chapman type equations with an integral operator. Far Eastern Math. J. **17**(2), 135–146 (2017) (In Russian)
9. Dynkin, E.B., Yushkevich, A.A.: Theorems and Problems about Markov Processes. Nauka, Moscow (1967). (In Russian)
10. Shiryaev, A.N.: Probability. Nauka, Moscow (1989). (In Russian)

Advantages of Data Mining for Digital Transformation of the Educational System

Kamelia Shoilekova(✉)

Angel Kanchev University of Ruse, 8 Studentska Str., 7000 Ruse, Bulgaria
kshoylekova@uni-ruse.bg

Abstract. Digitalization provides opportunities for developing skills and knowledge transfer between large groups of people that determines its application for the purposes of the educational system. Analysis of inhomogeneous data and the construction of interconnections and models is a prerequisite for resolving optimization tasks. In this case, the problem is related to raising the educational quality by using the advantages of the data mining process.

Keywords: Data mining · Digital transformation · Educational system

1 Digital Transformation in the Educational System

The wide usage of digital assets as well as the dynamics of the working and living environment reflect on the necessities of the contemporary learners. Nowadays, we face the situation of a continual search of possibilities for training options beyond traditional classroom approaches. Opposing the traditional methods, the educational institutions needed to react in a flash and offer alternative forms of training. All circumstances required to adapt the digital technologies to the educational ecosystem. The digital transformations within the educational system are capable of improving the traditional face-to-face training. Educators should make their best to gain the benefits from the distance-training situation and create pedagogical models for establishing better relationships with learners [2, 3, 7, 8].

Today, learners are connected to the virtual world by means of internet and smart phones, which absorb enormous amounts of data online. Information technologies provide learners with the opportunity to purposely collect and synthesize for a short time the data extracted from different sources. This circumstance by itself helps the educational institutions equip a widened audience of users with affordable models for education transformation.

While the course syllabus is a dynamically developing domain based on current and future trends, the teaching and learning methodologies lay in a specific area particularly in need of experiments and innovations. Technological interventions within that period can add value to the present pedagogical practices.

As a whole, digital transformation can effect the teaching process, on one hand, by passing the limits of the traditional training and introducing distance learning, thus

© The Author(s), under exclusive license to Springer Nature Switzerland AG 2021
R. Silhavy (Ed.): CSOC 2021, LNNS 229, pp. 450–454, 2021.
https://doi.org/10.1007/978-3-030-77445-5_42

benefitting from both traditional methods and modern information technologies. On the other hand, digitalization helps the automation of academic administrative duties related to student admittance, accommodation, curricular activities, etc.

Every academic year universities and colleges are required to complete a number of tasks connected with the preparation and delivery of institutional data to the accreditation and classification agencies. They sum up and present a huge amount of data that, in most cases, is manually prepared. The process can be relieved with the help of a data control system. Digitalizing the bigger part of the administrative services leads to a more effective control and monitoring over the processes and systems, which is a premise for optimizing the administrative services of the educational institutions [5].

The force majeure circumstances, which occurred this academic year, urged fundamental changes of the educational practices leading to speeded digital transformation of many processes related to the educational system, and society in general. The innovations, promptly introduced in the educational sphere, effect the teaching process and can be listed as follows:

- Online teaching processes (traditional classroom training switched to distance learning) (Fig. 1);
- Electronic register for recording students' marks, absences, as well as lesson topics for all subjects included in the curricula for students from 1st to 12th school grades, notifications to students and parents, etc. It is noted that the eRegister content can be duly modified for the purposes of higher education institutions (Fig. 2).

Fig. 1. Digitalization around us

The advantages, which can be counted in result of digital transformation, involve:

- The digital transformation effects not only the way a given technology is applied, but also, how the process is renewed in order to become more effective;
- The transformation of the existing processes;
- The digital culture of both learners and teachers;
- The transition to online teaching with regard to its positive effect for destructing the silos of mindset and improving the conventional models of learning.

Fig. 2. Cloud applications

The main issue emerging at this stage is not how to apply the new technologies in the existing educational system, but what changes are imposed by the digital revolution regarding processes and activities. It is possible to analyze billions of daily online transactions connecting processes, data, and users (learners and teachers). The analysis of data that can be extracted from the eRegister helps to detect hidden models, which may contribute to the improvement of the educational system [4, 6] (Fig. 3).

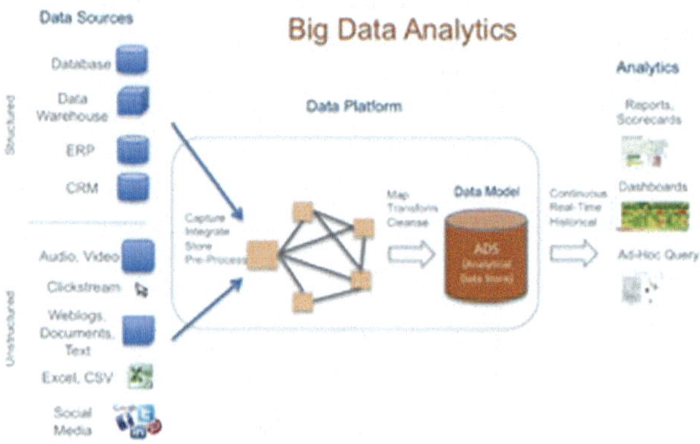

Fig. 3. Big data and analytics reporting

2 Basic Application Aspects of the Huge Data Sets in the Educational System

The process of teaching and learning in a digital environment, and the application of an electronic class Register provide the following opportunities:

- Parents can closely watch the children's progress and the acquired knowledge level.
- Teachers can send text messages to parents about student's involvement in the educational process. The Text Mining method, applied for analyzing the big data sets of the educational system, helps to detect models of unstructured data and to identify users who interfere with other users in the process of knowledge acquisition.

- Experts, who analyze the data of the electronic register, interpret and describe models in accordance with the recorded data.

In the context of the latter possibility, experts are provided with an option to regard the teaching process as an optimization problem. Optimization problems calculate the values of model's control parameters, which lead to the desired target state. These goal values of parameters can be found out through a process of iteration, known as "model learning". It can be carried out using two basic strategies: supervised learning and unsupervised learning. Furthermore, the classification model can be applied aiming to identify the repeating interconnections of the input parameters, which describe the objects belonging to the same class. Consequently, the detected interconnections are transformed into classification rules that are used for predicting the class of objects with known values of the predicting variables. The classification rules can be presented in various forms depending on the type of the applied model.

Figure 4 shows a process with three components - generator of observations, class learning and classification algorithm [1].

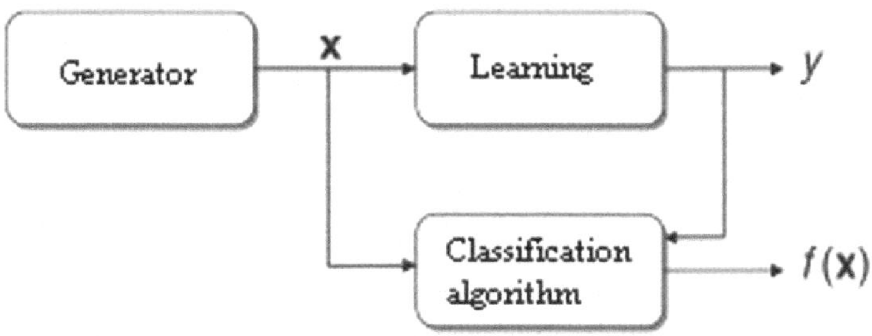

Fig. 4. Process of learning at applied classification [1]

Classification models are suitable for data description and interpretation, and for making predictions, both providing the possibility for generating regularities through learning. Easy-to-interpret classification rules can be generated with the help of simple models, while sophisticated models generate rules, which being more difficult to interpret are more accurate for predictions.

Based on the learning process at classification applied for the data stored in the electronic register, a comprehensive analysis can be made to obtain the regularity regarding:

- For which sequence of subject topics the students' results are higher and it may be an assumption for enhanced knowledge;
- Estimate whether gender influences the level of knowledge acquisition;
- Estimate whether students, which have chosen a given vocational class, achieve higher results of the professional subjects.

In conclusion, it may be affirmed that the Data Mining analysis is applicable for domains where statistical and analytical methods, respectively the built models, are not sufficient for obtaining the target results. Data Mining analysis is suitable for domains where the prevailing data is inhomogeneous, heterogeneous, non-stationary, variable and in huge volumes.

An effective digitalization is achieved when it is used as an instrument for stimulation of institutional and system changes related to the process of knowledge transfer and skills development, rather than to automation and quick inspection of the learning process.

References

1. Vercellis, C.: Business Intelligence: Data Mining and Optimization for Decision Making. Wiley, Chichester (2009)
2. Najmul Islam, A.K.M.: E-learning system use and its outcomes: Moderating role of perceived compatibility. Telematics Inform. **33**(1), 48–55 (2016). https://doi.org/10.1016/j.tele.2015.06.010
3. Ancheva, V., Voinohovska, V.: Integrating the stem methodology in robotics education in after school enrichment classes for students. In: 12th International Conference of Education, Research and Innovation, Seville, Spain, pp. 4610–4616 (2019)
4. Sclater, N., Peasgood, A., Mullan, J.: Learning Analytics in Higher Education. JISC, London (2016)
5. Kasemsap, K.: The role of learning analytics in global higher education. Developing effective educational experiences through learning analytics (2016)
6. Adam, K., Bakar, N.A.A., Fakhreldin, M., Majid, M.: Big Data and Learning Analytics: A Big Potential to Improve e-Learning. American Scientific Publishers, New York (2018)
7. Voinohovska, V., Goranova, E.: Student peer teaching – an innovative instructional method to foster student problem-solving and in-depth learning. In: 10th annual International Conference of Education, Research and Innovation, Seville, Spain, pp. 741–748 (2017)
8. Velikova, E., Mierlus-Mazilu, I., Vasileva-Ivanova, R., Georgieva, D.: About the stem education. In: Proceedings of University of Ruse, vol. 57, book 6.4, pp. 10–14, ISSN 1311-3321 (2018)

Matrix Product State Representations for Machine Learning

Eric Howard[1,2(✉)], Iftekher S. Chowdhury[1], and Ian Nagle[1]

[1] Department of Physics and Astronomy, Macquarie University, Sydney, NSW 2109, Australia
eric.howard@mq.edu.au

[2] Centre for Quantum Dynamics, Griffith University, Brisbane, QLD 4111, Australia

Abstract. Machine learning architectures provide a novel perspective on the study of quantum many body states. Restricted Boltzmann machines (RBM) tool is used to represent quantum many-body states in order to find connections to tensor network tools for studying quantum many-body physics, for a deeper understanding of the origin of entanglement entropy in quantum systems. Here, we seek the conditions for the optimal mapping of RBMs into Matrix Product States (MPS), with the aim to show that machine learning methods are a powerful tool for quantum state representations. We here showcase an efficient algorithm for translating RBMs into MPS, with a particular proof for Ising model. We also study the upper entropy bound condition and the entanglement properties of such mapping.

Keywords: Tensor networks · Matrix product states · Machine learning · Restricted Boltzmann machines · Deep learning

1 Introduction

Machine learning algorithms [1] have found a wide use in several applications in physics, engineering, astronomy [2] and computer science, including computer vision, image classification and recognition systems, speech recognition, natural language processing or recommender systems. Machine learning methods, many-body and condensed matter physics are separate disciplines but share a common need for highly efficient representations of multivariate function classes. Big data analytics require novel techniques to efficiently manage broad and diverse sets of data.

Classical information algorithms deal with automated pattern recognition and classification. Recently, machine learning techniques have been introduced to solve physical problems. Existing problems in machine learning, computational neuroscience, natural language processing, robotics, image recognition or signal processing generate vast amounts of high dimensional data. Tensors provide an efficient, natural and compact representation for such data via low-rank approximations. Machine learning and deep learning techniques are useful to compress

© The Author(s), under exclusive license to Springer Nature Switzerland AG 2021
R. Silhavy (Ed.): CSOC 2021, LNNS 229, pp. 455–468, 2021.
https://doi.org/10.1007/978-3-030-77445-5_43

high dimensional data into low dimensional sets and have been applied to find hidden patterns from various data input representations, with highly robust results against noise. Neural networks have become an effective tool for efficient encoding of classical and quantum correlations with iterative abstraction that can be learned from training data in higher dimensional spaces.

Meanwhile, Restricted Boltzmann machines (RBM) [3] are one of the most applied neural network algorithms and a cornerstone of deep learning architectures, used in greedy layer-by-layer pre-training of deep belief networks and multilayer deep Boltzmann machines. The unsupervised nonlinear features of RBMs can be useful in various applications with feature extraction or dimensional reduction methods and probability distribution modelling of diverse data input such as image, video or sound. More precisely, a Boltzmann machine is a stochastic neural network that contains a neuron-like network of nodes with undirected edges, where each neuron has a binary value with a probabilistic value, dependent on the neighbouring units to which it is connected. RBMs have recently attracted great interest due to their versatility in several unsupervised and supervised learning tasks.

Recently, RBM has also attracted significant attention in the field of quantum many-body physics [4], extending the bridge between the fields of deep learning and quantum physics in general. A study of the physical rules behind neural networks could provide an answer to the great success and development of deep learning in physics. Using neural network techniques, the properties of quantum many-body systems manifesting the underlying interactions, with exponentially large degrees of freedom can be successfully reduced to efficient simulations for learning the long-distance correlations encoded in the local Hamiltonian. The scale dependence of the correlations is not evident in the local Hamiltonian and it will be revealed by existing renormalization techniques via iterative coarse-graining of the local variables in the system. An RBM can be seen as a bipartite undirected model describing the generative process in which a data vector can be generated from a binary hidden vector. This bipartite structure allows fast data encoding and sampling-based inference, helping with processing a great amount of data using complex models.

Several schemes of machine learning, including supervised, unsupervised or reinforcement learning have been successfully applied to quantum systems of spins, bosons or fermions, in combination with gradient methods, Monte Carlo or others. Moreover, finding new methods to extract information from quantum devices has become a feasible and efficient method and important stage in the development of recent quantum simulators. Previously applied to mostly data science problems, machine learning methods started to address important questions from physics, with recent advances in quantum many-body systems for reliable classification and detection of phases of matter or enhancing the quantum systems simulators.

Finding efficient wavefunctions that usually parametrize physically relevant degrees of freedom in an exponentially large Hilbert space is a major tasks for the quantum many body theory [5]. The main issue is the exponential complexity of quantum many-body states with the number of constituents. Highly sophisticated computational models are today capable to describe the entanglement features of many-body wave-function representations in modern condensed matter physics.

The great success behind the strong connection and exceptional overlap [6] between deep learning and physics may be buried at a more fundamental level in the laws of physics, intricately responsible for symmetry, compositionality, locality and polynomial log-probability. An insight on quantum entanglement and its deep connection to tensor network states may deepen our understanding on symmetry, locality and causality in quantum physics.

The complexity of quantum systems is dependent on the size of the state space and grows exponentially with the number of degrees of freedom, a similar problem to the 'curse of dimensionality' found in machine learning. However, new machine learning methods with versatile abilities have been developed in order to recognize, classify or characterize datasets and may be used in the physical scenario under investigation.

Tensor network states have become an important tool in the study of critical systems and their dynamical features, even for long range interactions where the quantum correlations description falls beyond the area-law entanglement [7]. The search for compressed representations of entangled systems are a open problem in quantum physics. Tensor networks in general have been quite successful in approximating quantum states containing exponentially many parameters. A tensor network is a set of tensors with indices contracted according to a specific pattern.

Deep learning methods can be used to recognize and discover patterns in the input data. On the other hand, a study of entanglement for MPS may help with developing new deep learning algorithms [8]. Despite all achievements, there are still many challenges and concerns. For example, the underlying connection between well-known properties of classical data and entanglement properties of quantum states is still elusive.

We are here searching for a constructive connection between RBM and MPS methods for quantum many-body systems and develop an algorithm to translate one language into another [9]. We here present a constructive connection between RBM and MPS methods. The entanglement properties of a many-body quantum system should be revealed by the dependencies required for performing a deep learning tasks.

In a much broader context, there are several tensor networks that describe quantum many-body states as low-lying eigenstates of specific Hamiltonians, from matrix product states (MPS) and projected entangled pair states (PEPS) [10] to hierarchical structures like tree tensor networks (TTN), holographic renormalization [2] and multiscale entanglement renormalisation ansatz (MERA).

In particular, matrix product states (MPS) are a type of a tensor network where the tensors are arranged in a one dimensional geometry, same as tensor train decomposition from applied mathematics. Tensor networks and MPS share many common features in their intrinsic mathematical structure. In a MPS there is one tensor per site in the quantum many-body system.

MPS can describe any state of the quantum many-body Hilbert space by increasing sufficiently the value of the bond dimension. In order to contain all states of the Hilbert space, the bond dimension size has be exponentially large. However, for one dimensional critical systems, the bond dimension diverges polynomially in size.

The Matrix product states (MPS) Ansatz describes exponentially decaying correlations reflected in the area law entanglement of the wavefunction and efficiently captures the ground states of local gapped Hamiltonians.

MPS networks are an efficient tool [11] that has been successfully applied in many-body physics and it is based on the main assumption that physically relevant many-body states do not require a description that scales exponentially with the system size, but with a polynomial number of coefficients specified. MPS is widely considered as a key tool for modelling [12] low-entangled quantum states but also used successfully to model highly-entangled states, using deep quantum circuits or high-dimensional physical systems.

MPS tensor network method is very successful in describing quantum states with relatively low entanglement entropy following the area law. In general, the entanglement area law states that the entanglement entropy [13] of a system composed of two subsystems will scale linearly with the size of the boundary separating the two subsystems. The entanglement entropy is a measure of the information exchange between the two subsystems.

MPS have been also widely used for modelling time-evolving 1D systems, with particular applications in cold atom physics, stochastic modelling and optimization, time-periodic driving systems or quantum disordered systems. MPS methods are similar to tensor train decomposition methods already used in applied mathematics. We are interested in finding the general and optimal conditions for an MPS to efficiently translate into a specific RBM architecture as an efficient representation of quantum states.

The general goal here is to establish a proper connection between RBM and MPS methods in order to describe quantum states in statistical physics models and machine learning datasets. An RBM can be trained such that the probability distribution of the visible units can translate the probability distribution of the input data. The hidden units in the network could contain important information with significant physical meaning that can be used in pattern recognition to generate new samples from previously learned data.

The search for well defined wavefunctions is a key problem in quantum many-body physics. Here, the quantum-mechanical description of a system, including complex properties such as multi-qubit entanglement becomes mainly a data-driven problem. Machine learning architectures are able to identify phases or phase transitions in several condensed matter Hamiltonians.

Neural networks can be trained to find in raw datasets interesting order parameters or non-trivial states with no conventional order. An artificial neural network can be pictured [14] as a functional mapping of multiple variables that can be trained. In this way, many concepts and techniques from deep learning and quantum physics can be efficiently exchanged. For example, an insight into the entanglement entropy bound in MPS may quantify a number of complex features in RBM datasets.

In general, Boltzmann Machines are a particular form of log-linear Markov Random Field where the energy function is linear in its free parameters. For complex distributions, some of the variables are considered as never observed (hidden variables or units). The universal approximation theorem in machine learning states [15] that there exists an RBM to accurately describe a dataset if there is no limit on the number of hidden variables.

The modelling capacity of the Boltzmann Machine increases with the number of the hidden variables. RBMs further restrict generic Boltzmann Machines to networks without visible-visible and hidden-hidden connections. There are only connections between the visible and hidden units, not within them.

The RBMs are a specific case of energy-based models which associate a scalar energy to each configuration of the variables of interest. Learning is translated into modifying the energy function so that its shape has specific properties.

An energy-based probabilistic model defines a probability distribution via an energy function with a partition function as a normalizing factor, in analogy to physical systems [16]. The energy-based model is learnt by performing stochastic gradient descent on the negative log-likelihood of the input training data.

2 RBM Learning of MPS Networks

If we want to describe the probability distribution of an input dataset, based on an existing model, new samples can be generated from the learned probability distribution. The data distribution can be described by a quantum state. In a probabilistic setting, a dataset $\mathcal{T}$ can be represented by a repeated number of binary strings $v \in \mathcal{V} = \{0,1\}^{\otimes N}$. These strings are mapped to the basis vectors of a Hilbert space of dimension 2^N.

Nonetheless, if the probability distribution is described by the wavefunction $\Psi(\boldsymbol{v})$, the collapse produced by a measurement will give a final result $\boldsymbol{v} = (v_1, v_2, \cdots, v_N)$. The probability will be proportional to $|\Psi(\boldsymbol{v})|^2$.

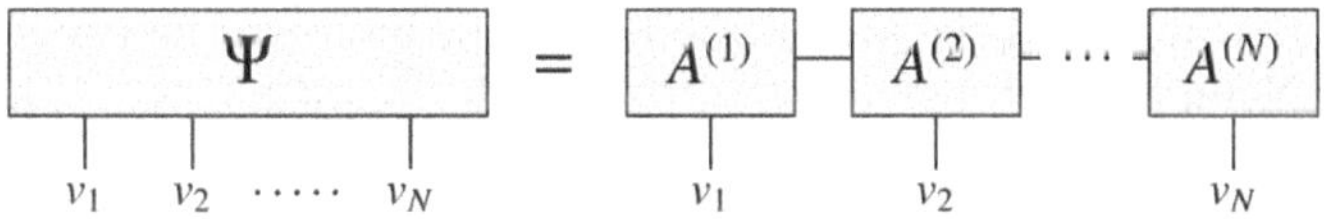

Fig. 1. The wavefunction in MPS representation

In a general context, MPS are states which are defined on a quantum chain with N sites, each component being a d-level system. The quantum wavefunction using MPS is

$$\Psi(v_1, v_2, \cdots, v_N) = -\text{Tr}\left(\prod_{i=1}^{N} A^{(i)v_i}\right) - \text{Tr}\left(A^{(1)v_1} A^{(2)v_2} \cdots A^{(N)v_N}\right), \tag{1}$$

where each $A^{(k)v_k}$ is a $\mathcal{D}_{k-1}$ by $\mathcal{D}_k$ matrix, and $\mathcal{D}_0 = \mathcal{D}_N$ closes the trace.

In Fig. 1 the blocks represent tensors while the connected lines are tensor contractions over virtual indices. The free vertical legs are physical indices. The number of operations to be done in order to obtain the final contraction of the tensor network depends on the order [17] in which the indices are contracted. This is a significant fact and a difficult problem, because in tensor network techniques, a great number of contractions are necessary and one of the goals is to make these contractions as efficiently as possible.

The variables are divided into two groups, we firstly have $\boldsymbol{v} = (\boldsymbol{v}_A, \boldsymbol{v}_B)$ and also $\rho_A = \sum_{\boldsymbol{v}_B} \Psi(\boldsymbol{v}_A, \boldsymbol{v}_B)\Psi(\boldsymbol{v}'_A, \boldsymbol{v}_B)$ as the reduced density matrix of one of the subsystems.

MPS can represent very efficiently a large number of quantum states [18] and accurately describe the ground states for one-dimensional gapped local Hamiltonians. The low-energy eigenstates of gapped Hamiltonians with local interactions obey the area-law for the entanglement entropy. This means that the entanglement entropy of a large system tends to scale, as the size of the boundary of the system and not as the volume.

In general, the low-energy states of realistic Hamiltonians are constrained by locality so that they must obey the area law. The entanglement and correlation properties of MPS networks play a significant role in simulating strongly correlated quantum chains and therefore a deeper understanding of the area law for entanglement entropy.

The area law bounds the entanglement that the ground states of a local Hamiltonian exhibit in a gapped system. More specifically, for a given subset S of particles, the area law bounds the entanglement entropy of the reduced density matrix of the ground state ρ_S that is restricted by the area of S, given by the number of local interactions between S and $\overline{S}$. MPS has a natural area law character and it is a good description for ground states of non-critical systems but a less efficient description for critical systems. In the valence-bond picture, for example, MPS obeys the area law by construction, where the entanglement entropy is based on individual valence bonds connecting the two subsystems.

Figure 2 shows how the entanglement entropy between two arbitrary subsystems A and B scales with the size of the boundary between the two subsystems.

The Von Neumann entanglement entropy of the quantum state is here defined as

$$S = -\text{Tr}(\rho_A \ln \rho_A), \tag{2}$$

The model probability distribution can be described as

$$\mathbb{P}(\boldsymbol{v}) = \frac{|\Psi(\boldsymbol{v})|^2}{Z}, \tag{3}$$

where $Z = \sum_{\boldsymbol{v} \in \mathcal{V}} |\Psi(\boldsymbol{v})|^2$ is the normalization factor.

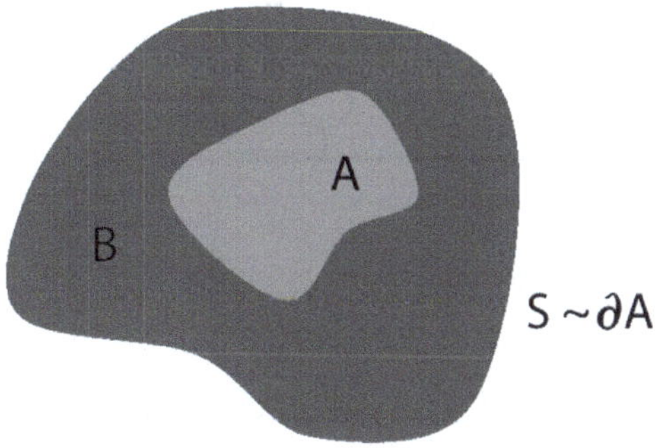

Fig. 2. The scaling of entanglement entropy in the area law

The RBM is composed of a set of visible $v = \{v_i\}$ and hidden $h = \{h_j\}$ binary variables coupled by the coupling matrix W_{ij} and satisfying a Boltzmann distribution with the energy functional

$$E(v, h) = -\sum_i a_i v_i - \sum_j b_j h_j - \sum_{i,j} v_i W_{ij} h_j, \tag{4}$$

It is important to mention the fact that while the hidden units interact one with another, there is no direct interaction between the visible units here.

The number of visible and hidden binary variables are n_v and n_h. The general ansatz of an RBM is shown in Fig. 3. If the number of hidden units (and interactions, consequently) is increased in the RBM architecture, the network will contain functions of higher complexity between the visible units. The RBM variational ansatz works here in any dimension. The biases applied to the visible and hidden units are a_i and b_j. The RBM can naturally generate an arbitrary probability distribution, if a large enough number of hidden nodes is present in the network.

Moreover, due to the RBM natural properties, the visible and hidden variables are conditionally independent given one-another. If the hidden units are integrated out, the training network becomes the marginal distribution of visible variables

$$\begin{aligned} \Psi_{\text{RBM}}(v) &= \sum_h e^{-E(v,h)} \\ &= \prod_i e^{a_i v_i} \prod_j \left(1 + e^{b_j + \sum_i v_i W_{ij}}\right). \end{aligned} \tag{5}$$

As a general rule, the number of parameters that can be learned has to be smaller than the number of input neurons of the neural network. As the number of parameters that can be learned is always polynomially dependent on the number of input neurons, the representation is efficient when the neural network represents the function or distribution in polynomial time.

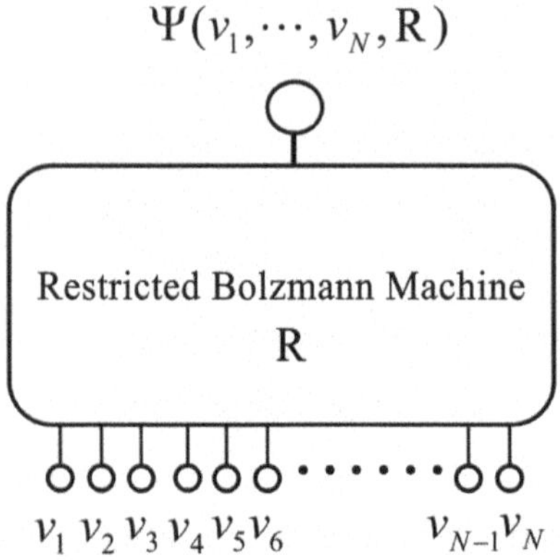

Fig. 3. Architecture of the RBM state

The network is equivalent to a system with N qubits (spin-1/2 subsystems) described by the local basis states $|v\rangle$, where $v \in \{0,1\}$ are the eigenstates of the z-Pauli operator $\hat{\sigma}^z |v\rangle = (-1)^v |v\rangle$. A many-body state of the system is

$$|\Psi\rangle = \sum_{\mathbf{v}} \Psi(\mathbf{v}) |\mathbf{v}\rangle , \tag{6}$$

where $\mathbf{v} = (v_1, v_2, \ldots, v_N)^{\mathrm{T}} \in \{0,1\}^N$ is a bit string with the configuration basis state $|\mathbf{v}\rangle$ and $\Psi(\mathbf{v})$ the associated complex amplitude.

After the MPS wavefunction is chosen, the learning process is done by adjusting the parameters of the wave function until the distribution becomes as close as possible to the data distribution. For example, the Maximum Likelihood Estimation algorithm describes a negative log-likelihood function and optimizes it by adjusting the parameters of the model. The log-likelihood in the logistic regression will have a loss function as being the negative log-likelihood. The task is to minimise the loss function, which is done using a gradient descent algorithm. By contracting the tensor network associated with the model, the gradient of the loss can be computed.

At zero temperature, the quantum many-body problem [19–21] performs in two stages: (i) find the ground state and low-lying excitations of the Hamiltonian $\hat{H}$, and (ii) time-evolve an initial state associated with the Hamiltonian (time dependent or not). The As the quantum state possesses 2^N amplitudes $\Psi(\mathbf{v})$, the quantum many-body problem reveals a 'curse of dimensionality'. The amplitudes $\Psi(\mathbf{v})$ are represented by an order-N tensor $\Psi_{v_1 v_2 \cdots v_N}$. The main idea behind tensor networks is the decomposition of this tensor $\Psi_{v_1 v_2 \cdots v_N}$ into a network of lower order tensors.

In an intuitive approach, the network is represented by a graph G where each vertex ν has an associated tensor $T^{(\nu)}$ with a small number of indices, each of dimension at most χ, with also possible physical indices v_j of dimension 2.

The edges of G describe how the internal legs of every tensor are contracted. The contraction process represents a generalisation of the matrix multiplication. A tensor network decomposition can be represented as

$$\Psi_{v_1 v_2 \cdots v_N} = \mathrm{tTr}\left[\otimes_{\nu \in G} T^{(\nu)}\right], \tag{7}$$

where tTr is the tensor trace performing the contractions of the internal indices described by G, with N open physical indices v_j. If χ scales exponentially with N, the tensor network decomposition describes any possible state $|\Psi\rangle$. The MPS network contains a series of tensors contracted together where the open vertical legs represent the physical indices corresponding to the lattice sites of the many-body system. The internal indices that are contracted between the tensors corresponds to the entanglement between different parts of the many-body system represented by the tensors.

Furthermore, the dimension χ of the internal indices describes how much entanglement can be modelled by the network and it is illustrated by the entanglement area-law. Even for a small χ many networks describe efficiently and accurately highly compressed physical states.

For MPS models, the network essentially retains the same geometry of the underlying physical system that tries to depict, for example a chain or lattice. If the system has a coordination number $\mathcal{Z}$, the network is built from tensors with $\mathcal{Z}$ internal indices and one physical index. The RBM can be translated into an MPS by representing visible and hidden variables as physical and virtual units in the same network structure.

The process of tensor network decomposition requires to variationally minimize the tensor elements, starting with finding the expectation values $\langle\psi|\,\hat{h}\,|\psi\rangle$, using the product operator $\hat{h}$ in the Hamiltonian $\hat{H}$. This demands a contraction of the tensor network for $\langle\psi|\,\hat{h}\,|\psi\rangle$. The 1D chain intrinsic geometry makes possible the exact contraction of an MPS.

A wavefunction of n_v physical variables is represented by the MPS as

$$\Psi_{\mathrm{MPS}}(v) = \mathrm{Tr}\prod_i A^{(i)}[v_i], \tag{8}$$

where $A^{(i)}$ is a three-index tensor. For any v_i, $A^{(i)}[v_i]$ is represented by a matrix, whose dimension is the virtual bond dimension of the MPS.

In a computationally efficient description of the quantum state of a many-body system, the final representation of the wavefunction in terms of a an MPS depends on a polynomial number of parameters. If the bond dimension is increased, the MPS can better describe complex multivariable functions.

In fact, the power of an RBM depends on finding an MPS representation [22–24] with the smallest possible bond dimensions. If the network is split into n_v sites containing at least a visible variable, the contraction of the network for every site translates into merging all hidden variables. The geometry of the hidden units with respect to the visible units is arbitrary. If the bond dimensions are trivial, then there is no entanglement present in the wavefunction, and the tensor network state is just a product state, as in mean field theory.

We introduce a diagonal tensor, $\Lambda_v^{(i)}$ or $\Lambda_h^{(j)}$, for every visible or hidden site and a 2×2 matrix $M^{(ij)}$ at every bond that connects each v_i to h_j

$$\Lambda_v^{(i)} = \mathrm{diag}\left(1, e^{a_i}\right), \tag{9}$$

$$\Lambda_h^{(j)} = \mathrm{diag}\left(1, e^{b_j}\right), \tag{10}$$

$$M^{(ij)} = \begin{pmatrix} 1 & 1 \\ 1 & e^{W_{ij}} \end{pmatrix}. \tag{11}$$

In this way, an MPS is generated by merging the links between sites onto virtual bonds. Overall, the MPS will posses a bond dimension that is fully dependent on the thickness of the virtual bonds and consequently on the total number of the merged links. If we have enough information about the MPS and its entanglement properties, the RBM can be efficiently generated.

In order to describe the entanglement entropy of the system, we need to divide its visible variable into two subsystems, X and Y. The entanglement entropy of a function Ψ between the two subsystems is

$$S = -\mathrm{Tr}(\rho \ln \rho), \tag{12}$$

where ρ is the reduced density matrix defined by

$$\rho = \sum_{v_Y} \Psi^* \left(v'_X, v_Y\right) \Psi \left(v_X, v_Y\right). \tag{13}$$

The entanglement entropy describing all the information from Ψ represents the short range correlations between the degrees of freedom from X and Y. v_Y contains all the visible variables in Y. The entanglement entropy depends only on the size of the boundary between the subsystems X and Y and it is bounded by the logarithm of the bond dimension, $\ln D$. If X and Y are not entangled, the entanglement entropy is zero. The maximal entanglement entropy between X and Y sites depends on the bond dimension of the MPS.

On the other hand, the RBM can be described [25,26] by the connections between visible units via the hidden units. The geometrical structure of the hidden variables with respect to the visible ones is arbitrarily chosen. In analogy with the MPS method [27,28], we introduce a boundary between the visible variables, splitting the system in two sites, X and $Y (= Y_1 \cup Y_2)$. The boundary Y_1 contains all visible variables connected to X, while Y_2 contains the rest of the variables.

The RBM function is

$$\Psi_{\mathrm{RBM}}(v) = \psi\left(v_X, v_{Y_1}\right) \phi\left(v_{Y_1}, v_{Y_2}\right). \tag{14}$$

If the visible variables from Y_1 are fixed, the RBM is a direct product of the visible variables from X and Y_2. The entanglement entropy between X and Y is given by the total number of visible variables at the boundary of Y_1, called $|Y_1|$. The MPS bond dimension becomes $D = 2^{|Y_1|}$.

We can also separate the region X into two sites $X = X_2 \cup X_1$ containing all variables connected to Y in X_1 and the rest of the variables in X_2. The entanglement entropy will be now bounded by the number of variables in X_1 and consequently by $S_{\max} = \min(|X_1|, |Y_1|) \ln 2$.

Most interestingly, any probability distribution of a N-bit system can be represented by an MPS as long as its bond dimensions are free from restrictions. As the bond dimension increases, the MPS improves its capacity to parametrize more complicated functions.

For a number of variables m, the entanglement entropy becomes

$$S_{\max} \sim mV^{(d-1)/d}, \tag{15}$$

where d is the spatial dimension, and V is the volume of the system.

More generally, the entanglement entropy of an RBM with sparse degrees of freedom follows the area law, while the entropy for dense RBMs scales with $m \sim V^{1/d}$ and consequently $S_{\max} \sim V$ grows linearly with the volume. As the number of parameters for a dense RBM scales polynomially with the full size of the system, these RBMs can describe highly entangled quantum states where the area law is not satisfied.

The connecting bond indices (or ancillary indices) describe the structure of the many-body entanglement in the quantum state. If the bond dimension is changed, the multiplicative factor of the area law also changes. If we want to modify the scaling with the boundary size, the geometric pattern of the tensor network has to change. The entanglement in the tensor network is a consequence of both of the bond dimension and the geometric pattern of the interconnection between bond indices. The number of the values that any of the bond indices can take is a measure of the amount of the quantum correlations encoded in the wavefunction.

The correlation functions of the MPS decay exponentially with the correlation distance. The correlation length of MPS states is finite, therefore MPS can't fully describe the properties of critical or scale-invariant systems, where the correlation length diverges.

The condition for an MPS to posses an RBM representation is that each tensor has the form

$$A_{\alpha\beta}[v] = L_{\alpha v} R_{v\beta}, \tag{16}$$

where L and R are two 2×2 matrices. The product of R and L is described by the coupling to the hidden RBM variables. The product RL will take the symmetric form

$$\begin{aligned} RL = &\begin{pmatrix} 1 & \\ & e^{a/2} \end{pmatrix} \begin{pmatrix} 1 & 1 \\ 1 & e^{W} \end{pmatrix} \begin{pmatrix} 1 & \\ & e^{b} \end{pmatrix} \\ &\begin{pmatrix} 1 & 1 \\ 1 & e^{W} \end{pmatrix} \begin{pmatrix} 1 & \\ & e^{a/2} \end{pmatrix}. \end{aligned} \tag{17}$$

As an MPS representation has gauge degrees of freedom, this this makes possible to restrict the tensors with canonical conditions, with direct benefits on computing an exact partition function. As an example, the partition function of the statistical Ising model with $s_i = \pm 1$ Ising spins is

$$Z = \sum_{\{s_i\}} \exp\left(K \sum_{\langle i,j\rangle} s_i s_j + H \sum_i s_i\right) \tag{18}$$

where K is the coupling constant, H the external field and the binary variables are defined as $v_i = (s_i + 1)/2$.

The wavefunction can be complex valued, but here we restrict it to be real valued. The partition function can be represented as a summation of the RBM function. In more physical terms, the RBM amplitudes can be translated into the effective action for physical spins which is obtained by tracing out a bath of hidden spin variables.

For the one dimensional Ising case, the partition function can be represented as an MPS. The matrix product on each bond is

$$RL = \begin{pmatrix} e^{K+H} & e^{-K} \\ e^{-K} & e^{K-H} \end{pmatrix}. \tag{19}$$

3 Conclusions

We analyze here the connection between unsupervised RBM methods and MPS tensor network tools, presenting an efficient algorithm to map an RBM into an MPS and provide an RBM variational ansatz for many-body quantum systems with a quantum entanglement view on learning of probabilistic models.

The organic connection between these parallel methods opens new doors into novel translating methods that can be developed and applied to quantum and many body physics in general, enabling learning of complex multi-variate information from raw data. We here find the conditions for implementing an RBM representation mapped to an MPS, and reveal an upper bound for the entanglement entropy of a quantum system represented as an parametrized RBM.

We studied the equivalence of several significant features and the entanglement properties for RBM and MPS, with the goald to quantify the strong potential of RBM techniques to help understanding the nature of entanglement of quantum many-body states in a classical context.

The significant crossover and great structural similarities between classically simulable tensor networks and quantum algorithms bridge several current numerical approaches, including tensor networks, the variational quantum eigensolver, quantum approximate optimization and quantum computation. RBM tensor networks generate quality results for large sets of learning tasks including classical resources and no further requirement of dedicated quantum hardware.

The bridge between RBMs and MPS models has the great potential to employ such tools in the near future for highly complex operations that may be efficiently executed at a classical level, and further accelerate the development of hybrid quantum/classical algorithms together with the use of classical algorithms on quantum hardware.

References

1. LeCun, Y., Bengio, Y., Hinton, G.: Deep learning. Nature **521**, 436–444 (2015)
2. Howard, E.M.: Machine learning algorithms in Astronomy. In: Astronomical Data Analysis Software and Systems XXV, Astronomical Data Analysis Software and Systems ADASS XXV, vol. 512, p. 245 (2017)
3. Le Roux, N., Bengio, Y.: Representational power of restricted Boltzmann machines and deep belief networks. Neural Comput. **20**, 1631–1649 (2008)
4. Carleo, G., Troyer, M.: Solving the quantum many-body problem with artificial neural networks. Science **355**, 602 (2017)
5. Nomura, Y., Darmawan, A.S., Yamaji, Y., Imada, M.: Restricted Boltzmann machine learning for solving strongly correlated quantum systems. Phys. Rev. B **96**, 205152 (2017)
6. Saito, H., Kato, M.: Machine learning technique to find quantum many-body ground states of bosons on a lattice. J. Phys. Soc. Jpn. **87**, 014001 (2018)
7. Amico, L., Fazio, R., Osterloh, A., Vedral, V.: Entanglement in many-body systems. Rev. Mod. Phys. **80**, 517 (2008)
8. Gao, X., Duan, L.-M.: Efficient representation of quantum many-body states with deep neural networks. Nat. Commun. **8**, 662 (2017)
9. Vidal, G.: Class of quantum many-body states that can be efficiently simulated. Phys. Rev. Lett. **101**, 110501 (2008)
10. Verstraete, F., Murg, V., Cirac, J.I.: Matrix product states, projected entangled pair states, and variational renormalization group methods for quantum spin systems. Adv. Phys. **57**(2), 143–224 (2008)
11. Howard, E.: Holographic renormalization with machine learning arXiv:1803.11056 [physics.gen-ph] (2018)
12. Tagliacozzo, L., Evenbly, G., Vidal, G.: Simulation of two-dimensional quantum systems using a tree tensor network that exploits the entropic area law. Phys. Rev. B **80**, 235127 (2009)
13. Hastings, M.B.: An area law for one-dimensional quantum systems. J. Stat. Mech. **2007**, P08024 (2007)
14. Vidal, G., Latorre, J.I., Rico, E., Kitaev, A.: Entanglement in quantum critical phenomena. Phys. Rev. Lett **90**, 227902 (2003)
15. Carleo, G., Nomura, Y., Imada, M.: Constructing exact representations of quantum many-body systems with deep neural networks. arXiv:1802.09558 (2018)
16. Chen, J., Cheng, S., Xie, H., Wang, L., Xiang, T.: Equivalence of restricted Boltzmann machines and tensor network states. Phys. Rev. B **97**, 085104 (2018)
17. Bottou, L.: Large-scale machine learning with stochastic gradient descent. In: Proceedings of the 19th International Conference on Computational Statistics, pp. 177–187 (2010)
18. Kingma, D.P., Ba, J.L.: Adam: a method for stochastic optimization arXiv:1412.6980 (2014)
19. Eisert, J., Cramer, M., Plenio, M.B.: Colloquium: area laws for the entanglement entropy. Rev. Mod. Phys. **82**, 277 (2010)
20. McCulloch, I.P.: Infinite size density matrix renormalization group, revisited, arXiv:0804.2509 (2008)
21. Vidal, G.: Efficient simulation of one-dimensional quantum many-body systems. Phys. Rev. Lett. **93**(4), 040502 (2004)
22. Jastrow, R.: Many-Body Problem with Strong Forces. Phys. Rev. **98**(5), 1479–1484 (1955)

23. Gutzwiller, M.C.: Correlation of electrons in a narrow s band. Phys. Rev. **137**(6A), A1726–A1735 (1965)
24. Huang, Y., Moore, J.E.: Neural network representation of tensor network and chiral states arXiv:1701.06246 (2017)
25. Clark, S.R.: Unifying neural-network quantum states and correlator product states via tensor networks. J. Phys. A: Math. Theor. **51**, 135301 (2018)
26. Bottou, L., Curtis, F.E., Nocedal, J.: Optimization methods for large-scale machine learning. Siam Rev. **60**(2), 223–311 (2018)
27. Carrasquilla, J., Melko, R.G.: Machine learning phases of matter. Nat. Phys. **13**, 431–434 (2017)
28. Wang, L.: Discovering phase transitions with unsupervised learning. Phys. Rev. B **94**, 195105 (2016)
29. Deng, D.-L., Li, X., Sarma, S.D.: Quantum entanglement in neural network states. Phys. Rev. X **7**, 021021 (2017)
30. Deng, D.-L., Li, X., Sarma, S.D.: Machine learning topological states. Phys. Rev. B **96**, 195145 (2017)

Improved Kangaroo Mob Optimization and Logistic Regression for Smart Grid Stability Classification

Dorin Moldovan(✉)

Faculty of Automation and Computer Science, Technical University of Cluj-Napoca, Cluj-Napoca, Romania
dorin.moldovan@cs.utcluj.ro

Abstract. The paper presents an improved version of the Kangaroo Mob Optimization (KMO) which is further called Improved Kangaroo Mob Optimization (IKMO). IKMO modifies the standard equations for the updating of the positions of the kangaroos to avoid the premature convergence and to explore the search space more. IKMO is validated using 9 objective functions and is compared to other bio-inspired algorithms such as Horse Optimization Algorithm (HOA), Chicken Swarm Optimization (CSO), Particle Swarm Optimization (PSO), Grey Wolf Optimizer (GWO), and Cat Swarm Optimization Algorithm (CSOA). Finally, the article applies IKMO in the tuning of the hyperparameters of a Logistic Regression (LR) model for the classification of the stability of a smart grid using the Electrical Grid Stability Simulated Dataset from the UCI Machine Learning Repository as experimental support.

Keywords: Improved Kangaroo Mob Optimization · Swarm intelligence · Smart grid · Logistic regression · Machine learning

1 Introduction

The bio-inspired computing research field evolved a lot in the last years and the sources of inspiration of the bio-inspired algorithms are various ranging from the bats behavior in nature like in the case of the Bat Algorithm (BA) [1] to the mating patterns of the naked mole-rats [2] like in the case of the Naked Mole-Rat (NMR) algorithm.

However, the original versions of the bio-inspired algorithms are sometimes not good enough for specific types of optimization problems. Possible solutions to address this challenge are the hybridization of two or more bio-inspired algorithms like in the case of the approach presented in [3] where the authors consider the hybridization of Genetic Algorithm (GA) [4] with Particle Swarm Optimization (PSO) [5] for the forecasting of the electricity demand, or the improvement of the original version of the bio-inspired algorithm as can be seen in [6] where the authors propose an improved version of Chicken Swarm Optimization (CSO) [7,8] for the prediction of the short-term wind power.

© The Author(s), under exclusive license to Springer Nature Switzerland AG 2021
R. Silhavy (Ed.): CSOC 2021, LNNS 229, pp. 469–487, 2021.
https://doi.org/10.1007/978-3-030-77445-5_44

Motivated by the previous research background in the application of bio-inspired algorithms for the improvement of the performance of the machine learning algorithms [9] considered in the classification of the smart grid stability, this paper proposes an improved version of the original Kangaroo Mob Optimization (KMO) algorithm [10] which targets the better exploration of the search space and the avoidance of the premature convergence.

The contributions of the paper are the following:

(1) The improvement of the KMO algorithm to a new form called IKMO;
(2) The introduction of a novel objective function for the evaluation of the bio-inspired algorithms;
(2) The comparison of the IKMO algorithm to other bio-inspired algorithms from the research literature using 9 objective functions;
(3) The application of IKMO in the tuning of a Logistic Regression (LR) model applied in the classification of the stability of a smart grid;

The paper has the following structure: Section 2 presents the related work, Sect. 3 presents the IKMO algorithm, Sect. 4 presents the IKMO evaluation, Sect. 5 presents the IKMO LR method for the classification of the smart grid stability, and Sect. 6 presents the conclusions.

2 Related Work

The first part presents bio-inspired algorithms for the tuning of machine learning models, while the second part presents approaches for the classification of the smart grid stability.

2.1 Bio-Inspired Algorithms for Machine Learning Models Tuning

Next are presented several approaches which consider bio-inspired algorithms for the tuning of the hyperparameters of the machine learning models.

The authors of [11] considered the application of a hybrid chaotic quantum BA for the tuning of three parameters of a SVMR considered in the electric load forecasting. The three tuned parameters are the penalty parameter (C), the Radial Basis Function (RBF) kernel parameter (σ) and the parameter of the ϵ-insensitive loss function (ϵ). The drawbacks of the BA algorithm approached by the authors are the premature convergence and the trapping in local optima.

Another approach, the one presented in [12], considers the application of a modified version of the Grey Wolf Optimizer (GWO) [13] called MGWO for the optimization of the kernel parameter (σ) and the kernel function (γ) of a Least Squares Support Vector Machine (LSSVM) which is used for predicting the dissolved gas concentration. Compared to other methods such as PSO-LSSVM and GWO-LSSVM, the MGWO-LSSVM shows better results in terms of performance metrics.

The authors of [14] considered the application of PSO for the optimization of the performance of a neural network (NN) considered in the prediction of the energy consumption. The PSO is used for the finding of the weights and biases that lead to the best classification results. The authors proposed also a re-generation based modification and a velocity boost-based modification in order to increase the prediction accuracy.

In [15] the GA was applied to optimize a Kernel Extreme Learning Machine (KELM) considered in the electrocardiogram (ECG) detection. The optimized parameters were the regularization parameter, the kernel parameter and the kernel type. The approach proposed in that article showed promising classification results for the arrhythmia database from the UCI repository.

2.2 Smart Grid Stability Classification

The classification of the smart grid stability using as experimental support the Electrical Grid Stability Simulated Data Data Set [16,17] was approached in the previous research work from different perspectives.

In the approach presented in [18], the features of the input data were extracted first using the Feature Extraction on basis of Scalable Hypothesis tests (FRESH) algorithm [19], then the features were selected using a binary version of the Horse Optimization Algorithm (HOA), and finally the data was classified using the Random Forest (RF) classification algorithm. A similar approach in which the features of the data were extracted prior to the application of the classifier was considered in [20], but in that case no features selection algorithm was applied after the extraction of the features and the dataset considered in the experiments was from another research domain.

A similar approach is the one presented in [9] where a methodology which integrates a features extraction a phase, a features selection phase, and a system stability prediction phase is presented. In that approach the KMO was used in the features selection phase in a binary version. The applied classifiers were LR, RF, Gradient Boosted Trees (GBT), and Multilayer Perceptron Classifier (MPC). Compared to other features selection techniques such as the one based on Multivariate Adaptive Regression Splines (MARS) [21] and a binary version of PSO, the binary version of KMO returned the best results in terms of accuracy for the LR and GBT classifiers.

The authors of [22] applied a Multidirectional LSTM model for the prediction of the smart grid stability. The model proposed by the authors resulted in a training accuracy equal to 99.07% and a testing accuracy equal to 99.07%. These values are much better than the values obtained in our previous research work. However, the LSTM model is more complex than the standard machine learning algorithms. Moreover, the experiments presented in that article were performed using an online Graphical Processing Unit (GPU) provided by Google.

3 IKMO

Figure 1 presents the flowchart of IKMO.

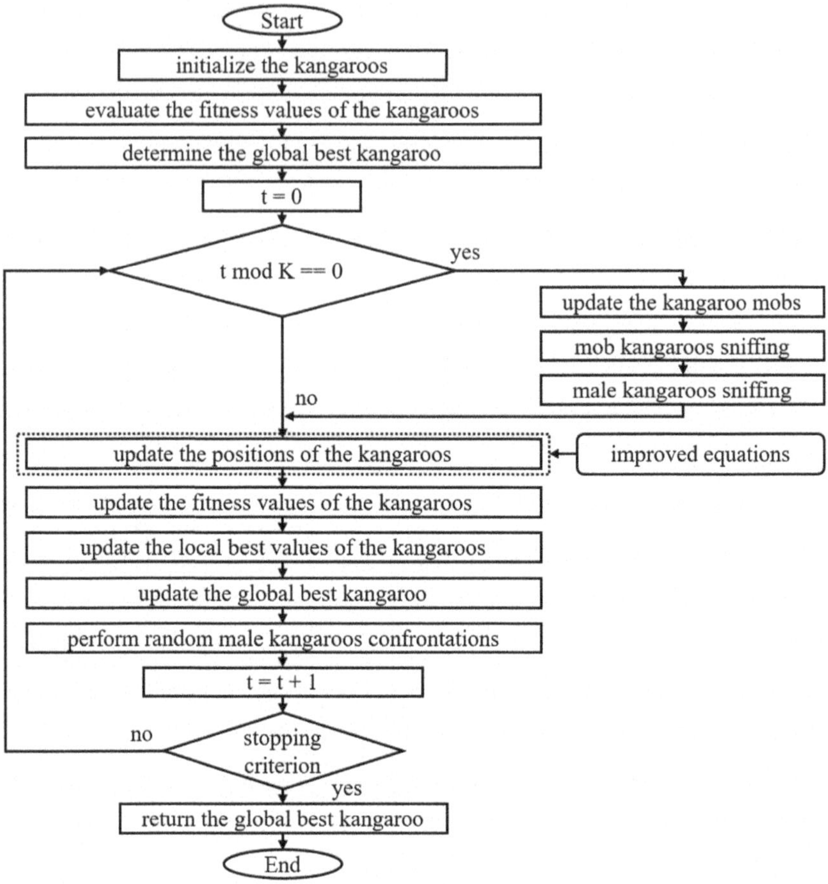

Fig. 1. Flowchart of IKMO

The steps of IKMO are presented next.

Step 1. A population of $N_{kangaroos}$ kangaroos is initialized in a search space with D dimensions. The characteristics of the kangaroos are initialized as follows:

- **position:** a D-dimensional array with random values from $[P_{min}, P_{max}]$ such that P_{min} is the minimum possible value of the position and P_{max} is the maximum possible value of the position;
- **velocity:** a D-dimensional array with random values from $\{1, 1.5, 2\}$;
- **gender:** a random value from $\{male, female\}$;
- **strength:** a random integer from $\{S_{min}, S_{min} + 1, ..., S_{max}\}$;
- **Lbest:** the local best value of the position of the kangaroo, a value equal to the initial position of the kangaroo;

Step 2. The fitness value of each kangaroo is evaluated using an Objective Function (OF).

Step 3. The global best kangaroo is selected as the one for which the fitness value is minimum.

Step 4. The value of the current iteration t is initialized to 0.

If the value of the current iteration t modulo K is equal to 0, where K is the mob updating frequency, then the algorithm continues with **Step 5**. Otherwise, the algorithm continues with **Step 8**.

Step 5. The kangaroos are distributed randomly in M mobs. The velocity of each kangaroo is updated to a new D-dimensional array with values from $\{1, 1.5, 2\}$ selected randomly.

The male kangaroos and the female kangaroos of each mob are sorted in decreasing order by strength value obtaining two sorted lists and are organized as follows:

- the male kangaroo with the highest male strength value is single;
- the female kangaroo with the highest female strength value is coupled with the male kangaroo with the second highest male strength value;
- the female with the second highest female strength value is coupled with the male kangaroo with the third highest male strength value;
- and so on while it is still possible to couple the kangaroos in pairs;
- the remaining kangaroos are left uncoupled;

Figure 2 presents an example of kangaroos mob which contains 3 kangaroo pairs and 2 single male kangaroos. For illustrative purposes, the strength of each kangaroo is proportional with its dimension.

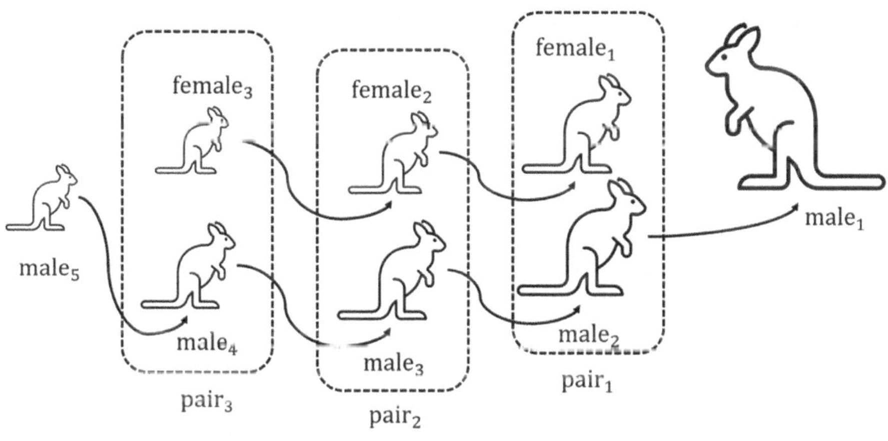

Fig. 2. Kangaroos mob illustrative example

Step 6. Mob kangaroos sniffing is performed. During this operation, the kangaroos of each mob update the *Lbest* value to the best *Lbest* value among the *Lbest* values of the kangaroos from the mob they belong to.

Figure 3 presents an example for the mob kangaroos sniffing operation for a mob with 3 kangaroos.

Fig. 3. Mob kangaroos sniffing operation illustrative example

Step 7. The algorithm continues with the male kangaroos sniffing operation in which the value of the strength of each male kangaroo is incremented with a value equal to the number of male kangaroos from the mob it belongs to and which have the strength value less than its strength value. If the new strength value is greater than S_{max} then the new strength value is updated to S_{max}.

Figure 4 shows an example of male kangaroos sniffing operation for a mob with 6 male kangaroos such that the number of plus symbols of each kangaroo describes the number of kangaroos with the strength less than its strength.

Fig. 4. Male kangaroos sniffing operation illustrative example

Step 8. The positions of the kangaroos are updated considering characteristics such as: the current position, the gender, the position of the partner if there exists one, the status (single or coupled), and the *Lbest* value.

IKMO algorithm introduces two modifications:

- **modification 1:** a random factor from $[0, 1]$ is included for the kangaroos that lost a confrontation in the previous iteration to explore the search space more;
- **modification 2:** a random factor from $[0, 1]$ is included in the formula for the updating of the positions of the kangaroos which corresponds to the part that describes the movement towards the *Lbest* value to avoid the premature convergence of the algorithm to a local optimum;

There are five equations for the updating of the positions of the kangaroos considering the following criteria: the gender (male/female), the status (single/coupled) and the losing of a confrontation in the previous iteration (busy/idle). The third criterion is applicable only in the case of the male kangaroos such that busy means that the kangaroo didn't lose a confrontation, while idle means that the kangaroo lost a confrontation in the previous iteration.

The modified equations for the updating of the positions of the kangaroos are presented next for each category of kangaroos.

- idle male kangaroos:

$$x^{t+1} = x^t \times r_1 \tag{1}$$

- single busy male kangaroos:

$$x^{t+1} = x^t + x^t \times r_1 \times \mathcal{N}(0, sd) + v^t \times r_2 \times \left(x^t_{female} - x^t\right) \tag{2}$$

- coupled busy male kangaroos:

$$x^{t+1} = x^t + x^t \times r_1 \times \mathcal{N}(0, sd) + v^t \times r_2 \times \left(x^t_{partner} - x^t\right) \tag{3}$$

- single female kangaroos:

$$x^{t+1} = x^t + x^t \times r_1 \times \mathcal{N}(0, sd) \tag{4}$$

- coupled female kangaroos:

$$x^{t+1} = x^t + x^t \times r_1 \times \mathcal{N}(0, sd) + v^t \times r_2 \times \left(x^t_{partner} - x^t\right) \tag{5}$$

such that x^{t+1} is the value of the position in the new iteration $t + 1$, x^t is the value of the position in the current iteration t, v^t is the value of the velocity in the current iteration t, r_1 and r_2 are random numbers from $[0, 1]$, $x^t_{partner}$ is the position of the partner in the iteration t, and x^t_{female} is the position of the female of the couple from the same mob as the kangaroo for which the position is updated such that the strength of the male partner is: (1) less than the strength of the kangaroo for which the position is updated and (2) better than the strength of the other coupled male kangaroos from the mob which have the strength less than the strength of the kangaroo for which the position is updated.

In Eqs. (2), (3), (4), and (5), $\mathcal{N}(0, sd)$ is a normal distribution with the mean 0 and the standard deviation sd defined by the equation:

$$sd = exp\left(-\frac{OF(x^t) - OF(Lbest)}{OF(x^t) + \epsilon}\right) \tag{6}$$

where ϵ is a constant used for avoiding division by 0.

The newly introduced random factor r_1 from the Eqs. (1), (2), (3), (4), (5) is underlined with two lines, while the movement towards the female is underlined with a dotted line, as can be seen in Eq. (2), because for a part of the kangaroos such a female might not exist.

Figure 5 presents an illustrative example for the updating of the positions of the kangaroos for the mob which was also presented as illustrative example in Fig. 2.

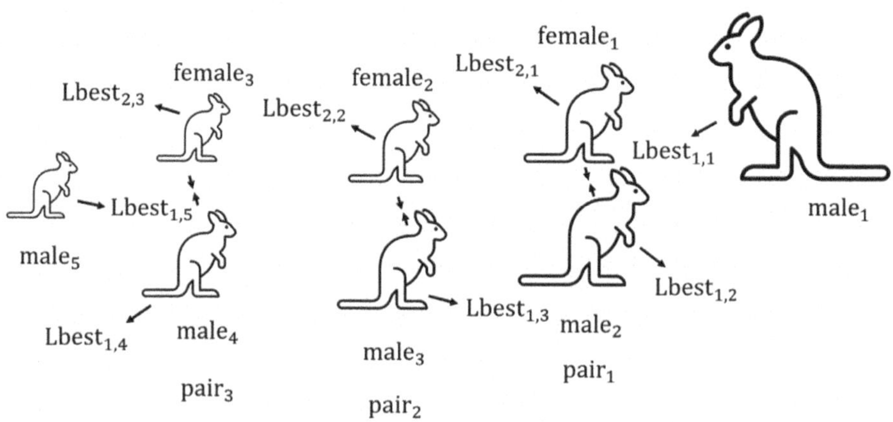

Fig. 5. Kangaroos positions updating inside a mob illustrative example

Step 9. The fitness values of the kangaroos are updated using the objective function OF.

Step 10. For each kangaroo, if the new fitness value is better than the old fitness value then its *Lbest* value is updated to the value of its new position.

Step 11. The global best kangaroo is updated to the kangaroo with the best fitness value.

Step 12. In each mob, two male kangaroos are selected randomly and if the absolute value of the difference between their strength values is less than or equal to a confrontation threshold (CT) then the strengths of the two kangaroos are updated as follows:

- the strength of the kangaroo with smallest strength value is decremented with 1 and is further updated to S_{min} if the new strength value is less than S_{min};

– the strength of the kangaroo with highest strength value is incremented with 1 and is further updated to S_{max} if the new strength value is greater than S_{max};

Figure 6 presents an illustrative example of kangaroos confrontation operation such that the strength of each kangaroo is represented by a certain number of stars. The previous strength values of the two kangaroos are 3 and 5 and new strength values are 2 and 6, respectively.

Fig. 6. Male kangaroos confrontation operation illustrative example

Step 13. The value of the current iteration t is incremented with 1.

If the stopping criterion is met, which in general happens when the current iteration t is equal to the number of iterations ($N_{iterations}$), then the algorithm continues with **Step 14**. Otherwise, the algorithm checks again if the current iteration t modulo K is 0 and continues either with **Step 5** or with **Step 8**.

Step 14. The algorithm returns the global best kangaroo.

4 IKMO Evaluation

Table 1 presents the objective functions considered in the evaluation and validation of IKMO. A part of the objective functions were used in [13], while the last objective function (OF_9) is a new objective function as far to our knowledge. For each objective function the table presents the mathematical formula, the range, the number of dimensions (Dim) and the best possible value.

The performance of IKMO is compared with performance of HOA, CSO, PSO, GWO, and Cat Swarm Optimization Algorithm (CSOA) [23,24], respectively. The number of individuals $N_{individuals}$ and the number of iterations $N_{iterations}$ were set to 100 and 2500 for each algorithm, respectively. For the other parameters, the following values were considered:

– IKMO: The number of kangaroo mobs (M) was set to 10. The frequency of the kangaroos mob updating (K) was set to 10. The values for the strength minimum value and strength maximum value (S_{min} and S_{max}) were set to 0 and 100, respectively. The value of the confrontation threshold parameter (CT) was set to 10, and the constant for avoiding division by zero (ϵ) was set to 10^{-9}.

Table 1. Objective functions for IKMO evaluation

OF	Mathematical Formula	Range	Dim	Best
OF_1	$\sum_{i=1}^{N} x_i^2$	$[-100, 100]$	30	0
OF_2	$\sum_{i=1}^{N} i \times x_i^4 + random[0, 1)$	$[-1.28, 1.28]$	30	0
OF_3	$max(\|x_i\|, 1 \leq i \leq N)$	$[-100, 100]$	30	0
OF_4	$\sum_{i=1}^{N-1} \left[100 \times \left(x_{i+1} - x_i^2\right)^2 + (x_i - 1)^2\right]$	$[-30, 30]$	30	0
OF_5	$\sum_{i=1}^{N} \left[x_i^2 - 10 \times cos(2\pi x_i) + 10\right]$	$[-5.12, 5.12]$	30	0
OF_6	$\sum_{i=1}^{N} \|x_i\| + \prod_{i=1}^{N} \|x_i\|$	$[-10, 10]$	30	0
OF_7	$\frac{1}{4000} \times \sum_{i=1}^{N} x_i^2 - \prod_{i=1}^{N} cos\left(\frac{x_i}{\sqrt{i}}\right) + 1$	$[-600, 600]$	30	0
OF_8	$\sum_{i=1}^{N} \left(\sum_{j=1}^{i} x_j\right)^2$	$[-100, 100]$	30	0
OF_9	$\frac{1}{2000} \times \sum_{i=1}^{N} x_i^2 - \prod_{i=1}^{N} cos^4\left(\frac{x_i^2}{i}\right) + 1$	$[-300, 300]$	30	0

- HOA: The default values from the original article which introduces HOA [18] were considered.
- CSO: The CSO configuration parameters values which were also used for the CSO algorithm considered in the testing of HOA [18] were used in experiments.
- PSO: The minimum and the maximum values of the velocity (V_{min} and V_{max}) were set to -10000 and 10000, respectively. The inertia value (w) was set to 0.5. The values for the cognitive acceleration coefficient (c_1) and the social acceleration coefficient (c_2) were both set to 2.
- GWO: The default value of the control parameter (a) from the original article which introduces GWO [13] was considered.
- CSOA: The count of dimensions to change (CDC) was set to 24 and the other configuration parameters values were set to the values used for the CSOA algorithm considered in the testing of HOA [18].

Each algorithm was run 10 times for each objective function. Table 2 summarizes the mean of the results, the standard deviation (std) of the results, the best result and the worst result for each objective function for each algorithm considered in the evaluation of IKMO.

As can be seen in the table, the performance of IKMO is comparable with the performance of other bio-inspired algorithms for a part of the objective functions. For example, IKMO returns the best result for the objective functions OF_5, OF_7, and OF_9. Another observation is that for each objective function there is at least another bio-inspired algorithm which returns worse results than IKMO.

Next, Fig. 7 compares the evolution of the best runs of each bio-inspired algorithm for each objective function.

As can be seen in the figure, IKMO, HOA, CSO and GWO converge to an optimal region of the search space in the initial iterations, CSOA explores the search space more in the initial iterations, and PSO converges to an optimal region of the search space in the final iterations.

OF	Result	IKMO	HOA	CSO	PSO	GWO	CSOA
OF_1	mean	2.7E–24	0.00692	0	2.0E–24	4E–207	0.00050
	std	6.0E–24	0.00137	0	6.1E–24	0	0.00020
	best	8E–136	0.00556	0	1.5E–26	5E–213	0.00019
	worst	1.7E–23	0.00987	0	1.9E–23	3E–206	0.00093
OF_2	mean	5.8E–05	3.8E–04	0	1.63631	3.8E–06	1.1E–09
	std	5.1E–05	2.0E–04	0	1.28136	2.4E–06	3.0E–10
	best	5.7E–06	1.5E–04	0	0.41164	7.7E–07	6.0E–10
	worst	0.00014	7.8E–04	0	4.46138	9.1E–06	1.5E–09
OF_3	mean	3.3E–10	0.01695	5.6E–45	1.56608	1.2E–50	0.00828
	std	1.0E–09	0.00416	0	0.87239	2.0E–50	0.00337
	best	1.5E–48	0.01016	5.6E–45	0.50907	4.7E–54	0.00428
	worst	3.2E–09	2.3E–02	5.6E–45	3.28612	6.1E–50	1.3E–02
OF_4	mean	28.7249	29.4095	28.9482	27138.3	27.0818	28.9483
	std	0.15958	0.08546	0.02141	43406.8	0.52290	0.02551
	best	28.3127	29.2779	28.9133	25.5375	26.2374	28.9120
	worst	28.8810	29.5454	28.9828	90072.9	28.0534	28.9990
OF_5	mean	0	1.02194	0	62.5209	0	0.00157
	std	0	0.20228	0	30.5344	0	0.00082
	best	0	0.77583	0	7.69815	0	0.00021
	worst	0	1.36581	0	118.541	0	2.8E–03
OF_6	mean	4.2E–13	0.31866	1.5E–43	22	2E–119	0.01497
	std	8.9E–13	0.01978	1.9E–44	13.9841	6E–119	0.00201
	best	4.5E–67	0.27817	1.0E–43	4.5E–17	1E–120	0.01250
	worst	2.3E–12	0.34159	1.6E–43	50	2E–118	1.8E–02
OF_7	mean	0	3.9E–04	0	0.00983	0	0.00047
	std	0	6.3E–05	0	0.02188	0	0.00075
	best	0	2.8E–04	0	0	0	5.5E–05
	worst	0	4.8E–04	0	0.06878	0	2.5E–03
OF_8	mean	1.6E–23	0.00282	0	13224.5	8.3E–66	0.00094
	std	5.2E–23	0.00111	0	8587.21	2.5E–65	0.00054
	best	2.1E–94	0.00139	0	99.7873	5.0E–73	0.00024
	worst	1.6E–22	4.6E–03	0	30000.2	8.1E–65	1.6E–03
OF_9	mean	0	2.7E–06	0	3.1E–04	0	2.7E–05
	std	0	5.6E–07	0	9.3E–04	0	8.4E–05
	best	0	1.8E–06	0	1.1E–16	0	3.4E–07
	worst	0	3.4E–06	0	0.00314	0	2.6E–04

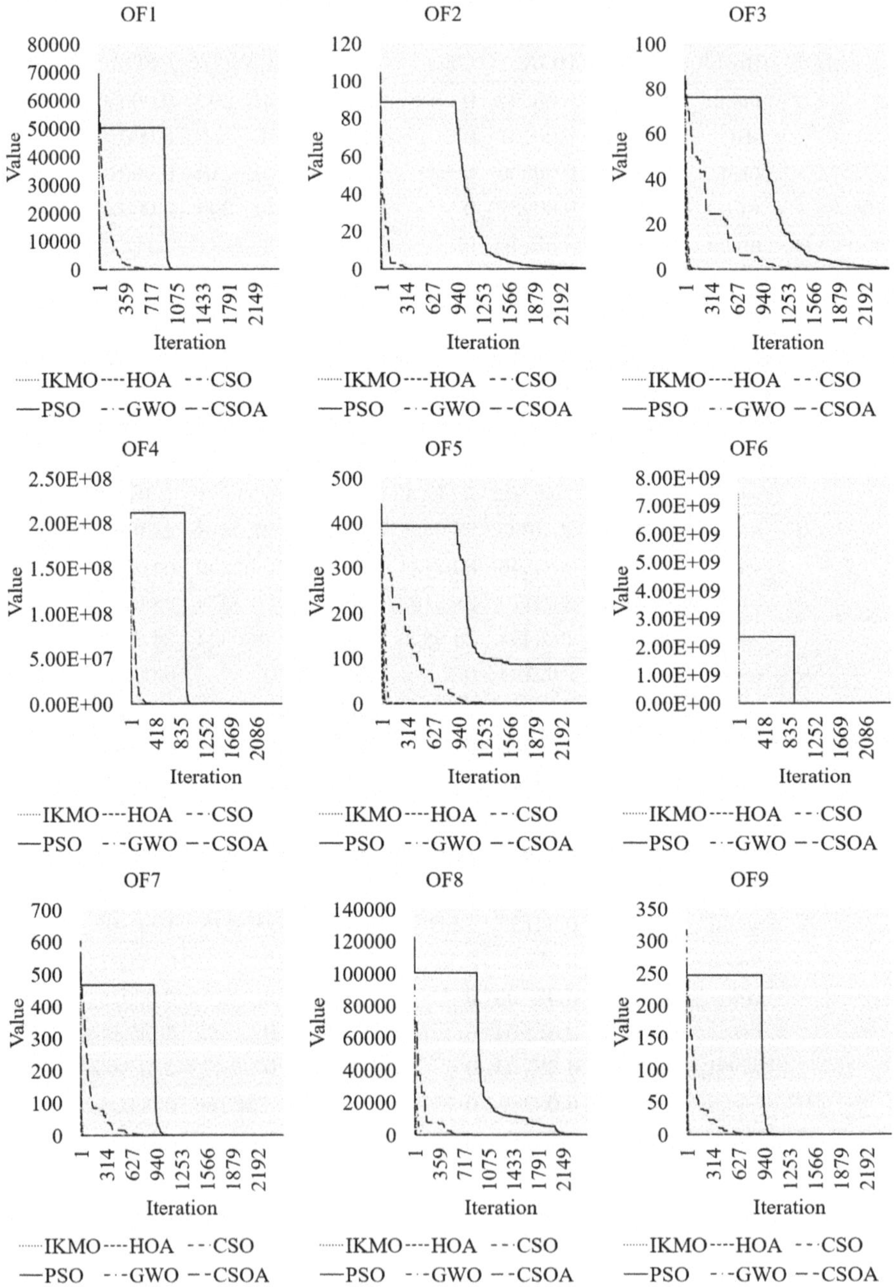

Fig. 7. Comparison of the evolution of the best runs for each bio-inspired algorithm for each objective function

As a general remark, the performance of the bio-inspired algorithms is highly influenced by the initial values of the configurable parameters and by the way in which the random numbers are generated both in the case of the results presented in Table 2 and in the case of the results presented in Fig. 7.

Therefore, the results presented in the paper are not directly comparable to the results obtained in the research literature for the same bio-inspired algorithms by other authors even if we tried to be as close as possible to the implementations of the original versions of the algorithms in our versions of the bio-inspired algorithms considered in the experiments.

5 IKMO LR Method for Smart Grid Stability Classification

5.1 High-Level View of the IKMO LR Method for Smart Grid Stability Classification

Figure 8 presents the high-level view of the IKMO method for the classification of the smart grid stability.

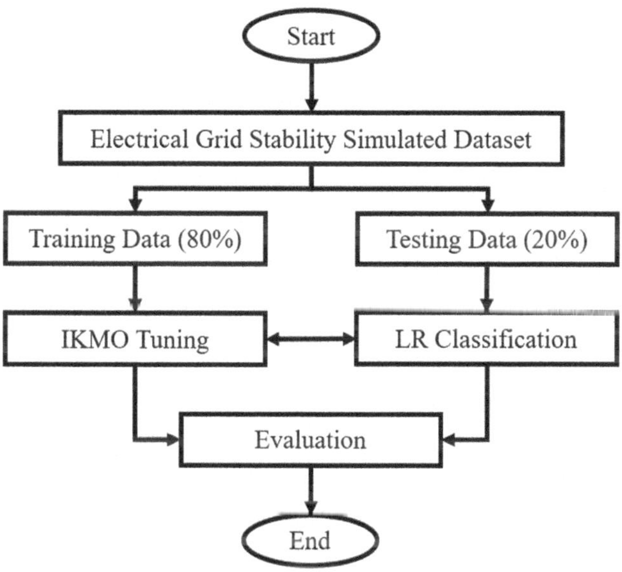

Fig. 8. IKMO LR method for smart grid stability classification high-level view

The input of the methodology is represented by the *Electrical Grid Stability Simulated Dataset*. The initial version of this dataset is characterized by 10000 data samples and 14 features such that 2 features out of the 14 features are labels. Since the approach presented in this paper considers the dataset from a classification perspective, the 13th column is ignored. The labels of the 14th column are converted as follows: the label *unstable* is replaced with 0 and the label *stable* is replaced with 1.

The data is further divided in *Training Data (80%)* and *Testing Data (20%)*.

In the *IKMO Tuning* phase, IKMO is adapted for the optimal selection of two hyperparameters of the LR model as follows:

- the tolerance parameter (tol): the tolerance for the stopping criteria;
- the regularization parameter (reg): a parameter which has control over the fitting parameters;

The number of dimensions of the search space (Dim) is set to 2 as each solution of the search space is represented by an array of the form (tol, reg).

The objective function has the formula:

$$OF = 1 - acc \tag{7}$$

where acc is the accuracy of the LR model for the *Testing Data* such that the LR model is trained using the *Training Data* and the values (tol, reg) are the ones which describe the position of the kangaroo.

In the *LR Classification* phase, the LR model is trained considering the position of the global best kangaroo and the *Training Data*. The LR model is further applied on the *Testing Data*.

Finally, in the *Evaluation* phase the IKMO algorithm is evaluated using the accuracy, the F1-Score, the weighted precision and the weighted recall for the *Testing Data* and the data predicted by the LR model such that the elements of the confusion matrix have the following significance:

- True Positive (TP): predicted stable and actual stable;
- False Positive (FP): predicted stable and actual unstable;
- False Negative (FN): predicted unstable and actual stable;
- True Negative (TN): predicted unstable and actual unstable;

5.2 Experiments Configuration

The experiments were performed on a machine with the following properties: Windows 10 Pro N operating system, 8.00 GB RAM, 64-bit operating system, x64-based processor, Intel(R) Core(TM) i7-7500 CPU @ 2.70 GHz 2.90 GHz.

The LR model was implemented using the Apache Spark [25] machine learning libraries from version 2.2.0. In the experiments performed in this section only the values of the *tol* and *reg* parameters were varied, while the values for the other hyperparameters were the default ones.

Table 3 summarizes the values of the IKMO parameters used in experiments.

The initial dataset is splitted in 5 folds such that each fold is characterized by an equal number of samples and 5 experiments are performed. In each experiment the *Testing Data* is represented by one different fold and the *Training Data* is represented by the remaining four folds.

Table 3. IKMO parameters values summary

Parameter	Value
number of dimensions (Dim)	2
number of iterations ($N_{iterations}$)	30
number of kangaroos ($N_{kangaroos}$)	10
number of mobs (M)	2
mobs updating frequency (K)	5
confrontation threshold (CT)	10
minimum strength value (S_{min})	0
maximum strength value (S_{max})	100
constant for avoiding division by zero (ϵ)	0.000000001
tol search space ($[tol_{min}, tol_{max}]$)	$[1E{-}6, 0.00003]$
reg search space ($[reg_{min}, reg_{max}]$)	$[1E{-}4, 0.3]$

5.3 Experimental Results

Figure 9 presents the improvement of the accuracy after 30 iterations of IKMO for each fold out of the 5 folds considered in the experiments.

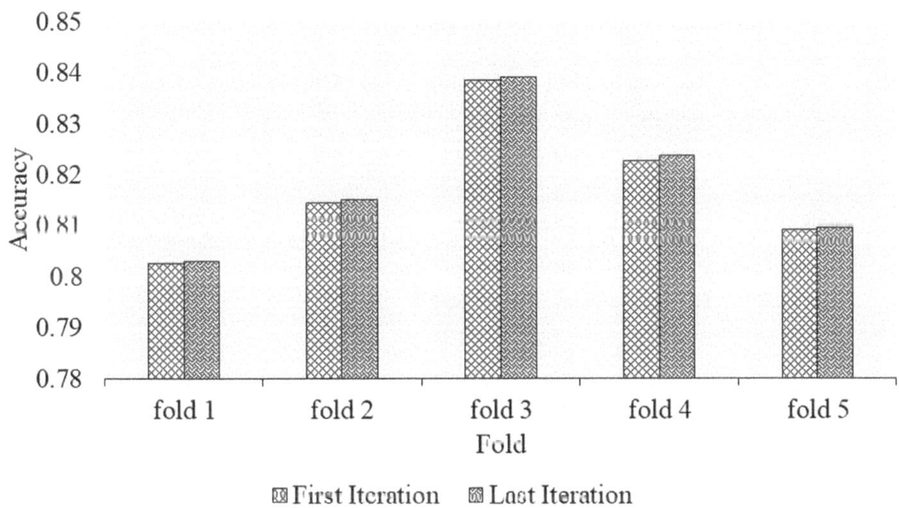

Fig. 9. Accuracy improvement using IKMO after 30 iterations for each fold

The results presented in the figure show an improvement of the accuracy value for each fold.

Next, Table 4 presents a comparison of the results obtained using the default values of tol and reg from Apache Spark with the results obtained after the running of the IKMO algorithm.

Table 4. Comparison of the results obtained using the IKMO returned parameters values with the results obtained using the default parameters values

Approach	Fold	Accuracy	F1-Score	Weighted Precision	Weighted Recall
default	fold 1	0.8005	0.7986	0.7981	0.8005
	fold 2	0.8145	0.8125	0.8122	0.8145
	fold 3	0.8365	0.8351	0.8347	0.8365
	fold 4	0.8210	0.8189	0.8190	0.8210
	fold 5	0.8085	0.8066	0.8063	0.8085
IKMO	fold 1	0.8030	0.8006	0.8004	0.8030
	fold 2	0.8150	0.8115	0.8127	0.8150
	fold 3	0.8390	0.8367	0.8369	0.8390
	fold 4	0.8235	0.8200	0.8220	0.8234
	fold 5	0.8095	0.8050	0.8078	0.8094

The application of IKMO results in an improvement of the accuracy in the case of all five folds.

Table 5 presents a comparison of the running time expressed in milliseconds for each approach.

Table 5. Comparison of the running time for each approach

Approach	Fold	Running Time (Milliseconds)
default	fold 1	15546
	fold 2	14792
	fold 3	18513
	fold 4	15877
	fold 5	15811
IKMO	fold 1	1580581
	fold 2	1388135
	fold 3	1769384
	fold 4	1654876
	fold 5	1482165

As expected, the running time for the IKMO is much greater than the running time for the cases when the default values are used.

Finally, Table 6 presents the characteristics of the global best kangaroo for each fold. For each kangaroo, the value of the position corresponds to the (tol, reg) values.

As can be seen in the table, the characteristics of the global best kangaroos differ significantly for each fold.

Table 6. Global best kangaroos characteristics for each fold

Fold	Gender	Single	Strength	Fitness	Position (tol, reg)	Velocity
fold 1	male	no	23	0.1969	(1.41E–6, 0.00403)	(2, 1)
fold 2	female	yes	85	0.1850	(4.96E–5, 0.01955)	(1, 1)
fold 3	female	yes	10	0.1610	(2.67E–5, 0.01859)	(2, 2)
fold 4	female	no	25	0.1765	(5.62E–6, 0.03064)	(2, 1)
fold 5	female	yes	36	0.1905	(2.14E–5, 0.03837)	(2, 1)

5.4 Comparison with Literature Results

Compared to the approach presented in [18], the approach presented in this article did not return better results. However, the results are not directly comparable because that approach was based on features extraction, features selection and classification, and a different classifier was used.

Compared to the approach from [9] where the classifiers LR, RF, GBT, and MPC were applied, the results obtained in this article are better than the results obtained by the LR classifier, but not better than the results obtained by the other classifiers.

Finally, the results presented in [22] are also better than the results presented in this approach but that approach uses a GPU and a more complex model.

6 Conclusions

This paper presented a new version of the KMO algorithm called IKMO in which the equations for the updating of the positions of the kangaroos were improved using a random factor in order to explore the search space more and to avoid the premature convergence.

The IKMO was evaluated using 9 objective functions and showed results comparable to the ones obtained using other bio-inspired algorithms.

Finally, IKMO was applied in the tuning of the hyperparameters of a LR model applied in the classification of the smart grid stability and showed superiority compared to the case when the default values of the hyperparameters of the LR from Apache Spark were considered.

As future research work, the following directions are proposed: (1) further improvements of the KMO algorithm, (2) the evaluation and the validation using other objective functions and (3) the application of the new improved versions of KMO in other research areas.

References

1. Yang, X.-S.: A new metaheuristic bat-inspired algorithm. In: Gonzalez, J.R., Pelta, D.A., Cruz, C., Terrazas, G., Krasnogor, N. (eds.) Nature Inspired Cooperative Strategies for Optimization (NICSO 2010). Studies in Computational Intelligence, vol. 284, pp. 65–74. Springer, Berlin, Heidelberg (2010). https://doi.org/10.1007/978-3-642-12538-6_6
2. Salgotra, R., Singh, U.: The naked mole-rat algorithm. Neural Comput. Appl. **31**, 8837–8857 (2019). https://doi.org/10.1007/s00521-019-04464-7
3. Anand, A., Suganthi, L.: Hybrid GA-PSO optimization of artificial neural network for forecasting electricity demand. Energies **11**(4), 728 (2018). https://doi.org/10.3390/en11040728
4. Holland, J.H.: Adaptation in Natural and Artificial Systems. MIT Press, Cambridge (1975)
5. Kennedy, J., Eberhart, R.: Particle swarm optimization. In: Proceedings of ICNN 1995 - International Conference on Neural Networks, Perth, WA, Australia, vol. 4, pp. 1942–1948 (1995). https://doi.org/10.1109/ICNN.1995.488968
6. Xue, H.-R., Li, L.-L., Chao, K.-H., Fu, C.: Short-term wind power prediction based on improved chicken algorithm and support vector machine. In: Proceedings of the 2018 International Symposium on Computer, Consumer and Control (IS3C), Taichung, Taiwan, pp. 137–140 (2018). https://doi.org/10.1109/IS3C.2018.00042
7. Meng, X., Liu, Y., Gao, X., Zhang, H.: A new bio-inspired algorithm: chicken swarm optimization. In: Tan, Y., Shi, Y., Coello, C.A.C. (eds.) Advances in Swarm Intelligence. ICSI 2014. Lecture Notes in Computer Science, vol. 8794, pp. 86–94. Springer, Cham (2014). https://doi.org/10.1007/978-3-319-11857-4_10
8. Moldovan, D., Slowik, A.: Chicken swarm optimization. In: Swarm intelligence algorithms. A tutorial. Slowik, A. (eds.), pp. 71–84. Taylor & Francis Group(CRC Press), Boca Raton, USA (2020). https://doi.org/10.1201/9780429422614-6
9. Moldovan, D., Salomie, I.: Detection of sources of instability in smart grids using machine learning techniques. In: Proceedings of the 2019 IEEE 15th International Conference on Intelligent Computer Communication and Processing (ICCP), Cluj-Napoca, Romania, pp. 175–182 (2019). https://doi.org/10.1109/ICCP48234.2019.8959649
10. Moldovan, D., Anghel, I., Cioara, T., Salomie, I., Chifu, V., Pop, C.: Kangaroo mob heuristic for optimizing features selection in learning the daily living activities of people with Alzheimer's. In: Proceedings of the 2019 22nd International Conference on Control Systems and Computer Science (CSCS), Bucharest, Romania, pp. 236–243 (2019). https://doi.org/10.1109/CSCS.2019.00046
11. Li, M.-V., Geng, J., Wang, S., Hong, W.-C.: Hybrid chaotic quantum bat algorithm with SVR in electric load forecasting. Energies **10**(12), 2180 (2017). https://doi.org/10.3390/en10122180
12. Zeng, B., et al.: Prediction model for dissolved Gas concentration in transformer oil based on modified grey wolf optimizer and LSSVM with grey relational analysis and empirical mode decomposition. Energies **13**(2), 422 (2020). https://doi.org/10.3390/en13020422
13. Mirjalili, S., Mirjalili, S.M., Lewis, A.: Grey wolf optimizer. Adv. Eng. Softw. **69**, 46–61 (2014). https://doi.org/10.1016/j.advengsoft.2013.12.007
14. Malik, S., Kim, D.: Prediction-learning algorithm for efficient energy consumption in smart buildings based on particle regeneration and velocity boost in particle swarm optimization neural networks. Energies **11**(5), 1289 (2018). https://doi.org/10.3390/en11051289

15. Dalal, S., Vishwakarma, V.P.: GA based KELM optimization for ECG classification. Procedia Comput. Sci. **167**, 580–588 (2020). https://doi.org/10.1016/j.procs.2020.03.322
16. Arzamasov, V., Bohm, K., Jochem, P.: Towards concise models of grid stability. In: Proceedings of the 2018 IEEE International Conference on Communications, Control, and Computing Technologies for Smart Grids (SmartGridComm), Aalborg, Denmark, pp. 1–6 (2018). https://doi.org/10.1109/SmartGridComm.2018.8587498
17. Dua, D., Graff, C.: UCI Machine Learning Repository. Irvine, CA: University of California, School of Information and Computer Science (2019). [http://archive.ics.uci.edu/ml]
18. Moldovan, D.: Horse optimization algorithm: a novel bio-inspired algorithm for solving global optimization problems. In: Silhavy, R. (eds.) Artificial Intelligence and Bioinspired Computational Methods. CSOC 2020. Advances in Intelligent Systems and Computing, vol. 1225, pp. 195–209. Springer, Cham (2020). https://doi.org/10.1007/978-3-030-51971-1_16
19. Christ, M., Braun, N., Neuffer, J., Kempa-Liehr, A.W.: Time series FeatuRe extraction on basis of scalable hypothesis tests (tsfresh - A Python package). Neurocomputing **307**(1), 72–77 (2018). https://doi.org/10.1016/j.neucom.2018.03.067
20. Moldovan, D.: Scalable hypothesis tests for detection of epileptic seizures. In: Silhavy, R., Silhavy, P., Prokopova, Z. (eds.) Computational Statistics and Mathematical Modeling Methods in Intelligent Systems. CoMeSySo 2019 2019. Advances in Intelligent Systems and Computing, vol. 1047, pp. 157–166. Springer, Cham (2019). https://doi.org/10.1007/978-3-030-31362-3_16
21. Koc, E.K., Bozdogan, H.: Model selection in multivariate adaptive regression splines (MARS) using information complexity as the fitness function. Mach. Learn. **101**, 35–58 (2015). https://doi.org/10.1007/s10994-014-5440-5
22. Alazab, M., Khan, S., Krishnan, S.S.R., Pham, Q.-V., Reddy, M.P.K., Gadekallu, T.R.: A multidirectional LSTM model for predicting the stability of a smart grid. IEEE Access **8**, 85454–85463 (2020). https://doi.org/10.1109/ACCESS.2020.2991067
23. Chu S.-C., Tsai P., Pan J.-S.: Cat swarm optimization. In: Yang, Q., Webb, G. (eds.) PRICAI 2006: Trends in Artificial Intelligence. PRICAI 2006. Lecture Notes in Computer Science, vol. 4099, pp. 854–858. Springer, Berlin, Heidelberg (2006). https://doi.org/10.1007/978-3-540-36668-3_94
24. Moldovan, D., Chifu, V., Salomie, I., Slowik, A.: Cat swarm optimization. In: Swarm intelligence algorithms. A tutorial. Slowik, A. (eds.), pp. 55–70. Taylor & Francis Group (CRC Press), Boca Raton, USA (2020). https://doi.org/10.1201/9780429422614-5
25. Fu, J., Sun, J., Wang, K.: SPARK - a big data processing platform for machine learning. In: Proceedings of the 2016 International Conference on Industrial Informatics - Computing Technology, Intelligent Technology, Industrial Information Integration (ICIICII), Wuhan, China, pp. 48–51 (2016). https://doi.org/10.1109/ICIICII.2016.0023

Design and Development of the Optimal Technique for the Enhancement of Contrast and Brightness Quality of Hand Capture Images Using Wavelet Transformation

B. S. Mahalakshmi(✉) and S. V. Sheela

Department of Information Science and Engineering, BMS College of Engineering, Bengaluru, India
bsmahalakshmi.ise@bmsce.ac.in

Abstract. Owing to the unique traits of palm, palm-based biometric is one of the increasing trends of authenticating the user in most commercial applications. A review of existing literature shows significant work being carried out towards authentication of the palm; however, the existing studies are witnessed to exclude some practical impediments towards authentication measures. The proposed study identifies two essential practical impediments, i.e., illumination condition and variance in palm texture during database building and authentication, leading to the unreliable authentication system. Therefore, the proposed system introduces a simplified and novel preprocessing approach where multi-variant discrete algorithms address these practical challenges, where poor signal quality poses issues in the authentication. The study outcome shows the beneficial aspect of all the algorithms under varied conditions.

Keywords: Authentication · Pre-processing · Palm image · Biometric · Security · Identification and recognition

1 Introduction

Authentication is an essential operation for any organization or application to secure their network and resources by granting access to only the legitimate user for their privileged resources [1]. At present, there are various types of authentication mechanisms, viz. password-based [2], multi-factor-based [3], certificate-based [4], biometric-based [5], token-based [6], etc. All these existing forms of authentication mechanism has their beneficial aspect as well as limitations too. At present, there is various work towards biometric security being carried out using various approaches [7, 8], while it is noticed that palm-based biometrics are simpler and more frequently adopted [9, 10]. The signal

© The Author(s), under exclusive license to Springer Nature Switzerland AG 2021
R. Silhavy (Ed.): CSOC 2021, LNNS 229, pp. 488–498, 2021.
https://doi.org/10.1007/978-3-030-77445-5_45

quality of palm image plays a significant role in offering reliable authentication modelling; however, there are various problems in this operation. The primary problem is associated with the illumination condition, which results in an unreliable and error-prone authentication system. Such consideration is less adopted in existing approaches. The secondary problem is associated with the palm's texture, which may vary from the registration and authentication stage due to genuine reason. Hence, the presence of such factors is inevitable and will significantly affect the authentication process. Therefore, the proposed system considers three practical problems associated with an authentication system using palm image. The study introduces three algorithms that offer to preprocess towards palm image to address the palm image's illumination and texture problem while attempting to register in the dataset or performing query during the actual authentication process. The paper's organization is as follows: Sect. 2 briefs about existing authentication schemes using palm image, while its associated research problem is discussed in Sect. 3. Discussion of the research methodology is briefed in Sect. 4, while algorithm discussion is carried out in Sect. 5. The result obtained is discussed in Sect. 6, while the conclusion is discussed in Sect. 7.

2 Related Work

This section discusses the existing approaches where authentication work is carried out using palm image as the biometric. The work carried out by Chen et al. [11] has presented an identification of palm vein using complex wavelet and convolution neural network for feature extraction. Another work carried out by Palma et al. [12] have used a recursive verification approach for palm image, and it uses a dynamical system of linear order. Wang et al. [13] have used a photoacoustic imaging technique to perform the palm vessel's verification as a biometric. The work carried out by Ahmad et al. [14] has used a transformed-based mechanism using wave atom for privacy preservation of palm-vein image during the recognition process. The work of Palma et al. [15] has presented an authentication mechanism for the pattern of palm vascular using a recursive approach. The adoption of the deep learning concept was implemented by Obayya et al. [16], where the region of interest is used while the Bayesian algorithm optimizes the process. Kuzu et al. [17] have used a convolution autoencoder for carrying out the verification of the palm vein with an emphasis on extracting patterns of the hand vein. Agbinya [18] has presented a geometry modelling of the human palm to construct a secure authentication system. A study using a convolution neural network towards an identification-based authentication system is discussed by Jhong et al. [19]. The study has also used adaptive filtering of the background. The study carried out by Martin et al. [20] has presented a discussion about recognition f palm vein using an experimental based approach. Apart from this, there are also other unique variants of authentication approach using palm image viz. feature extraction based authentication [21], multimodal authentication [22], matching-of vein based authentication [23], integrated authentication using the wrist, dorsal, and palmer [24], total-variation decomposition-based intrusion resistance for finger vein [25], deep residual network-based authentication [26], Gabor filter based encryption of biometric [27], phase-based identification [28], generation of the secret key from palm image [29], and generation of a pseudonym for safeguarding dataset [30]. The next section discusses the research problems associated with the existing approach.

3 Problem Description

From the prior section, it is noticed that the existing authentication mechanism using palm image is claimed for certain advantages and is also linked with certain inherent issues. The identified research problems are as follows: i) existing authentication does not emphasize on demands of various illumination condition on palm when placed over palm scanner, ii) the surface of palm scanner is affected by various illumination condition as well as dust, which will affect even the creation of dataset, iii) The inevitable problem associated with the texture of a palm surface could significantly affect the authentication system which is not considered in existing techniques, and iv) majority of the existing authentication scheme is carried out over dataset which is assumed to be free from any error causing outliers during authentication process when queried with real-time palm image. The research mentioned above problems is addressed in the proposed study, as discussed in the next section.

4 Proposed Methodology

The proposed study's prime research goal is to obtain an optimal quality image of the palm, addressing various practical problems that act as an impediment towards authentication using the palm as biometrics. The diagrammatic representation of the proposed implementation is shown in Fig. 1.

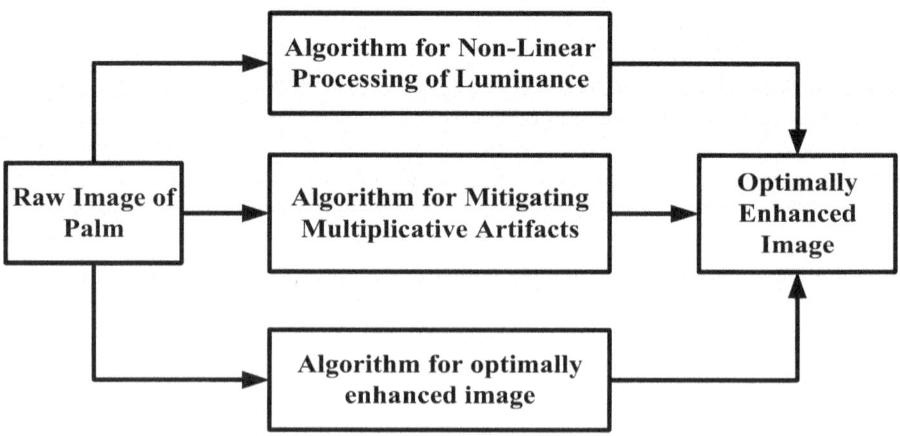

Fig. 1. Proposed implementation flow

Figure 1 highlights that the proposed system introduces multi-variants processing techniques over the same image, which offers a highly customizable and highly flexible form of the image enhancement process. Considering three sets of problems that can arise from illumination conditions, the proposed system constructs three preprocessing approaches that help extract the significant image with optimal quality. This implementation will act as a potential supportive system to extract precisive features in the future direction of work towards authentication using a precisive feature from palm image. The next section discusses algorithm implementation.

5 Algorithm Implementation

The proposed algorithm's core target is to obtain the highest possible image quality by applying multi-variant preprocessing over the palm image. The study introduces three discrete algorithms with three different objectives viz. i) To carry out non-linear processing of luminance, ii) For mitigating the multiplicative noise, and iii) Optimize the quality of palm image. The discussion of the proposed algorithms are as follows:

5.1 Algorithm for Non-linear Processing of Luminance

Luminance carries the essential properties of the hand geometry, and often it gets affected when subjected to a variable intensity of illumination when the hand is placed on a hand scanner. Therefore, this algorithm offers a primary enhancement level to ensure effective encoding and decoding processes carried out over the luminance factor. The steps of the algorithm are as follows:

Algorithm for Non-Linear Processing of Luminance
Input:*I* (palm image), *h*(encoding value)
Output: Img3 (Enhanced image)
Start
1. **If** s(I,3) = =3
2. $Img_1 = f_1(I)$
3. $Img_2 = f_2(Img_1, h)$
4. $Img_3 = f_3(Img_2)$
5. **End**
End

The algorithm takes the raw palm image I input and assesses a colored image (Line-1). In such a case, it applies a function $f_1(x)$ that can transform it into greyscale image Img_1 (Line-1). It also changes the precision of the image to double. The algorithm then applies function $f_2(x)$, which carries out non-linear processing of luminance (Line-2), and it considers h as encoding value, and its value is always less than 1. It also contributes towards compression and expansion. If the value of h < 1, it can be used for encoding considering power-law nonlinearity resulting in compression, while if h > 1, it can be used for decoding operation resulting in expansion. Hence, just by minor changes in the value of h, the palm image's complete contrast factor can be controlled in a highly customized manner. This leads to the generation of primary enhanced image

Img_2 (Line-3). Finally, normalization $f_3(x)$ is applied considering an unsigned integer of 8 bits to obtain an enhanced image. This completes the algorithm operation resulting in an effective contrast image, which mitigates primary illumination variation over the scanner.

5.2 Algorithm for Mitigating Multiplicative Artifacts

The hand scanner's authentication process may also fail if there is a texture difference between the palm geometry of the subject in the database and at the time of authentication. Due to various reasons, there are possible changes in the hand surface texture, which significantly affect the brightness factor during authentication. However, palm lines or their geometry is inherent properties, and it is assumed that it cannot change. However, the roughness of skin and differences in texture may significantly result in uneven brightness and contrast generation, and this problem is results in multiplicative noise, which occurs while capturing or processing the palm image. This algorithm is responsible for addressing this problem, which finally results in an enhanced image. The steps of this algorithm are as follows:

Algorithm for Mitigating Multiplicative Artifacts

Input: *nr* (number of rows), *nc* (number of columns), *n* (degree of filter)

Output:Img_5 (filtered image)

Start

1. **For** i=1: *nr*
2.   **For** j=1:*nc*
3. $\alpha(i, j)=f_4(i, j, nr, nc)$
4. $\beta(i, j)=f_5(\alpha, n)$
5. $\beta=f_6(\Delta q. \beta)$
6. Img_4→obtain inverse log of Img_3
7. $Img_5=norm(Img_4)$
8.   **End**
9. **End**

End

Considering all rows (Line-1) and all columns (Line-2), the algorithm applies a function $f_4(x)$, which obtains α signal using the expression as showcased in the Line-3

$$\alpha(i,j) = \sqrt{(i - nr/2)^2 + (j - nc/2)^2} \tag{1}$$

It further applies another mathematical expression to obtain the enhanced image β as shown in Line-4 as,

$$\beta(i,j) = \left(1/1 + 10/\alpha(i,j)\right)^{2n} \tag{2}$$

Finally, the algorithm applies another function, $f_6(x)$, which uses enhanced image β in the previous step and Δq, where Δq represents a difference of high and low frequency components (Line-5). The next step obtains an inverse log of the absolute value of DFT for the filtered image that is obtained by applying DFT over the log of the original image (Line-6). This results in another version of enhanced image Img4 followed by further applying it to normalization (Line-7). The outcome is an image with a balanced brightness and contrast factor associated with hand palm, and it ensures that when a dataset is created, this factor is considered. So that in case of any variation of texture, brightness and contrast factor is not significantly affected during authentication or querying process.

5.3 Algorithm for Optimally Enhanced Image

Authentication of palm image could also go wrong if the hand's shape and size are differently oriented apart from the illumination factor. Simultaneously, the system may be required to perform authentication of the massive number of palm images of different users, affecting the computational operation, especially when the palm image is compared with the feature stored in the dataset in decomposed form. Therefore, this algorithm is responsible for addressing this issue with a complete focus on the optimization performance for obtaining the highly enhanced image. The steps of the algorithm are as follow:

Algorithm for optimally enhanced image

Input: r (level of decomposition)

Output:

Start

1. $D_{img}=f_7(I, \text{method})$

2. $Img6=f_l(D_{img}, r)$

End

This algorithm takes the input of the level of decomposition r and applies a function f7(x), which carries out the following operation to yield decomposed image D_{img} (Line-1). The algorithm applies two-dimensional discrete wavelet transform using various methods, e.g., Haar, Daubechies, Symlet, etc. It also carries out a very simplified normalization operation. The method looks for an effective thresholding scheme that computes the global thresholding level that could be exploited to transform image intensity to a binary image with its normalized value of intensity lying within the probability limits of 0 and 1. A similar function is applied again to the decomposed image over the low pass filter considering a similar decomposition level (Line-2) to obtain finally optimized image Img6. The complete process performs optimization in that threshold value is selected to reduce the possible occurrences of intraclass variance associated with the white and black labelled pixels. The operation is carried out in simplified steps. It performs autonomous configuration of the thresholding where the contrast ratio of foreground and background is usually good; however, in case of any glare due to illumination condition, the algorithm can suitably enhance the image contrast level. Hence, the algorithm results in a highly optimized palm image.

6 Results Discussion

The proposed study's analysis is carried out considering the standard dataset of CASIA [31] and real-time palm images. The scripting of the proposed algorithm is carried out in MATLAB, where all three algorithms have been studied concerning 4 performance parameters, i.e., i) Perceptive contrast, ii) Mean difference of intensity, iii) Contrast, and iv) Sharpness. The visual and numerical outcomes are briefed in this section.

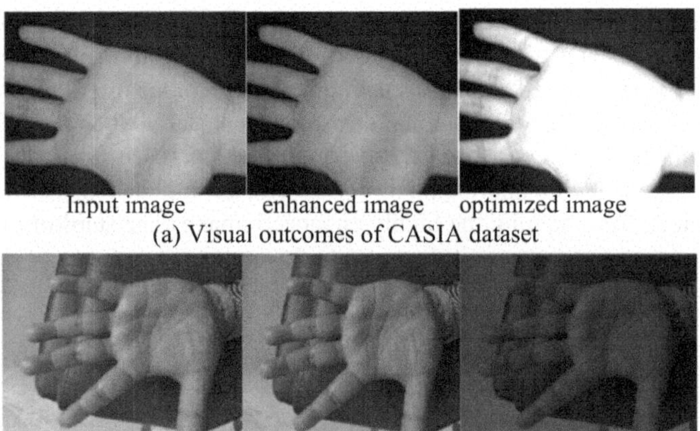

Fig. 2. Visual outcome of non-linear processing of luminance

Fig. 3. Visual outcome of mitigating multiplicative artifacts

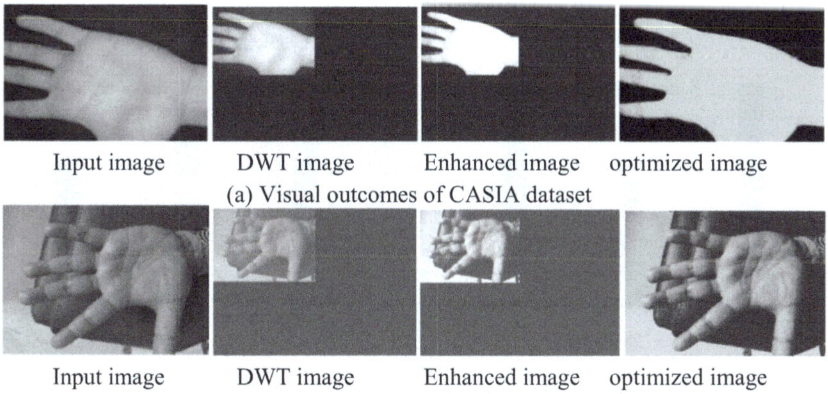

(a) Visual outcomes of CASIA dataset

(a) Visual outcomes of CASIA dataset

Fig. 4. Visual outcome of optimally enhanced image

Table 1. Comparative analysis of algorithms of preprocessing

	Perceptive contrast		Mean difference of intensity		Contrast		Sharpness	
	Dataset	Real	Dataset	Real	Dataset	Real	Dataset	Real
ALG1	7.8752	3.0875	19.1441	19.1504	4.9669	1.4484	2.7663	6.8801
ALG2	3.9628	0.98907	23.9739	24.7796	2.8461	0.50993	3.044	3.546
ALG3	8.1835	3.3244	14.8586	12.8922	4.3203	1.2844	18564	5.0311

The visual outcomes shown in Fig. 2, Fig. 3, and Fig. 4 show that the third algorithm offers a better resolution for the sample test image. However, this conclusive remark could vary from different palm images, which means that the proposed system offers a highly flexible scheme where different variants of quality resolution of the image can be obtained. From the numerical outcome shown in Table 1 and the corresponding graphical outcome shown in Fig. 5, it can be said that the third algorithm offers better perceptive contrast (Fig. 5(a)), while the second algorithm offers a better mean difference in intensity (Fig. 5(b)), while the first algorithm offers better contrast (Fig. 5(c)) as well as sharpness (Fig. 5(d)). Hence, the outcome suggests that the proposed system offers multiple variants of preprocessing algorithms to enrich the signal quality of palm image, which further contributes towards precise feature extraction, thereby facilitating better authentication mechanisms in biometrics.

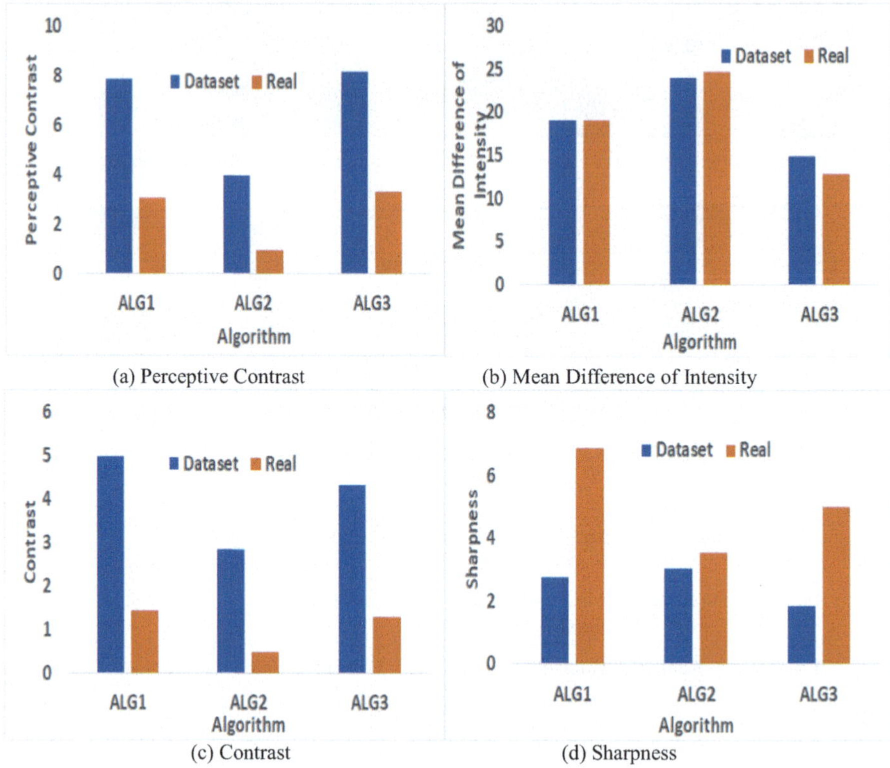

Fig. 5. Graphical Outcome of Comparative Analysis

7 Conclusion

There is no doubt that palm-based biometric has been a recent topic of interest among researchers in authentication. However, from the review of literature, it is found that existing approaches are mostly reported to extract the feature of palm vein and use them as prime authentication There is no consideration of practical problems which distort both in the dataset registration process as well as authentication process where illumination and texture of palm is the prime issue. Hence, the proposed system addresses the problem by introducing a multi-variant algorithm, which addresses three practical scenarios of authentication issues caused due to illumination and texture of palm. The contribution of the study are i) The study introduces three different algorithms of preprocessing which addresses three discrete problems both during registration and querying process using palm image, ii) The complete operation offers the highest range of flexibility to the user to not only obtain the enhanced image but also obtain the optimized image, iii) The performance outcome shows that three introduced algorithms offer discrete performance pattern with a different set of image and hence better granularity in precision is obtained, iv) The complete process in order to execute does not require any specialized skill or dependency of any explicit software, and hence it is a quite cost-effective preprocessing framework in present time.

References

1. Xiong, L., Zhong, X., Yang, C.-N.: DWT-SISA: a secure and effective discrete wavelet transform-based secret image sharing with authentication. Signal Processing **173**, 107571 (2020)
2. Rao, K., Avadhani, P.S.: Authentication based on wavelet transformations. J. Discret. Math. Sci. Cryptogr. **9**(3), 513–521 (2006)
3. Al-Assam, H., Sellahewa, H., Jassim, S.: On the security of multi-factor biometric authentication. In: 2010 International Conference for Internet Technology and Secured Transactions, pp. 1–6. IEEE (2010)
4. Xia, C.-L.: Examination certificate based on two-dimensional bar code technology. In: 2008 International Symposium on Computer Science and Computational Technology, vol. 1, pp. 232–234. IEEE (2008)
5. Wang, H., Fang, H., Xing, L., Chen, M.: An integrated biometric-based security framework using wavelet-domain HMM in wireless body area networks (WBAN). In: 2011 IEEE International Conference on Communications (ICC), pp. 1–5. IEEE (2011)
6. Meng, F., Qi, Z., Chen, Z., Wang, B., Shi, Y.: Token based crack detection. J. Intell. Fuzzy Syst. **38**(3), 3501–3513 (2020)
7. Bhatnagar, G., Jonathan, W., Raman, B.: Fractional dual tree complex wavelet transform and its application to biometric security during communication and transmission. Futur. Gener. Comput. Syst. **28**(1), 254–267 (2012)
8. Sanchez-Avila, C., Sanchez-Reillo, R., de Martin-Roche, D.: Iris-based biometric recognition using dyadic wavelet transform. IEEE Aerosp. Electron. Syst. Mag. **17**(10), 3–6 (2002)
9. Saranraj, S., Padmapriya, V., Sudharsan, S., Piruthiha, D., Venkateswaran, N.: Palm print biometric recognition based on scattering wavelet transform. In: 2016 International Conference on Wireless Communications, Signal Processing and Networking (WiSPNET), pp. 490–495. IEEE (2016)
10. Belgacem, N., Fournier, R., Nait-Ali, A., Bereksi-Reguig, F.: A novel biometric authentication approach using ECG and EMG signals. J. Med. Eng. Technol. **39**(4), 226–238 (2015)
11. Chen, P., et al.: Design of low-cost personal identification system that uses combined palm vein and palmprint biometric features. IEEE Access **7**, 15922–15931 (2019). https://doi.org/10.1109/ACCESS.2019.2894393
12. Palma, D., Montessoro, P.L., Giordano, G., Blanchini, F.: Biometric palmprint verification: a dynamical system approach. IEEE Trans. Syst. Man Cybern. Syst. **49**(12), 2676–2687 (2019). https://doi.org/10.1109/TSMC.2017.2771232
13. Wang, Y., et al.: A robust and secure palm vessel biometric sensing system based on photoacoustics. IEEE Sens. J. **18**(14), 5993–6000 (2018). https://doi.org/10.1109/JSEN.2018.2843119
14. Ahmad, F., Cheng, L., Khan, A.: Lightweight and privacy-preserving template generation for palm-vein-based human recognition. IEEE Trans. Inf. Forensics Secur. **15**, 184–194 (2020). https://doi.org/10.1109/TIFS.2019.2917156
15. Palma, D., Blanchini, F., Giordano, G., Montessoro, P.L.: A dynamic biometric authentication algorithm for near-infrared palm vascular patterns. IEEE Access **8**, 118978–118988 (2020)
16. Obayya, M.I., El-Ghandour, M., Alrowais, F.: Contactless palm vein authentication using deep learning with Bayesian optimization. IEEE Access **9**, 1940–1957 (2021). https://doi.org/10.1109/ACCESS.2020.3045424
17. Kuzu, R.S., Maiorana, E., Campisi, P.: Vein-based biometric verification using densely-connected convolutional autoencoder. IEEE Signal Process. Lett. **27**, 1869–1873 (2020). https://doi.org/10.1109/LSP.2020.3030533

18. Agbinya, J. I.: Human palm geometry modelling for biometric security systems. In: 2019 Cybersecurity and Cyberforensics Conference (CCC), Melbourne, Australia, pp. 160–164 (2019). https://doi.org/10.1109/CCC.2019.00011
19. Jhong, S.-Y.: et al.: An automated biometric identification system using CNN-based palm vein recognition. In: 2020 International Conference on Advanced Robotics and Intelligent Systems (ARIS), Taipei, Taiwan, pp. 1–6 (2020). https://doi.org/10.1109/ARIS50834.2020.9205778
20. Garcia-Martin, R., Sanchez-Reillo, R.: Vein biometric recognition on a smartphone. IEEE Access **8**, 104801–104813 (2020). https://doi.org/10.1109/ACCESS.2020.3000044
21. Doi, J., Yamanaka, M.: Discrete finger and palmar feature extraction for personal authentication. IEEE Trans. Instrum. Measur. **54**(6), 2213–2219 (2005). https://doi.org/10.1109/TIM.2005.858820
22. Bianco, S., Napoletano, P.: Biometric recognition using multimodal physiological signals. IEEE Access **7**, 83581–83588 (2019). https://doi.org/10.1109/ACCESS.2019.2923856
23. Alejo, W., Rodriguez, D., Kemper, G.: A biometric method based on the matching of dilated and skeletonized IR images of the veins map of the dorsum of the hand. IEEE Lat. Am. Trans. **13**(5), 1438–1445 (2015). https://doi.org/10.1109/TLA.2015.7112000
24. Toygar, O., Babalola, F., Bitirim, Y.: FYO: a novel multimodal vein database with palmar, dorsal and wrist biometrics. IEEE Access **8**, 82461–82470 (2020). https://doi.org/10.1109/ACCESS.2020.2991475
25. Qiu, X., Kang, W., Tian, S., Jia, W., Huang, Z.: Finger vein presentation attack detection using total variation decomposition. IEEE Trans. Inf. Forensics Secur. **13**(2), 465–477 (2018). https://doi.org/10.1109/TIFS.2017.2756598
26. Kim, C.S., Cho, N.S., Park, K.R.: Deep residual network-based recognition of finger wrinkles using smartphone camera. IEEE Access **7**, 71270–71285 (2019). https://doi.org/10.1109/ACCESS.2019.2920391
27. Zainon, N.A.F.M., Razak, S.A.: Region of interest extraction for biometric cryptosystem. In: 2018 IEEE Conference on Application, Information and Network Security (AINS), Langkawi, Malaysia, pp. 33–37 (2018). https://doi.org/10.1109/AINS.2018.8631495
28. Uçan, O.N., Bayat, O., Çoşkun, M.B.: Development and evaluation of the authentication systems by using phase-only correlation palm print identification methods. In: 2017 International Conference on Engineering and Technology (ICET), Antalya, pp. 1–4 (2017). https://doi.org/10.1109/ICEngTechnol.2017.8308200
29. Zainon, N.A.F.M., Razak, S.A.: Master and child key generation from palm vein. In: 2017 IEEE Conference on Application, Information and Network Security (AINS), Miri, pp. 37–41 (2017). https://doi.org/10.1109/AINS.2017.8270421
30. Mohd Nazari, N.H., Razak, S.A.: Palm vein pseudonym for anonymous database record. In: 2018 IEEE Conference on Application, Information and Network Security (AINS), Langkawi, Malaysia, pp. 38–42 (2018). https://doi.org/10.1109/AINS.2018.8631463
31. Liu, C.-L., Yin, F., Wang, D.-H., Wang, Q.-F.: CASIA online and offline Chinese handwriting databases. In: 2011 International Conference on Document, pp. 37–41. IEEE (2011)

Distance Learning Advantages and Disadvantages: Teaching Experience Analysis at the University with the Basis on Different Informational-Communicative Technologies

Yu. I. Lobanova(✉)

Saint Petersburg State University of Architecture and Civil Engineering (SPbGASU), 4, 2nd Krasnoarmeyskaya, St. Petersburg 190005, Russia

Abstract. The transition to distance learning during the coronavirus pandemic is seen as a natural pedagogical experiment, which became more manageable for all participants of educational process in autumn 2020. Comprehending work experience in spring, as well as training in the field of informational and communication technologies (ICT), allows the teacher not only to expand their capabilities when working online, but also to experiment with using different digital tools (Teams, Zoom, Moodle, Kahoot), analyzing their application success and the reasons for difficulties occurred. The article presents different courses' students survey results and forms of education regarding distance learning features which they note. The analysis of attitudes towards certain distance learning aspects of by students of different ages (and, accordingly, different generations - Z and Y) is carried out. It is shown that the hypotheses regarding more positive perception of online learning and greater satisfaction with the use of digital tools by the teacher that ensure the interactive classes nature (holding meetings in subgroups in separate channels, organizing joint work using the One Note tool, quizzes by discipline using the Kahoot digital tool) by junior students was not confirmed: senior students note more positive aspects of distance learning and, with the same satisfaction as generation Z students, perceive the use of ICT by the teacher to organize interaction within the educational process. The paper analyzes the results obtained, as well as outlines the prospects for further research in the field of distance learning: overcoming the problems of insufficient (loss) information) when using ICT, overcoming limitations in the transfer of personal knowledge, studying the influence of group dynamics when organizing interaction using ICT.

Keywords: Distance learning · Pros and cons · Informational-communicative technologies · Online · Offline · Teams · Kahoot · Moodle · One note · Webinar · Webcam · Chat · Microphone · Generations Y and Z · Monotony · Eye contact · Need for communication · Time · Personal knowledge

© The Author(s), under exclusive license to Springer Nature Switzerland AG 2021
R. Silhavy (Ed.): CSOC 2021, LNNS 229, pp. 499–506, 2021.
https://doi.org/10.1007/978-3-030-77445-5_46

1 Introduction

The coronavirus pandemic has put everyone in the conditions of some kind of experiment… In the case of education - pedagogical: suddenly and instantly, teaching methods and means have radically changed. Experts had seen future behind digital learning [1], but it seemed that it was still very far away. However, the self-isolation and quarantine regimes introduced in different countries have made distant things close.

Unlike pre-COVID reality, when the teacher-researcher had the opportunity to plan an experiment, and in the event of problems with the introduction of new technologies and teaching methods, immediately return to the traditional ones [10, 13–15], the new reality forced everyone to turn to the study of informational-communicative technologies at a new level and adaptation in their use [9, 11, 12, 16].

How this affected the learning outcomes - it will be clear little later, but the impressions that have arisen and arise among the trainees involved in the processes of change can be studied now.

In June last year, the article's author conducted a comparative analysis of student behavior when learning offline (within the traditional classroom-lesson system) and at the very beginning of a natural experiment to change the format and means of teaching due to the coronavirus pandemic: in terms of class attendance, the peculiarities of using feedback in work online, according to the activity of using various technical means (equipment) when connecting to webinars, according to the activity and discipline of completing the tasks received in the classroom. There was a slight decrease in classes' attendance by all students (when working online compared to offline), differences in the activity of using microphones and chats during communication with the teacher, senior and junior students (and in general, older and younger students). Less technically advanced older students were happy to master not only the new teaching format and the opportunities that it provides. The junior students were better oriented in the technologies themselves, but during the classes they behaved more restrained and even passively [11].

But if in spring 2020 the situation in education resembled a plunge into virtual reality, then in autumn the participants in the educational process have already somewhat adapted to new realities, and when working with them, it is already possible to use both retrospective and situational reflection [4, p. 33] to analyze the situation. This work is devoted to this analysis.

2 Methods

1. The survey (in free form) regarding distance learning peculiarities, which are noticed by its participants (pilot study) - 55 people (students of two study groups: first and fourth years).
2. Assessment of selected features significance (focus group survey): total 80, 43 (1st year) and 37 (4th year)
3. Assessment of satisfaction with the educational process when using ICT, allowing to organize interaction in teams and group work: 60 - 30 and 30 in total (1st year and masters)

2.1 Object of Study

Undergraduate students (representatives of generation Z) - 101 people.

Senior undergraduate and graduate students (generation Y) - 94 people.

2.2 Subject of Study

Perception features of positive and negative distance learning aspects, as well as individual teaching methods using informational and communicative technologies by students of different ages, training levels and forms of education.

The study was conducted at a technical college. Levels of preparation: bachelor's and master's degrees. Forms of study: full-time and part-time. Used environments, programs, technologies:

- for training: Moodle; Mail; WhatsApp; Teams (in spring only lectures were used, in autumn they added case solving in separate groups (in separate channels), discussions and even elements of game design with the presentation of their intermediate results using the One Note tool, and the final ones with demonstration of the presentation on the screen in a common channel.
- to test knowledge: tests in Moodle; quiz on Kahoot.

3 Results

The study was carried out in two stages: the first one revealed the features that the students noticed when switching from offline learning to online, the second one assessed and compared the significance of individual features for junior and senior students.

The results of the survey regarding the features of distance learning (advantages and disadvantages) are presented in Table 1.

Table 1. Features of distance learning through students' eyes.

Disadvantages of distance learning	Advantages of distance learning
Presence of technical difficulties	Having a recording
Monotony	Saving time
Sensory hunger	Visibility (visualization availability)
Lack of kinesthetic sensation	Ergonomics
Lack of communication (during lectures)	
Noise disturbance	
Lectures are like watching a movie	

Note that separation of distance learning trainees' marked features into positive and negative was carried out by the author of the article on objective grounds (the negative aspects included the moments of deprivation of resources, opportunities, deprivation of needs, the occurrence of interference in activities - those conditions that are considered the reasons for tension or stress states development, frustrations, etc.

Before presenting individual features of distance learning for assessment by students, a hypothesis was formulated that first-year students (due to their attribution to generation Z [17]) would generally perceive distance learning format more positively and focus more on its positive aspects than on the negative.

In groups of students of different ages (and belonging to different generations - Y and Z), an open vote was held, the results of which are presented in Tables 2 and 3. The teacher named one or another feature of distance learning, and the students raised their hand if they subjectively felt its influence.

Table 2. Percentage of students who report negative aspects of distance learning

Negative features	First year (reaction percentage)	Fourth year and masters
Technical difficulties	23.8	4.5
Monotony, sensory hunger, lack of kinesthetic sensation	26.2	7.7
Noise disturbance	50	50

Table 3. Percentage of students who report positive aspects of distance learning

Positive features	First year (reaction percentage)	Fourth year and masters
Saving time (for sleep)	80	0
Saving time for other activities	75	50
Having a recording	90	75
Visibility	55	50
Ergonomics	90	90

It should be noted that the hypothesis put forward was not confirmed - the opposite tendency was found: senior students and masters reacted more positively to distance learning in general and were less fixed on its negative features: they complained less about monotony and sensory deprivation (perhaps just because they are more actively used feedback, and thereby provide themselves contact with the teacher alike to communication in the classroom).

4 Discussion

One of the main factors that determined a positive attitude towards distance learning was a key feature of adults - they always have a shortage of time [6]. Even within the framework of study, it should be noted that elder students spent saved time on other things, and the younger ones also on sleep. But in any case, it is obvious that for adults, saving a resource - particularly so valuable and irreplaceable as time - more than compensated all the negative aspects of distance learning.

As for the first = year students, most of them "rushed" into the audience (unlike senior students), and among the factors - motivators were named the following:

1. New impressions (familiarity with a new environment and fellow students) (20%).
2. Curiosity (differences between higher education and secondary education) (25%).
3. The need for eye contact with the teacher (as an important condition of the learning process) - 55%.

The last point needs to be given special attention. The children have repeatedly emphasized the importance of eye contact with the teacher. What can be considered as signs, if not understanding, then at least involvement in the process and a positive attitude towards learning? Some knowledge from the field of neurolinguistic programming NLP can come to the rescue [2, p. 17–27] (Table 4 shows the characteristics of the client's behavior, by which it is possible to determine the client's involvement in the process of interaction with the sales manager, for example, and his intention to answer "yes" or "no"):

Table 4. Non-verbal signs of involvement in the interaction

Yes	No
Keeping eye contact (at least according to Russian culture tradition)	Absence of eye contact – withdraw eyes from dialog partner
Body tilt forward	Body tilt back
Head nod	Head shake from side to side
Decrease the distance with dialog partner	Increase the distance
Increased pupil size	Shoulder tightness and lifting
Relaxation of the shoulder girdle	Decreased pupil size
Relaxed lips	Tight and compressed lips

Unfortunately, most of indicated signs cannot be fully used to recognize the psychoemotional state and attitude to the situation by the participants in the educational process when teaching a large group of people online.

The main problem with distance learning (even when using the most modern teaching tools) is that it is not possible to provide meaningful contact through non-verbal means (and even through webcams) when working with a group. Individual training can provide a certain illusion of contact, but part of the information is inevitably lost: a change in communication distance is often impossible for participants in the educational process entangled with wires, headphones and a microphone; using the microphone are distorted.

And looking eye to eye becomes an illusion - people are looking at the screen, not at each other's faces.

As for the opportunities for feedback that exist during lectures-webinars, namely, turning on / off webcams and chats, the tendency in terms of interaction with teachers remains the same as in spring: senior students and masters actively use every possible means of communication with teachers, and not only when conducting a synchronous online lecture, but also at other times: they use personal chats in Teams, in Moodle, e-mail, etc. (those are from 10 to 20% of the group). As for first-year students, they use this kind of channels much less often and rather at the teacher's request (about 5%).

In autumn 2021, discussions in channels (in small groups), remote design (group work), polls using Kahoot were introduced in the classroom with students. The hypothesis was the same as in the first stage of the study: senior students will perceive new tools more "coolly", showing a tendency towards learning offline.

Students' survey regarding learning satisfaction with ICTs' inclusion that provide interactive cooperation showed that the hypothesis was not confirmed: positive emotions from the introduction of tools into the educational process that provide classes' interactive nature (holding meetings in separate channels, working together using One Note, conducting surveys in the form of a quiz on Kahoot) were equally related to students of all courses and different ages: 90–95% of respondents expressed satisfaction with conducting surveys and learning (solving cases) in interactive format.

The reasons for these emotions were different: first-year students were primarily happy to have the opportunity to communicate with each other, while senior students perceived knowledge control tools (Kahoot) as "playful" and "motivating".

5 Conclusion

The use of interactive technologies in distance learning not only corresponds to the ideas of modern didactics, but also during pandemic, with a sharp decrease in the number of contacts, with a pronounced lack of communication, contributes to the closure of the deprived needs of students. It is no coincidence that one of the most common reactions of the trainees was the words "Well, at least to communicate with each other in this way…".

The use of the above-mentioned digital tools after completing even a small refresher course does not present a problem for the teacher. Their use brings the interaction of trainees closer to interaction in the classroom. Conclusion: a good teacher even in new conditions (distance learning conditions can work well, laying the students' knowledge base, skills and abilities in their discipline.

Remaining problems are those that cannot be overcome due to the actual human factor (with a plus sign) - the qualifications of the teacher. These are problems caused by the specifics of modern informational and communicative technologies - the lack of information about each other by participants in the educational process, information that cannot be transmitted through the auditory and visual channels.

Lack of this kind of information can become (with the final and irreversible departure in the educational process to distance learning) a factor of professional burnout of teachers and instructors (just as burnout develops in psychologists working on the helpline [3].

In addition, in the role-based communication of a teacher and a student, not only educational information is transmitted: there is also an emotional and social-perceptual side of communication. A certain emotional background can influence the process and learning outcomes in different ways, but this has always been a given of the educational process, and how the obvious change in the conditions of its manifestation and influence, respectively, will affect remains to be studied. After all, what determines the transfer of not just information by a teacher, but personal meanings [5], personal knowledge is thoroughly unknown, and this despite the fact that it is generally accepted that knowledge is not only of a social nature [8], and that the role of tacit knowledge is no less, and perhaps even more important than explicit knowledge, which can be transmitted through educational texts, presentations, etc. The ways and means of implicit knowledge transfer are yet to be studied - just in connection with solving the problems of transferring it and with the help of electronic learning tools. The same largely applies to group dynamics (group effects): a separate factor, which is not yet fully implemented in the conditions of distance learning (and in general, the possibilities and limitations of its influence on learners in the format of distance learning are still only remains to be seen).

Modern ICT (Teams, Zoom, Miro, Trello, Padlett, Kahoot, etc.) and e-learning itself are the step into the future, including the future of the educational process. However, it is impossible to build this future all at once, hurriedly - the pandemic educational experiment has clearly shown this. It is necessary to carefully study the psychological and socio-psychological factors that determine the effectiveness of the educational process in "contact" work and to look for and create digital tools to preserve the positive influence of the "human factor" on the process and learning outcomes. Only a synthesis of the best that is in a person-teacher and the opportunities provided by modern informational and communicative technologies can ensure progress in the educational sphere, making distance learning not only affordable, but also truly effective.

References

1. Vinogradova, V.V.: Problems and prospects of modern education in the framework of the development of digital technologies. In: Proceedings of the V International Scientific and Practical Conference "Pedagogical parallels", pp. 114–120 (2018)
2. Goryainova, O.V.: Learn to Manipulate People! Trainings and Role-Playing Games, 3rd edn. Phenox, Rostov-on-Don (2006)
3. Zhilina, E.V.: Professional features of the client's image, formed on the basis of auditory information: on the example of the activities of counseling psychologists of the helpline. Russian state pedagogical university, St. Petersburg, Russia (2007)
4. Karpov, A.V.: Reflexive determination of activity and personality. RAO, Moscow, Russia (2012)
5. Leontiev, D.A.: The psychology of meaning: nature, structure and dynamics of meaningful reality. Smysl, Moscow, Russia (1999)
6. Lobanova, Yu.I.: Adult education: problems and solutions. In: Proceedings of the V International Scientific and Practical Conference "Pedagogical parallels", pp. 375–381 (2018)
7. Nosenko, E.L.: An attempt at a systematic approach to the analysis of speech in a state of emotional tension. Psychol. J. **6**, 54–61 (1980)

8. Polani, M.: Personal knowledge. Progress, Moscow, Russia (1985)
9. Dzhangarov, A.I., Hanmurzaev, H.E., Potapova, N.V.: Digital education in the coronavirus era. In: Journal of Physics: Conference Series, vol. 1691, p. 012133 (2020).
10. Lobanova, Y.: Basic guidelines, principles and psychological-pedagogical technologies of creation of the engineer of the future. In: Anikina, Zhanna (ed.) IEEHGIP 2020. LNNS, vol. 131, pp. 621–631. Springer, Cham (2020). https://doi.org/10.1007/978-3-030-47415-7_66
11. Lobanova, Yu.I.: Distant studying experience reflection during pandemic of covid-19 (On the example of teaching in the technical university). In: Journal of Physics: Conference Series, vol. 1691, p. 012152 (2020)
12. Makarenya, T.A., Stash, S.V., Nikashina, P.O.: Modern educational technologies in the context of distance learning. Journal of Physics: Conference Series 1691, 012117 (2020).
13. Harri-Augstein, S., Thomas, L.: Self-organized learning and the relativity of knowing: towards a conversational methodology. In: Stringer P., Bannister, D. (eds.) Constructs of Sociality and Individuality. Academic Press, London (1979)
14. Thomas, L.: Learning and meaning. In: Fransella (ed.) Personal Construct Psychology. Academic press, London (1978)
15. Thomas, L.: Nothing more theoretical than good practice: teaching for self-organized learning. In: Bannister, D. (ed.) Issues and Approaches in Personal Construct Theory. Academic press, London (1985)
16. Yarychev, N.U., Mentsiev, A.U.: New methods and ways of organizing the educational process in the context of digitalization. In: Journal of Physics: Conference Series, vol. 1691, p. 012128 (2020)
17. Theory of generations. https://psixologiya.org/razdely/biznes/2155.html?showall=1. Accessed 16 Dec 2021

Heat Load Forecasting: Comparison Between Neuro-Fuzzy and Neural Network Approaches, a Real World Case Study

Szabolcs Kováč(✉), Igor Halenár, German Michaľčonok, and Andrea Némethová

Institute of Applied Informatics, Automation and Mechatronics, Faculty of Materials and Science and Technology in Trnava, Slovak University of Technology in Bratislava, Trnava, Slovakia
{szabolcs.kovac,igor.halenar,german.michalconok, andrea.nemethova}@stuba.sk

Abstract. An important task in each field of energetics is an accurate forecasting model because by accurate forecasting significant economical, social and technical improvements can be achieved. The aim of this paper was short-term heat load forecasting for the one of the largest heat plants in Slovakia. In this research we studied historical load and outdoor temperature from periods of autumn, winter and spring. Thorough data pre-processing and data analysis yield important findings that during the modeling process were included as an input variables. This paper presents and compares two machine learning approaches – artificial neural networks and neuro-fuzzy. Accuracy of proposed models were validated on a dataset from a newer period where ANFIS model have outperforms in every statistical metrics the conventional ANN model. In general, both proposed models can forecast day ahead heat demand relatively accurately.

Keywords: Neural network · Neuro-fuzzy · Forecasting · District heating · ANFIS

1 Introduction

Nowadays, forecasting plays a key role in heat load forecasting as well in various areas of energetics. It determines the future direction of the company, implementing new technologies, production planning and so on. Based on the time perspective, forecasting can be divided into three groups – a long term (yearly), medium term (monthly) and short term (daily, hourly) [1]. Heat plants often face a problem of production planning for a certain period. An accurate forecasting is a crucial especially for the field of heating. Overestimated production causes a surplus which later needs to be stored also it may cause a startup of new boilers. The opposite case is underestimation that means the required demand is not ensured. These mentioned everyday problems confirm the importance of the short term load prediction (STLF). During the years several approaches have been proposed. The load forecasting can be divided into two main categories – parametrical approach and non-parametrical (data-driven) approach [2]. The parametrical

© The Author(s), under exclusive license to Springer Nature Switzerland AG 2021
R. Silhavy (Ed.): CSOC 2021, LNNS 229, pp. 507–517, 2021.
https://doi.org/10.1007/978-3-030-77445-5_47

model describes the physical behavior of the system [3]. Related to non-parametrical (data-driven) approach various statistical models were proposed such as linear regression [4], ARMA [5] and several artificial intelligence models like artificial neural networks (ANN) [6], support vector machines (SVM) [7] and other.

2 Related Works

The scientific field of predicting the use of energy resources is one of the most current topics. As a confirmation of this is the number of publications available around the world which deal with issues similar to the topic of this article. As mentioned above there are more kinds of mathematical procedures to realize STLF. In available literary sources are not only but often used fuzzy systems [8], genetic algorithms [9], neural networks [10].

The core of this work is focused into energetic industry field. Similar prediction problems are solved on range of type of systems and involves almost every scientific discipline including for example medicine [11], weather forecast [12–14], traffic density [15, 16], earthquakes [17, 18].

Short term load forecasting is most used in energy field such a power grids, smart power distribution networks, power generators. Problem in area of predictive planning power grids load solves work of authors Ye Jianhua and Li Yang [19]. They propose creation of hybrid STLF model via joining the technology of empirical mode decomposition together with harmony search and optimized support vector regression.

The output of this combination is model for prediction of electricity load in power grid. Suchlike problem is solved by Inteha et al. in [20]. They are using more complex predictive model to solve problem of prediction of generation, transmission and distribution network of smart power management system. Their short-term load forecasting prediction technique is based on artificial intelligence and an updated variant of RNN family called long short-term memory (LSTM) model. Outputs of predictive model are compared and verified to simpler model identified as Gated recurrent unit (GRU), but both LSTM and GRU model are providing similar forecasting values.

Authors Nadeem and Naveed in [21] describe in their work preliminary results of applying STLF techniques on dataset from local electricity distribution smart grid for power consumption prediction. As the result is interest finding, that in this case Multiple Linear Regression and Support Vector Regression methods perform better outputs than used artificial neural network.

Same goal (problem of forecast power demand in house grid) is solved in work of Bonetto et al. [22]. They considered to use another type of neural network - autoregressive (NAR) neural network and compare outputs to ARMA and ARIMA models. The result is that besides smaller prediction errors in the mean and the variance, the proposed NAR architecture provides reliable indications of rises in the power demand while standard ARMA and ARIMA models entirely fail. Huang et al. in their work [10] show possibility of using different approach for predictive control of grid-connected power converters. They use novel data-enabled predictive control (DeePC) algorithm which uses an N-step ARMA model for. The ARMA model is identified via an N-step prediction error method in a recursive way. The suggested model can serve as an alternative control system that can stabilize an unstable system by performing optimal and constrained receding-horizon control. Another work is pretty similar to heat prediction. Jiandong et al. [23] propose combined prediction algorithm for electricity consumption prediction. Suggested hybrid system consists of two predictive procedures. After data normalization and preprocessing is applied the Least Squares Support Vector Machine based on Maximum Correlation Entropy Criterion and parallelly are data processed using Auto-Regressive Moving Average Model (ARMA). Final predicted value is generated as superposition from two procedures. As a conclusion is clear, that the established algorithm has good prediction accuracy and ability.

Different problem, but similar solutions are used in [24]. Article deals with prediction of indoor temperature via two different approaches. First model is created using the support vector machine (SVM) theory and second is based on back propagation neural network (BPNN). The experimental results show that the prediction error of the SVM model were less than the prediction error of the BPNN model and SVM model has the better prediction accuracy than the BPNN model. A different methodology for the issue of STLF procedure is used in work of Rafi et al. [25]. Published article describe using convolutional neural network (CNN) for state prediction. In conclusion, the forecasting outcomes of the proposed CNN model is compared with the outcomes of autoregressive integrated moving average (ARIMA) model that are most frequently used in time series forecasting arena. The conclusion is that every validation test of predicted load values using CNN model outperforms ARIMA model. This article illustrates the predominance of the artificial neural networks such a technology that can precisely handle a long sequence of time series load data and predict the future load in different periods.

2.1 Methodology

Artificial neural networks are, for their characteristics, widely used in various study fields, such as feature extraction, clustering, classification and prediction. It is clear, that ANNs can represent quite successful results, especially about predicting the values of design parameters in handled problems [26].

Artificial Neural Networks (ANN)

The domain of ANN creation and implementation is comprehensive and well developed. ANN used in our problem is created via MATLAB. Generally, every neural network consist of a series of layers, neurons from every layer are interconnected between other neurons from higher and lower layers. The final layer produces the network's output. In practical science, there are used many types of neural networks, more or less complicated. Figure 1 shows the basic structure of a feed-forward neural network. Here, the current input neuron layer by i, the related input neuron by j, and the output neuron in next layer by k are represented respectively, and input neuron by X_i, weight values of connections by w_{ijk}, and the bias value by b_{ij} are expressed respectively as i, j, k = 1, 2, ..., n.

Dependent variables expressed via neurons in the output layer are predicted by means of neurons in the input layer in this network structure and corresponding to independent variables. Each input value is multiplied via interconnection weights related to the output node. The output value is obtained via processing all obtained multiplication results by means of an activation function [26].

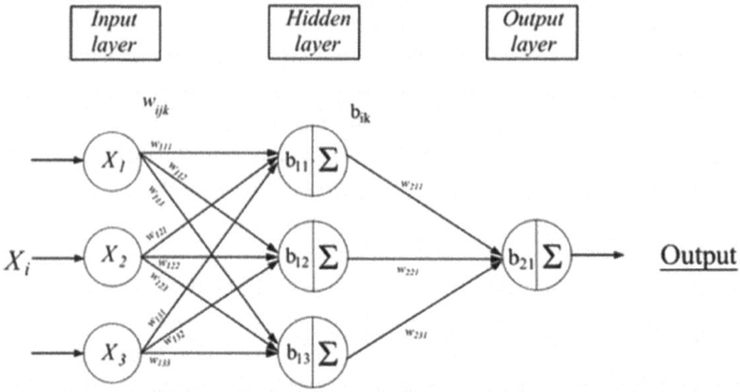

Fig. 1. Multilayer feed-forward neural network [26]

We are using simply Feedforward neural network, composed of three layers with seventeen neurons in input layer.

Artificial Neural Fuzzy Inference System (ANFIS)
The name and architecture of ANFIS itself evokes that it is a combination of a fuzzy logic and neural networks. ANFIS uses Takagi-Sugeno type fuzzy inference system (FIS). Basic architecture of ANFIS consists of five different layers. The first layer is known as a fuzzification layer where membership values are assigned to the inputs. Layer 2 is known as a rule layer where incoming signals are multiplied, and the product is sent to the layer 3 where the strength (weight) of the rules are calculated. Layer 4 consists of adaptive nodes which is named as a defuzzification layer. In this layer the weighted results are calculated for each rule. The fifth layer sums up outputs from the layer 4 which results to the real (crisp) output. Similarly, like ANN, ANFIS uses backpropagation learning rule or hybrid method which is a combination of a least squares and backpropagation rule to estimate the membership function parameters. Several membership function can be chosen i.e. gaussian, triangular, bell shaped and so on [27].

3 Data Set

The input data were provided by local heat plant which consists of three separate heat supply units. Our research is based on real data analysis from the heating process environment from periods of October and March from year 2014 up to 2017. This data set contains various heating load profiles from early autumn, winter and spring seasons. It is convenient that during this long period of data monitoring various weather changes occurred. The total load consumption consists of domestic hot water data and logs of hot water used for central heating, these samples were taken every hour. In addition to the data recorded from heat production, data from local weather station are also included in the data set. The units for load are in MW and for temperature in °C (see Fig. 2).

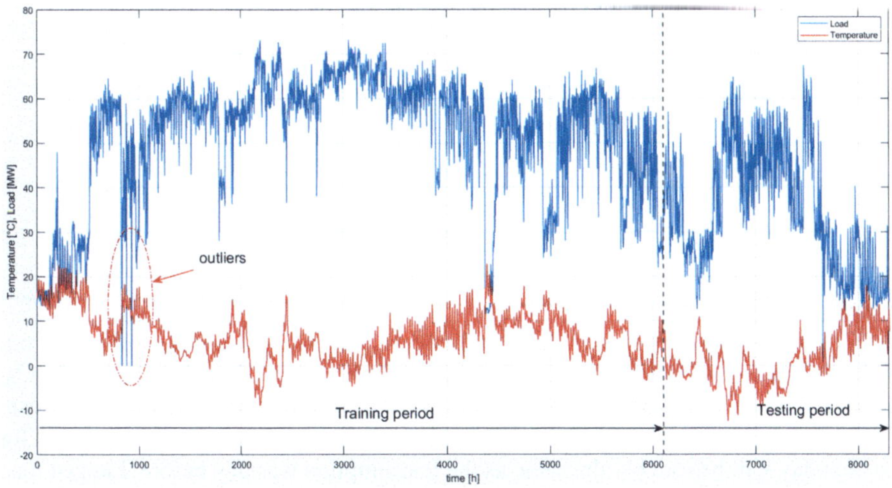

Fig. 2. Raw data set

3.1 Data Sets Normalization and Preprocessing

Data acquired from technological processes are often full of noise, measurements errors which are caused by equipment failure or incorrect calibration. These disturbing observations need to be removed from the data set in order to have the most realistic model. From depicted raw data set Fig. 2. Raw data set we can clearly see especially at the beginning of the heating season that our dataset contains several outliers which need to be detected and later replaced. In our research Z-Score standardization (1) was employed to detect the outliers. Detected outliers were replaced by linear interpolation.

$$x_s = \frac{(x - \mu)}{\sigma} \tag{1}$$

Where x – data, μ – data mean, σ – data standard deviation.

The next step was to filter out the noisy data by a simple moving average filter (2), which is frequently used among researchers due their simplicity. Large number of N could cause smoother output but it leads to lose the original heat load curve. In our case N was set to 3 to respect the original shape of the heat load curve. Statistical characteristics of the data are presented in Table 1.

$$x_{MA} = \frac{1}{N}\sum_{i=0}^{n-1} x_n - i \tag{2}$$

Where N – number of points used in moving average.

3.2 Feature Selection

Short Term Heat Load Forecasting is a complex task which is affected by several exogenous and endogenous factors. It is well known that consumption is mainly influenced by

Table 1. Statistical characteristics of input data set

	Raw set	After pre-processing
Min	0	12.06
Max	73.23	70.53
Std	16.63	16.07
Mean	45.83	45.81

meteorological especially by outdoor temperature, but other variables need to be considered during the modelling process. Calendar variables (day of week, hour, weekends) can improve the forecasting accuracy these variables were also included. Significant influence on consumption has the past heat load. It is likely that the consumption for the next day will be roughly the same as the consumption was day before. Lagged load variables were determined by autocorrelation and Fourier analysis where large peaks were detected at 24 h and 48 h. Selected inputs are presented in Table 2.

Table 2. List of input variables

No.	Name/Abbreviation	Representation
1	Hours (h)	0–23
2	Day of week (DoW)	1–7
3	Month (m)	10, 11, 12, 1, 2, 3
4	Weekend (w)	Binary
5–12	Average Temperature of T_{t-18}, T_{t-12}, T_{t-6}, Temperature T_{t-24h}, T_{t-2}, T_{t-1}, T_t,	Various
12–18	Average Load P_{t-24}, P_{t-18}, P_{t-12}, P_{t-6} Past Load P_{t-48}, P_{t-24}	Various

Above mentioned inputs are used to forecast P_{t+24} load with hourly sampling interval. Finally, the selected inputs were scaled into the range $[-1, 1]$ to avoid problems regarded to the network training (stuck in local minima) and computational problems. The data set was divided into training and testing set later the generalization ability of proposed models were validated on a holdout set from the period of 1.1.2018 to 31.3. 2018.

3.3 ANN and ANFIS Modelling

Firstly a single hidden layer feedforward neural network was created with Levenberg-Marquardt training algorithm. Number of neurons in the hidden layer were determined iteratively. The initial number of neurons were set to 10 each cycle the number of neurons was increased by 5. After performing several experiments, the networks with N neurons that produced the lowest error on the test set were used as lower and upper bound. In this part the number of neurons were increased by one. The best architecture was reached when number of hidden neurons were set to 34. Hyperbolic tangent sigmoid activation function was used in the hidden layer and purelin in the output layer. Both models were trained and tested on same period (Fig. 2) with the same input parameters. Trial and error approach was used to determine the best ANFIS architecture. Gaussian membership function was used in the fuzzification layer and subtractive clustering had to been used due large amount of data to determine the number of fuzzy rules and membership functions. Hybrid optimization was used to tune input/output mapped relationships.

The proposed methodology from raw inputs to prediction is depicted in Fig. 3.

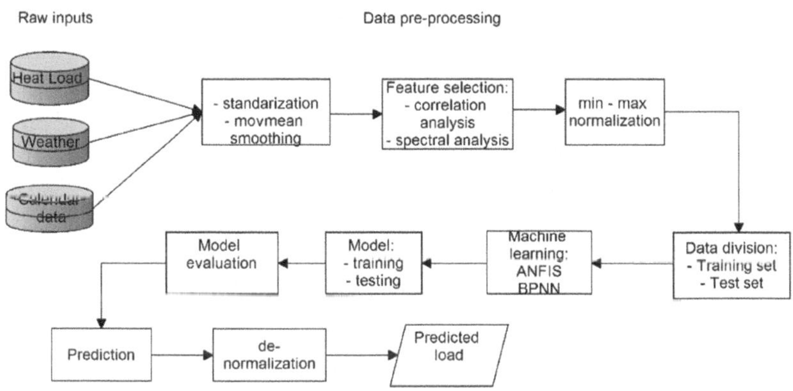

Fig. 3. The proposed flowchart of ANN and ANFIS modelling

4 Results

MATLAB 2020b environment was used to implement ANN and ANFIS models.

Three periods from validation set have been randomly selected to validate our proposed models: January 19, 2018 February 21, 2018, and March 22, 2018. The predicted horizon was 24 h with one-hour sampling interval.

4.1 Numerical Results

The prediction accuracy was evaluated with the following statistical metrics Mean Average Error (MAE), Mean Absolute Percentage Error (MAPE) and Root Mean Square Error (RMSE). The mathematical representation of metrics is described below:

1. Mean Average Error

$$MAE = \frac{1}{n}\sum_{i=1}^{n}\left|Y_i - \bar{Y}_i\right| \tag{3}$$

2. Mean Absolute Percentage Error

$$MAPE = \frac{1}{n}\sum_{i=1}^{n}\left|\frac{Y_i - \bar{Y}_i}{Y_i}\right| \tag{4}$$

3. Root Mean Square Error

$$RMSE = \sqrt{\frac{1}{n}\sum_{i=1}^{n}(Y_i - \bar{Y}_i)^2} \tag{5}$$

Where n – number of samples, Y_i – actual value, $\bar{Y}_i$ – predicted value. The smaller results represent better prediction accuracy. Validation numerical results are presented in Table 3 and Table 4. As seen in presented tables the ANFIS model have better results in every criterion compared to the ANN model.

Table 3. ANN results

Validation set	MAE [MW]	MAPE [%]	RMSE [MW]
19 January	1.38	6.04	1.42
21 February	1.26	3.78	1.28
22 March	0.96	2.09	0.98

Table 4. ANFIS results

Validation set	MAE [MW]	MAPE [%]	RMSE [MW]
19 January	0.68	2.93	0.73
21 February	0.34	1.01	0.43
22 March	0.50	1.07	0.62

Results obtained from models are depicted in Fig. 4. Black dashed line corresponds to the original heating curve for a target day, blue line represents the ANFIS output and red line corresponds to the output from the ANN. By visual checking we can state that our proposed models have good generalization ability. Each case the daily load curve tendency is captured. Some findings are clearly noticeable e.g. peaks for ANFIS prediction at approximately 5 h which is considered as input and ANN has much smoother curve compared to the ANN model. Also, we can state that ANN underestimates the daily load and ANFIS have tendency overestimate the load. In general, the results are acceptable.

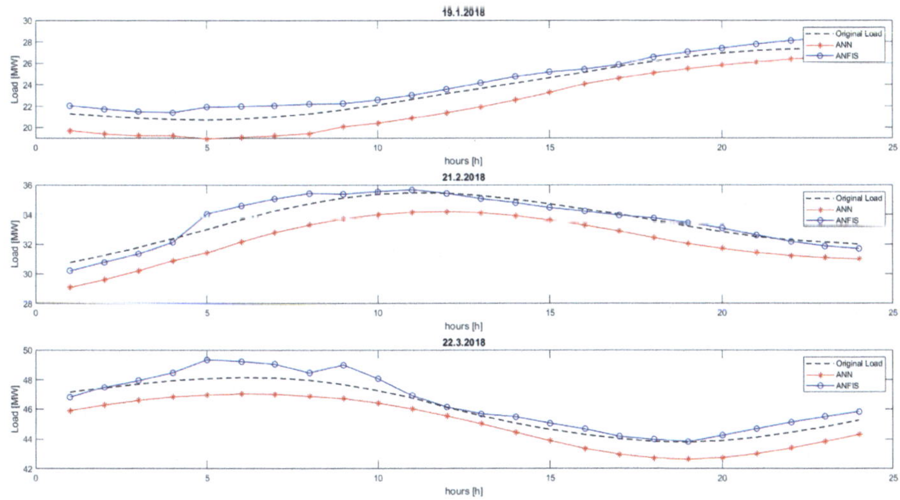

Fig. 4. ANN and ANFIS results

5 Conclusion

In this research two machine learning methods namely ANN and ANFIS were employed to predict day ahead heat load. Firstly, data pre-processing techniques were applied on raw data set to remove outliers. Spectral and correlation analysis was performed to detect important inputs. Main feature of used black-box modelling is absence of precise guidelines which makes it hard to obtain the best internal structure of the system. Number of hidden neurons was determined iteratively. Similarly, several membership functions were tested to obtain the best configuration for ANFIS. The performance of proposed models was measured by different statistical metrics. For validation of models three sets from three random days have been selected. In each case ANFIS model performed better outcomes than ANN model. ANN had tendency to underestimate the amount of heat load. However, the differences between models are not significant.

Overall, from outputs of proposed models it is clear that each model proved good generalization abilities and the obtained results are acceptable. These results encourage us for further research in this area. To obtain more accurate models in next research several changes are required. This means not only implementation of more advanced data-preprocessing is required, like wavelet analysis, empirical mode decomposition and alternatively others but also use of more accurate data patterns.

Acknowledgement. This publication is the result of implementation of the project KEGA 20/1769, 005STU-4/2020: "Innovation and new learning opportunities in industrial process management with PLC" supported by the KEGA.

References

1. Soliman, S.A.H., Al-Kandari, A.M.: Electrical Load Forecasting – Modeling and Model Construction, 1st edn. Butterworh-Heinemann, Oxford (2010)

2. Baliyan, A., Gaurav, K., Mishra, S.K.: A review of short term load forecasting using artificial neural network models. In: International Conference on Intelligent Computing, Communication & Convergence, ICCC 2015, vol. 48, pp. 121–125. Elsevier, Netherlands (2015)
3. Idowu, S., et al.: Forecasting heat load for smart district heating systems: a machine learning approach. In: IEEE International Conference on Smart Grid Communications 2014, SmartGridComm, Italy, pp. 554–559 (2014)
4. Fang, T., Lahdelma, R.: Evaluation of a multiple linear regression model and SARIMA model in forecasting heat demand for district heating system. Appl. Energy **179**, 544–552 (2016)
5. Chramcov, B.: Utilization of mathematica environment for designing the forecast model of heat demand. WSEAS Trans. Heat Mass Transf. **6**(1), 21–30 (2011)
6. Wojdyga, K.: Predicting heat demand for a district heating systems. Int. J. Energy Power Eng. **3**(5), 237–244 (2014)
7. Al-Sammari, E.I., et al.: Prediction of heat load in district heating systems by support vector machine with firefly searching algorithm. Energy **95**, 266–273 (2016)
8. Ahmmed, S., Khan, M.A.A., Hasan, M.K., Saber, A.Y., Huda, M.N., Rahman, M.Z.: STLF using neural networks and fuzzy for anomalous load scenarios - a case study for Hajj. In: International Conference on Electrical & Computer Engineering (ICECE 2010), Dhaka, pp. 722–725 (2010). https://doi.org/10.1109/ICELCE.2010.5700794
9. Panda, S.K., Ray, P., Mishra, D.P.: Effectiveness of GA on short term load forecasting. In: 2019 International Conference on Information Technology (ICIT), Bhubaneswar, India, pp. 27–32 (2019). https://doi.org/10.1109/ICIT48102.2019.00011
10. Huang, L., Coulson, J., Lygeros, J., Dörfler, F.: Data-enabled predictive control for grid-connected power converters. In: 2019 IEEE 58th Conference on Decision and Control (CDC), Nice, France, pp. 8130–8135 (2019). https://doi.org/10.1109/CDC40024.2019.9029522
11. Wanga, H., Lia, Y., Khan, S.A., Luoa, Y.: Prediction of breast cancer distant recurrence using natural language processing and knowledge-guided convolutional neural network. Artif. Intell. Med. **110** (2020). https://doi.org/10.1016/j.artmed.2020.101977
12. Yonekura, K., Hattori, H., Suzuki, T.: Short-term local weather forecast using dense weather station by deep neural network. In: 2018 IEEE International Conference on Big Data (Big Data), Seattle, WA, USA, pp. 1683–1690 (2018). https://doi.org/10.1109/BigData.2018.8622195
13. Naveen, L., Mohan, H.S.: Atmospheric weather prediction using various machine learning techniques: a survey. In: 2019 3rd International Conference on Computing Methodologies and Communication (ICCMC), Erode, India, pp. 422–428 (2019). https://doi.org/10.1109/ICCMC.2019.8819643
14. Fente, D.N., Singh, D.K.: Weather forecasting using artificial neural network. In: 2018 Second International Conference on Inventive Communication and Computational Technologies (ICICCT), Coimbatore, pp. 1757–1761 (2018). https://doi.org/10.1109/ICICCT.2018.8473167
15. Poonia, P., Jain, V.K.: Short-term traffic flow prediction: using LSTM. In: 2020 International Conference on Emerging Trends in Communication, Control and Computing (ICONC3), Lakshmangarh, Sikar, India, pp. 1–4 (2020). https://doi.org/10.1109/ICONC345789.2020.9117329
16. Mackenzie, J., Roddick, J.F., Zito, R.: An evaluation of HTM and LSTM for short-term arterial traffic flow prediction. IEEE Trans. Intell. Transp. Syst. **20**(5), 1847–1857 (2019). https://doi.org/10.1109/TITS.2018.2843349
17. Asim, K.M., Idris, A., Martínez-Álvarez, F., Iqbal, T.: Short term earthquake prediction in Hindukush region using tree based ensemble learning. In: 2016 International Conference on Frontiers of Information Technology (FIT), Islamabad, pp. 365–370 (2016). https://doi.org/10.1109/FIT.2016.073

18. Li, R., Lu, X., Li, S., Yang, H., Qiu, J., Zhang, L.: DLEP: a deep learning model for earthquake prediction. In: 2020 International Joint Conference on Neural Networks (IJCNN), Glasgow, UK, pp. 1–8 (2020). https://doi.org/10.1109/IJCNN48605.2020.9207621
19. Ye, J., Yang, L.: Short-term load forecasting using ensemble empirical mode decomposition and harmony search optimized support vector regression. In: 2019 14th IEEE Conference on Industrial Electronics and Applications (ICIEA), Xi'an, China, pp. 851–855 (2019). https://doi.org/10.1109/ICIEA.2019.8834259
20. Inteha, A.,Nahid-Al-Masood, Deeba, S.R.: An Artificial intelligence based day lag technique for day ahead short term load forecasting. In: 2020 IEEE Region 10 Symposium (TENSYMP), Dhaka, Bangladesh, pp. 626–629 (2020). https://doi.org/10.1109/TENSYMP50017.2020.9230805
21. Nadeem, A., Arshad, N.: Short term load forecasting on PRECON dataset. In: 2019 International Conference on Advances in the Emerging Computing Technologies (AECT), Al Madinah Al Munawwarah, Saudi Arabia, pp. 1–6 (2020). https://doi.org/10.1109/AECT47998.2020.9194176
22. Bonetto, R., Rossi, M.: Parallel multi-step ahead power demand forecasting through NAR neural networks. In: 2016 IEEE International Conference on Smart Grid Communications (SmartGridComm), Sydney, NSW, pp. 314–319 (2016). https://doi.org/10.1109/SmartGridComm.2016.7778780
23. Jiandong, D., Peng, W., Wentao, M., Xuan, T., Xinyu, Q.: Daily electricity consumption combined forecast based on variational mode decomposition and sample entropy. In: 2019 IEEE Sustainable Power and Energy Conference (iSPEC), Beijing, China, pp. 762–767 (2019). https://doi.org/10.1109/iSPEC48194.2019.8975040. Author, F.: Article title. Journal 2(5), 99–110 (2016)
24. Qi, C., Wenbiao, W., Siyuan, W.: Application of indoor temperature prediction based on SVM and BPNN. In: The 27th Chinese Control and Decision Conference (2015 CCDC), Qingdao, pp. 2883–2887 (2015). https://doi.org/10.1109/CCDC.2015.7162418.Bb
25. Rafi, S.H., Nahid-Al-Masood: Short term electric load forecasting using high precision convolutional neural network. In: 2020 International Conference on Computer, Electrical & Communication Engineering (ICCECE), Kolkata, India, pp. 1–7 (2020). https://doi.org/10.1109/ICCECE48148.2020.9223102
26. Bekdaş, G., Nigdeli, S.M., Yücel, M.: Artificial Intelligence and Machine Learning Applications in Civil, Mechanical, and Industrial Engineering. IGI Global (2020). https://app.knovel.com/hotlink/toc/id:kpAIMLACM8/artificial-intelligence/artificial-intelligence
27. Jang, J.R.: Input selection for ANFIS learning. In: 1996 Proceedings of IEEE 5th International Fuzzy Systems, New Orleans, LA, USA, pp. 1493–1499 (1996)

Emergency Locksmith: A Restricted Dynamic Device Authentication Design Pattern for the Healthcare IoT

Alper Ugur(✉)

Pamukkale University, 20020 Denizli, Turkey
augur@pau.edu.tr

Abstract. Authentication is one of the emerging mechanisms required for a reliable IoT system. In health care environment, this became critical due to the nature of data carried and tasks of the system. There are various proposed schemes for authentication. Password based, cryptography based, hardware based, and behavioral are the extensive approaches with different characteristics. However, as in other IoT environments, the devices in healthcare are vary widely with their hardware limitations. The sensitive data they are responsible for is another case. Therefore, accomplishing the desired cooperation between diverse devices is limited in applications. In emergency, the healthcare IoT devices must be available and coordinate to share vital data to the emergency systems in need. Software design patterns are known to enable integration of software objects and modules in a reusable way. In this paper we described emergency authentication cases and proposed a design pattern for dynamic changing environment of the IoT. As on software paradigm, the pattern is convenient for heterogeneity of healthcare IoT system to adapt and implement its authentication task and priority of human health. The security of the data access is provided with applied restrictions. Achievements and shortcomings are also discussed.

Keywords: Healthcare · Emergency · IoT · Authentication · Design patterns

1 Introduction

Health systems are rapidly adopted make use of current technologies, such as artificial intelligence and robotics on operations, sensor networks and Internet of Things (IoT) on monitoring patients health status, real-time diagnosis etc. [1]. IoT solutions aims health care to be much reasonable and accessible. An efficient healthcare ser-vice is about prevention, early detection and homecare which expected to minimize high-priced operations in hospitals as clinical care [2].

Healthcare IoT applications enable online interaction with patients, identification and tracking patients and their records, determination locations of paramedics, tracing medical supply chain, logistics, even drug management and drug counterfeit detection [3]. The leading event of a healthcare application is the patient's body monitoring as capturing ECG, body temperature, heartrate and blood pressure data [4]. Ambient monitoring

© The Author(s), under exclusive license to Springer Nature Switzerland AG 2021
R. Silhavy (Ed.): CSOC 2021, LNNS 229, pp. 518–526, 2021.
https://doi.org/10.1007/978-3-030-77445-5_48

is another main concept to capture date of patient's environment such as temperature, light, humidity, location, body position, motion data, CO2, etc. The data acquired from IoT sources can be used to be a feedback for a smart diagnosis system or a smart home for comfortable environment adjustments even tele-robotic surgery [3–6].

Heterogeneity of different integration domains causes problem in finding a suitable cooperated IoT platforms. Although, IoT platforms employ similar tasks, their implementation and the core mechanisms may differ [7]. In communication protocols the diversity spans from 6LoWPAN to NFC, and REST to CoAP in application protocols [4].

In the environment it was designed for, most healthcare appliances work like a clockwork. Authenticates itself for instance to the smart home data collection inspector, mutual authenticate with other appliances for communication and analysis etc. Yet, outside of its safe domain, the security precautions as authentication and re-striction on sharing private data access can cause vital problems.

In case of public transportation accidents, building collapses due to the nature causes etc., identification of the injured person during emergency intervention is important for obtaining the health history. If the person is conscious the health-related query would be quick and may be effective. Otherwise, any identifier like medical alert bracelet or credential to query in the health system is useful. In above cata-strophic cases, these are expectations that were hoped to be found but often not with-in range. The healthcare IoTs may still working but prevents access due to authentication restrictions.

Numerous works on emergency healthcare IoT models were focused on identification of patient by discovering related IoT provider [8] or authenticating implanted sensors for health condition measurements [9]. The current approaches utilize limited healthcare IoT devices in an emergency. Through widespread of new usable technologies wearables as in [10] will provide status of a patient more than remote assistant systems. We propose an emergency authentication model to enable access to the existing healthcare IoT on the casualty. The model involves restricted emergency data definition and a dynamic authentication pattern enabling to operate in heterogeneity of IoT environment.

The paper is arranged as follows. Section 2 consists of related work on healthcare IoT and patterns used in IoT. In Sect. 3, the emergency authentication pattern is presented. The results obtained are given in Sect. 4. Finally, in Sect. 5, the paper is concluded with discussion and future directions.

2 Background Research

2.1 Current Emergency Oriented Healthcare IoT Works

Emergency oriented systems aimed to ensure the survival of the patient or casualty in a quick and efficient way. The critical data in most cases are the identification, the location - if it is an emergency call-, and the situation of the patient. The patient condition may not be stable, or the psychological status of the witness waste vital time through verbal reporting [8], that even might not that sufficient at the end.

In [8], the system suggests a biometric identification system. The fingerprint of patient is achieved by a mobile application and send to identification middleware to inform local hospital for patient preparation. A mobile first aid protocol through mobile phone is activated until paramedics' arrival.

In [9], patient's fingerprints are used to authenticate with the implantable device, for ECG readings on patient. This enables access data for emergency medical teams.

In [11] a connection oriented mHealth system is proposed to achieve data from a patient even in noisy environments by using close mobile resources. In [12], the privacy and security of communication between patient and the help center is supported.

2.2 Design Patterns in IoT

The above approaches deal with network and perception layer. The cross-domain resource sharing, and collaborations are still a research challenge for IoT access control [13]. This work presents an introductory design pattern for a sufficiently flexible framework for heterogeneity characteristic of IoT.

Various design patterns were presented for IoT architecture. These design patterns enable forming an abstract model of data flow used by different applications. Several architectural, general and domain specific patterns were submitted in recent years. In [14], Industrial IoT design patterns were summarized. In [15], self-adaptive IoT design patterns were presented. In [16], classification of recent design patterns on IoT was given.

The most similar patterns to the proposed work are given as follows. In [17], strategy design pattern is used with device-dependent class which implements the behavior interface, where the agent core act as a proxy. Authentication restrictions were not implicated in the study. In [18], authorization and authentication patterns serve for remote access of database records. The proposed pattern aimed real-time data through healthcare IoT devices. In [19], in an energy system case an actor with an observer pattern is presented as a middleware platform.

3 Emergency Locksmith: An Authentication Design Pattern

As stated in prior sections, most works in healthcare IoT are focused on achieving remote electronic health records (EHR). In an emergency, authentication phase is accomplished with the patient's biometrics as in [8, 9]. In public transportation accidents or disasters, it is likely that mobile phones even bodies are thrown around and away. Authentication with persons own key device as mobile phone may fail in the absence of it. The authentication of devices with the biometrics becomes dysfunctional where the casualty is not in close-range.

Even the devices on the patient can emit signal, they turn out to be inaccessible. Without a proper authentication, crucial data is unavailable for emergency response team (ERT). Likewise, authentication of the ERT to the patient healthcare IoT devices as a trusted entity will provide direct and real-time status reports without the need to solve any routing problems.

In IoT environment, high diversity of authentication methods exists. Password based, cryptography based, hardware based, and behavioral are the wide-ranging approaches with distinct characteristics [20]. As in other IoT environments, the devices in healthcare

are vary widely by their hardware limitations. This suppresses implementation of any scheme on desired device.

We propose a solution for the case with an ERT device designed as multi-functional authentication entity. This requires implementation design pattern for emergency ready devices in healthcare IoT solutions.

The pattern is described with the document format used in [21]. The document sections are as follows.

- *Pattern Name:* This name is used to refer and identify the design challenge. In this case the pattern is named as Emergency Locksmith to be abstract-enough to reflect the pattern's intended design.
- *Problem:* This is a short summary of the pattern.
- *Context:* This describes the setting in which the problem arises.
- *Motivation Forces:* A use-case scenario, in which the problem is likely to happen and therefore the pattern can be used effectively.
- *Solution Details:* This is the statement about how the pattern solves the problem.
- *Sketch:* The resulting architecture following utilization of the pattern.

Problem. How can healthcare IoT devices dynamically authenticate with an ERT device in emergency? How ERT device adapt its authentication mechanism according to requesting patient device?

Context. The healthcare IoT device need to authenticate with an unknown device. This may be triggered by an event. The heterogeneity of devices and authentication schemes require large scale support on authentication. This demands multiple authentication implementation interface. The data flow will be one way and restricted due to security. This yields to the need of defining safe to share data in an emergency.

Motivation Forces. Suppose in an emergency the patient is away from reach at the time of arrival to the accident scene and he is unconscious. The scenario is illustrated in Fig. 1. If paramedics can learn his credentials, they may be able to plan next step, arranging the medicine or hospital. The healthcare IoT devices on his body or smart car may help his current status for a vital decision.

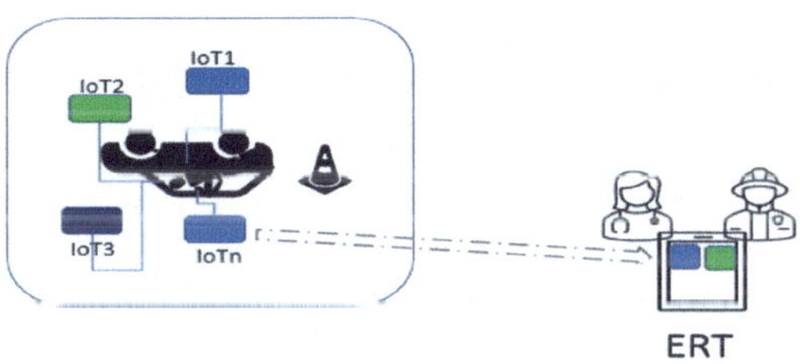

Fig. 1. A dynamic authentication case of ERT device

Devices protect sensitive data with authentication mechanism where an authenticated device such as personal mobile phone has access but any malicious device are rejected.

You may implement an ERT device which is able to authenticate with the healthcare device. It may provide broad set of authentication methods of device vendors. And your implementation will interchange routinely to the matched method and parameters. As in middleware systems the structure can support distinctive protocols. To restrict unauthorized access, the sensitive data flow will not be in two way like common middleware operations [16].

Solution Details. Initial phase is event-driven, when healthcare IoT devices measure unusual values such as high or low heartbeat rate, heat or cold they activate emergency mode as in common sense. Unlike the usual, the devices will use a temporal emergency subroutine that enables the ERT authentication. A notification and recover procedure are activated for false positive interpretations. The user interaction will end emergency state.

The solution is based on multiple utilization of software design patterns. The different protocols on ERT can be employed in the design with strategy pattern (Fig. 2a). The selection is associated to the preferred healthcare IoT device. If exists health care devices on or near patient will implement observer pattern (Fig. 2b). This enables ERT is identified by all devices with a single device authentication. The authenticator changes its state to notify its dependents that the ERT is authentic as an emergency entity. Henceforth, they push restricted and momentarily data to ERT.

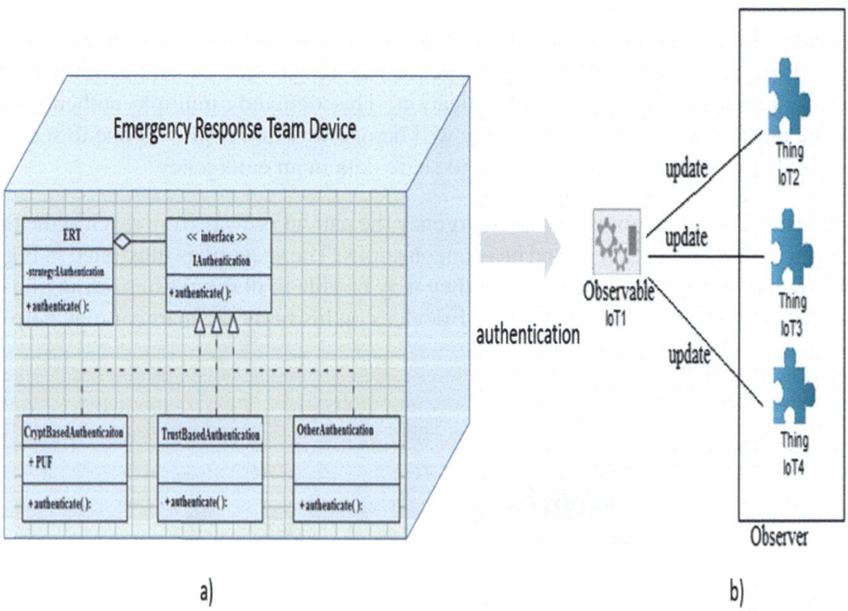

Fig. 2. Patterns in design a) Strategy pattern on ERT, b) Observer pattern on healthcare IoT

The ERT device has a physically unclonable function (PUF) code [22] and use it for authentication as an emergency device. The PUF data must be available in observable device.

If the healthcare devices have restricted hardware capability, ERT device changes its authentication mode to trust based one. The anomaly sensing healthcare devices share data to the ERT and use recommendation bias by changing conditions. If unusual measurements follow normalization curve after contribution, the ERT device authenticated as trusted entity. Else, after a threshold time data share will stop and the device is added on blacklist as a security precaution. This also provides protection to passing by malicious devices in normal conditions.

A colocation parameter is determined between ERT and the healthcare devices for emergency subroutine to prevent malicious nodes. The data shared with the device that introduces itself as ERT is limited with anomaly sensing vital measurements. The historical data is bound to the trust level. The UML sequence diagram of the design is given in Fig. 3.

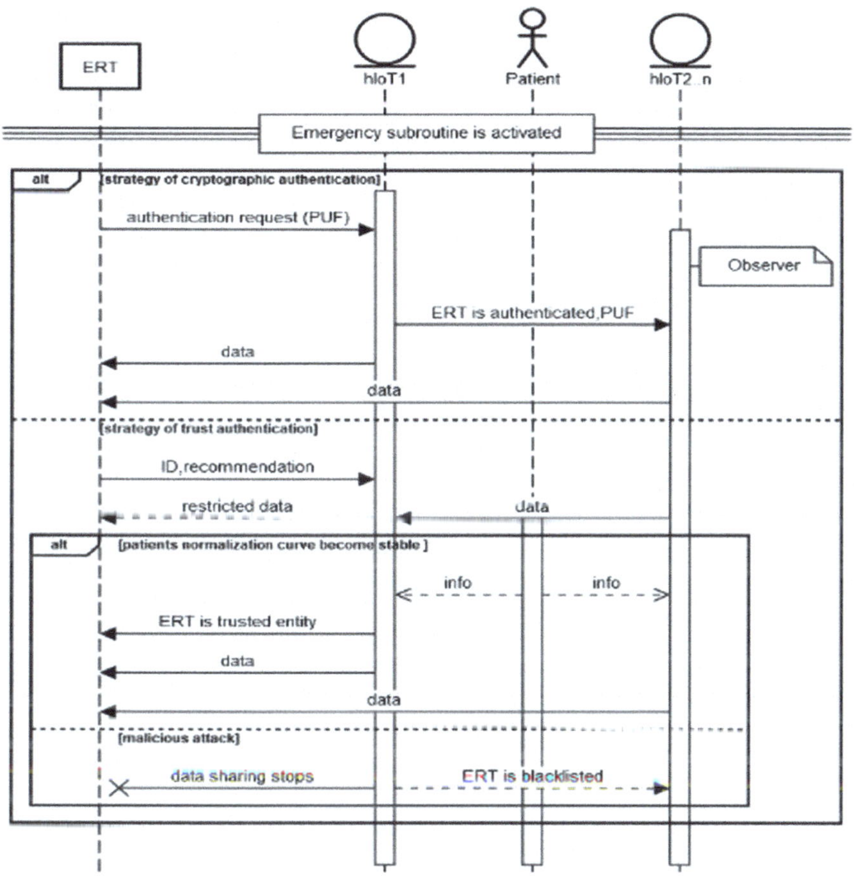

Fig. 3. Sequence diagram of the proposed pattern.

4 Results

The characteristics of the proposed design is summarized with the mentioned recent works in Table 1. The main advantage of the proposed scheme is its decentralized access mechanism. The status of the patient can be acquired without propagation delay. The design does not require physical contact as in biometric authentication models and has ability to access collocated IoT devices. To acquire heterogeneity design creates density but direct communication avoids the complexity of routing. Patient biometric data is not shared with post authorized reporters as in [8] but data sharing is restricted to circumvent the emergency. The security level is marked as low due to the lack of a comprehensive attack analysis. It also depends on reliability of the authentication mechanisms contained in strategy pattern.

Table 1. Comparison of the characteristics of the proposed design with the recent healthcare IoT authentication schemes

Features	[8]	[9]	[12]	Proposed
Healthcare IoT data access type	Indirect	Indirect/Direct	Indirect	Direct
Require Central server	Yes	Yes	Yes	No
Require physical connection to the patient	Yes	Yes	N/A	No
Device	Mobile phone	Mobile Phone and implant	Multiple devices	ERT device
Design complexity	Low	Medium	High	Medium
Heterogeneity of IoT devices	No	No	N/A	Yes
Authentication mechanism	Biometric	Biometric and cryptographic	Cryptographic	Multiple
Multiple IoT device access	Single	Single	Multiple (for routing)	Multiple
Privacy of sensitive data	Protected	Protected	Protected	N/A
Security level	Low	Medium	High	Low

5 Discussion and Future Directions

A restricted dynamic authentication design pattern is presented for healthcare IoT devices. An ERT device that is capable of multiple authentication scheme is proposed. The pattern does not need central data server to access healthcare vital data.

It employs direct and real-time measurement after authentication phase. Design differs among healthcare models with this efficient feature for emergency situations. The pattern supports heterogeneity of the IoT devices and adapts to their service.

The strategy pattern requires parameters for each specific authentication scheme used by healthcare IoT devices. Consequently, the data storage and process resources of ERT device must be sufficient.

IoT authentication is one of the widely studied topic and there are numerous attack scenarios for specific use cases. There are security safeguards proposed for the malicious devices such as PUF, emergency threshold etc. Hence functionality is priority of the design, security is not fully supported against entire authentication attacks.

Potential solutions in literature are evaluated as the security-functionality tradeoff decision. For instance, employing cryptographic countermeasures may cause negative impact on heterogeneity where for instance resource restricted healthcare devices should be excluded. The security improvements and detailed attack analysis of the design is left for further works.

References

1. Jara, A.J., Zamora-Izquierdo, M.A., Skarmeta, A.F.: Interconnection framework for mHealth and remote monitoring based on the internet of things. IEEE J. Sel. Areas Commun. **31**(9), 47–65 (2013)
2. Kodali, R., Swamy, G., Boppana, L.: An implementation of IoT for healthcare. In: 2015 IEEE Recent Advances in Intelligent Computational Systems (RAICS), Trivandrum, India, pp. 411–416 (2015)
3. Tyagi, S., Agarwal, A., Maheshwari, P.: A conceptual framework for IoT-based healthcare system using cloud computing. In: 2016 6th International Conference - Cloud System and Big Data Engineering (Confluence), Noida, India, pp. 503–507 (2016)
4. de Morais Barroca Filho, I., de Aquino Junior, G.S.: IoT-based healthcare applications: a review. In: Gervasi, O., et al. (eds.) ICCSA 2017. LNCS, vol. 10409, pp. 47–62. Springer, Cham (2017). https://doi.org/10.1007/978-3-319-62407-5_4
5. Periyasamy, K., Alagar, V., Wan, K.: Dependable design for elderly health care (2017)
6. Saha, R., Kumar, G., Rai, M.K., Thomas, R., Lim, S.J.: Privacy ensured e-healthcare for fog-enhanced IoT based applications. IEEE Access **7**, 44536–44543 (2019)
7. Guth, J., Breitenbücher, U., Falkenthal, M., Leymann, F., Reinfurt, L.: Comparison of IoT platform architectures: a field study based on a reference architecture. In: 2016 Cloudification of the Internet of Things (CIoT), pp. 1–6. IEEE, Paris (2016)
8. Yun, S., Jung, H., Yun, S.: Smart emergency rescue system based on biometric authentication. In: Park, J., Stojmenovic, I., Jeong, H., Yi, G. (eds.) Computer Science and its Applications. LNEE, vol. 330, pp. 1043–1048. Springer, Berlin (2015). https://doi.org/10.1007/978-3-662-45402-2_148
9. Belkhouja, T., Du, X., Mohamed, A., Abdulla, K., Guizani, M.: Biometric-based authentication scheme for implantable medical devices during emergency situations. Futur. Gener. Comput. Syst. **98**, 109–119 (2019)
10. Chen, M., Ma, Y., Li, Y., Wu, D., Zhang, Y., Youn, C.: Wearable 2.0: enabling human-cloud integration in next generation healthcare systems. IEEE Commun. Mag. **55**(1), 54–61 (2017)
11. Lu, R., Lin, X., Shen, X.: SPOC: a secure and privacy-preserving opportunistic computing framework for mobile-healthcare emergency. IEEE Trans. Parallel Distrib. Syst. **24**(3), 614–624 (2013)

12. Meshram, C., Lee, C.-C., Meshram, S.G., Ramteke, R.J., Meshram, A.: An efficient mobile-healthcare emergency framework. J. Med. Syst. **44**(3), 1–14 (2020). https://doi.org/10.1007/s10916-019-1458-3
13. Bertin, E., Hussein, D., Sengul, C., Frey, V.: Access control in the internet of things: a survey of existing approaches and open research questions. Ann. Telecommun. **74**(7–8), 375–388 (2019). https://doi.org/10.1007/s12243-019-00709-7
14. Bloom, G., Alsulami, B., Nwafor, E., Bertolotti, I.C.: Design patterns for the industrial internet of things. In: 2018 14th IEEE International Workshop on Factory Communication Systems (WFCS), Imperia, pp. 1–10 (2018)
15. Krupitzer, C., Temizer, T., Prantl, T., Raibulet, C.: An overview of design patterns for self-adaptive systems in the context of the internet of things. IEEE Access **8**, 187384–187399 (2020)
16. Washizaki, H., Ogata, S., Hazeyama, A., Okubo, T., Fernandez, E.B., Yoshioka, N.: Landscape of architecture and design patterns for IoT systems. IEEE Internet Things J. **7**(10), 10091–10101 (2020)
17. Jung, E., Cho, I., Kang, S.M.: An agent modeling for overcoming the heterogeneity in the IoT with design patterns. In: Park, J., Adeli, H., Park, N., Woungang, I. (eds.) Mobile, Ubiquitous, and Intelligent Computing. LNEE, vol. 274, pp. 69–74. Springer, Heidelberg (2014). https://doi.org/10.1007/978-3-642-40675-1_11
18. Ali, I., Asif, M.: Applying security patterns for authorization of users in IoT based applications. In: 2018 International Conference on Engineering and Emerging Technologies (ICEET), Lahore, pp. 1–5 (2018)
19. Walker, M.A., Dubey, A., Laszka, A., Schmidt, D.C.: Platibart: a platform for transactive IoT blockchain applications with repeatable testing. In: Proceedings of the 4th Workshop on Middleware and Applications for the Internet of Things, pp. 17–22 (2017)
20. El-Hajj, M., Fadlallah, A., Chamoun, M., Serhrouchni, A.: A survey of internet of things (IoT) authentication schemes. Sensors **19**(5), 1141 (2019)
21. Qanbari, S., Pezeshki, S., Raisi, R., et al.: IoT design patterns: computational constructs to design, build and engineer edge applications. In: 2016 IEEE First International Conference on Internet-of-Things Design and Implementation (IoTDI), pp. 277–282. IEEE (2016)
22. Garg, S., Kaur, K., Kaddoum, G., Choo, K.R.: Toward secure and provable authentication for internet of things: realizing industry 4.0. IEEE Internet Things J. **7**(5), 4598–4606 (2020)

Fuzzy Model with Meteorological Variables for the Determination of the THSW Index and the Electric Field in the Area of East Lima, Peru

Juan J. Soria[1(✉)], Orlando Poma[2], David A. Sumire[2], and Joel Hugo Fernandez Rojas[2]

[1] Escuela Profesional de Ingeniería de Sistemas, Universidad Peruana Unión, Lima, Peru
jesussoria@upeu.edu.pe

[2] Escuela Profesional de Ingeniería Ambiental, Universidad Peruana Unión, Lima, Peru

Abstract. The prediction of environmental variables is important because it facilitates understanding the processes in real time. The objective of this study was to formulate a fuzzy model of four environmental input variables, namely temperature, UV radiation, humidity and wind speed, and two environmental output variables, namely THSW index and electric field. The Mandani fuzzy method was used with 80 fuzzy rules concatenated with the "and" conector defuzzifying the THSW index and the electric field with 19 654 records. The fuzzy model was based on a maximum temperature for 2019 of 29.8 °C with an average of 24.595 °C and a maximum UV radiation of 11.7 and an average of 1.9506 nm, as well as an average humidity of 6.317%, in addition to a wind speed of 12.90 m/s resulting lower in spring with an average of 1.2727 m/s. Likewise, an average THSW index of 25.169 was obtained as output, measured with four meteorological variables with a minimum of 21.055 in spring and a maximum of 25.169 in summer. In addition, an average electric field of −1.6218 kV/m was found in winter and in summer 2019, a value of 0.1682 kV/m was obtained with a minimum value of −6.6993 kV/m and a maximum value of 3.6323 kV/m.

Robustness of the Fuzzy model was determined with the Friedman test, where the THSW index had a value of 0.0014 and the electric field had a value of 0.0021, which shows a good performance of the proposed model.

Keywords: Fuzzy model · Meteorological variables · Electric field · THSW index · Artificial intelligence · Mandani method

1 Introduction

The earth is a system that interacts with other subsystems, lithosphere, hydrosphere, biosphere and atmosphere, each of these spheres interact with each other. Elizalde (2009) states that between the outermost layers of the earth, seen from the outside in the sequence of atmosphere, hydrosphere, biosphere, pedosphere, lithosphere, there are constant and important exchanges of matter and energy. In this study, the emphasis is on the process

© The Author(s), under exclusive license to Springer Nature Switzerland AG 2021
R. Silhavy (Ed.): CSOC 2021, LNNS 229, pp. 527–544, 2021.
https://doi.org/10.1007/978-3-030-77445-5_49

of the electrical circuit, predominantly capacitive, where the earth at a global level can be seen as a large capacitor, in which the plates are the upper layer of the ionosphere and the earth; both behave as a capacitor separated by a variable dielectric which is the air. In this dielectric, therefore, there is an atmospheric electric field, whenever there is a wide or partial voltage difference between its plates. (Martinez 2014).

Nicoll et al. (2019) have established that the earth has a global atmospheric electric circuit, which can be seen in areas that are not electrically disturbed by weather. Therefore, studies have been conducted in this area for more than a century and have been used in atmospheric sciences to understand the quasi-static atmospheric electric field.

Antunes de Sá et al. (2020), indicate that the global electric circuit is maintained even in the absence of electrified clouds with a near-surface electric field in fair weather conditions of around −100 V/m.

The electric field near the Earth's surface in the absence of electrified clouds is maintained by the global electric circuit.

Harrison et al. (2020), indicate that the atmospheric positive charge distribution of the weather is analyzed in more detail in the context of layer cloud effects, considering their complexity, which is a key element in modeling the global electric field, the complexity and variability of the processes occurring within the planetary boundary layer.

For this reason, we agree with Marcos et al. (2021), who emphasize the need for tools to identify and study the characteristics of the atmosphere. The atmosphere is a complex system and an important aspect for its understanding is the meteorological parameters.

Mirbagheri et al. (2020) studied the propagation of the GSM frequency band and its dependence on propagating waves in the troposphere. They suggest that the intensity values of electromagnetic waves can be used for different conditions. Romero et al. (2020) further consider that the meteorological parameters that influence particle concentration are surface air temperature, relative surface air humidity, total daily precipitation, total column cloud cover, surface wind speed and wind speed direction.

Likewise, Azad et al. (2020) suggest that the deterministic part is explained by the motion between the Sun and the Earth, and follows a diurnal cycle, with the most significant and unexpected fluctuations in photovoltaic production due to the stochastic behavior of the atmosphere, uncertainty that is due to multiple meteorological variables, such as, temperature, humidity level, visibility, air pressure, wind speed and cloud cover.

In Peru, Romero et al. (2020) conducted studies on the effects of meteorological parameters on air quality with respect to fine particulate matter with diameters of 1, 2, 5 and 10 μm.

To understand the complex atmospheric system, various models of artificial intelligence, including neuro-fuzzy (NF) techniques by Shiri (2019) and Azad et al. (2020) have been applied in recent years, applying computational models such as artificial neural networks (ANN) and adaptive neuro-fuzzy inference system (ANFIS), as fuzzy rules can provide a transparent linguistic language to describe the model. Fuzzy systems provide the possibility to integrate information processing with the attractive mathematical properties of general membership function approximators such as triangular, trapezoidal, Gaussian and sigmoidal functions like the one used by Saboya et al. (2019) in their fuzzy model to determine the graduate profile of Peruvian university students.

In this study, fuzzy models under rules were applied to analyze the behavior of meteorological variables such as temperature, relative humidity, wind speed, ultraviolet (UV) radiation and THSW index associated with the electric field, determining the characteristic values in the different seasons of the year 2019.

2 Literature Review

2.1 Fuzzy Logic

The membership function of a fuzzy set A over a universe X is defined by $\mu_A : X \to [0, 1]$, where $\mu_A(x) = r$ if r is the degree to which x belongs to A. If the set is crisp, the membership function takes the values of {0, 1}, while if it is fuzzy, it will take them within the closed interval [0, 1] (Sivanandam et al. 2007).

Matute and Bernal Suárez (2017) mention that the most used membership functions are the triangular, trapezoidal and Gaussian, which are defined as follows:

The triangular function, defined by the lower bound **a**, the upper bound **b** and the modal value **m**, such that **a < m < b**, whose correspondence rule is shown in Eq. (1).

$$\mu_A(x, a, m, b) = \begin{cases} 0, & Si\, x < a \\ \dfrac{x-a}{m-a}, & Si\, a < x \leq m \\ \dfrac{b-x}{b-m}, & Si\, m < x < b \\ 0, & Si\, x \geq b \end{cases} \tag{1}$$

The trapezoidal membership function, defined by its lower bound **a**, upper bound **d**, and the lower support limit **b** and upper support limit **c**, such that **a < b < c < d**, whose correspondence rule is shown in Eq. (2).

$$\mu_A(x, a, b, c, d) = \begin{cases} 0, & Si\, x < a \\ \dfrac{x-a}{b-a}, & Si\, a \leq x \leq b \\ 1, & Si\, b \leq x \leq c \\ \dfrac{d-x}{d-c}, & Si\, c \leq x \leq d \\ 0, & Si\, x > d \end{cases} \tag{2}$$

The Gaussian function defined by two parameters {σ, c}, which denote the center and width of the function whose correspondence rule is shown in Eq. (3).

$$\mu_A = (x, \sigma, c) = e^{-\frac{1}{2}\left(\frac{x-c}{\sigma}\right)^2} \tag{3}$$

Fuzzification is the process where the clear quantities become fuzzy, identifying some of the uncertainties present in the clear values, forming fuzzy values. The conversion of fuzzy values is represented by the membership functions. (Hutter et al. 2019).

Defuzzification is the process of converting to a quantifiable value into classical logic, given fuzzy sets and their corresponding degrees of membership, in other words, it transforms a fuzzy set into a classical set. The function reduces the collection of values of membership functions into a single accumulated amount called a centroid. Shan et al. (2019) mention that the output in the fuzzy union can be expressed mathematically with the Eq. (4), represented in Fig. 1.

$$C_n = \sum\nolimits_{i=1}^{R} C_i = C \tag{4}$$

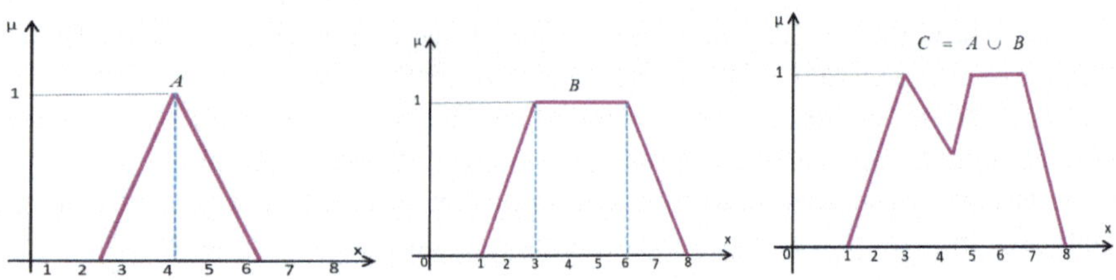

Fig. 1. Union of fuzzy sets

Centroid Method

Rick et al. (2000) argue that the use of the centroid method as shown in Eq. (5) is the most used, which is defined in algebraic form by y as represented in Fig. 2.

$$x^* = \int \frac{\mu_{\tilde{C}}(x).x.dx}{\mu_{\tilde{C}}(x)} \tag{5}$$

Where the integral symbol is used for algebraic integration.

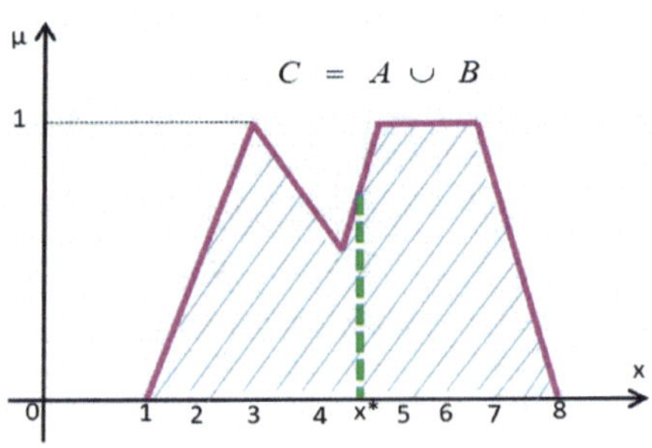

Fig. 2. Centroid method

3 Methodology Selection: THSW Index and Electric Field

In the study of the fuzzy model in meteorological variables to determine the THSW index and the electric field, Matlab software was used as a technological tool to obtain the best predictive model with the lowest error and optimal accuracy (Mathworks 2020).

In the measurement of meteorological variables, the Vantage Pro2 console equipment was used, as shown in Fig. 3, recording the measurements of the meteorological variables under study, such as temperature, ultraviolet radiation, humidity, wind speed and THSW index. The Vantage Pro2 console operates in five modes distributed in configuration, current weather, maximum graph mode, minimum graph mode and final graph (Davis 2019a).

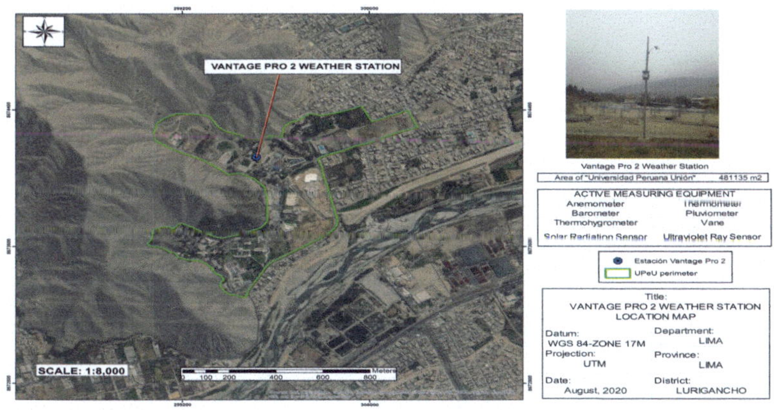

Fig. 3. Vantage Pro2 located on the main campus of the Universidad Peruana Unión.

WeatherLink software was used to connect the Vantage Pro2 station to the computer located in the laboratories of the School of Environmental Engineering of the Universidad Peruana Unión, Lima.

The Weatherlink database log was connected to the Vantage Pro2 console and stored data on the weather variables under study, which were recorded on the computer through WeatherLink.

The weather station is located in the laboratories of the Universidad Peruana Unión, which transmits data from the external Integrated Sensor Suite (ISS) sensor to the console via a low power radio frequency system.

The EFM 100 sensor monitored the electric field during 2019, and the data was then related to the measurements of meteorological variables within the fuzzy model with the Mandani technique, allowing for a classification of the electric field in each season of the year (Silhavy 2020).

3.1 Meteorological Variables

This study has considered the temperature, which refers to the ambient temperature expressed in degrees Celsius, as well as the UV radiation, which is the amount of solar radiation that reaches a horizontal surface. This irradiation includes the direct component from the sun and the component reflected from the rest of the sky. The solar radiation readings give a measure of the amount of solar radiation which reaches the solar radiation sensor at any given time, expressed in Watts/m^2. In addition, the relative humidity of the air expressed in percentage and the wind speed expressed in m/s were included. The THSW index uses temperature and humidity to calculate an apparent temperature. This THSW incorporates the thermal effects of direct solar radiation and the cooling effects of the wind in its perception. On the other hand, the electric field is a force field created by the attraction and repulsion of electric charges expressed in V/m (Davis 2019b).

4 Results

4.1 Fuzzy Model Design for Meteorological Variables

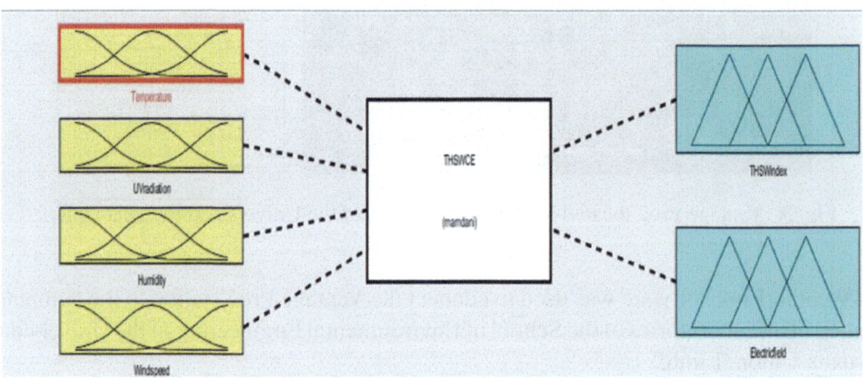

Fig. 4. Fuzzy model of meteorological variables (Mathworks 2020)

In Fig. 5 we observe the four input variables of the fuzzy model, temperature (a), UV radiation (b), humidity (c), wind speed (d), and the two output variables, THSW index (e) and electric field (f) with their respective membership functions and graphs made with MatLab software (Fig. 4).

Figure 6 shows eighty fuzzy rules used with their different combinations connected with the "and" connector which yielded accurate results of the output variables which were then corroborated with experts in the research topic.

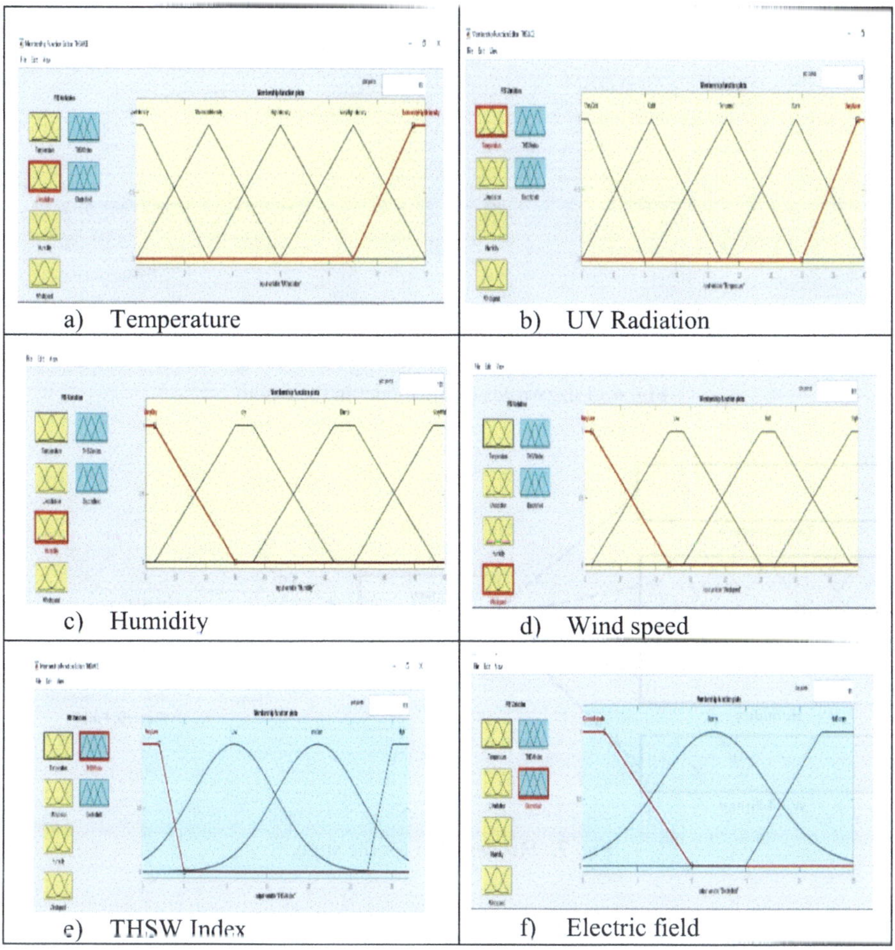

Fig. 5. Meteorological variables

4.2 Results of the Fuzzy Logic Model

The fuzzy model has four input variables; temperature, UV radiation, relative humidity, wind speed and two output variables; THSW index and electric field.

The fuzzy engine processes the inputs through the Mandani method. This method first inputs the input values, processes them through If-Then rules and finds a single value in the output (Fig. 7).

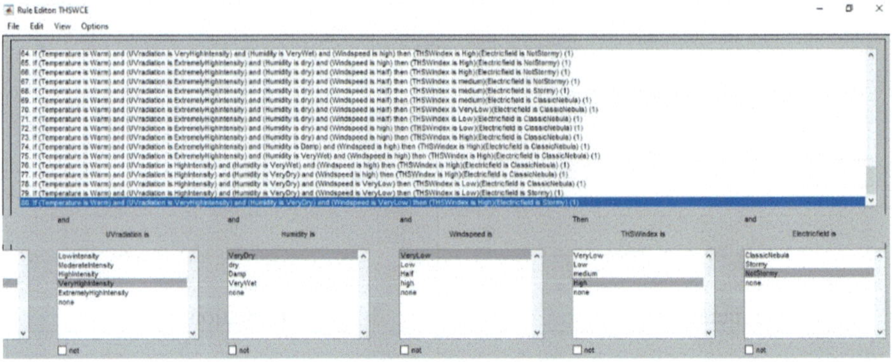

Fig. 6. Fuzzy rules of the model in the studio

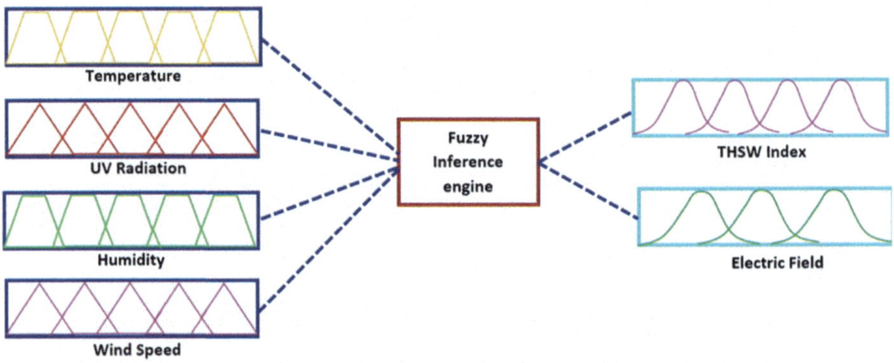

Fig. 7. Fuzzy model for the study

The evaluation of the THSW index and electric field is very important to analyze the state of the atmospheric zone within a region, as well as to determine the electric field to analyze, if there is cloudiness or not in the different seasons of the year. This study closes the gap by applying a Fuzzy Logic approach to analyze the THSW index and electric field with input and output variables, with their corresponding membership functions, linguistic values and parameters defined in Table 1.

Table 1 shows the four input variables and two output variables, with their respective ranges within a domain, with parameters and membership functions to be used in the proposed fuzzy model.

Table 2 shows the mathematical functions of Fuzzy membership of a meteorological variable, in this case temperature, where the very cold (a) and very warm (e) temperatures are represented by trapezoidal functions; while the cold (b), temperate (c) and warm (d) temperatures are triangular functions.

Table 1. Input/output of the fuzzy logic model

Input/output	Domain	Linguistic value	Parameters	Membership
Temperature	[–5–40]	Very cold	[–15, –6, –4, 5]	Trapezoidal
		Cold	[–5, 6, 17]	Triangular
		Temperate	[6, 18, 29]	Triangular
		Warm	[18, 29, 40]	Triangular
		Very warm	[30, 39, 41, 50]	Trapezoidal
UV radiation	[0–12]	Low intensity	[–3, –0.3, 0.3, 3]	Trapezoidal
		Moderate intensity	[0, 3, 6]	Triangular
		High intensity	[3, 6, 9]	Triangular
		Very high intensity	[6, 9, 12]	Triangular
		Extremely high intensity	[9 11.5, 12, 14]	Trapezoidal
Humidity	[0–100]	Very dry	[–30, –3, 3, 30]	Trapezoidal
		Dry	[3, 30, 36, 63]	Trapezoidal
		Humid	[36,63,70,97]	Trapezoidal
		Very humid	[70, 96, 105, 120]	Trapezoidal
Wind speed	[0–39]	Very low	[–11, –1, 1, 12]	Trapezoidal
		Low	[1, 12, 14, 25]	Trapezoidal
		Medium	[14, 25, 27, 38]	Trapezoidal
		High	[27, 38, 40, 50]	Trapezoidal
Output THSW index	[0–32]	Very low	[–9, –2, 2, 5]	Trapezoidal
		Low	[5, –11]	Gaussian
		Medium	[5, 21]	Gaussian
		High	[27, 30, 34, 40]	Trapezoidal
Electric field	[0–25]	Classical nebulosity	[–10, –2, 2, 10]	Trapezoidal
		Stormy	[5, 12]	Gaussian
		Not stormy	[15, 22, 27, 35]	Trapezoidal

4.3 Evaluation of the Proposed Solution

To validate the effectiveness of the constructed fuzzy model and evaluate the impact of the results of the THSW index and the electric field, a panel of specialists in the area was consulted, consisting of seven professionals with experience and knowledge in the research topic, to support the reliability of the results.

Table 2. Fuzzy membership functions for temperature

$\mu_A(x,-15,-6,-4,5)=\begin{cases}0 & ,Si\ x<-15\\ \frac{x+15}{9} & ,Si\ -15\leq x\leq -6\\ 1 & ,Si\ -6\leq x\leq -4\\ \frac{5-x}{9} & ,Si\ -4\leq x\leq 5\\ 0 & ,Si\ x>5\end{cases}$ a) Trapezoidal: Very Cold	$\mu_A(x,18,29,40)=\begin{cases}0 & ,Si\ x\leq 18\\ \frac{x-18}{11} & ,Si\ 18<x\leq 29\\ \frac{40-x}{11} & ,Si\ 29<x<40\\ 0 & ,Si\ x\geq 40\end{cases}$ d) Triangular: Warm
$\mu_A(x,-5,6,17)=\begin{cases}0 & ,Si\ x\leq -5\\ \frac{x+5}{11} & ,Si\ -5<x\leq 6\\ \frac{17-x}{11} & ,Si\ 6<x<17\\ 0 & ,Si\ x\geq 17\end{cases}$ b) Triangular: Cold	$\mu_A(x,30,39,41,50)=\begin{cases}0 & ,Si\ x<30\\ \frac{x-30}{9} & ,Si\ 30\leq x\leq 39\\ 1 & ,Si\ 39\leq x\leq 41\\ \frac{50-x}{9} & ,Si\ 41\leq x\leq 50\\ 0 & ,Si\ x>50\end{cases}$ e) Trapezoidal: Very Warm
$\mu_A(x,6,18,29)=\begin{cases}0 & ,Si\ x\leq 6\\ \frac{x-6}{12} & ,Si\ 6<x\leq 18\\ \frac{29-x}{11} & ,Si\ 18<x<29\\ 0 & ,Si\ x\geq 29\end{cases}$ c) Triangular: Temperate	

To reach a group agreement on the opinions of the results of the output variables, the Delphi method by rounds proposed in (Esquivel Garcia 2013) was applied to ensure the reliability of the statements.

The selection criterion used was a concordance greater than 60%. The final rules were determined by the characteristics of this process and were also configured to meet the following standards (Pareja 2003):

- **Completeness**, where the rules have to comply with all possible combinations of inputs to the controller, so that there are no loopholes in which no action is taken.
- **Consistency**, here two control actions cannot coexist for the same situation, which would produce a contradiction.
- **Interaction**, this corresponds to the weight that a rule has among the remaining rules, for which the effect of the rules on the control action must be analyzed.
- **Robustness**, a measure of the controller's reaction to disturbances of the inputs.

Part of the knowledge base of the fuzzy inference model is provided by the fuzzy rules shown in Table 3.

Table 3. Knowledge base for the knowledge of the output variables

Knowledge base			
THSW index and electric field			
Yes		Then	
Temperature	Very cold	THSW Index	
	Cold		
	Temperate		
	Warm		Very Low
	Very warm		Low
UV radiation	Low intensity		Medium
	Moderate intensity		High
	High intensity		
	Intensity very high		
	Extremely high intensity		
Humidity	Very dry	Electric field	
	Dry		
	Wet		Classic nebulosity
	Very humid		Stormy
Wind speed	Very low		Not stormy
	Low		
	Medium		
	High		

After analyzing the ranges (expressed in short phrases such as low and warm) with their respective combinations by means of fuzzy rules, the centroid method shown earlier in Eq. (5) was used to defuzzify the THSW index and the electric field.

4.4 Experimental Evaluation of the Fuzzy Model

To validate the fuzzy model experimentally, 80 cases were formulated with the meteorological input variables and an analysis was carried out with the output result given by the MatLab software with those given by the panel of experts. When comparing the analyzed data, it was observed that the experiment yielded a high coincidence of the output values with the criteria generated by the expert specialists. The robustness of the fuzzy model in answering the different questions of the meteorological input variables was tested using SPSS (ver 27) by applying the Friedman test (Haroldo 2008) and the ROC curves (Navidi 2006), to validate the certainty of each fuzzy rule which corresponds to the output variables.

Table 4. Fuzzy model validation indicators

Variables	Friedman sig.	ROC curve VLow	ROC curve low	ROC curve medium	ROC curve high	% of concordant cases
THSW index	0.0014	0.862	0.874	0.864	0.882	78.63
Electric field	0.0021	0.932	0.824	0.810	0.883	81.42

In Table 4 we observe that the THSW Index has a Friedman's test value of 0.0014 which, since it is less than the significance level $\alpha = 5\%$, validates the output results of the Matlab software with the central tendency of the results of the experts in the research topic. Likewise, it is observed that the electric field has a pvalue of 0.0021 which, since it is less than the significance level $\alpha = 5\%$, validates the results and guarantees the robustness of the fuzzy model in which the four ROC curves have an area significantly greater than 0.5, showing that the fuzzy model has a good level of perfomance.

4.5 Results of the Meteorological Research Variables

In Table 5, the values of the temperature input variable in its different seasons are observed in which it is seen that in summer 2019 reached a maximum temperature of 29.8 °C and an average of 24.595 °C, which is reflected in Fig. 8.

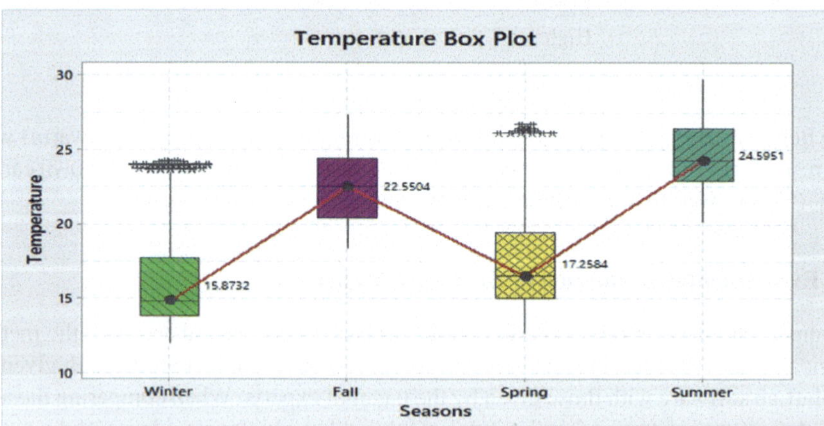

Fig. 8. Temperature box plot

Table 6 shows the results of UV radiation in its different seasons in which it is seen that in summer 2019 a maximum UV radiation of 11.7 nm and a mean of 1.9506 nm was reached, as shown in Fig. 9.

Table 5. Temperature results (°C)

Season	Count	Mean	Stan. Dev.	Minimum	Maximum
Winter	6681	15.873	2.651	11.900	24.200
Fall	3179	22.550	2.341	18.300	27.400
Spring	5570	17.258	2.901	12.600	26.700
Summer	4134	**24.595**	2.112	20.100	**29.800**

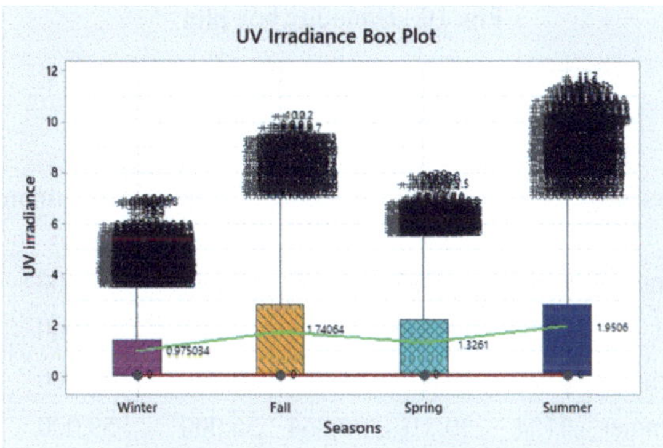

Fig. 9. UV radiation box plot

Table 6. Results for UV radiation (nm)

Season	Count	Mean	Stan. Dev.	Minimum	Maximum
Winter	6681	0.9750	1.6578	0.0000	6.9000
Fall	3179	1.7406	2.8103	0.0000	10.2000
Spring	5570	1.3261	2.1128	0.0000	7.9000
Summer	4134	1.9506	3.2487	0.0000	11.7000

Table 7 shows the results of humidity in the different seasons, in which it can be seen that in the winter of 2019 there was a minimum humidity of 0%, highlighting that in the winter it had the lowest average of 6.317%, which affected the area of eastern Lima, causing viral diseases such as flu and allergies, which is represented in the box diagram in Fig. 10.

Table 8 shows the results of the wind speed in the different seasons in which it can be seen that in winter and autumn of 2019 there was a maximum wind speed of 12.90 m/s, highlighting that in spring it had the lowest with an average of 1.2727 m/s, this is represented in Fig. 11 with a box plot.

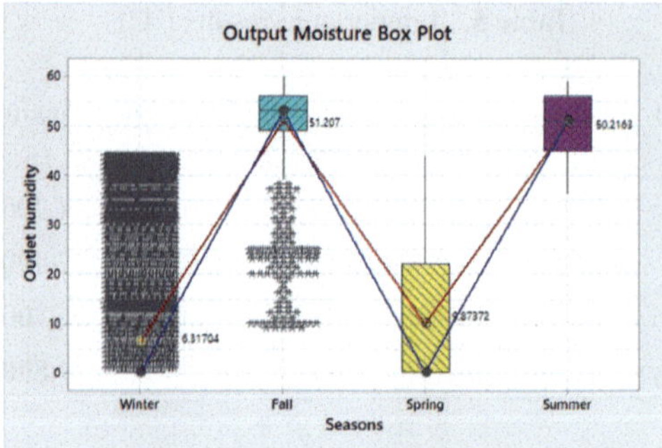

Fig. 10. Humidity box plot

Table 7. Results for humidity (%)

Season	Count	Mean	Stan. Dev.	Minimum	Maximum
Winter	6681	6.317	12.839	0.000	44.000
Fall	3179	51.207	7.251	9.000	60.000
Spring	5570	9.874	15.605	0.000	44.000
Summer	4134	50.216	5.864	36.000	59.000

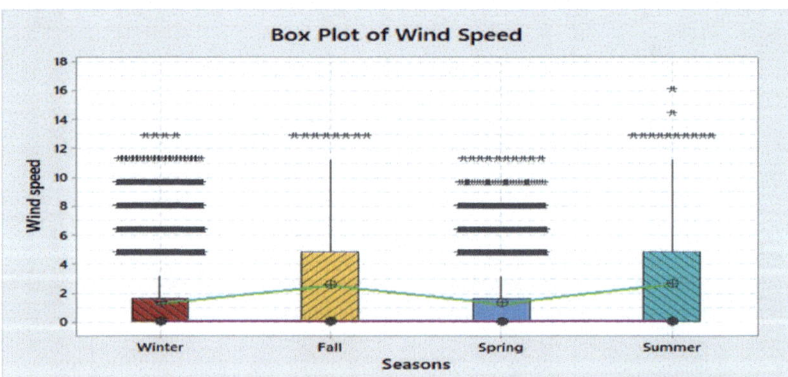

Fig. 11. Wind speed box plot

Table 9 shows the results of the THSW index in its different seasons in which it can be seen that in the summer of 2019 the index reached 36.2, with an average of 25.169, as shown in the box plot in Fig. 12.

Table 10 shows the results of the electric field in its different seasons, in which it can be seen that in winter 2019 there was a maximum electric field of 11.8697 kV/m with

Table 8. Results for wind speed (m/s)

Season	Count	Mean	Stan. Dev.	Minimum	Maximum
Winter	6681	1.2801	2.4288	0.0000	12.9000
Fall	3179	2.5320	3.2303	0.0000	12.9000
Spring	5570	1.2727	2.2716	0.0000	11.3000
Summer	4134	2.6339	3.2683	0.0000	16.1000

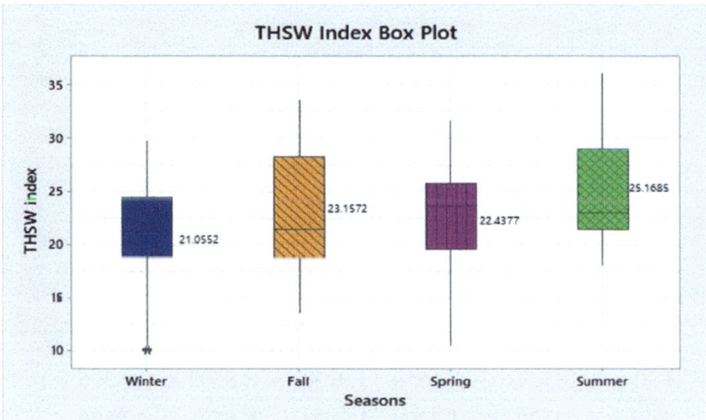

Fig. 12. THSW index box plot

Table 9. Results for the THSW index

Season	Count	Mean	Stan. Dev.	Minimum	Maximum
Winter	6681	**21.055**	4.445	9.900	29.800
Fall	3179	23.157	5.431	13.500	33.700
Spring	5570	22.438	4.893	10.400	31.700
Summer	4134	**25.169**	4.932	17.900	36.200

an average of −1.6218 kV/m and a minimum electric field of −20.48 kV/m, as shown in the box plot in Fig. 13.

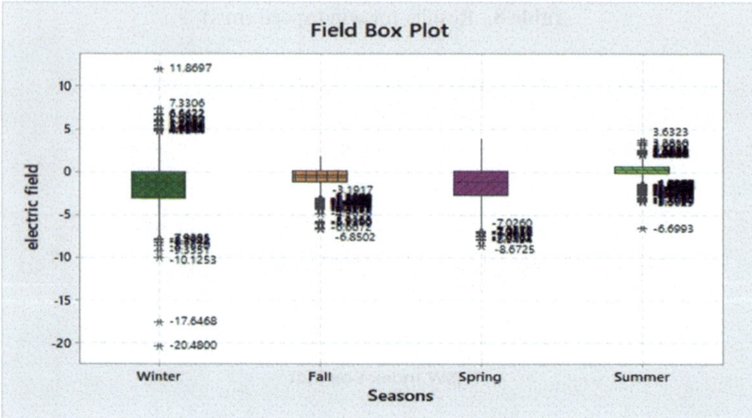

Fig. 13. Electric field box plot

Table 10. Results for the electric field

Season	Count	Mean	Stan. Dev.	Minimum	Maximum
Winter	6681	**–1.6218**	2.1174	**–20.480**	**11.8697**
Fall	3179	– 0.6550	1.0233	– 6.8502	1.8796
Spring	5570	– 1.4309	1.9024	– 8.6725	3.8296
Summer	4134	**0.1682**	0.7111	**-6.6993**	**3.6323**

5 Conclusions

The fuzzy model used meteorological variables as its input variables, where the maximum temperature in summer 2019 was 29.8 °C with an average of 24.595 °C, likewise the UV radiation was 11.7 nm as a maximum with an average of 1.9506 nm, also the average humidity was 6.317%, as well as the wind speed of 12.90 m/s which was lower in spring with an average of 1.2727 m/s.

The fuzzy model's outputs showed that the THSW index measured the four input meteorological variables with an average of 25.169, with a minimum in spring of 21.055 and a maximum in summer with 25.169, as well as measuring the electric field obtaining in winter an average of −1.6218 kV/m with a minimum value of −20.480 kV/m and a maximum value of 11.8697 kV/m, which means that there was a lot of cloudiness and it was cold in the area of eastern Lima; as well as in the summer of 2019 an average of 0.1682 kV/m was obtained with a minimum value of -6.6993 kV/m and a maximum value of 3.6323 kV/m, which means that the sky was clear and it was hot in the study area.

The input variables of the fuzzy model for a temperate of 22 °C, with a moderate radiation of 4 nm, a dry humidity of 40% and an average speed of 29 m/s, the fuzzy model yielded an average THSW index of 19 and a classical cloud electric field of 8 kV/m, which reflects the accuracy and performance of the fuzzy model on the output variables. Considering 80 fuzzy rules, it was possible to defuzzify the THSW index and the electric field, obtaining an accuracy at a 95% confidence level.

The fuzzy model was validated with the output results of the MatLab software and the results of the experts with a good robustness using the Friedman test, for which the THSW index had a pvalue of 0.0014 and the electric field a pvalue of 0.0021 which are both less than the significance level $\alpha = 5\%$. The performance of the fuzzy model was determined regarding the accuracy results in which all ROC curves have an area less than 0.5, showing the reliability of the model.

References

Antunes de Sá, A., Marshall, R., Sousa, A., Viets, A., Deierling, W.: An array of low-cost, high-speed, autonomous electric field mills for thunderstorm research. Earth Space Sci. **7**(11) (2020). https://doi.org/10.1029/2020EA001309

Azad, A., et al.: Novel approaches for air temperature prediction: a comparison of four hybrid evolutionary fuzzy models. Meteorol. Appl. **27**(1) (2020). https://doi.org/10.1002/met.1817

Davis: Vantage Pro2 Manuel de la console (2019a). https://www.davisnet.com/legal.

Davis, I.: Options de collecte de données (2019b). www.weatherlink.com

Elizalde, G.: julio-diciembre. **14**(2), 265–292 (2009)

Esquivel Garcia, R.: Educación Superior y formación de directivos View project Evaluación del impacto de la capacitación view project (2013). https://www.researchgate.net/publication/324278460

Haroldo, P.: Haroldo Elorza Pérez-Tejada. CERGAGE Learning (ed.) Tercera (2008)

Harrison, R.G., Nicoll, K.A., Mareev, E., Slyunyaev, N., Rycroft, M.J.: Extensive layer clouds in the global electric circuit: their effects on vertical charge distribution and storage. Proc. Roy. Soc. A: Math. Phys. Eng. Sci. **476**(2238), 20190758 (2020). https://doi.org/10.1098/rspa.2019.0758

Hutter, F., Kotthoff, L., Vanschoren, J.: Automated Machine Learning Methods, Systems, Challenges. The Springer Series on Challenges in Machine Learning (2019). https://doi.org/10.1007/978-3-030-05318-5. http://www.springer.com/series/15602

Marcos, J.L., Sánchez, J.L., Merino, A., Melcón, P., Mérida, G., García-Ortega, E.: Spatial and temporal variability of hail falls and estimation of maximum diameter from meteorological variables. Atmos. Res. **247** (2021). https://doi.org/10.1016/j.atmosres.2020.105142

Martinez, M.: Medición del campo eléctrico atmosférico en la ciudad de León. Establecimiento de límites para prevención ante la ocurrencia de descargas atomosféricas (2014)

Mathworks, C.: Statistics and machine learning toolbox™ user's guide R2020a (2020). www.mathworks.com

Matute, A., Bernal Suárez, W.: Fuzzy logic techniques in control engineering. Rev. Ciencia Innov. Tecnol. (RCIYT) **III**, 125–134 (2017)

Mirbagheri, S.A., Mohammadi, M., Bazrehkar, S.: Assessing the weight of weather parameters on MW signal strength by descriptive data mining. SN Appl. Sci. **2**(2), 1–5 (2020). https://doi.org/10.1007/s42452-020-2093-9

Navidi, W.: Estadística para ingenieros y científicos. Pablo Eduardo Roig Vázquez (ed.) Primera (2006)

Nicoll, K.A., et al.: A global atmospheric electricity monitoring network for climate and geophysical research. J. Atmos. Solar-Terrestrial Phys. **184**, 18–29 (2019). https://doi.org/10.1016/j.jastp.2019.01.003

Pareja, I.V.: El método Delphi El metodo delphi (2003)
Rick, A., Bothorel, S., Bouchon-Meunier, B., Muller, S., Rifqi, M.: Fuzzy techniques in mammographic image processing. In: Kerre, E.E., Nachtegael, M. (eds.) vol. 52, pp. 308–336. Physica-Verlag HD (2000). https://doi.org/10.1007/978-3-7908-1847-5_12
Romero, Y., Diaz, C., Meldrum, I., Arias Velasquez, R., Noel, J.: Temporal and spatial analysis of traffic – related pollutant under the influence of the seasonality and meteorological variables over an urban city in Peru. Heliyon **6**(6) (2020). https://doi.org/10.1016/j.heliyon.2020.e04029
Saboya, N., Loaiza, O.L., Soria, J.J., Bustamante, J.: Fuzzy logic model for the selection of applicants to university study programs according to enrollment profile. In: Antipova, T., Rocha, A. (eds.) DSIC18 2018. AISC, vol. 850, pp. 121–133. Springer, Cham (2019). https://doi.org/10.1007/978-3-030-02351-5_16
Shan, H., Liu, Y., Stefanov, T.: Ensemble of convolutional neural networks for P300 speller in brain computer interface. In: Tetko, I.V., Kůrková, V., Karpov, P., Theis, F. (eds.) ICANN 2019. LNCS, vol. 11730, pp. 376–394. Springer, Cham (2019). https://doi.org/10.1007/978-3-030-30490-4_31
Shiri, J.: Evaluation of a neuro-fuzzy technique in estimating pan evaporation values in low-altitude locations. Meteorol. Appl. **26**(2), 204–212 (2019). https://doi.org/10.1002/met.1753
Silhavy, R.: Advances in Intelligent Systems and Computing 1225. In: Janusz Kacprzyk, P. (ed.) vol. 2, Systems Research Institute, Polish Academy of Sciences, Warsaw & Advisory (2020). http://www.springer.com/series/11156
Sivanandam, S.N., Sumathi, S., Deepa, S.N.: Introduction to fuzzy logic using MATLAB. Springer, Heidelberg (2007). https://doi.org/10.1007/978-3-540-35781-0

Hyper-heuristics: Method of Differential Evolution and Bat Method for Selecting Classification Features

Sergey Rodzin and Lada Rodzina(✉)

Southern Federal University, Taganrog 344090, Russia
srodzin@sfedu.ru

Abstract. The paper proposes a hyper-heuristic GEBA algorithm for selecting significant classification features and evaluating the classification accuracy. The GEBA algorithm combines the bat echolocation algorithm and a modified version of the differential evolution algorithm. The experiments used the well-known k nearest neighbor classifier (KNN) and the decision tree-based classifier (CART). The effectiveness of the proposed GEBA algorithm was evaluated using a set of reference data sets from the UCI world library. The performance of the GEBA algorithm for solving the problem of selecting significant classification features was compared with the competing algorithms of bats, crows, and gray wolves, and the hybrid algorithm of particle swarm and gravity search. The proposed algorithm leads in the average and best values of the fitness function for most of the tested data sets and is also statistically significant compared to competing algorithms for the Wilcoxon T-test and the Friedman test with a confidence level of 0.95. The stability and accuracy of the GEBA algorithm are higher than that of competing algorithms on most reference datasets. A comparative analysis of the convergence rate of algorithms on a set of reference data sets for a given number of iterations showed that the GEBA algorithm converges faster to global optima than competing algorithms. The algorithm is planned to be used for solving traveling salesman problems, integer programming and minimax problems, medical image segmentation problems, and multidimensional optimization problems.

Keywords: Hyper-heuristics · Classifier · Feature selection · Bat algorithm · Differential evolution algorithm · Data set · Statistical test

1 Introduction

1.1 Background

The problem of feature selection for data classification and analysis consists in detecting the most significant (essential) features and eliminating irrelevant and redundant data. The correct selection of features can improve the quality of supervised machine learning models, reduce training time and reduce the required computing power, and in the case of multidimensional input data, weaken the "curse of dimension". Feature selection is often used in areas where there are many features, and the data samples are relatively

© The Author(s), under exclusive license to Springer Nature Switzerland AG 2021
R. Silhavy (Ed.): CSOC 2021, LNNS 229, pp. 545–556, 2021.
https://doi.org/10.1007/978-3-030-77445-5_50

small. In various applications, excessive feature space is often used when presenting data, so it is necessary to select the most significant features and eliminate redundant or insignificant features. This is an NP-hard optimization problem and its exact solution by standard optimization methods in the worst case may require the construction of a search tree for solutions of exponential size.

The object of research in this work is the features that characterize the data set, as well as methods that allow the selection of significant features for subsequent classification. The current class of optimization methods that allow finding solutions for a wide range of *NP*-hard problems from various applications are bioinspired methods that can solve complex problems without knowing the search space, which is why these methods make it possible to solve intractable optimization problems. Bioinspired methods include (but are not limited to) evolutionary and swarm algorithms for individual and collective agent behavior; heuristics based on physical processes in nature, as well as heuristics based on cognitive processes and human activities. The current classification of bio-heuristics includes evolutionary algorithms such as differential evolution (DE), evolutionary strategies (ES), genetic algorithms (GA); swarm bio-heuristics such as ant colony (ACO) and bee (ABC) colony algorithms, bats (BA), crows (CSA), cuckoos (CS), fireflies (FA), gray wolves (GWO), krill school (KHA), light-flying moths (MFO), flocks of birds (PSO), spider colonies (SSO), groups of whales (WOA); bio-heuristics based on physical processes such as charged particle motion (EM), gravitational interaction (GSA), simulated annealing (SA), transition of matter from one thermal state to another (SMS), water droplet algorithm (ADW); bio-heuristics based on cognitive processes and human activities such as fireworks algorithm (FWA), memetics (MA), biogeography (BBO), search for harmony (HS), neo-colonialism algorithm (ICA) [1]. In particular, to solve the problem of choosing significant classification features, the bats algorithm (BA) [2], the crow search algorithm (CSA) [3], the gray wolf algorithm (GWO) [4], and the hybrid particle swarm and gravity search algorithm were proposed. (PSO-GSA) [5]. Also, hybridization of these algorithms attracted many researchers to solve the problem of feature selection for classification and data analysis, for example, [6, 7].

In this paper, we propose a hyper-heuristic algorithm for selecting significant classification features and assessing the classification accuracy. The main idea of the proposed hyper-heuristic algorithm is to combine two heuristics – the algorithm of bats and the algorithm of differential evolution, each of which has its own strengths and weaknesses – and then, the construction of a mechanism for their selection depending on the current state of the solution. The choice of the parameters of the BA and DE heuristics is based on experimental data of high dimension. The main purpose of the proposed algorithm is to increase the efficiency of the method for selecting significant classification features. The resulting algorithm should be fast and convenient for implementation, but at the same time, it should be sufficiently reliable for solving the specified problem and should have advantages in comparison with known competing algorithms.

2 Methods

2.1 Proposed Methodology

The key success factor for many industries is the ability to efficiently process large volumes and flows of information. The search for optimal methods for processing large amounts of data is becoming important for promising areas of science and industry; it is one of the key areas of development of problem-oriented systems in the field of presentation, analysis, and extraction of useful patterns from large amounts of data of various nature.

The search for effective and efficient methods for choosing the optimal set of significant classification features is one of the main components of data mining [8]. The methodology adopted in this paper includes the following elements. Firstly, the largest repository of data and model machine learning problems UCI is used as data [9]. The repository contains real data on applied problems in biology, medicine, physics, technology, sociology, etc. Data sets of this particular repository are most often used by the scientific community for empirical analysis of machine learning algorithms. Next, a method for selecting significant features for classification is discussed and an assessment of the results obtained for several types of data sets from the UCI is presented.

2.2 Theoretical Basis

Bat Algorithm. The canonical Bat Algorithm (BA) was proposed by Young [10] based on the echolocation behavior of bats. Echolocation consists of two stages: emitting loud frequency-modulated sound pulses and receiving (listening) echoes that are reflected from surrounding objects, which allows you to build a three-dimensional scenario of the immediate environment. Typically, bats use their specialized sonar system to locate prey, avoid obstacles and locate an overnight stay.

The bat uses echolocation to achieve some goal, defined as the global best solution. The BA algorithm uses the following parameters, the values of which are constantly adjusted during the search: frequency, volume, and speed of the pulse emission. The frequency is modeled by a set of d-dimensional vectors, each of which is associated with the ith bat. The values of these vectors are randomly adjusted at each kth iteration by the formula:

$$\mathcal{F}_{\text{i}}^{\text{k}} = \mathcal{F}_{\text{min}} + \beta(\mathcal{F}_{\text{max}} - \mathcal{F}_{\text{min}}) \tag{1}$$

where the parameter $\mathcal{F}_{min}$ and $\mathcal{F}_{max}$ denote the minimum and maximum frequencies, respectively; β is a vector of random numbers, each of which is in the interval [0, 1].

The parameters of the loudness A_i and the speed of the pulsed radiation r_i, the initial values of which $A_i{}^0$ and $r_i{}^0$ are determined during the initialization of the algorithm. As the process of finding the optimal solution develops, the values of these parameters change in accordance with the following expression:

$$A_i^{k+1} = \alpha A_i^k,\ r_i^0(1 - \exp(-\gamma k)) \tag{2}$$

Where $\alpha < 1$ and $\gamma < 1$ are some constants.

Regarding updating the location of the ith bat at the kth iteration, the BA-algorithm uses the following operator:

$$\boldsymbol{x}_i^k = \boldsymbol{x}_i^{k-1} + \boldsymbol{v}_i^k \tag{3}$$

where $\boldsymbol{x}_i^{k-1}$ is the position for i^{th} bat in the previous iteration ($k - 1$), at the same time $\boldsymbol{v}_i^k$ is the speed of ith bat, which, in turn, is calculated as follows:

$$\boldsymbol{v}_i^k = \boldsymbol{v}_i^{k-1} + \mathcal{F}_i^k \cdot \left(x_i^{k-1} - \boldsymbol{x}_{best}\right) \tag{4}$$

where $\boldsymbol{x}_{best}$ is the current global optimal solution, and $\mathcal{F}_i^k$ represents the frequency vector defined according to (1).

In addition, BA also proposes to include a local search scheme, according to which at each iteration, a randomly selected individual among the current best solutions is additionally refined by performing a random walk as follows:

$$\boldsymbol{x}_*^k = \begin{cases} \boldsymbol{x}_i^k + \varepsilon A_i^k, & if\left(rand > r_i^k\right) \\ \boldsymbol{x}_i^k, & in\ the\ opposite \end{cases} \tag{5}$$

where *rand* is a random number, obtained from uniformly spaced interval [0, 1].

Moreover, the new solution $\boldsymbol{x}_*{}^k$ is accepted as the location of the ith bat only if certain conditions are met, in particular:

$$\boldsymbol{x}_i^k = \begin{cases} \boldsymbol{x}_*^k, & if\left(rand < A_i^k \& x_i^k < x_*^k\right) \\ \boldsymbol{x}_i^k, & in\ the\ opposite \end{cases} \tag{6}$$

Differential Evolution Algorithm. The differential evolution (DE) algorithm was proposed by R. Storn and K. Price in 1996 in [11] and, along with genetic algorithms (GA), is one of the most popular optimization approaches inspired by evolutionary processes.

At each k-step of the DE algorithm, the operators of mutation, crossing over, and selection are applied to allow the population of solutions $\boldsymbol{X} = \{\boldsymbol{x}_1, \boldsymbol{x}_2, \ldots, \boldsymbol{x}_N\}$ to "evolve" towards the optimal solution.

When performing the mutation operation, new solutions $\boldsymbol{m}_i{}^k = [m^k{}_{i,1}, m^k{}_{i,2}, \ldots, m^k{}_{i,d}]$ are generated for each individual xi by the formula:

$$\boldsymbol{m}_i^k = \boldsymbol{x}_{r3}^k + F\left(\boldsymbol{x}_{r1}^k - \boldsymbol{x}_{r2}^k\right) \tag{7}$$

where $r_1, r_2, r_3 \in \{1, 2, \ldots, N\}$ $(r_1 \neq r_2 \neq r_3 \neq i)$ is a random index, $F \in [0, 2]$ – differential weight used to control the difference in values $\left(\boldsymbol{x}_{r1}^k - \boldsymbol{x}_{r2}^k\right)$.

When the crossing over operator is executed, a trial solution vector is generated $\boldsymbol{u}_i{}^k = [u^k{}_{i,1}, u^k{}_{i,2}, \ldots, u^k{}_{i,d}]$ for i^{th} individual population solution. Components $u^k{}_{i,j}$ in the trial vector are formed from the solution $\boldsymbol{x}^k{}_i$ and the corresponding decision $\boldsymbol{m}^k{}_i$ after performing the mutation operation as follows:

$$u_{i,j}^k = \begin{cases} m_{i,j}^k, & if\,(rand \le CR)\, or\, (n = n^*) \\ x_{i,j}^k, & if\,(rand > CR) \end{cases} \tag{8}$$

where $n^* \in \{1, 2,\ldots, d\}$) is a random index, *rand* is a random number from the interval [0, 1], $CR \in [0, 1]$, which is used in the DE algorithm when performing crossing over to form elements $u^k{}_{i,j}$ from elements $x^k{}_{i,j}$ or $m^k{}_{i,j}$.

Finally, when executing the selection operator, the fitness function of the trial solution $\boldsymbol{u}_i{}^k$ is compared with the fitness function of the candidate solution $\boldsymbol{x}_i{}^k$ using the greedy test.

2.3 Proposed Hyper-heuristic Algorithm

Let us consider a new hyper-heuristic bioinspired algorithm (GEBA) that combines the advantages of the BA and DE algorithms.

If n is the number of features characterizing the data, then the best solution is the one with the minimum number of features and high classification accuracy.

The algorithm starts with generating a population of random solutions (a subset of classification features). The fitness function is used to evaluate decisions. The $\boldsymbol{x}_{best}$ denotes the best solution in the population. The main loop of the algorithm is repeated many times. In each generation, for the VA algorithm, the main action is to update the position of the ith bat at the kth iteration, according to formula (3), and for the DE algorithm, the execution of the mutation operator according to (7). In BA, the operator of the pulse radiation velocity r_i is applied based on the V-shaped transfer function according to (2), and in DE – on the basis of the S-shaped transfer function.

In the hyper-heuristic algorithm GEBA, the probability of choosing a solution according to the VA algorithm is equal to 0.6, according to the DE algorithm – 0.4. The V-shaped transfer function in VA is used to establish the position of the bat according to (3) as follows:

$$f = \arctan\left(\boldsymbol{x}_i^k\right) \tag{9}$$

The S-shaped transfer function is used in DE to obtain a mutant vector according to (7) as follows:

$$f = \frac{1}{1 - \exp\left(10\left(m_i^k - 0,5\right)\right)}. \tag{10}$$

At the next *(k + 1)* step of the algorithm, the new binary position in the solution is determined as follows:

$$x_i^{k+1} = \begin{cases} 0, & rand < f; \\ 1, & \text{в противном случае,} \end{cases} \tag{11}$$

where *rand* () random number of uniform distributions on the interval [0, 1], and $x_i{}^{k+1}$ indicates a new binary solution in the $(k + 1)^{th}$ generation.

After updating the binary position in the solution, it is necessary to maintain a balance between the convergence rate of the algorithm and the diversification of the solution search space. For this purpose, the VA uses the speed of the pulsed radiation of a bat r_i with a generated random vector in the range [0, 1]. If the random value is less than r_i, then the search for the set of significant classification parameters continues by storing the updated binary positions according to the formula (11). Next, the fitness function is calculated and the transition to a new population of solutions is carried out. This process is repeated until the stopping conditions are met, in particular, the specified maximum number of iterations k_{max}. The pseudocode of the GEBA algorithm is as follows.

Procedure GEBA:

1: Set the following parameters for algorithm: $\mathcal{F}_{min} = 0{,}1$; $\mathcal{F}_{max} = 2$; $r_i{}^0 = 0{,}2$, k_{max}.
2: Initialization of a set of randomly selected solutions in the form of binary vectors
3: $k = 1$
4: ***for*** (each element of the vector $i = 1..d$)
5: ***begin***
6: ***if*** (*rand* > 0,4) ***then****Procedure BA* ***else****Procedure DE*
7: *Procedure BA*
8: ***begin***
9: $\mathcal{F}_i^k = \mathcal{F}_{min} + \beta(\mathcal{F}_{max} - \mathcal{F}_{min})$; $\beta = rand()$
10: $\boldsymbol{v}_i^k = \boldsymbol{v}_i^{k-1} + \mathcal{F}_i^k \cdot \left(x_i^{k-1} - \boldsymbol{x}_{best}\right)$,
11: $\boldsymbol{x}_i^k = \boldsymbol{x}_i^{k-1} + \boldsymbol{v}_i^k$,
12: $f = \arctan(\boldsymbol{x}_i^k)$
13: ***end***
14: *Procedure DE*
15: ***begin***
16: Choose $\boldsymbol{x}_{r1}^k, \boldsymbol{x}_{r2}^k, \boldsymbol{x}_{r3}^k$ for $\boldsymbol{X}$
17: $\boldsymbol{m}_i^k = \boldsymbol{x}_{r3}^k + F\left(\boldsymbol{x}_{r1}^k - \boldsymbol{x}_{r2}^k\right)$, $F = rand()$
18: $f = \frac{1}{1 - \exp(10(m_i^k - 0{,}5))}$
19: ***end***
20: ***if*** $(rand() > f)$ ***then*** $(x_i{}^{k+1} = 1)$ ***else*** $(x_i{}^{k+1} = 0)$
21: ***if*** $(rand() > r_i)$ ***then*** $(x_i{}^{k+1} = \boldsymbol{x}_{best})$ ***else*** (*go to* 6)
22: $k = k + 1$
23: ***if*** $(k \leq k_{max})$ ***then*** (*go to 4*) ***else*** (*end*)
24: ***end***

3 Experimental Results and Discussion

To test the performance of the proposed algorithm, 10 *UCI*'s reference datasets were used [12]. Table 1 provides a brief description of each dataset.

Each solution is evaluated using a fitness function at each iteration of the algorithm. The data is randomly divided into two separate parts: training and test datasets. The first part is for training the classifier and the second is for testing. The function of losses (classification errors) for linear classification models is used as a criterion for evaluating the classification. The classification accuracy is determined by the ratio of the number of correctly classified specimens to the total number of specimens.

The work uses two popular classifiers: KNN (*k* nearest neighbors) and CART (classification and regression by building a decision tree) [3]. According to KNN, to classify each of the objects of the test sample, it is necessary to perform the following operations

Table 1. Description of the used UCI datasets.

Dataset ID	Dataset name	Number of features	Number of copies	Number of classes	Data area
S1	Bank loan	24	1000	2	Business
S2	Sonar	60	20	2	Physics
S3	Diabetes	6	144	3	Medicine
S4	Oncology diagnostics	32	596	2	Medicine
S5	Zoo	18	101	2	Animal world
S6	Ionosphere	34	351	2	Physics
S7	Lymphography	18	148	4	Medicine
S8	Rehabilitation	309	126	2	Life
S9	Heart	13	270	2	Medicine
S10	Chess	36	3196	2	Game

sequentially: calculate the distance to each of the objects of the training sample; select *k*-objects of the training sample, the distance to which is minimal; determine the class of the classified object as the most common among the k nearest neighbors, depending on the distance between the new and training instances. Another classifier is CART, which builds a binary decision tree, where each node of the tree, when split, has only two children. At each step of building a tree, a rule formed at a node divides a given set of data sets into two parts – a part in which the rule is executed and a part in which the rule is not executed. The scoring function used by the CART algorithm is based on the intuitive idea of reducing the uncertainty in a node. In the CART algorithm, the idea of uncertainty is formalized in the *Gini* index. There are several modifications of CART. We use the modification of CART presented in [13]. The goal of any classifier is to obtain the highest accuracy, that is, the best solution is the one that minimizes the classification error, minimizes the number of selected classification features, and maximizes the classification accuracy. The classification process according to the GEBA algorithm stops when the maximum number of iterations is reached. For comparison with competing algorithms, the maximum number of iterations in all experiments was set equal to 50.

The efficiency of the proposed GEBA algorithm was evaluated using 10 reference datasets. The experiments were carried out using MATLAB on a Core i5 processor with 8 GB RAM. The parameter value for the maximum number of iterations is $k_{max} = 50$, the population size parameter is 20.

The following criteria were used to compare the GEBA algorithm with modern competing algorithms:

- classification error after applying the classifier to the test dataset (the maximum, average and minimum values of the fitness function are compared;
- average sample size;

- *T*-test for the sum of Wilcoxon's ranks to check the differences between samples by the level of any quantitative trait (a statistical test allows you to establish not only the direction of changes, but also their severity);
- Friedman's test is a nonparametric statistical test with a 5% significance level [14].

Statistical tests using Wilcoxon and Friedman's criteria are needed to demonstrate that the proposed algorithm is significantly improved over competing algorithms. A T-value < 0.05 is considered adequate evidence against the null hypothesis. The value $h = 1$ for the Wilcoxon test shows that the null hypothesis is rejected, and the proposed algorithm is superior to the competing algorithm.

In the course of the experiments, the performance of the GEBA algorithm for solving the problem of choosing significant classification features was compared with competing algorithms: the binary BA [2] algorithm, the CSA binary crow algorithm [3], the GWO binary gray wolf algorithm [4] and the binary hybrid particle swarm and gravity search algorithm PSO-GSA [5].

The following parameters indicated in the original works were used as settings for competing algorithms. For the BA algorithm, the pulse rate r is 0.1; the volume of the sound A emitted by the bat is 0.25; parameter $F_{min} = 0$, $F_{max} = 2$. For the CSA algorithm, the parameter $fl = 2$. For the hybrid PSO-GSA algorithm, the initial value of the gravitational constant is $g_0 = 1$, and the factors $c_1^{'} = -2\frac{t^3}{T^3} + 2$, $c_1^{'} = -2\frac{t^3}{T^3}$, where T is the maximum number of iterations, t is the current iteration, the coefficient α is 20. For the GEBA algorithm, the parameter $F_{mi} = 0.1$; $F_{max} = 2$. The GWO algorithm does not imply parameter settings.

Computational experiments were carried out on 10 reference datasets from various fields and were aimed at testing the efficiency of the proposed GEBA hyper-heuristic algorithm for the problem of choosing significant classification features.

Table 2 shows the best, average and worst fitness function values obtained by competing algorithms (BA, CSA, GWO, PSO-GSA) and the GEBA algorithm for 50 iterations.

Table 2. Best, average and worst fitness function values obtained by the BA, CSA, GWO, PSO-GSA and GEBA algorithms.

Dataset number	*k*-classifier – nearest neighbors					*CART*-classifier				
	BA	*CSA*	*GWO*	*PSO-GSA*	*GEBA*	*BA*	*CSA*	*GWO*	*PSO-GSA*	*GEBA*
S1	**0,251**	0,278	0,259	0,255	0,266	0,237	0,246	0,236	0,229	**0,226**
	0,277	0,300	0,280	**0,271**	0,275	0,251	0,260	0,247	0,241	**0,238**
	0,299	**0,328**	0,299	0,293	0,291	0,268	**0,272**	0,263	0,266	0,256
S2	0,048	0,082	0,048	**0,034**	0,048	0,135	0,135	0,139	0,125	**0,120**
	0,075	0,098	0,075	**0,057**	0,063	0,163	0,163	0,175	**0,157**	0,165
	0,101	**0,120**	0,096	0,091	0,091	0,188	0,188	0,197	0,178	**0,202**
S3	**0,021**	0,028	**0,021**	0,035	**0,021**	**0,021**	**0,021**	**0,021**	0,028	**0,021**
	0,032	0,040	0,038	0,044	**0,030**	0,026	0,027	0,027	0,033	**0,025**
	0,049	0,056	0,056	**0,076**	0,049	0,063	0,035	0,035	**0,069**	0,028
S4	**0,000**	**0,000**	**0,000**	**0,000**	**0,000**	**0,012**	0,013	**0,012**	0,013	0,013

(continued)

Table 2. (*continued*)

Dataset number	*k*-classifier – nearest neighbors					*CART*-classifier				
	BA	*CSA*	*GWO*	*PSO-GSA*	*GEBA*	*BA*	*CSA*	*GWO*	*PSO-GSA*	*GEBA*
	0,000	**0,000**	**0,000**	**0,000**	**0,000**	**0,015**	0,016	**0,015**	0,016	**0,015**
	0,001	0,000	0,000	**0,004**	0,000	0,021	0,020	**0,018**	0,020	0,018
S5	**0,000**	0,010	**0,000**	**0,000**	**0,000**	**0,020**	0,040	0,030	**0,020**	**0,020**
	0,007	0,022	0,012	0,009	**0,000**	0,029	0,054	0,044	0,035	**0,025**
	0,010	**0,040**	0,020	**0,040**	0,000	0,050	**0,069**	0,059	0,050	0,040
S6	0,046	0,074	0,063	**0,046**	0,054	**0,048**	0,060	**0,048**	**0,048**	0,051
	0,066	0,085	0,074	**0,059**	0,062	0,061	0,066	0,062	**0,057**	0,060
	0,080	**0,094**	0,083	0,074	0,074	0,071	0,071	**0,074**	0,071	0,071
S7	0,108	0,122	0,108	**0,101**	0,115	0,108	0,128	**0,101**	0,108	0,115
	0,127	0,162	0,128	**0,120**	0,123	0,139	0,148	0,139	0,133	**0,132**
	0,149	**0,196**	0,162	0,149	0,135	**0,176**	0,169	0,155	0,155	0,162
S8	0,294	0,317	0,333	0,230	**0,222**	**0,111**	0,119	0,127	**0,111**	0,119
	0,341	0,344	0,348	0,318	**0,310**	0,138	0,137	0,141	**0,133**	0,137
	0,357	**0,365**	**0,365**	**0,365**	**0,365**	**0,159**	0,151	**0,159**	**0,159**	0,151
S9	**0,156**	**0,156**	**0,156**	**0,156**	**0,156**	0,143	0,146	0,143	**0,139**	**0,139**
	0,169	0,198	0,178	0,172	**0,163**	0,159	0,166	0,157	0,157	**0,155**
	0,190	**0,231**	0,221	0,224	0,190	0,173	**0,187**	0,173	0,180	0,170
S10	**0,022**	0,052	**0,022**	0,023	**0,022**	**0,007**	0,018	**0,007**	**0,007**	**0,007**
	0,026	0,069	0,030	0,027	**0,024**	**0,008**	0,038	**0,008**	0,009	**0,008**
	0,041	**0,096**	0,043	0,032	0,027	0,014	**0,059**	0,009	0,014	0,009

Analysis of the results shows the following. For the KNN classifier, the GEBA algorithm obtains the best fitness mean for 6 out of 10 datasets tested (S3, S4, S5, S8 S9 and S10), while the rival PSO-GSA algorithm leads for 5 out of 10 datasets (S7, S1, S4, S6 and S2). The BA, CSA, and GWO algorithms obtained the best fitness mean for 1 in 10 datasets tested (S10). For the KNN classifier, the GEBA algorithm obtains maximum fitness values for 6 out of 10 tested datasets (S3, S4, S5, S8, S9 and S10), as well as the BA and PSO-GSA algorithms. The CSA algorithm gives the worst results.

For the CART classifier, the GEBA algorithm obtains the best fitness mean for 7 out of 10 datasets tested (S1, S3, S4, S5, S7, S9, and S10), and the best competing PSO-GSA algorithm gets 3 out of 10. The GEBA algorithm gets the highest fitness scores for 6 out of 10 datasets tested, and the worst results on most datasets are given by CSA.

The Wilcoxon rank-sum *T*-test and the Friedman test with a confidence level of 0.95 show that the GEBA algorithm is statistically significant compared to the competing algorithms BA, CSA, and GWO. The GEBA algorithm takes a leading position, obtained in 50 independent runs for each of 10 datasets, in comparison with the PSO-GSA algorithm, both for the KNN classifier and for the CART. Table 3 illustrates the robustness of the compared algorithms.

The robustness of the GEBA algorithm is higher than that of competing algorithms on 7 out of 10 datasets (S1, S3, S4, S5, S7, S9 and S10) when using the KNN classifier, and on 6 out of 10 datasets (S3, S4, S5, S7, S9, and S10) when using the CART classifier. The main advantage of the GEBA algorithm is that it has fewer adjustable variables, which definitely affects the performance of any optimization algorithm.

Table 3. Standard deviation for BA, CSA, GWO, PSO-GSA, and GEBA algorithms.

Dataset number	*k*-classifier – nearest neighbors					*CART*-classifier				
	BA	*CSA*	*GWO*	*PSO-GSA*	*GEBA*	*BA*	*CSA*	*GWO*	*PSO-GSA*	*GEBA*
S1	0,011	0,014	0,010	0,008	**0,006**	0,010	**0,007**	0,008	0,010	0,008
S2	0,014	**0,009**	0,014	0,014	0,011	0,016	**0,012**	0,014	0,013	0,019
S3	0,008	0,011	0,011	0,011	**0,005**	0,009	0,005	**0,003**	0,009	**0,003**
S4	**0,000**	**0,000**	**0,000**	0,001	**0,000**	0,002	0,002	**0,001**	0,002	**0,001**
S5	0,005	0,008	0,006	0,009	**0,000**	0,011	0,009	**0,008**	0,010	**0,008**
S6	0,011	0,007	**0,005**	0,007	0,006	0,005	**0,004**	0,006	0,007	0,005
S7	0,013	0,017	0,016	0,011	**0,007**	0,016	**0,012**	0,014	0,015	**0,012**
S8	0,016	0,012	**0,009**	0,046	0,050	0,013	0,009	**0,007**	0,012	0,009
S9	0,010	0,023	0,019	0,019	**0,008**	**0,008**	0,011	0,009	0,011	**0,008**
S10	0,005	0,011	0,005	0,003	**0,001**	0,001	0,011	0,001	0,002	**0,000**

The experimental results obtained allow us to assert that the hyper-heuristic bioinspired algorithm GEBA allows increasing the productivity and stability of its work when solving the problem of choosing significant classification features, in comparison with competing for bioinspired algorithms.

The results shown in Tables 2 and 3 demonstrate the best average accuracy of the GEBA algorithm for the KNN and CART classifiers for most datasets used, including the largest dataset S1.

Comparative analysis of the rate of convergence of algorithms on 10 reference datasets and 50 iterations shows that the GEBA algorithm converges faster to global optima than competing algorithms BA, CSA, GWO, and PSO-GSA. We believe that the reason here is a combination of the Bat Algorithm and the differential evolution algorithm using a mechanism to select them based on the current state of the solution. If a balance is maintained between the rate of convergence of the algorithm and the diversification of the solution search space. The lowest convergence rate is demonstrated by the CSA algorithm. This is consistent with the results presented in Table 2.

4 Conclusion

The article proposes the GEBA hyper-heuristic algorithm for selecting significant classification features and assessing the classification accuracy. The GEBA algorithm combines the Bat Echolocation algorithm and a modified version of the Differential Evolution algorithm. The experiments used the well-known *k* nearest neighbors (KNN) classifier and a decision tree-based classifier (CART). The efficiency of the proposed GEBA algorithm was evaluated using a variety of reference datasets from the world UCI library. The performance of the GEBA algorithm for solving the problem of selecting meaningful classification features was compared with competing algorithms for bats, crows, gray wolves, and a hybrid particle swarm and gravity search algorithm. The proposed algorithm leads in the average and best fitness function values for most of the tested datasets and is also statistically significant compared to competing algorithms according to the Wilcoxon *T*-test and Friedman's test with a confidence level of 0.95. The robustness and accuracy of the GEBA algorithm are higher than that of competing algorithms on most reference datasets. Comparative analysis of the rate of convergence of algorithms on a set of reference datasets for a given number of iterations showed that the GEBA algorithm converges faster to global optima than competing algorithms.

The algorithm is planned to be used to solve traveling salesman problems [15, 16], integer programming [17, 18], medical image segmentation problems [19], multidimensional problems [20].

Acknowledgements. The study was carried out with the financial support of the Russian Foundation for Basic Research within the framework of research projects No. 19-07-00570 and No. 18-29-22019.

References

1. Cuevas, E., Fausto, F., Gonzales, A.: New Advancements in Swarm Algorithms: Operators and Applications. Intelligent Systems Reference Library, vol. 160, 296 p. Springer, Heidelberg (2020). https://doi.org/10.1007/978-3-030-16339-6
2. Mirjalili, S.M., Yang, X.-S.: Binary bat algorithm. Neural Comput. Appl. **25**, 663–681 (2014)
3. Sayed, G.I., Hassanien, A.E., Azar, A.T.: Feature selection via a novel chaotic crow search algorithm. Neural Comput. Appl. **31**, 171–188 (2019)
4. Emary, E., Zawbaa, H.M., Hassanien, A.E.: Binary grey wolf optimization approaches for feature selection. Neurocomputing **172**, 371–381 (2016)
5. Mirjalili, S., Wang, G.-G., Coelho, L.D.: Binary optimization using hybrid particle swarm optimization and gravitational search algorithm. Neural Comput. Appl. **25**, 1423–1430 (2014)
6. Tawhid, M.A., Ibrahim, A.M.: Hybrid binary particle swarm optimization and flower pollination algorithm based on rough set approach for feature selection problem. In: Yang, X.-S., He, X.-S. (eds.) Nature-Inspired Computation in Data Mining and Machine Learning. SCI, vol. 855, pp. 249–273. Springer, Cham (2020). https://doi.org/10.1007/978-3-030-28553-1_12
7. Tawhid, M.A., Dsouza, K.B.: Hybrid binary dragonfly enhanced particle swarm optimization algorithm for solving feature selection problems. Math. Found. Comput. **1**(2), 181–200 (2018)
8. Hedar, A.R., Ibrahim, A.M., Abdel Hakim, A.E., Sewisy, A.A.: Modulated clustering using integrated rough sets and scatter search attribute reduction. In: Proceedings of the Genetic and Evolutionary Computation Conference Companion (GECCO 2018)., pp. 1394–1401. ACM, New York (2018)
9. Asuncion, A., Newman, D.J.: UCI machine learning repository. University of California, School of Information and Computer Science, Irvine (2007)
10. Yang, X.S.: A new metaheuristic bat-inspired algorithm. In: González, J.R., Pelta, D.A., Cruz, C., Terrazas, G., Krasnogor, N. (eds.) Nature Inspired Cooperative Strategies for Optimization (NICSO 2010), vol. 284, pp. 65–74. Springer, Heidelberg (2010). https://doi.org/10.1007/978-3-642-12538-6_6
11. Storn, R., Price, K.: Differential evolution – a simple and efficient heuristic for global optimization over continuous spaces. J. Glob. Optim. **11**, 341–359 (1997)
12. Dua, D., Graff, C.: UCI machine learning repository (2017). http://archive.ics.uci.edu/ml
13. Gupta, B., Rawat, A., Jain, A., Arora, A., Dhami, N.: Analysis of various decision tree algorithms for classification in data mining. Int. J. Comput. Appl. **163**(8), 15–19 (2017)
14. Derrac, J., García, S., Molina, D., Herrera, F.: A practical tutorial on the use of nonparametric statistical tests as a methodology for comparing evolutionary and swarm intelligence algorithms. Swarm Evol. Comput. **1**(1), 3–18 (2011)

15. Rodzin, S., Rodzina, O.: Metaheuristics memes and biogeography for trans computational combinatorial optimization problems. In: Proceedings of the 6th International Conference - Cloud System and Big Data Engineering, pp. 1–5 (2016)
16. Tawhid, M.A., Savsani, P.: Discrete sine-cosine algorithm (DSCA) with local search for solving traveling salesman problem. Arab. J. Sci. Eng. **44**(4), 3669–3679 (2019)
17. Rodzin, S.I., Rodzina, O.N.: Effectiveness evaluation of memetics and biogeography algorithms using benchmark and trans computational tasks of combinatorial optimization. In: Abraham, A., Kovalev, S., Tarassov, V., Snášel, V. (eds.) Proceedings of the First International Scientific Conference "Intelligent Information Technologies for Industry" (IITI'16). AISC, vol. 450, pp. 463–473. Springer, Cham (2016). https://doi.org/10.1007/978-3-319-33609-1_42
18. Tawhid, M.A., Ali, A.F.: Multidirectional harmony search algorithm for solving integer programming and minimax problems. Int. J. Bio-Inspired. Comput. **13**(3), 141–158 (2019)
19. El-Khatib, S., Rodzin, S.I., Skobtcov, Y.A.: Investigation of optimal heuristical parameters for mixed ACO-k-means segmentation algorithm for MRI images. In: Proceedings of the Conf. on Information Technologies in Science, Management, Social Sphere and Medicine (ITSMSSM 2016), pp. 216–221 (2016)
20. Rodzin, S., Rodzina, L., Rodzina, O.: Optimization of multi-extreme multidimensional functions: a population-based nature-inspired algorithm. In: Proceedings of the International Russian Automation Conference, 6–12 September 2020, pp. 595–599. IEEE, Sochi (2020)

Method for the Synthesis of Fuzzy Controllers from Experimental Data Based on Clustering

Vladimir Vladimirovich Ignatyev(✉), Viktor Vladimirovich Soloviev, Denis Aleksandrovich Beloglazov, and Andrey Vladimirovich Kovalev

Southern Federal University, Bolshaya Sadovaya Str., 105/42, 344006 Rostov-on-Don, Russia
vvignatev@sfedu.ru

Abstract. The purpose of this work is to develop a method for the synthesis of fuzzy controllers based on experimental data based on clustering. To achieve this goal, it is proposed to take experimental data on the input and output signals of the control system of a technical object with a classical controller. On the basis of experimental data, a data clustering method is developed that allows determining the term-sets of input and output linguistic variables of a fuzzy controller that implements the Mamdani fuzzy inference algorithm. Clustering is performed by estimating the boundaries of the intervals of change in experimental data, evenly dividing them into clusters depending on the required power of the linguistic variables term-sets, and determining whether the data belongs to one or another cluster. Since the experimental data are related, that is, for each moment of time, data on both the input and output signals of the classical controller is stored and their belonging to clusters is determined, then the rule base of a fuzzy controller is easily compiled. According to the results of experiments, it was found that in most cases the rule base is formed with duplicate rules, that is, there are rules with the same antecedent, but different consequent, so the authors conducted additional research related to the reduction of the rule base and assigning redundant rules of different weights. To simplify the research, the authors developed software in the MatLab environment that allows both to obtain experimental data, and to synthesize a fuzzy controller and check its performance. The research results will be useful for developers of fuzzy control models. #CSOC1120.

Keywords: Fuzzy controller · PID controller · Data clustering · Rule weight · Rule base reduction · Automatic rule base synthesis

1 Introduction

To date, the problems of synthesis of control systems for technical objects with fuzzy controllers attract the attention of researchers from all over the world, since they allow obtaining better control quality indicators (duration of the transient process, over-control) in comparison with systems with classical controllers. In this work, the authors understand classical controllers as PID controllers and their various modifications. Consider publications devoted to research related to the structures of fuzzy controllers and the synthesis of fuzzy rule bases.

© The Author(s), under exclusive license to Springer Nature Switzerland AG 2021
R. Silhavy (Ed.): CSOC 2021, LNNS 229, pp. 557–572, 2021.
https://doi.org/10.1007/978-3-030-77445-5_51

In work [1], the authors proposed the structure of a fuzzy controller in the form of a parallel connection of PI and PD fuzzy controllers with two parametric membership functions for each of the input signals. This made it possible to synthesize simple rule bases containing 4 rules each. The authors have determined the stability conditions for a system with the proposed fuzzy controller under constraints on input and output signals. The results of model experiments showed the best indicators of the quality of control with a fuzzy controller in comparison with a classical PID controller.

The work [2] presents a method for synthesizing a fuzzy PID controller using evolutionary algorithms. A study of a fuzzy PID controller with three independent rule bases for P-, I- and D- components is carried out. A method for setting the parameters of the controller for a dynamic plant based on a genetic algorithm is proposed, which allows obtaining the coefficients of each channel of the controller and forming nonlinear transform functions. In experimental studies, an oscillatory link was used as a model of the control object and it was found that, in comparison with a classical PID controller, a fuzzy controller can significantly improve control quality indicators.

In work [3], the authors reviewed various ways of representing PID controllers and their modifications in the form of fuzzy models with one, two or three inputs. As elements of the term-set of fuzzy variables, uniformly distributed triangular membership functions with overlapping are used. The results of the analysis of the possibilities of adapting the parameters of fuzzy controllers with variable gains are presented. The authors note that fuzzy controllers with decoupled control channels contain fewer rules compared to controllers that have two or three inputs and one output. The results of theoretical calculations are not supported by experiments.

The work [4] presents an overview of possible structural implementations of fuzzy PID controllers and ways to adjust their parameters. The results show that fuzzy PID controllers can contain from one to three input signals. In this case, the rule bases of individual channels can be both related and independent. Separately, the structural diagrams of systems in which I- or D- components are implemented as separate from the PD and PI fuzzy models are given. As methods of tuning, a variant with adaptive tuning of the input and output gains of fuzzy PID controllers using observers of the input and output signals and a parametric function is considered. A separate class of controllers is considered, in which fuzzy models are used to adjust the gain of a classical PID controller, as well as hybrid implementations of fuzzy and classical controllers.

In work [5], methods for adapting a fuzzy controller are considered. The authors point out that when developing fuzzy controllers, even for the same control objects, they make an individual contribution to the form of membership functions, the power of term-sets of linguistic variables, rule bases and create suboptimal controllers. A model of the desired behavior of the control object is proposed, which, in contrast to the reference model, uses information about the time derivatives of the state signals and the control action. In the developed adaptation method, the overlap value of adjacent terms of fuzzy variables is adjusted, the terms are absorbed and split, and the consequents of the rules are changed. The results of theoretical calculations are not supported by experiments.

In work [6] presents a method of parametric synthesis of a digital fuzzy controller of automatic control systems for non-stationary control objects. Parametric synthesis is carried out by determining the dependence of the optimal tuning parameters of the fuzzy controller on the change in the parameters of the control object. As configurable parameters, the error change interval, its first and second derivatives, and the control action change interval are considered, which are configured for each of the sub-ranges of the control object state change. According to the author, the proposed method allows build an optimal tunable fuzzy controller according to a given quality criterion.

In work [7], the authors consider the problem of synthesizing a control law in the presence of observation noise. It is assumed that the state assessment and control synthesis are solved sequentially. Uncertainty was formalized using fuzzy intervals and the Takagi-Sugeno fuzzy model describing the dynamics of the system. The simulation results are presented, which showed a decrease in energy costs for controls by 11% and an increase in control accuracy by 6%.

In work [8], the authors consider a procedure for the synthesis of a fuzzy feedback controller for nonlinear neutral systems with state delay with impulse effects. The authors carried out theoretical studies and proposed a method for analyzing the stability of a system with a fuzzy Takagi-Sugeno model. Conditions for the stability of a closed-loop system with a fuzzy controller are formulated and the results of model experiments are presented.

The work [9] presents a nonlinear mathematical model of a quadrocopter, which was linearized in the vicinity of the operating point. PID controllers with angular and linear values of the quadrocopter have been developed. The results of model experiments that demonstrate a large overshoot (30%) in flight altitude are presented. Based on the results of model experiments, data on the magnitude of errors for linear and angular flight parameters were recorded, on the basis of which fuzzy controllers for a quadcopter were synthesized. The authors have manually defined the parameters of fuzzy variables and developed fuzzy rule bases. Based on the results of comparison of fuzzy and classical PID controllers, it was found that fuzzy controllers provide the best indicators of the quality of flight control of a quadcopter.

The work [10] presents a method for constructing regions of guaranteed stability of nonstationary nonlinear systems using fuzzy PID controllers. The method makes it possible to determine the stability areas that are larger in area than the areas determined by classical methods. These results were achieved due to the development of an algorithm for constructing Lyapunov spline functions. Examples of the synthesis of fuzzy P-, PI-, PID- controllers for a non-stationary third-order control object are given. Based on the results of model experiments, it was found that the expansion of the stability boundaries allows improving the control accuracy in the steady state and reducing the duration of the transient process. The authors note that the further purpose of the study is to identify areas of stability for self-learning fuzzy controllers.

In work [11], fuzzy logic is used to formalize the uncertainty of the wood drying process. Based on the theory and practice of wood drying, the authors determined the input linguistic variables of temperature and humidity, and the drying time was chosen as the output variable. Based on the reference data, the intervals for changing the input and output parameters are determined. Term-sets of linguistic variables containing uniformly distributed membership functions of triangular and trapezoidal types are proposed. The rule base in the MatLab environment is developed and the results of comparison with the model based on heat and mass transfer are presented, showing the adequacy of the developed fuzzy model.

The work [12] presents the results of the development of a fuzzy controller for a wind power plant and presents the results of modeling in comparison with the classical PI controller. The fuzzy controller synthesis procedure is based on a self-organization algorithm that consistently forms a base of fuzzy rules. It is found that the system with a fuzzy controller provides a shorter duration of the transition process by 53% during acceleration and by 79% during deceleration, and reduces the deviation from the set value by 23%.

In work [13], an algorithm for the synthesis of a fuzzy controller for a steam boiler is presented. The synthesis mechanism is based on a regression equation with the input parameters of the steam temperature of the primary superheater, the condensation temperature, the valve opening angle, and the output parameter steam temperature at the output of the secondary superheater. The primary superheater steam temperature and condensation temperature were considered as the input parameters of the fuzzy controller, and the valve opening value was the output parameter. On the basis of the experimental data obtained by clustering, term-sets of fuzzy variables with triangular membership functions were obtained and a base of fuzzy rules was compiled. The authors note the problem of duplication of rule antecedents and solve it by assigning weight coefficients to the rules.

In work [14], a method for the development of fuzzy models based on experimental data is proposed, which makes it possible to exclude from the base rules with the same antecedents and different consequents using numerical estimates of the degree of importance of the rules. A method for optimizing the rule base is presented, based on generating a base containing all possible combinations of membership functions of input and output variables and its further correction using heuristic methods. To check the results, a nonlinear function was chosen, with which the experimental data were taken and a fuzzy model was synthesized, which showed a high accuracy of its approximation.

In the early works of the authors of this article [15, 16], some issues of the synthesis of fuzzy models based on the Mamdani and Sugeno algorithms based on experimental data were already considered. Clustering and learning algorithms were proposed that allow obtaining fuzzy controllers that exceed the classical ones in terms of characteristics, which are confirmed by the simulation results.

Based on the results of the literature review, the following problems can be noted:

- methods for the synthesis of fuzzy models based on experimental data are not yet sufficiently developed;
- there remains the problem of eliminating uncertainty in fuzzy rule bases, in which some rules have the same antecedent, but different consequent;
- the problem of fully automatic synthesis of a fuzzy model without the involvement of expert knowledge based on the results of experimental studies has not been solved;
- studies of the optimal structures of fuzzy models for the cases of multiple input signals and one output signal have not been conducted, since it is possible to build separate or related controller rule bases for the same control object;
- the structure of a software application that allows synthesizing fuzzy controllers based on experimental data in a fully automatic mode has not been proposed.

In this work, the authors focused on solving these problems.

2 Proposed Method

This work is a development of the results obtained by the authors as a result of research on the development of theoretical foundations and methods for automatically configuring the parameters of intelligent control systems for technical objects with a priori uncertainty [15, 16]. Let us consider the stages of implementation of the proposed method for the synthesis of fuzzy controllers, the sequence of which is shown in Fig. 1.

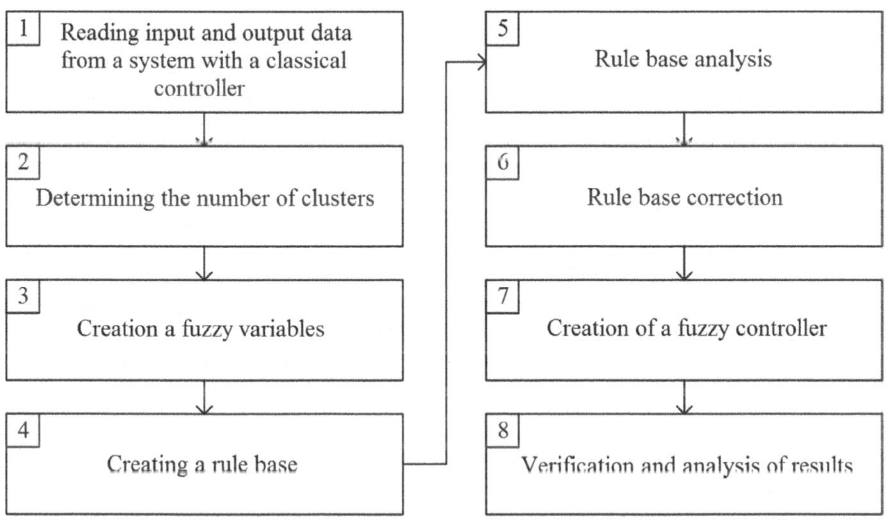

Fig. 1. Stages of the fuzzy controller synthesis method.

1. **Reading input and output data from a system with a classical controller.** This data can either be taken from a real object in the course of its operation, or a mathematical model of the control system can be developed and obtained in the course of modeling. In this work, the research was carried out using a mathematical model of the control system developed in the MatLab Simulink environment. The simulation scheme is shown in Fig. 2.

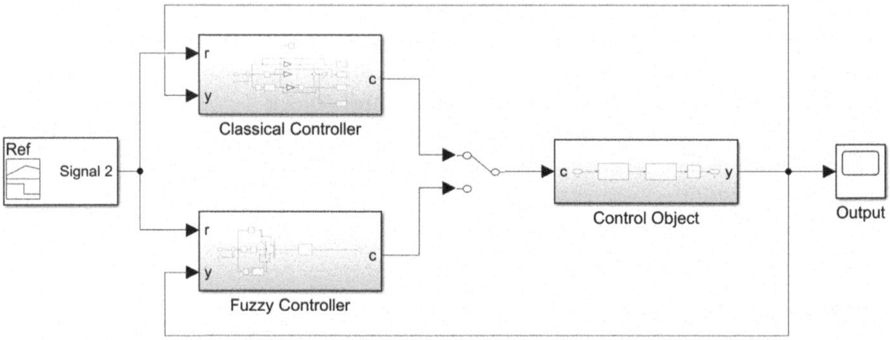

Fig. 2. Simulation scheme.

The simulation scheme contains a single control object that can be controlled by either a classical or fuzzy controller via a switch. The input of the system receives a setting effect. The input and output data of the classical controller and the simulation time intervals are written to the DataPoints.mat file. It should be noted that it is necessary to read data from a real object or from a model with a fixed time step.

2. **Determining the number of clusters.** When developing fuzzy models with the involvement of expert knowledge, usually try to use no more than 7 ± 2 membership functions for each fuzzy variable [17]. Indeed, if a fuzzy model has three input variables with seven membership functions each, then the complete base must contain $7 * 7 * 7 = 343$ fuzzy rules. In the case of creating fuzzy models without the involvement of experts, this problem does not arise and the number of membership functions of linguistic variables is determined by computational resources, as well as the difference between measurements in a series of experimental data. This work considers a uniform distribution of clusters in the range of variation of experimental data, which made it possible to formulate conditions for determining the number of clusters. For example, let the signal data be recorded in a control system with a classical controller, as shown in Fig. 3.

Then the minimum cluster size will be equal to δ. If you select a smaller cluster, then after determining whether the experimental clusters belong to these clusters, there will be clusters that do not have any measurements. The value of δ can be determined in a loop, going through the data of each measured signal. In this regard, the maximum number of signal clusters can be found by the formula:

$$N_{i\max} = \left\lceil \frac{s_{i\max} - s_{i\min}}{\delta_i} \right\rceil \quad (1)$$

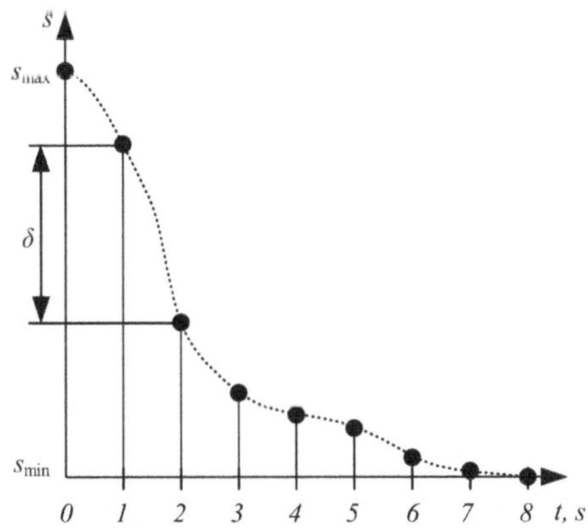

Fig. 3. Signal data in a control system with a classic controller.

where $N_{i\max}$ – the maximum clusters number of the i-th signal; $s_{i\max}$, $s_{i\min}$ – the maximum and minimum signal value, respectively; δ_i – the maximum deviation between two adjacent measurements; ⌈⌉ – the rounding to a larger integer.

Then the cluster size can be calculated by the formula:

$$pc_i = \frac{s_{i\max} - s_{i\min}}{N_{i\max}} \tag{2}$$

where pc_i – the cluster size of the i-th signal.

It is also necessary to take into account that the clusters number of output signal data satisfies the condition:

$$N_{\text{out}} \leq \prod_{i=1}^{m} N_{i\text{in}} \tag{3}$$

where $N_{i\text{in}}$ – the clusters number of the i-th input signal of the fuzzy model; m – the number of input signals.

If condition (3) is met, all the membership functions of the fuzzy variable of the output signal will be taken into account in the rule base. If condition (3) is not met, it is necessary to reduce the clusters number of output signal data and recalculate their sizes according to the formula (2). The minimum number of clusters for signals is 2. Therefore, the number of clusters for each of the variables is selected from the $[2,\ N_{i\max}]$ interval.

3. **Creation of fuzzy variables.** According to the results of the second stage, the number of clusters N, the interval of change $[s_{\min},\ s_{\max}]$ and the size pc are determined for each variable. The boundaries and centers of clusters can be determined by the formulas [18]:

$$c_1 = [s_{i\min},\ s_{i\min} + 0.5 \cdot pc),\ ck_1 = c_{11},$$

$$c_j = [\max(c_{j-1}),\ \max(c_{j-1}) + pc),\ ck_j = 0.5\big(c_{j1} + c_{j2}\big), (j = 2,\ 3, \ldots k-1), \tag{4}$$

$$c_k = [\max(c_{j-1}),\ \max(c_{j-1}) + 0.5 \cdot pc],\ ck_k = c_{k2}.$$

where k – the number of variable clusters, ci – the cluster intervals, cki – the cluster centers.

The second index in the cij parameters in expression (4) means: 1 – the left boundary of the interval (ci1), 2 – the right boundary of the interval (ci2). It should be noted that in (4) the cluster center is not its geometric center, but the point to which the neighboring membership functions are attached.

Expressions (4) allow for the series of experimental data to match the numbers of the clusters in which they fall. Table 1 shows an example of determining whether data points belong to clusters.

Table 1. Result of determining whether data points belong to clusters.

Input 1	Cluster number	…	…	Input N	Cluster number	Output	Cluster number
0.15	2	…	…	0.34	4	0.95	9
0.28	3	…	…	0.01	1	0.84	8
0.82	6	…	…	0.18	2	0.62	8
…	…	…	…	…	…	…	…

For each experimental signal, a linguistic variable L is introduced, which is characterized by the set:

$$L = \langle N, T, X \rangle \tag{6}$$

where N – the name of the linguistic variable; T = {"j"} (j = (1, 2, …, k)) – term-set of a linguistic variable; X = [smin, smax] – domain of definition of elements from T (base set).

Define the elements of the term-set T in the form [19]:

– z-type membership function:

$$\mu_1(x; a, b) = \begin{cases} 1, & x \le a, \\ 1 - 2\left(\frac{x-a}{b-a}\right)^2, & a \le x \le \frac{a+b}{2}, \\ 2\left(\frac{x-a}{b-a}\right)^2, & \frac{a+b}{2} \le x \le b, \\ 0, & x \ge b, \end{cases} \tag{6}$$

где $a = c_{11}$; $b = ck_2$; x – variable.

– triangular membership function:

$$\mu_i(x, a, b, d) = \begin{cases} 0, & x \le a; \\ \dfrac{x-a}{d-a}, & a < x \le d; \\ \dfrac{b-x}{b-d}, & d < x \le b; \\ 0, & x > b. \end{cases} \tag{7}$$

где $a = c_{k-1}$; $b = c_{k+1}$; $i = 2, 3, \ldots, k-1$.

– s-type membership function:

$$\mu_1(x; a, b) = \begin{cases} 0, & x \le a, \\ 2\left(\frac{x-a}{b-a}\right)^2, & a \le x \le \frac{a+b}{2}, \\ 1 - 2\left(\frac{x-a}{b-a}\right)^2, & \frac{a+b}{2} \le x \le b, \\ 0, & x \ge b, \end{cases} \tag{8}$$

where $a = c_{k-1}$; $b = c_{k2}$.

Note that, for example, in [3, 13, 14] and in the early work of the authors [18], only triangular membership functions were used. But according to the results of further research, it was found that after the synthesis of a fuzzy controller and its approbation with a control object, collisions may occur due to signals going beyond the universal sets and, as a consequence, incorrect fuzzy inference. In addition, in expressions (6)–(8), the coefficients a, b depend on the parameters of neighboring clusters, which makes it possible to automate the process of determining term-sets of linguistic variables of the fuzzy controller and create fuzzy variables as shown in Fig. 4.

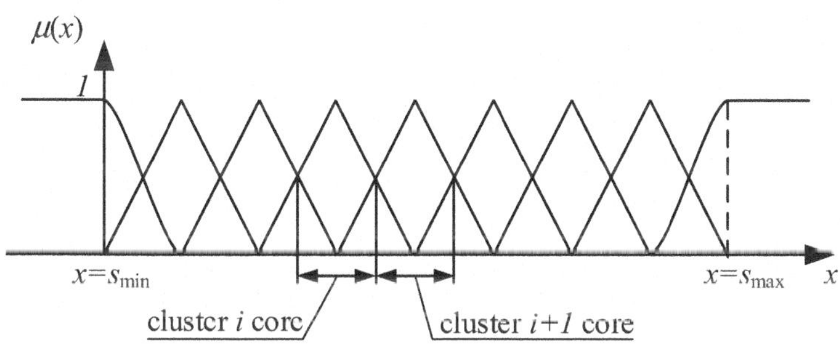

Fig. 4. Creating a fuzzy variable based on clusters.

As can be seen from Fig. 4, the number of membership functions is equal to the number of clusters. To account for the uncertainty, each membership function exceeds the size of its cluster, that is, according to (7): $d - a > pc$.

4. **Creating a rule base.** The fuzzy controller rule base is formed automatically after determining the belonging of experimental data points to clusters, as shown in Table 2.

Table 2. Fuzzy rule base.

Rule number	*Tinp1*	T_{inp2}	…	T_{inpM}	T_{out}	W
1	2	4	…	6	2	18
2	3	1	…	11	8	35
3	6	2	…	4	1	4
…	…	…	…	…	…	…

In Table 2, the following notation is introduced: Tinp – the elements of the term-sets of the input linguistic variables, Tout – the elements of the term-sets of the output linguistic variables, W – the number of experimental data points that fell into the cluster, M – the number of input variables. The parameter W will be needed in the next step to analyze and correct the rule base.

5. **Analysis of the rule base.** When analyzing the rules, the following characteristics of the rule base should be determined: completeness and consistency.

The complete base must contain $N_{inp1} \cdot N_{inp2} \cdot \ldots \cdot N_{inpM} = R_F$ of fuzzy rules. If the number of rules is less than RF, then the rule base is incomplete. If the number of rules is greater than R_F, then the rule base is redundant. It should be noted that when creating fuzzy controllers based on experimental data, obtaining a complete rule base is not guaranteed, since not all combinations of elements of term-sets of input linguistic variables can be included in the base.

A rule base is inconsistent if it contains rules with the same antecedent and different consequent:

……………………………………………………………………

Rule i: If x_{inp1} is T_{inp2} and … x_{inpM} is T_{inp1} then x_{out} is $\boldsymbol{T_{out4}}$;

……………………………………………………………………

Rule j: If x_{inp1} is T_{inp2} and … x_{inpM} is T_{inp1} then x_{out} is $\boldsymbol{T_{out2}}$;

……………………………………………………………………

Checking for completeness and consistency is easily performed algorithmically.

6. **Correction of the rule base.** If there are conflicting rules, the rule base needs to be adjusted. It is possible either to collapse the conflicting rules and synthesize one rule with the generalizing element of the term set in the consequent, or exclude from the conflicting rules those that have fewer experimental points W (see Table 2).

The convolution of the rules will be performed according to the method from the work [18]. It is necessary to select the cluster centers of consequent *ck* redundant rules and find the arithmetic mean by the formula:

$$s = \frac{\sum_{j=1}^{p} ck_j}{p} \tag{9}$$

where p – the number of redundant rules with the same antecedent.

Then it is necessary to determine the belonging of s to one of the clusters from T_{out} and assign its number to the consequent in one of the redundant rules, and delete the rest. It should be noted here that if there are two redundant rules and their consequents correspond to adjacent terms, then the use of expression (9) leads to uncertainty. For this case, it is necessary to introduce an additional term with parameters into the term-set of the output linguistic variable (see Fig. 5)

$$a_j = c_{d(j-1)}, \quad b_j = s, \quad d_j = c_{d(j+1)} \tag{10}$$

where $j - 1, j + 1$ – the clusters centers that form the membership functions of the sequents of redundant rules.

Redundand rules

№	Inp1	Inp2	Inp3	Out
...	...	...	...	...
l	1	2	2	4
...	...	...	...	...
k	1	2	2	5
...	...	...	...	...

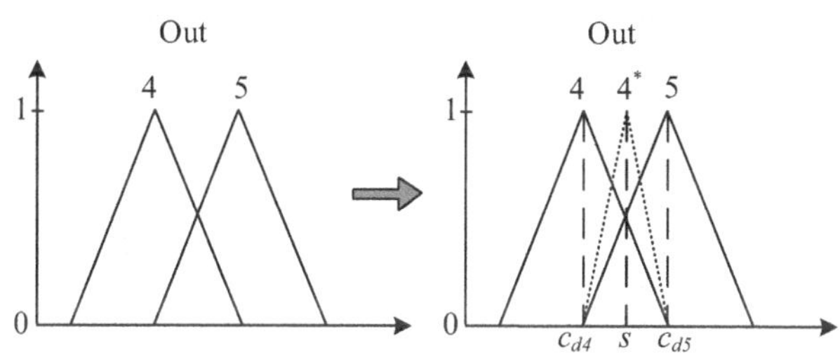

Fig. 5. Combining redundant rules in a rule base.

Performing this procedure for all sets of redundant rules, obtains a reduced rule base.

7. **Creating a fuzzy controller.** By the beginning of this stage, term-sets of input and output linguistic variables have already been defined, and the rule base has been obtained and reduced. The authors have developed a software module in the MatLab 2018b environment to create a fuzzy controller. The input parameters of the program module are input and output linguistic variables with specified universal and term-sets, and the rule base in the form of an array, the columns of which define the terms of the input variables, the output variable and the weight of the rule. An empty fis structure [20] is created into which the data on input and output variables are sequentially entered with the addvar command, terms – with the addmf command, the rule base - with the addrule command.
8. **Verification and analysis of the result.** The generated fis-structure is saved to a file and then exported to the Fuzzy Controller block in the Simulink environment

for verification and analysis, as shown in the simulation scheme in Fig. 2. First, the output signals of the classical and fuzzy controllers are compared, and then the output signals from the control object are compared with both controllers.

3 Results

The mathematical model of coupled capacities from work [1], which is described by the transfer function, was chosen as the control object:

$$G(s) = \frac{0.75e^{-8.067\,s}}{(1 + 61.45\,s)^2}. \tag{11}$$

As a classical controller, a PD controller is used with the parameters: $K_p = 10.45$, $K_d = 293.34$ and a limitation on the signal amplitude of the derivative error in the interval [−1, 1].

The signal shown in Fig. 6 was applied to the input of the system.

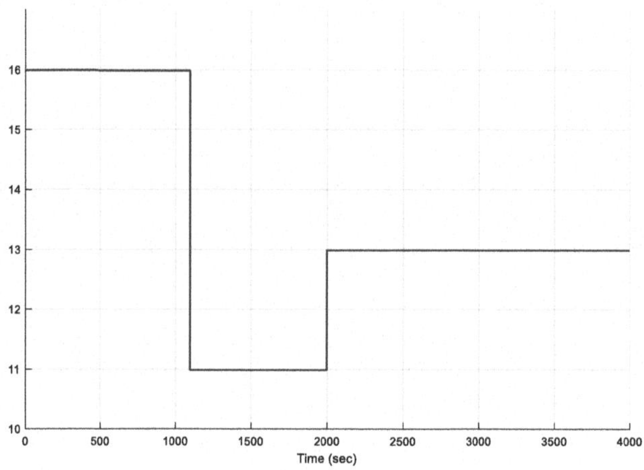

Fig. 6. Reference signal.

Based on the results of the simulation the classical controller, the error signals, its derivative and control were recorded in the amount of 40,000 points with a sampling step of 0.1 s. In accordance with steps 1–7, a fuzzy controller containing input variables with 13 terms and an output variable with 7 terms was synthesized, as shown in Fig. 7.

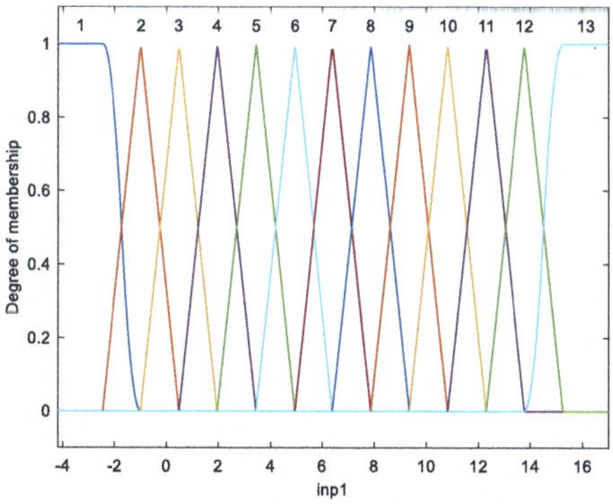

a – Input 1 (Error)

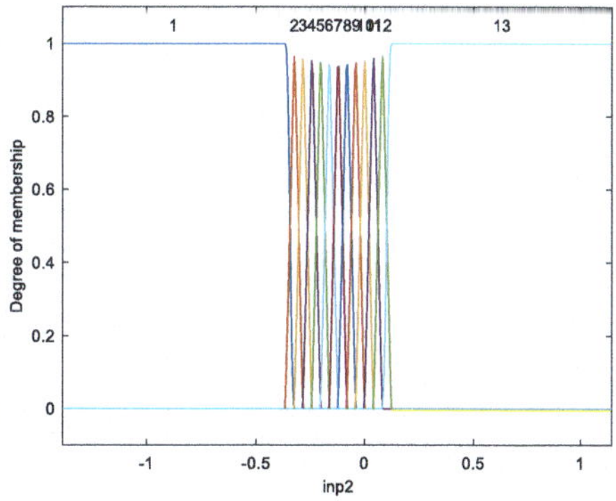

b – Input 2 (Derived error)

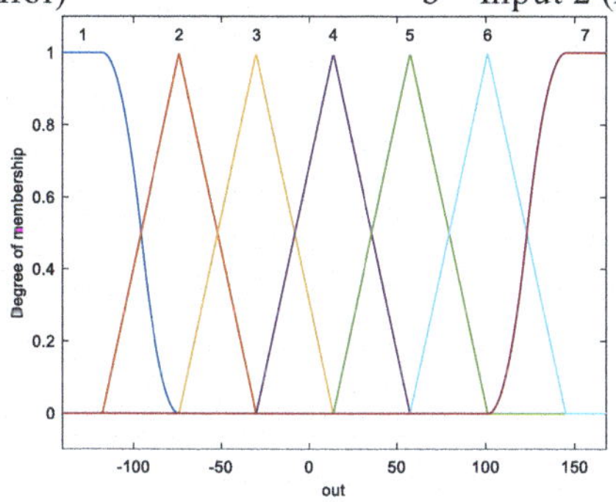

c – Output (Control signal)

Fig. 7. Fuzzy controller parameters.

The comparison results of the classical and fuzzy controllers are shown in Fig. 8.

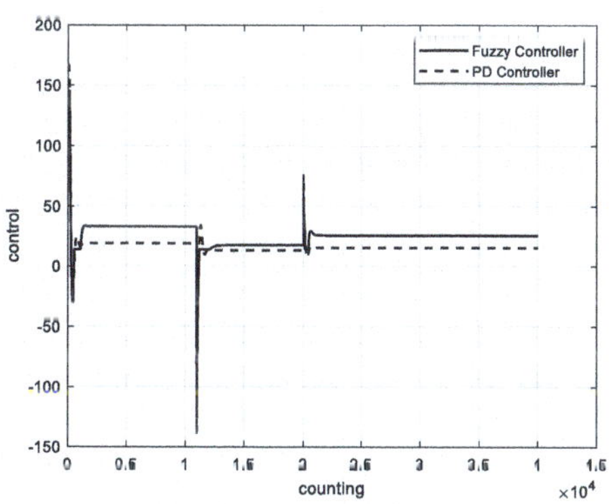

a – the output signals of the controllers

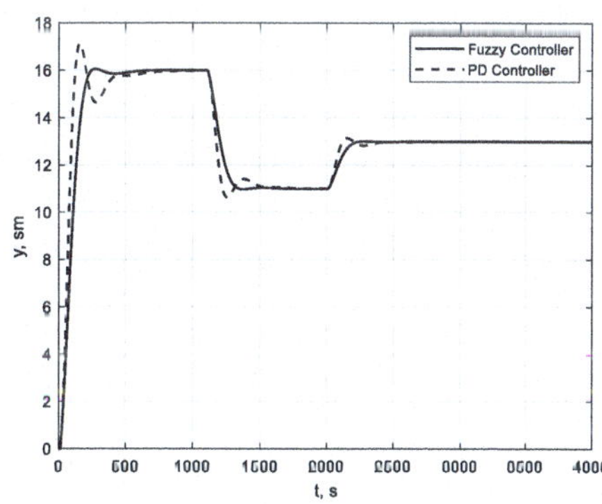

b – the output signals of the object

Fig. 8. Comparison results of classical and fuzzy controllers.

Generated base containing 59 fuzzy rules, which was reduced to 53 rules due to conflicting rules.

4 Discussion

Based on the simulation results, a fuzzy controller was synthesized for a second-order control plant with delay, for which the input linguistic variable "error" contained 13 terms on the universal set [−4, 16], the linguistic variable "error derivative" contained 13 terms on the universal set [−1, 1], the linguistic variable "control signal" contained 7 terms on the universal set [−150, 150]. Universal sets are determined automatically, and the number of linguistic variables terms was set manually. It was automatically determined that for the "error" variable the maximum possible number of terms is 79, for the "error derivative" variable – 34, for the "control signal" variable – 56.

Besides the fact that a controller with such a number of terms will be computationally expensive, it will be approximate the classical controller with high accuracy and, therefore, will not lead to an improvement in the quality of control.

For a reduced number of linguistic variables terms, a discrepancy is seen in the signals of the classical and fuzzy controllers (Fig. 8, a), but as a result, transient processes in an object with a fuzzy controller have a shorter duration and overshoot (Fig. 8, b).

It should be noted that the resulting base of 53 fuzzy rules is incomplete and in other conditions of the control object functioning, such combinations of terms of input linguistic variables may occur that are not present in the rule base. To eliminate this uncertainty, if possible, it is necessary to remove the data in all modes of operation of the object, so that they are sufficient to synthesize a complete rule base.

In addition, the use of z- and s- membership functions on the boundaries of universal sets of linguistic variables allows the fuzzy controller to function correctly, even if the signals exceed their values.

5 Conclusion

In contrast to the previously developed method for the synthesis of fuzzy controllers [18], this method adds stages of automatic analysis of the boundaries of changes in input and output signals, suggests conditions for determining the limiting number of terms of linguistic variables, allowing automatic determination of these parameters. Experiments have shown that an increase in the number of terms does not always have a positive effect on the quality of control, since the synthesized fuzzy controller in its characteristics is very close to the classical. The use of non-triangular membership functions at the boundaries of the variable change intervals has a positive effect and allows the control system to function when signals go beyond the specified ranges. In contrast to work [18], the procedure for creating clusters and determining the belonging of experimental data points has become more formalized, which makes it possible to synthesize fuzzy controllers virtually without human participation.

In addition, the approach to the synthesis of fuzzy controllers in this work is presented in a generalized manner, without reference to the characteristics of classical controllers and, therefore, can be further developed for various structures of controllers: with one input and one output or with several inputs and one output, without essential changes.

The authors have developed a software application in the MatLab environment that implements stages 1–8 in automatic mode. Individual stages of the method can also be of interest to researchers, as they can find their application in various fields of science.

Further work of the authors is aimed at creating a procedure for finding the optimal number of terms of linguistic variables and considering the issues of stability of control systems with synthesized controllers.

Acknowledgements. Scientific research was carried out as part of the project "Creating a high-tech production of hardware and software systems for processing agricultural raw materials based on microwave radiation" (Agreement with the Ministry of Education and Science of the Russian Federation № 075-11-2019-083 dated 20.12.2019, Agreement South Federal University № 18 dated 20.09.2019, number of work in South Federal University № HD/19-25-RT).

References

1. Xu, J.X., Hang, C., Liu, C.: Parallel structure and tuning of a fuzzy PID controller. Automatic **36**, 673–684 (2000)
2. Burakov, M.V., Konovalov, A.S., Yakovets, O.B.: Evolyutsionnyy sintez nechetkikh regulyatorov. [Evolutionary synthesis of fuzzy controllers]. Inf. Control Syst. **6**, 28–33 (2015). (in Russian)
3. Mann, G.K.I., Hu, B.-G., Gosine, R.G.: Analysis of direct action fuzzy PID controller structures. IEEE Trans. Syst. Man Cybern. Part B (Cybern.) **29**(3), 371–388 (1999)
4. Yesil, E., Guzelkaya, M., Eksin, I.: Fuzzy PID controllers: an overview. In: Conference: The 3rd Triennial ETAI International Conference on Applied Automatic (2003)
5. Soloviev, K.A., Muravyova, E.A., Sultanov, R.G.: Adaptatsiya nechetkogo regulyatora. [Adaptation of a fuzzy regulator]. Oil Gas Bus. **12**(3), 123–128 (2014). (in Russian)
6. Kiryanov, D.V.: Metodika parametricheskogo sinteza tsifrovogo nechetkogo regulyatora SAU nestatsionarnymi ob'yektami upravleniya. [Methods of parametric synthesis of digital fuzzy ACS controller by non stationary control objects]. Inf. Process. Syst. **5**(54), 53–57 (2006). (in Ukrainian)
7. Kostoglotov, A.A., Lazarenko, S.V., Pugachev, I.V.: Method of synthesis of multi-mode control under the expected uncertainty using the analysis of the phase-space decomposition on the basis of the generalized power maximum condition. In: AIP Conference Proceedings, vol. 2188, p. 030005 (2019)
8. Zhang, J., Chen, W.H., Lu, X., Wen, J.: Fuzzy controller synthesis for nonlinear neutral state-delayed systems with impulsive effects. Inf. Sci. **555**, 293–313 (2020)
9. Alsafadi, L.A., Chulin, N.A., Mironova, I.V.: Synthesis of fuzzy controller based on simple PID controller. Procedia Computer Science **150**, 28–38 (2019). 13th International Symposium "Intelligent Systems" (INTELS 2018)
10. Berdnikov, V., Lokhin, V.: Synthesis of guaranteed stability regions of a nonstationary nonlinear system with a fuzzy controller. Civ. Eng. J. **5**(1), 107–116 (2019)
11. Pobedinsky, V.V., Hertz, E.F., Gorokhovsky, A.G., Shishkina, E.E.: Synthesis of a fuzzy model in matlab for automatic control of the drying process. Adv. Eng. Res. **151**, 243–248 (2018)
12. Uskov, A., Shchokin, V., Mykhailenko, O., Kryvenko, O.: The fuzzy logic controllers synthesis method in the vector control system of the wind turbine doubly-fed induction generator. In: E3S Web of Conferences, vol. 166, p. 04006 (2020)

13. Muravyova, E.A., Radakina, D.S.: Using the fuzzy controller to control process parameters. Int. Sci. J. Industry 4.0. **3**(5), 236–239 (2018)
14. Dutu, L.C., Mauris, G., Bolon, Ph.: A fast and accurate rule-base generation method for mamdani fuzzy systems. IEEE Trans. Fuzzy Syst. Inst. Electr. Electron. Eng. **26**, 715-733 (2018)
15. Ignatyev, V., Kovalev, A., Spiridonov, O., Kureychik, V., Soloviev, V., Ignatyeva, A.: A method of optimizing the rule base in the Sugeno fuzzy inference system using fuzzy cluster analysis. Int. Rev. Electr. Eng. (I.R.E.E.) **15**(4), 316–327 (2020). ISSN 1827-6660
16. Ignatyev, V.V., Soloviev, V.V., Beloglazov, D.A., Boldyreff, A.S.: Development of a method for automatic generation and optimization of fuzzy controller parameters using genetic algorithm. In: Proceedings of SPIE. Artificial Intelligence and Machine Learning in Defense Applications II, vol. 11543, p. 115430A (2020)
17. Volosencu, C.: Properties of fuzzy systems. WSEAS Trans. Syst. **8**, 210–228 (2009)
18. Ignatyev, V., Soloviev, V., Beloglazov, D., Kureychik, V., Ignatyeva, A., Vorotova, A.: System for automatic adjustment of intelligent controller parameters. In: Kravets, A.G., Groumpos, P.P., Shcherbakov, M., Kultsova, M. (eds.) CIT&DS 2019. CCIS, vol. 1084, pp. 226–242. Springer, Cham (2019). https://doi.org/10.1007/978-3-030-29750-3_18
19. Medasani, S., Kim, J., Krishnapuram, R.: An overview of membership function generation techniques for pattern recognition. Int. J. Approx. Reason. **19**, 391–417 (1998)
20. Ojha, V., Abraham, A., Snasel, V.: Heuristic design of fuzzy inference systems: a review of three decades of research. Eng. Appl. Artif. Intell. **85**, 845–864 (2019)

Heuristic Backward Chaining Based on Predicate Tensorization

Alexander Sakharov(✉)

Synstretch, Framingham, MA, USA
mail@sakharov.net

Abstract. Inference methods for first-order logic or its fragments are inherently slow. Neural networks make it possible to rapidly approximate the truth values of ground atoms but the results are not explainable and not necessarily accurate. A hybrid neural-symbolic inference method is proposed in this paper. It is a framework for incorporating predicate tensorization techniques into backward chaining and its extensions. Heuristic functions for best-first search strategies for backward chaining are defined via tensor representations of predicates. These strategies speed up inference by reducing backtracking.

Keywords: Knowledge base · First-order logic · Heuristic search · Neural-symbolic computing

1 Introduction

The facts and rules of knowledge bases (KB) are usually expressible in first-order logic (FOL) [17]. Typically, KB facts are literals. Quantifier-free implications $A \Leftarrow A_1 \wedge ... \wedge A_k$, where $A, A_1, ..., A_k$ are literals, are arguably the most common form of rules in KBs. All these literals are positive in Prolog rules. These rules are equivalent to disjunctions of literals in classical FOL. These disjunctions are known as non-Horn clauses [23], and the disjunctions corresponding to Prolog rules are called Horn clauses. Their source rules are also called non-Horn and Horn, respectively. Any FOL formula can be expressed by such set of non-Horn clauses that their conjunction is equisatisfiable with this formula.

Backward chaining (BC) is a complete inference method for Horn clauses [17]. BC can be applied to non-Horn clauses as well, but it is not complete for FOL. Complete FOL inference methods are much slower than BC. Due to efficiency considerations, incomplete inference procedures are common for KBs. Prolog also utilizes an incomplete inference procedure [23]. Even BC is not sufficiently fast for some applications. The inefficiency of inference in FOL and even in its Horn fragment prompted attempts to replace inference with neural networks (NN) [5,12,16,18,21,25]. Most commonly, it is done via predicate tensorization. Predicates are represented by one or more tensors of various ranks. These tensor representations are learned.

© The Author(s), under exclusive license to Springer Nature Switzerland AG 2021
R. Silhavy (Ed.): CSOC 2021, LNNS 229, pp. 573–586, 2021.
https://doi.org/10.1007/978-3-030-77445-5_52

Objects are embedded as real-valued vectors of a fixed length for the use in NNs. The truth values of ground atoms of any predicate P are approximated by applying a symbolically differentiable function σ to an algebraic expression. The range of σ is the interval $[0, 1]$. One corresponds to *true*, and zero corresponds to *false*. The expression is composed of the tensors representing P, embeddings of the constants that are P arguments, tensor contraction operations, and other symbolically differentiable functions.

One key advantage of this machine learning approach over inference is that approximation of truth values of ground atoms is fast if σ and other functions from the aforementioned expression are efficiently implemented. Unfortunately, there are serious cons to this approach. This approach is limited to ground atoms. If the result of an approximation is around 0.5, it is not possible to draw any conclusion about the truth value of an atom. Approximation results may not be reliable. Usually, their accuracy is unknown. The truth values yielded by NNs are not explainable because machine learning does not provide any justification for the results. The aim of some inference tasks is to bind variables from goals to constants. In many AI tasks such as automatic code generation, robotic planning, etc., the aim is actually a derivation itself not the mere knowledge of the truth value. The approximation of truth values based on NNs does not contribute to these tasks.

Hybrid approaches that combine symbolic reasoning and machine learning models are considered the most promising [11]. This paper presents a hybrid method that retains the accuracy of inference, produces derivations, and takes advantage of learned predicate representations. Our method gives a framework for incorporating various predicate tensorization techniques into inference. Heuristic functions are defined via neural predicate representations. The application of heuristic search to BC inference helps make better choices at every derivation step, and thus, helps reduce backtracking which usually consumes the vast portion of time during inference. The idea to use neural predicate representation for guiding inference is introduced in [20].

An algorithm presented here can be used in the implementation of heuristic functions for best-first search strategies for BC. This algorithm is parameterized by tensor representations of predicates. These representations are automatically generated. In contrast to this automated approach, crafting heuristic functions is an inventive task. It is conjectured that best-first search strategies employing existing predicate tensorization techniques produce results that are comparable or better than those of traditional heuristic search. This method is also applicable to the BC extensions described in this paper.

2 Backward Chaining and Heuristic Search

We consider KBs in which facts are literals and rules are non-Horn, goals are literals. BC is a form of goal-directed inference based on Generalized Modus Ponens [17]. Given rule $A \Leftarrow A_1 \wedge ... \wedge A_k$, goal list $\Gamma = \{...G...\}$, and substitution θ that is the most general unifier [17] of A and G, every step of BC replaces G with $A_1\theta, ..., A_k\theta$ in Γ and also applies θ to the other goals in Γ.

Non-Horn rules can be transformed to equivalent disjunctions of literals and goal lists can be replaced by disjunctions of the complements of the goals. In this case, BC can be treated as a resolution strategy [1]. In this resolution strategy, one of any two resolved literals is a rule head or a fact [19], i.e. this strategy is a restricted form of input resolution [1]. This strategy is also called SLD resolution if all rules are Horn. Given this similarity, we will use the resolution calculus terminology in application to BC.

BC is particularly suitable for KBs and logic programs. Complete inference procedures for FOL and its extensions are more suitable for theorem provers. Inference in KBs is supposed to be faster. Unlike theorem provers, the number of facts and rules involved in KB inference may be huge, which slows down the inference. The use of incomplete strategies is also justified by the fact that KBs are almost always incomplete.

Various search strategies can be used in implementations of BC. Search strategies determine the order in which goals, rules, or facts are resolved. These strategies include depth-first, breadth-first, iterative deepening, and others. Prolog relies on the depth-first strategy. It is incomplete but efficient. Unit preference [17] is one well-known best-first search strategy for resolution. Facts are resolved before rules under unit preference. Otter [13] conducts heuristic search on the basis of rule weight, i.e. the number of symbols in a rule. Lighter rules are preferred. BC should include some form of loop detection [22]. There also exist optimization techniques that make resolution implementations more efficient. One notable example of these techniques is tabling [24].

Non-Horn clauses may contain Skolem functions or constants. We refer to both of them as Skolem functions for short. They are introduced in the process of eliminating existential quantifiers from FOL formulas [1]. Skolem functions are not evaluable because they are unknown. It is fair to assume that all other functions occurring in KB rules and facts are evaluable. They are implemented as recursive functions. These functions are assumed to be total but may yield undefined values. Ground terms composed of evaluable functions are reduced to constants during inference. Usually, the majority of rules and facts do not contain Skolem functions.

Extensional predicates are the predicates that do not occur in rule heads, they are wholly defined by the set of their facts. A KB may have evaluable predicates [13]. They are implemented as boolean recursive functions by means that are external to rules. Equality is an example of an evaluable predicate. The truth value of any evaluable predicate with an undefined argument is false.

Literals of evaluable predicates without Skolem functions are not resolved. Instead, the truth values of these literals with constant arguments are calculated. For any goal list containing such literals, this calculation is done before other goals are resolved. Other literals of evaluable predicates are not processed until a resolution step produces a substitution that binds all their arguments to constants. If a goal list contains only non-ground literals of evaluable predicates, then an iteration over variable values is performed until true values are encountered or a limit on the number of steps is reached.

Many AI search problems can be formulated as inference tasks. Normally, it is not done despite the simplicity of such formulation because heuristic search methods are necessary for solving these problems. A specific heuristic function is devised for every problem. Some AI search problems are specified by a fixed number of possible state transformation functions $f_1, ..., f_n$ and the respective predicates $P_1, ..., P_n$ guarding the application of these functions generating new states. These problems can be specified as inference problems by fact $S(x, x)$, the following rules, and goals $S(a, b)$ for constants a, b:

$S(x, y) \Leftarrow P_i(x) \wedge S(f_i(x), y) \quad$ (for $i = 1...n$)

For other AI search problems, the number of possible state transformations varies. In this case, the relation between a given state and its expansions can be specified by a predicate. If P is such predicate, then the search problem is specified as follows:

$S(x, x) \qquad\qquad S(x, y) \Leftarrow P(x, z) \wedge S(z, y)$

Constraint satisfaction problems can be specified by rules as well. For instance, the boolean satisfiability problem (aka SAT) can be specified by fact $S(1)$ and these rules:

$S(x) \Leftarrow n \leq d(x) \wedge S(s(x, n, 1)) \qquad\qquad S(x) \Leftarrow n \leq d(x) \wedge S(s(x, n, 0))$

Function d yields the number of variables in a boolean expression. Function $s(x, n, b)$ substitutes boolean value b for the n-th variable in boolean expression x and simplifies the result via standard transformations.

BC is sufficient for solving these search problems specified by rules. These formulations by rules leave out specific heuristic functions for the respective problems. Nonetheless, more general heuristic functions can be incorporated into inference engines. Otter is an example of that. Our method also falls into this category of general informed (heuristic) search methods for BC.

3 Heuristic Backward Chaining

BC is a typical and-or search problem [17]. A boolean function will be defined later. This function yields 1 for goal lists whose all elements are likely derivable. Otherwise, it yields 0. The complement of this function can be used as a heuristic function in informed search, and thus it can be used in a variety of best-first search strategies. Greedy search and AO* search are examples of such strategies [17]. This function will be called core function because it can serve as the core of various heuristic functions. For instance, a heuristic function can assign lower values to goal lists with fewer predicates and Skolem functions.

A best-first strategy based on our core function may also work as a supplement to other inference search strategies which will be called base strategies in this case. If a base strategy leaves multiple choices, then the best-first strategy is invoked. For example, this best-first strategy can be used to select individual facts in addition to employing unit preference. Our core function relies on the following property of BC derivations:

Proposition 1. *Every goal in any successful BC derivation is derivable or is a fact.*

Since every step of BC is interpreted as an application of Generalized Modus Ponens, the proof of Proposition 1 is a straightforward induction on the depth of derivations.

Suppose $\mathcal{D}(v)$ denotes the domain of variable v and $\mathcal{D}(P_n)$ denotes the domain of the n-th argument of predicate P. For any ground literal B, let $[B]$ denote its truth value, i.e. 0 (*false*) or 1 (*true*). For any set of goals $\{B_1, ..., B_b\}$ with free variables $u_1...u_w$, its truth value $[B_1, ..., B_b]$ can be expressed as:

$$[B_1, ..., B_b] = \max_{u_1 \in \mathcal{D}(u_1)...u_w \in \mathcal{D}(u_w)} \min_{i=1...b} [B_i] \tag{1}$$

The values of ground terms without Skolem functions are calculated, and so are the truth values of the respective ground atoms of evaluable predicates. The truth values of literals present in the KB and of their complementary literals are known. The truth value of any such literal of an extensional predicate with constant arguments that neither this literal nor its complement is in the KB is approximated as 0.5. If subsymbolic representations of non-evaluable non-extensional predicates are available, then the truth values of their ground atoms without Skolem functions can be approximated by real numbers from interval $[0, 1]$. The arguments of these atoms are calculated. Tensors representing predicates and object embeddings are used to calculate truth value approximations [5,12,16,18,21,25].

Subsymbolic representations of Skolem functions, or other non-evaluable functions for that matter, can be learned like subsymbolic representations of predicates [18]. Like predicates, Skolem functions are represented by one or more tensors of various ranks. These representations of Skolem functions map embeddings of function arguments into embeddings of function results. Subsymbolic representations of non-evaluabele predicates and Skolem functions make it is possible to approximate the truth value of any ground atom.

If r is the truth value (or its approximation) of ground atom A, then the truth value of $\neg A$ is approximated as $1 - r$. We use the same notation $[B_i]$ for the approximate truth value of ground literal B_i given by a NN, or multiple NNs in the presence of Skolem functions. It is expected that predicate tensorization methods are accurate, that is, there is such interval $[t_1, t_2]$, where $0.5 < t_1$ and $t_2 < 1$, that the set of ground literals satisfying $[B] > t$ and the set of true ground literals do not differ much for $t \in [t_1, t_2]$. Due to Proposition 1, Formula 1 can be used for approximating the truth values of arbitrary sets of goals. Our heuristic function draws on this formula.

We do not specify a particular predicate tensorization method producing NNs that represent predicates. One advantage of our core function is that it can be used with any predicate tensorization method including [5,12,16,18,21,25]. These papers describe how tensor representations are learned, and some of them contain experimental evaluations of the accuracy of truth value approximations. The heuristic function can also be used with other neural methods such as [3, 26]. The only requirement for such method is that it maps ground atoms to approximate truth values from interval $[0, 1]$.

The evaluation of the truth value of any goal list R is possible only if the domains of all variables from R are finite. Even in this case, the evaluation may be too costly. Our heuristic function iterates over finite subsets of the domains of variables from R. In order to minimize the time of iterating over object domains, we rely on pre-built linked lists which are similar to graph adjacency lists. The effect of these linked lists on the time of calculating the heuristic function is also similar to the effect of using adjacency lists in graph search algorithms.

3.1 Tabling

Tabling enables an efficient approximate evaluation of the heuristic function. In contrast to tabling for resolution [24], values $[B_i]$ for ground literals B_i are tabled. These values are calculated in advance and persisted for use in all derivations. Since the aim of search strategies based on our core function is to find goal lists with the highest values of $[B_i]$, only these values are tabled. For atom $P(c_1, ..., c_p)$ where $c_1, ..., c_p$ are constants, the values $[P(c_1, ..., c_p)] > t$ and the values $[\neg P(c_1, ..., c_p)] > t$ are tabled. Here, t is a constant from the interval $[t_1, t_2]$ mentioned earlier. The selection of an optimal t value requires experimentation, its value may vary for different predicate tensorization methods and even for different domains.

To guarantee a fast access to tabled values, they are stored in hash tables. Every entry in such hash table $\mathcal{H}_P$ for predicate P is a key-value pair of one of the two following forms: $P(c_1, ..., c_p) \rightarrow [P(c_1, ..., c_p)]$, $\neg P(c_1, ..., c_p) \rightarrow [\neg P(c_1, ..., c_p)]$. These key-value pairs are also organized as linked lists for predicates having more than one argument. One linked list is created for every subset $c_1, ..., c_{i-1}, c_{i+1}, ..., c_p$ of P arguments. This list is comprised of all possible values of the i-th argument of P yielding truth value approximations greater than t. These linked lists become values in another hash table $\mathcal{L}_P$ with keys $P(c_1, ..., c_{i-1}, \_, c_{i+1}, ..., c_p)$ and $\neg P(c_1, ..., c_{i-1}, \_, c_{i+1}, ..., c_p)$. Algorithm 1 populates $\mathcal{H}_P$ and $\mathcal{L}_P$. Function $aug(E\ (c_1, ..., c_p), i)$ adds c_i to the linked list that is the value of $\mathcal{L}_P$ for key $E(c_1, ..., c_{i-1}, \_, c_{i+1}, ..., c_p)$ if one exists. Otherwise, it creates such linked list with single element c_i. The total number of iterations in this algorithm is limited by constant $\overline{n}$.

It is fair to say that all computer data types are countable (enumerable). This applies even to the types that represent continuous sets such as real numbers because the number of bits representing these values on a computer is limited by a constant. For any computer data type, there exists a recursive function that is a bijection of natural numbers into the set of values of this type. We reuse the notation $\mathcal{D}(...)$ for denoting this bijection for the respective domain. Let p be the number of arguments of predicate P. As usual, $|A|$ denotes the cardinality of set A.

Functions $\mathcal{D}(...)$ can be viewed as sampling functions. It is beneficial if they are consistent with the probability distributions. For the majority of finite domains, Algorithm 1 iterates over entire domains. Countable domains are recursively defined. Functions $\mathcal{D}(...)$ are expected to count the values constructed from other values after the values from which they are built, and the former are

Algorithm 1

$\mathcal{H}_P := \emptyset$; $\mathcal{L}_P := \emptyset$; $n_1 := min\{\overline{n}, |\mathcal{D}(P_1)|\}$; ... $n_p := min\{\overline{n}, |\mathcal{D}(P_p)|\}$;
for $i_1 := 1...n_1; ...; i_p := 1...n_p$ **do**
  if $[P(\mathcal{D}(P_1, i_1), ..., \mathcal{D}(P_p, i_p))] > t$ **then**
    $\mathcal{H}_P := \mathcal{H}_P \cup P(\mathcal{D}(P_1, i_1), ..., \mathcal{D}(P_p, i_p)) \rightarrow [P(\mathcal{D}(P_1, i_1), ..., \mathcal{D}(P_p, i_p))]$
    for $k = 1...p$ **do** $\mathcal{L}_P.aug(P(\mathcal{D}(P_1, i_1), ..., \mathcal{D}(P_p, i_p)), k)$ **end for**
  end if
  if $[P(\mathcal{D}(P_1, i_1), ..., \mathcal{D}(P_p, i_p))] < 1 - t$ **then**
    $\mathcal{H}_P := \mathcal{H}_P \cup \neg P(\mathcal{D}(P_1, i_1), ..., \mathcal{D}(P_p, i_p)) \rightarrow 1 - [P(\mathcal{D}(P_1, i_1), ..., \mathcal{D}(P_p, i_p))]$
    for $k = 1...p$ **do** $\mathcal{L}_P.aug(\neg P(\mathcal{D}(P_1, i_1), ..., \mathcal{D}(P_p, i_p)), k)$ **end for**
  end if
end for

expected to have lower probabilities than the latter. Real numbers are expected to have a Gaussian distribution centered at zero. And functions $\mathcal{D}(...)$ are supposed to comply with that [8, chapter 3.4.1].

3.2 Core Function

Algorithm 2 implements the core function. Goal list $\{R_1, ..., R_r\}$ is its input. Let $u_1, ..., u_w$ be all the variables from literals $R_1, ..., R_r$. Note that w is usually small, it is zero or one for the majority of goal lists in derivations of ground literals. Every $[R_i\{u_1/i_1, ..., u_w/i_w\}]$, where $i_1, ..., i_w$ are constants, can be evaluated, approximated, or retrieved from $\mathcal{H}$. Function *get* retrieves a value from the hash table. If the complement of its argument is in the hash table, *get* returns one minus the retrieved value. Otherwise, *get* returns 0.5.

We say that argument c_i and its literal are outer if $c_i = \mathcal{D}(P_i, k)$ and $k > \overline{n}$. Constant $\underline{n}$ limits the number of iterations over outer arguments, $\underline{n} < \overline{n}$. Function $lit(u)$ returns the index of the first such non-outer literal L that u is the only variable in L and u is not a part of any term. Otherwise, $lit(u) = 0$ if u occurs in an outer literal, and $lit(u) = -1$ if it does not.

NN calculations are avoided when possible. These are relatively fast compared to derivations but picking up values from a hash table is much faster than NN calculations. Initially, outer literals do not occur as keys in $\mathcal{H}$ tables. The approximated truth value of any ground literal is calculated once during a derivation, and then it is added to the respective hash table. These new hash table entities are not reused in other derivations.

Suppose x is the maximum time of the neural approximation of the truth value of a ground atom. The time complexity of the first loop is $O(xr)$. The time complexity of the second loop of Algorithm 2 is $O(rw)$. The total number of executions of the last innermost loop is no more than $r\overline{n}^w$, and at most $r\underline{n}^w$ of them involve NN approximations. Hence, the time complexity of Algorithm 2 is $O(\max\{r\overline{n}^w, xr\underline{n}^w\})$. This estimate is based on the assumption that a single operation on a hash table takes a constant time.

The execution of Algorithm 2 is less expensive for goal lists with no variables or fewer variables. Given that, it may be worth limiting the calculation of the core function to goal lists with no more than one variable or even with no variables for some KBs. A KB may contain a gigantic number of facts for some predicates, and this leads to huge branching factors in relevant derivation steps. Therefore, branching reduction for derivation steps involving facts is crucial. Fortunately, goal lists obtained from facts usually have no more than one variable.

Algorithm 2

```
if w = 0 then
    for j = 1...r do
        h := H_{R_j}.get(R_j)
        if h = 0.5 ∧ outer(R_j) then
            h := [R_j];  H_{R_j} := H_{R_j} ∪ R_j → h
        end if
        if h < t then a := 0; break else a := 1 end if
    end for
else
    for i = 1...w do   j := lit(u_i)
        if j = 0 then  T_i := {D(u_i, 1), ..., D(u_i, min{n̲, |D(u_i)|})
        else if j = −1 then  T_i := {D(u_i, 1), ..., D(u_i, min{n̄, |D(u_i)|})
        else
            T_i := L_{R_j}.get(R_j{u_v/_})
            if T_i = nil then return 1 end if
        end if
    end for
    for i_1 ∈ T_1; ...; i_w ∈ T_w do
        for j = 1...r do
            h := H_{R_j}.get(R_j{u_1/i_1, ..., u_w/i_w})
            if h = 0.5 ∧ outer(R_j) then
                h := [R_j{u_1/i_1, ..., u_w/i_w}];  H_{R_j} := H_{R_j} ∪ R_j{u_1/i_1, ..., u_w/i_w} → h
            end if
            if h < t then a := 0; break else a := 1 end if
        end for
        if a = 1 then break end if
    end for
end if
return a
```

3.3 Examples

The base strategy in these examples is such depth-first search that literals are resolved according to their order in rule bodies, and preference is given to facts. Greedy search is conducted, i.e. such rule or fact matching the current goal is picked that the value of our core function for the resulting goal list is 1.

Transitive Closure. Suppose binary extensional predicate E specifies edges of a directed graph. Our task is to find such paths connecting pairs of nodes in the graph that all interim nodes on these paths have property P which is also dependent on the target node. Graph path search algorithms can be adopted to fulfill this task, but the ultimate goal of automatic programming is to have a declarative specification of this problem and to expect that solutions are derived without coding an algorithm. The following two rules define predicate T as the transitive closure of E with an additional constraint given by P. Derivations of goals $T(a, b)$, where a and b are nodes, give desired paths. Arguably, it is the simplest possible specification of the task. This approach works for infinite graphs too.

$T(x, y) \Leftarrow E(x, y) \qquad T(x, y) \Leftarrow E(x, z) \wedge P(z, y) \wedge T(z, y)$

If $T(a, b)$ is derivable but $E(a, b)$ is not a fact, then the second rule is the only choice for goal $T(a, b)$. For goal list $(E(a, z), P(z, b), T(z, b))$, the base strategy picks up the first literal. If the neural approximations of P and T are accurate, then such fact $E(a, c)$ is selected that c is on a desired path because both $[P(c, b)]$ and $[T(c, b)]$ are expected to be close to one. If $P(z, b)$ derivations for z on this path do not involve backtracking, then $T(a, b)$ derivation should not involve backtracking either.

Research aimed at finding good heuristics for graph path search has been going on for decades [2]. These hard-earned heuristics are not directly applicable to modified tasks incorporating additional conditions such as the one given by predicate P. In contrast, neural approximations are not limited to specific tasks. Tensor representations of predicates are tailored to individual tasks because exact rules are input to training in predicate tensorization methods.

$\mathbf{n^2}$–1 Puzzles. These puzzles such as the 8-puzzle are typically dealt with by heuristic search. Fact $G(x, x)$ and the rules given below specify solutions for the n^2–1 puzzles. Function p yields the horizontal and vertical coordinates of the blank tile. Function m moves the blank tile in a given direction.

$$G(x, y) \Leftarrow p(x, 0) > 1 \wedge G(m(x, 'l'), y)$$
$$G(x, y) \Leftarrow p(x, 0) < n \wedge G(m(x, 'r'), y)$$
$$G(x, y) \Leftarrow p(x, 1) > 1 \wedge G(m(x, 'u'), y)$$
$$G(x, y) \Leftarrow p(x, 1) < n \wedge G(m(x, 'd'), y)$$

With an accurate neural representation of G, derivation of goals $G(a, b)$ is expected to proceed without backtracking for the majority of constants a, b. Heuristic-function calculation is fast for these derivations because goal lists do not contain variables. BC derivations based on the neural representation of G are conjectured to be shorter than search paths obtained with using standard heuristic functions for the n^2–1 puzzles such as the Manhattan distance. As examples from [15] show, search using the Manhattan distance may involve sizable backtracking even for the simplest 8-puzzles.

Minimization Operator. Consider the following rules:

$M(x, '') \Leftarrow P(x, '') \qquad M(x, y) \Leftarrow \neg(y = '') \wedge P(x, y) \wedge N(x, p(y))$

$N(x, '') \Leftarrow \neg P(x, '') \qquad N(x, y) \Leftarrow \neg(y = '') \wedge \neg P(x, y) \wedge N(x, p(y))$

$P(x, y) \Leftarrow Q(x, y)$

$\neg P(x, y) \Leftarrow S(x, y)$ $\neg P(x, y) \Leftarrow \neg Q(x, '') \wedge R(x, y)$

Suppose R is an evaluable predicate and Q is an extensional predicate, function $p(y)$ trims the last character off string y. Derivation of $M(a, z)$, where a is a constant, binds variable z with a minimal string c satisfying $P(a, c)$ unlike c prefixes. Suppose $\neg Q(a, '')$ and $Q(a, c)$ are facts, $Q(a, d)$ is not a fact, $S(a, d)$ is not derivable, and $R(a, d)$ holds for every d that is a prefix of c. If the neural representation of N is accurate, then $[N(a, c)]$ is close to zero whereas $[N(a, d)]$ is close to one for these d.

After exercising the second M rule, the first P rule is applied. $Q(a, z)$ is resolved with fact $Q(a, c)$ because c is the only value of z for which $[Q(a, z)] = 1$, $[\neg(z = '')] = 1$, and $[N(a, p(z))]$ is close to one. After that, $\neg(c = '')$ is evaluated. If the neural representation of S is relatively accurate, then derivations of $N(a, p(c)), ..., N(a, '')$ are not expected to involve backtracking. Due to the use of the heuristic function, the second P rule is avoided. Depending on the rules related to S, the second P rule could be a source of extensive backtracking while attempting to derive $\neg P(a, e)$ for multiple values of e.

These rules specify a concrete problem, they do not correspond to a well-known class of problems, and hence there are no readily available traditional heuristic methods for the problem. The Otter [13] search strategy based on rule weight would be counterproductive for this example. It would always select the second P rule.

4 Backward Chaining Extensions

Inference procedures that are more powerful than BC may be necessary for some KBs, especially for KBs with a large number of negative literals. Our core function will not work well for complete resolution methods such as linear resolution because not every goal in those derivations is derivable. For example, goal P is derivable from rules $P \Leftarrow Q$, $P \Leftarrow \neg Q$ in FOL. Inevitably, either Q or $\neg Q$ is a resolvent in any resolution refutation of $\neg P$ and the rules. None of the two respective goals is derivable.

For any rule $A \Leftarrow A_1 \wedge ... \wedge A_k$, the implications $\sim A_i \Leftarrow \sim A \wedge A_1 \wedge ... A_{i-1} \wedge A_{i+1} ... \wedge A_n$ for $i - 1...k$ are called contrapositives to this rule [23]. $\sim A$ denotes the complement of literal A, i.e. it is $\neg A$ if A is positive, and it is the atom of A otherwise. One of two resolved literals at any derivation step of BC with contrapositives is a fact, rule head, or contrapositive head. This step can be viewed as an application of the resolution rule [1] to a fact or a KB rule represented as a disjunction of literals. Conversely, if KB facts and rules along with an initial goal list constitute a resolution input, then every application of the resolution rule to the input clauses except for the initial goal list maps to a BC step. Hence, BC with contrapositives is almost identical to input resolution without the factoring rule [1]. The only difference is that the initial goal list may occur in the first resolution step only.

Rules imply contrapositives in FOL. Clearly, Proposition 1 holds for this extended BC provided that it refers to derivability in FOL. This warrants the applicability of our heuristic function to BC with contrapositives. The use of contrapositives may significantly increase the range of derivable literals. For instance, if literals A and $\neg D$ are derivable, then literal $\neg B$ is derivable by BC from rules $C \Leftarrow A \wedge B$, $D \Leftarrow C$ and their contrapositives. This derivation is not possible without contrapositives. Still, BC with contrapositives is an incomplete inference method. Proofs by contradiction are not possible with it [23]. For instance, literal P is not derivable by BC with contrapositives from rules $P \Leftarrow Q \wedge R$, $P \Leftarrow \neg Q \wedge S$ and facts R, S.

The power of BC with contrapositives could be further enhanced. It is accomplished by targeted KB completion, i.e. by addition of rules and contrapositives that are derived from the original set of KB rules and facts in FOL. Proposition 1 will still hold in this case. Only such rules that their heads are not derivable from the literals of their bodies by BC with contrapositives affect its power. The set of literals that are not derivable by BC with contrapositives can be characterized in terms of the PTTP model elimination procedure [23]. Clearly, any derivation of such literal includes an application of the reduction rule. The reduction rule eliminates the literals that unify with the complements of previously resolved goals. Please refer to [23] for details.

The targeted KB completion is done as follows. We generate derivative rules by applying the PTTP procedure to the head G of a rule or contrapositive, that is, G is an original goal. If $G_1, ..., G_n$ is a goal list after a number of steps of this procedure, then $G \Leftarrow G_1 \wedge ... \wedge G_n$ can be viewed as a derivative rule due to the deduction theorem. A derivative rule is added to the KB if it does not contain ground literals and the last step in its derivation is such application of the reduction rule that the reduced literal is unified with $\sim G$. As a result, only the shortest derivations that cannot be obtained by BC with contrapositives produce new rules. This minimizes the number of new rules. Contrapositives of these derivative rules are also added to the KB. It makes sense to set up a limit on the number of literals in the derivative rules. In order to guarantee the termination of this targeted KB completion, a limit on the size of intermediate goal lists should be set as well.

As an example of the targeted KB completion, consider these rules:

$A(x, y) \Leftarrow B(x, y) \wedge C(x, y)$ $\quad$ $A(x, y) \Leftarrow E(x, y) \wedge F(x, y)$

$B(x, 1) \Leftarrow \neg D(x, 1)$ $\quad$ $F(x, y) \Leftarrow D(x, y) \wedge H(y + 1)$

If $H(2)$ is a fact, then rule $A(x, 1) \Leftarrow C(x, 1) \wedge E(x, 1)$ is added to the KB because the last step in its derivation is the reduction of $\neg A(x, 1)$.

BC extensions should be exercised with caution because they may degrade KB inference performance. Increasing the number of rules increases branching factors. Contrapositives may be redundant for such rules as the n^2-1 puzzle rules.

5 Related Work

Paper [6] is a survey of research in the area of neural-symbolic computing. The aforementioned works [5, 16, 21, 25] use rules or FOL formulas to train NNs based on subsymbolic predicate representations. These NNs are utilized to approximate the truth values of ground formulas. The methods from these papers do not incorporate inference. Function-free rules with unary and binary predicates are used for training NNs in [16]. Neural Logic Machines [5] are capable of representing non-cyclic Horn clauses over finite object domains. Differentiable Reasoning introduced in [25] employs function-free FOL formulas. Logic Tensor Networks [21] accept all FOL formulas.

A simple hybrid algorithm combining logical reasoning with the approximation of truth values is suggested in [18]. This algorithm searches for partial derivations. Once such ground literal occurs in a derivation that its approximated truth value is close to 1, this literal is considered a fact not requiring further derivation. The results of this hybrid algorithm may not be explainable. The reliability of the results is dependent of the accuracy of the approximation of the truth values of ground atoms. Paper [20] is a preliminary report about a particular search strategy for BC using predicate tensorization. A relation between that strategy and heuristic search is not addressed there. The algorithm implementing that strategy is not optimized. Extensions of BC are not considered in that paper.

A method of differentiable reasoning for the function-free fragment of FOL is suggested in [14]. Unification is done by finding the nearest neighbors in the Euclidean space of predicate embeddings. An approximate proof is constructed by solving an optimization problem. In contrast to symbolic reasoning, differentiable reasoning may not be reliable but it is appropriate for areas with imprecise knowledge such as natural language. Extending the method of [14] onto rules containing atoms with functions is problematic due to unification of terms.

In other research on the neural representation of knowledge, individual predicates are not directly represented via NNs but are rather blended into a graph NN that represents multiple predicates along with rules [26]. Nonetheless, this graph NN can be used to approximate the truth values of ground literals. Potential advantages of graph NNs for the neural representation of knowledge are advocated in [9]. Function-free normal logic programs are represented by tensors in DeepLogic [3], and ground atoms are represented by vectors. These representations are used together to approximate the truth values of ground atoms.

Deep logic models proposed in [12] are two-layer NNs. The first layer represents predicates, and its output is approximate truth values of atoms. The second layer represents constraints on the outputs of the first layer. These constraints are expressed by FOL formulas over predicates of the first layer.

DeepProbLog [10] is a hybrid system that combines probabilistic logic programming with NNs. The latter are used to calculate probabilities associated with ground atoms in rules annotated with probabilities. These annotated rules are transformed into probabilistic logic programs. Exact inference for these programs is intractable, and approximate inference for them is more complicated than FOL inference [7]. TensorLog [4] is another system integrating probabilistic logical reasoning with NNs. TensorLog programs are function-free, they contain unary or binary predicates, and probabilities are associated with facts only.

6 Conclusion

Our method is neural-symbolic but its results are derivations, and they are precise and explainable. If some neural predicate representations are not reliable, this has effect on search speed only. As the examples have shown, best-first search based on our core function may significantly reduce backtracking for typical derivation tasks. It has the potential to outperform best-first search based on heuristic functions designed for specific tasks. Our method eliminates the need to design heuristic functions for individual AI tasks.

Further experiments comparing performance of our method with traditional heuristic search would be interesting. Note that these experiments would actually compare the quality of neural approximations and specific heuristic search methods. Such comparisons for multiple predicate tensorization methods are beyond the scope of this paper. The value of our technique is the ability to convert unexplainable and inaccurate results from a variety of predicate tensorization methods into precise and explainable derivations. The power of BC can be substantially extended without compromising the legitimacy of the core function. It done by adding rules, which does not void the property of Proposition 1. Only proofs by contradiction are beyond the reach of extended BC but such proofs are more relevant to theorem proving rather than to KB inference.

References

1. Chang, C.L., Lee, R.C.T.: Symbolic Logic and Mechanical Theorem Proving. Academic press, Cambridge (1973)
2. Chen, R., Gotsman, C.: Efficient fastest-path computations in road maps. arXiv preprint arXiv:1810.01776 (2018)
3. Cingillioglu, N., Russo, A.: DeepLogic: towards end-to-end differentiable logical reasoning. In: AAAI 2019 Spring Symposium on Combining Machine Learning with Knowledge Engineering, vol. 2350. CEUR-WS.org (2019)
4. Cohen, W., Yang, F., Mazaitis, K.R.: Tensorlog: a probabilistic database implemented using deep-learning infrastructure. J. Artif. Intell. Res. **67**, 285–325 (2020)
5. Dong, H., Mao, J., Lin, T., Wang, C., Li, L., Zhou, D.: Neural logic machines. In: International Conference on Learning Representations (2019)
6. d'Avila Garcez, A.S., Gori, M., Lamb, L.C., Serafini, L., Spranger, M., Tran, S.N.: Neural-symbolic computing: an effective methodology for principled integration of machine learning and reasoning. J. Appl. Logics **6**(4), 611–632 (2019)
7. Kimmig, A., Demoen, B., De Raedt, L., Costa, V.S., Rocha, R.: On the implementation of the probabilistic logic programming language ProbLog. arXiv preprint arXiv:1006.4442 (2010)
8. Knuth, D.E.: Art of Programming, vol. 2: Seminumerical Algorithms, vol. 2. Addison-Wesley, Boston (1998)
9. Lamb, L.C., d'Avila Garcez, A.S., Gori, M., Prates, M.O.R., Avelar, P.H.C., Vardi, M.Y.: Graph neural networks meet neural-symbolic computing: a survey and perspective. In: Proceedings of the Twenty-Ninth International Joint Conference on Artificial Intelligence, pp. 4877–4884 (2020)

10. Manhaeve, R., Dumancic, S., Kimmig, A., Demeester, T., De Raedt, L.: DeepProbLog: neural probabilistic logic programming. In: Advances in Neural Information Processing Systems, pp. 3749–3759 (2018)
11. Marcus, G.: The next decade in AI: four steps towards robust artificial intelligence. arXiv preprint arXiv:2002.06177 (2020)
12. Marra, G., Giannini, F., Diligenti, M., Gori, M.: Integrating learning and reasoning with deep logic models. In: Joint European Conference on Machine Learning and Knowledge Discovery in Databases, pp. 517–532 (2019)
13. McCune, W.: Otter 3.3 reference manual and guide. Technical report, Argonne National Laboratory (2003)
14. Minervini, P., Bosnjak, M., Rocktäschel, T., Riedel, S., Grefenstette, E.: Differentiable reasoning on large knowledge bases and natural language. In: Knowledge Graphs for eXplainable Artificial Intelligence: Foundations, Applications and Challenges, pp. 125–142. IOS Press (2020)
15. Nilsson, N.J.: Principles of Artificial Intelligence. Morgan Kaufmann, Massachusetts (2014)
16. Rocktäschel, T.: Combining representation learning with logic for language processing. CoRR (2017). http://arxiv.org/abs/1712.09687
17. Russell, S., Norvig, P.: Artificial Intelligence: A Modern Approach. Prentice Hall Press, 3rd edn. (2009)
18. Sakharov, A.: Hierarchical rules for knowledge representation and learning. In: IEEE 2nd International Conference on Artificial Intelligence and Knowledge Engineering, pp. 167–171 (2019)
19. Sakharov, A.: Hierarchical resolution for structured predicate definitions. In: 11th Hellenic Conference on Artificial Intelligence, pp. 202–210 (2020)
20. Sakharov, A.: A best-first backward-chaining search strategy based on learned predicate representations. In: 13th International Conference on Agents and Artificial Intelligence, vol. 2, pp. 982–989 (2021)
21. Serafini, L., d'Avila Garcez, A.S.: Logic tensor networks: deep learning and logical reasoning from data and knowledge. In: 11th International Workshop on Neural-Symbolic Learning and Reasoning, vol. 1768. CEUR-WS.org (2016)
22. Shen, Y.D., Yuan, L.Y., You, J.H.: Loop checks for logic programs with functions. Theor. Comput. Sci. **266**(1–2), 441–461 (2001)
23. Stickel, M.E.: A Prolog technology theorem prover: a new exposition and implementation in Prolog. Theor. Comput. Sci. **104**(1), 109–128 (1992)
24. Swift, T.: Design patterns for tabled logic programming. In: Abreu, S., Seipel, D. (eds.) Applications of Declarative Programming and Knowledge Management. INAP 2009. Lecture Notes in Computer Science, vol. 6547, pp. 1–19. Springer, Heidelberg (2009). https://doi.org/10.1007/978-3-642-20589-7_1
25. Van Krieken, E., Acar, E., Van Harmelen, F.: Semi-supervised learning using differentiable reasoning. J. Appl. Logics **6**(4), 633–651 (2019)
26. Zhang, Y., et al.: Efficient probabilistic logic reasoning with graph neural networks. In: International Conference on Learning Representations (2020)

Deep Learning for the Classification of Charge Density Plots of Lithium-Ion Cathode Materials

Oleg O. Kartashov(✉), Maria A. Butakova, Andrey V. Chernov, and Konstantin A. Namavir

The Smart Materials Research Institute, Southern Federal University, 178/24 Sladkova, 344090 Rostov-on-Don, Russia
{okartashov,mbutakova,cherno,namavir}@sfedu.ru

Abstract. This paper proposes a novel approach to the classification of charge density plots of several types of lithium-ion cathode materials. Charge density plots visualize essential materials properties and are obtained as a result of the ab-initio atomic-scale materials supercomputer modeling. The paper presents a novel computational procedure for slicing the ab-initio simulated 3D data along each lattice vector and stepped on the grid points of computed material. Image datasets formed since the slicing procedure are classified with a deep learning convolution neural network, which is a pre-trained one, customized and fine-tuned after transfer learning. Computation experiments with the proposed deep learning classifier are performed and estimated.

Keywords: Deep learning · Transfer learning · Image classification · Charge density · Lithium-Ion batteries

1 Introduction

Lithium-ion cathode materials have been attracted much research attention in recent decades [1–3]. The traditional advantages of lithium-ion batteries include superior energy density compared to other rechargeable battery materials, high reliability and round trip efficiency, relatively low self-discharge rate, and fast rechargeable rate [4]. At the same time, lithium-ion energy cells have well-known weaknesses such as the high cost of initial material, suffer from aging, have transportation and safety restrictions, need advanced overcharging and discharging protection circuits with reliable fault diagnostic algorithms [5]. Improvements in material electrochemical performance, high-current cycling stability, and safety-related thermal stability have resulted in a wide variety of lithium-ion cathode materials. Despite the cathode materials diversity, there are several structure arrangement models and dimensionality of the Li^+ ions transportation [6]. They include 1-D framework (olivines $LiFePO_4$, $LiCoPO_4$, and some polyanion compounds [7]), the most prominent 2-D layered framework $LiCoO_2$ [8] (with promising layered oxide materials such as $LiNi_{1-x-y}Co_xMn_yO_2$, $LiNi_{1-x-y}Co_xAl_yO_2$, Li-rich layered oxides) and 3-D spinel framework $LiMn_2O_4$ with various modifications [9].

© The Author(s), under exclusive license to Springer Nature Switzerland AG 2021
R. Silhavy (Ed.): CSOC 2021, LNNS 229, pp. 587–593, 2021.
https://doi.org/10.1007/978-3-030-77445-5_53

Hands-on work in a chemistry laboratory for the discovery of new lithium cathode materials requires not only a long period of research experience. Still, it comes along with high-cost equipment, substances, and compounds. Computational simulations at the atomic and molecular dynamics level have become essential and valuable tools in the field of cathode materials improvement [10]. Computational approaches designed for computer modeling of crystal structures based on their chemical composition, as well as predicting their properties by solving quantum mechanical equations, are often called first-principle or ab-initio calculations. Being an extremely attractive tool that calculations are developing in several directions and one of the main is the solution of many-electron problems in the framework of the Kohn–Sham density functional theory (DFT). With over fifty years of research history and a huge number of publications, including exhaustive monographs [11], DFT remains a "subtle, seductive, provocative business" [12]. However, the results of DFT calculations reflect the vast majority of experimental findings, and moreover, DFT and its implementation in the supercomputer software (for example, Vienna Ab-initio Simulation Package, VASP) is possibly the adequate primary simulation for solid-state matters. Calculation results after VASP applying are useful for the estimation of many material characteristics and one of the fundamental properties of condensed phases of matter, namely electron density. Charge density plots allow visualizing (for example, by VESTA software) the probability of finding electrons at a given point within the designed lithium-ion cathode cell. Consequently, determining charge density difference analysis and further the Bader charge analysis is possible.

Despite the relatively simple idea behind the charge density plot visualizing and analyzing for perspective cathode materials investigations, there is a lack of enough data to making appropriate machine learning classifiers for the plots mentioned above and corresponding images. This paper proposes a novel approach for collecting the charge density plot data of several types of lithium-ion cathode materials using VASP first-principle calculations combined with calculation results slicing procedure to making image sets and then applying transfer learning approach with deep learning convolutional neural network for classification formed image sets. The paper is structured as follows. Section 2 reveals the materials and methods utilized to solve the problem of charge density plots obtaining and deep learning classifier designing. Section 3 presents the computational experiments, results estimation, and their discussion. Section 4 concludes the paper.

2 Materials and Methods

There are several approaches to obtaining electron density, which provide the ability to plot the charge density of the particular (and cathode) material. Some of them, such as high-resolution diffraction, also named X-ray charge density analysis, or by means of high-resolution transmission electron microscopy, are out of the scope of this paper because they are related to chemical experiments in the central part. The area of interest of this work includes techniques that are related to supercomputer simulation. It should be noted that such modeling is also costly, but this already refers to the relatively long waiting for the results of computational experiments. And one of the critical issues of the simulated charge density remains the statical state of atoms during the simulation procedure that gives less opportunity for real-time electron charge analysis. Due to these facts and the need to reduce exploration cost and time, machine learning (ML) models and techniques are becoming more attractive for simulations on the atomistic level [13–15], including transferrable ML model for electron density [16].

The proposed method consists of several steps. The first step is performing VASP (vasp.at) simulation of the electronic density of chosen lithium-ion cathode material. As a result, VASP CHGCAR files were produced. These files have essential information about interatomic interactions, including material structure descriptors (lattice, symbols, atomcount, selective dynamics, coordinates, cell volume), gridsize [X Y Z] for fast Fourier transform, 3D array (X*Y*Z) of points according to chosen gridsize and 3D array (X*Y*Z) points of magnetization. The second step is applying the novel 3D "slicing" procedure to the simulated CHGCAR data files. The idea of 3D slicing is shown in Fig. 1. Creating CD (charge density plots) has been performed using the following lattice planes (Miller indices): (0, 1, 2); (1, 0, 2), (2, 0, 1). An image data set with CD plots have been obtained and then used for deep learning classifier training. Implementation of that idea is shown in **Algorithm 1** pseudocode. There are several steps in **Algorithm 1**: reading of CHGCAR file and decomposing its format; determining density and atoms characteristics; determining cell values, matrix dimensions, and lattice vectors; multi-index cycle through each axis for calculation projections containing CD; save CDs to file.

Algorithm 1 is implemented in *Python* scripting language with external extensions and modules. *Numpy* extension was used to perform many math operations included matrix transpose and diagonalization (for example, pseudocode for transp_diag, transp), vector operations, and many others. We created charge density plots with the *Pylab* module from *Matplotlib*. From the *ASE* (*Atomic Simulation Environment)* toolset, we used the *VaspChargeDensity* module for processing of specialized structure of VASP CHGCAR and calculations of charge density, atoms, matrix dimension, and unit cell.

The next step in the proposed method is designing the architecture of ML deep learning classifier. From the numerous DL techniques and artificial neural network (ANN) architectures, we have chosen the Transfer Learning (TL) technique. TL one of the research areas in machine learning that studies the possibility of applying the knowledge gained from solving one problem to another. Nowadays, TL is a powerful artificial intelligence tool with over ten pre-trained deep networks. A pre-trained model is a model that was created and trained previously to solve a problem like the one we want to investigate. In practice, researchers, having large computing resources, construct a large ANN to solve a specific problem, train it on Big Data, for example, ImageNet or Wikipedia Corpus. Some of the most significant and widely used resources with pre-trained models include *TensorFlow Hub*, *Keras Applications*, *PyTorch Hub*. *Mathworks Matlab* software comprises different well-known pre-trained ANN: SqueezeNet, AlexNet, Inception v3, GoogleNet, variants of VGG and DarkNet and ResNet, and others.

Choosing the appropriate TL ANN architecture for the particular problem is not a trivial task. It strongly depends on the quality of datasets. For example, in their previous work [17], the authors used an Inception v3 ANN successfully. In DL architectures, the initial layers learn general information, while the layers at the last level are more specific features, so this process in TL is called retraining. Instead, it is more useful to supplement

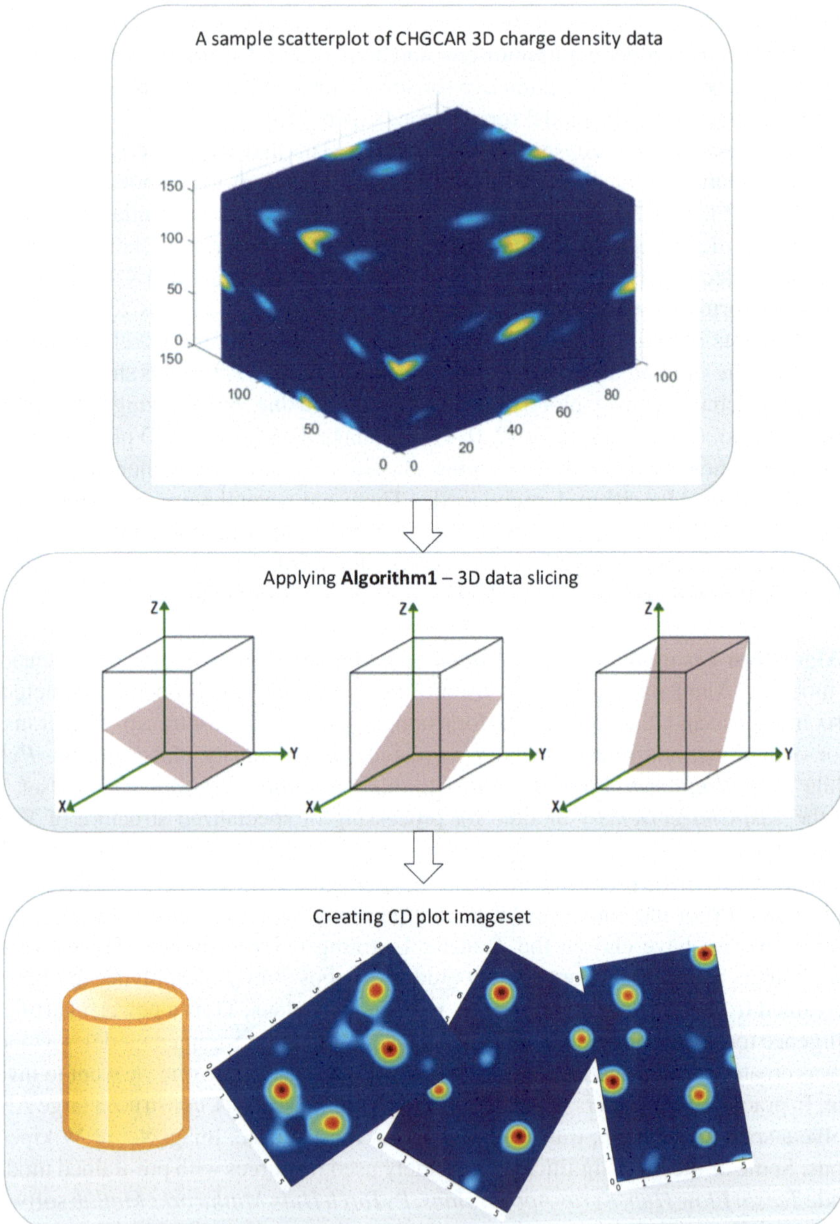

Fig. 1. Creating the image set from 3D simulated data

it with new specific features, retraining only the last layers to repurpose it for your own needs, by touching more layers for retraining, the risk of overfitting increases. Another task in TL is feature extraction. For this, the representations obtained by the previous model are used, and to extract features from new samples, they are passed through a new classifier. This method simply adds a classifier that will train from scratch on top of the pre-trained model to solve the objective function.

3 Results and Discussion

Experiments included two types of calculations for three types of materials. The first experiment is the creation of a CHGCAR VASP 5.2 file with a designated charge density. The second experiment was the generation of datasets with charge density potential plots. The check was carried out for three types of materials with different volumetric structures. Investigations were performed with different types of cathode materials: 1D - ($LiFePO_4$ sample), 2D - ($LiCoO_2$ sample), and 3D - ($LiMnO_2$ sample), and essential parameters of materials were determined. All calculations were performed on a 354-core high-performance computing cluster "Blokhin" (nano.sfedu.ru), with a peak computing power of 20 TFlops and 31 working nodes with different performance. The optimization and run of supercomputer resource management have been implemented with SLURM. Creating CHGCAR file for 1D material, a node of class b7 worked (2X20-core Intel Xeon Gold 6230 processors with 384 GB RAM, 6 physical cores are involved) took 17 min and 30 s. 2D material was calculated node b6 of the cluster (2X20-core Intel Xeon Gold 6230 processors with 512 GB RAM, 6 physical cores are involved) for 1 min 14 s. For 3D material, it took 27 min 45 s there. The second type of calculations was the creation datasets of CD plots file $LiFePO_4$ included 328 images produced in approximately 7 h on h13 (based on 6-cores processors Intel Original Core i7 X6 5930K @ 3.5 GHz (Haswell), 64 Gb DDR4). $LiCoO_2$ was rated for 312 measurements with about 4 h of use b6. It took 09 h 27 min to generate 384 charge density plots of $LiMnO_2$ using the computational node h11 (equivalent to h13).

After CD plots image data sets creation TL approach was applied. Mathworks Matlab pre-trained model SqueezeNet has been used. Two layers (conv2layer and output classification layer) undergo necessary modifications according to the number of classes. Then the model was retrained, customized, and fine-tuned. Results of ANN accuracy and loss estimation are shown in Fig. 2 and Fig. 3, accordingly.

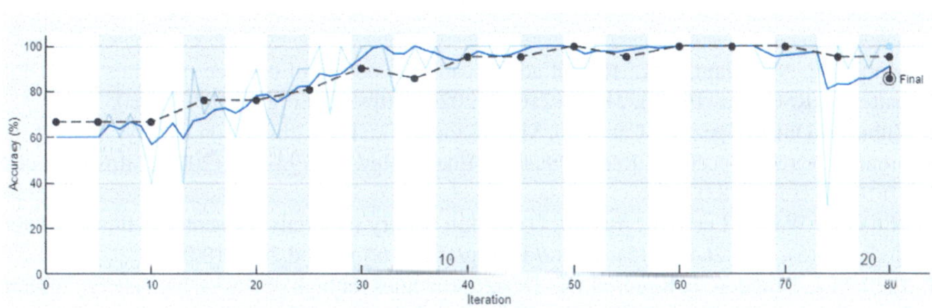

Fig. 2. ANN (retrained SqueezeNet) accuracy estimation

Model retraining process was performed with sgdm solver, and initial learning rate equals 0.0001. Retraining of ANN took approximately one hour using Matlab 2020b on

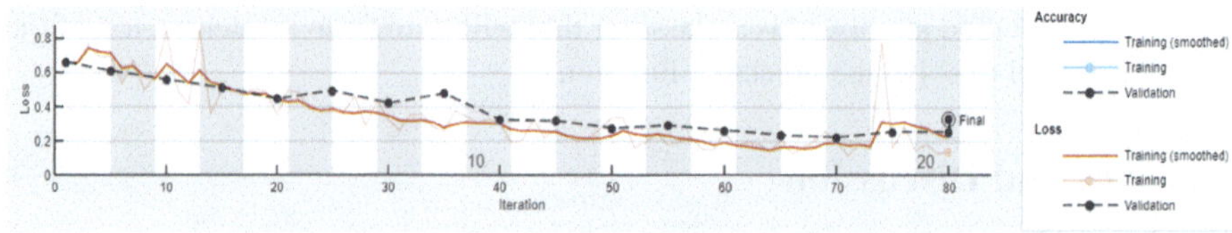

Fig. 3. ANN (retrained SqueezeNet) loss estimation

GPU Nvidia GTX1060 (6 Gb) with CPU Core i7 9700k 32Gb RAM OS Windows 10 personal computer. The final accuracy of ANN is approximately 87.6%.

4 Conclusion

The paper proposes a novel approach to DL classification of CD plots of several common cathode materials for batteries having 1D, 2D and 3D framework. The problems and approaches to material charge density estimation using DTF and supercomputer simulation on the atomistic level are defined. As a central part of the paper, the novel idea for CD image sets has been described. Based on that idea, a novel approach and algorithm for 3D slicing have been developed. A variant of DL learning technique is TL with ANN customizing, and retraining has been used for ML classifier. Experiments for simulation of three types of lithium-ion cathode materials have been performed by means of VASP software and 354-core high-performance computing cluster “Blokhin”. Experiments with created CD image set showed the adequate accuracy of the developed DL classifier.

Acknowledgment. The research was financially supported by the Ministry of Science and Higher Education of the Russian Federation (State assignment in the field of scientific activity, № 0852-2020-0019).

References

1. Mohamed, N., Allam, N.K.: Recent advances in the design of cathode materials for Li-Ion batteries. RSC Adv. **10**(37), 21662–21685 (2020). https://doi.org/10.1039/d0ra03314f
2. Zubi, G., Dufo-López, R., Carvalho, M., Pasaoglu, G.: The Lithium-ion battery: State of the art and future perspectives. Renew. Sustain. Energy Rev. **89**, 292–308 (2018). https://doi.org/10.1016/j.rser.2018.03.002
3. Nitta, N., Wu, F., Lee, J.T., Yushin, G.: Li-Ion battery materials: present and future. Mater. Today **18**(5), 252–264 (2015). https://doi.org/10.1016/j.mattod.2014.10.040
4. Lain, M.J., Brandon, J., Kendrick, E.: Design strategies for high power vs. high energy lithium ion cells. Batteries **5**, 64 (2019). https://doi.org/10.3390/batteries5040064
5. Tran, M.-K., Fowler, M.A.: Review of Lithium-ion battery fault diagnostic algorithms: current progress and future challenges. Algorithms **13**, 62 (2020). https://doi.org/10.3390/a13030062
6. Julien, C.M., Mauger, A., Zaghib, K., Groult, H.: Comparative issues of cathode materials for Li-Ion batteries. Inorganics **2**, 132–154 (2014). https://doi.org/10.3390/inorganics2010132

7. Wu, X.B., et al.: Polyanion compounds as cathode materials for Li-Ion batteries. In: Zhang, Z., Zhang, S.S. (eds.) Rechargeable Batteries. GET, pp. 93–134. Springer, Cham (2015). https://doi.org/10.1007/978-3-319-15458-9_4
8. Daniel, C., Mohanty, D., Li, J., Wood D.L.: Cathode materials review. In: AIP Conference Proceedings, vol. 1597,p. 26 (2014). https://doi.org/10.1063/1.4878478
9. Zhang, Y., et al.: Research status of spinel LiMn2O4Cathode materials for Lithium ion batteries. In: IOP Conference Series: Earth and Environmental Science, vol. 603, no. 1 (2020). https://doi.org/10.1088/1755-1315/603/1/012051
10. Li, G., Monroe C.W.: Multiscale Lithium-battery modeling from materials to cells. Ann. Rev. Chem. Biomol. Eng. **11**(1), 277–310 (2020). https://doi.org/10.1146/annurev-chembioeng-012120-083016
11. Engel, E., Dreizler R.M.: Density Functional Theory. An Advanced Course. Springer, Heidelberg (2011). 531 p. https://doi.org/10.1007/978-3-642-14090-7
12. Becke, A.D.: Perspective: Fifty years of density-functional theory in chemical physics. J. Chem. Phys. **140** (18) (2014). https://doi.org/10.1063/1.4869598
13. Nagai, R., Akashi, R., Sugino, O.: Completing density functional theory by machine learning hidden messages from molecules. NPJ Comput. Mater. **6**(1) (2020). https://doi.org/10.1038/s41524-020-0310-0
14. Dick, S., Fernandez-Serra, M.: Machine learning accurate exchange and correlation functionals of the electronic density. Nat. Commun. **11**(1). (2020). https://doi.org/10.1038/s41467-020-17265-7
15. Deringer, V.L.: Modelling and understanding battery materials with machine-learning-driven atomistic simulations. J. Phys. Energy **2**, 041003 (2020). https://doi.org/10.1088/2515-7655/abb011
16. Grisafi, A., Fabrizio, A., Meyer, B., Wilkins, D.M., Corminboeuf, C., Ceriotti, M.: Transferable machine-learning model of the electron density. ACS Cent. Sci. **5**(1), 57–64 (2019). https://doi.org/10.1021/acscentsci.8b00551
17. Chernov, A., Butakova, M., Guda, A., Shevchuk, P.: Development of intelligent obstacle detection system on railway tracks for yard locomotives using CNN. In: Bernardi, S., et al. (eds.) EDCC 2020. CCIS, vol. 1279, pp. 33–43. Springer, Cham (2020). https://doi.org/10.1007/978-3-030-58462-7_3

Circular Encryption Algorithms to Provide a Secure Channel for Storing and Transmitting Information in Mobile Robot Groups

Shabanov Vladimir[1(✉)], Kapustjan Sergej[2], Prakapovich Ryhor[3], and Sycjov Uladzislau[3]

[1] Southern Federal University, 2 Chehova St., 3479328 Taganrog, Russia
[2] Southern Scientific Centre of the Russian Academy of Sciences, SSC RAS, 41 Chekhov av., 344006 Rostov on Don, Russia
[3] State Scientific Institution "United Institute of Informatics Problems of the National Academy of Sciences of Belarus, 220012 Minsk, Belarus

Abstract. Information interaction in teams of robots, is currently a hot topic in group robotics. The high relevance of the problem was achieved by using robot collectives in solving technically complex tasks, they found themselves in solving tasks of monitoring, working with radioactive substances, search operations, moving large-sized bodies, etc. Fulfillment of all these purposes requires a high level of team interaction, taking into account dynamically changing environment and limited radius of data transfer, as a consequence, the question of team information interaction, storage and protection of transmitted information from unauthorized access becomes acute. In article the approach of realization of information interaction by means of anonymous cryptoprotected algorithms which use method of ciphering of the sender of the information from unauthorized access, allowing to receive data on an environment group without personalization of a source is considered. The existing algorithms of anonymous decentralized data transfer based on blockchain are also briefly considered.

Keywords: Micro robot · Blockchain · Cryptography · Robot group · Decentralized data storage · Information exchange

1 Introduction

Groups of mobile robots have long found their fans and methods of application, such as: mapping, search operations [1], work with radioactive waste [2], surveillance. Indeed, the application of group interaction allows to expand the range of tasks to be solved, by distributing goals and tasks between team members.

Team management requires the creation of an externally protected information storage channel, which allows to achieve a high level of team interaction in a changing environment. Ideally, if the system will be unified with the knowledge base and tasks, forming a protected shell of the system. The realization of such a system is possible through the use of blockchain technology and circular encryption algorithms, at the

© The Author(s), under exclusive license to Springer Nature Switzerland AG 2021
R. Silhavy (Ed.): CSOC 2021, LNNS 229, pp. 594–600, 2021.
https://doi.org/10.1007/978-3-030-77445-5_54

moment this algorithm is used only for the implementation of anonymous cryptocurrencies [3, 4], with regard to group management, the algorithm will create a decentralized cryptoprotected data storage.

2 Principle of Function Hash Operation

The convolution function or so-called hash function converts a data sequence to a bit string of fixed length, converted using special algorithms. The main requirement for transformations is uniformity of value distribution when the argument is chosen arbitrarily. In a distributed registry, information in a block, such as: data structure, transaction, and timestamps, must be compressed into short messages that can be checked later.

Cryptographic hash functions are used everywhere [3–5]. Their main purpose is to convert the textual information of any length into fixed messages, secured from the eyes of common man. The converted result has the property of constancy when the same data is entered. The process of calculating a hash must not be too simple to protect against a malicious user attempting to artificially create collisions by brute force methods.

At present there are many hashing algorithms, such as SHA algorithms [6–8] developed in 1995 and recommended for use in the U.S. government agencies; Ethash [9, 10] using modified functions SHA3–256 [11] allows to implement smart contracts, aimed at forming lists and controlling property of something.

3 Anonymous Hashing Algorithms

The aforementioned algorithms use an open blockchain as their basis, in relation to group robotics it is more appropriate to use a closed blockchain, to protect incoming information from agents. Popular algorithms with data anonymization include algorithms X11 [14] - developed in 2014, which uses a hybrid technology pos/pow for data processing, having at its core master nodes that allow voting. Master nodes provide anonymity by blending transactions, allowing data to be hidden from ordinary ordinary people. The working principle is divided into several points:

- Splitting all the transactions equally, to hide the sender's address
- If transactions match, an additional mixing process takes place which is done by another master node participating in the network.

The use of master nodes in the algorithm, allowed to achieve a traffic distribution as well as to modernize the consensus, all conflicting transactions are blocked and rejected, if the consensus process is resolved by master nodes, the mechanism is passed to the participants of the algorithm and is solved by block confirmation;

Another well-known anonymization algorithm is an algorithm that works on a proof protocol with zero solution [15, 16], which allows one to verify the correctness of the solution without any additional information from the second party. The algorithm is functional if the verifier can convince the checked in the correctness of the solution, variants with an incorrect solution are unlikely in view of the high complexity of the problem being solved. The principle of operation of the protocol was most vividly formulated by Jean-Jacques Kiskater [17], showing the principle of operation of the algorithm on the example of a cave in the form of a circle (Fig. 1 As soon as it happens, the inspector moves to the point Y, informing the inspected where to go "right" or "left". During the primary inspection, the probability of the inspected leaving the cave is 50%, in case of increasing n times, the probability decreases to 1/2^n, increasing the knowledge of the inspected person about the exit.

One of the most famous blockchain projects using this algorithm is ZCash developed in 2016 [18]. ZCash organizes confidentiality and controlled transaction transparency, realized by generating random values used as initial parameters and key fragments. The blockchain reflects only the transaction, the source and the receiver are not disclosed unless they have agreed to reveal this information.

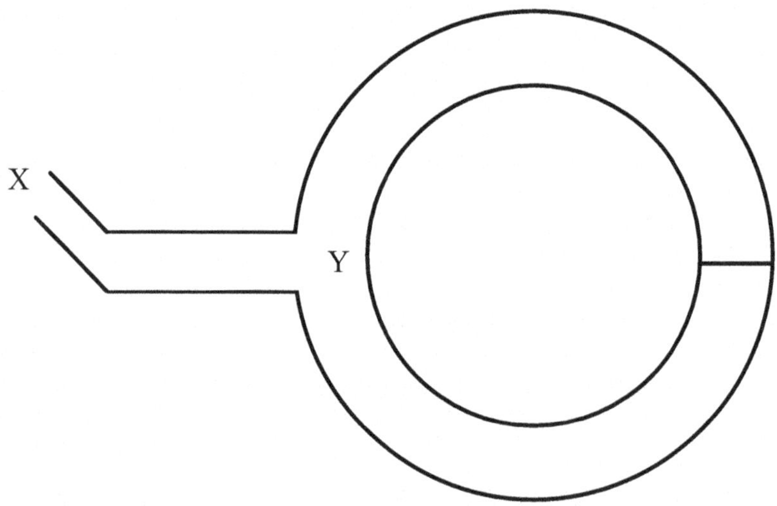

Fig. 1. The principle of zero-disclosure protocol

One of the promising algorithms can be considered - a ring encryption algorithm based on a simplified version of group signatures. Developed in 2012, by a man under the pseudonym Nicholas van Saberhagen, it is interesting to use circular signatures encrypting the sender and one-time addresses hiding the recipient.

The process of finding the correct hash in the algorithm does not depend on frequent calls to the miner, as it does in algorithms like SHA-256 [19], it is stored in the memory of the device. This allows to build a network protected from the so-called "51% attack" [20], when more than half of the network computing power belongs to a single device.

The procedure of anonymization of the user takes place by generating the public key of the sender and several other keys, making impossible the procedure of identification of the sender (Fig. 2). On the recipient side a new key is generated for each transfer, including if the sender is the same.

A distinctive feature of the algorithm is an arbitrary number of validator signatures, including the author of the transaction. Computations are performed on the transmitter initiator base using the keys of the collective, and the collective members may not even be aware of the use of keys for transaction signatures. These features effectively ensure the anonymity of transmitted messages [3].

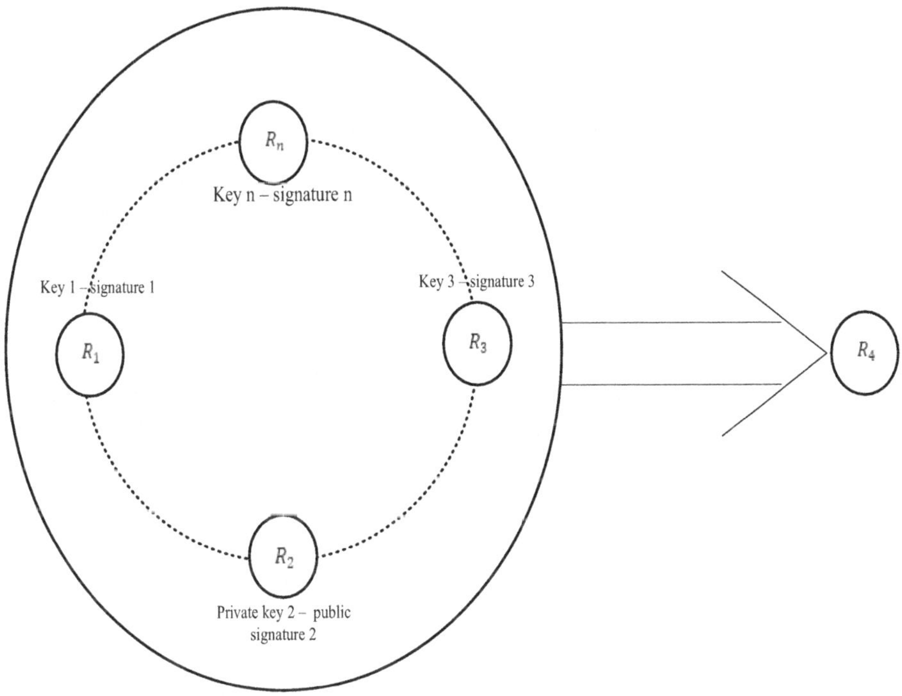

Fig. 2. Diagram of the circular signature with solution elaboration

4 Cryptonote for Information Interaction of the Robot

The algorithm has not yet been used for group robotics and decentralized data storage systems. The significant advantages of its use can be considered:

- Complete anonymity of the message sender. Unlike more known blockchains like Ethereum, Bitcoin, Litecoin in which it is possible to find the sender by analyzing the chain of blocks, CryptoNote using a method of circular signatures allows to protect a device that sends data in the chain of blocks. This advantage makes it impossible to trace the sender to analyze the data it sends and possibly to determine its location;
- No coded constants, CryptoNote allows block size to be adjusted based on historical network data;
- recalculation of block complexity after each mined block.

The scheme of algorithm's work in mobile robot groups is shown on the figure (Fig. 3). It shows the process that allows to achieve the anonymity of the source of transmitted information. The data comes from the whole group, and it is not known for sure, which member of the team transmitted it. In mobile robot group control systems, this allows for a secure data storage system without the risk of revealing the source of information in the unit, which in turn protects team members from being purposefully disrupted. To increase the reliability of the system, according to the algorithm, one-time addresses are used as recipient addresses indicating the right to dispose of one of the inputs. When viewing the system from the outside, the transaction can be used as an entry point to any other transaction from the referenced ones. It is also worth noting the possibility of expanding the circle of muxing by increasing the transaction fee.

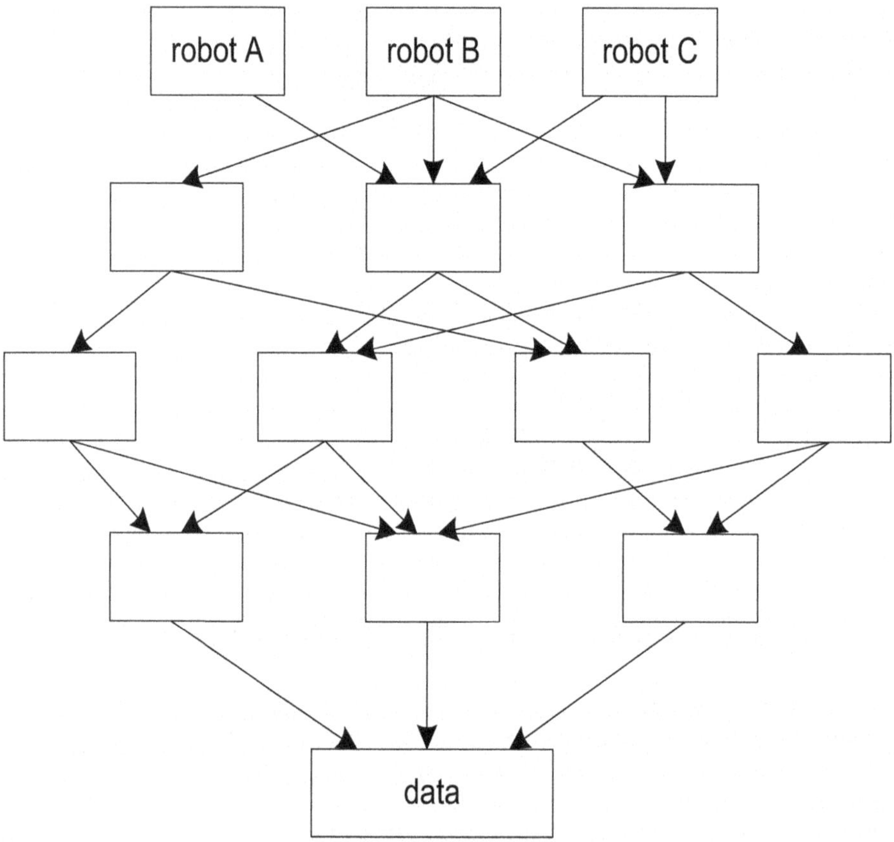

Fig. 3. Diagram of information transmission between units

5 Conclusions

When examining the existing solutions that use blockchain in robot groups, one cannot help but notice the trend of these projects. Most of the existing solutions use Ethereum smart contracts, which allows to reach a block creation speed of 15 s.

Blockchain allows to solve the problem of lack of knowledge of the collective, regardless of the possible expansion of the collective, dispersion or diversity of robots. All this is achieved through protocol consistency and network replication. In this case, the use of anonymous protocols increases the security of the system and team members.

Among the disadvantages and vulnerabilities of the blockchain, it is worth mentioning the possibility of "51% attack" [20], which in the case of using circular encryption algorithms will make false data in blocks, but not the sources of previously transmitted data. On the other hand, to carry out such an attack requires direct intervention in a supposedly closed network of a collective of robots and obtaining computing power over 51%, which in practice would be difficult to implement due to the lack of data on the protocols used.

It is worth noting a side feature of using blockchain technology in collective robotics - audit logging. Sensor data can be recorded and act as a "black box" that is stored in a decentralized network.

Acknowledgements. The publication was prepared as part of GZ SSC RAS project №AAAA-A19-119011190173-6 and RFBR projects 18-58-00051.

References

1. Bernard, M., et al.: Autonomous transportation and deployment with aerial robots for search and rescue missions. J. Field Robot. **28**(6), 914–931 (2011)
2. Bashyal, S., Venayagamoorthy, G.K.: Human swarm interaction for radiation source search and localization. In: Swarm Intelligence Symposium, 2008. SIS 2008, pp. 1–8. IEEE (2008)
3. Noether, S.: Ring SIgnature confidential transactions for Monero. IACR Cryptol. ePrint Arch. **2015,** 1098 (2015)
4. Van Saberhagen, N.: CryptoNote v 2.0. (2013)
5. Debiez, J., Hughes, J.P., Apvrille, A.: Data integrity check method using cumulative hash function. Google Patents (2003)
6. Sun, D., Kleiner, A., Schindelhauer, C.: Decentralized hash tables for mobile robot teams solving intra-logistics tasks. International Foundation for Autonomous Agents and (2010)
7. Mitzenmacher, M., Vadhan, S.: Why simple hash functions work: exploiting the entropy in a data stream. In: Proceedings of the Nineteenth Annual ACM-SIAM Symposium on Discrete Algorithms, pp. 746–755 (2008)
8. Housley, R.: A 224-bit one-way hash function: SHA-224 (2004)
9. Eastlake, D., Jones, P.: US secure hash algorithm 1 (SHA1). RFC 3174, September 2001
10. Dang, Q.H.: Secure hash standard (2015)
11. Wood, G., et al.: Ethereum: a secure decentralised generalised transaction ledger. In: Ethereum Proj. Yellow Pap. vol. 151, pp. 1–32 (2014)
12. Dannen, C.: Introducing Ethereum and Solidity. Springer (2017)
13. Dworkin, M.J.: SHA-3 standard: Permutation-based hash and extendable-output functions (2015)
14. Duffield, E., Diaz, D. Dash: A privacycentric cryptocurrency. Self-published (2015)
15. Goldreich, O., Krawczyk, H.: On the composition of zero-knowledge proof systems. JSIAM J. Comput. SIAM **25**(1), 169–192 (1996)
16. Goldreich, O., Oren, Y.: Definitions and properties of zero-knowledge proof systems. J. Cryptol. **7**(1), 1–32 (1994)

17. Quisquater, J.-J., et al.: How to explain zero-knowledge protocols to your children. In: Conference on the Theory and Application of Cryptology, pp. 628–631 (1989)
18. Ben-Sasson, E., et al.: Zerocash: Decentralized Anonymous Payments from Bitcoin (extended version) [Electronic resource]. 2014, pp. 1–56. http://zerocash-project.org/media/pdf/zerocash-extended-20140518.pdf
19. Nakamoto, S., et al.: Bitcoin: A peer-to-peer electronic cash system. Working Paper (2008)
20. Bastiaan, M.: Preventing the 51%-attack: a stochastic analysis of two phase proof of work in bitcoin. http://referaat.cs.utwente.nl/conference/22/paper/7473/preventingthe-51-attack-astochastic-analysis-of-two-phase-proof-of-work-in-bitcoin.pdf (2015)

Utilisation of EU Employment Data in Lecturing Data Mining Course

Jan Masopust, Zdena Dobesova(✉), and Karel Macků

Department of Geoinformatics, Faculty of Science, Palacky University, 17. listopadu 50, 779 00 Olomouc, Czech Republic
{jan.masopust01,zdena.dobesova,karel.macku}@upol.cz

Abstract. This article describes the utilisation of Eurostat employment data in the Data Mining course. The course is the obligatory course for a master degree Geoinformatics and Cartograhy study program at Palacký University in Olomouc. The article shows an example of the implementation of several methods like correlation, principal components analysis, k-means and hierarchical clustering on the same dataset in the course's teaching. The processing data in the Orange software and following interpretation of results gained by these methods are explained to students. Moreover, students create the MS PowerBI dashboard based on the same data. Teacher final finding is that the use of the current European data is for students more illustrative and increases their awareness of the status of employment in European countries within economic activities categorised by NACE. Practical processing of real data brings a deeper understanding of the lectured topics. Presented outputs, such as clustering, discover similar countries according to the same sector of employment.

Keywords: Education · Data Mining · European Union · Eurostat · NACE · GIS · MS PowerBI · UrbanDM

1 Introduction

The educational aspect of teaching data science-oriented courses faces many challenges. One of them lies in selecting relevant sample data that are sufficiently illustrative and contribute to an easier understanding of the presented methods. The authors of the article lecture the course Data Mining in the first grade of master study program Geoinformatics and Cartography at Palacký University in Olomouc, Czech Republic. The Data Mining course syllabus consists of basic topics like data pre-processing, multi-dimensional statistics methods (such as clustering or principal component analysis), decision trees, association rules and time series analysis. The student's evaluation of previous courses brought criticism of repetitive using of the sample dataset Iris. Iris flower dataset is probably one of the best-known multivariate sample datasets for many statistical techniques such as classification or clustering. The dataset consists of 150 samples from three species of Iris with four measured characteristics. Many manuals and teaching materials for data mining software frequently use this dataset as a transparent example.

© The Author(s), under exclusive license to Springer Nature Switzerland AG 2021
R. Silhavy (Ed.): CSOC 2021, LNNS 229, pp. 601–616, 2021.
https://doi.org/10.1007/978-3-030-77445-5_55

Nevertheless, in the field of geoinformatics, students would prefer more geographically oriented data. As a response to student evaluation, the new practical examples were prepared using the EU (European Union) data. There are dozens of interesting themes in the Eurostat database [1], so we decided to employ this data source to bring a more European perspective into the Data Mining course. The data about employment in various sectors in the EU states were selected for practical exercises to make taught data mining methods easier to understand. Simultaneously, students' subconsciousness about the structure of employment on a European scale is raising.

The finding of new ways of education is a topical theme for pedagogy at universities. Real data and a combination of new methods are mentioned in the literature [2, 3]. The article's authors have a positive experience using a practical example to boost students' knowledge in other courses. Contributive experience is mentioned in the article about the relational database design for the botanical garden plant database under the BotanGIS project [4]. Another example is from the same Data Ming course about constructing a decision tree by self-gathered data [5].

In the practical example demonstrated in this paper, the main goal was to confirm differences between west and middle/east European countries using employment data. From a historical perspective, the separation after the Second World War has influenced industries' development in two parts of Europe for near 45 years in the 20th century. East Europe has been oriented more to agriculture and heavy industry and (primary and secondary sector); west Europe has emphasised the public services, science and research (tertiary and quaternary sector). Based on this historical development, the research question arises: how have the following 30 years (since the 90s of the 20th century and the beginning of the 21st century) changed employment structure in economy sectors? Do the disparities remain or have they disappeared?

This article presents practical examples using data about employment in different sectors of economic activities in Europe. Related to the Data Mining course, several sample tasks were presented as they are taught in the course, including interpreting results.

2 Data and Methods

For successful lecturing, it is crucial to choose software and training data wisely. The next chapters briefly explain used data source Eurostat database, Statistical Classification of Economic Activities NACE and practical examples in ORANGE software.

2.1 Eurostat Database

Eurostat is the main statistical body of the European Union. Its main task is to provide harmonised statistics at the EU level at both national and regional levels to compare different areas in selected topics. All data published by Eurostat are obtained from the statistical authorities of individual member states (national statistical offices). Eurostat itself is primarily concerned with the collection of these data and their harmonisation into a comparable form. Unfortunately, this approach is also the main weakness, as member states are only required to supply a limited amount of statistical data. The remaining part is not legally enforceable, so their availability depends only on the member state's willingness.

For this reason, much of the data is regionally or temporally unavailable. Most of the European statistics is freely disseminated in the Eurostat Database. Eurostat also provides microdata, such as data from sample survey EU-SILC (European Union Survey on Income and Living Condition). This data is only available on request, consisting of verification of the applicant entity and submission of a research plan in which the data is planned to be used.

2.2 Statistical Classification of Economic Activities - NACE

The Statistical Classification of Economic Activities in the European Community, commonly referred to as NACE, is the industry-standard classification system used in the EU. The Regulation establishing NACE Rev. 2 was adopted in December 2006 [6].

NACE uses four hierarchical levels:

- Level 1: 21 sections identified by alphabetical letters A to U;
- Level 2: 88 divisions identified by two-digit numerical codes (01 to 99);
- Level 3: 272 groups identified by three-digit numerical codes (01.1 to 99.0);
- Level 4: 629 classes identified by four-digit numerical codes (01.11 to 99.00).

Eurostat offers data under *National accounts employment data by industry* section [7], online data code NAMA_10_A64_E. Two levels of detail - Level 1 and Level 2 are reported. For a practical example, Level 1 with 21 categories was selected. This level was downloaded in the format for MS Excel spreadsheet. The reported unit of measure is a thousand persons. This unit is incomparable for searching for similar countries and unsuitable for particular methods like clustering. The purpose was to find similar countries with a similar or equal structure of employee by economic activities. Students recalculate the original values from a thousand persons to the relative value - the percentage of each country's total number of employed persons in lectures.

Some statistics are available for countries outside the EU, but they suffer from a certain level of incompleteness. In total, countries Montenegro, Serbia, Liechtenstein, and Malta were excluded from the processed dataset, and 30 countries remain for father processing. The reported year 2018 was selected as a reference point for a practical example. The data for 2019 is also accessible, but many values are labelled as *(p) provisional*. The detection of the incompleteness of data is the first practical experience for students. Recalculation data and solving incompleteness belongs to the introductory lesson about pre-processing data in the Data Mining course.

Table1 shows the classification and description of economic activities at Level 1.

Table 1. NACE classification at Level 1

Code	Description
A	AGRICULTURE, FORESTRY AND FISHING
B	MINING AND QUARRYING
C	MANUFACTURING
D	ELECTRICITY, GAS, STEAM AND AIR CONDITIONING SUPPLY
E	WATER SUPPLY; SEWERAGE, WASTE MANAGEMENT AND REMEDIATION ACTIVITIES
F	CONSTRUCTION
G	WHOLESALE AND RETAIL TRADE; REPAIR OF MOTOR VEHICLES AND MOTORCYCLES
H	TRANSPORTATION AND STORAGE
I	ACCOMMODATION AND FOOD SERVICE ACTIVITIES
J	INFORMATION AND COMMUNICATION
K	FINANCIAL AND INSURANCE ACTIVITIES
L	REAL ESTATE ACTIVITIES
M	PROFESSIONAL, SCIENTIFIC AND TECHNICAL ACTIVITIES
N	ADMINISTRATIVE AND SUPPORT SERVICE ACTIVITIES
O	PUBLIC ADMINISTRATION AND DEFENCE; COMPULSORY SOCIAL SECURITY
P	EDUCATION
Q	HUMAN HEALTH AND SOCIAL WORK ACTIVITIES
R	ARTS, ENTERTAINMENT AND RECREATION
S	OTHER SERVICE ACTIVITIES
T	ACTIVITIES OF HOUSEHOLDS AS EMPLOYERS; UNDIFFERENTIATED GOODS- AND SERVICES-PRODUCING ACTIVITIES OF HOUSEHOLDS FOR OWN USE
U	ACTIVITIES OF EXTRATERRITORIAL ORGANISATIONS AND BODIES

3 Practical Example

The introductory chapters and theoretical presentation for the Data Ming course's lecturing are based on books by P. Pavel and J. Šarmanová [8, 9]. The examples presented in these lecture-books are mostly oriented on medical care, insurance, and the financial sector. Therefore, the theme of employment by economic activities in the EU, which belongs to economic geography, is more suitable for Geoinformatics students.

3.1 Example EU Employment in Software ORANGE

The WEKA software has been used in the Data Mining course for five years due to its free access. Alternatively, software R is practised. Starting at last academic year, 2019/2020, Orange software has been implemented into practical lessons. Orange is freely available [10, 11]. The interface is user friendly and easy to handle using the drag and drop approach. Figure 1 shows an example of a workflow containing several data mining methods. This workflow presents all practised methods that are described in the following section of this article.

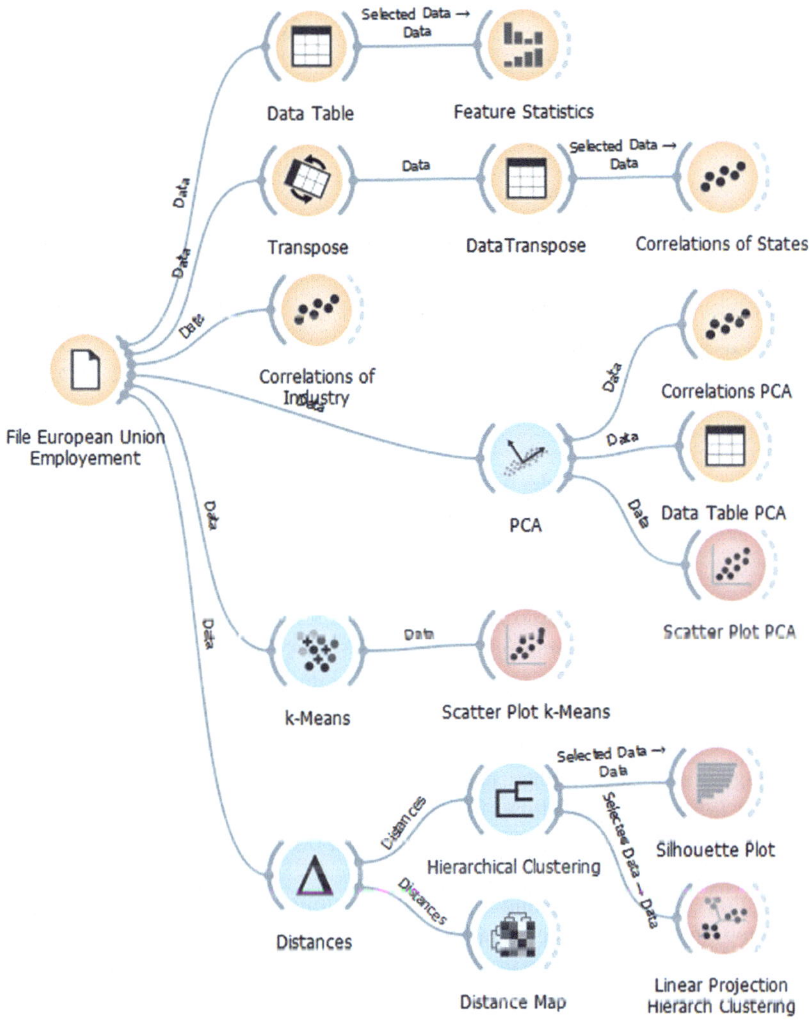

Fig. 1. The workflow in Orange software for processing data by several methods. Methods are widgets represented by colour circular nodes.

The first widget, *File European Union Employment* at the left, connects the source XLSX data in the workflow. The first top branch from the *File* widget connects the *Data Table* widget. This widget shows the input table content about employments in sectors. Each row represents one country as an instance. Columns contain the type of economic activity in percentage (Fig. 2).

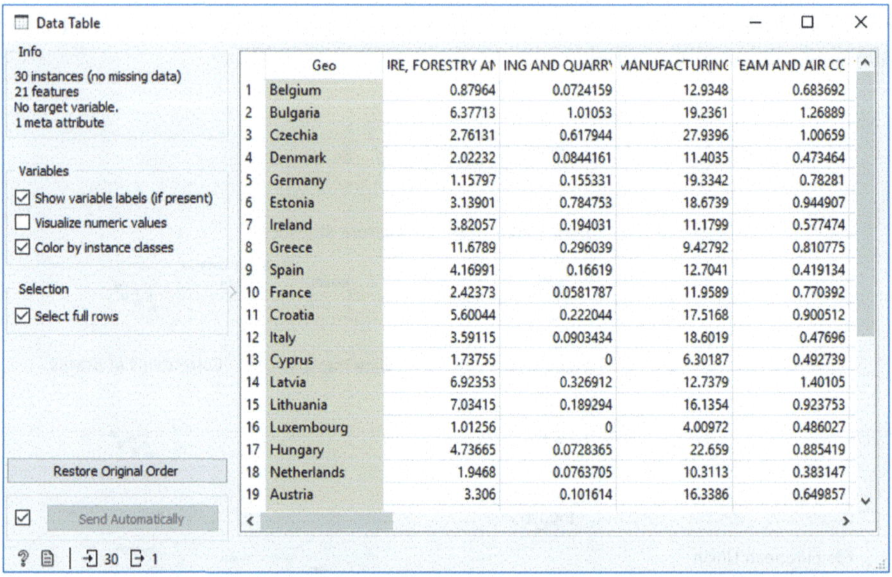

	Geo	IRE, FORESTRY AN	ING AND QUARRY	MANUFACTURING	EAM AND AIR CC
1	Belgium	0.87964	0.0724159	12.9348	0.683692
2	Bulgaria	6.37713	1.01053	19.2361	1.26889
3	Czechia	2.76131	0.617944	27.9396	1.00659
4	Denmark	2.02232	0.0844161	11.4035	0.473464
5	Germany	1.15797	0.155331	19.3342	0.78281
6	Estonia	3.13901	0.784753	18.6739	0.944907
7	Ireland	3.82057	0.194031	11.1799	0.577474
8	Greece	11.6789	0.296039	9.42792	0.810775
9	Spain	4.16991	0.16619	12.7041	0.419134
10	France	2.42373	0.0581787	11.9589	0.770392
11	Croatia	5.60044	0.222044	17.5168	0.900512
12	Italy	3.59115	0.0903434	18.6019	0.47696
13	Cyprus	1.73755	0	6.30187	0.492739
14	Latvia	6.92353	0.326912	12.7379	1.40105
15	Lithuania	7.03415	0.189294	16.1354	0.923753
16	Luxembourg	1.01256	0	4.00972	0.486027
17	Hungary	4.73665	0.0728365	22.659	0.885419
18	Netherlands	1.9468	0.0763705	10.3113	0.383147
19	Austria	3.306	0.101614	16.3386	0.649857

Fig. 2. Source data with EU states as records with the percentage of employment in specific sectors of economic activities NACE. Widget Data Table depicts it in Orange software.

Feature Statistics

Students are taught that exploratory data analysis (EDA) should always be the first step in the data processing process. EDA includes, for example, basic descriptive statistics of centre and variability, which provide an elementary overview about the nature of data. Basic feature statistics (minimum, maximum and mean values) has been calculated in the Orange software by widget *Feature Statistics* (Fig. 3). This widget follows the widget *Data Table* in the workflow (Fig. 1). It is possible to order rows interactively by maximum or minimum values in columns. Data is ordered by the maximum value in Fig. 3. The maximum value in Manufacturing (near 28% employee) was observed in Czechia, followed by Slovenia and Slovakia. The simple graph with the distribution of values (histogram) is also depicted in the *Feature Statistics* window. The distribution is very variable in different economic activities in Europe. The widget brings a quick and complete first overview of data.

Correlation Between Sector and Correlation Between Countries

The next basic tasks is the examination of correlation using the Pearson correlation coefficient. The widget node *Correlation* shows the list of Pearson correlation coefficients between pairs sorted form the highest absolute value to zero (Fig. 4a). The values are accompanied by colour lines (green positive, blue negative correlation) where length expresses the value. The interesting thing is the investigation of the correlation between partial economic activities. Very high correlation (above + 0.8) are between *Activities of extraterritorial organisations* and *Financial and insurance activities; Electricity, gas and steam supply* correlates with *transportation and storage.* Surprisingly, negative correlation (-0.7) of *Manufacturing* with *Professional, scientific and technical activities.* The second-highest negative correlation is between *Professional, scientific and technical activities* and *Water supply and sewerage* (0.652).

After transposing the data table (widget *Transpose* in Fig. 1), it is possible to calculate the correlation coefficient for countries (Fig. 4b). The strongest correlation is between *Hungary* and *Slovakia* (+0.989*), Czechia* and *Slovenia* (+0.988) and *Belgium* and *France* (0.982). Significant correlations are evident among the east (former socialist) countries, and the western countries also correlate.

No negative correlation was observed, but several combinations report low, insignificant correlations: *Luxembourg – Romania* (0.07), *Czechia* (0.252), *Bulgaria* (0.277). Especially *Luxembourg* can be marked as a very specific country form the point of employment by economic activities.

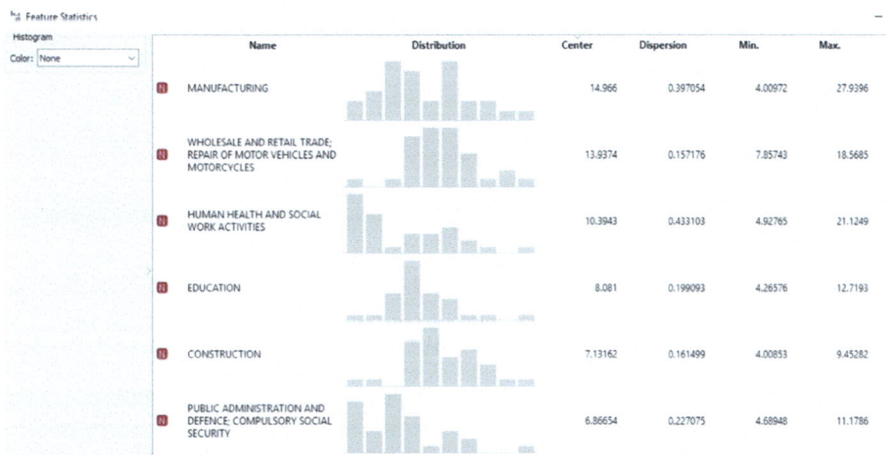

Fig. 3. Basic description by Feature Statistics widget.

Principal Component Analysis and Scatter Plot in PC1 a PC2

A principal component analysis (PCA) is a useful technique for revealing patterns when dealing with multi-dimensional correlated attributes. PCA can reduce the processed file's dimension and reveal new, hidden properties - components [12]. The main components are linear combinations of the original attributes, which in descending order capture as much of the input data variance as possible. PCA can be used widely - not only for simple data size reduction but also for checking the data file's consistency or calculating indexes (see example in [13]).

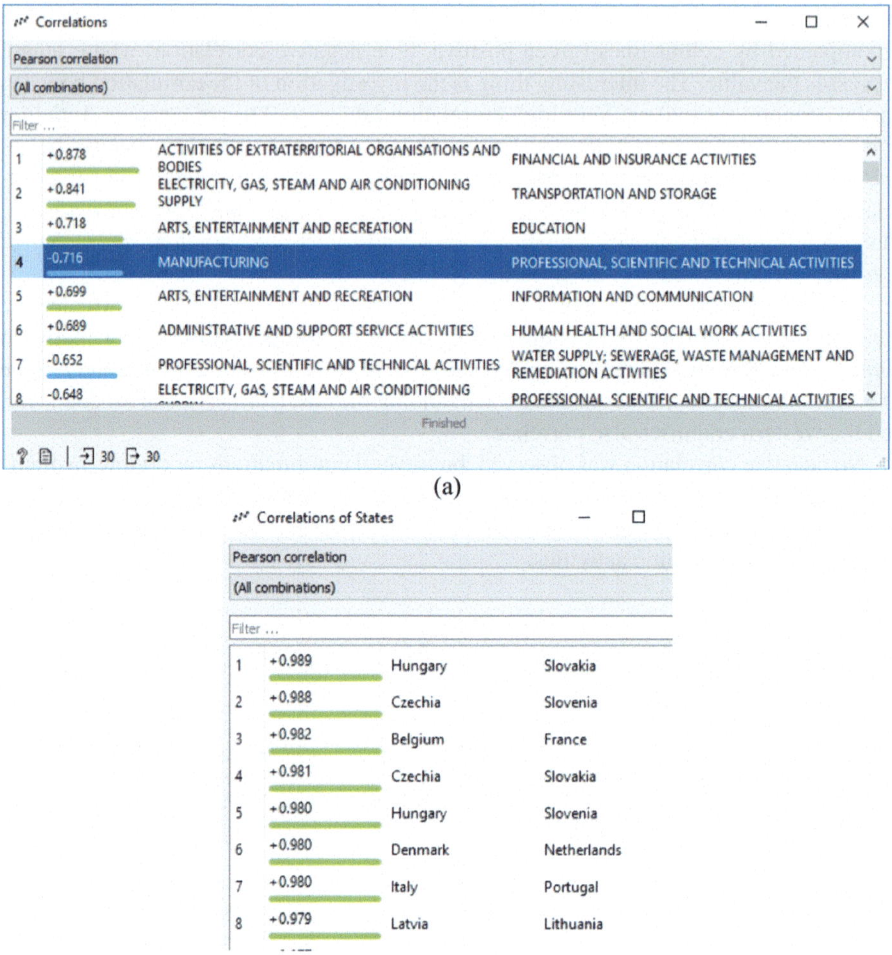

Fig. 4. (a) Correlation of economic sectors, (b) Correlation of countries.

First, the number of new significant principal components is determined using the Kaiser criterion. If the component's eigenvalue is greater than one, the component brings new information, and it is advisable to include it for analysis. In our case, five components were labelled as significant, and they together describe 92% of the variance in original input data (Fig. 5). The widget PCA shows the red scree graph where the user can interactively move the black vertical line to select the final number of components. The graph also shows a green curve of cumulative variance labelled by explained variance [14]. The interactive control is very user–friendly and helps student in the selection of the final count of a new component.

Several graphical outputs can analyse the results of PCA. The graph of instances has the form of the two-dimensional scatterplot, showing the position of individual inputs projected into two-dimensional space defined by the scores of two selected components (one graph for each pairwise combination of components is possible). In Fig. 6, the input instances (countries) are displayed in the context of the first and second component, PC1 and PC2. The similarities among countries can be then observed concerning the newly created components. The mutual position of individual states describes their similarity in the context of the components of the score. For example, *Hungary, Slovakia, Slovenia* and *Czechia* are close to each other (yellow group), which means they have quite similar components scores. The green group in the middle consists of Lithuania, Croatia, Latvia, Austria, Portugal, and Italy. There is another significant group of similar countries consisting of the *United Kingdom, Netherlands, Sweden, Denmark, France, Switzerland* and *Belgium* (blue group).

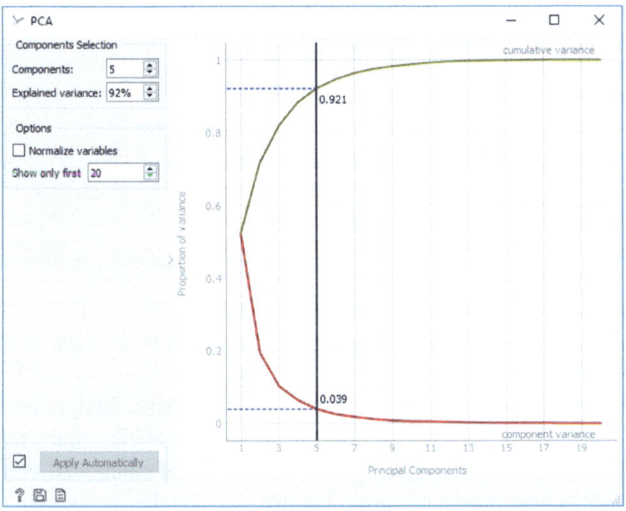

Fig. 5. The widget PCA with interactive determining of number of components in the scree graph.

Similarities within only one component can be observed as well (Fig. 6).: *Luxembourg and Norway* have an almost identical score on the first component, on the second component *Greece* has the highest component score, while the first component has a score of around zero, which points to the typical profile of *Greece* on the second component.

Depending on the PCA's efficiency to the data size reduction, all of the components might be deeper interpreted and named. This task is often extremely challenging when there are many correlated indicators, which form a particular component. The similarity of countries can be explained by detecting the correlation between the original attributes and new components PC1 and PC2. The following node *Correlation PCA* represents this step after node *PCA* (Fig. 1). In our example, attributes *Manufacturing* (0.91), *Water supply* (0.77), *Electricity supply* (0.62), and *Agriculture* (0.59) have the highest contribution for PC1. This contribution value of original attributes helps with the labelling and interpretation of components. With this finding, we can label the PC1 as *Industry and production. Agriculture* (correlation 0.62), *Whole retail trade* (0.57) and *Accommodation and food service* (0.48) have the highest contribution for PC2. We can assign the

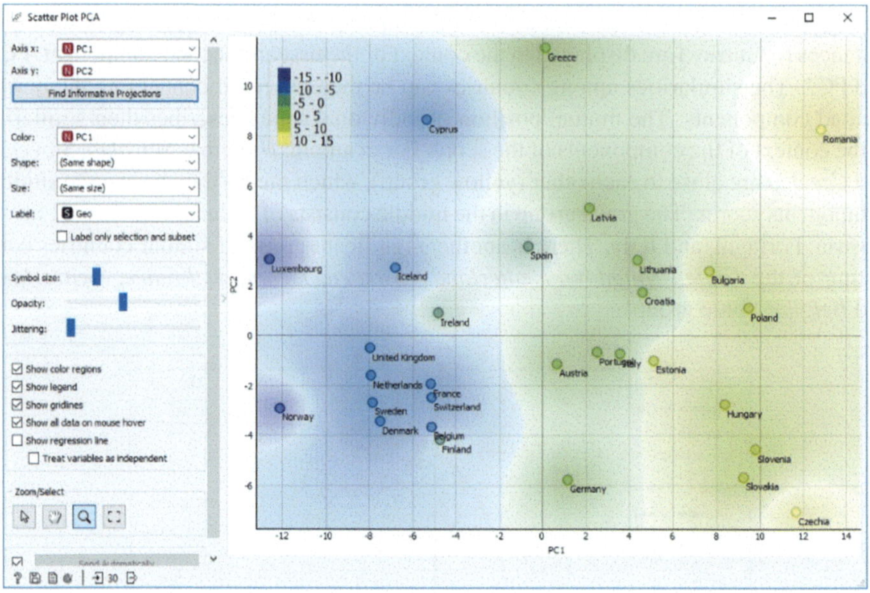

Fig. 6. The result PCA scatter plot of instances in PC1 and PC2 coordinate system.

label PC2 as *Trading and services. Financial and insurance activities* have the highest contribution to PC3.

There are several graphical tools, which can provide deeper insight into PCA results: namely *correlation circle*, which visualise the relation between input indicators in the coordinates of PCA; *scree plot* showing the importance of calculated components; *bar plots* presenting the contribution of particular attributes on the component (or the quality of their representation on the component) or *graph of individuals* showing the component score of every record included in the analysis. All of these outputs are the subject of detective work, aiming at a better interpretation of the PCA and understanding the revealed links.

One of the benefits of a *Scatter plot* implemented for PCA is the possibility of changing pairs of the axis (PCs), forming the plot's coordinates system and distinguishing the potential clusters (based on PC score) by colour. The colouring depends on the selected component (PC1 in case Fig. 6), so the different results and cluster arrangement could be deduced. E.g., Greece creates a single element cluster due to many people working at accommodation and food services. Luxembourg and Cyprus have many employees in financial and insurance activities.

Clustering Methods

Another group of typical data mining method is cluster analysis. In general, it is an objective numerical approach aiming at searching for the similarities in n-dimensional space. Meloun & Militký [15] describe cluster analysis as a method to investigate the similarity of multi-dimensional objects and their classification into classes. The essence of cluster analysis is calculating distances between individual objects in n-dimensional space, which represents the input data. Subsequently, clusters (groups of objects similar to each other and simultaneously differ as much as possible from the objects of other clusters) are defined using various algorithms.

In the literature, many authors point out the importance of selecting a suitable clustering algorithm, but not to the selection of a suitable degree of similarity [16]. According to Hastie, Tibshirani, & Friedman [17], the specification of an appropriate degree of similarity (distance) is often more important than a particular algorithm's choice. At least the elementary examples of the distances, such as Euclidean, Mahalanobis, Manhattan or Cosine measures, are introduced to students in the Data Mining course. It is always emphasised that it is necessary to play and experiment with the metrics and choose how they make sense for the given data (e.g. use the Mahalanobis distance for strongly correlated data). The Orange offer several metrics in the widget *Distances*. The visualisation of the used metrics in the form of a heatmap helps to understand the mutual relations between the objects analysed. The widget *Distance Map* shows a matrix of cosine distances with a dendrogram (Fig. 7). In Fig. 7, two significant blue rectangles (couples close to each other, i.e. potential clusters) can be distinguished. Luxembourg and Romania are both at the edges of the matrix, which determines they are dissimilar to other countries. The finding about Luxembourg correspond to the previous findings outputting from correlation analysis; the separation of Romania was also visible in the

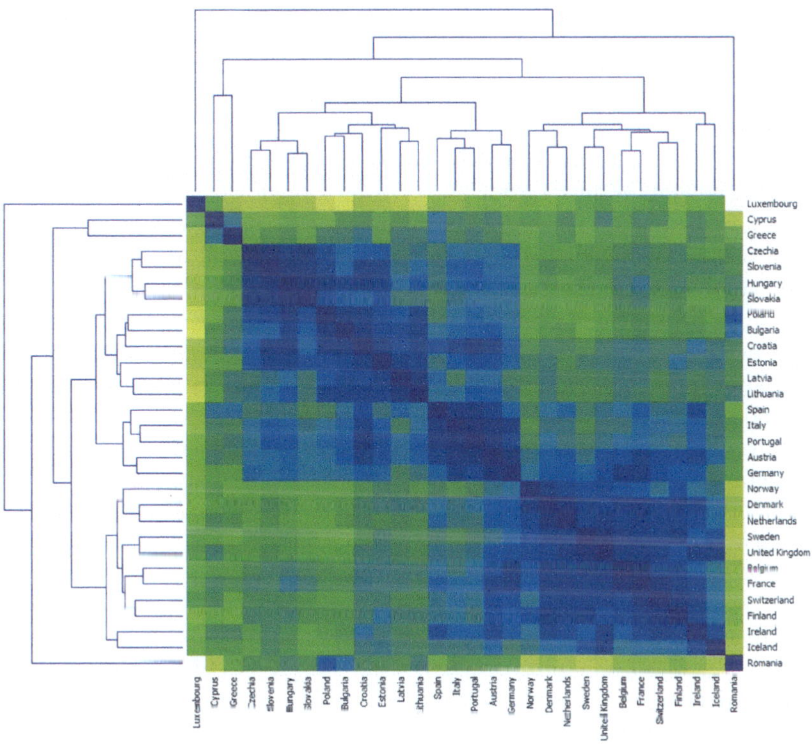

Fig. 7. Visualisation of distance matrix based on cosine distance for economic activities

result of PCA. The explanation lies in the great portion of employees in *agriculture, forestry and fishing* (19.8%). That belongs to the primary sector. The second country with a high amount of employees in agriculture is Greece (11.7%).

Hierarchical Clustering with Ward's Method and Silhouette Plot
There are many possibilities of clustering approaches and algorithms – see, e.g. [18–20]. In the course, the main difference between the hierarchical and non-hierarchical are introduced and demonstrated in the examples of Ward's method (hierarchical group) and K-means (non-hierarchical). These methods were also applied to find similar enterprises from the C sector – Manufacturing in previous research. The source data was about various types of innovations [21]. Ward's method was selected due to the best results in the presented example of economic activities.

The widget *Hierarchical Clustering* computes dendrogram from a matrix of distances. The level of selection could be stated manually, by height ratio or by top N nodes. The big advantage is the interactive manual placing of the vertical dot line of selection and automatic colouring clusters (Fig. 8a). Clustering was processed on a cosine distance matrix, using Ward's method in the presented dendrogram in Fig. 8a.

The widget *Silhouette Plot* offers a graphical representation of consistency within clusters. User can visually assess cluster quality. The silhouette score close to 1 indicates that data instances are close to the centre of the cluster. Instances possessing a score close to 0 are on the border between two clusters. Figure 8b verifies the consistency of the selected level of clustering. Students can experiment with selecting the number of clusters and assessing the consistency of clusters via the widget *Silhouette Plot*. The colours are assigned automatically, and they correspond between the dendrogram and silhouette plot.

They are seven result clusters in the dendrogram. The arrangement of clusters corresponds to the previous results. Luxembourg creates a single-element entity - cluster C1, since it is very dissimilar to other European countries. The next widget, *Linear Projection,* helps with an examination of other clusters and similarities. Clusters of countries are placed in multiple axes of source attributes. Cluster C2 consists of the western countries of Europe. These have more employees in *Human health and social activities, Educations, Professional, scientific and technical activities, Arts, entertainment and recreation.* Four countries from cluster C3: Czechia, Slovakia, Slovenia and Hungary, are similar due high portion of employees in manufacturing and water supply.

Moreover, they are neighbouring countries (except Slovenia). Cluster C4 is a single-element entity of Romania. Cluster C5 is consists of Lithuania, Latvia, Bulgaria, Croatia, Estonia and Poland. They have a similar portion of employees in electricity, gas, steam supply and transportation and agriculture. An interesting cluster is C6 which consists of Austria, Portugal, Italy, Spain and Germany. They are similar in *Accommodation and food services, trading* and *transportation.* Finally, cluster C7 consists of Cyprus and Greece that were also found similar by PCA (sectors like accommodation and trading).

K-Means Clustering
The last presented method is k-Means clustering. It belongs to unsupervised machine learning as well as PCA and hierarchical clustering [8]. The disadvantage of k-Means is the need of choosing the number of clusters at the start. This widget helps to solve this

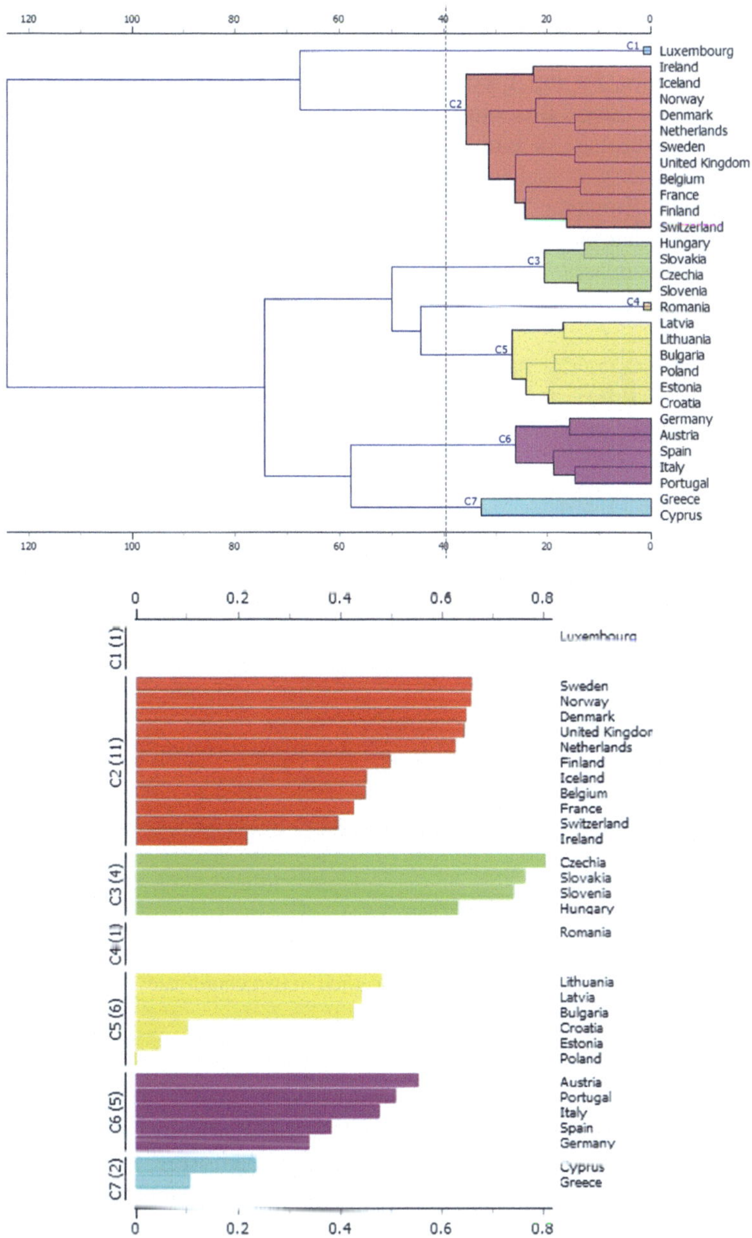

Fig. 8. Hierarchical clustering (a) dendrogram, (b) corresponding silhouette plot.

problem by finding the number of clusters using silhouette scores. In the case of NACE economic activities, the suggested number of clusters is two.

Moreover, there are two options for determining the starting centroids of clusters: *Random initialisation* or *Initialisation with KMeans++* in Orange. The followed widget

Scatter Plot shows the result of K-Means in automatically selected the best projection (Fig. 9). For economic activities, the suggested projection is for the X-axis *Human health and social activities*, for the Y-axis *Manufacturing*. The result red and blue clusters confirm previous findings. Blue cluster with Norway, Denmark, Netherlands and others have a higher percentage of employees in *Human health and social activities.* Red clusters are countries with higher manufacturing numbers (Czechia, Slovenia, Slovakia, Hungary, Poland, Romania, Italy, Bulgaria, Germany). Germany and Austria belong to the red cluster, but they are near the clusters' border because they also have many employees in *Human health and social activities.*

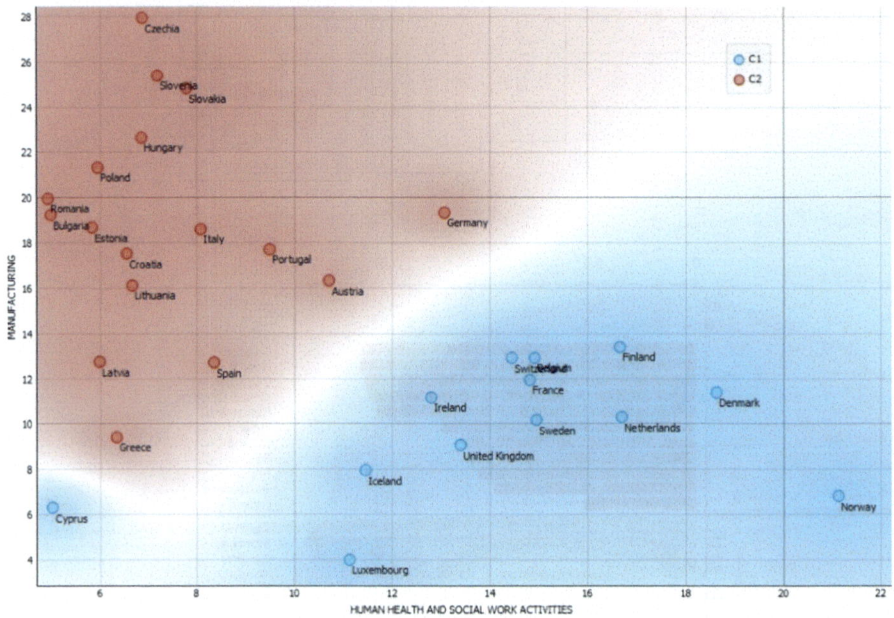

Fig. 9. Clusters determined by the K-Means method in two coordinates (economic sectors).

PowerBI Dashboard

After statistical analysing of the dataset in the Orange software, we had visualised data using online interactive business intelligence (BI) dashboards (Fig. 10). A dashboard is a basic tool of the visual analytics technique. They are one of the most common use cases for data visualisation, and their design and contexts of use are considerably different from exploratory visualisation tools [22]. We had chosen the Microsoft Power BI Desktop. It is free for use. Students had uploaded data into the application. Then they used some visuals suitable for the dataset. We used the matrix, bar chart, choropleth map, filters, or slicers in this tutorial. The dashboard allows interactive selecting and filtering of data by clicking into any visual. Students can click on the YES or NO button in the EURO section to filter out only countries within the Eurozone. They can click anywhere in the table, map, or chart also. The given country will be highlighted in every visual. The final dashboard from our course is available online on https://bit.ly/3dpSV9w.

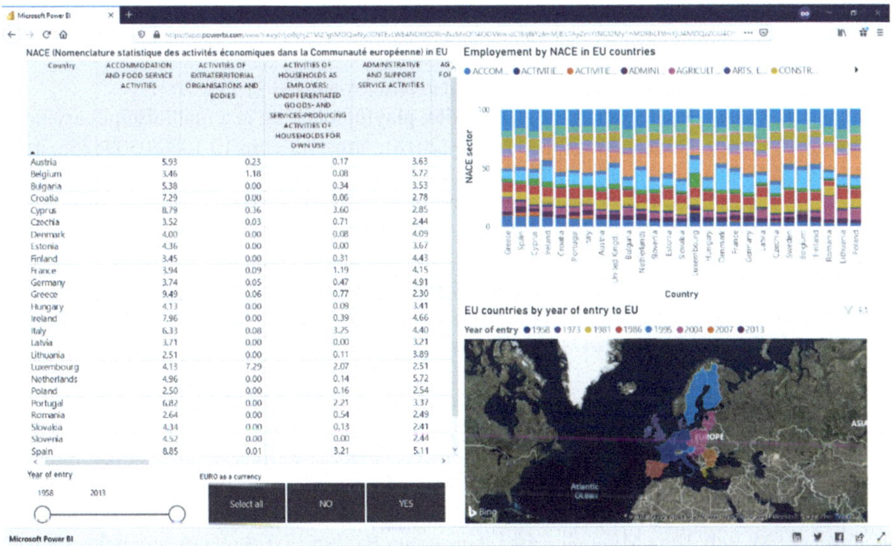

Fig. 10. MS PowerBI Dashboard built with EU NACE employment dataset.

4 Conclusion

The use of real Eurostat data is very valuable for students. The best way is a combination of several data mining methods. Some methods bring new findings, and also the combination of a method brings confirmation of partial finding. This article discussed the correlation, PCA, clustering, and interactive visualisation using Orange and MS PowerBI software.

The result is that some differences in the west and middle/east European countries exist. West countries are more oriented to a tertiary and quaternary economic sector (human health, science, services), and east countries are oriented to agriculture, manufacturing and industry (primary and secondary sector). Some outliers are identified, like Romania, Luxembourg and Greece.

Moreover, there is a very good experience with Orange software for the Data Mining course. Orange software is very intuitive and allows the quick design of workflow, especially for novice students in analysing multi-dimensional data. The advantage is interactive exploration results in multiple windows, compare, combine and verify the results in the interpretation phase.

Acknowledgement. This article was created with the support of the Erasmus+ Programme of the European Union, Jean Monnet Module (Project No. 620791-EPP-1-2020-1-CZ-EPPJMO-MODULE, UrbanDM - Data mining and analysing of urban structures as a contribution to European Union studies). The second supporting project is the Application of geospatial technologies for spatial analysis, modelling, and visualisation of spatial phenomena (IGA_PrF_2021_020) of the Internal Grant Agency of Palacký University Olomouc.

References

1. Eurostat. https://ec.europa.eu/eurostat/data/database. Accessed 31 Dec 2021
2. Pánek, J., Pászto, V., Perkins, C.: Flying a kite: playful mapping in a multidisciplinary field-course. J. Geogr. High. Educ. **42**, 317–336 (2018). https://doi.org/10.1080/03098265.2018.1463975
3. Egiebor, E.E., Foster, E.J.: Students' perceptions of their engagement using GIS-story maps. J. Geogr. **118**, 51–65 (2019). https://doi.org/10.1080/00221341.2018.1515975
4. Dobesova, Z.: Teaching database systems using a practical example. Earth Sci. Inf. **9**(2), 215–224 (2015). https://doi.org/10.1007/s12145-015-0241-3
5. Dobesova, Z.: Teaching decision tree using a practical example. In: Silhavy, R. (ed.) CSOC 2020. AISC, vol. 1226, pp. 247–256. Springer, Cham (2020). https://doi.org/10.1007/978-3-030-51974-2_23
6. NACE Rev. 2 - Statistical Classification of Economic Activities. https://ec.europa.eu/eurostat/web/nace-rev2/overview. Accessed 10 Feb 2021
7. National accounts employment data by industry. https://ec.europa.eu/eurostat/databrowser/view/NAMA_10_A64_E__custom_221911/default/table?lang=en. Accessed 12 Feb 2021
8. Šarmanová, J.: Metody analýzy dat. VŠB – Technická univerzita, Ostrava (2012)
9. Pavel, P.: Metody Data Miningu, část I. Univerzita Pardubice, Fakulta ekonomicko-správní, Pardubice (2014)
10. University of Ljubljana: Orange. https://orange.biolab.si/. Accessed 04 May 2019
11. Orange visual programming documentation, Release 3. https://buildmedia.readthedocs.org/media/pdf/orange-visual-programming/latest/orange-visual-programming.pdf. Accessed 24 Sep 2020
12. Jolliffe, I.: Principal Component Analysis. Springer, New York (2002)
13. Macků, K., Voženílek, V.: Statistická syntéza indikátorů kvality života – návrh tvorby indexu v evropských regionech. Geographia Cassoviensis 13 (2019). https://doi.org/10.33542/GC2019-2-06
14. Getting Started with Orange 09: Principal Component Analysis. https://www.youtube.com/watch?v=OmaAC8a52YI&t=2s. Accessed 10 Dec 2019
15. Meloun, M., Militký, J.: Statistical Data Analysis. Woodhead Publishing India (2011)
16. Mimmack, G.M., Mason, S.J., Galpin, J.S.: Choice of distance matrices in cluster analysis: defining regions. J. Clim. **14**, 2790 (2001). https://doi.org/10.1175/1520-0442(2001)014%3c2790:codmic%3e2.0.co;2
17. Hastie, T., Tibshirani, R., Friedman, J.: The Elements of Statistical Learning. Springer New York Inc., New York (2001)
18. Hartigan, J.A., Wong, M.A.: Algorithm AS 136: a K-means clustering algorithm. J. Roy. Statist. Soc. Ser. C (Appl. Stat.) **28**, 100–108 (1979). https://doi.org/10.2307/2346830
19. Ward, J.H.: Hierarchical grouping to optimise an objective function. J. Am. Stat. Assoc. **58**, 236–244 (1963). https://doi.org/10.1080/01621459.1963.10500845
20. Kaufman, L., Rousseeuw, P.J.: Finding Groups in Data: An Introduction to Cluster Analysis. John Wiley (1990)
21. Dobesova, Z., Paszto, V., Macku, K.: Analysis of similarities in context of enterprise innovations. In: Slavíčková, P. (ed.) Knowledge for Market Use, p. 8, Olomouc (2017)
22. Sarikaya, A., Correll, M., Bartram, L., Tory, M., Fisher, D.: What do we talk about when we talk about dashboards? IEEE Trans. Visual Comput. Graphics **25**, 682–692 (2019). https://doi.org/10.1109/TVCG.2018.2864903

A Decentralized Ontology Versioning Model Designed for Inter-operability and Multi-organizational Data Exchange

Andrei Tara[1,2(✉)], Nicolae Taban[2], Cristina Vasiu[2], and Constantin Zamfirescu[1]

[1] Lucian Blaga University of Sibiu, 10 Victoriei, 550024 Sibiu, Romania
research@andreitara.com, zbc@acm.org
[2] Openfabric Network SRL, 16, Iosif Velceanu, 550057 Sibiu, Romania
{nicu,cristina}@openfabric.ai

Abstract. Current information systems operated in a multi-organization context use centralized or hierarchical data models proven inefficient as the network's scale rapidly increases. The introduction of blockchain and decentralized specific mechanisms stimulate the use of ontology-based data models. While the adoption of blockchain as a communication medium facilitates the communication and data exchange in a more reliably and securely fashion, this opens new challenges regarding the evolution and versioning of the ontology. This paper proposes a versioning mechanism tailored to the decentralized environment. It provides the conceptual building blocks allowing an ontology to evolve under the supervision of a community of individuals and organizations proposing progressive refinements to the ontology elements. Also, it details the aggregation mechanism used to select the changes for the new version.

Keywords: Decentralization · Distributed ontology · Blockchain

1 Introduction

Interoperability remains a significant challenge in the world of information technology [26]; however, there are efforts in this direction that endeavour to uniformize the interaction between multiple parties [21, 28]. In order to promote multi-organizational communication, the systems should adhere to the same data representation. While maintaining a common data format for a reduced number of participants is relatively effortless [18], this becomes a real struggle as the number of participants increases [14, 18]. Various endeavours involving the use of standards [19] attempt to close the interoperability gap by involving the shared effort of domain experts and company groups [13]. Once parties begin to comply to the same standard, then communication ceases to be a limiting factor [23]; however, the current standardization process is slow, due to the lack of a mechanism for promoting inter-organizational communication at scale. The current paper proposes a practical design which empowers groups of individuals and

© The Author(s), under exclusive license to Springer Nature Switzerland AG 2021
R. Silhavy (Ed.): CSOC 2021, LNNS 229, pp. 617–628, 2021.
https://doi.org/10.1007/978-3-030-77445-5_56

organizations to design ontology-based data formats by introducing a practical versioning mechanism nurtured by distributed ledger technology innovations.

2 Related Work

A seminal work enquiring the versioning of ontologies, in the context of the Semantic Web [4], is exhibited on [17] and proposes multiple approaches on how ontology elements may be versioned. In more recent work [16], the author provides an improvement by introducing the concept negotiation strategy. A different approach is expressed in Memento protocol [29], which introduces temporal versioning of a dataset and access to the operational version through the use of a timestamp. The presented approaches rely on monolithic models in which onology components are mapped using compact structures, a design choice which undermines vertical fragmentation and increases the amount of data transferred in the network [31]. In contrast, a layered ontology that is split into multiple independent semantic structures offers a more flexible way to transfer in a network. Additionally, even though there are mentions that various participants can modify an ontology, the process of ensuring consensus among participants is not detailed.

A more formal model of the connections between ontologies [15] provides mathematical descriptions of the ontologies, information sources, and a subset of the relationships between them, such as ontology extension, the relationships between versions of the same ontology, and the dependencies between data and the ontology used to interpret it. The paper also draws parallels between the proposed formal model and the OWL [2] and RDF [20] languages.

In contrast to this current design, the paper provides a more granular versioning system tat is centered around the concepts in an ontology. Versioning and evolving an ontology can be achieved in centralized environments, and there are tools such as Protégé [22] that are leveraged for developing ontologies collaboratively. The functionality of these tools can be extended with plugins, like the one proposed in [24], that ensure versioning facilities. The disadvantages of this approach lie within the centralized client-server architecture, and in the fact that the development process involves only the developers, who are not always the end-users of the ontology.

Maintaining a particular domain's ontological classifications in a decentralized environment can be accomplished through the use of peer-to-peer networks [11,30] by following various approaches. The approach followed by [5] and [10] papers argues that peers (which represent the organizations or the individuals in the network) should form their local understanding of the domain, and that algorithmic solutions should be applied in order to align the peer ontologies when they need to communicate. Alternatively, the solution proposed in [27,32] is based on a "core ontology" deployed by a central board which observes how organizations extend the ontology locally, and the board then gradually refines it to meet the needs of the community.

With the advent of blockchain technology, the second approach can be improved-upon by removing the central board and replacing it with all of the interested organizations' decision power, which can determine the evolution of the ontology through a democratic vote. In this approach, the blockchain, being an unforgeable source of truth, can serve as the shared ontology model's repository [8].

3 Proposed Model

3.1 Overall Architecture

The distributed ontology model from [31], which this paper expands upon, is a multi-layered architecture, with each layer presenting contexts for a particular dimension of the ontology. The benefits of the decentralized approach, compared to the classic approach (centralized), rely on the data structures' accessibility. The data structures can be created and updated in a decentralized environment, which brings with it the benefits of high availability, resilience, and redundancy.

While providing some immediate benefits, decentralized ontologies come with a set of new challenges related to distributed storage [33], ensuring finality [1], and consensus involved in the versioning process. From a practical perspective, the versioning process should be transparent and reflect the multi-organizational environment needs.

By considering the design of distributed ledgers (DLT) [7] and the associated challenges, the current design proposes an ontology model that is fragmented both horizontally and vertically [31]. This feature has the beneficial effect that users can request layers of an ontology concept, instead of requesting the entire concept, diminishing the distributed network's load.

The ontology architecture is composed of the following layers:

- Structural layer - the basic layer, which consists of classifications, concepts, properties, and relations;
- Connection layer - contains information about concept mapping and ontology location;
- Encoding layer - specifies the encoding format;
- Defaults layer - stores the default values for particular properties;
- Validation layer - defines property validation rules;
- Restriction layer - holds properties restriction in the concept context;
- Naming layer - human-readable text for multi-language properties;
- Instruction layer - provides guidelines on how concept properties can be used;
- Versioning layer - represents the sets of future release updates;
- Template layer - allows for partitioning the initial schema for different contexts.

3.2 Ontology Versioning and Evolution

To ensure the evolution of an ontology in a multi-organizational context, the multilayered ontology model provides a mechanism which serves to form the core of the versioning layer. Each organization in the decentralized environment can leverage this layer to request a change to the ontology, which can consist of a single modification or a set of changes. Any ontology layer can be the subject of the change request; and additionally, these modification proposals can be applied to both concept [12] and classification [9] spaces. Apart from the demanded adjustments, the author of the proposal should submit - whenever possible - the direct and inverse mappings specifying how one can navigate between the current state of the ontology and the state, which includes the proposed adjustments.

In this context, the blockchain (or, more generally, the Distributed Ledger [7]) technology helps in achieving a trusted and highly reliable platform. It can serve as a storage medium for the ontology, or - if the size becomes unmanageable - the storage of the ontology can otherwise be delegated to a content addressable storage system such as IPFS [3], which guarantees that the contents of the retrieved files have not been tampered. By leveraging the DLT technology, a voting system can be constructed (using a smart contract) so that any participant can trust that their will is being equitably represented. In addition, all participants have access to submitted requests, even conflicting requests, so that the entire multi-organizational environment decides the future of the ontology. After the request proposals have been submitted, the community can then vote on the changes; some modifications are grouped in the same volume, and they should thus be voted on and processed as a sole entity.

There will be cases when some change requests are conflicting. In these cases, the participants should reach a consensus regarding which updates will be added to the next ontology version.

3.3 Formal Definition

Considering the following notations:

- CRS - represents an aggregation containing collections of *change requests* and *change mappings*;
- $\Lambda = \{\lambda | \lambda - layer\ type : "structural", "connection", ...\}$ - represents target layer impacted by the CRS;
- $A = \{a | a - action type : "create", "update", "delete"\}$ is the set of all possible actions that can be applied to a change request; a change request can perform only one of these actions.

In this setting, the following elements of the versioning layer model are defined:

- **Change mappings**: considering the concept c, the change request mapping is defined as a set of transformation functions that are required in order to transport property value from the current version to a future one. This is formally defined as:

$$CM_c = \bigcup_{p \in P_c} \Upsilon_{p,p'} \quad (1)$$

where $\Upsilon_{p,p'}$ is the mapping between two versions of a property belonging to concept c and p denotes the state of the property in the current version of the ontology, while p' is the state of the property in the future version of the ontology.

- **Change request**: a change request for a designated concept c is defined as a tuple:

$$CR_c = \{< r, a, \lambda > | a \in A; \lambda \in \Lambda\} \tag{2}$$

where r stands for the request payload data structure containing all the primary attributes that are required in order to update the specified layer λ, the payload can be fully-specified (for the "create" case) or partially specified (for the "update" and "delete" actions).
- **Change request set**: a change request set represents a union of change requests and mappings:

$$CRS = \bigcup_{c \in C_s} < \Upsilon_{p,p'}, < r, a, \lambda >>, \quad where\, C_s \subset C_o \tag{3}$$

In the above formula, C_o represents the set of all concepts belonging to ontology O.
The creator of the change request set decides whether or not the set should be executed atomically. An atomic change request set contains requests and mappings that are interdependent and should be executed together. A non-atomic change request set contains multiple change requests over different concepts that can be approved and executed independently.
Let *request* R be an entity that will be voted on and approved during the ontology release process. If a changeset is marked as atomic, then R becomes the whole change request set; otherwise, R is regarded as any simple mapping (Υ) or change request (CR_c) that belongs to a non-atomic changeset. Each request R is voted on and applied independently; further sections offer more details about this.

3.4 Data Model

Evolving from one version of an ontology to another requires two pieces of information. Firstly, one must describe the changes that are to be performed by using *change requests*; and secondly, one must associate forwards and backwards *mappings* which are needed to transition between the old and new versions.

In the multilayered ontology model, the versioning layer contains a collection of *change request sets* targeting the classification space and the concept space. From a design perspective, the change request set embeds change mappings and change requests. A change request set has an attribute called *atomic* that defines how the set is executed.

Mappings are usually added when there are modifications to the structural layer of a concept. A change request provides the following essential attributes: *space-type*, *action*, *layer* and *target*.

In Fig. 1, there is a full list of change request attributes. Space type can have the following values: "concept" and "classification", and it is used to discriminate between concept and classification spaces. Using the action field, one can define what kind of action will be applied; the allowed actions are: "create", "update", "delete".

Specifying which ontology layer is affected can be accomplished via the *layer* field. In the concept space, the versioning layer cannot be the scope of a change request, whereas in the classification space, only two layers are relevant: structure and naming. The identifier from the *target* field specifies which entity is going to be modified in the next version.

Additional information regarding why the modification is necessary can be given in the *comment* field; the voters and approvers use this information in order to understand the purpose of the changes.

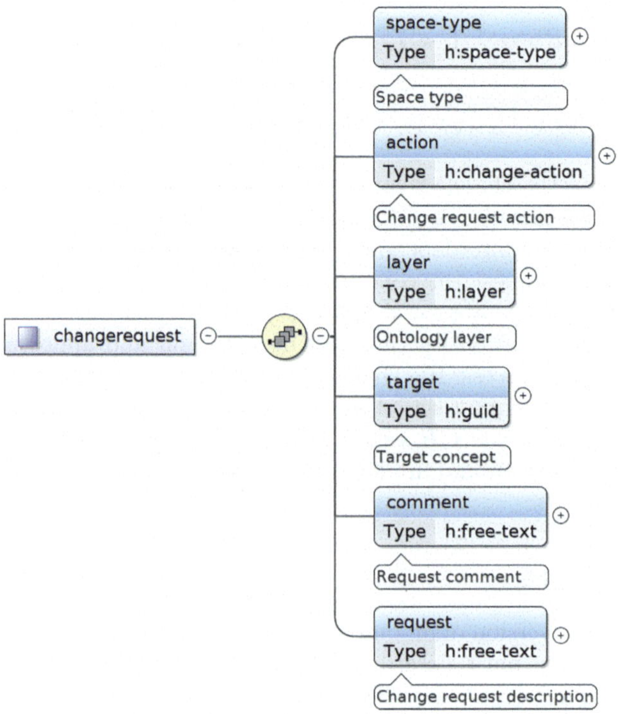

Fig. 1. Versioning layer - change request schema

The *request* field contains the change request payload, which holds the concept's modifications, and is coded in a text field. A possible implementation of the payload is a text that corresponds to the serialized representation of the concept for that particular layer. For a create action, the payload is the full concept representation for that specific layer, whereas in contrast, for an update or delete action, the full definition is not required, but only the parts that will be modified.

A mapping between two properties is a pair of forward and backward transformation functions. Figure 2 represents a transformation mapping, and contains

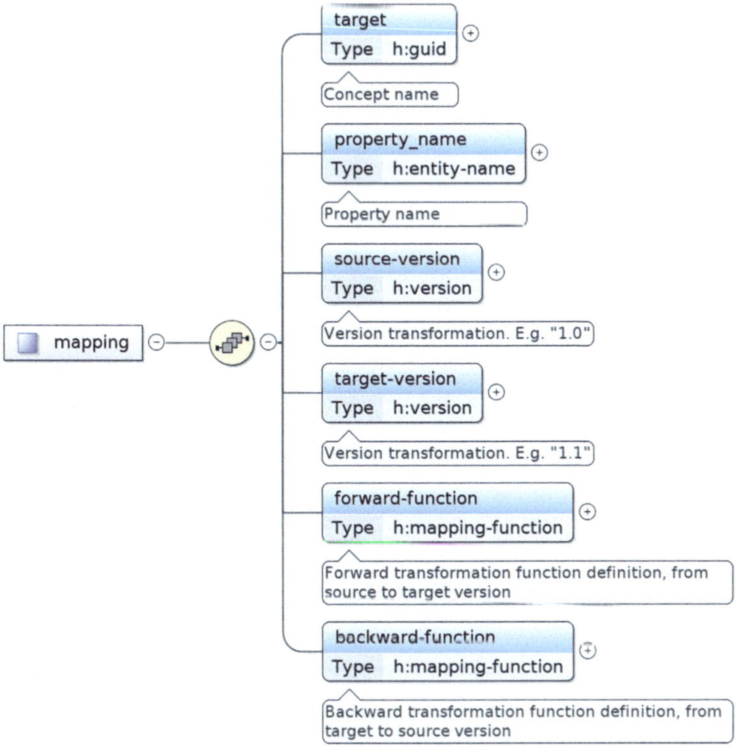

Fig. 2. Versioning layer - change mapping schema

information regarding property name, source and target version, and the mapping functions. The target field stores the link to the concept that mapping schema applies to.

The forward function maps source-version to target-version, while the backward function represents the inverse mapping. The content of any mapping function is a function definition that has the following format: "function($source, $target){ $function-body }". In the case of the forward-function, $source represents the concept from the source_version ontology, and $target is the concept from the target_version, while for the backward-function, the mapping is reversed.

The format of the $function-body is highly correlated with the ontology implementation, and a Turing complete language [25] is recommended for it, so that it may perform any transformation. In addition, for performance reasons, mapping functions should be as simple as possible.

3.5 Voting System

Deploying ontologies in a decentralized environment brings forward a set of challenges, one of which is ensuring that the ontologies undergo a process of gradual

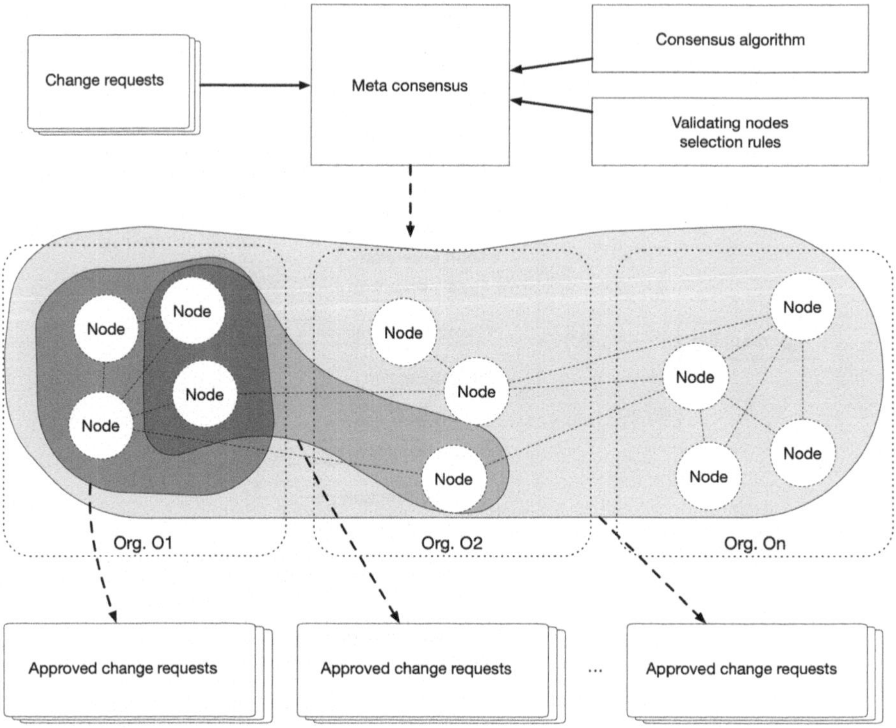

Fig. 3. Versioning layer - consensus mechanism

refinement which reflects the will of the participants. However, this hurdle can be overcome by implementing the use of a voting mechanism. The voting system is also necessary because the ontology is a means of establishing communication among heterogeneous organisations; and so, the maintenance of this communication medium should be an effort undertaken by the whole community.

Blockchain technology ensures a shared view of the network state, and can be leveraged to facilitate the voting process. One way in which the blockchain could prove beneficial is through the use of smart contracts [34]. For example, a smart contract can be developed to hold the change request sets. As all data on the ledger is publicly-available, members of the community will be able to query for the latest sets and decide on their adequacy.

In the same way, another smart contract should tally the votes of the proposition under debate and provide the result of the vote after a specified period. The following formula describes an individual voter u that casts a vote w_i for request R.

$$v_{R_i} = \begin{cases} +1/p, \; u \, agrees \, with \, R \\ -1/(1-p), \; u \, disagrees \, with \, R \end{cases} \tag{4}$$

where p is the required percentage of positive votes needed to accept the request, a percentage that is convened-upon in advance by the community. The final result of the voting employs the utilization of a simple summation of individual votes:

$$w_R = \sum_{u \in Co} v_{R_i} \tag{5}$$

In the above formula, C_0 represents the entire community, and u is any individual/organization in the community. In order to decide which propositions have the appropriate community support, a limit function λ is introduced, which allows for the filtering out of the requests that do not have a minimum number of participants - but the organizations can agree to overwrite it. The default version is:

$$\lambda = \frac{t_{w_R}}{log_2(a)} > 1 \tag{6}$$

where t_{w_R} is the total number of votes for request R and a is the number of members. By introducing a timestamp t, at which the request R is evaluated, a binary function can describe the results of the vote as follows:

$$Result_R{}^t = \begin{cases} passed, w_R \geq 0 \, and \, \lambda = true \\ rejected, w_R < 0 \, or \, \lambda = false \end{cases} \tag{7}$$

3.6 Consensus Mechanism

The voting mechanism does not ensure that the result of an accepted request R (utilizing voting) will be materialized in the next version of the ontology. There are situations in which multiple change requests meet the voting requirements, but they propose conflicting views of the ontology. This use-case requires a decision from a high-level perspective, and needs to involve all of the affected participants in order to reach an agreement regarding the conflict.

Additionally, there are multiple circumstances where not all of the participants are directly interested/impacted by a particular change on the ontology structure. In these particular situations, there is no need for full participation to reach a consensus, and complementary, different (not necessarily disjointed) groups of experts and interested users can take part in the maintenance and evolution processes.

These scenarios would benefit from using a dynamic consensus mechanism [6], where users can make multiple decisions over different parts of an ontology in parallel.

Figure 3 presents the multi-organization decentralized environment in which each organization has its nodes involved in the network, each storing parts of the ontology. Distinct levels of grey represent distinct communication channels that are topic oriented. So, if an organization wants to have a say in a particular section of an ontology, it should subscribe its peer to the face detection topic. The system's governance is ensured by a meta consensus logic responsible for selecting the nodes and the consensus algorithm. The input change requests are distributed across the topic-oriented channels. The result of the process will create approved change requests used as input for the new ontology version.

When a conflicting request R appears in a section of the ontology, those who are interested in the topic mapped to that particular section will follow the rules of a consensus mechanism that has been agreed-upon by all of them, as this process ensures that a viable decision is reached. Two properties emerge from the process described above. Firstly, different channels can use different consensus mechanisms (as the participants in the channels see fit), and secondly, nodes which subscribe to distinct topics of the ontology can make multiple decisions in parallel, thus expediting the versioning procedure.

4 Conclusion

Efficient communication in a multi-organizational environment is achievable only if the participants share a common understanding of the data, processes, and concepts exchanged through the business interactions. The shared information vision acts as a mechanism aimed to ensure interoperability and cross-organization integrations. While semantic models achieved some encouraging results, their practical application remains limited because of a lacking efficient means of communication. The current architecture combines the blockchain specific features with the ontology formalism to empower data exchange. Organization data formats are evolving in time; a core feature required by the before-mentioned systems represents the possibility to decide about ontology progress jointly.

The current paper a decentralized ontology versioning model with the following unique characteristics:

- The formal framework for the elements of the versioning layer of a multilayered decentralized ontology
- An accompanying data model to meet the requirements of the formal definitions
- A new approach that utilizes the advantages of blockchain technology to democratize the evolution process of an ontology
- A mechanism resolving conflicting modification requests employing blockchain consensus mechanisms

Future iterations of this project aim to provide a practical implementation of the described model, including comparing the various blockchain alternatives and deploying a full-fledged ontology model.

References

1. Anceaume, E., Pozzo, A., Rieutord, T., Tucci-Piergiovanni, S.: On finality in blockchains. arXiv preprint arXiv:2012.10172 (2020)
2. Bechhofer, S., et al.: Owl web ontology language reference. W3C Recommendation **10**(02) (2004)

3. Benet, J.: Ipfs-content addressed, versioned, p2p file system. arXiv preprint arXiv:1407.3561 (2014)
4. Berners-Lee, T., Hendler, J., Lassila, O.: The semantic web. Scientific american **284**(5), 34–43 (2001)
5. Bonifacio, M., Bouquet, P., Mameli, G., Nori, M.: Peer-mediated distributed knowledge management. In: van Elst, L., Dignum, V., Abecker, A. (eds.) International Symposium on Agent-Mediated Knowledge Management. Lecture Notes in Computer Science, pp. 31–47. Springer, Berlin, Heidelberg (2003). https://doi.org/10.1007/978-3-540-24612-1_2
6. Butean, A., Pournaras, E., Tara, A., Turesson, H., Ivkushkin, K.: Dynamic consensus: Increasing blockchain adaptability to enterprise applications. In: Computer Science On-line Conference, pp. 433–442. Springer, Cham (2020). https://doi.org/10.1007/978-3-030-51974-2_41
7. Casino, F., Dasaklis, T.K., Patsakis, C.: A systematic literature review of blockchain-based applications: current status, classification and open issues. Telematics Inform. **36**, 55–81 (2019)
8. De Kruijff, J., Weigand, H.: Understanding the blockchain using enterprise ontology. In: International Conference on Advanced Information Systems Engineering, pp. 29–43. Springer (2017). https://doi.org/10.1007/978-3-319-59536-8_3
9. Dumais, S., Chen, H.: Hierarchical classification of web content. In: Proceedings of the 23rd Annual International ACM SIGIR Conference on Research and Development in Information Retrieval, pp. 256–263 (2000)
10. Ehrig, M., et al.: The swap data and metadata model for semantics-based peer-to-peer systems. In: German Conference on Multiagent System Technologies, pp. 144–155. Springer (2003). https://doi.org/10.1007/978-3-540-39869-1_13
11. Fang, Q., Zhao, Y., Yang, G., Zheng, W.: Scalable distributed ontology reasoning using dht-based partitioning. In: Asian Semantic Web Conference. pp. 91–105. Springer (2008). https://doi.org/10.1007/978-3-540-89704-0_7
12. Fudholi, D.H., Rahayu, W., Pardede, E., et al.: A data-driven approach toward building dynamic ontology. In: Information and Communication Technology-EurAsia Conference. pp. 223–232. Springer (2013). https://doi.org/10.1007/978-3-642-36818-9_23
13. Grangel-González, I., et al.: The industry 4.0 standards landscape from a semantic integration perspective. In: 2017 22nd IEEE International Conference on Emerging Technologies and Factory Automation (ETFA), pp. 1–8. IEEE (2017)
14. Hayes, J.G., Peyrovian, E., Sarin, S., Schmidt, M.T., Swenson, K.D., Weber, R.: Workflow interoperability standards for the internet. IEEE Internet Comput. **4**(3), 37–45 (2000)
15. Heflin, J., Pan, Z.: A model theoretic semantics for ontology versioning. In: International Semantic Web Conference. pp. 62–76. Springer (2004). https://doi.org/10.1007/978-3-540-30475-3_6
16. Hyland, B., Atemezing, G., Villazón-Terrazas, B.: Best practices for publishing linked data, (2014). http://www.w3.org/TR/ld-bp
17. Klein, M.C., Fensel, D.: Ontology versioning on the semantic web. In: SWWS, pp. 75–91 (2001)
18. Kosanke, K.: Iso standards for interoperability: a comparison. In: Interoperability of enterprise software and applications, pp. 55–64. Springer (2006). https://doi.org/10.1007/1-84628-152-0_6

19. Kosanke, K., Nell, J.G.: Enterprise Engineering and Integration: Building International Consensus: Proceedings of ICEIMT 1997, International Conference on Enterprise Integration and Modeling Technology, Torino, Italy, 28–30 October 1997. Springer Science & Business Media (2012)
20. Lassila, O., Swick, R.R., et al.: Resource description framework (rdf) model and syntax specification (1998)
21. Lu, Y., Asghar, M.R.: Semantic communications between distributed cyber-physical systems towards collaborative automation for smart manufacturing. J. Manuf. Syst. **55**, 348–359 (2020)
22. Musen, M.A.: The protégé project: a look back and a look forward. AI matters **1**(4), 4–12 (2015)
23. Mykkänen, J.A., Tuomainen, M.P.: An evaluation and selection framework for interoperability standards. Inf. Softw. Technol. **50**(3), 176–197 (2008)
24. Noy, N.F., Chugh, A., Liu, W., Musen, M.A.: A framework for ontology evolution in collaborative environments. In: International semantic web conference. pp. 544–558. Springer (2006). https://doi.org/10.1007/11926078_39
25. Ofria, C., Adami, C., Collier, T.C.: Design of evolvable computer languages. IEEE Trans. Evol. Comput. **6**(4), 420–424 (2002)
26. Ouksel, A.M., Sheth, A.: Semantic interoperability in global information systems. ACM SIGMOD Rec. **28**(1), 5–12 (1999)
27. Pinto, H.S., Tempich, C., Staab, S.: Ontology engineering and evolution in a distributed world using diligent. In: Handbook on ontologies, pp. 153–176. Springer (2009). https://doi.org/10.1007/978-3-540-92673-3_7
28. Sheth, A.P.: Changing focus on interoperability in information systems: from system, syntax, structure to semantics. In: Interoperating geographic information systems, pp. 5–29. Springer (1999). https://doi.org/10.1007/978-1-4615-5189-8_2
29. Van de Sompel, H., Nelson, M.L., Sanderson, R.: Http framework for time-based access to resource states-memento. Rfc 7089, IETF (2013)
30. Tao, Y., Jin, H., Wu, S., Shi, X.: Scalable dht-and ontology-based information service for large-scale grids. Futur. Gener. Comput. Syst. **26**(5), 729–739 (2010)
31. Tara, A., Butean, A., Zamfirescu, C., Learney, R.: An ontology model for interoperability and multi-organization data exchange. In: Computer Science On-line Conference, pp. 284–296. Springer (2020). https://doi.org/10.1007/978-3-030-51971-1_23
32. Tempich, C., Pinto, S., Staab, S., Sure, Y.: A case study in supporting distributed, loosely-controlled and evolving engineering of ontologies (diligent). In: Proceedings of the 4th International Conference on Knowledge Management (I-KNOW 2004), pp. 225–232 (2004)
33. Uriarte, R.B., DeNicola, R.: Blockchain-based decentralized cloud/fog solutions: Challenges, opportunities, and standards. IEEE Commun. Stan. Mag. **2**(3), 22–28 (2018)
34. Wang, S., Yuan, Y., Wang, X., Li, J., Qin, R., Wang, F.Y.: An overview of smart contract: architecture, applications, and future trends. In: 2018 IEEE Intelligent Vehicles Symposium (IV), pp. 108–113. IEEE (2018)

Digital Transformation in HR Management

Elena Nikolaevna Sochneva[1(✉)], Anna Andreevna Malakhova[1], Dmitry Ivanovitch Kravtsov[1], Kolesnikova Natalya Anatolyevna[2], Anastasiya Valer'evna Kutuzova[3], Dmitry Valeryevitch Zyablikov[1], and Olga Valeryevna Starova[1]

[1] Siberian Federal University, Krasnoyarsk, Russia
[2] The Academy of Labor and Social Relations, Moscow, Russia
[3] Krasnoyarsk Institute of Railway Transport - Branch of the Irkutsk University of Railway Transport, Krasnoyarsk, Russia

Abstract. This article examines the analysis of the introduction of teleworking in Russia, which has accelerated under the influence of the digitalization of the economy. Remote work in Russia is developing in the same directions as in foreign countries. The study showed that in Russia there are similar legal structures in telecommuting. Remote work can be for a certain period, as well as with the possibility of alternating remote and traditional forms. Today, both Russian and foreign legislation is developing in the direction of clarifying the rights and obligations of teleworkers, as well as the legal possibilities of employers in relation to them. When introducing remote work, a number of problems arose. In particular, an economic mechanism is needed to compensate for the cost of equipment owned by a remote worker. It is also advisable to substantiate the working hours of remote workers and their labor standards. Taking into account the introduction of electronic document management and other attributes of digitalization, remote work will be greatly developed in the near future.

Keywords: Digitalization · Personnel management · Teleworking · Electronic document management · Compensation for equipment cost · Labor rationing

1 Introduction

The past coronavirus pandemic has significantly changed many processes in the economy, and many activities have been filled with new meaning and content. Global economies have changed, in many countries a new institutional environment has begun to be created in connection with the changes. First of all, the coronavirus pandemic has given impetus to the development of digitalization. The processes of digitalization of the economy were already inevitable, in fact, they represent the basis of a new stage in the development of the economy. The coronavirus epidemic has accelerated and catalyzed the inevitable processes of global economic changes. It touched to one degree or another all kinds of activities.

This article presents the most important changes in traditional work within the framework of the type of HR management activity in Russia in connection with digitalization, and shows the problems and prospects for the development of these transformations.

© The Author(s), under exclusive license to Springer Nature Switzerland AG 2021
R. Silhavy (Ed.): CSOC 2021, LNNS 229, pp. 629–643, 2021.
https://doi.org/10.1007/978-3-030-77445-5_57

So, digitalization has given an impetus to the transition of organizations to electronic document management, uncertificated employment of workers, and, even, the possibility of exercising control and supervision over the social and labor sphere through contactless electronic technologies.

During the coronavirus pandemic, it became necessary to transfer part of the workers and business processes to a remote format. Today, this way of working has already largely shown its economic efficiency, thanks to cost savings. This is due to savings in floor space, transportation of workers to and from the workplace, workplace equipment, safety training, and other cost savings in doing business. Thus, telecommuting has significant prospects, and it will develop.

In foreign countries, the definition of "telework" is most often used for remote work. The European Teleworking Organization has proposed a classification in which four types of teleworking have been identified.

1. Home-based telework (telework at home). This type of telecommuting involves workers who work from home one day a week rather than at their workplace, using a computer and telecommunications links to interact with colleagues from home.
2. Mobile telework (mobile telework). This type of telecommuting involves employees staying outside their workplace and away from home for at least 10 h a week (usually on business trips), as well as the use of computer communication connections during this time.
3. Telecommuting in telecottages, which provide people in the community with learning opportunities, as well as high-performance information and computer technologies.
4. Telework by self-employed in SOHOs (telework for self-employed in small offices - home offices), which is a combination of self-employment and telework (in Russia, it is analogous to home work) [1].

2 Methods

2.1 Comparison of Telework in Russia and Abroad

It should be noted that in Russia there are legal structures similar to foreign ones, only they are called "flexible forms of employment" and are scattered in different parts of the Labor Code. This means that all countries are moving along the path of transforming the labor process, and, if possible, separating it from a rigidly fixed workplace (Table 1).

Nowadays it is generally accepted that telecommuting (teleworking) is more suitable for intellectual work. But in the context of the development of information technology, robotics and technical cybernetics, as well as the emergence of "smart" systems to support business processes, a time will follow when all work processes will be conducted through remote control of a person.

The authors of the article conducted a survey among students of the Siberian Federal University, studying in the direction of Management, as to how they see the processes of organizing work in the future after the coronavirus pandemic. 90 people took part in the survey, the period is April–May 2020, that is, just in the midst of the coronavirus pandemic. The following answer was received on average.

The most general characteristics in terms of work processes include the following.

Table 1. Comparative analysis of telework forms of employment in Russia and abroad

№	Telework forms in European countries	Russian legal structure, similar to foreign
1	Home-based telework (telework at home). This type of telecommuting involves workers who work from home one day a week rather than at their workplace, using a computer and telecommunications links to interact with colleagues from home	Remote work (Ch. 49.1 of the Labor Code of the Russian Federation, Art. 312.1 of the Labor Code of the Russian Federation. The employee's performance of the labor function remotely periodically, provided that the periods of the employee's performance of the labor function remotely and the periods of performance of the labor function at the stationary workplace are alternated
2	Telecommuting in telecottages that provide community-based learning opportunities as well as high-performance information and communication technologies	Remote work (Ch. 49.1 of the Labor Code of the Russian Federation, Art. 312.1 of the Labor Code of the Russian Federation. The employee's performance of the labor function remotely on an ongoing basis (during the term of the employment contract) or temporarily (continuously during the period specified by the employment contract or supplementary agreement to the employment contract, not exceeding six months
3	Mobile telework (mobile telework). This type of telecommuting involves employees staying outside their workplace and away from home for at least 10 h a week (usually on business trips), as well as using computer communication connections during this time	Irregular working hours, traveling nature of work, in Russia the exact number of hours is not regulated
4	Telework by self-employed in SOHOs (telework for the self-employed in small offices - home offices) which is a combination of self-employment and telecommuting	Home work, Article 310 Labor Code

1. A certain part of the employees will now constantly work remotely, and the main criterion for the effectiveness of their activities will be the result, not the time spent. Consequently, the questions of how to properly arrange a remote worker are being actualized. How to provide him with the necessary funds to carry out work (if necessary), how to create working conditions for such workers, how to carry out safety and labor protection measures with this employee, etc. Transferring employees to a remote form will significantly reduce the costs of employers. They will not have to pay for rent and maintenance of premises, delivery of workers to the place of work, security, utility bills, etc. Now employers have understood and adapted their workers to teleworking and many will not leave it.

2. Development of online commerce. Merchants will have to reformat themselves for e-commerce. For this, it will be necessary to develop online fitting rooms, retrain employees, and develop a new sales business model.
3. Reformatting shopping centers. They will be irrelevant after the pandemic, and they will have to organize coworking areas or recreation areas in order to survive.
4. Development of online learning. The pandemic has given impetus to distance learning in all areas, from professional to continuing education. The existing experience has shown the effectiveness of distance forms and for many they seemed very convenient, because now you can improve your educational status and expand competencies without leaving your home. The pandemic will give an impetus to the development of distance education and improve the forms and methods of its implementation, and will give an impetus to the development of digitalization in this area.
5. Instead of regular couriers delivery of food. drugs and other goods will be handled by drones. For this, it will be necessary to develop legislation on the legality of the work of drones in Russia. A similar experience of using drones has already been in the United States.
6. Since the entire banking system is under threat during the crisis, in the foreseeable future cryptocurrencies, that is, monetary units that exist only in electronic form, will become the main payment units.
7. Finances should be invested in IT companies, as with the transition to remote work and training, the demand for them has become even greater.
8. Robotics and smart technologies will develop.

Thus, in the future, the transfer of all types of activities to remote modes is obvious. There will be very few activities that will require "manual" execution. Working hands will take control of robotic machines.

In connection with the development of telecommuting in Russia, the question has been raised: how to legalize it. In particular, what should be written in the supplementary agreement (employment contract)? How to calculate compensation to an employee for the cost of equipment, materials and other necessary funds used? How to protect the rights of an employee in a teleworking environment? And much more.

Government agencies began to respond to these economic needs. Thus, draft law No. 966659-7 "On Amendments to Article 57 of the Labor Code of the Russian Federation" has been prepared. It says about the need, as a prerequisite, when concluding a contract with a remote worker, to indicate the period during which the employee temporarily performs work in a remote form, or a schedule according to which the employee partially performs work in a remote form. The conditions for performing work in a remote form are also necessary, including the working hours, methods and means of communication between the employee and the employer, the procedure for providing the employee with equipment, tools, technical documentation and other means and items necessary for the performance of his job duties, or providing appropriate compensation, as well as the procedure for the employee's presentation of the results of the work performed and their acceptance by the employer [2].

In addition, in December 2020, Russia issued a federal law "On Amendments to the Labor Code of the Russian Federation regarding the regulation of remote (remote) work and temporary transfer of an employee to remote (remote) work on the initiative of the employer in exceptional cases", which entered into effective from January 1 of the current year [3]. This federal law amends Chapter 49.1 of the Labor Code of the Russian Federation, which regulates the work of remote workers.

Now remote work (telework) in Russia, by analogy with foreign countries, is divided into forms. In Russia, there are three of them: 1) remote work on a permanent basis; 2) temporary remote work for up to six months; 3) periodic remote work, involving the alternation of periods of performing work remotely and at a stationary workplace.

It also specifies the ways of interaction between a remote worker and an employer when concluding an employment contract and performing remote work, the grounds for terminating an employment contract with such an employee.

Federal law also establishes guarantees for remuneration of a remote worker, payment of travel expenses when sending him to another locality to carry out an official assignment and providing him with equipment, software, hardware and other means necessary for performing remote work at the employer's expense.

Thus, now in the local regulatory legal acts of the organization will be spelled out the conditions that must be included in the employment contract (additional agreement) when transferring an employee to remote work. It will also indicate the period for which the employee was transferred to a remote form, the mode of work, the conditions for the exchange of information between the employee and the employer, methods and forms of providing labor results, criteria for assessing the quality of labor, a mechanism for compensating the employee for the equipment, materials used, etc.

In addition, the law under consideration determines the procedure for the temporary transfer of an employee to remote work at the initiative of the employer (without the consent of the employee) in the event of a natural or man-made disaster, industrial accident, industrial accident, fire, flood, earthquake, epidemic or epizootic and in any exceptional in cases that endanger the life or normal living conditions of the entire population or its part, as well as in the event that an appropriate decision is made by a public authority and (or) a local self-government body. At the same time, if the specifics of the work do not allow the temporary transfer of the employee to remote work or the employer cannot provide the employee with the necessary equipment and funds, the time during which such an employee does not perform his job is paid according to the rules for payment of downtime for reasons beyond the control of employer and employee, if a larger amount of payment is not provided for by collective agreements, agreements, local regulations.

Foreign legislation has also taken the path of transformation due to the inevitable transition to remote forms of work.

Thus, the German government plans to secure the human right to work remotely, but at the same time it will not give the employer the right to force an employee to perform tasks remotely. This should be the employee's own choice. The draft law, which the minister lobbies and proposes to submit for approval this fall, provides for freedom of choice for the employee.

He can stay in the office and perform his work according to the requirements of the company, or he can switch to a remote type of employment, if the conditions for fulfilling the employer's tasks allow it.

In addition, a temporary transition to a remote type of work or with performing work at home from one to several days is assumed.

Note that in Russian legislation the conditions for working remotely or in full-time format are determined by the employer. Also, the employer decides to transfer existing employees to a remote format.

In theory, an employee may not accept this. But, then, according to Russian law, the employment contract with him will be terminated.

At the same time, in Germany, they intend to establish restrictions in the legislation so that work does not absorb a person's private life.

To do this, it is necessary to clearly define the time frame for performing work with a remote mode of employment. In addition, the law will explicitly prohibit forced labor remotely in a private house or apartment of a worker.

They positively assess the economy's ability to distribute resources so that people can work remotely. This form of employment is very convenient and comfortable for some workers. It not only creates benefits for business, but also reduces pressure on urban infrastructure, allows a person to more freely manage their personal time [4].

In order to assess how effective a remote worker is and how to manage it, it is necessary to develop operating modes and calculate scientifically based norms of time for each type of work. The norms can be of a recommendatory nature, but they should be at the level of benchmarks for employers to understand the efficiency of work and the degree of workload of the employee.

The possibility of freedom of choice for the employee: to work remotely or in the office, certainly deserves attention in the framework of labor relations in modern Russia. It's no secret that not everyone in Russia has conditions for working remotely. Let's hope that Russian labor legislation will provide for freedom of choice for the employee in this matter.

The disadvantage of the existing amended legislation in modern Russia is that the lists of conditions under which the employer can unilaterally terminate the employment contract with the employee are not stipulated. Those lists of reasons for the termination of labor relations on the part of the employer unilaterally, which today exist in labor legislation (Article 81 of the Labor Code of the Russian Federation), are not suitable for teleworkers.

2.2 Methods for Calculating Compensation for Equipment Used by a Remote Worker

The most important question from the point of view of the economy, which arises in connection with these changes: how to properly compensate the employee for the costs of communications, materials and depreciation of his equipment. In this case, one can be guided by sound logic and practice prevailing in economics.

It is already known that in accordance with Art. 188 of the Labor Code of the Russian Federation, when an employee uses personal property with the consent or knowledge of the employer and in his interests, the employee is paid compensation for the use, depreciation (depreciation) of tools, personal vehicles, equipment and other technical means and materials belonging to the employee, as well as reimbursed the costs associated with their use. The amount of reimbursement of expenses is determined by the agreement of the parties to the employment contract, expressed in writing.

Also, in accordance with article 312.9 of the Labor Code of Russia, in certain cases related to the category of emergency, the employer can unilaterally transfer the employee to remote work, but then it is necessary to provide the latter with all the necessary engineering and technical means and equipment, or pay compensation to the employee in case. if he uses his equipment and the necessary facilities. The mechanism for calculating such compensation is left to the discretion of the employer.

From the point of view of economic feasibility, the amount of monthly compensation for the use of personal property should be established on the basis of the estimated amount of depreciation on it, that is, taking into account its value and useful life (in accordance with the Classification of fixed assets included in depreciation groups, approved by the Decree of the Government of Russia dated 01.01.2002 No. 1) [5]. For example, let's calculate the amount of compensation that is due to a remote worker for using a personal laptop for work purposes.

The useful life of computer equipment and printing devices is over two years up to three years inclusive (OKOF code 330.28.23.23 "Other office machines"[6].

Obviously, the laptop is used not only for work, but also for personal purposes, so only working hours should be included in the calculation of compensation for its use.

For example, in April - this is 24% of the total time (175 working hours/(30 cal. Days * 24 h) = 24%), in May - this is 18% of the total time (135 working hours/(31 cal days * 24 h) = 18%).

Thus, compensation for the use of a personal laptop for work purposes worth 56,000 rubles. in April will amount to 373.33 rubles. (56,000/36 * 24% = 373.33 rubles), in May - 280.00 rubles. (56,000/36 * 18% = 280.00 rubles).

In a formalized form, the calculation of the employee's working time ratio can be presented as:

$$Kwork = \frac{t_{work}}{tgen}, \tag{1}$$

где K_{work} - working time utilization;
t_{work} - he time that the employee performs labor functions in accordance with the work schedule;
tgen - general fund of working time.

It is known that along with the wear and tear of the technical means used in the work, the employee bears the costs of mobile communications, Internet, electricity, cartridges and paper for printing (if required by the workflow), office supplies, and other materials.

In the costs of telephony and the Internet, as well as when using technology, one should allocate the amount that falls on working hours.

So, with a tariff plan worth 500 rubles. per month the employee will be compensated in April 120 rubles. (500 * 24% = 120 rubles), in May - 90 rubles. (500 * 18% = 90 rubles).

Expenses for cartridges, paper, stationery can be compensated in a fixed amount (there must also be a reasonable calculation) or according to the checks presented by the employee.

Since compensation to a remote worker for the use of personal property for business purposes is a requirement of the legislation of the Russian Federation and the condition for its payment must be spelled out in the employment contract (additional agreement) with the employee, it is taken into account in labor costs on the basis of paragraph 25 of Art. 255 of the Tax Code of the Russian Federation, like other types of payments provided for by an employment contract and made in favor of the employee. At the same time, it should be noted that the Ministry of Finance of Russia allows such compensation to be accounted for as other expenses on the basis of paragraphs 49 p. 1 of art. 264 Tax Code (letter of the Ministry of Finance of Russia dated 08.12.2006 No. 03-03-04/4/186).

In both of these cases, it is important to understand that one way or another compensation payments can be included in the expenses of the organization and affect the lowering of the tax base for corporate income tax.

Amounts of compensation for the use of personal property by employees for work purposes and other costs for remote work:

- are not subject to personal income tax (PIT) on the basis of par. 10 p. 1 art. 217 of the Tax Code of the Russian Federation (see also the letter of the Ministry of Finance of Russia dated November 27, 2015 No. 03-04-06/69014); –
- are not subject to contributions to insurance funds from the amount of wages on the basis of par. 10 pp. 2 p. 1 art. 422 of the Tax Code of the Russian Federation and para. 13 pp. 2 p. 1 art. 20.2 of the Federal Law of 24.07.1998 No. 125-FZ.

At the same time, the Ministry of Finance of Russia believes that in order to exempt the amount of compensation for the use of personal property of employees from personal income tax and insurance premiums, the compensation condition must be spelled out in the employment contract (supplementary agreement). You also need documents confirming the employee's ownership of the property used, documents confirming the actual costs incurred, calculations of compensation (letters of the Ministry of Finance of Russia dated 12.09.2018 No. 03-04-06/65168, dated 30.04.2015 No. 03-04-05/25434, dated 17.06.2009 No. 03-04-06-01/138).

It is important that the condition that the employee owns the property used in the work is mandatory only if compensation is paid on the basis of Art. 188 of the Labor Code of Russia. For teleworkers, Art. 312.3 of the Labor Code of Russia expressly stipulates that they are paid compensation for the use of not only their own, but also rented property. Therefore, it will not be subject to personal income tax and insurance premiums even in the case when the property used in the work is not personal. But then it is necessary to confirm that it is rented. In any case, it may be a problem for employees to confirm ownership of the equipment used or the fact of purchase for rent. Not all citizens keep receipts and settlement documents for purchased equipment. In addition, how then to be with the equipment that was generally taken from relatives or friends and no documents, of course, were not issued. In any case, the legislation has not been finalized. In this case, it would be more logical to proceed not from the message who is the owner of this equipment, but about the fact that the employee has this equipment and is willing to do work on it. The compensation can be set based on the average market price of the given equipment, which is not difficult to do by analyzing the market.

To obtain the average market price, it is necessary to obtain at least three quotations for such equipment (or be guided by the concepts of identity and homogeneity from the Tax Code). Further, it is advisable to process these price offers using the coefficient of variation in order for the selected data set to be homogeneous, and then calculate the arithmetic mean.

The price variation coefficient is determined by the following formula:

$$V = \frac{\sigma}{[P]} \times 100\%, \quad (2)$$

where:

V - is the coefficient of variation;
σ- standard deviation;σ
[P] - average price level.

$$\sigma = \sqrt{\frac{\sum_{i=1}^{n} (P_i - [P])^2}{n}}, \quad (3)$$

P_i - is the price of a unit of goods, work, services indicated in the source with the number i;
[P] - the arithmetic mean of the price of a unit of goods, work, services;
n - is the number of values used in the calculation.

The coefficient of variation can be calculated using the standard functions of the spreadsheet editors. The set of values used in the calculation is considered heterogeneous in determining the NMCK if the price variation coefficient exceeds 33%. If the coefficient of variation exceeds 33%, it is advisable to conduct additional research in order to increase the amount of price information used in the calculations.

The average market price by the method of comparable market prices (market analysis) is determined by the formula:

$$P_{av.mark} = \frac{v}{n} \times \sum_{i=1}^{n} P_i, \quad (4)$$

where: $P_{av.work.}$ - average market price determined by market analysis;
v - is the quantity (volume) of the purchased goods (work, services);
n - is the quantity (volume) of the purchased goods (work, services);
i - number of the source of price information;
P_i - the price of a unit of goods, work, services, presented in the source with the number i.

Thus, you can always calculate the initial market value of any equipment, in the absence of financial documents confirming its purchase. However, this raises another important practical question: how to repair the equipment on which the remote worker performs his work? Of course, in local regulatory legal acts, this issue should be worked out at the organization level. But it is much more rational to consolidate this position in federal legislation, namely, the rights and obligations of the employee and the employer when conducting remote work.

There is a similar experience abroad. In March 2020, Chile passed a law regulating telecommuting, which spelled out the rights and obligations of an employee and an employer. The obligation to bear the costs of organizing production and ensuring the functioning, maintenance and repair of equipment was imputed to the employer [7].

It is also important to note that if, when determining depreciation costs, price information is used that is a rock of prescription more than six months ago, then it is necessary to use the indexation mechanism according to Rosstat price indices. It is advisable to use producer price indices for indexing.

$$P_{real} = \frac{P_n}{I_{pp}}, \tag{5}$$

where Preal is the price of the product, adjusted for the producer price index;

Ipp - producer price index, calculated according to the data of the Russian statistical service for the required period.

If statistical data is presented with a monthly breakdown, then the final price index is calculated using the formula:

$$I_{pp} = \prod_{i=1}^{n} I_i, \tag{6}$$

where Ii is the price index for the corresponding billing period;

n - is the number of periods for which statistical reporting data are used.

If statistical data are presented for a year, then the calculation of the price index, according to which it is necessary to bring the price of goods to a comparable form, will be carried out according to the formula:

$$I_{pm} = \sqrt[12]{I_{py}}, \tag{7}$$

where Ipm - is the average monthly price increase;

Ipy - price index for the year according to statistical reporting data.

Then the final price index required to calculate the real value of the goods will be calculated using the formula:

$$I_{pp} = I_{pm}^{n}, \tag{8}$$

2.3 Labor Safety of Remote Workers

Thus, the problem of compensating for the costs of a remote worker can be solved within the framework of modern legislation. But another global problem is being actualized, which has not yet been resolved. We are talking about the protection of the employee within the framework of labor protection and social insurance of the employee from accidents at work and occupational diseases. How can a remote worker be instructed on labor protection? Already now in the legal literature there is a dispute about the need for such instructions. Thus, a methodological study of such issues is needed.

According to the position of the Ministry of Labor and Social Protection of the Russian Federation, teleworkers and their employers are subject to all provisions of the Labor Code of the Russian Federation regulating labor protection.

According to the International Labor Office, a subdivision of the International Labor Organization, employers have a responsibility to take care of all their workers and, to the extent practicable, to provide a safe work environment that does not pose a risk to physical and mental health. This includes analyzing risks and taking control and counteraction measures in locations outside of the normal production facility, such as the home of a remote worker [8].

The remote worker, like any other worker, needs to be trained in the safe ways of working on production equipment. With him, you need to conduct, inter alia, introductory and primary briefings, allowing the employee to safely carry out production activities in the remote work mode [9].

The next question is how to qualify a production accident for a teleworker. The federal law in Russia "On compulsory social insurance against industrial accidents and occupational diseases" of July 24, 1998 N 125 does not yet contain any clarifications in this regard [10].

It would be most logical to fix in the specified law in the list of insurance situations that are subject to compensation from the Social Insurance Fund, the work of a remote worker within the framework of his established working hours. Recall that the working hours are now clearly defined by local regulations and the employment contract of the employee himself.

2.4 Labor Rationing

Labor rationing is the establishment of reasonable rates of production, time, service, manageability, number, etc. It is also a rationale for rationed tasks for workers working on a time-based wage system.

In order to justify the remuneration of any employee, it is necessary to determine the time frame and quantitative parameters of his work. This is of particular relevance for a teleworker. Since the employer is deprived of the possibility of direct control over the progress of his work. This means that in this situation, reasonable labor standards are needed. In the context of remote work, the most important are standardized tasks. At the moment, remote forms of employment, first of all, include forms of intellectual labor, therefore, it is impossible to determine the norms of production.

The standardized tasks, which are most appropriate for the rationing of the labor of remote workers, must be developed and approved in each organization where the labor of remote workers is used. An effective method of setting standards in this case will be the photograph of working hours, as a method of setting labor standards.

When analyzing and processing photographs, it is necessary to assess the effectiveness of the use of working time by employees.

1. Coefficient of use of working time. It shows: what percentage of the working time was used productively by the employee.

$$Tuse = \frac{T_{pf} + T_m + T_s + T_r}{T_g}, \quad (9)$$

где T_{use} - is the working time utilization factor;
T_{pf} - preparatory and final time;
T_m - operational (main) time of work;
T_s - is the service time of the workplace;
T_r - time for rest and personal needs;
T_g - is the total shiftable working time fund.

2. Working time loss coefficient. It characterizes the percentage of time that is associated with production losses that must be minimized and which are inevitable. For example, in the case of a remote worker: the lights were turned off, a communal accident at home, the arrival of unexpected visitors, etc. The time spent on all these unforeseen, but real-life situations must be included in the time norm of a remote worker as a percentage of the operational time. A similar percentage can be set on average for the organization based on the rationing of workers.

$$K_l = \frac{T_l}{T_g}, \quad (10)$$

где T_l - loss of time due to organizational and technical reasons.
3. Coefficient of losses of working time due to violations of labor discipline. This coefficient shows: what percentage of the working time the employee was absent from the workplace without good reason. In the case of a remote worker, it can be calculated only if a clear working time schedule is assigned to the latter. However, since in the case of a teleworker it is difficult or impossible to do this, this indicator does not need to be laid down in time norms. And besides, the scope of the working time of a remote worker is blurred, therefore, in the event of unproductive losses of working time, he will perform the full amount of work outside the established working time.
4. Possible increase in labor productivity with the complete elimination of direct losses of working time. This is the additional amount of work that we can charge workers if they remove unproductive waste of time. In the case of teleworkers, it is important to take into account the time for self-organization of workers, which they will spend in the absence of control from the employer over the working hours.

$$P_g = \frac{[T_l + T_o + (T_{rf} - T_{rn})]}{T_m} \times 100\%, \quad (11)$$

где T_l - loss of time due to organizational and technical reasons;
T_o - loss of time for self-organization;
T_{rf} - actual time for rest and personal needs during working hours;
T_{rn} - time for rest and personal needs during working hours is standard.

3 Results

Now let's consider the prospects for such changes. It is clear that the transition to the electronic format of employment will lead to the fact that a number of potential employees will "drop out" from the "supply" on the labor market. After all, it is obvious that this innovation will be accepted by the youngest and most, as they say, "advanced". What about the rest? Their chances of finding a job in the face of rising unemployment will be greatly reduced. How to involve such people in digital employment processes? Many of them will find themselves in employment centers registered as unemployed, where they will be provided with comprehensive assistance in this direction. But it is no secret that not all unemployed are registered at employment centers, many are looking for work on their own. For such workers, the situation will be much worse compared to those who own Internet technologies.

In terms of control over the social and labor sphere of enterprises and organizations, it is known that the supervisory authorities have suspended all scheduled inspections until the end of the year in connection with the coronavirus pandemic. This follows from the Decree of the Government of 03.04.2020 No. 440 "On the extension of the validity of permits and other features in relation to licensing activities in 2020". Also, on the basis of this Resolution, methods of electronic supervision are being developed and the validity of all permits in the Russian Federation has been extended. This weakens control from the outside over the social and labor sphere of enterprises and organizations, which especially actualizes internal or trade union control.

4 Discussions

The conclusions proposed in the work are substantiated by the irreversibility of digitalization processes in the economy. The economy is at the beginning of a new upward wave that will result in a digital economic recovery. HR processes will be no exception in global transformation.

So, as part of the development of digitalization, the Russian Government proposed Federal Law No. 122-FZ dated 24.04.2020 "On conducting an experiment on the use of electronic documents related to work." The organization can enter in itself an electronic personnel document flow. This can be done at will, while the organization needs to fix the list of local documents that can be in electronic form. As a rule, these are documents related to internal personnel events in the organization. The employer must notify the employee two months in advance of the transition to electronic document management and the employee's disagreement cannot be the reason for his dismissal. Trade unions should oversee the observance of workers' rights.

It is important to remember that during the period of an emergency it was allowed to use electronic copies of educational documents for employment. This is stated in the Federal Law of 08.06.2020 No. 164. And the Federal Law "On the Protection of Population and Territories from Natural and Technogenic Emergencies" dated 21.12.1994 N 68-FZ was amended on 01.04.2020, according to which an emergency the situation referred to the spread of the disease, which poses a danger to others. This innovation will simplify the procedures for employment and confirmation of their qualifications.

Also in the Labor Code of Russia they plan to introduce a norm on the procedure for the exchange of legally significant messages. This is intended to ensure the legal regulation of the transmission of information in electronic form, including when using remote forms of labor relations.

It is assumed that the exchange of legally significant messages can take place not only in the personal presence in one place of the employee and the employer or when sending a message to the party to the employment contract in the form of a paper document by regular mail, but also using electronic or other technical means. The employer is obliged by the draft law to ensure the storage of all legally significant messages received from the employee and copies of legally significant messages sent to the employee in the manner prescribed by the local act.

The moment from which a legally significant message entails legally significant consequences is determined by the moment the message is delivered to the addressee, that is, when the addressee has the opportunity to get acquainted with the content of the message. The message will be deemed delivered even if it was received by the party to the employment contract to which it was sent, but due to circumstances depending on it, it was not delivered to it, or the party did not familiarize itself with it.

In addition, it is proposed to establish the sender of the message in any way that allows you to reliably identify the sender of the message, for example, preliminary indication in the employment contract of the addresses of the parties (including email addresses) for the exchange of messages. Also, the way to identify the sender can be the use of an electronic digital signature (simple or enhanced).

In addition, it is planned to consolidate in Art. 67 of the Labor Code of the Russian Federation, the norm on concluding an employment contract by exchanging documents and information through electronic and other technical means, since, in fact, the exchange by e-mail of scanned documents containing images of the parties' signatures is already often used. However, in practice, in the absence of a direct indication in the law, the developers note, there is a difference in approaches to the recognition of such agreements as concluded. The existing norm in the Civil Code on legally significant messages (Art. 165.1 of the Civil Code of Russia) makes it possible to exclude civil disputes about the moment of occurrence, change and termination of a person's respective rights and obligations. There is no such norm in the Labor Code of the Russian Federation, but its necessity is justified by its no less importance for labor legal relations, both at the stage of concluding an employment contract and during the period of its validity. The authors also noted that the development of electronic forms of information transfer within the framework of the digitalization of the economy is especially relevant when using remote forms of labor relations in a pandemic and other emergencies. This means that the employee can notify the employer by means of electronic messages about his will on basic personnel issues. For example, to agree to relocation, to quit, to withdraw the application, etc. The foregoing transfers trade union control to the sphere of electronic labor relations, which is new for trade unions, but they are ready for such challenges.

The tendencies listed above indicate that all personnel processes will soon be converted to electronic format.

5 Conclusion

1. It is necessary to consolidate the methodology for accounting for the costs of a remote worker for equipment belonging to him when performing work. The proposed algorithm can serve as a basis. It consists of several stages. At the first, the cost of the employee's own equipment is determined. On the second, depreciation is calculated in a monthly equivalent. At the third stage, the share of working time in the general calendar fund is estimated. Finally, the amount of compensation for the equipment is calculated.
2. It is advisable to methodologically study the issues of rationing the labor of remote workers. To establish labor standards for remote workers, it is effective to take photographs of working hours. It is important to take into account the normal time for self-organization of a remote worker.
3. It is important to raise the question of the labor safety of a remote worker. The corresponding solutions should be worked out in national legislation.

References

1. Kosmach, U.B.: Legal framework for labor regulation of teleworkers in Russia and foreign countries. Russia - Asia - Africa - Latin America: The economy of mutual trust. Materials of the X Eurasian Economic Youth Forum. In 3 volumes. Responsible for the release of Ya.P. Silin, R.V. Krasnov, E.B. Dvoryadkin Publisher: Ural State Economic University (Yekaterinburg). S. 79–81 (2019)
2. Draft Federal Law N 966659-7 "On Amendments to Article 57 of the Labor Code of the Russian Federation". http://www.consultant.ru/cons/cgi/online.cgi?req=doc&base=PRJ&n=195779#08837607700048757
3. Labor Code of the R ussian Federation of December 30, 2001 N 197-FZ (as amended on December 29, 2020)
4. Goloborodko Viktor Remote work in Germany will be included in the legislation and limited. https://aussiedlerbote.de/2020/04/udalennuyu-rabotu-v-germanii-vklyuchat-v-zakonodatelstvo-i-ogranichat/
5. Resolution of the Government of the Russian Federation of 01.01.2002 N 1 (as amended on 27.12.2019) "On the Classification of Fixed Assets Included in Depreciation Groups"
6. "OK 013-2014 (SNA 2008). All-Russian Classifier of Fixed Assets" (adopted and put into effect by Rosstandart Order of 12.12.2014 N 2018-st) (as amended on 08/05/2018)
7. Dentons, "Chile Telework law," April 2020; Remote Working in the COVID-19 Pandemic: A Guide for Employers Geneva: International Labor Office, May 2020, p. 7. https://www.ilo.org/
8. Employers' Guide Geneva: International Labor Office, May 2020, p. 8. https://www.ilo.org/.
9. Overview of topical issues from workers and employers for June 2020 © Material from the Help system "Labor protection". https://vip.1otruda.ru/#/document/99/565298354/ZAP2DQC3HA/?of=copy-513da410ec
10. Federal Law "On Compulsory Social Insurance against Industrial Accidents and Occupational Diseases" of 24.07.1998 N 125-FZ

A Time Series Forecasting of Electricity Demand by ARIMA and ReLU Based ANN

Karin Kandananond(✉)

Valaya Alongkorn Rajabhat University, Pathum Thani 13180, Thailand
Karin@vru.ac.th

Abstract. The accurate forecasting of electricity demand is an important issue. The objective of this study is the application of two forecasting methods, Box-Jenkin's autoregressive integrated moving average (ARIMA) and artificial neural network (ANN), to fit a time series data of electricity demand. Three models of ARIMA, AR (1), ARMA (1, 1), and IMA (1, 1), were utilized to fit the demand data. Another method, ANN, was based on the activation function of rectified linear unit (ReLU) and the training algorithm of resilient backpropagation with weight-backtracking. The performance of these models was compared by considering the root mean square error (RMSE) which was used to measure the errors when each model was applied to fit the data.

Keywords: Autoregressive integrated moving average (ARIMA) · Artificial neural network (ANN) · Rectified linear unit (ReLU) · Root mean square error (RMSE)

1 Introduction

The ability to forecast the electricity demand is important since it will assist the relat-ed personnel to efficiently allocate the organization budget for the utility bill. It is also a useful information for the promotion of the energy conservation scheme in an organization. Moreover, it might be used as a baseline to create the future-plan of the power consumption. If the data is accurately forecasted, the efficient plan will be drafted. As a result, this study focuses on the application of different forecasting methods to fit the available data of electricity demand in an organization. The errors of each forecasting models were compared and the model with the lowest error was indicated.

2 Literature Review

Tan et al. (2010) proposed the combination model of wavelet transform, autoregressive integrated moving average (ARIMA), and autoregressive conditional heteroskedasticity (GARCH) to predict the electricity price. Kandananond (2011) applied three different methods, ANN, ARIMA, and multiple linear regression to forecast the electricity demand in Thailand. The amount of population, gross domestic product (GDP), stock index, and

© The Author(s), under exclusive license to Springer Nature Switzerland AG 2021
R. Silhavy (Ed.): CSOC 2021, LNNS 229, pp. 644–649, 2021.
https://doi.org/10.1007/978-3-030-77445-5_58

revenue from exporting industrial products, were used as the input factors for forecasting. Zhang et al. (2012) proposed a hybrid model based on ARIMA, wavelet transform and least squares support vector machine, to forecast electricity price. The new hybrid method was optimized by using the particle swarm optimization (PSO). Gao and Fan (2017) compared the performance of ARIMA and ANN models to forecast the electricity price in United Kingdom (UK). The hybrid model between ANN and PSO was utilized by Anand and Suganthi (2017) to forecast the electricity demand and its forecasting performance was benchmarked with ANN-Genetic algorithm (GA), ANN-BP, and ARIMA models. Kim et al. (2019) applied ARIMA, ARIMA-GARCH, multiple seasonal exponential smoothing, and ANN models to forecast the peak load demand in buildings.

3 Methods

Two methods were applied to forecast the data of time series and these are ARIMA and ANN. For ARIMA, the general model (Hyndman and Athanasopoulos 2018) is shown in (1).

$$\left(1 - \emptyset_1 B - \ldots - \emptyset_p B^p\right)\left(1 - B)^d (y_t - \frac{\mu t^d}{d!}\right) = \left(1 + \theta_1 B + \ldots + \theta_q B^q\right)\varepsilon_t \quad (1)$$

where y_t is the observation at time t.

ϕ_i is the autoregressive (AR) coefficient at lag i
θ_i is the moving average (MA) coefficient at lag i
μ is the mean
B is the backshift operator
ε_t is the error at time t
p is the order of AR terms
d is the order of differencing
q is the order of MA terms.

Basically, the values of AR and MA parameters are between 1 and 1. On the other hand, ANN is one of the machine-learning methods. It is based on the neural network training and composed of input, hidden and output layers. The example of ANN structure is illustrated in Fig. 1. According to Fig. 1, the observations at time t − 1, t − 2, and t − 3 were utilized as the inputs (y_{t-1}, y_{t-2}, and y_{t-3}) and the output is the observation at time t. In this study, the activation function used is the rectified linear unit or ReLU which is shown in (2) (Nair and Hinton 2010).

$$f(x) = \max(0, x) \quad (2)$$

According to (2), $f(x) = x$ if $x \geq 0$ and $f(x) = 0$ if $x < 0$. Therefore, the output of the ReLU function is graphically depicted in Fig. 2.

The error measurement of each forecasting method is performed in the form of root mean square error (RMSE) as shown in (3).

$$RMSE = \sqrt{\frac{1}{n}\sum\nolimits_{t=1}^{n}(\hat{y}_t - y_t)^2} \quad (3)$$

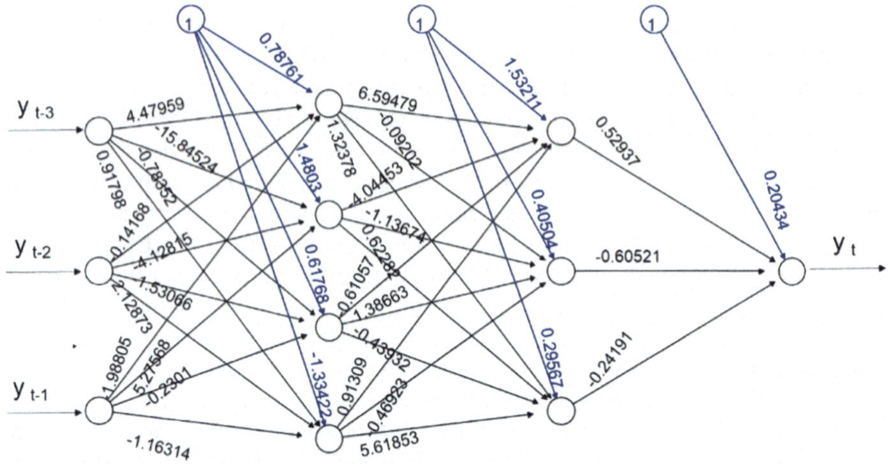

Fig. 1. Example of ANN.

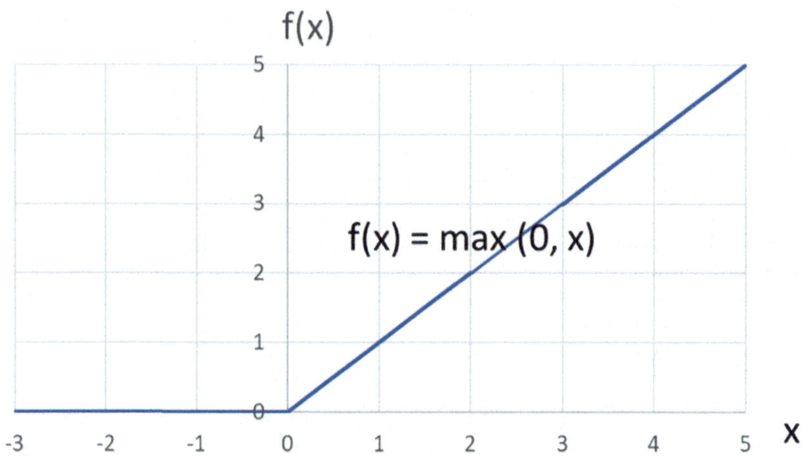

Fig. 2. ReLU output (Nair and Hinton 2010).

4 Results

The data set used is the monthly electricity demand of an organization from January 2015 to December 2020 and the time series plot is shown in Fig. 3.

Afterwards, different forecasting models were applied to fit the time series data of electricity demand. R program was run to calculate the coefficients of ARIMA models as well as the parameters of ANN.

4.1 ARIMA Model

Three ARIMA models were utilized to fit the data. These models were ARIMA (1, 0, 0) or AR (1), ARIMA (0, 1, 1) or IMA (1, 1), and ARIMA (1, 0, 1) or ARMA (1, 1).

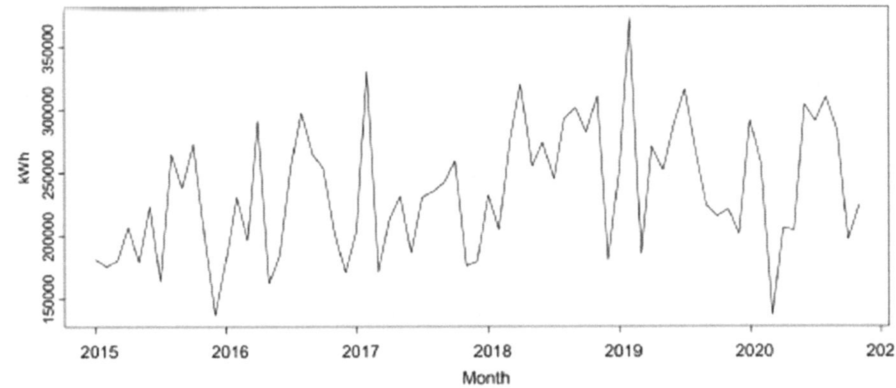

Fig. 3. Time series plot.

4.1.1 AR (1)

The first model is AR (1) which is selected to fit the data and the error measurement in the form of RMSE equals 49417.94. The equation of AR (1) model is shown in (4).

$$y_t = 184775.1 + 0.214y_{t-1} + \varepsilon_t \tag{4}$$

4.1.2 ARMA (1, 1)

For ARMA (1, 1), after fitting data with the selected model, the calculated RMSE is equal to 49402.74 and the ARMA (1, 1) model is listed as follows:

$$y_t = 156009.2 + 0.3361y_{t-1} - 0.1278\varepsilon_{t-1} + \varepsilon_t \tag{5}$$

4.1.3 IMA (1, 1)

When IMA (1, 1) was chosen to fit the data, the RMSE equals 49366.54. The equation of IMA (1, 1) model is shown in (6)

$$y_t = y_{t-1} - 0.8888\varepsilon_{t-1} + \varepsilon_t \tag{6}$$

4.2 ANN

The activation function used in this study was ReLU and the algorithm utilized to train the network was resilient backpropagation with weight-backtracking. The observation at time t − 1, t − 2, and t − 3 were applied as the inputs, y_{t-1}, y_{t-2}, and y_{t-3}, and the output was the observation at time t (y_t). In this study, the number of hidden layers were set at 2. Let S contain the number of nodes at each layer, so S = [number of nodes in the input layer, number of nodes in the 1st hidden layer, number of nodes in the 2nd layer, number of nodes in the output layer]. Two networks were trained by the available data. For the first network, the number of nodes in the 1st and 2nd hidden layer were 5 and 4, respectively. On the other hand, another network has 3 and 3 nodes in the 1st and 2nd hidden layer. Therefore, the sets of S were [3, 5, 4, 1] and [3, 3, 3, 1]. The ANN structures are illustrated in Fig. 4 and 5 and they were based on the feed-forward network that nodes in each layer were linked to other nodes in the next layer (input layer → hidden layers → output layer) in the forward direction. Networks in Fig. 4 and 5 show weights and biases from one layer to another. The RMSE of proposed ANN with S = [3, 5, 4, 1] is equal to 46711.71 while that of ANN with S = [3, 3, 3, 1] equals 37040.58.

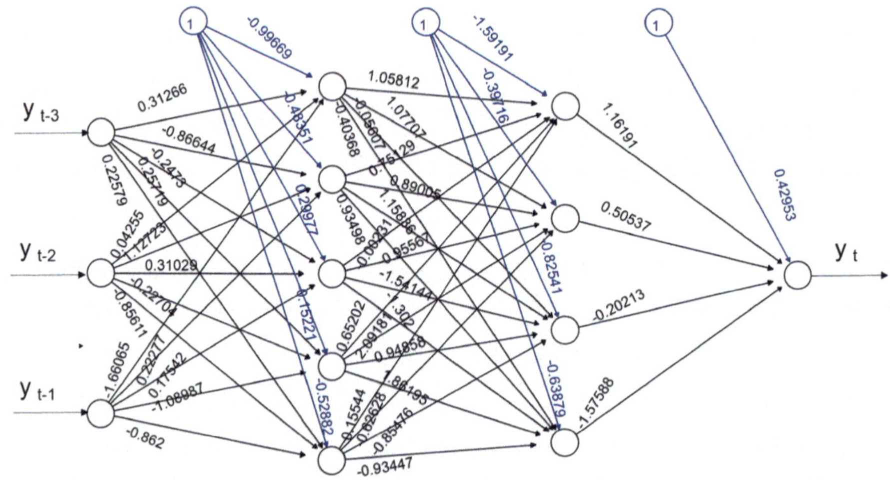

Fig. 4. ANN with S = [3, 5, 4, 1].

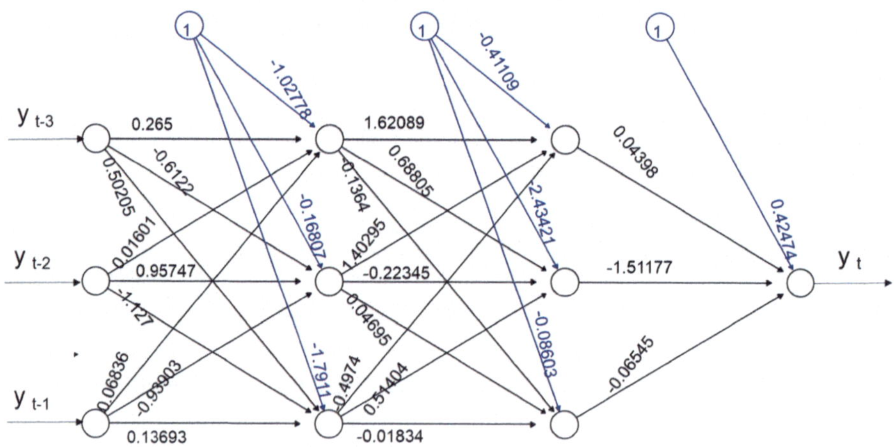

Fig. 5. ANN with S = [3, 3, 3,1].

5 Conclusions

The error measures of each model were concluded in Table 1.

Table 1. RMSE of each model.

	ARIMA			ANN	
	AR (1)	ARMA (1, 1)	IMA (1, 1)	[3, 5, 4, 1]	[3, 3, 3, 1]
RMSE	49417.94	49402.74	49366.54	46711.71	37040.58

The results in Table 1 show that the performance of different ARIMA models is not significantly different. However, ANN seems to outperform ARIMA models since their RMSEs are lower than those of ARIMA models. For the ANN, the RMSE of ANN with S = [3, 3, 3, 1] is lower than the one of ANN with S = [3, 5, 4, 1].

References

Nair, V., Hinton, G.E.: Rectified linear units improve restricted Boltzmann machines. In: Proceedings of the 27th International Conference on International Conference on Machine Learning, pp. 807–814. Omnipress, Haifa (2010)

Tan, Z., Zhang, J., Wang, J., Xu, J.: Day-ahead electricity price forecasting using wavelet transform combined with ARIMA and GARCH models. Appl. Energy **87**(11), 3606–3610 (2010)

Kandananond, K.: Forecasting electricity demand in Thailand with an artificial neural network approach. Energies **4**(9), 1246–1257 (2011)

Zhang, J., Tan, Z., Yang, S.: Day-ahead electricity price forecasting by a new hybrid method. Comput. Ind. Eng. **63**, 695–701 (2012)

Gao, G., Fan, F.: Comparison of ARIMA and ANN Models used in electricity price forecasting for power market. Energy Power Eng. **9**, 120–126 (2017)

Anand, A., Suganthi, L.: Forecasting of electricity demand by hybrid ANN-PSO models. Int. J. Energy Optim. Eng. **6**(4), 66–83 (2017)

Hyndman, R.J., Athanasopoulos, G.: Forecasting: Principles and Practice, 2nd edition, OTexts, Melbourne, Australia (2018). OTexts.com/fpp2. Accessed 26 Feb 2021

Kim, Y., Son, H., Kin, S.: Short term electricity load forecasting for institutional buildings. Energy Rep. **5**, 1270–1280 (2019)

Measurement of Transmission Speed in a Wireless Network with Encrypted Communication Using Various Security Protocols

Hung Bui Duc, Martin Koppl, Milos Orgon(✉), and Stefan Pocarovsky

Faculty of Electrical Engineering and Information Technology, Slovak University of Technology in Bratislava, Bratislava, Slovakia
orgon@ktl.elf.stuba.sk

Abstract. A necessary condition for measuring data flow in networks is the availability of network resources and the transparency of network routes. The basic parameters influencing the transmission quality in the network are delay, jitter fluctuation, and packet loss. The speed of receiving and delaying the data of a given data connection in wireless networks is significantly affected by the use of encryption. Determining the impact of the used security protocols used in wireless networks on the reduction of transmission speed is the content of this article.

Keywords: Cybersecurity · Encryption · Wireless · Transmission speed · Wi-Fi

1 Introduction

The wireless networks are one of the most used type for many different areas such as smart city, smart home, smart factory, smart grid, smart health, smart transportation, smart farming, smart building, internet of things, and other application areas comparing with Power Line Communication, Ethernet or even optic fiber [1–6]. However, the wireless technologies, IEEE 802.11 primarily, has many cybersecurity shortcomings, which in praxis creates insecure networks [7–11]. Moreover, there is no clear way how to select between current security protocols used in IEEE 802.11 networks in case of performance, which might be critical in networks for industry or other critical applications [12–15].

This paper focuses on determining the performance of the most common security protocols used in a family of standards IEEE 802.11 via experimental measurement of a selected scenario. This should help with a better understanding

Hung Bui Duc—This article is created with the support of the KEGA agency project - 034STU-4/2021 "Utilization of Web-based Training and Learning Systems at the Development of New Educational Programs in the Area of Optical Wireless Technologies".

© The Author(s), under exclusive license to Springer Nature Switzerland AG 2021
R. Silhavy (Ed.): CSOC 2021, LNNS 229, pp. 650–657, 2021.
https://doi.org/10.1007/978-3-030-77445-5_59

of the current state of IEEE 802.11 and the selection of the right security protocol for a given application.

The rest of the paper is organized as follows. Section 2 provides basic information about the used network and setup of the client/server. Section 3 provides experimental measurements of IEEE 802.11n standard followed by experimental measurements of IEEE 802.11ac in Sect. 4. Finally, Sect. 5 comes with a brief description of the work done, summarizing our results and conclusion of our findings.

2 Measuring the Transmission Speed in a Wireless Network

The transmission speed in a wireless network is one of the main factors for the user. This article compares individual security protocols and their impact on the throughput. Transmission speed measurements were performed according to pre-prepared test scenarios and will take place in "open" communication (without the use of security protocol) and in the use of security protocols WEP, WPA, WPA2. There are several software tools that can be used to measure the throughput. One of them is the "iperf", that we chose for our experiments. This tool allows you to manually set the size of the send socket buffer and the size of the TCP RWND on both sides (client and server). It is necessary to take into account the performance of both sides (end devices to avoid degradation of the result. Figure 1 shows the wiring diagram used. From a large group of IEEE 802.11 standards, we selected two standards: currently the most used – IEEE 802.11n standard to date released for the commercial operation – IEEE 802.11ac.

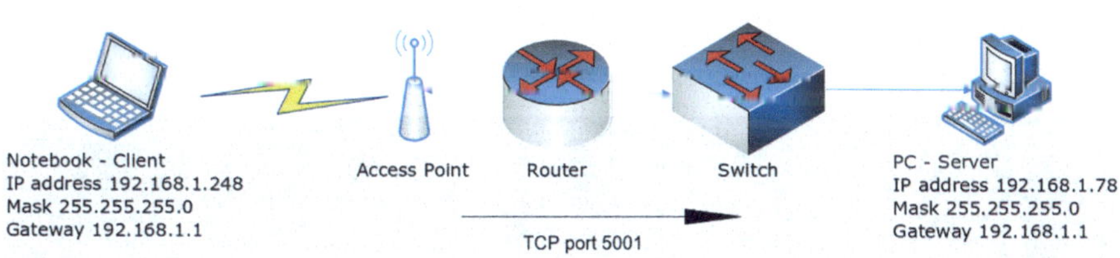

Fig. 1. Wiring diagram.

The connection configuration was as follows [16]:

- For Client:

```
./iperf3.exe -c 192.168.1.78 -w 2M -p 1  i -p 5001 -f m
   -t 10
```

- For server:

```
./iperf3.exe -s
```

The connection setup parameters were as follows [16]:

- `-r` to measure speed in both directions, upload first from the client and then download.
- `-d` to measure speed in both directions simultaneously.
- `-w` to resize TCP window (2 - 65,535 bytes).
- `-t` to specifies the duration of the measurement in seconds.
- `-i` for periodic bandwidth interval in seconds.
- `-f` to display the results in the desired form.
- `-b` to allocate the required bandwidth.
- `-m` for maximum segment size - the largest amount of data, in bytes, that a computer can support in a single TCP segment.
- `-u` for UDP testing including information on the amount of delay jitter and lost packets.
- `-P` for parallel testing.

3 Transmission Rate Measurements Using the IEEE 802.11n Standard

Transmission speed measurements using the IEEE 802.11n standard were first performed with an "open" unsecured wireless WiFi network and with various security protocols. The throughputs were taken by the client sending a request to the server and the server then responding ten times at regular one-second intervals. Furthermore, various other parameters have been defined, such as the size of the TCP window, the bandwidth interval, and the server. With an unsecured open WiFi network, the highest throughput was expected because the transmitted data did not have to be encrypted. The transmission speed reached an average value of 96.8 Mbit/s (Fig. 2).

```
[ ID] Interval           Transfer     Bandwidth
[  4]   0.00-1.00   sec  13.2 MBytes   111 Mbits/sec
[  4]   1.00-2.01   sec  13.0 MBytes   108 Mbits/sec
[  4]   2.01-3.00   sec  12.8 MBytes   108 Mbits/sec
[  4]   3.00-4.01   sec  11.4 MBytes  95.0 Mbits/sec
[  4]   4.01-5.00   sec  7.38 MBytes  62.0 Mbits/sec
[  4]   5.00-6.00   sec  10.5 MBytes  88.1 Mbits/sec
[  4]   6.00-7.00   sec  12.4 MBytes   104 Mbits/sec
[  4]   7.00-8.01   sec  12.2 MBytes   102 Mbits/sec
[  4]   8.01-9.00   sec  11.5 MBytes  97.0 Mbits/sec
[  4]   9.00-10.01  sec  11.1 MBytes  92.7 Mbits/sec
- - - - - - - - - - - - - - - - - - - - - - - - -
[ ID] Interval           Transfer     Bandwidth
[  4]   0.00-10.01  sec   116 MBytes  96.8 Mbits/sec                  sender
[  4]   0.00-10.01  sec   115 MBytes  96.8 Mbits/sec                  receiver
```

Fig. 2. Measured values of transmission speed with open (unsecured) WiFi network.

The second measurement used the WEP security protocol with a 64-bit key. Here, the transmission speed was significantly reduced compared to an unsecured

```
[ ID] Interval           Transfer     Bandwidth
[  4]   0.00-10.01  sec  26.4 MBytes  22.1 Mbits/sec                  sender
[  4]   0.00-10.01  sec  26.4 MBytes  22.1 Mbits/sec                  receiver
```

Fig. 3. Throughput using WEP 64 bit key.

WiFi network - up to 74.7 Mbit/s (Fig. 3). This reduction was expected due to the encryption used, which uses the WEP security protocol.

Using the WEP security protocol with a 128-bit key, the transmission speed was even 2 Mbit/s lower than the WEP protocol with a 64-bit key (Fig. 4). Thus, there was no further significant slowdown in throughput using a stronger key.

```
[ ID] Interval           Transfer     Bandwidth
[  4]   0.00-10.02  sec  24.0 MBytes  20.1 Mbits/sec                  sender
[  4]   0.00-10.02  sec  24.0 MBytes  20.1 Mbits/sec                  receiver
```

Fig. 4. Throughput using WEP 128 bit key.

When using the WPA-PSK + (TKIP) (Temporal Key Integrity Protocol) security protocol, the average value of the transmission rate in this measurement increased by 28.6 Mbit/s compared to the measurement with the WEP protocol. The average transfer rate reached 50.7 Mbit/s (Fig. 5).

```
[ ID] Interval           Transfer     Bandwidth
[  4]   0.00-10.00  sec  60.5 MBytes  50.7 Mbits/sec                  sender
[  4]   0.00-10.00  sec  60.5 MBytes  50.7 Mbits/sec                  receiver
```

Fig. 5. Throughput using WPA-PSK + (TKIP).

The WPA2-PSK + (AES) security protocol was used for the last measurement. It achieved the highest average value of the transmission speed. This speed achieved of 82.6 Mbit/s (Fig. 6), even though it used the most complex AES encryption. Interestingly, the most complex encryption has the least impact on reducing the transmission rate compared to using previous older encryption algorithms (see Table 1 and Fig. 7).

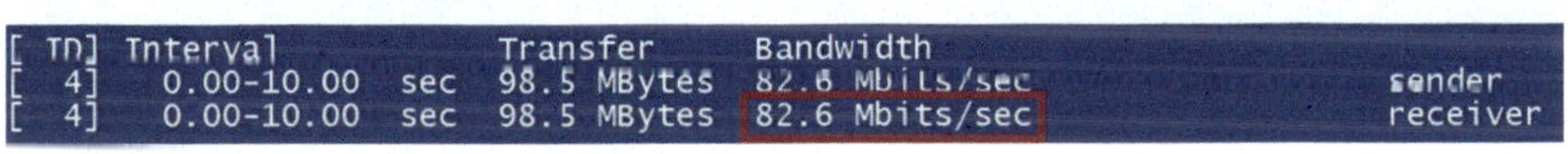

```
[ ID] Interval           Transfer     Bandwidth
[  4]   0.00-10.00  sec  98.5 MBytes  82.6 Mbits/sec                  sender
[  4]   0.00-10.00  sec  98.5 MBytes  82.6 Mbits/sec                  receiver
```

Fig. 6. Throughput using WPA2-PSK + (AES).

Table 1. Average values of throughputs using different security protocols.

Security protocol	Protocol N/A	WEP 64b	WEP 128b	WPA-PSK TKIP	WPA2-PSK AES
Average bit					
Rate [Mbit/s]	96.8	22.1	20.1	50.7	82.6

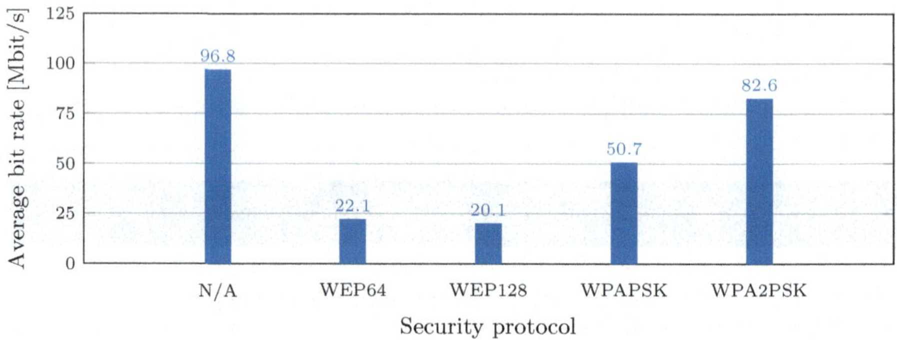

Fig. 7. Average values of throughputs for different security measures.

4 Transmission Rate Measurement with the IEEE 802.11ac Standard

When measuring the transmission speed in the WiFi network using the IEEE 802.11ac standard (Fig. PC.2), the same wiring diagram and the same set parameters were used as for the IEEE 802.11n standard. The advantage of the IEEE 802.11ac standard is that it uses a 5 GHz frequency band that is not as interfered with as the 2.4 GHz frequency band. These measurements were expected to increase the transmission speed in a secure wireless network. The first measurement took place in an unsecured WiFi network, where the transmission speed reached 577 Mbit/s (Fig. 8). The huge difference is visible in comparison to the older IEEE 802.11n standard, where the average value of the transmission speed was lower by 480.2 Mbit/s.

Fig. 8. Average transmission speed for unsecured WiFi network.

When testing the throughput with WPA-PSK + (TKIP) security, the value of the average throughput was 471 Mbit/s (Fig. 9). Using this security protocol

Fig. 9. Average throughput for WPA-PSK + (TKIP) security protocol.

significantly reduced the average throughput, as was the case when using the IEEE 802.11n standard.

In the last throughput measurement, using the WPA2-PSK + (AES) security protocol, when a large amount of data (1.23 GB) was transmitted, the average throughput was 529 Mbit/s (Fig. 10 and Table 2). WPA2-PSK with AES block cipher proved to be the best security protocol for wireless networks.

Fig. 10. Average throughput for WPA2-PSK + (AES) security protocol.

Table 2. Average transmission speed of different security settings.

Security protocol	Protocol N/A	WPA-PSK TKIP	WPA2-PSK AES
Average bit rate [Mbit/s]	577	471	529

In Fig. 11 is a comparison of average throughputs using different security protocols.

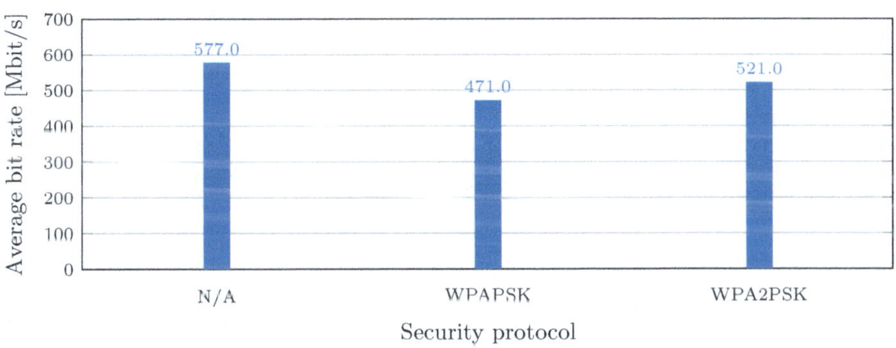

Fig. 11. Comparison of measured values of transmission speed depending on the used security protocol WPA-PSK + TKIP and WPA2-PSK + AES.

5 Conclusion

This paper was focusing on the impact of security protocols in wireless networks. When measuring the throughput of the network running on 802.11n standard a WEP security protocol, the average value of the throughput was 22.1 Mbit/s (Fig. 7). A significant decrease compared to the unsecured wireless network (the average value of the transmission speed without the use of the security protocol was 96.8 Mbit/s) was caused by the RC4 stream cipher used in the WEP protocol.

When measuring the throughput in a wireless network using the IEEE 802.11n standard using the WPA2-PSK TKIP security protocol, the average value of the throughput was 50.7 Mbit/s (Table 1), which was a less significant decrease compared to the unsecured 96.8 Mbit/s network. It is also interesting that when using higher quality ("stronger") WPA2-PSK AES encryption, the average value of the transmission speed was even higher - 82.6 Mbit/s (Table 1), which means that even though a stronger AES cipher, the effect on the throughput was the smallest. Using the IEEE 802.11ac standard, the average throughput was 529 Mbit/s. The significantly higher transmission speed of the IEEE 802.11ac standard was caused, among other things, by the network bus architecture supporting the increase of the transmission speed in the wireless network and the use of the 5 GHz frequency band. The performed transmission rate measurements show that the use of the WPA2 AES security protocol is currently the most advantageous, especially in terms of increasing the transmission speeds and less impact of the WPA2-PSK AES encryption used in data transmission in the network.

References

1. Mlynek, P., Fujdiak, R., Misurec, J., Slacik, J.: Experimental measurements of noise influence on narrowband power line communication. In: 8th International Congress on Ultra Modern Telecommunications and Control Systems and Workshops (ICUMT). IEEE **2016**, 94–100 (2016)
2. Raso, O., Mlynek, P., Fujdiak, R., Pospichal, L., Kubicek, P.: Implementation of elliptic curve diffie hellman in ultra-low power microcontroller. In: 2015 38th International Conference on Telecommunications and Signal Processing (TSP). IEEE, pp. 662–666 (2015)
3. Mlynek, P., Misurec, J., Toman, P., Silhavy, P., Fujdiak, R., Slacik, J., Hasirci, Z., Samouylov, K.: Performance testing and methodology for evaluation of power line communication. Elektronika ir Elektrotechnika **24**(3), 88–95 (2018)
4. Mlynek, P., Misurec, J., Kolka, Z., Slacik, J., Fujdiak, R.: Narrowband power line communication for smart metering and street lighting control. IFAC-PapersOnLine **48**(4), 215–219 (2015)
5. Fujdiak, R., Mlynek, P., Misurec, J., Masek, P.: Design of low-power random number generator using signal quantization error in smart grid. In: 2016 39th International Conference on Telecommunications and Signal Processing (TSP). IEEE, pp. 7–10 (2016)

6. Mlynek, P., Hasirci, Z., Misurec, J., Fujdiak, R.: Analysis of channel transfer functions in power line communication system for smart metering and home area network. Adv. Electr. Comput. Eng. **16**(4), 51–56 (2016)
7. Fujdiak, R., Misurec, J., Mlynek, P., Janer, L.: Cryptograph key distribution with elliptic curve diffie-hellman algorithm in low-power devices for power grids. Rev. Roum. des Sci. Tech. pp. 84–88 (2016)
8. Krivtsova, I., et al.: Implementing a broadcast storm attack on a mission-critical wireless sensor network. In: International Conference on Wired/Wireless Internet Communication. Springer, pp. 297–308 (2016). https://doi.org/10.1007/978-3-319-33936-8_23
9. Fujdiak, R., Mlynek, P.,Blazek, P., Barabas, M., Mrnustik, P.: Seeking the relationbetween performance and security in modern systems: metrics and measures. In: 2018 41st International Conference on Telecommunications and Signal Processing (TSP). IEEE, pp. 1–5 (2018)
10. Fujdiak, R., Masek, P., Hosek, J., Mlynek, P., Misurec, J.: Efficiency evaluation of different types of cryptography curves on low-power devices. In: 7th International Congress on Ultra Modern Telecommunications and Control Systems and Workshops (ICUMT). IEEE, **2015**, 269–274 (2015)
11. Fujdiak, R., Mlynek, P., Misurec, J., Slacik, J.: Simulation of intelligent public light system in smart city. In: Progress In Electromagnetics Research Symposium-Spring (PIERS). IEEE **2017**, 2515–2519 (2017)
12. Baghaei, N., Hunt, R.: Ieee 802.11 wireless lan securityperformance using multiple clients.In: Proceedings of the 2004 12th IEEE International Conference on Networks(ICON 2004)(IEEE Cat. No. 04EX955), vol. 1, pp. 299–303. IEEE (2004)
13. Leon, M., Aldeco, R., Merino, S.: Performance analysis of the confidentiality security service in the ieee 802.11 using wep, aes-ccm, and ecc. In: 2005 2nd International Conference on Electrical and Electronics Engineering. IEEE, pp. 52–55 (2005)
14. Sari, A.: Security approaches in ieee 802.11 manet–performance evaluation of usm and ras. Int'l J. Commun. Network Syst. Sci. **2014** (2014)
15. Mitchell, C., He,C.: Security analysis and improvements for ieee 802.11 i. In: The 12th Annual Network and Distributed System Security Symposium (NDSS 2005) Stanford University, Stanford. Citeseer, pp. 90–110 (2005)
16. IPERF: Tool. Jperf, graphical frontend for iperf–network performance measurement graphical tool (2014). https://github.com/AgilData/jperf

Severity Estimation of Potholes in Imagery Using Convolutional Neural Networks

Vernon Kok, Nontokozo Mpofu, and Micheal Olusanya(✉)

Sol Platje University, Kimberley, Northern Cape, South Africa
{201605330,nontokozo.mpofu,michael.olusanya}@spu.ac.za
http://www.spu.ac.za

Abstract. Potholes are holes in the surface of the road caused by water entering into cracks in the road and expanding over time. Pothole remediation is a time-consuming process undergone by every municipality to ensure quality of roads. Holes in the road vary in severity and can prove life-threatening, especially in cases where they are on roads which experience high traffic volumes. This paper aims to demonstrate how deep learning processing architectures and smartphone technology can be used to localize potholes in images to estimate their severity for automated reporting and remediation. Localization refers to locating a pothole in an image. Determining its size can then help decide which potholes need to be prioritized. The approach taken in this paper is model-based as opposed to a hardware-based approach, where the underlying algorithm is a deep learning model as opposed to sensors attached to a microcontroller. We use a pre-trained object detection model to extract the region of an image containing a pothole and feed the extracted region of interest (ROI) to a convolutional neural network for estimation of pothole depth. The researchers make use of ordinary smartphone sensor technology to construct a processing architecture that can capture pothole depth and location. This information can be used subsequently to guide decision making with regards to fixing road damage. The proposed architecture, obtained through a Random Search in the hyper-parameter space, is able to identify the bounding box of a pothole contained in an image and estimate its depth.

Keywords: Deep learning · Potholes · Object detection · Convolutional neural networks

1 Introduction

Roads and highways are free to use for all users, and as a result, attract traffic of varying shapes and sizes ranging from daily commuters from home to work, vehicles carrying goods to be delivered, to pedestrians. Road networks are essential as major transportation and communication enablers in modern civilizations [1].

© The Author(s), under exclusive license to Springer Nature Switzerland AG 2021
R. Silhavy (Ed.): CSOC 2021, LNNS 229, pp. 658–672, 2021.
https://doi.org/10.1007/978-3-030-77445-5_60

Municipalities dedicate time and resources to designing and maintaining road and traffic networks. Traditional methods for road quality management such as manually looking for holes, documenting their location and returning to fix, are time-consuming, expensive, labor-intensive, and error-prone [2]. Maintenance of road networks is time-consuming and can never be complete as potholes occur as a result of road usage. According to [1], weather conditions and vehicle loads contribute to road deterioration over time. Thus, road quality assurance is an ongoing effort. Road damage can range from simple cracks on road surfaces which later develop into potholes when water enters into the soil, to fading road markings and even damaged signs. Potholes are one of the major types of road defects/damages and can prove life-threatening in some cases. As such, the recent advances in the field of computer vision have been applied to automate the detection and severity estimation of potholes. Automated road quality management systems that leverage the automatic feature engineering exhibited by deep learning algorithms can be used to replace the manual process of pothole detection and assist in remediation.

Knowledge of pothole location and severity can assist road agencies in planning repairs as highlighted by [2] and [4], and knowledge of pothole dimensions (surface area and depth) can be used to gauge severity to prioritize fixing of potholes. Previous model-based approaches emphasize the localization of the pothole in an image (object detection). However, the identification of a pothole in an image may assist in avoidance but does not really contribute towards remediation. Pothole depth information would allow for efficient distribution of resources for pothole remediation. The road agencies can now allocate resources more efficiently, as they have a view of which potholes require more time and resources to fix.

This article is aimed at designing an efficient road damage detection system which can be used to automate damage examination and streamline allocation of resources for fixing of roads. The question is, "How can an efficient severity estimation model be constructed with limited amounts of data, without the use of specialized equipment?" Locating a pothole in an image before passing to an image regression model will allow the model to learn. Therefore, the contribution of this paper is to demonstrate how a pothole severity estimation pipeline, leveraging automatic feature extraction demonstrated by convolutional neural networks, can be developed using data collected with a smartphone camera.

The paper is structure as follows: Sect. 2 briefly reviews related works. Section 3 describe the proposed method, data collection, model training and model testing. The results of the study are discussed in Sect. 4, followed by concluding remarks in Sect. 5 with future research direction.

2 Related Works

Road damage inspection is usually performed by humans using expensive and time-consuming methods [4,5]. In their paper, Avellaneda and Lopez-Parra [6] make use of a smartphone-based GIS system that uses the sensors available, namely GPS, accelerometer and compass to capture information about roads and produces output in the form of a map visualization and automated report generating. The approach they take is to place the smartphone on the inside of a vehicle. The data collected and stored by the system is latitude, longitude, GPS accuracy, speed of the vehicle in which the device is placed, accelerometer readings, date-time and compass azimuth. The researchers limit themselves to the detection of potholes and categorization using the Superficial Deterioration Index which ranks the potholes into three mutually exclusive groups. The potholes are ranked as either 'good', 'fair' or 'deficient'. The researchers do not catalogue any other road damage.

According to [2] the damages range from a simple tire puncture to costlier ones such as damage to vehicle suspension and vehicle steering alignment. The researchers model the interaction between the wheels of a car and the pothole encountered. In their paper, the researchers make use of a multi-phase dynamic model which decomposes the interaction between the pothole and the wheel of a car into three separate models. Namely, a model for when the car is rolling on flat ground, a model for interaction of the front wheel/s with the edge of the pothole, and a model for the interaction between the rear wheels of the car and the edge of the hole. In their paper, [3] take a combination of image processing methods and machine learning methods to detect the presence of potholes in an image. An input image is first preprocessed and fed to a decision tree classifier, K-nearest neighbor classifier and a support vector machine classifier.

Departing from the sensor-based approach taken previously, [5] take a computer-vision based approach to road-damage detection. They highlight that no benchmark dataset exists for road-damage detection, and as such, construct their own using a dashboard-mounted smartphone. The dataset constructed consists of 163–664 road images, each 600×600 pixels. They then make use of object-detection algorithms to predict the bounding boxes of areas that contain road-damage. The model constructed is able to predict the bounding boxes of eight kinds of damage categories.

The system proposed by [5] is able to accurately predict bounding boxes for eight kinds of road damage but does not provide an estimate of the severity. The system proposed by [7] makes use of thermal imagery to detect the presence of potholes but the need for a specialized camera means that only individuals who have such specialized imagery hardware would be able to contribute to the fixing of roads.

3 Proposed Method

Modern smartphones are equipped with a wide variety of sensors ranging from location, luminous intensity and even gyroscopes. Using this data, combined with state-of-the-art computer vision algorithms, the researchers propose a road-damage detection system that accepts as input an image containing a pothole and estimates the depth of the pothole.

A dataset of 399 images of potholes is collected from the internet. The dataset (see Fig. 1) is annotated with the bounding boxes of potholes contained in each image. The images vary in size and as a result, the researchers first resize all images to 300 × 300 before annotating. Annotating images first and then resizing results in loss of information as the ground truth bounding boxes no longer refer to the same image. The dataset was annotated using 'LabelImg', an open-source annotation tool. The image annotations are in the PASCAL-VOC format. The resizing of the images is done for ease of processing.

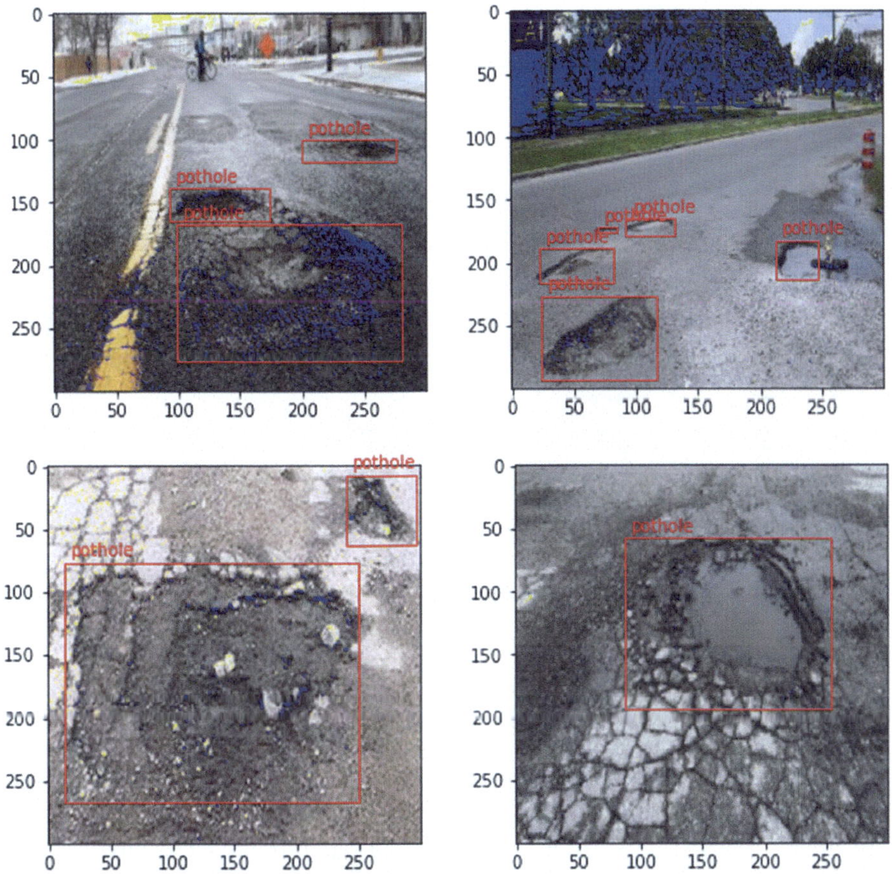

Fig. 1. Images from dataset 1

A second dataset of pothole images along with their corresponding dimensions is collected and used to train a convolutional neural network for prediction of pothole depth. The dataset used in stage two consists of pothole images collected manually by using a smartphone to capture the image of the pothole. The pothole dataset consists of images collected in the central business district of Johannesburg, South Africa. The researchers make use of squares to record pothole length and breadth, a second smaller square is used to capture pothole depth. A dataset of images (see Fig. 2) containing potholes is used to train an object detector. The object detector (ROI extraction network) accepts as input an N × M image containing a pothole and outputs the bounding box coordinates. The coordinates are used to extract the region of an image containing the pothole. A second smaller dataset of pothole images labelled with their corresponding depth is used to train a convolutional neural network to estimate pothole depth. The first network aims to reduce the dimensionality of the input to the depth estimation network. The authors are not aware of any dataset that consists of potholes labeled with their corresponding depth and as such one is collected using a smartphone and measuring instruments.

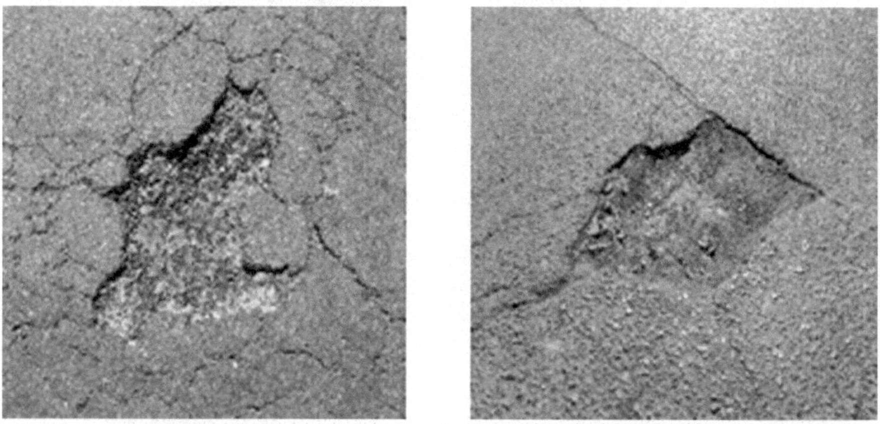

Fig. 2. Images from dataset used to train depth estimator

All measurement are in millimeters and rounded to the nearest whole number because potholes deteriorate further over time. A Faster R-CNN [8] with a Resnet50 frame proposal network pretrained on the PASCAL-VOC dataset and fine-tuned on dataset 1 (see Table 1), is used as an initial ROI extraction network. The ROI extraction network is evaluated using the mean average precision (mAP). The extracted ROI then feeds to a separate model for prediction of pothole depth. Prediction of pothole severity from an image is broken down into two separate models. The first model predicts a bounding box. The bounding box is the models estimate on where a pothole is in an image. The bounding box coordinates are used to extract a ROI. The extraction of an ROI also reduces the complexity of the input to the depth-estimation model by getting rid of any background noise that could be present in the image. The background noise is removed by cropping the ROI from the rest of the image, and the depth estimator is able to focus solely on the pothole for depth prediction.

The feature extraction network as shown in Fig. 3 accepts as input an N × N image and outputs a M × Z bounding box which contains the pothole obtained from the input image. The ROI is resized to (300, 300) and passed

Table 1. Sample of the dataset used for training of depth estimation network

Filename	Depth	Length	Breadth
20200811_130957.jpg	35	600	570
20200811_130951.jpg	35	600	570
20200811_130945.jpg	35	600	570
20200811_130941.jpg	35	600	570
20200811_131824.jpg	25	250	190
20200811_131820.jpg	25	250	190
20200811_131817.jpg	25	250	190
20200811_132452.jpg	40	370	150
20200811_132440.jpg	40	370	150
20200811_132435.jpg	40	370	150
20200811_134733.jpg	45	560	420
20200811_134727.jpg	45	560	420
20200811_134722.jpg	45	560	420

to a convolutional neural network pre-trained on ImageNet which is modified to estimate pothole depth by adding a combination of convolutional layers and fully connected layers. Resnet50, VGG16 and VGG19 are trained for 10 epochs using a batch size of 4 and Stochastic Gradient Descent (SGD) with a learning rate of 0.001. The layers of Resnet50, VGG16 and VGG19 are frozen so that only the weights of the convolutional and fully connected layers are updated during training. The learning rate is kept small while fine-tuning to avoid overwriting parameter values too much.

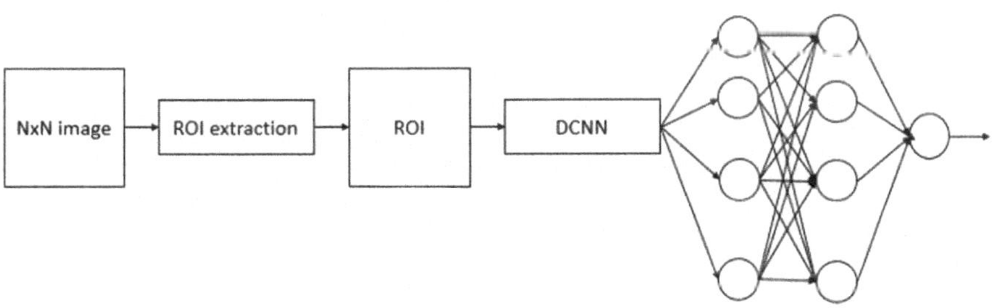

Fig. 3. The proposed model

The extraction of the ROI (see Fig. 4) allows for a reduction in input complexity to the second model. A two-stage architecture requires two separate training phases and two separate training datasets. The dataset used to train the Faster R-CNN network is not in the correct structure to predict pothole depth, see Fig. 5.

Fig. 4. Example output from the ROI extraction network

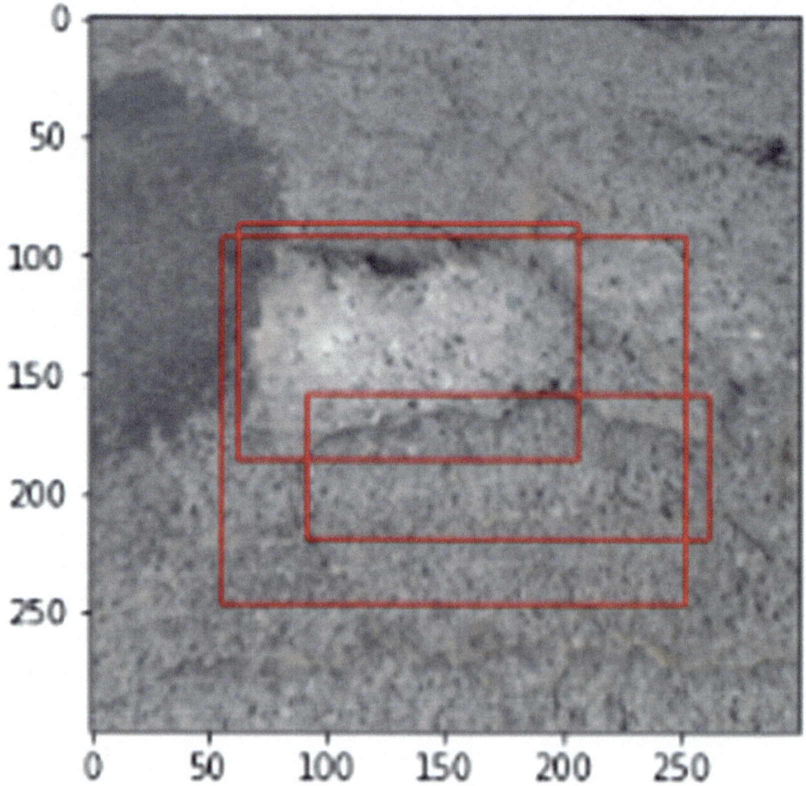

Fig. 5. Classifying depth of this example will be difficult and illustrates that the accuracy of the depth estimate is largely dependent on the accuracy of the ROI extraction network

```
_________________________________________________________________
Layer (type)                 Output Shape              Param #
=================================================================
vgg16 (Functional)           (None, 9, 9, 512)         14714688
_________________________________________________________________
conv2d (Conv2D)              (None, 9, 9, 108)         497772
_________________________________________________________________
conv2d_1 (Conv2D)            (None, 9, 9, 92)          89516
_________________________________________________________________
conv2d_2 (Conv2D)            (None, 9, 9, 4)           3316
_________________________________________________________________
flatten (Flatten)            (None, 324)               0
_________________________________________________________________
dense (Dense)                (None, 42)                13650
_________________________________________________________________
dropout (Dropout)            (None, 42)                0
_________________________________________________________________
dense_1 (Dense)              (None, 1)                 43
=================================================================
Total params: 15,318,985
Trainable params: 604,297
Non-trainable params: 14,714,688
_________________________________________________________________
```

Fig. 6. Summary VGG16-based CNN architecture

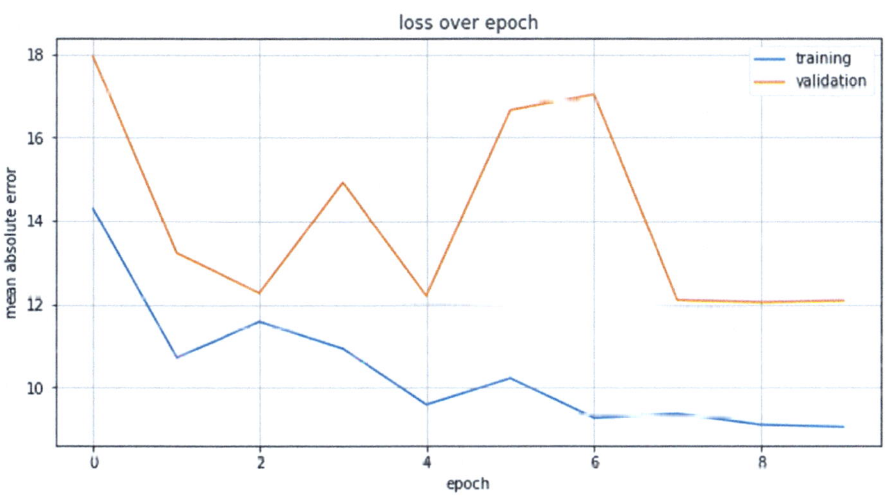

Fig. 7. VGG16-based CNN learning curve

```
_________________________________________________________________
Layer (type)                 Output Shape              Param #
=================================================================
vgg19 (Functional)           (None, 9, 9, 512)         20024384
_________________________________________________________________
conv2d (Conv2D)              (None, 9, 9, 100)         460900
_________________________________________________________________
conv2d_1 (Conv2D)            (None, 9, 9, 64)          57664
_________________________________________________________________
flatten (Flatten)            (None, 5184)              0
_________________________________________________________________
dense (Dense)                (None, 362)               1876970
_________________________________________________________________
dropout (Dropout)            (None, 362)               0
_________________________________________________________________
dense_1 (Dense)              (None, 1)                 363
=================================================================
Total params: 22,420,281
Trainable params: 2,395,897
Non-trainable params: 20,024,384
_________________________________________________________________
```

Fig. 8. Summary VGG19-based CNN architecture

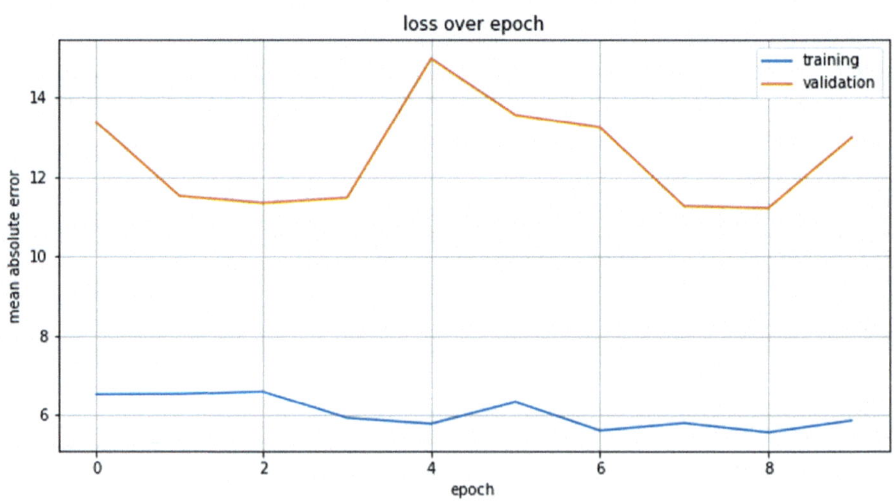

Fig. 9. VGG19-based CNN learning curve

```
_________________________________________________________________
Layer (type)                 Output Shape              Param #
=================================================================
input_1 (InputLayer)         [(None, 300, 300, 3)]     0
_________________________________________________________________
cast_to_float32 (CastToFloat (None, 300, 300, 3)       0
_________________________________________________________________
resnet50 (Functional)        (None, None, None, 2048)  23587712
_________________________________________________________________
flatten (Flatten)            (None, 204800)            0
_________________________________________________________________
regression_head_1 (Dense)    (None, 1)                 204801
=================================================================
Total params: 23,792,513
Trainable params: 23,739,393
Non-trainable params: 53,120
_________________________________________________________________
```

Fig. 10. Summary RESNET50-based CNN architecture

4 Results

The metric used for gauging the performance of the ROI extraction network is the intersection-over-union (IOU)/mAP. The IOU measures the 'tightness' of the predicted bounding box with the ground truth bounding box. The IOU measure the overlap between the two by dividing the area where the two boxes overlap by the total area of both. First a pre-trained faster- RCNN network with a RESNET 50 backbone is trained, when training the network for 10 epochs with a batch size of 4, the model obtains a mAP of 23.3.

For comparison of the Convolutional Neural Network models, the Mean Absolute Error (MAE) is used. The MAE measures the absolute difference between the predicted depth and ground truth. First, a VGG16-base is trained for 10 epochs using the network architecture obtained from a Random Search. Random Search takes a range of values for a specific hyper-parameter as input and selects random combinations between them to obtain the best performing model. The search is performed to suggest the number of convolutional layers, the number of units in the fully connected layer and the dropout ratio. The network architecture consists of VGG16 as a base model which produces 512 9×9 'VGG16' feature maps. These feature maps feed into three convolutional layers which produce 108, 92 and 4 feature maps each of size 9×9. The images are zero-padded to avoid losing information that could prove valuable for the fully connected layers and prediction layers. The fully connected layer consists of 42 units which then feed to a single output neuron for prediction of pothole depth. The network architecture is further summarized in Fig. 6 and 7.

When tested on a 0.2 portion of the training set (80 images), the VGG16-based CNN achieves a MAE of 32.2852. From the learning curve above it can be

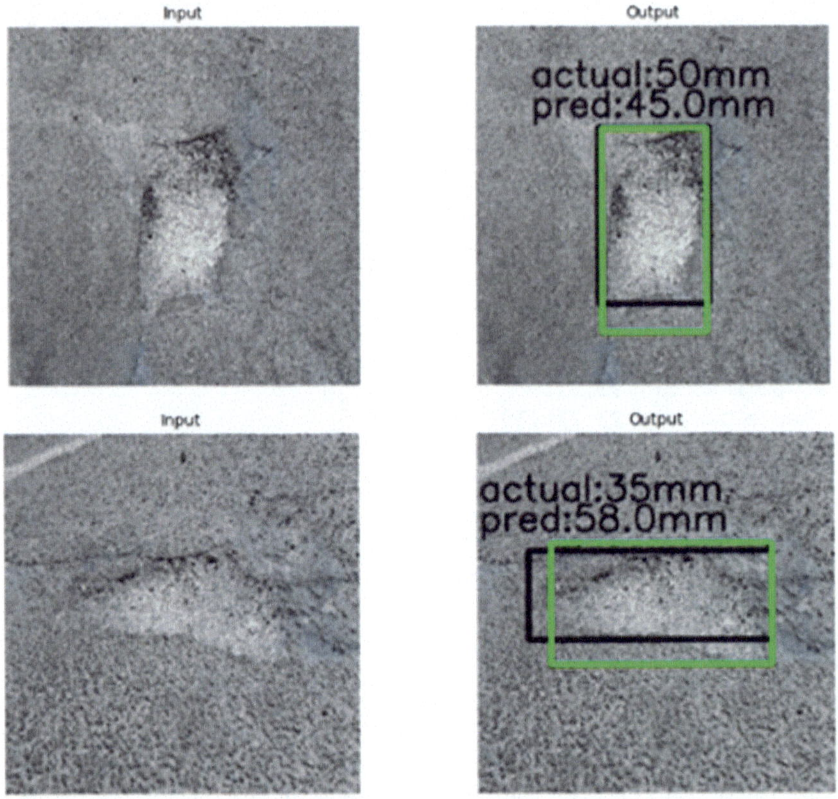

Fig. 11. Example output when using VGG16-based DCNN as depth estimator

seen that the training loss shows a downward trend. Both training and validation loss start to stabilize after 8 epochs. The VGG16 base model is replaced with VGG19. The RandomSearch search space is kept the same as for the VGG16 base model. The network architecture consists of a VGG19 base model that produces 512 9×9 'VGG19' feature maps. The main difference is that the VGG19-based network contains two convolutional layers which produce 100 and 64 9×9 feature maps respectively. The features are fed to a fully-connected layer with 362 units which in turn feeds to a single unit for prediction of pothole depth. The network is summarized in Fig. 8 and 9. When trained for 10 epochs and tested on a 0.2 split of the training data, the VGG19-based CNN model achieves a MAE of 98.59481. The final network tested in this study is Resnet50. The researchers make use of a customized Neural Architecture Search (NAS) approach, in which the neural network architecture is built iteratively, and obtain a neural network consisting of resnet50 (layers frozen). The output of the resnet50 layers feed directly into a single output neuron. The network is summarized in Fig. 10.

When trained for 10 epochs and tested on a 0.2 split of the training data, the Resnet50-based CNN achieves a MAE of 21.6045. The Resnet50-based network

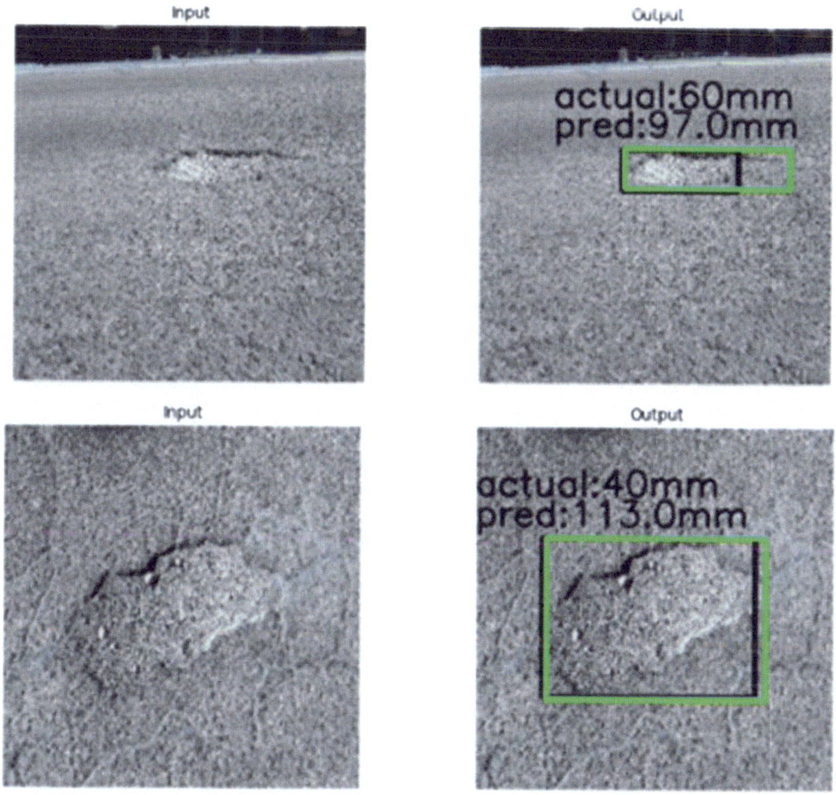

Fig. 12. Example output when using VGG19-based DCNN as depth estimator

shows the greatest reduction in error on the validation set. It achieves the best results of all networks under consideration in this study with a MAE of 21.6045 on the validation set.

Examples output from using the various CNN architectures are shown in Figs. 11, 12, 13 and 14. The predicted bounding box is shown in black with the ground-truth bounding box shown in green. The predicted depth from the depth estimator is labelled as 'pred' with the ground truth labelled as 'actual' for comparison of results. The VGG16-based CNN is selected to form the second stage of the two-model pipeline as it produces the most accurate results even though it performs worse than the Resnet50-based CNN network on training and validation datasets.

Figure 11 shows example outputs from the proposed model for depth estimation. The input to the ROI extraction network is shown on the left, labelled 'Input'. Once a ROI has been found in the input, it is extracted from the original image, resized and fed to a VGG16-based CNN model which estimates the depth (see Fig. 12 and 13). In cases where multiple bounding boxes are detected, the researchers select the bounding box corresponding to the highest confidence

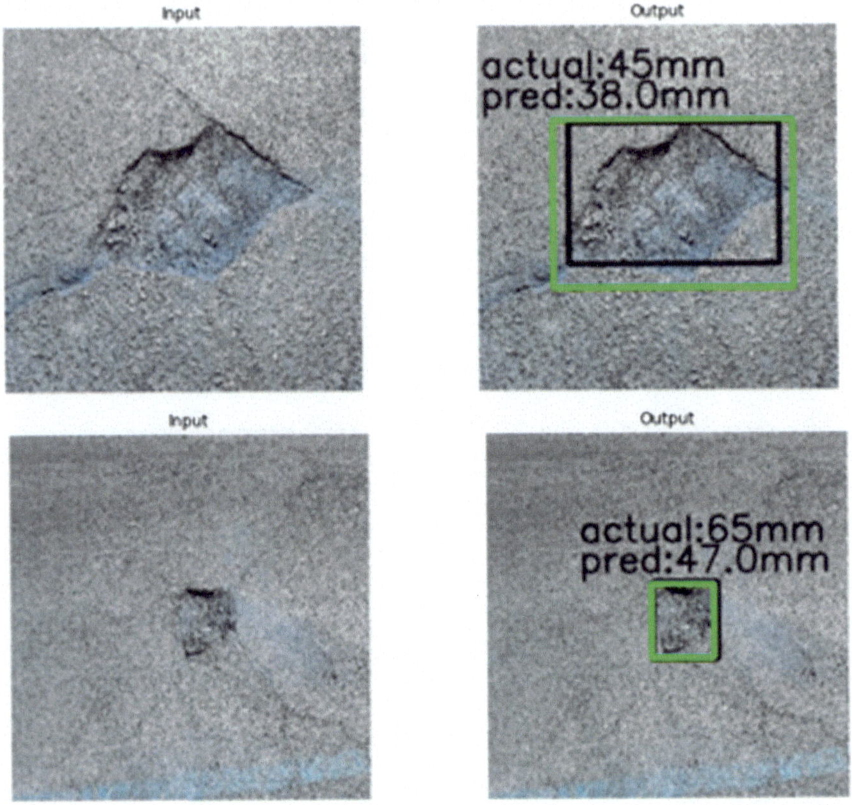

Fig. 13. Example output when using the resnet50-based DCNN as depth estimator

score. Found by ranking predictions by confidence score and select the first prediction in the list. In cases where the predicted bounding box and ground truth box are very similar in terms of IOU, the models estimate of depth is relatively close to the ground truth. An example of this is shown in Fig. 14. In cases where the pothole in image is not clearly visible or distinguishable from its background, the bounding box is not easily found, and the depth estimate is performed on random noise contained in the image as shown in Fig. 14.

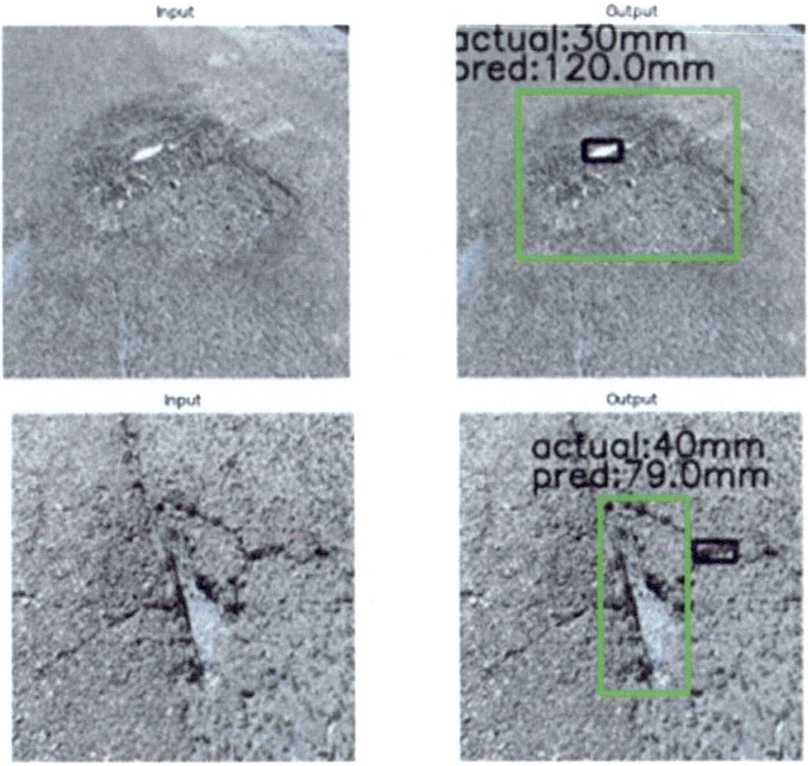

Fig. 14. Example outputs from the proposed model (VGG16) when bounding boxes are not found correctly.

5 Conclusion

Roads and traffic networks have become the standard mechanism used for transportation of goods and passengers. Roads connect provinces and even cities together, and this has been an enabler for ever-expanding metropolitan areas. The importance of roads is rivaled only by the need to maintain road and traffic networks. The manual maintenance of these roads is infeasible, considering the size and coverage of traffic networks

This article has highlighted some of the previous hardware-based and model-based approaches employed to detect road damage. Recent advances in the field of deep learning have led to more model-based methods where features are automatically learnt from the training data and less effort is spent on the manual construction of features. Knowledge of the presence of a pothole in an image is not sufficient for remediation efforts - knowledge/estimates of pothole dimensions is also needed for effectively distributing time and resources.

The study also proposed is the use of two separate models to estimate the depth of a pothole from an image. The first model is an object detection model which acts as a ROI extraction network. The ROI is the model's estimate of where a pothole is in an image. Once the pothole has been identified in an image, it is removed from the rest of the input image, reducing the complexity of the image input to the second model. This extraction of the ROI also helps to remove background noise from the input which make depth estimation harder. Consider an image taken of a pothole which contains humans in it. When passed as-is for prediction of depth, the model may have difficulty producing accurate output. Extracting the area of the image containing the pothole first solves this problem. The second stage of the two-model pipeline consists of a VGG16-based CNN. The architecture of the second network was found by performing a Random Search in a parameter search space. The search space can be kept constant by the introduction of a seed value.

Finally, the proposed method is able to provide the relevant authorities with information vital to the distribution of resources for road remediation. Knowing the location of a pothole has been helpful for avoidance but not for remediation. This article has highlighted a computer vision approach to pothole remediation. A study for the future will focus on potholes remediation web application and combining the two proposed models into a single model.

References

1. Tiong, P., Mustaffar, M., Hainin, M.: Road Syrface assessment of pothole severity by close range digital photogrammetry method. World Appl. Sci. J. **19**(6), 867–873 (2012)
2. Li, Z., Kolmanovsky, I., Atkins, E., Lu, J., Filev, D.: Road anomaly estimation: model based pothole detection. In: American Control Conference, pp. 1315–1320 (2015)
3. Karmel, A., Adhithiyan, M., Kumar, P.: Machine learning based approach for pothole detection. Int. J. Civ. Eng. Technol. **9**(5), 882–888 (2018)
4. Gouwan, L.: Arduino-based pothole detection, logging, and analysis (2019)
5. Maeda, H., Sekimoto, Y., Seto, T., Kashiyama, T., Omata, H.: Road damage detection and classification using deep neural networks with smartphone images. In: Computer-Aided Civil and Infrastructure Engineering, pp. 1–15 (2018)
6. Avellaneda, D., Lopez-Parra, J.: Detection and localization of potholes in roadways using smartphones. DYNA **83**, 156–162 (2016)
7. Arjapure, S., Kalbande, D.: Review on analysis techniques for road pot-hole detection. In: Pant, M., Sharma, T., Verma, O., Singla, R., Sikander, A. (eds.) Soft Computing: Theories and Applications, vol. 1053, pp. 1189–1197. Springer, Singapore (2020). https://doi.org/10.1007/978-981-15-0751-9_109
8. Ren, S., He, K., Girshack, R., Sun, J.: Faster R-CNN: towards real-time object detection with region proposal networks. IEEE Trans. Pattern Anal. Mach. Intell. **39**, 1137–1149 (2019)

Portfolio Optimization Model for Asset Allocation Problem Based on Alternative Risk Measures

Anna Andreevna Malakhova[1(✉)], Elena Nikolaevna Sochneva[1], Svetlana Anatolyevna Yarkova[2], Anastasiya Vladimirovna Yarkova[3], Olga Valeryevna Starova[1], Dmitry Ivanovitch Kravtsov[1], and Dmitry Valeryevitch Zyablikov[1]

[1] Siberian Federal University, Krasnoyarsk, Russia
amalakhova@sfu-kras.ru

[2] Krasnoyarsk Institute of Railway Transport - Branch of the Irkutsk State Transport University, Krasnoyarsk, Russia

[3] Peter the Great Saint Petersburg Polytechnic University, Saint Petersburg, Russia

Abstract. In this paper new approaches to synthesize investment portfolios are proposed based on the introduced alternative risk measures for portfolio. It is shown that the difference between the portfolio mean return and 'low-mean' return, the difference between the portfolio mean return and 'value-at-risk' return may be considered as alternative risk-measures. The algorithm for portfolio 'low-mean' return and 'value-at-risk' return evaluation is formulated. The experimental study of the traditional models and of the alternative risk measures has been performed on the basis of common stocks traded via Moscow Exchange. The feasible sets for the proposed models have been constructed. It has been shown that the feasible sets for the developed investment portfolio models are conceptually similar to the 'mean – variance' set of the Markowitz model. The efficient frontiers for the Markowitz model, for the Sharpe index model and for the proposed alternative risk measures have been constructed; the optimal weight distributions have been obtained. The efficiency of the portfolios based on traditional risk measures and of the portfolios with alternative risk measures have been evaluated and compared. It has been shown that models based on the 'left-side' risk measures have produced better results for the market under consideration. These findings confirm the validity of application of alternative measures for risk assessment under financial market conditions.

Keywords: Efficient frontier · Markowitz model · Market model · Investment portfolio · Asset allocation problem · Value-at-Risk

1 Introduction

With the securities market development the investment portfolio analysis and the asset allocation problem have been recently well studied as is demonstrated by many

© The Author(s), under exclusive license to Springer Nature Switzerland AG 2021
R. Silhavy (Ed.): CSOC 2021, LNNS 229, pp. 673–693, 2021.
https://doi.org/10.1007/978-3-030-77445-5_61

publications on this topic. There are different formulations and models developed to synthesize investment portfolios (see e.g. [1–3]). This problem is often defined as a problem of optimization (risk minimizing or return-maximizing problem). One of the discussed problems is the way to measure risk or volatility of an asset and of the portfolio. Among the most common used and efficient approaches are the Markowitz model and the Market model (the Sharpe index model) [1, 4]. In the Markowitz model the risk of the portfolio is measured by the variance of the portfolio return and depends on the correlation between the assets, measured by the covariance [5]. According to the Market model the risk of portfolio depends on the correlation between the assets and the market portfolio, which is measured by the beta-coefficients [1]. Nevertheless the way to assess portfolio risk remains a discussed question.

The present state of the art of portfolio risk concept [1, 6, 7] demonstrates that none of the existing risk measures is a universal way of assessing it: traditional risk assessment measures can underestimate negative deviations from the expected return and thus be conducive to incorrect conclusions. An important problem that hinders the use of the studied models is the failure to fulfill in practice the assumption of the normal distribution of the return on the portfolio and the assets included in it [1, 8]. Considering the aforesaid it seems necessary to develop alternative 'down-side' risk indicators taking into account the deviations of the observed returns down from the expected mean return [1, 2, 9].

The aim of this paper is to develop investment portfolio models for asset allocation based on the alternative risk measures and to evaluate their efficiency.

The paper is organized as follows. The classical formulation of the Markowitz problem for portfolio optimization is formulated in the second section. The modification of the Markowitz problem based on the Sharpe index model is formulated. Alternative risk measures for portfolio are proposed and an algorithm for defining the initial parameters of the investment portfolio optimization problem is developed. A modified Markowitz model is proposed on the basis of the introduced 'left-side' risk measures. In the third section an illustrative experimental case-study is performed on the basis of common stocks traded via 'Moscow Exchange'. The investment portfolio is optimized on the basis of the Markowitz model, the Market model and the modified Markowitz model for the introduced alternative risk measures. The efficiency of the portfolios based on traditional risk measures and of the portfolios with alternative risk measures have been evaluated and compared. The results of model verification are discussed in the forth section. The main findings of the research are summarized in the conclusion.

2 Methods

2.1 The Markowitz Model for Portfolio Optimization

The Markowitz model reveals the trade-off between return and variance of the portfolio. The model is based on the principle of 'effective diversification' and allows constructing an efficient frontier on which the optimal portfolios are located, that is, the efficient set provides the investor with the best combinations of portfolio risk and return [5]. According to the efficient set theorem, an investor will choose an optimal portfolio from a set of portfolios, each of which provides the maximum expected return for a certain level of risk (minimum risk for a certain value of the expected return). This approach is known as 'mean-variance' optimization [1, 4, 6].

Initial premises of the classical Markowitz problem are as follows:

1. Investors evaluate investment portfolios based on expected returns and standard deviations over the holding period.
2. Investors are never jaded, that is, other conditions being equal, they prefer a portfolio with a higher expected return.
3. Investors are rational, that is, other conditions being equal, they prefer a portfolio with less risk.
4. The portfolio consists only of risky assets, the risk measure of which is the standard deviation of return.
5. Investors assume a normal probability distribution of the expected return on assets. The portfolio return aggregates the return on the securities included in it.
6. Private assets are infinitely divisible, that is, an investor can buy part of a share.
7. Taxes and transaction costs are immaterial.

The mean return of a portfolio of n assets is obtained as:

$$R_p = \sum_{i=1}^{n} w_i \cdot \bar{R}_i, \tag{1}$$

where w_i is the weight of the asset i in the portfolio, $\bar{R}_i$ is the mean return on the asset i.

The variance of a portfolio return is defined as [4]:

$$\sigma_p^2 = \sum_{ij} w_i \cdot w_j \cdot Cov_{ij} \tag{2}$$

where w_i and w_j are the weights of the assets i and j in the portfolio, respectively, Cov_{ij} is the covariance between two assets i and j.

The points of the efficient frontier are obtained by solving two optimization problems that yield the same result: to minimize (maximize) the variance (the mean return) of (on) the portfolio under fixed value R (σ^2) of the mean return (variance). That is [5, 9]:

$$J_{Markowitz} = \sum_{ij} w_i \cdot w_j \cdot Cov_{ij} \rightarrow min \tag{3}$$

subject to

$$\sum_i w_i \cdot R_i = R \tag{4}$$

$$\sum_i w_i = 1$$

$$w_i \geq 0,\ i = \overline{1, n}$$

or

$$J_{Markowitz} = \sum w_i \cdot \bar{R}_i \rightarrow max \tag{5}$$

subject to

$$\sum_{ij} w_i \cdot w_j \cdot Cov_{ij} = \sigma^2,\ i = \overline{1, n} \tag{6}$$

$$\sum_{i} w_i = 1,\ i = \overline{1, n}$$

$$w_i \geq 0,\ i = \overline{1, n}$$

The solution of the problem yields the optimal weight coefficients w_i^* for the assets in the portfolio.

This problem refers to the class of quadratic programming problems, since the objective function is quadratic and the constraints are linear [4]. This problem can be solved, for example, in MatLab and Excel.

The primary processing of statistical data can be carried out in the DataAnalysis Excel software package using the following Analysis Tools: Correlation, Covariance, Moving Average, as well as the Average, Covar, Frequency functions.

Solver Tool can be used to optimize portfolio in Excel. We will enter the portfolio variance into the target cell, and we will set the asset weights as the changing cells. In the constraints, we will set the sum of the asset weights, the expected portfolio return, and constraints on the specific asset weights in the portfolio. To get an efficient set of portfolios, we will use the Visual Basic for Application add-in through recording a macros.

To estimate the expected returns and the covariance matrix, MatLab offers the Financial Toolbox, the Analyzing Portfolios function block, the ewstats function, Syntax: [ExpReturn, ExpCovariance, NumEffObs] = ewstats (RetSeries, DecayFactor, WindowLength).

To obtain the weight distribution and parameters of portfolios lying on the efficient frontier, the function portopt, Syntax will be used: [PortRisk, PortReturn, PortWts] = portopt (ExpReturn, ExpCovariance, NumPorts, PortReturn, ConSet). To set constraints related to the weights of specific assets, we will use the pcalims, Syntax function: [A, b] = pcalims (AssetMin, AssetMax, NumAssets).

2.2 The Market Model (the Sharpe Index Model) for Investment Portfolio Optimization

The basic assumption of the Market model is the existence of correlation between the stock price and the market index (for example, *S&P 500* – Standard & Poor's Index, *DJIA* – Dow Jones Industrial Average, *NYA* – New York Stock Exchange Composite etc.), which is described by the 'beta' coefficient [1]. The expected return on an asset is described by a linear relationship to the return on the market index, and the expected return on a portfolio is the weighted return on the assets included in the portfolio. The risk of an asset is associated with the variance of the market index through the beta coefficient, and also includes the intrinsic risk of the asset, expressed by the variance of the random error. Random error arises due to the fact that the asset return may not be fully explained by the return on the market index, that is, there are other factors explaining the fluctuations of the return.

Thus, the Market model implies a special structural property for the return of the asset. The expected return on the asset i, $i = \overline{1, n}$, is assumed in the form [1]:

$$R_i = \alpha_{iI} + \beta_{iI} \cdot R_I + \varepsilon_{iI}, \tag{7}$$

where α_{iI} is the shift coefficient, β_{iI} is the slope coefficient, ε_{iI} is a random variable, R_I is the return on the market index. One has:

$$\alpha_{iI} = \bar{R}_i - \beta_{iI} \cdot \bar{R}_I, \tag{8}$$

where $\bar{R}_i$ and $\bar{R}_I$ are the mean returns on the asset and on the index, respectively.

The coefficient β_{iI} represents the sensitivity of asset i to market movement. 'Beta' shows how much the asset performance moves when the market moves:

$$\beta_{iI} = \frac{Cov_{iI}}{\sigma_I^2}, \tag{9}$$

where Cov_{iI} is the covariance between the return on the asset and the return on the index; σ_I^2 is the variance of the return on the index.

For the further analysis we consider the Index I_t of the Trading Organiser 'Moscow Exchange' – MOEX Russia Index *(IMOEX)* [8], that is defined as follows [4]:

$$I_t = I_0 \cdot \frac{MC_t}{MC_0}, \tag{10}$$

where MC_t and MC_0 are the total market capitalization of Index component stocks on the current date t and on the initial date *0*, and I_0 is the Index value on the initial date. Thus,

$$MC_t = \sum\nolimits_{i=1}^{k} P_{ti} \cdot Q_{ti}, \; t = \overline{1, N}, \tag{11}$$

where Q_{ti} is the number of stocks i existing on the current date t, P_{ti} is the price of unit stock i on the time t, k is the total number of component stocks used in the Index calculation.

The risk of the asset i is measured by the variance:

$$\sigma_i^2 = \beta_{iI}^2 \cdot \sigma_I^2 + \sigma_{\varepsilon i}^2, \tag{12}$$

where $\sigma_{\varepsilon i}^2$ is the variance of the random variable ε_{iI} of the asset i.

The return on the portfolio of n assets is defined as in (1).

The risk of the portfolio is defined as [1, 4]:

$$\sigma_p^2 = \beta_{pI}^2 \cdot \sigma_I^2 + \sigma_{\varepsilon p}^2, \tag{13}$$

where $\sigma_{\varepsilon p}^2$ is the variance of the random variable ε_{iI} of the portfolio, β_{iI} represents the sensitivity of the portfolio to market movement, σ_I^2 is the variance of the market index.

$$\beta_{pI}^2 = \left(\sum\nolimits_{i=1}^{n} w_i \cdot \beta_{iI}\right)^2 \tag{14}$$

$$\sigma_{\varepsilon p}^2 = \sum_{i=1}^{n} w_i^2 \cdot \sigma_{\varepsilon i}^2 \tag{15}$$

The first term of the sum in (13) represents the systematic risk of the portfolio, which cannot be reduced by diversification, because every asset with nonzero beta contains this risk. The second part of the sum in (13) is termed specific risk. This risk is uncorrelated with the market and can be reduced by diversification.

Taking into account the initial assumptions of the model and the decomposition of portfolio risk into intrinsic and market risk according to the Sharpe index model, the modified Markowitz problem is formulated as follows:

The optimization problem is formulated as:

$$J_{Market} = \beta_{pI}^2 \cdot \sigma_I^2 + \sigma_{\varepsilon p}^2 \rightarrow min \tag{16}$$

subject to

$$\sum_i w_i \cdot \overline{R_i} = R \tag{17}$$

$$\sum_i w_i = 1$$

$$\alpha_k \leq w_k \leq \beta_k,\ 0 \leq \alpha_k < 1,\ 0 < \beta_k \leq 1,\ k \in \overline{1, n}$$

The solution of the problem yields the optimal weight coefficients w_{Market}^* for the assets in the portfolio.

To estimate the input parameters of the model, we will use the Data Analysis Toolbox Excel, the Regression function, which estimates the parameters of the linear regression model: Intercept, X Variable1, as well as the standard error and coefficient value. The problem is related to nonlinear programming problems. Solver ToolBox of Excel can be used to solve the problem. To solve problems based on the application of the one-factor Sharpe model, a ready-made package in SOLVSAMP of Excel is also offered, namely the Portfolio of Securities block.

2.3 Modification of the Markowitz Model Based on the $(R_p - R_{plow})$- and $(R_p - R_{pVaR})$-Risk Measures for Portfolio

To characterize the full structure of the left-side deviations of the portfolio, we introduce a parameter called the 'left-side expected return' on the portfolio p, defined as:

$$R_{plow} = \sum_{k \in Z^-} R_{pk} \cdot p'_{pk}, \tag{18}$$

where Z^- is the set of indices k such for which the relation $R_{pk} < R_p$ holds, R_p is the expected return on the portfolio p, p'_{pk} is the normalized 'left-side' frequency.

The value R_{plow} measures the average return on the left ('unwanted') part of the distribution of the return on portfolio p, that is, the mathematical expectation of those returns that lie to the left of the expected portfolio return.

If the number of observations is not enough to divide the series of returns into intervals, then the left-side expected return can be defined as:

$$R_{plow} = \frac{1}{cardZ^-} \sum\nolimits_{t \in Z^-} R_{pt}, \tag{19}$$

where Z is the set of indices t such that $R_{pt} \langle R_p$, $cardZ^-$ denotes the number of elements Z^-, R_{pt} is the moving portfolio return.

Based on the introduced measure R_{plow}, we propose to evaluate the portfolio risk as the difference between the expected return and the left-side expected portfolio return:

$$R_p - R_{plow} = \sum\nolimits_i w_i \cdot \bar{R}_i - \sum\nolimits_{k \in Z^-} R_{pk} \cdot p'_{pk} \tag{20}$$

If the portfolio risk is estimated as a $(R_p - R_{plow})$-value, then, taking into account (20), the portfolio synthesis problem will be formulated as follows:

$$J_{(R-R_{low})} = \sum\nolimits_i w_i \cdot \bar{R}_i - \sum\nolimits_{k \in Z^-} R_{pk} \cdot p'_{pk} \rightarrow min \tag{21}$$

subject to

$$\sum\nolimits_i w_i \cdot \overline{R_i} = R \tag{22}$$

$$\sum_i w_i = 1$$

$$\alpha_k \le w_k \le \beta_k,\ 0 \le \alpha_k < 1,\ 0 < \beta_k \le 1,\ k \in \overline{1, n}$$

The solution of the problem yields the optimal weight coefficients $w^*_{(R-Rlow)}$ for the assets in the portfolio. The objective function is non-linear from w_i. We do not need to describe a nonlinear function, since there is an algorithmic scheme sufficient for determining $J_{(R-R_{low})}$ and for solving an optimization problem. We do not know the form of the function from w_i, but we know its values, which we need to find the efficient set of portfolios. To solve the problem, we propose the following algorithm:

1. To generate portfolio weight distributions for all possible asset weights combinations w_i with a given step (0.01, 0.05, or 0.1), $\sum_i w_i = 1$ provided.
2. To construct a series of moving portfolio returns R_{pt} for the weight distribution for each generated portfolio.
3. To determine the expected return R_p, 'left-side expected return' R_{plow}, the $(R_p - R_{plow})$-risk measure for each series of moving returns corresponding to its generated portfolio.
4. To construct a feasible set of portfolios in terms of 'R_p -$(R_p - R_{plow})$' and to determine efficient portfolios with minimum $(R_p - R_{plow})$- values for the given values of return.

We introduce an alternative risk measure for assessment of the portfolio risk, the distribution of returns of which is different from the symmetric one. It is based on the portfolio value-at-risk measure, determined by the expression:

$$VaR_p = R_{pVaR} : \left[P\{R_p > R_{pVaR}\} = 1 - \alpha\right], \tag{23}$$

where R_{pVaR} is the return on portfolio p, α is the given level of significance.

We denote the portfolio risk as as the difference between the expected portfolio return and the VaR_p- return. Portfolio risk for discrete distribution can be formalized by the expression:

$$R_p - VaR_p = \sum\nolimits_i w_i \cdot \bar{R}_i - R_{pk},\ k = k_{VaR} \tag{24}$$

Using $(R_p - VaR_p)$-value as a measure of portfolio risk, taking into account (24), we formulate the portfolio synthesis problem as follows:

$$J_{(R-VaR)} = \sum\nolimits_i w_i \cdot \bar{R}_i - \underset{k=k_{VaR}}{R_{pk}} \rightarrow min \tag{25}$$

subject to

$$\sum\nolimits_i w_i \cdot \overline{R_i} = R \tag{26}$$

$$\sum_i w_i = 1$$

$$\alpha_k \le w_k \le \beta_k,\ 0 \le \alpha_k < 1,\ 0 < \beta_k \le 1,\ k \in \overline{1, n}$$

The solution of the problem yields the optimal weight coefficients $w^*_{(R-VaR)}$ for the assets in the portfolio. The objective function is non-linear from w_i. We do not need to describe the nonlinearity, since there is an algorithmic scheme sufficient for determining $J_{(R-VaR)}$ and solving the optimization problem. To obtain a solution to the problem, Solver Toolbox Excel will be used, the VaR_p-value will be set using the *Small* function under the appropriate significance level. We propose the following algorithm to find an efficient set of portfolios:

1. To generate portfolio weight distributions for all possible asset weights combinations w_i with a given step (0.01, 0.05, or 0.1), $\sum_i w_i = 1$ provided.
2. To construct a series of moving portfolio returns R_{pt} for the weight distribution of each generated portfolio.
3. To determine the expected return R_p, VaR_p-value and $\left(R_p - VaR_p\right)$-risk measure for each series of moving return corresponding to its generated portfolio;
4. To construct a feasible set of portfolios on in terms of 'R_p - $\left(R_p - VaR_p\right)$' and to determine efficient portfolios with minimum $\left(R_p - VaR_p\right)$- values for the given values of return.

3 Results

3.1 The Markowitz Model and the Market Model for Portfolio Optimization

The 'Moscow Exchange' was chosen as the information basis for conducting this part of the experimental study as the oldest organizer and the leading trading platform for the securities market in Russia [10]. A crucial factor in the choice of this trading platform was the open access to information on trading for the entire historical period of its existence including the former 'Russian Trading System', which had been affiliated further to 'Moscow Exchange', namely the period of 1995–2021. The prices that are formed in Moscow Exchange are a generally recognized guideline for investors who transact with Russian shares and depositary receipts on other Russian stock exchanges, on European and US stock exchanges. Information about trading on the Moscow Exchange is broadcast to a huge number of consumers in financial market in Russia and abroad and is the basis for calculating the main indicator of the Russian stock market – the RTS Index and Moscow Exchange Index IMOEX [10].

For the analysis and synthesis of the portfolio, diverse assets have been be selected that represent various sectors of the economy according to the chosen principle of branch diversification: electricity, oil and gas, financial sectors, communications and telecommunications, metallurgical, air transportation, consumer sector. Preference is to be given to securities with good growth potential and liquidity.

Thus, six assets were selected that meet the listed characteristics of the following issuers: Public Joint stock company "ROSSETI" (ordinary share *RSTI*), Public Joint-Stock Company VTB Bank (ordinary share *VTBR*), Public Joint Stock Company "Mining and Metallurgical Company "NORILSK NICKEL" (ordinary share *GMKN*), Public Limited Liability Company Yandex N.V., shares of a foreign issuer (ordinary share *YNDX*), Mobile TeleSystems Public Joint Stock Company (ordinary share *MTSS*) [10].

For the further analysis we selected one-year period, namely the period December, 01, 2019 – December 31, 2020. The price series of assets were transformed into series of moving returns with a smoothing window of one month. Monthly moving returns have been obtained for the one-year period, namely January, 01, 2020 – December 31, 2020. The initial parameters of the assets are presented in Table 1.

Table 1. Statistic parameters for the Markowitz model

Asset	Asset parameters			
	$\bar{R}_i$	σ_i	σ_i^2	Coefficient of variation
GMKN	1,023189	0,096822	0,009375	0,094628
MTSS	1,006183	0,066658	0,004443	0,066249
YNDX	1,062315	0,107592	0,011576	0,10128
VTBR	1,058802	0,832589	0,693204	0,78635
RSTI	1,03565	0,151427	0,02293	0,146215

The estimated asset parameters from Table 1 show that the *VTBR*-asset is the most risky, as is demonstrated by the high value of the coefficient of variation (78.6%). The least variable are *MTSS*- and *GMKN*-assets (6.62% and 9.46%, respectively).

Variance-covariance matrix is given in Table 2.

Table 2. Covariance matrix for the selected assets

Asset	Cov_{ij}				
	GMKN	MTSS	YNDX	VTBR	RSTI
GMKN	0,00933704	0,00413593	0,00232696	0,0115345	0,00986401
MTSS	0,004135931	0,00442554	0,00393745	0,0097648	0,00733078
YNDX	0,00232696	0,00393745	0,01152967	0,0071513	0,00528334
VTBR	0,011534529	0,00976478	0,00715132	0,6904312	0,02447735
RSTI	0,009864008	0,00733078	0,00528334	0,0244773	0,02283856

Note that during the period under study, it was not possible to find negatively correlated assets, as is shown in Table 2, which can be explained by the overall decline of the market in 2020 as a whole due to pandemia COVID-19

In order to evaluate the location of the Markowitz feasible set we got different return-risk combinations letting the weight-coefficients w_i range over all possible combinations using a 0.1 step, so that $\sum w_i = 1$. The feasible set for the Markowitz model is presented in Fig. 1. The optimal portfolios are located on the efficient frontier of plotted 'mean-variance' set.

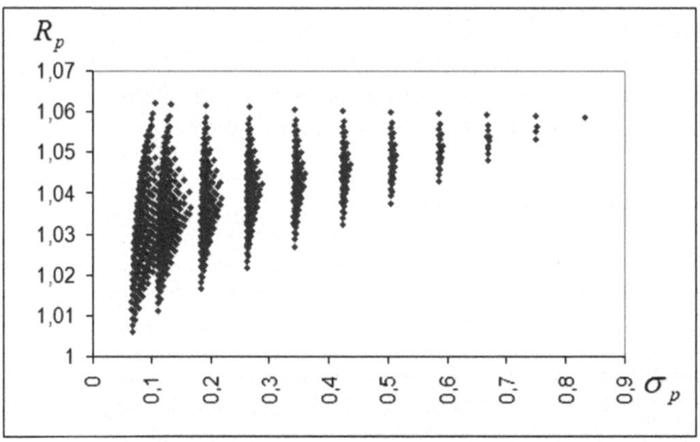

Fig. 1. Feasible set for the Markowitz model

We solve the asset allocation problem as a variance-minimizing problem for the parameters of the Table 1, using (3), (4). The arbitrary values R are chosen using a 0.5% step. The solution of the problem yields the optimal weight coefficients $w^*_{iMarkowitz}$ for efficient frontier portfolios, presented in Table 3.

Table 3. Portfolio parameters and weight distribution for optimal portfolios $w^*_{iMarkowitz}$

Portfolio return	Standard deviation	Weights of the assets				
		GMKN	MTSS	YNDX	VTBR	RSTI
1,0623	0,107376	0,00000100	0,00000000	1,00000100	0,00000000	0,00000000
1,0600	0,102399	0,05880663	0,00000000	0,93705812	0,00413525	0,00000000
1,0550	0,093020	0,18680083	0,00000000	0,81132568	0,00187348	0,00000000
1,0500	0,085690	0,31473468	0,00000000	0,68526532	0,00000000	0,00000000
1,0450	0,080875	0,39074852	0,03610852	0,57314296	0,00000000	0,00000000
1,0400	0,077089	0,34301222	0,15845785	0,49852993	0,00000000	0,00000000
1,0350	0,073762	0,29527592	0,28080716	0,42391692	0,00000000	0,00000000
1,0300	0,070958	0,24753961	0,40315651	0,34930388	0,00000000	0,00000000
1,0250	0,068743	0,19980331	0,52550584	0,27469085	0,00000000	0,00000000
1,0200	0,067173	0,15206700	0,64785517	0,20007782	0,00000000	0,00000000
1,0150	0,066295	0,10433069	0,77020453	0,12546478	0,00000000	0,00000000
1,0010	0,066136	0,05660396	0,89252827	0,05086677	0,00000000	0,00000000

The minimum-variance set for the Markowitz model is shown in Fig. 2.

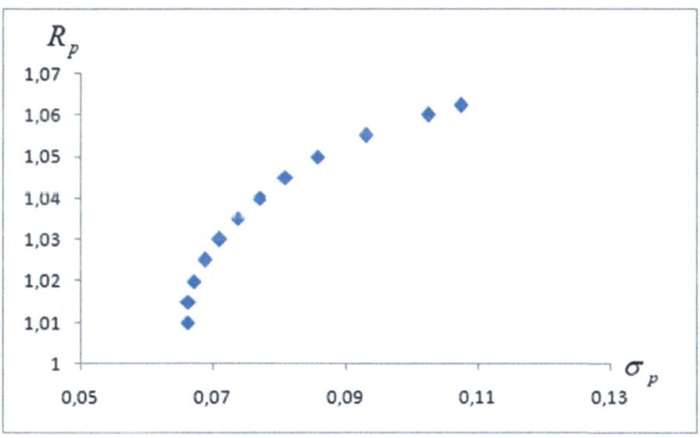

Fig. 2. Efficient frontier for the Markowitz model

To evaluate the parameters for the Market model the Regression analysis of Data Analysis Toolbox of Excel has been used. Table 4 presents the ‘beta’-coefficients of the assets, and regression statistics: the correlation coefficient R, R square and the standard error (variance of the random variable ε_{it}).

The high values of the R-coefficient indicate that the funds’ fluctuations are explained by performance fluctuations of the index [1, 2, 4]. Among the selected assets the *VTBR*-asset may be considered to be affected by factors other than the market to the highest

Table 4. Statistic parameters for the Market model

Asset	Asset parameters					
	$\bar{R}_i$	β_{iI}	Multiple R	R^2	Standard error	Observations
GMKN	1,021932	1,012719	0,997856	0,995717	0,067399	250
MTSS	1,003563	0,994515	0,998735	0,997471	0,050814	250
YNDX	1,060220	1,050662	0,996472	0,992957	0,089789	250
VTBR	1,062192	1,052617	0,791981	0,627234	0,823392	250
RSTI	1,038655	1,050662	0,995894	0,991805	0,094938	250

degree (the lowest value of R-square). Note that the most risky from the market point of view are *YNDX*-, *VTBR*- and *RSTI*-assets, since their 'beta'-coefficient exceeds unitary value significantly, while *MTSS*-asset is less sensitive to movements in the market index ($\beta_{iI} < 1$). Note that during the period under study, it was not possible to find assets negatively correlated with the market portfolio measured by the market index (no negative 'beta' found), as is shown in Table 4, which can be explained by mutual trend of market decline.

It can be seen that the obtained values of assets returns and 'beta'-factor confirm the basic statement of risk theory that a higher value of return $\bar{R}_i$ is associated with an increased value of risk (measured as 'beta'-factor in the Market model).

The return and the variance of the *IMOEX* Index are as follows: $R_I = 1.009097$; $\sigma_I^2 = 0.007228$.

We solve the risk-minimizing problem (16), (17) to find the efficient frontier of a feasible set for the Market model. The fixed values R were chosen using a 0.5% step. The obtained minimum-variance set is presented in Fig. 3.

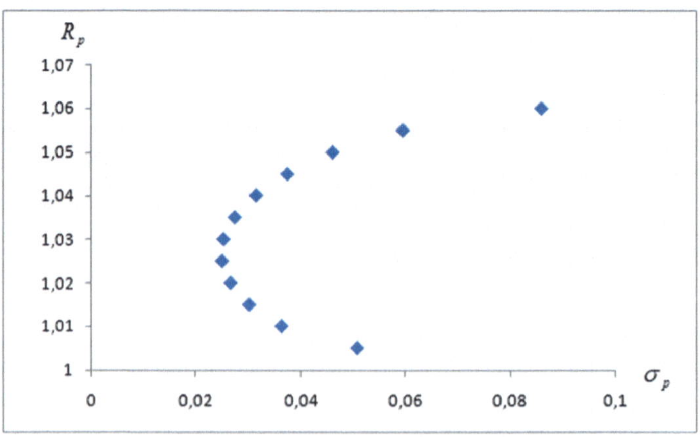

Fig. 3. Minimum-variance set for the Market model

The optimal solution yields the weight distribution $w^*_{iMarket}$, given in Table 5. Note that with regard to asset allocation it differs from the Markowitz weight distribution $w^*_{iMarkowitz}$ (see Table 3).

Table 5. Portfolio parameters and weight distribution $w^*_{iMarket}$

Portfolio return	Portfolio risk σ_p^2	Weights of the assets				
		GMKN	MTSS	YNDX	VTBR	RSTI
1,0600	0,085872	0,0000000	0,0000000	0,0000000	1,0000000	0,0000000
1,0550	0,059450	0,0000000	0,0000000	0,8762388	0,1040844	0,0196766
1,0500	0,046006	0,0000000	0,0000000	0,6733336	0,0775141	0,2491522
1,0450	0,037460	0,1163735	0,0000000	0,5482456	0,0623743	0,2730065
1,0400	0,031512	0,2073631	0,0284949	0,4464739	0,0503208	0,2673471
1,0350	0,027457	0,2215488	0,1157329	0,3764130	0,0422274	0,2440777
1,0300	0,025296	0,2357346	0,2029710	0,3063520	0,0341340	0,2208082
1,0250	0,025027	0,2499203	0,2902090	0,2362911	0,0260406	0,1975388

The optimal solutions $w^*_{iMarkowitz}$ and $w^*_{iMarket}$ presented in Table 3 and Table 5, respectively, have been obtained using the Optimization Toolbox of Excel and the Visual Basic Editor.

3.2 Case Study for $(R_p - R_{plow})$- and $(R_p - R_{pVaR})$-Minimizing Problem

In this part of experimental study the constructed portfolios for the introduced alternative risk measures. Letting the weight-coefficients w_i range over all possible combinations using a 0.1 step, so that $\sum w_i = 1$, we get different return-risk combinations. It order to obtain all possible weight combinations for asset allocation we used html-code for the considered number of assets (5 assets), setting the step to 10%.

```
<html>

<body>
<script>

i=0;
j=0;
k=0;

for(i=0;i<101;i+=10){
 for(j=0;j<101;j+=10){
  for(k=0;k<101;k+=10){
    for(m=0;m<101;m+=10){
  for(n=0;n<101;n+=10){
  if(i+j+k+m+n==100){
       document.write(i);
       document.write(" - ");
       document.write(j);
       document.write(" - ");
       document.write(k);
  document.write(" - ");
       document.write(m);
       document.write(" - ");
       document.write(n);
  document.write("<br>");
      }
    }
   }
  }
}
}
</script>
</body>
</html>
```

The obtained feasible set for $(R_p - R_{plow})$-minimizing problem is shown in Fig. 4. Note, that the obtained 'R_p - $(R_p - R_{plow})$' set conceptually is similar to that of the Markowitz model (see Fig. 1).

The optimal portfolios are on the efficient frontier of the minimum-$(R_p - R_{plow})$ set. The solution of the problem yields the weight distribution $w^*_{i(R-R_{low})}$ given in Table 6.

Note that the obtained weight distribution $w^*_{i(R-R_{low})}$ is closer to $w^*_{iMarket}$ (see Table 5) then to $w^*_{iMarkowitz}$ shown in Table 3

In order to analyze the introduced alternative $(R_p - R_{pVaR})$-risk measure we consider how the named risk estimates change for the entire set of portfolios. We estimated the risk for all possible combinations of asset weights, setting a 10% step in the weight change using the aforesaid html-code. For each combination of weights, using the values R_{it}, the series of moving portfolio returns R_{pt} were evaluated as well as the expected portfolio

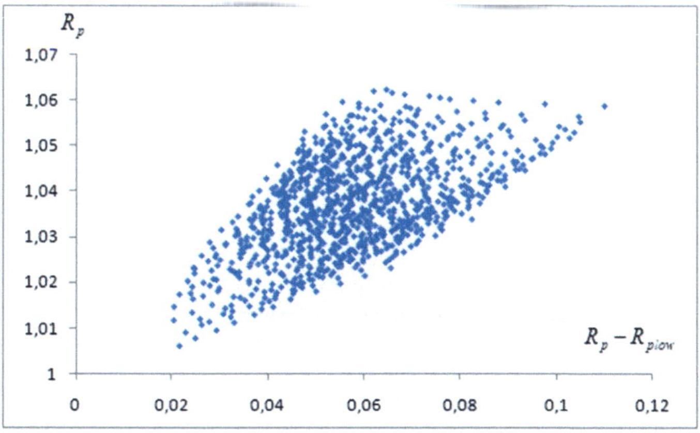

Fig. 4. Feasible set for $(R_p - R_{plow})$-minimizing problem

Table 6. Portfolio parameters and weight distribution for optimal portfolios $w^*_{i(R-R_{low})}$

Portfolio return	Portfolio risk $(R_p - R_{plow})$	Weights of the assets				
		GMKN	MTSS	YNDX	VTBR	RSTI
1,061964	0,062055	0	0	0,9	0,1	0
1,059649	0,055532	0	0	0,9	0	0,1
1,056982	0,052101	0	0	0,8	0	0,2
1,051369	0,047273	0	0,1	0,7	0	0,2
1,045756	0,041942	0	0,2	0,6	0	0,2
1,037476	0,036426	0	0,3	0,4	0	0,3
1,042809	0,039656	0	0,3	0,6	0	0,1
1,034529	0,03377	0	0,4	0,4	0	0,2
1,028916	0,027876	0	0,5	0,3	0	0,2
1,031583	0,030311	0	0,5	0,4	0	0,1
1,023303	0,024553	0	0,6	0,2	0	0,2
1,025969	0,026161	0	0,6	0,3	0	0,1
1,020356	0,023308	0	0,7	0,2	0	0,1
1,014743	0,020155	0	0,8	0,1	0	0,1
1,01741	0,021572	0	0,8	0,2	0	0
1,011796	0,020427	0	0,9	0,1	0	0

return. The $(R_p - R_{pVaR})$-risk measure was assessed for significance levels $\alpha = 0.1$ and $\alpha = 0.05$. The feasible sets of portfolios were obtained, which in shape turned out to be close to the Markowitz feasible set. The feasible sets 'R_p - $(R_p - R_{pVaR})$' for the significance levels $\alpha = 0.05$ and $\alpha = 0.1$ are shown in Fig. 5 and Fig. 6, respectively.

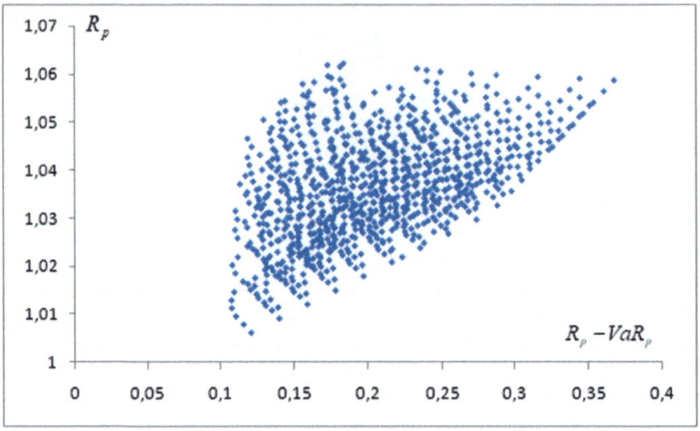

Fig. 5. Feasible set for $(R_p - R_{pVaR})$-minimizing problem, $\alpha = 0.05$

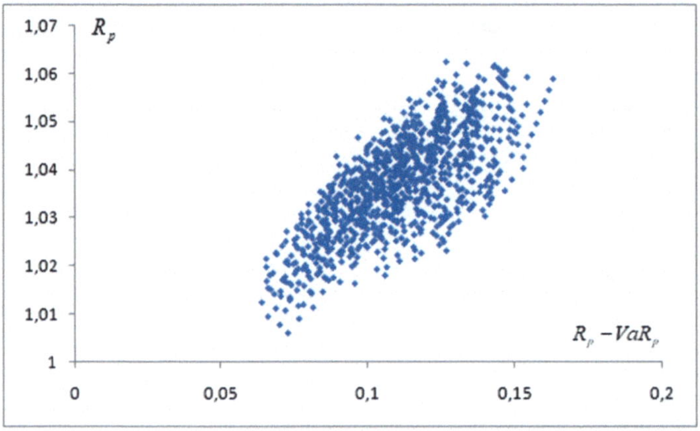

Fig. 6. Feasible set for $(R_p - R_{pVaR})$-minimizing problem, $\alpha = 0.1$

Weight distributions for asset allocation and parameters of efficient portfolios for $(R_p - R_{pVaR})$-minimizing problem for significance levels $\alpha = 0.05$ and $\alpha = 0.1$ are given in Table 7 and Table 8, respectively.

The efficient frontiers for the obtained weight distributions $w^*_{i(R-R_{low})}$ and $w^*_{i(R-VaR)}$ for the significance levels $\alpha = 0.05$ and $\alpha = 0.1$ are shown in Fig. 7.

Figure 7 demonstrates that the efficient frontier corresponding to the $(R_p - R_{plow})$-risk measure is located to the left of all. The efficient frontier for the $(R_p - R_{pVaR})$-risk measure at the significance level $\alpha = 0.1$ is located more to the left than at the significance level $\alpha = 0.05$. That is, efficient frontier lies the more to the left, the higher the level of significance and the narrower the confidence interval, and vice versa, the efficient frontier

Table 7. Portfolio parameters and weight distribution for optimal portfolios $w^*_{i(R-VaR)}$, $\alpha = 0.05$

Portfolio return	Portfolio risk $(R_p - R_{pVaR})$	Weights of the assets				
		GMKN	MTSS	YNDX	VTBR	RSTI
1,061964	0,172208	0	0	0,9	0,1	0
1,058051	0,155459	0,1	0	0,8	0,1	0
1,054139	0,14059	0,2	0	0,7	0,1	0
1,050578	0,128621	0,3	0	0,7	0	0
1,046665	0,117745	0,4	0	0,6	0	0
1,012986	0,106712	0,4	0,6	0	0	0
1,037139	0,112675	0,5	0,1	0,4	0	0
1,031526	0,109558	0,5	0,2	0,3	0	0
1,020299	0,10722	0,5	0,4	0,1	0	0
1,027613	0,109116	0,6	0,2	0,2	0	0

Table 8. Portfolio parameters and weight distribution for optimal portfolios $w^*_{i(R-VaR)}$, $\alpha = 0.1$

Portfolio return	Portfolio risk $(R_p - R_{pVaR})$	Weights of the assets				
		GMKN	MTSS	YNDX	VTBR	RSTI
1,062315	0,126379	0	0	1	0	0
1,05449	0,113975	0,2	0	0,8	0	0
1,012531	0,064113	0,2	0,7	0	0	0,1
1,050578	0,105457	0,3	0	0,7	0	0
1,03072	0,077458	0,3	0,3	0,2	0,1	0,1
1,046665	0,096808	0,4	0	0,6	0	0
1,032772	0,079972	0,4	0,2	0,3	0	0,1
1,027159	0,072556	0,4	0,3	0,2	0	0,1
1,021545	0,065632	0,4	0,4	0,1	0	0,1
1,042752	0,089355	0,5	0	0,5	0	0

is located the more to the right, the wider (more reliable) the confidence interval is set to be. Thus we may conclude that setting the significance level for the $(R_p \quad R_{pVaR})$ risk measure depends on the investor"s subjective risk expectations: the more the investor is 'reinsured', the more risky portfolios on the efficient set seem (with the same return).

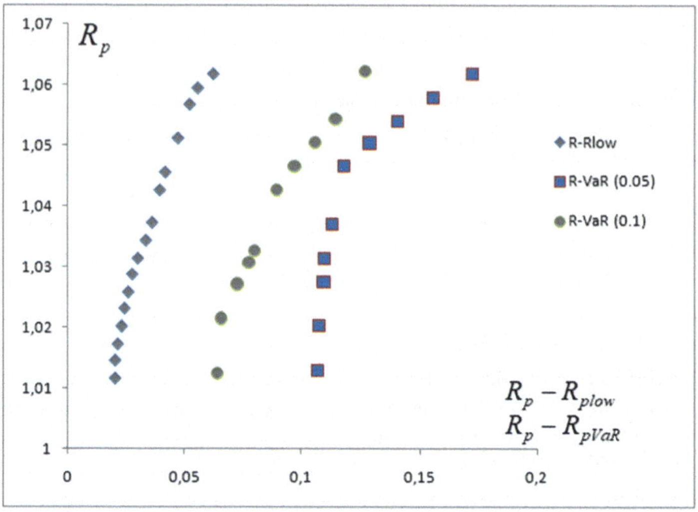

Fig. 7. Efficient frontiers for $(R_p - VaR_p)$- and $(R_p - R_{plow})$-risk measures

3.3 Portfolio Efficiency Assessment in Real Time

In this part of experimental study the constructed portfolios from the efficient frontiers with the weight distributions $w^*_{iMarkowitz}$, $w^*_{iMarket}$, $w^*_{i(R-R_{low})}$ and $w^*_{i(R-VaR)}$ for $\alpha = 0.05$, $\alpha = 0.1$ synthesized on the basis of the $(w_i w_j Cov_{ij})$-Markowitz model, of the $\left(\beta^2_{pI} \cdot \sigma^2_I + \sigma^2_{\varepsilon p}\right)$-Sharpe index model and on the basis of the introduced alternative risk $(R_p - R_{plow})$- and $(R_p - R_{pVaR})$-measures were verified in real time in a subsequent period, namely January 2021.

On the basis of series of the assets' moving returns we calculated portfolio moving returns for the mentioned period. The actual parameters of the portfolios for the named period were evaluated. The values of R_p and σ_p, corresponding to weight distributions $w^*_{iMarkowitz}$, $w^*_{iMarket}$, $w^*_{i(R-R_{low})}$ and $w^*_{i(R-VaR)}$ for $\alpha = 0.05$, $\alpha = 0.1$, vary within the intervals, presented in the Table 9.

It has been found that portfolios with $w^*_{i(R-VaR)}$ weight distribution in real time produced better results than portfolios with traditional risk assessment: higher mean return values were obtained at a lower risk.

Portfolios with distribution $w^*_{iMarkowitz}$ and $w^*_{i(R-R_{low})}$ have yielded the lowest return at the same risk-level.

The most efficient portfolios in this experiment were produced by $(R_p - VaR_p)$-model for the significance level $\alpha = 0.05$. This finding confirms the assumption that the level of significance depends on the investor's assessment of risk. Setting a more reliable confidence interval increases the reliability of the portfolio, which explains higher efficiency of such a portfolio in real time.

The 'mean – standard deviation' diagrams for portfolios with weight distributions $w^*_{iMarkowitz}$, $w^*_{iMarket}$, $w^*_{i(R-R_{low})}$, $w^*_{i(R-VaR)}$ in real time are presented in Fig. 8.

Table 9. Interval values of R_p and $\boldsymbol{\sigma_p}$ for weight distributions $w^*_{iMarkowitz}$, $w^*_{iMarket}$, $w^*_{i(R-R_{low})}$, $w^*_{i(R-VaR)}$

Model formulation	Portfolio parameters	
	R_p	$\boldsymbol{\sigma_p}$
$(w_i w_j Cov_{ij})$	[0,990781; 1,034413]	[0,017696; 0,026457]
$\left(\beta^2_{pI} \cdot \sigma^2_I + \sigma^2_{\varepsilon p}\right)$	[0,993840; 1,049437]	[0,017189; 0,029276]
$(R_p - VaR_p), \alpha = 0.05$	[0,993455; 1,062886]	[0,017237; 0,025501]
$(R_p - VaR_p), \alpha = 0.1$	[0,990780; 1,051493]	[0,018258; 0,026457]
$(R_p - R_{plow})$	[0,992181; 1,028722]	[0,018735; 0,027372]

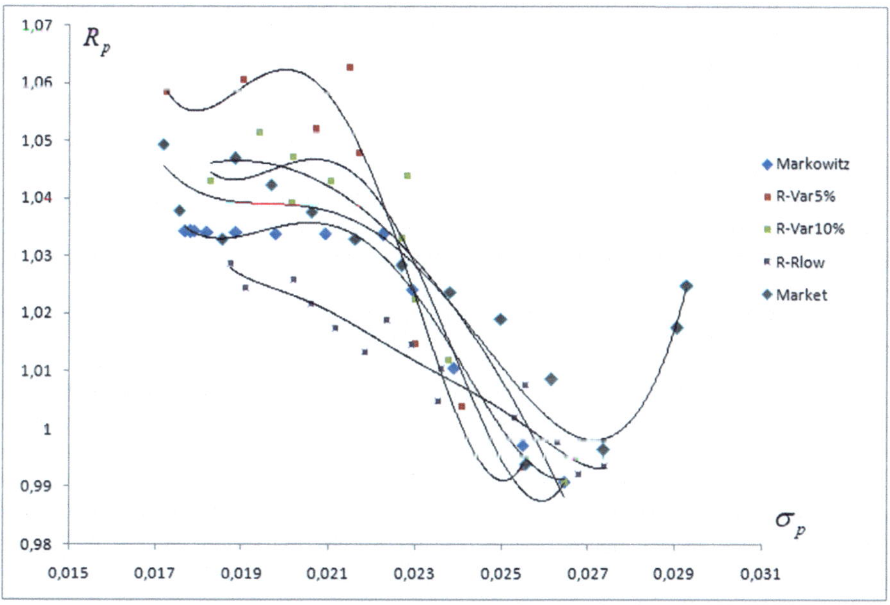

Fig. 8. 'Mean – standard deviation' diagrams for portfolios with weight distributions $w^*_{iMarkowitz}$, $w^*_{iMarket}$, $w^*_{i(R-R_{low})}$, $w^*_{i(R-VaR)}$ in real time.

4 Discussions

The feasible sets of portfolios for the introduced alternative risk measures turned out to be close to the Markowitz feasible set. That confirms the initial assumption of the essence of the risk assessment by the introduced $(R_p - R_{plow})$- and $(R_p - VaR_p)$-risk measures.

Comparative analysis of the efficient frontiers revealed that the efficient frontier corresponding to the $(R_p - R_{plow})$-risk measure is located to the left of all. The efficient frontier for the $(R_p - VaR_p)$-risk measure at the significance level $\alpha = 0.1$ is located more to the left than at $\alpha = 0.05$. That is, efficient frontier lies the more to the left, the higher the level of significance and the narrower the confidence interval, and vice versa, the efficient frontier is located the more to the right, the wider (more reliable) the confidence interval is set to be. Thus we may conclude that setting the significance level for the $(R_p - VaR_p)$-risk measure depends on the investor's subjective risk expectations: the more the investor is "reinsured", the more risky portfolios on the efficient set seem (for the same rate of return).

The differences in the obtained weight distributions $w^*_{iMarkowitz}$, $w^*_{iMarket}$, $w^*_{i(R-R_{low})}$, $w^*_{i(R-VaR)}$ for $\alpha = 0.05$, $\alpha = 0.1$ shown in Table 3, Table 5, Table 6, Table 7 and Table 8, respectively, confirm that the solution of the asset allocation problem is sensitive to the way of risk assessment.

The optimal solution $w^*_{iMarket}$ differs from the Markowitz weight distribution $w^*_{iMarkowitz}$. It may be explained by the differences in their approaches to measure the risk. As mentioned before, in the Markowitz model the risk of the portfolio depends on the covariance of the assets, while in the Market model the risk of portfolio depends on the correlation between the assets and the market portfolio.

The obtained weight distributions for the alternative risk measures turned out to be closer to each other than to the weight distributions for the traditional risk measures.

Note that portfolios with $w^*_{i(R-VaR)}$ weight distribution in real time produced better results than portfolios with traditional risk assessment: higher mean return values were obtained at a lower risk. The $(R_p - R_{plow})$-model has not proved its efficiency under experimental conditions and needs further research.

The most efficient portfolios in this experiment were produced by $(R_p - VaR_p)$-model for $\alpha = 0.05$. This finding confirms the assumption that the level of significance depends on the investor's assessment of risk. Setting a more reliable confidence interval increases the reliability of the portfolio, which explains higher efficiency of such a portfolio in real time.

To the best of our knowledge the conservative formulation of the $(R_p - VaR_p)$-model based on the more reliable confidence interval for $\alpha = 0.05$ yields better results relatively to $(R_p - VaR_p)$-model for $\alpha = 0.1$, to the Markowitz model and to the Market portfolio as well. This finding confirms the need of using the 'left-side' measures for risk assessments under financial market conditions.

5 Conclusion

In this paper we have proposed new approaches to synthesize investment portfolios: the $(R_p - R_{plow})$- and $(R_p - VaR_p)$-minimizing models. The performance of the proposed models has been compared with the common used Markowitz model and Market model, on the base of common stocks traded on the Moscow Exchange.

The difference between the portfolio mean return and 'low-mean' return, between the portfolio mean return and 'value-at-risk' return may be considered as alternative risk-measures. The 'mean- $(R_p - R_{plow})$' and the 'mean- $(R_p - VaR_p)$' sets are conceptually similar to the 'mean – variance' set of the Markowitz model.

The differences in the obtained weight distributions $w^*_{iMarkowitz}$, $w^*_{iMarket}$, $w^*_{i(R-R_{low})}$, $w^*_{i(R-VaR)}$ confirm that the solution of the asset allocation problem is sensitive to the way of risk assessment.

It has been shown that the $(R_p - VaR_p)$-model has produced the most efficient portfolios for the market under consideration: higher mean return values were obtained at a lower risk. It has been found out that setting a more reliable confidence interval increases the reliability of the portfolio, which is conducive to higher efficiency of such a portfolio in real time.

These findings confirm the validity of application and need for further development and research of alternative measures for risk assessments under financial market conditions.

References

1. Sharpe, W.F., Alexander, G.J., Beiley, J.V.: Investments. Prentice Hall International Inc., Upper Saddle River (1995)
2. Bodie, Z., Kane, A., Marcus, A.J.: Essentials of Investments. McGraw-Hill, New York (2001)
3. Fabozzi, F.J.: Quantitative strategies: multi-asset. J. Portfolio Manag. Multi-Asset Spec. Issue **46**(6), 1–3 (2020)
4. Luenberger, D.G.: Investment science. Oxford University Press, New York (1998)
5. Markowitz, H.M.: Portfolio selection. J. Financ. **7**(1), 77–91 (1952)
6. Chen, Z.-P., Zhao, C.-E.: Is the MV efficient portfolio really that sensitive to estimation errors. Asia-Pac. J. Oper. Res. **19**, 149–168 (2002)
7. Chopra, V.K., Ziemba, W.T.: The effect and errors in means, variances and covariances in optimal portfolio choice. J. Portfolio Manag. Winter **19**(2), 6–11 (1993)
8. Malakhova, A.A., et al.: Mean-variance portfolio optimization under parametric uncertainty. In: Silhavy, R., Silhavy, P., Prokopova, Z. (eds.) CoMeSySo 2020. AISC, vol. 1295, pp. 797–812. Springer, Cham (2020). https://doi.org/10.1007/978-3-030-63319-6_74
9. Malakhova, A.A., et al.: Reward-to-variability ratio as a key performance indicator in financial manager efficiency assessment. In: Silhavy, R. (ed.) CSOC 2020. AISC, vol. 1225, pp. 598–613. Springer, Cham (2020). https://doi.org/10.1007/978-3-030-51971-1_49
10. Moscow Exchange. https://www.moex.com/. Accessed 21 Feb 2021

The Assembly Processes Automation Methodology According to the IEC 61499 Standard on the LEGO Example

Artem Voinov[1(✉)] and Ilya Senokosov[2]

[1] Penza State University, Penza, Russia
[2] Cryptosoft, Penza, Russia

Abstract. This paper proposes a new approach to the 3D structures automating the assembly process. The approach that is designed allows you to verify assembly procedures by using formal predicate models. This uses Prolog's logical output language, which can be self-deposed during execution. The proposed approach consists of four stages: processing the source data and building an ontological model, building a connection diagram and building a formal predicate model, converting the model into the Prolog program, and translating the program into a function block. The technologist, developing the assembly process technological map, forms the raw data. This data is converted into a connection scheme. The connection scheme is described in the Prolog as a fact set. Prolog also develops rules that allow you to verify and upgrade the 3D structures assembly. The facts resulting set is converted into function blocks that are embedded in the assembly management system. The proposed approach is tested using the LEGO assembly process example. #CSOC1120

Keywords: Function blocks · LEGO-designs · CFS · Prolog · Ontology · Assembly · IEC61499 · Prolog · nxtStudio

1 Introduction

Industry 4.0 [1] is the cyber-physical systems (CPS) [2] widespread introduction into the world around us. The CPS is the physical part union with the intellectual component that allows this system to improve. This paper examines the algorithm for building the CPS for an automated 3D construction system.

In production, it is quite difficult to avoid errors in the complex 3D constructions assembly management system design. Errors in the assembly algorithm lead to the entire production downtime, which leads to significant time and material costs. The methodology proposed in this paper will avoid these errors by introducing an automated verification and modernization phase once the designer has provided a technological map.

Currently, there are many works devoted to the generation, assembly automated process and 3D structures processing. In the work of the 3D input model [3], the parts and connectors needed to build the corresponding physical object are automatically

© The Author(s), under exclusive license to Springer Nature Switzerland AG 2021
R. Silhavy (Ed.): CSOC 2021, LNNS 229, pp. 694–702, 2021.
https://doi.org/10.1007/978-3-030-77445-5_62

generated. The work [4] proposes a designing 3D models method, based on templates from the database. The paper [5] describes a method for creating complex 3D objects from flat 2D elements by making incisions in them and connecting them at certain angles. The article [6] presents an assembly design system from interconnected quadrangle ranges. 3D objects are assembled using 3D-printed and laser-cut 3D parts [7]. The article [8] presents a self-carrying structures optimal construction method. The work [9] describes various algorithms for constructing a given 3D figure from LEGO parts.

The main difference between this work and the previous ones is that during the command generation phase, the robot-builder uses intelligent assembly processing using the Prolog language [10].

The Prolog is a logical programming language based on the logic of the first-order predicate. This language main advantage over the rest is that it allows being self-indices directly during the execution. In the working with 3D structures context, this allows to process the input technology map and change it in two main areas:

1) The elements used number optimization. When designing 3D designs, a designer can create a design from several elements that can be replaced with one element. A similar algorithm is used in the work [11], but there is the entire LEGO design restructuring, and in this work, there is a point optimization where possible.
2) Refining the 3D design to improve by adding or removing blocks. The designer develops a technological map. It can contain errors that are difficult to identify during the design phase, and they can only prove themselves once the builder has started assembling.

2 Methodology for Assembling 3D Structures

The 3D construction process is the CPS, because any assembly algorithm has an intelligent device (cybernetic part) that controls the assembly process (physical part). The general principles of building a 3D design are reflected in the ontological model [12], which is the CPS ontology private case and is represented in Fig. 1.

2.1 Connection Scheme

The connection diagram is onto a graph, where the tops are parts, and bridges - directly connections. To transform from a technology map to a connection diagram, you must use a translator that allows you to present the assembly as oriented onto the graph. This ontograph is necessary to monitor the build sequence. The next step is the ontograph is transformed into the facts' description in Prolog.

2.2 Prolog

Prolog processing is the methodology presented in the most important part. The main goal is to optimize the assembly process and the 3D design itself. This uses the rules' ready-made sets that change the build process. The description in Prolog has two parts:

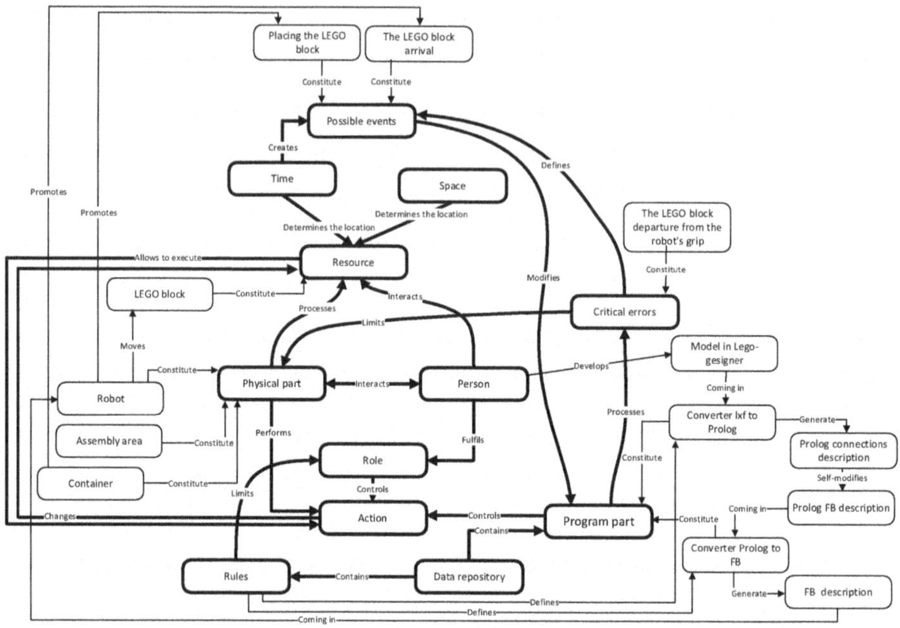

Fig. 1. The 3D constructions ontology based on the CPS high-level ontology.

1) Facts set derived from a compound scheme that fully describes all the details and connections used in the design.
2) A rule set for controlling and optimizing the build process.

After launch, the build process is modernized and optimized, at which a result two files will be created:

a. A file with a parts sequence coming from a conveyor to an assembly robot. The parts set may change after optimization and may not match the original set from the designer. The acceptable parts set definition for an assembly is determined by a separate rule in Prolog. For the parts sequence on the conveyor to correspond to the order specified in this file, the part sampling system described in [13] is used.
b. A file describing the commands sequence for the assembly robot

In this file, each command is encoded into a representation that can be read by the assembly robot. The valid commands set is determined by rules in the Prolog language.

2.3 FB

To control the assembly robot, a control system based on function blocks (FB) of IEC 61499 standard is used [14]. The IEC 61499 standard defines the distributed and reconfigurable control systems software structure, which is well suited for solving the automating the assembly process problem. The nxtStudio tool [15] can be used to design systems based on the IEC 61499 standard.

Given an abstract assembly robot existence based on the proposed control system, it becomes necessary to add an FB that generates an instructions sequence. A control scheme example of an assembly robot is shown in Fig. 2.

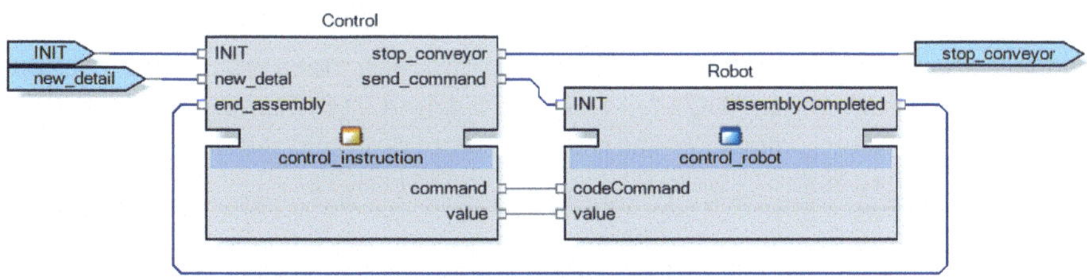

Fig. 2. FB control instruction and FB control robot.

In this view, FB *control_robot* is responsible for controlling the robot. The information input is supplied with a command code (*codeCommand*) and a numeric value (*value*). Event input *INIT* means that a new command has come for the robot. The *assembly completed event exit* reports that the part attached to the assembly is complete.

FB *control_instruction* is what we will create from Prolog. This FB has:

three event inputs:

1) *INIT* - starting the conveyer (starting the build system),
2) *new_detail* - a new part has arrived,
3) *end_assembly* - the previous part is attached to the assembly;

two event outputs:

1) *stop_conveyer* - an error occurred and the conveyor needs to be stopped,
2) *send_command* - a new command has been sent for building;

2 information outputs:

1) *command* - command code,
2) *value* is a numeric value.

For example: An instruction of the form rotate by 90^0 in Z will take the following form: *command* = *0xF0000012* and *value* = *90*, where *0xF0000012* is the command code Rotation along the + Z axis.

Any FB can be represented as a facts and rules set in Prolog language [16]. A part of the *control_instruction* Prolog representation is shown below:

```
event_input (control_insruction, new_detal).
event_output (control_insruction, stop_conveyor).
data_output (control_insruction, command, "UINT").
ec_state (control_insruction, getFromList).
ec_tran (control_insruction, "START", eil, _, getFromList).
ec_tran (control_insruction, getFromList, 1, _, "START").
eil :-( event_input (control_insruction, new_detal), event_input
(control_insruction, "INIT")); (event_input (control_insruction,
new_detal), event_input (control_insruction, end_assembly)).
ec_action    (control_insruction,    getFromList,    1,    alg1,
send_command).
algorithm (control_insruction, alg1, "Algorithm").
```

In this view, the string "Algorithm" is replaced with the generated instructions sequence for the robot.

The last step is to translate the obtained FB representation into *control_instruction* and embed it into the general scheme.

3 Implementation Example

LEGO is a prefabricated 3D construction prime example. Initially, the design is carried out using a special editor on the computer - the LEGO Designer [17]. This program creates a file with a description of the parts used (which are encrypted by id) and the affine transformations' matrix for the part correct orientation in space. Thus, based on the file created in LEGO Designer, it is impossible to understand the connection diagram between the blocks.

A primitive LEGO project, shown in Fig. 3, was chosen as its functionality demonstration.

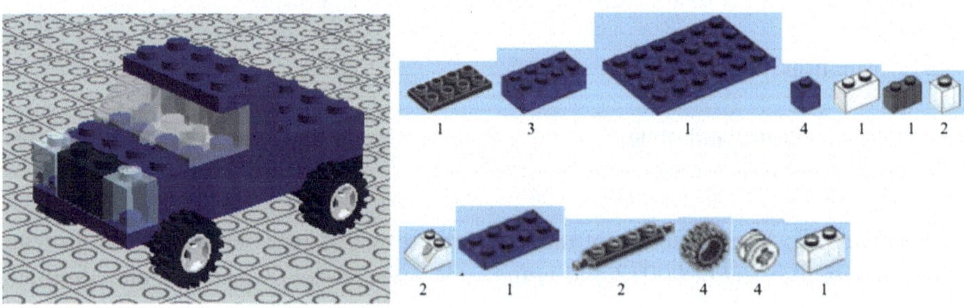

a) **b)**

Fig. 3. The model developed by the designer (a) and the details used in the model (b).

The part instances number is labelled under the part image in Fig. 3b. This model consists of 27 parts (13 types of parts). The LEGO pieces standard dimensions can be seen in [18].

Table 1. LEGO parts encoding.

id	Detail	Material	Assembly number	Thickness	Shape
102	4 × 2 Fiol	Plastics	3	9.6 mm	Rectangle
105	6 × 4 Fiol	Plastics	1	3,2 mm	Rectangle
112	2 × 1 Black	Plastics	1	9.6 mm	Rectangle
115	2 × 2 Clear	Glass	1	9.6 mm	Square

For further processing convenience, we will assign a unique identifier to each part. A part identification example is shown in Table 1.

The output from LEGO Designer is a file in the lxf format (a special LEGO Designer format), which is translated into a schematic represented in an owl representation form (owl is an ontology representation format). To work with this format you can use Protégé [19]. This view is necessary to highlight parallel assembly elements. an assembly ontology example is shown in Fig. 4.

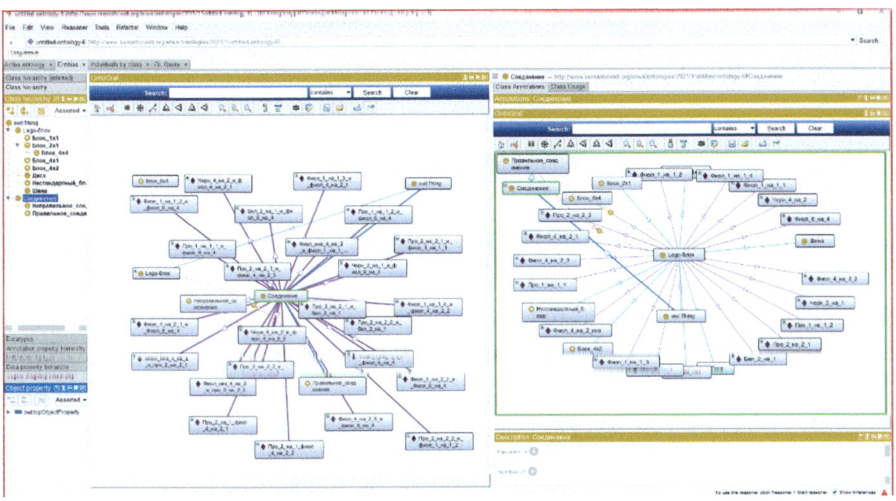

Fig. 4. An assembly ontology piece example in Protégé.

The next step is to convert the selected owl representation into a Prolog representation. A Prolog view example for a designed LEGO assembly:

```
first_detail (id_105).,
detail (id_105,8,4).
detail (id_102,4,2).
detail (id_112,2,1).
detail (id_115,2,2).
connection (number6, id_102,1, id_105).
coordinates (number6, id_102,1,1,4,2).
coordinates (number6, id_105,2,1,3,4).
connection (number7, id_112,1, id_105).
coordinates (number7, id_112,1,1,2,1).
coordinates (number7, id_105,1,2,1,3).
connection (number10, id_115,0, id_102).
coordinates (number10, id_115,1,1,2,2).
coordinates (number10, id_102,1,1,2,1).
```

Let's define the command identification rules for our example in Table 2. This table reflects some of the abstract assembly robot commands.

Table 2. Assembly robot commands.

id	Detail
0xF0000001	Hook detail
0xF0000008	Move along axis+X
0xF000000A	Move along axis+Y
0xF000000D	Move along axis−Z
0xF0000012	Rotation+Z
0xF0000014	To fix the position
0xF0000015	Return to fixed position
0xF0000016	Move to assemblage point
0xF0000017	Release detail
0xF0000018	Increase assembly height

For a given LEGO model, the assembly rules apply (which are loaded into the Prolog representation of the assembly). The example shows only a few rules.

```
 sub (X, Y, Res): - Res is Y - X.
 add_join: - connection (X, _, _, _), assert (unused_connection
(X)), not (unused_connection (X)), write (unused_connection
(X)), nl; true.
 rot (X): - X == 1, write ('0xF0000012'), write (' '), write
('90'), nl; true.
 second (Where): - connection (Join, That, Rot, Where),
unused_connection (Join), coordinates (Join, Where, W_X1, W_Y1,
W_X2, W_Y2), coordinates (Join, That, T_X1, T_Y1, T_X2, T_Y2 ),
detail (Where, WC_X, WC_Y), detail (That, TC_X, TC_Y), write
('0xF0000001'), nl, write ('0xF0000018'), nl, write (''), write
('10 ') , nl, write ('0xF0000016'), nl, sub (1, W_X1, Mov_X),
sub (1, W_Y1, Mov_Y), rot (Rot), write ('0xF0000008'), write
(''), write ( Mov_X), nl, write ('0xF000000A'), write (''),
write (Mov_Y), nl, write ('0xF000000D'), nl, write
('0xF0000017'), nl, write ('0xF0000015'), nl, retract
((unused_connection (Join))), assert (built (That)), second
(Where); true.
 four (X): - second (X), X \ == Y, (retract (built (X)); true),
built (Y), four (Y); true.
 first: -add_join, write ('0xF0000014'), nl, write
('0xF0000001'), nl, write ('0xF0000016'), nl, write
('0xF0000017'), nl, write ('0xF0000015'), nl, second (X).
```

The Prolog self-modification process initiation is carried out by the rule *first*. The output is a Prolog representation of the FB *control_instruction*, which is translated into the fbt format (the format for recording the FB in nxtStudio) and embedded in nxtStudio.

As the methodology all stages implementation result, a verified FB was created for controlling the robot-assembler. This FB does not contain primitive design errors. As a work emulation results in the nxtStudio environment, the FB worked correctly for the test case.

4 Conclusion

The developed methodology applied to the 3D structures assembly production will allow avoiding design errors, which contributes to the reduction in the production process cost. Finding errors at the loading commands stage into the robot will force the 3D structure design almost from the beginning, which will result in the entire production long-term downtime.

In the future, it is planned to develop the generated FB several variants, depending on a particular robot control scheme (CAT block, composite block, etc.).

Acknowledgements. The SS processing technological process development was carried out with the financial support of the Russian Federation President Council on Grants (the Russian Federation President scholarship for young scientists and graduate students) (project No. SP-2511.2021.5).

References

1. Lasi, H., Fettke, P., Kemper, H., Feld, T., Hoffmann, M.: Industry 4.0. Bus. Inf. Syst. Eng. **6**(4), 239–242 (2014)
2. Iarovyi, S., Mohammed, W., Lobov, A., Ferrer, B., Lastra, J.: Cyber-physical systems for open-knowledge-driven manufacturing execution systems. Proc. IEEE **104**(5), 1142–1154 (2016)
3. Lau, M., Ohgawara, A., Mitani, J., Igarashi, T.: Converting 3D furniture models to fabricate parts and connectors. ACM Trans. Gr. (TOG) **30**(4), 1–6 (2011)
4. Schulz, A., Shamir, A., Levin, D.I., Sitthi-Amorn, P., Matusik, W.: Design and fabrication by example. ACM Trans. Gr. (TOG) **33**(4), 1–11 (2014)
5. Cignoni, P., Pietroni, N., Malomo, L., Scopigno, R.: Field-aligned mesh joinery. ACM Trans. Gr. (TOG) **33**(1), 1–12 (2014)
6. Skouras, M., Coros, S., Grinspun, E., Thomaszewski, B.: Interactive surface design with interlocking elements. ACM Trans. Gr. (TOG) **34**(6), 1–7 (2015)
7. Song, P., et al.: CofiFab: coarse-to-fine fabrication of large 3D objects. ACM Trans. Gr. (TOG) **35**(4), 1–11 (2016)
8. Deuss, M., et al.: Assembling self-supporting structures. ACM Trans. Graph. **33**(6), 214–221 (2014)
9. Kim, J., Kang, K., Lee, J.: Survey on automated LEGO assembly construction. In: 22nd International Conference in Central European Computer Graphics, pp. 89–96 (2014)
10. Clocksin, WF, Mellish, C.S.: Programming in Prolog: Using the ISO standard. In: Springer Science + Business Media, Berlin (2012)
11. Stephenson, B.: A multi-phase search approach to the LEGO construction problem. In: Ninth Annual Symposium on Combinatorial Search (2016)
12. Guarino, N., Oberle, D., Staab, S.: What is an ontology? In: Staab, S., Studer, R. (eds.) Handbook on ontologies, pp. 1–17. Springer, Heidelberg (2009). https://doi.org/10.1007/978-3-540-92673-3_0
13. Dubinin, V., Senokosov, I., Vyatkin, V.: Auto-generation of distributed automation software based on formal product line specification. In: Mařík, V., Wahlster, W., Strasser, T., Kadera, P. (eds.) Industrial Applications of Holonic and Multi-Agent Systems. HoloMAS 2017. Lecture Notes in Computer Science, vol. 10444, pp 80–91. Springer, Cham (2017). https://doi.org/10.1007/978-3-319-64635-0_7
14. International Electrotechnical Commission 2005 IEC 61499–1: Function Blocks-Part 1 (Architecture. Int. Stand. First Ed. Geneva 1)
15. nxtStudio (nxtControl). http://www.nxtcontrol.com/
16. Dubinin, V., Vyatkin, V., Hanisch, H.: Modeling and verification of IEC 61499 applications using prolog. In: 2006 IEEE Conference on Emerging Technologies and Factory Automation, pp. 774–781. IEEE (2006)
17. LEGO Designer. https://www.lego.com/en-us/ldd/
18. LEGO - 01 Basic Dimensions & Bricks Explained. https://grabcad.com/tutorials/lego-01-basic-dimensions-bricks-explained/
19. Protege. https://protege.stanford.edu/

Conceptual Approach to Implementation of Adaptive Protection of Operational Cybersecurity Centers

M. M. Putyato(✉), A. S. Makaryan, and M. V. Evsyukov

Kuban State Technological University, Moskovskaya Str., 2, Krasnodar 350072, Russian Federation

Abstract. The article approaches conceptual aspects of analyzing and modeling adaptive protection systems of operational cybersecurity centers. The conditions of global information environment are constantly changing which provides security systems an opportunity to automatically adapt to requirements of modern software. The problem of integrating a complex security system into an operational cybersecurity center and defining interaction rules between the functional components consists of several tasks. The main of them are analysis and implementation of various formalized procedures for ensuring integrated security. An adaptive integrated security system is considered as an object that either implements the ability to control and promptly build a protection model, depending on the problem being solved, or assists in preventing or eliminating an incident.

The approach proposed is conditioned by two primary reasons. The first one is increasing functionality of modern technologies. The second one is strict requirements for adaptive integrated solutions that must have a scalable architecture to provide protection against a large variety of threats. The use of a complex security system within the framework of the approach proposed is expected to allow for prompt responds to incidents of both external and internal nature. Therefore, the amount of time spent on neutralization of the incident consequences will be reduced.

The adaptive integrated security system, as a common platform, will provide timely monitoring, context and control capabilities in various situations. The development of such an integration platform will increase automation degree and improve the quality of information provided by information security products.

Keywords: Complex security · Operational cybersecurity center · Information security · Complex adaptive systems · Full overlap model · Mathematical model

© The Author(s), under exclusive license to Springer Nature Switzerland AG 2021
R. Silhavy (Ed.): CSOC 2021, LNNS 229, pp. 703–713, 2021.
https://doi.org/10.1007/978-3-030-77445-5_63

1 Introduction

The rapid development of various economic sectors is imposing such new requirements for information systems as:

- minimization of the human factor;
- increasing automation of information processing and management processes;
- intellectualization of solutions;
- security of information technologies, customers and organizations;
- adaptivity of the system to constantly changing conditions.

A special place in this paradigm is occupied by security of IT systems, especially critical information infrastructures [1], regulated by the normative and legislative acts of Russian Federation.

The conditions of global information environment are constantly changing which provides security systems an opportunity to automatically adapt to requirements of modern software.

Operational cybersecurity center can be considered as a combination of multidimensional system that organizes all stages of decision-making and a set of subsystems which support hardware, technical and software types of work. Therefore, it is necessary to formulate tasks that are solved by the means of interaction between a complex security system (CSS) and structural elements of the operational cybersecurity center (OCC).

The functional features of decision support systems [2] (DSS), which are used within OCC, allow for constructing a multipurpose algorithm based on relationships among factors describing the task and the target state of a control object [3, 4].

The adaptive system for ensuring integrated security as component of an OCC has access to all the capabilities available to the main elements of the distributed structure. In particular, the integral stages of intellectual support, while implementing the adaptability of the security state, are situation modeling, expert support and intellectual Internet search.

The problem of integrating a CSS into an OCC and defining interaction rules between the functional components consists of several tasks. The main of them are analysis and implementation of various formalized procedures for ensuring integrated security. In various works particular cases of using both existing means of countering unauthorized access to critical data [5, 6] and new approaches to implementation of modern protection methods [7] were considered.

According to modern research [8], an adaptive security system should apply management and adaptation methods from other areas of scientific knowledge for information security issues. The use of generalized principles of adaptation of an applied nature reflect the specifics of a particular area. Studies [9] show that most of adaptive control systems are built on the basis of regulated and well-established models of adaptation of safety management systems. Such models offer either a parametric adaptation approach, which allows to adjust the system parameters without changing its operation principles, or structural adaptation, which means changing the model's structure but preserving the system parameters.

The use of multipurpose information security systems allows to perform a wide range of such functions as monitoring, processing and control of heterogeneous information with a sufficiently large degree of uncertainty. Processing information and using it for a specific situation takes a lot of time, informational and analytical resources [10].

The functioning of various types of systems [8], within the framework of a single adaptive integrated security system, will ensure the implementation of a larger number of tasks with minimal involvement of resources. Moreover the use of mathematical and intellectual apparatus will provide the methodological component for such systems.

Taking into consideration that an adaptive system with intelligent control is a necessary solution that affects the speed of response to information security incidents, we will analyze the set of artificial intelligence methods that have already been proposed for use in the distributed structure of OCC [11]:

1. Artificial intelligence techniques for monitoring:
   - Computer vision;
   - Intellectual agents;
   - Natural language processing;
   - Expert systems;
   - Cognitive modeling.
2. Artificial intelligence techniques for forecasting:
   - Neural networks;
   - Fuzzy logic;
   - Expert systems.
3. Artificial intelligence techniques for planning:
   - Neural networks;
   - Fuzzy logic;
   - Expert systems.

Interaction with Information Subsystems of the OCC. The organization of a complex security system (CSS) functioning is based on active inclusion of information systems without disrupting their stability and performance. CSS is an integrated platform for rapid detection and response to threats with automatic data collection and mining. Such a platform constantly monitors the security of the system and both notifies about violations and suspicious events and creates a model of events. CSS supports the processes of monitoring, forecasting, planning and decision-making from the point of view of information and integrated security at all stages of the functioning of the OCC [9].

Application of Smart Technologies for Ensuring Security. The use of mathematical methods and artificial intelligence technologies is able to minimize participation of incompetent specialists in ensuring security. Synergetic combination of knowledge and experience in certain areas of security makes it possible to solve the issue of ensuring an object protection by combining security areas via a certain set of markers or a system of criteria into a complex security system of the OCC.

Monitoring OCC Functioning in Various Modes. OCC is able to switch to various modes to create an optimal structure designed for effective interaction and use of all system components. In this case, a CSS provides structural, parametric and situational safety during the functioning of the integrated complex both in real time and offline [9].

Formation of New Safety Models Based on Modeling Systems and Expert Opinion. When considering multidimensional problems and the development of the situation, the systems of simulation, agent-based and discrete-event modeling can be actively used. Enabling expert support procedures will provide additional information about security issues. Using such procedures if effective so it will provide an opportunity to reduce the amount of time spent on making decisions [10] and to apply the heuristic apparatus of experts to accumulate a corresponding knowledge base.

Organizational and Legal Support. According to the results of a study conducted by Ponemon Institute and Fasoo [12], carelessness of employees is the main cause of data leaks. It was the cause of more than half of the incidents (56%). In 37% of cases, a leak occurred as a result of loss or theft of the device. 35% of incidents were attributed to third party error. Only 22% of confidential information disappeared due to the fault of intruders not associated with the company [13].

A CSS should also include functions of regulatory and organizational security support, especially in relation to OCC employees who have a sufficient amount of knowledge and information.

Data store. DB and DS security. Cloud technologies. Distributed data transmission and storage technologies. Visualization subsystem. A group of users for work in the protected perimeter of the closed part of the OCC is considered. Identification systems using combined access methods: biometric, hardware and software, organizational, passwords and access keys.

An adaptive integrated security system is considered as an object that either implements the ability to control and promptly build a protection model for the OCC, depending on the problem being solved, or assists in preventing or eliminating an incident. Let us review in detail interrelations and interaction of a CSS with all elements of an OCC.

1. Communication of a CSS with the information and analytical system for monitoring, planning, forecasting [9] and decision-making in an OCC:

   - intrusion prevention;
   - enforcing security policies;
   - prompt warning of violations and incidents;
   - visualization of key indicators based on the results of safety monitoring;
   - software and hardware protection;
   - technical protection.

2. Communication of a CSS with the universal data warehouse reflects the following functions:

- data backup;
- security of information bases;
- lossless data transmission;
- constant system availability.

3. Communication of a CSS with the structure and methodology software and artificial intelligence of OCC provides:

- complex security models and policies;
- artificial intelligence security methods;
- principles of complex security implementation;
- forecasting new threats and vulnerabilities.

4. Communication of a CSS with OCC staff:

- monitoring performance indicators of CSS;
- enforcing access control to the physical object;
- ensuring the regulatory aspect of security.

2 Mathematical Model of a Complex Security System

A mathematical model is a formalized description of scenarios in the form of logical algorithms represented by a sequence of violators' actions and responses. The quantitative characteristics of the model determine the main functional dependencies that are used to assess the vulnerability of an object, build a protection system and then consider its effectiveness.

There are several types of models distinguished by factors:

- risk minimization models [14];
- models based on system threat analysis [15];
- finite state models [16];
- models based on discrete components [17].

In order to select the required model of the protection system, it is necessary to compare them in terms of the parameters that determine the functioning of the protection system and permit or prohibit operations on the protected object (Table 1).

It can be seen from the presented table, that the best models for organizing efficient work of an intelligent OCC are models based on system threat analysis. The full overlap model [18] is one of the most popular and efficient methods of analysis. However, it should be noted that it must have at least one mean (a security subject) on every possible path of entry into the system. In order to take quantitative security factors of the security system into consideration, a system model with complete overlap has been defined, which belongs to the type of threat analysis models in the system. This model allows for continuous monitoring of quantitative indicators that enable their users to track a set of protected objects. Each of the objects must be associated with a certain amount of

Table 1. Comparison of models for creating a complex information security system.

Model criterias	Implementation mechanism	Dynamics of system change	Financial expenses	Assessment of vulnerability and damage degree
Risk minimization models	Middle	Middle	High	Middle
Models based on system threat analysis	Difficult	High	Low	High
Finite state models	Difficult	Middle	High	Middle
Models based on discrete components	Easy	Low	Low	Low

damage and a number of actions that an attacker can resort to gain unauthorized access to an object.

The full overlap model is considered in direct interaction of the "threat area", "defense system" and "protected area". Such a model can be described by the following sets:

- $T = \{t_i\}$ – set of threats;
- $O = \{o_j\}$ – set of protected system objects;
- $M = \{m_k\}$ – set of OCC security mechanisms.

The protection system is described by relations between elements of the sets.

In such a case, a graph model is usually used to describe the protection system, in which the set of threat-object relations form a bipartite graph. The goal of the protection is to cover all possible edges of the graph. This is achieved by introducing the third set called M. The standard full overlap model is shown in Fig. 1.equations are centered and set on a separate line.

A model which refines the general model with full overlap, where a set of vulnerabilities – V and a set of barriers – B are introduced, has shown to be the most effective one recently.

Vulnerability of the protection system is a possibility of the implementation of a threat t_i against an object o_j. The set of vulnerabilities – V is determined by the subset of the Cartesian products T*O: $v_r = \langle t_i, o_j \rangle$.

Barriers are determined by the Cartesian product V*M: $b = \langle t_i, o_j, m_k \rangle$. Barriers are ways of implementing security threats, blocked by means of protection.

For a full-coverage system, there is a barrier to any vulnerability. In other words, for all security threats, there must be protection mechanisms that prevent implementation of the threats.

Protection mechanisms provide certain resistance to threats. Therefore, barrier characteristics can be presented as a set $\langle P_i, D_i, R_i \rangle$, where P_i is probability of occurrence

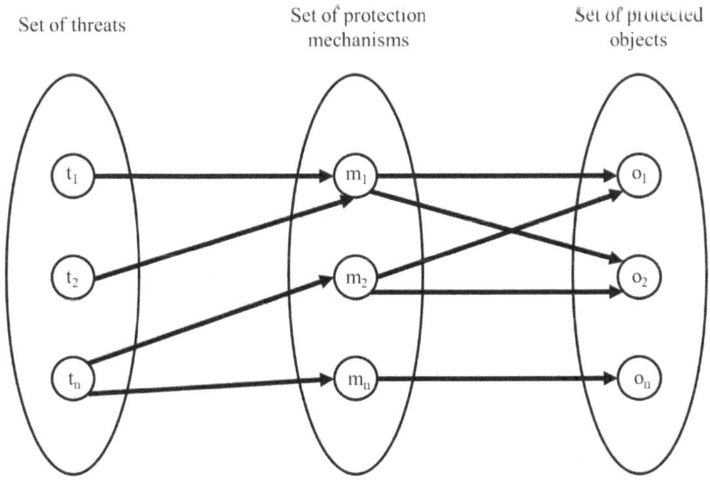

Fig. 1. Standard model of a complex security system with full overlap in the form of a three-part graph.

for the i-th threat, D_i – potential damage caused by implementation of the i-th threat and R_i – the degree of resistance of the protection mechanism.

The strength of the barrier is described by the value of the residual risk X_i, associated with the possibility of a threat t_i against an object o_j while using a protection mechanism s_k.

$$X_i = P_k * D_k * (1 - R_k) \tag{1}$$

In order to determine the value of security, the following ratio is used:

$$Z = \frac{1}{\sum_{\forall b_k \subset B} P_k * D_k (1 - R_k)} \tag{2}$$

In practice, obtaining the exact values of the above characteristics of barriers is difficult, since the concepts of threat, damage and resistance of the protection mechanism are difficult to formalize.

We consider that cost methods of assessing the effectiveness of protection means are relevant for defending information that can be assessed. For these methods, the characteristics of barriers are complemented by the quantity F_i - cost of creating a barrier bi. In this case, the choice of the optimal set of information security tools means minimizing total cost $A = \{a_i\}$, which consists of the cost of creating a barrier $F = \{f_i\}$ and losses caused by the threat $N = \{n_i\}$ if it occurs.

A modified model of an integrated security system with full coverage is presented in Fig. 2.

This model is general and does not fully reflect the conceptual structure of the CSS within the OCC. It is proposed to use an adaptive modified full overlap model. The proposed model will allow, depending on the goals and tasks to be solved, many mathematical models can be organized as part of an integrated protection system, which even at the design stage of the system make it possible to evaluate its efficiency (Fig. 3).

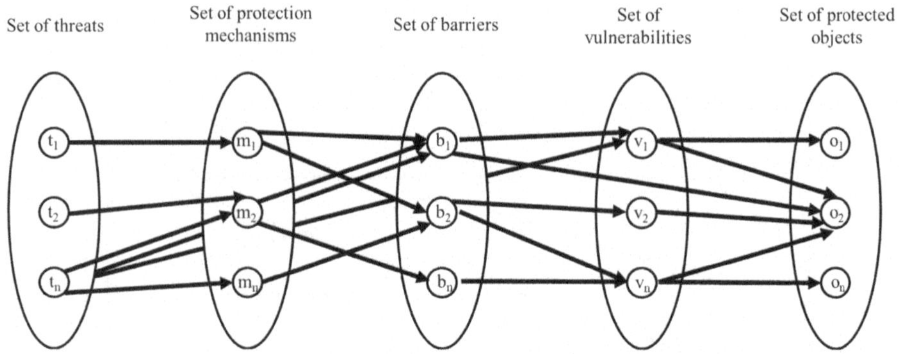

Fig. 2. Modified model of a complex security system with complete overlap in the form of a five-part graph.

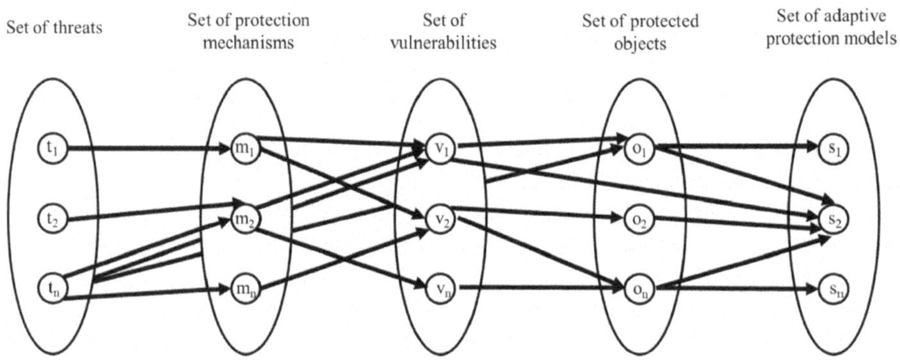

Fig. 3. Modified model of a complex security system with full overlap in the form of a five-part graph with the inclusion of a set of adaptive models.

It is proposed to describe the adaptive security management model by the formula *Security = Monitoring + Audit + Risk Analysis + Security Policy + Protections + Implementation of Countermeasures + Response.* The process of determining the efficiency of integrated security systems consists of substantiating indicators (criteria) for evaluating the efficiency and methods for calculating these indicators. We suggest choosing an optimization approach for assessing a complex security system, which means maximizing a certain function under given constraints. In the case of adaptive systems, it is proposed to expand the specified methodology with the parameter k, which reflects stages of the adaptive process.

We introduce the following notation:

$T = \{t_j\}$ – set of security threats, j = 1, … n;

$M^k = \{a_i^k\}$ – set of security mechanisms used on the k-th stage of the adaptive process, k = 0…R.

$V = \{v_i\}$ – set of vulnerabilities, i = 1…n.

$C^k = \{c_i^k\}$ – acceptable costs of creating object protection (the amount of costs for maintaining the system, taking into account the implementation of the k-th stage of the adaptive process), and ci are the costs of initial development, training and initial launch of the adaptive system;

$d^k(i,j)$ – neutralization efficiency of the j-th threat by the i-th mechanism on the k-th stage.

The variable p(i,j) is introduced to construct a mathematical model. It equals 1, if the j-th threat is eliminated by the i-th mechanism and 0 – otherwise. We also introduce the variable q such that:

$$q(i,j) = \begin{cases} 1 - \text{if}, \ i - \text{th security mechanism} \\ \text{if used to eliminate tht } j - \text{th threat} \\ 0 - \text{otherwise} \end{cases} \quad (3)$$

If information threats are not related to each other, then it is required to find the maximum effect from neutralizing the set of information threats T using the protection means S declared in the system under restrictions on the total amount of costs C.

$$\sum_{k=1}^{r} \sum_{j=1}^{m} \sum_{i=1}^{n} d^k(i,j)p(i,j) \to max \quad (4)$$

with restrictions

$$\sum_{i=1}^{n} c(i)^* sign \sum_{u_j \in U}^{n} p(i,j) \le C \quad (5)$$

$$p(i,j) \in [1, 0]$$

After slightly modifying the problem statement and introducing the concept of a membership function $\mu^S(x_i)$, we get an option for assessing the effectiveness of the information security system in fuzzy indicators. Let W– countable set of indicators $W = \{w_i\}$, i = 1..h, where h-number of indicators. Belonging to a certain security level is determined at a given interval [0, T]. Then the set of values of the model S that determine the fulfillment of security requirements is defined as: $S = \sum_{i-1}^{n} \frac{\mu^A(x_i)}{x_i}$, where $\frac{\mu^A(x_i)}{x_i}$ – pair «membership function/element».

A conceptual approach to the implementation of adaptive protection of the operational cybersecurity center can be represented using a structural model (Fig. 4).

Various security states of the system are described in the form of subsets of the fuzzy set, and the probability of breaking the evaluated system corresponds to the cardinal number (power) of the fuzzy set. This approach means assessing efficiency of an adaptive complex security system so it is necessary to gather information about the security requirements and to which extent they are met. Such an approach allows to establish continuous monitoring of the OCC security status, to predict the possibility of attacks and to change the requirements for security variables. Thus security staff gets an ability to control newly emerging threats, eliminate vulnerabilities that can lead to implementation of a threat, analyze the conditions leading to the emergence of vulnerabilities.

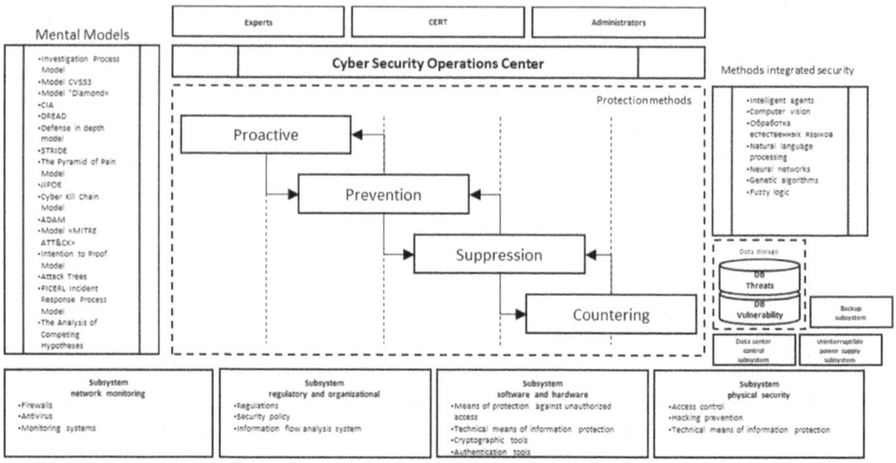

Fig. 4. Structural model for the implementation of adaptive protection of the operational cybersecurity center.

3 Conclusion

The analysis of the main functions and characteristics of the adaptive complex security systems based on OCC has led to obtaining the following results:

The approach proposed is conditioned by two primary reasons. The first one is increasing functionality of modern technologies. The second one is strict requirements for adaptive integrated solutions that must have a scalable architecture to provide protection against a large variety of threats.

The use of a complex security system within the framework of the approach proposed is expected to allow for prompt responds to incidents of both external and internal nature. Therefore, the amount of time spent on neutralization of the incident consequences will be reduced.

An adaptive complex security system, as a common platform, will provide timely monitoring, context and control capabilities in various situations. The development of such an integration platform will improve the automation process and significantly improve the quality of information provided by information security products.

References

1. Federal Law No. 187-FZ of July 26, 2017 «On the security of the critical information infrastructure of the Russian Federation» (2017)
2. Kurejchik, V.M.: Peculiarities of creating decision support systems. Bulletin of the Southern Federal University. Technical sciences. Thematic issue, pp. 92–98 (2012)
3. Putyato, M.M., Makaryan, A.S.: Cybersecurity as an essential attribute of a multilevel protected cyber space. Caspian J. Control High Technol. **3**(51), 94–102 (2020)
4. Putyato, M.M., Makaryan, A.S.: Study of the possibility of improving the cybersecurity of the Internet of Things infrastructure based on the integration of biometric authentication methods. In: Information systems and technologies in modeling and management, pp. 267–270. LNCS, Yalta (2020)

5. Vlasenko, A.V., Putyato, M.M., Makaryan, A.S.: Possibilities of improving the cyber security of mobile devices based on the integration of dynamic biometric methods. In: Taran, V.N. (eds.) Proceedings of the V International Scientific and Practical Conference, pp. 369–372. Remote Educational Technologies, Yalta (2020)
6. Makaryan, A.S., Putyato, M.M.: Classification of messengers based on analysis of the security level of stored data. Caspian J. Control High Technol. **4**(48), 135–143 (2019)
7. Ocheredko, A.R., Gerasimenko, V.S., Putyato, M.M., Makaryan, A.S.: Research of SIEM systems based on the analysis of mechanisms for detecting cyber attacks. The Bulletin of the Adyghe State University, the series "Natural-Mathematical and Technical Sciences", vol. 2, pp. 25–31 (2020)
8. Tyukin, Y.A., Terekhov, V.A.: Adaptation in nonlinear dynamic systems. LKI, Saint-Petersburg (2008)
9. Masaev, S.N., Dorrer, M.G.: Assessment of the company's management system based on the method of adaptation correlation to the external environment. Manag. Prob.**3**, 45–50 (2010)
10. Putyato, M.M., Makaryan, A.S.: Adaptive integrated security assurance system as an element of the infrastructure of the situation center. Caspian J. Control High Technol. **4**(52), 75–84 (2020)
11. Fedotova, Z.A., Martynov, V.: Intelligent systems in information and analytical (situational) centers in the context of the transition to the digital economy. Moscow Econo. J. **1**, (2019)
12. Ponemon Institute and Fasoo analyzed the causes of corporate data leaks. https://securenews.ru/ponemon_institute_fasoo_report/. Accessed 27 February 2021
13. Evolution of threats and defense strategies. https://habr.com/ru/company/smart_soft/blog/304764/. Accessed 27 February 2021
14. Barankova, I.I., Mikhailova, U.V., Afanasyeva, M.V.: Minimization of information security risks based on modeling security threats. Dyn. Syst. Mech. Mach. **7**(4), 060–066 (2019). https://doi.org/10.25206/2310-9793-7-4-60-66
15. Lyubimov, A.V.: Threat models in the common criteria methodology. Sci. Tech. J. Inf. Technol. Mech. Optics **8**, 75–81 (2006)
16. Shushlyapin, E.A., Kanoe, L.N.: End-state models for continuous-discrete systems. Radio Electronics, Computer Science, Control, pp. 129–132 (1999)
17. Gutova, G., Kartashov, Y.: Peculiarities of discrete modeling of dynamic objects. Bull. Kemerovo State Univ. **1**(41), 63–67 (2010)
18. Maksimenko, V.N., Yasyuk, E.V.: Main approaches to the analysis and estimation of risks of information security. Econ. Q. Commun. Syst. **2**, 42–48 (2017)

The Method of Context-Dependent Annotation Based on the Spectral Language Model in Information Systems for Monitoring Emergencies

Irina Pozharkova[1], Oleslav Antamoshkin[2,3], Kwasi Adu-Boahen Opare[4], Alexey Knyazkov[2], Vladislav Kukartsev[5], Viacheslav Shestakov[2], Valeriy Pomazan[2], and Roman Tsarev[2,6](✉)

[1] FSBEE HE Siberian Fire and Rescue Academy, EMERCOM of Russia, Zheleznogorsk, Russia
[2] Siberian Federal University, Krasnoyarsk, Russia
[3] Krasnoyarsk State Agrarian University, Krasnoyarsk, Russia
[4] Kwame Nkrumah University of Science and Technology, Kumasi, Ghana
[5] Reshetnev Siberian State University of Science and Technology, Krasnoyarsk, Russia
[6] V.A. Trapeznikov Institute of Control Sciences of Russian Academy of Sciences, Moscow, Russia

Abstract. The article describes the method of context-dependent annotation based on the spectral language model for solving problems of emergency monitoring based on information systems. A functional model of the emergency monitoring task in the IDEF0 notation is presented. The task of context-dependent annotation of operational reports as a basis for generating preliminary reports is formulated. An algorithm for solving this problem is presented, taking into account the specifics of the subject area. A comparative analysis of the quality of constructing annotations manually and on the basis of the presented method using assessor assessments is carried out. #CSOC1120

Keywords: Context-dependent annotation · Spectral language model · Information system · Emergency monitoring

1 Introduction

Emergency monitoring is associated with solving a wide range of tasks for processing a large amount of data of various structures [1–3]. One of these tasks is to generate preliminary overview reports, which are used to generate a final report on the results of monitoring, and can also be used to make operational decisions on the prevention of emergencies and the elimination of their consequences.

The main difficulty in solving the problem of generating preliminary reports is that the initial data is often presented in a fragmented form: monitoring results for various objects, operational reports from the territorial and functional subsystems of the unified state system of emergency prevention and response (RSPRES), etc. At the same time,

© The Author(s), under exclusive license to Springer Nature Switzerland AG 2021
R. Silhavy (Ed.): CSOC 2021, LNNS 229, pp. 714–721, 2021.
https://doi.org/10.1007/978-3-030-77445-5_64

if the solution of the problem of generating a report on clearly structured monitoring results based on an information system does not pose significant problems [1], then in the case of operational reports there are a number of difficulties, namely:

- the source data is presented in the form of text;
- the volume of operational reports on large monitoring objects can be significant.

At the same time, the selection of key (relative to the purpose of generating a preliminary report) fragments of summaries manually on large volumes of source data requires a significant amount of time, which, for instance, in a rapidly developing emergency situation can be critical. One of the ways to solve this problem is the context-dependent annotation of text summaries, which allows selecting fragments from the source document that has the maximum relevance to the target query [4]. A functional model for solving the problem of generating a preliminary report in the IDEF0 notation is shown in Fig. 1, 2.

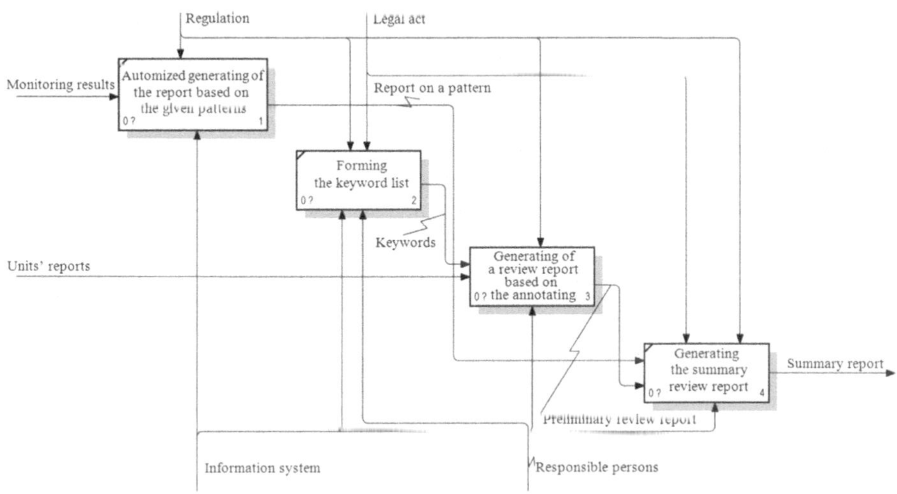

Fig. 1. Decomposition of the functional model of the task of generating a preliminary report.

2 Method

The basis of most annotation methods is the selection of such fragments of the source text that best meet the target criterion for the quality of the annotation construction. Accordingly, this task can be presented in the following form.

Basic sets of the task:

Target query – the set of tokens (keywords) of the target query:

$$Q = \langle q_i \rangle.$$

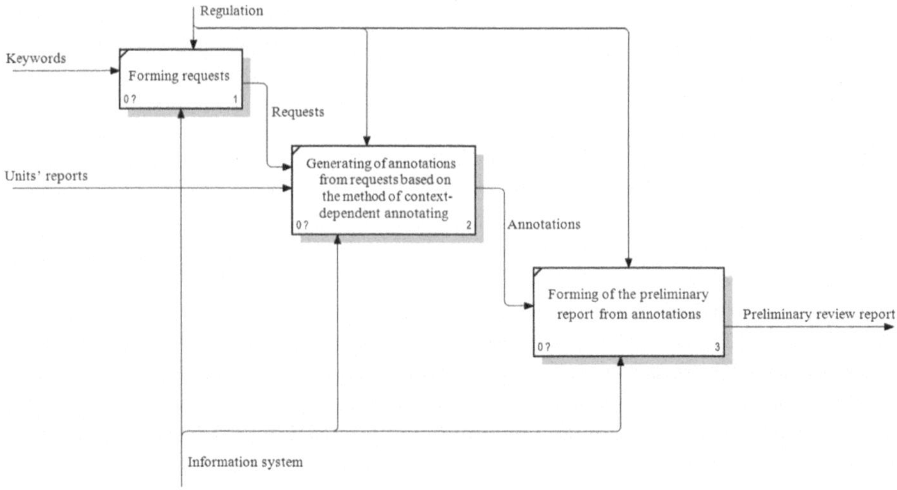

Fig. 2. Decomposition of the functional model of the task of generating an overview report based on annotation.

Annotated document – set of document tokens:

$$D = \langle d_i \rangle.$$

Annotation – set of tokens of the text obtained as a result of annotation

$$A = \langle a_i \rangle.$$

Task constraints:

The maximum length of the annotation can be used as a constraint – *MaxLen*:

$$\sum_{A} len(a_i) \leq MaxLen \quad (1)$$

where $len(a_i)$ is the length of the token.

Then the annotation problem can be formulated as generating from the tokens of the set *D* of the set *A* satisfying the constraint (1), such that:

$$rel(A, Q) = \max_{A \subseteq D},$$

where *rel*(*A*, *Q*) is a measure of the relevance of the annotation relative to the target query.

In the presented work, the task of annotating is solved by the following algorithm (Fig. 3).

1. Decomposition of the set *D* into fragments *F* by the method of sliding-window, which length is specified by the number of sentences *PropLen*:

$$F_i = \langle P_k \rangle, k = i, i + 1..i + PropLen - 1,$$

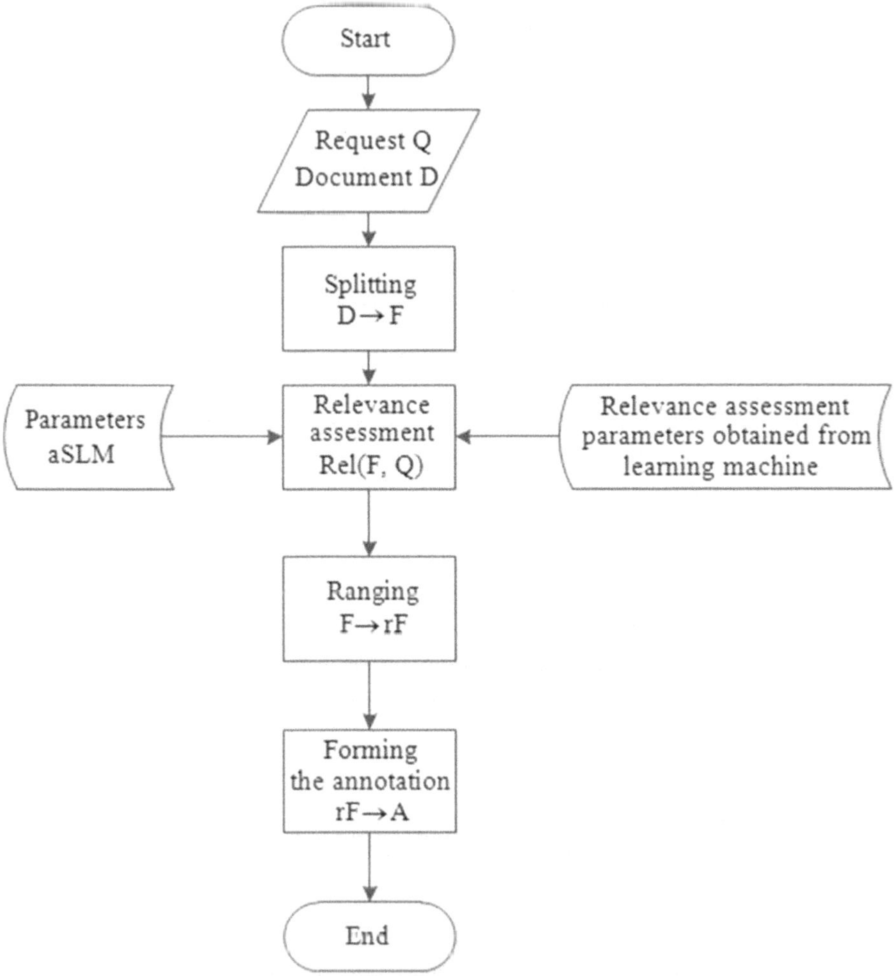

Fig. 3. The algorithm of context-dependent annotation.

$$\sum_{F_i} len(P_k) \leq MaxLen.$$

2. Calculation of the relevance of the obtained fragments on the basis of spectral language model (SLM) [5], in which the "importance" of a lexeme L of the assessed fragment when annotating a document D is calculated by the form:

$$aSLM(L, D) = a(L) + b(L) \cdot \log(\frac{TF(L, D)}{len(D)}), \qquad (2)$$

where $a(L)$, $b(L)$ are the parameters of the spectral language model of the lexeme L [5], determined from the entire set of documents D (collection of operational summaries).

$TF(L, D)$ is the frequency with which the lexeme L occurs in document D.
Two basic characteristics are used to evaluate the relevance of the fragment F relative to the query Q:
Assessment of the completeness of the fragment F with respect to query Q:

$$FQF(F, Q) = \sum_{q \in Q, q \notin F} aSLM\,(q, D).$$

Assessment of the completeness of a query Q with respect to fragment F:

$$FQF(F, Q) = \sum_{f \in F, f \notin Q} aSLM\,(q, D).$$

Use the second assessment is due to the feasibility of forming concise, readable annotations.
Then the relevance of the fragment F relative to the query Q is estimated by the formula:

$$rel(F, Q) = c_1 \cdot FFQ(F, Q)^{p_1} + c_2 \cdot FQF(F, Q)^{p_2}.$$

where c_1, c_2, p_1, p_2 are parameters determined using machine learning based on a sample of annotations, the quality of which is evaluated by experts.

3. Ranking fragments F in descending order of their relevance to the query Q:

$$F \overset{rel(F,Q)}{\rightarrow} rF,$$

$$rF = \langle rf : rel(rf_i, Q) \geq rel(rf_{i+1}, Q) \rangle.$$

4. Forming an annotation A by sequentially selecting fragments from the ranked set rF as long as the constraint (1) is satisfied:

$$A = \left\langle rf_i : \sum_i len(rf_i) \leq MaxLen \right\rangle.$$

The annotations obtained on the basis of this algorithm are then used to generate a preliminary summary report (Fig. 2). The collection of documents, on the basis of which the spectral language model is built and the parameters $a(L)$, $b(L)$ (2) are calculated, is formed from the operational summaries used for generating reports. This approach takes into account the specific features of the documents created in accordance with the current regulations.

3 Results

To assess the effectiveness of the presented method of context-dependent annotation, assessor ratings were used [6]:

Precision:

$$Prec = \frac{RFP}{RFP + RWP}$$

where *RFP*, *RWP* – the number, respectively, of relevant and irrelevant relative to the request for proposals in the annotation.
Accuracy:

$$Acc = \frac{RFP + WNP}{FF},$$

where *WNP* is the number of irrelevant suggestions relative to the request that was not included in the annotations.
FF – the total number of sentences in the documents.
Error:

$$Err = \frac{RWP + RNP}{FF},$$

where *RNP* is the number of suggestions relevant to the request that was not included in the annotations.

The evaluation was performed for annotations generated from the documents of the test sample, which does not overlap with the training sample. The results of the evaluation are presented in Table 1.

Table 1. Comparative analysis of the quality of annotations built on the basis of the context-dependent annotation method and manually

Assessment/Method	The method of context-dependent annotation	Manual method
Precision	0,882353	1
Accuracy	0,956693	1
Error	0,043307	0

Table 2 presents a comparative analysis of the average time of annotation formation for a single document.

Table 2. Comparative analysis of annotation construction time, based on the context-dependent annotation method and manually.

Method	The method of context-dependent annotation	Manual method
The time of building the annotations, sec	Less than 0,0001	148,2

4 Discussion

The obtained estimates of the quality of the context-dependent annotation method within the framework of the problem solved on its basis have acceptable values. At the same time, the speed of generating annotations is many times higher compared to the manual method. When processing a large number of documents manually in rapidly developing emergency situations that require urgent decision-making, the low speed of generating preliminary reports can be critical; therefore, in such scenarios, an acceptable reduction in the quality of generated documents will significantly reduce the time for generating a report and, accordingly, making decisions based on it, which indicates the effectiveness of the method under consideration.

5 Conclusion

A special feature of this method is the use of a spectral language model, which allows us to assess the" importance " of tokens in comparison with the classical metrics *BM25* [7], *DFR* more accurately [8]. In addition, to build the model used a collection of documents consisting solely of operative reports, which are annotation, which allowed, firstly, to take into account their specifics, which is largely determined by the relevant rules and regulations, and secondly, to reduce the amount of data that needs be stored and processed. It is worth noting that the models of the same words built on the basis of such a highly specialized collection and common vocabulary differ significantly.

The use of machine learning made it possible to achieve a high quality of the generated annotations, while significantly reducing the time of their formation in comparison with the manual method and, accordingly, increasing the efficiency of solving the problem of synthesizing a preliminary report.

References

1. Pozharkova, I., Pupkov, A., Tsarev, R., Yamskikh, T., Zhigalov, K., Scurtu, I.C.: Improvements in the functioning of the information system for monitoring and forecasting emergency situations in the Republic of Altai. Int. J. Eng. Res. Technol. **13**(12), 4861–4865 (2021)
2. Antamoshkin, O.A., Smirnov, N.A., Antamoshkina, O.A.: Multi-agent automation system for monitoring, forecasting and managing emergency situations. In: IOP Conference Series: Materials Science and Engineering, vol. 122, p. 012003 (2016)
3. Trofimova, N.V., Antamoshkin, O.A., Antamoshkina, O.A., Nicheporchuk, V.V.: Decision support system for responding to emergencies and accidents at hazardous production facilities. Civil Secur. Technol. **4**(30), 64–71 (2011)
4. Volk, M., Ripplinger, B., Vintar, Š, Buitelaar, P., et al.: Semantic annotation for concept-based cross-language medical information retrieval. Int. J. Med. Inform. **1–3**, 97–112 (2002)
5. Zyabrev, I., Pozharkov, O., Pozharkova, I.: Spectral characteristics in problems of text information processing. In: Proceedings of the 13th All-Russian Scientific Conference "Digital libraries: Advanced Methods and Technologies, Digital Collections", pp. 191–194. CEUR, Voronezh, Russia (2011)
6. Audeh, B., et al.: Improving exploratory information retrieval for neophytes: machine learning approach with feature analysis. ACM SIGAPP Appl. Comput. Rev. **20**(4), 50–64 (2021)

7. Kadhim, A.I.: Term weighting for feature extraction on Twitter: a comparison between BM25 and TF-IDF. In: 2019 International Conference on Advanced Science and Engineering (ICOASE), pp. 124–128. IEEE, Zakho-Duhok, Iraq (2019)
8. Thakur, N., et al.: Analysis and implementation of the bray–Curtis distance-based similarity measure for retrieving information from the medical repository. In: Bhattacharyya, S., Hassanien, A.E., Gupta, D., Khanna, A., Pan, I. (eds.) International Conference on Innovative Computing and Communications. LNNS, vol. 56, pp. 117–125. Springer, Singapore (2019). https://doi.org/10.1007/978-981-13-2354-6_14

Cased Based Reasoning in Business Process Management Design

Philipp Tueschen(✉) and Vítor Duarte dos Santos

Information Management Systems School, Universidade Nova de Lisboa, Lisbon, Portugal

Abstract. Case-based reasoning (CBR), another form of artificial intelligence, stores and retrieves past cases that can be adapted to find a solution to a current problem. The new solution can then be retained and made available to solve other future problems. Business Process Management analyzes and optimizes business processes to make them more effective and efficient for an organization's strategy to ultimately increasing shareholder value. CBR can help to support BPM, making better decisions with existing knowledge when solving process problems. This study investigates effectively store, retrieve, and adapt Business Process Management Notation (BPMN) solutions that best fit the underlying BPM problem using CBR as a tool. Therefore, a theoretical model was proposed, containing each CBR live cycle phase with different possible tools applied to BPMN diagrams, which was validated by expert interviews. This study concludes that a whole CBR life cycle can be applied to BPMN diagram problems with the need for human intervention. The objective was not to solve the whole problem but to contribute to a possible solution by using CBR through a theoretical model.

Keywords: Artificial intelligence · Business process management · Case-based reasoning · Graph-edit distance · Learning by analogy · XPDL

1 Introduction

Since the 90s artificial intelligence (AI) is becoming more and more useful; automating more and more tasks makes human life easy by copying the human way of learning, thinking, and reasoning. AI can conduct more and more complicated tasks, such as autonomous driving. This imitating of the human brain is accomplished based on the three basic concepts within the AI area: machine learning, deep learning, and neural networks [16].

Case-based reasoning (CBR) is part of the AI area as CBR learns by storing initial knowledge and retrieving it when needed. Further, CBR reuses this knowledge as a possible solution, which is then revised to solve a current problem. In the end, a successful solution to a problem is retained in the case database adding new knowledge to the CBR system, which is considered "learning" [15].

© The Author(s), under exclusive license to Springer Nature Switzerland AG 2021
R. Silhavy (Ed.): CSOC 2021, LNNS 229, pp. 722–741, 2021.
https://doi.org/10.1007/978-3-030-77445-5_65

Business process management (BPM) problems rely on expert's knowledge to resolve them. As BPM focuses on finding improvement opportunities and managing processes' consistency, experts rely on their past experiences to tackle the present problems.

CBR is imitating the expert's knowledge acquisition by storing knowledge from the past and using applicable knowledge to solve a present problem, so it is a good system to manage BPM problems more efficiently. In fact, Pichler [18] found that using CBR implemented in a BPM system allows them to execute processes that have clearly defined and standardized process models and react to events outside the limitations. Pichler [18] used Petri nets as a similarity function within the CBR system to find the most suitable cases to the underlying problem. However, this report will work with Carbonell's [6] learning by analogy, which compares semantics and considers the decisions' justifications of positive and negative outcomes. Further, Carbonell's "learning by analogy" is incorporated into the CBR system, supporting the retrieval and adaptation process.

Although Pichler [18] establishes that CBR can help to find solutions for BPM problems by using a similarity function establishing the distance between the attributes of the business process management notation (BPMN) diagrams, the article does not talk about the storing and retrieving of BPMN diagrams, which is an essential part of CBR in providing the best fit solution.

Existing literature established the possibility of translating BPMN models into XPDL language and that CBR can be an effective method for improving BPMN solutions [18,22]. However, there is a gap between how CBR could be used in comparing BPMN models and how CBR used with BPMN could be useful with different models then proposed by Pichler [18]. Hence, this thesis aims to find a solution to how BPMN can be stored effectively in XDPL and how the different stored XPDL files can be analyzed and retrieved effectively. This thesis also considers the BPMN adaptation, providing the best fit solution the user is looking for. This thesis's outcome is abstract knowledge; however, it could contribute to a more physical solution in the future [10].

2 Literature Review

Case-Based Reasoning. CBR uses and adapts solutions that have been successful in the past to solve new problems. Case histories are collected for CBR, and their most important features are described. New solutions of cases are added as new knowledge to the CBR databases. Thus, as databases enable the management of larger volumes of cases, it makes the need for explicit models obsolete [15]. All in all, CBR uses successful past cases to suggest a solution to a current, new to the system, and similar problem. Kolodner [13] came up with four assumptions. After the current problem case could be described with its features, the most similar case stored can be allocated. Although this stored solution may not directly fit the current problem case's differences, it can be adjusted. The derived and authenticated new solution to the current problem can then be stored as a new case solution to solve a similar problem in the future

[15]. This process builds the ground for the CBR working Cycle, that according to Aamodt and Plaza [2], is defined as the four Res:

1. Retrieve: After a current problem has been defined the most applicable solution is searched for and obtained from the case database
2. Reuse: The solutions obtained and used to solve a current problem
3. Revise: The obtained solution has to be adapted to the current problem to build a possible solution. If the solution is undesirable further adopting is required.
4. Retain: If the solution is desirable and verified as such it can be added as a new case to the database.

Derivational Analogy. Solving a problem with derivational analogy means to search for related problems in the database and adapt their solutions to be possibly relevant to the current problem, which is why Carbonell [6, p. 3] defines it as the following: "Analogical problem solving consist of transferring knowledge from the past problem-solving episodes to new problems that share significant aspects with the corresponding past experience - and using the transferred knowledge to construct solutions to the new problems." Carbonell points out that it is not enough to find a solution similar to the current problem but rather about the knowledge, it contains, which can be retrieved and interpreted. It implies the necessity for a very detailed adaptation model to obtain a desired adapted solution from the past. As this way of problem-solving is a fundamental part of human cognition, the derivational analogy is undoubtedly part of AI [6].

Building a model to adapt past case problems to the current situation requires specificity. The model needs to know what it means for both cases to have meaningful aspects in common. Further, what knowledge from these aspects is transmitted from the past solution to the current problem and how it occurs. Finally, it is important to know how these analogically associated aspects are selected from the case base [6].

Transformational Analogy. Carbonell [1983, as cited in 14] depicts transformational analogy as taking part in the CBR adaptation where a solution and its sequence of actions from a previous problem is being modified into the solution of the new problem. The modification includes removing, adding, and/ or changing the sequence's actions. Hence, after the closest solution has been found, the derivational and transformational analogy can be used as a CBR strategy to form a reasoning process that helps adapt the old solution to the current problem [14].

Business Process Management

BPM analyses operational activities performed in organizations and identifies advantages of improving processes, guaranteeing their consistency. Thus, BPM helps operating processes faster, more accurately, at lower costs, and with reduced assets while increasing their flexibility. Focusing on End-to-End processes across organizational boundaries can reduce nonvalue-adding tasks, which exist due to departmental boundaries. Further, processes are continuously monitored and improved if they no longer meet the customer's needs [12]. There are several methodologies with tools and techniques that help to identify redundant, inefficient, and ineffective processes or highlight specific processes for improvement [20]. Identifying business processes, their discovery, analysis, redesign, implementation, controlling and monitoring, is called the BPM Life Cycle [9].

The BPM as a management discipline can be divided into two approaches. Process improvement aims to analyze existing processes and consciously improve them while process reengineering questions existing processes from end-to-end and redesigns them from scratch. Therefore, BPM does not create business value by merely using IT or information systems to automate processes, but BPM uses IT and IS to enable process change, creating business value [9,12].

Business Process Management Notation 2.0. After processes are identified and discovered, they can be modeled and represented as a BPMN 2.0, which is a graphical notation in the form of a diagram. It provides an intuitive notation used by both technical and business users to represent complicated end-to-end process semantics, especially because many tools support this notation. It can also be identified as the most popular language for BPM [1]. Further, BPMN 2.0's Business Process Diagram (BPD) is the graphical notation and can be divided into groups, pools, and lanes. Pools can represent a business entity or a business role, while a lane is a subpartition of a pool representing specific business roles [17]. Within these divisions, BPMN uses four different essential elements. Flow objects consist of events, activities, and gateways. These are connected with Connecting Objects, divided by a Swimlane, and enriched with information through Artefacts. Thereby, actives are the work that is being performed, which can be further broken down into processes, subprocesses, and tasks. Events happen without any action and can be start events, intermediate events, or end events.

From BPMN to XPDL. XPDL is an XML-Process Definition language being used by a variety of workflow applications. Since BPMN is only a diagram, in order to be executable, it has to be translated into XPDL [22]. Both BPMN and XPDL are flow-chart structures enabling a direct, simple, and complete translation from BPMN into the semantically identical and compatible XPDL. According to the WfMC [21], XPDL writes out the BPMN diagram providing a file format. It enables other software to read and recreate the same process.

As many elements serve as graphical representations, once BPMN has been translated into XPDL, the elements' order changes. Pool and Lane names graphically describe what each participant does in the form of containers. Thus, they hold information about activities, events, and gateways that are executed by the participant. However, these containers are not present in the XPDL format and cannot hold the participant information. Instead, the task level within the activity is the only entity that can hold the participant information. To associate the Pool or Lane with a task, XPDL utilizes the elements <participant> and <performer>, which carry the actor executing the activity. Therefore, each <Task> holds a hidden <performer> [7].

Graph-Edit Distance. A distance measure has to be defined to calculate the graph similarity of two labeled graphs. Graph-Edit Distance (GED), first reported by Sanfeliu et al. [19], is a widely accepted and used distance measure for labeled graphs due to its flexibility and sensitivity. The definition of GED is the minimum cost of an edit path between two graphs. The edit path is a sequence of edit operations that change a graph to an isomorphic graph of B by adding, deleting, and substituting a node or an edge. These three operations are linked to a non-negative cost, which as a sum defines the cost of an edit path. Further, the edit costs transfer the metric properties to GED [4]. However, computing GED is complicated since it is NP-hard also if edit costs are uniform. Just using GED in matching algorithms would lead to computational explosions. Consequently, there needs to be a tradeoff between computational complexity and precision [8].

Therefore, only a few algorithms can handle large graphs that would allow, in principle, to design a GED algorithm performing more accurately than currently possible [5]. Abu-Aisheh et al. [3] mentioned that there are currently no reliable algorithms to compute GED for big graphs having more than 16 nodes within an acceptable time frame. Thus, using pure GED is not be enough for a similarity search of process models, given that real-life process models often exceed 20 nodes. Therefore, heuristic algorithms have been proposed that are based on graph similarity. The difference between graph edit similarity and GED is that "[GES] of two graphs is the maximum possible similarity induced by a mapping between graphs [while the] GED of two graphs is the minimal possible distance induced by some mapping" [8, p. 53]. The best performing algorithm using GES was a greedy algorithm that maps between a pair of process graphs [8].

Previous Related Work. Pichler [18] builds a business process management system using supply chain event management to identify and manage events that occur during process execution. It monitors processes, tracks, and traces the whole process down to its activities' progress and sends out an alert if there is a disturbance in the process, such as activities that do not finish on time. CBR is used to find similar past cases from the CBR database, which could be applied to the current event. The process and its model where the event occurred function as a basis for CBR systems search. The current and the old case have to be compared to find a suitable case. In other words, the similarity between the cases has to be calculated. The higher the similarity, the more compatible are the two cases. Therefore, Pichler uses Petri Nets and a process description. The current case is characterized through the process instance, its execution status, and its description, which is then compared to the case base. Although

Pichler [18] states the critical part needed to compare the old with new cases, it lags an explanation of how each case's descriptions are being translated to be comparable, since different people use different semantics in describing a case or a process. Aagesen and Krogstie [1]. Found that people working on process design tend to use text annotation extensively to work around design issues. Hence, much text would have to be analyzed instead of the process itself. Having a diagram in the first step to visualize and identify the process' problem would help to search for a solution with specific keywords in the case base.

3 Storing BPMN Diagrams

Using the CBR life cycle, the first cases have to be stored to build a case database that can be later used to optimize BPMN diagrams' problems (see Fig. 1). Before any BPMN diagram can be stored first, the key information must be retrieved for proper labeling to index each diagram correctly. To be able to retrieve information, the BPMN diagram has to be first converted to an XPDL file, which keeps the diagram's structure. Thus, the indexing system is only working with the BPMN's respective XPDL files, which can later be translated back into a BPMN diagram. Therefore, only the XPDL file of each BPMN diagram has to be stored. There are specific sections within XPDL that contain information needed to build the labels for the BPMN diagram's index. Creating the BPMN diagram index at the storing stage helps reduce the computing time later in the retrieving phase. Further, the BPMN information used is limited to the diagram itself. No additional information, such as activity time, is used to analyze this thesis for simplicity reasons.

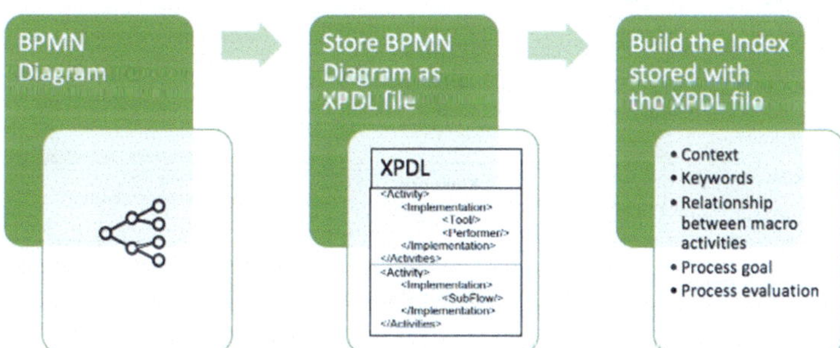

Fig. 1. From a BPMN diagram to its index

Building the Index

The index's labels are only effective if they adequately capture and describe the process's content. Hence, each XPDL file's labels can be broken down into the following five different categories, namely context, keywords, the relationship between macro activities, process goal, and process evaluation. Together they capture the actors involved in the process, the tasks they fulfill, and how they relate to each other. Further, they contain the goal the process is supposed to achieve and its success or failure. This information will help later in the retrieving process to faster allocate processes with similar goals and use positive and negative outcomes to understand better how the right solution should look. Therefore, it strengthens the understanding of how processes are structured in detail and what professions are involved in getting to the desired goal.

Information in the Original Diagram. The context label consists of the process and participants' names for retrieving, which a few steps have to be taken. From where and what information is obtained from the XPDL file can be seen in Fig. 2. First, the workflow process tag provides the name of the process. Next, as previously mentioned, BPMN's pool- and lane -names contain the information about the process participants, which are not directly represented within XPDL. Since only a task or activity can hold the participants' information, the participants in XPDL associated with the pool and lane names of the BPMN diagram have to be retrieved from the activity itself. The context can show if a process needs different actors or if the current actor setup is a good fit for the current process problem. In the rest of this paper, the current problem refers to the problem case that is tried to be solved with the herein developed artifact.

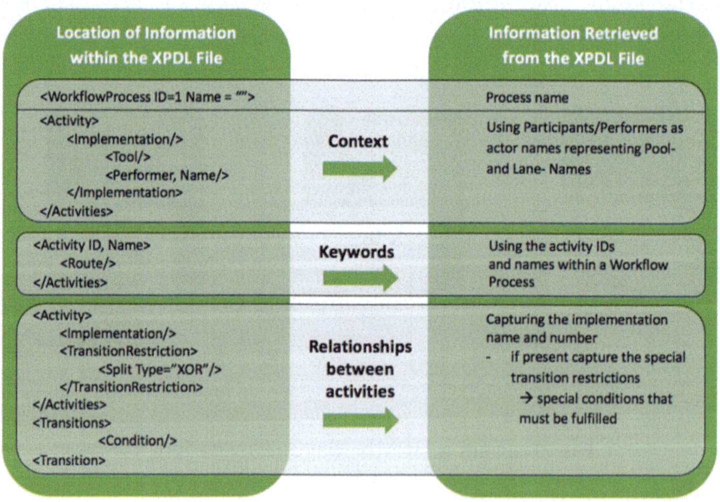

Fig. 2. XPDL information extraction

Further, the macro activities are used as keywords. Therefore, the keyword label summarizes the XPDL activity tags within a workflow process, which pro-

```
(<PackageHeader>
   <XPDLDescription> Goal = {}>
 </PackageHeader>
 <WorkflowProcess ID>)
```

List. 1. XPDL Header adding the Goal Label

```
(<PackageHeader>
   <XPDLDescription> Goal = {}, Success = {True or False}>
 </PackageHeader>
 <WorkflowProcess ID>)
```

List. 2. XPDL Header adding the Success Label

vides information about what happens during each activity. One process's activities can then be compared to other process's activities on similarity and if the current problem process might have to change, add or delete activities.

Next, the relations between macro activities have to be captured. Here the implementations, transitions, and conditions explain what happens with the token after each activity. In other words, as they are like the BPMN's gateway, they describe how two activities are connected, for example, through an exclusive or parallel gateway. This gives additional information on how complex the flow of the diagram is structured. When compared to other processes, it can show the differences in structure and flow to reach the goal of the process. Although having the same goal, they can be vastly different and thus can show opportunities for improvement of the current problem process. As a process can involve multiple participants, activities, or relationships, it is favorable to store them all together in a list. These lists can then be compared to each other position by position.

Information Not in the Original Diagram. The following information is not directly translated from a BPMN diagram into XPDL; however, it has beneficial effects when added to XPDL as it makes comparisons easier. Knowing the goal of a process at the beginning will help later to compare processes better. Processes with the same or similar goal can be grouped and thus limit the search space for similar solutions.

These similar solutions can the compared based on the previously explained labels context, keywords, and relationship between macro activities. Hence, if similar goals are available in the database, they can be an additional supportive label and can be added as a name into XPDL's package header (see List. 1). Further, it is essential to have a process evaluation knowing whether the process was a success or failure (see List. 2).

Negative examples support building the new models by showing how the process's goal could not have been reached. However, a more effective method is to separate the negative examples since this binary label would strongly polarize a later similarity search. Another possible method is leaving successful and unsuccessful cases in the similarity search without having the current problem's success labeled, which would provide more freedom of search.

Since the other cases in the case database would still possess this label, the other labels' similarity would determine the current problem's success or failure. Thus, there would be negative and positive examples retrieved, providing broader knowledge that would support a more accurate solution in the end. On the other hand, successful models possibly provide an instant alternative solution to the current problem. Having established all five labels, they can be used to describe and classify each XPDL file process. Combined, they form the index for each BPMN process consisting of the process ID and the five labels. This construct can then be stored together with each XPDL file in the case database and be queried against the before-mentioned index's labels. However, as the previous arguments show, there are many possibilities to combine and restrict labels during the similarity search to obtain one or multiple solutions to a current problem. These combinations and restrictions have to be chosen by the user's preferences and tailored to the knowledge required to find a solution to the current problem.

4 Retrieving BPMN Diagrams

In the second phase of CBR matching cases to a current BPMN problem are being retrieved. Each XPDL file belonging to a BPMN diagram has been stored with its own index consisting of five different labels in the form of lists. The retrieving phase utilizes Carbonells suggested difference function in the form of a semantic search engine and a similarity metric to compare the lists with each other. The closest matches are then being returned for the adaptation step.

Difficulties of Retrieving BPMN Diagram

Retrieving analogical BPMN diagrams means searching for some degree of similarity related to content or structure. The words and language each BPM specialist used can be different within the process schema, for example, naming the activities. Hence, two different BPM specialist can solve the same topic or problem using different words which are semantically similar. However, a normal search would only be able to find similar words with tokens or an edit-distance-based system. Therefore, semantics play a significant role when searching for similar BPMN diagrams.

Simultaneously, solutions in the case database may solve the same or a similar problem while having a completely different structure. In another scenario, the solutions may solve different problems with very similar structures. Thus, the structure is also important and can be different for each situation.

Consequently, two underlying problems have to be solved. The different diagram structures that could be beneficial and the used words that could be synonymous. While the different usage of words could mean the problem is covering a different problem, a different structure could just implicate another creative way of solving the current problem.

Semantic Search Model

A semantic search model queries the database's index for its keywords. There are many search engines that can retrieve information based on the similarity of words calculated by different edit-distances, but that is not the goal of this engine. Retrieving the same words limits the search to a strong match of similarly spelled words leaving just the possibility of a different diagram structure. This, however, would disregard the other powerful impact of semantically similar words. Therefore, the semantic search model uses a semantic engine that also finds semantically similar keywords or a list of keywords.

As previously established, the index's information is obtained through the original XPDL file of the BPMN diagram (context, keywords, and the relationships of the activities) and some additional information (goals and success), which has to be cleaned first. A preprocessing stage has to take place, normalizing and tokenizing the data.

The semantic search model then uses the preprocessed index and calculates a semantic similarity score for each label for each diagram in the database. In order to calculate the similarity score, the semantic search model has to build a vector space, grouping together semantically similar words using constructs like unsupervised learning models. Afterward, the semantic search model calculates an overall similarity score based on the weight the user assigned to each label (different levels of importance for different labels) and returns the most similar solution. Hence, priorities have to be set on which labels are most important to the user. As a result, this paper proposes that the semantic model is trained by the whole database (see Fig. 3).

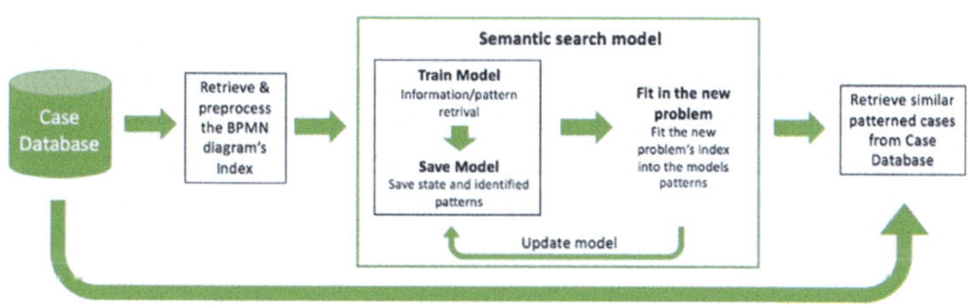

Fig. 3. Sematic search model used for retrieving similar cases

It breaks down the index and learns what phrases and word patterns, which are not necessarily linked to a specific label, exist in the database. Once a problem diagram is introduced to the model, it will compare the problem diagram's words to the database's pre-identified patterns. The space of patterns the new problem fits most can then be assumed to be semantically similar and retrieved. Hence, the model categorizes the new problem according to pre-identified patterns. For example, there could be a pattern of how activities are being named around a particular process goal. Thus, the model remembers that these names are used when a particular process goal has to be achieved. If the words of the current problem are using words associated with this pattern, the model will retrieve the diagrams with the same word patterns from the database. As only word patterns are being analyzed, the index's context or goal labels can vary from the current problem giving a possible solution more freedom and more knowledge. Here the current problem's index must not contain the label of the process' success because the semantic engine will bring into a context space around other processes sorted around the label success or failure, which is not desirable when looking for knowledge that helps to find a successful solution to the current problem. The success label will help sort the most similar cases at a later stage. It is also important to notice that this analysis does not consider the order of activities in the diagram, as it just looks at the similarity of word patterns in the index, giving the search engine further freedom.

Further, the model's training time could be very time-consuming depending on the database's size; however, the training does not have to be repeated for every search process because, without new cases, the patterns and phrases will not change. Instead, after a threshold of new diagrams has been stored on the database, the training process can be conducted during system downtime.

Nevertheless, the built index, which the model uses for the analysis, can be categorized as short text. Short text analysis has shortness and sparsity, which is a critical challenge for traditional text mining tools [11]. Traditional text mining tools require cohesive text to train the model to learn patterns and other information, which can later be found in other texts/documents. In the model's case, the patterns can be seen as semantic similarity clusters. Although much research studied the application of text mining tools for whole documents and more extended text parts, other papers suggest much current research about short text analysis, for example, using discovering topic representative terms or self-teaching convolutional neural networks [23,24]. Hence, the semantic search model is still a theoretical construct, which practical development is out of scope for this thesis and is left to future research. Therefore, future tests are necessary to prove the viability of the above suggest construct and if the indexes are long enough and with adequate information to find and retrieve semantically similar indexes.

Graph-Edit Distance

Before, the semantic search model looked at the words used in different diagrams to find some similarities. This section looks at the BPMN diagrams' structure that visualizes the process flow from beginning to end. As previously established, the diagram structure holds much additional, important information. If a process is solved similarly to another process, it means that the structures coincide. The more the processes differ in the solution, so do their structures. Hence, when comparing two diagram structures for similarity, there should always be some threshold that allows for some structure differences providing some alternative structure solutions.

The need to compare two diagrams requires some metrics measuring their similarity, for which usually some distance has to be calculated. Therefore, this thesis proposes GED as a measure since the BPMN diagram's process structure can be translated into a graph [8]. An example of the mapping from an XPDL to a graph can be seen in Fig. 4.

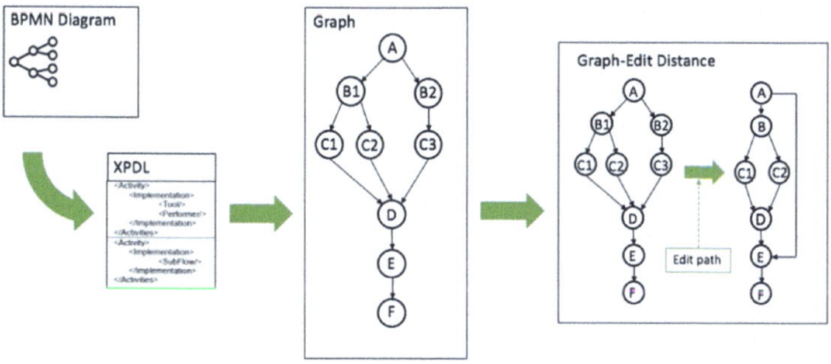

Fig. 4. From XPDL to graph-edit distance

Once the BPMN diagram is translated into a graph, its activities are now called vertexes, and its connecting objects are called edges. All other parts of the diagram are not considered in this step. GED can be applied to any graph where it calculates the minimum operations needed to adjust one graph to the other. Therefore, each vertex/edge from the first graph has to be mapped to one particular vertex/edge of the second graph or a dummy vertex/edge, which can absorb structural transformations.

Changing one graph to the other is achieved through the edit operations insert, delete, and substitute, which sequence is called edit-path. This edit-path is applied to the graph's vertexes and edges. Next, the extent of the transformations applied to a graph by the edit-path is represented by a cost function. The cost function can measure the magnitude of distortions since each edit operation has a predefined cost. Hence, the edit-distance between two graphs is defined by the edit path with the lowest cost function indicating the highest structural similarity. Since there is a lot of research on calculating graph similarity, there are already a few algorithms using the GES, a GED form. According to the research, GES can be better handled by a greedy algorithm, which is currently the leading algorithm, especially when graphs get as big as a BPMN diagram with more than 18 nodes. However, the test and possible implementation of this GED form and algorithm is out of this thesis's scope. That is why this paper will use GED, which can later be translated into GES.

Retrieving Using a Semantic Search Model and Graph-Edit Distance

The previous sections covered the retrieving phase analyzing the written and structural parts of a BPMN diagram. Now, as a final step, the retrieving phase has to coordinate these two analysis parts. First, both analyses need a threshold of the degree of similarity to retrieve similar cases, which has to be done by the user. The model needs the thresholds to find some similarity that narrows down the possible solutions because, without a degree of similarity, the system does not know what cases to retrieve. Once the thresholds have been set, the order of the two parts has to be set up. Since semantics can better describe a solution's content and whether it has dealt with a similar or the same problem, the most logical order is the semantical analysis followed by the structural analysis. Accordingly, the system must first learn the semantical context of possible solutions to the problem, which it will then retrieve. After, it can look at the structures of the same context, possible solutions. According to the threshold, based on the graph-edit distance or the GES and a greedy algorithm, it can then retrieve the solutions with the highest similarity. These solutions can then be used in the next step to adapt the current problem to the successful retrieved solution.

Retrieving possible solutions based on similarity always poses a knowledge barrier problem. As the system looks for some similarity, it is impossible to retrieve all knowledge present in the database. There might be a solution with no semantic or structural similarity to the current problem but still offers a viable solution. This solution might even be from a completely different domain. However, since there is no similarity, the system will not be able to detect this kind of solution. Thus, these solutions also cannot be considered by the system.

5 Adapting and Retaining BPMN Diagrams

The CBR live cycle's adaptation phase is about the knowledge transfer from the past solutions, retrieved from the case base to the current problem, which is not just about the similarity between the past solution and the current problem. According to Carbonell (1983), it is important to look at what kind of knowledge is transmitted for adapting the old solution to the current problem. Ideally, the adaptation model consists of positive and negative problem solutions, which share the same issues. Additionally, past solutions and the current problem should also have similar reasons of decisions taken available for a thorough understanding of the two problems at hand. If the reasons for decisions taken are congruent, the past solution could solve the current structure's problem.

In the current construct, the adoption phase can have positive and negative solutions available, the amount of which is based on the similarity threshold set in the retrieving phase. Additionally, the ratio of positive and negative examples can vary randomly for each CBR life cycle depending on the similarity score.

However, not all information Carbonell suggests is available in the BPMN diagrams. For example, the reasons for decisions taken are often not written down in the diagram and thus are difficult to obtain. If information about the decisions taken were present in the diagram, it would be in the form of process descriptions used to explain some parts of the diagram. Since not all parts of the diagram are always explained, the reasons for decisions taken are often incomplete and would make the text analysis even more complicated and time-consuming. Hence, due to incompleteness and increased complexity of the retrieving phase, the reasons for decisions taken are not considered in this adapting approach.

In the end, there are only two points of knowledge available for the system to derive a valuable solution for the current problem, namely semantical context and a degree of similarity between the process models' segments. Consequently, there is not enough information to implement derivational analogy since the system cannot reuse decisions taken in the retrieved solutions and build a new solution from null. Therefore, this report suggests a transformational analogy for the system to build the solutions to the current problem based on past solutions, which can deal with the context and structure of retrieved solutions and the current problem. Additionally, next to the transformational analogy, the adaptation system takes into account positive and negative solutions. Now that the basic concepts of the adaptation system are set, detailed procedures can be developed.

The Adaptation System

The adaptation system (see Fig. 5) consists of multiple steps, from applying old knowledge to validating the new solution and its functionality. Since the semantic engine already retrieves the contextually most similar cases in the previous retrieving step and the GED distance further narrows down the search to the structurally closest solutions, all these solutions can be temporarily stored in a separated pool (Step 1 of Fig. 5). Next, the adaptation process (2) applies the changes (Fig. 6). The GED has already calculated all changes that have to be made between the vertices and edges of the old solution and the current problem in the retrieving phase. In this case, the vertices are the process's activities, and the edges are their connections. Thus, the adaptation process implements the most straightforward change calculated by the GED, making the current problem's process successful.

In order to validate the successful change from the current problem to the new viable solution, the changes within the BPMN process's XPDL file have to be temporarily stored (3). Since the transformational analogy applies the best successful solution to the current problem, the negative examples have not been considered yet. Therefore, the negative retrieved solutions can be separated into an extra pool (4) so that in the next step, the temporary new solution can be compared to the negative examples (5).

Since the system is now dealing with negative examples, the desired outcome is a low similarity between the new temporary solution and the case base's negative solutions. However, they will have some congruities, given all cases have been retrieved beforehand based on a high similarity score. Hence, the comparison looks at whether there is less similarity to the negative examples than

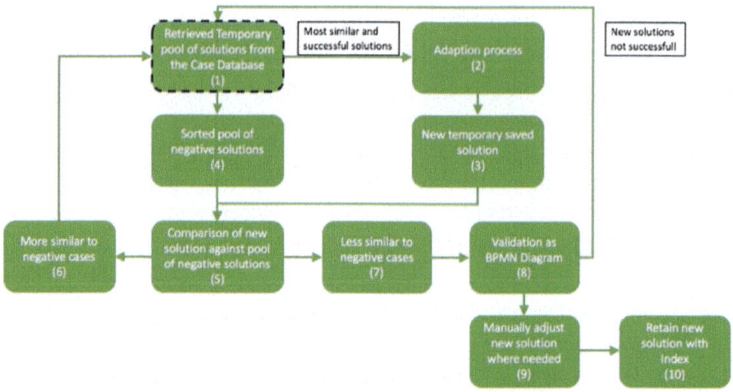

Fig. 5. Adaptation process overview

before in both semantics and structure. Therefore, the comparison uses the previous similarity score that has been established during the retrieving phase and calculates a new overall similarity score. The assumption is that if the old similarity score of the current problem and the negative solutions are similar in the beginning, the newly calculated score should be lower after the adaptation of a successful solution. Thus, according to the previous assumption, if the new solution is not less similar to the negative examples or if the new solution is even more similar than before, it would indicate a negative, unsuccessful new solution (6). As a result, the next best retrieved successful solution would have to be retrieved from the temporary pool (1) to go through the adaptation and comparison process (2–5) again. Once the similarity is less than before (7), the assumption is that the new solution to the current problem must be more successful than before.

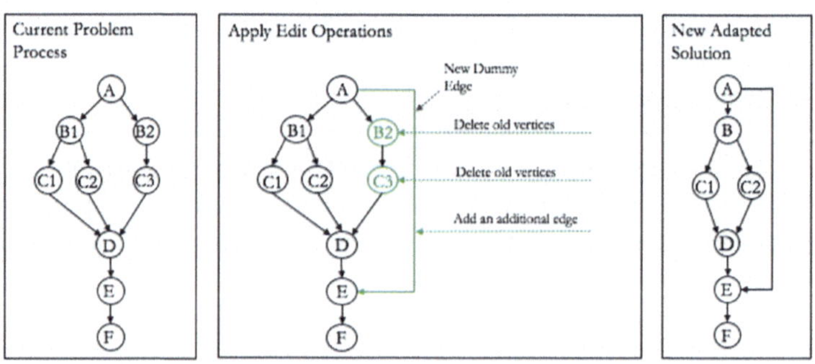

Fig. 6. Adaptation process using transformational analogy

For validation purposes, the new solution, transformed into a BPMN Diagram, is then tested for a correctly running process (8). If it is also successful at this stage, the current problem can be solved with the new solution. However, this new solution was built based on the structure disregarding the context as it is assumed to be similar. Thus, there is the need to check for misinterpreted activities by the system, which contextually and structurally do not fit precisely the place the system put it (9). For example, there is always the possibility of two times the same activity at different places in the process diagram. It can result from the two processes having the same activity before the adaptation but at different places. In addition, due to the double copy of one activity, there exists the possibility of a missing activity. In this case, it is required to manually adjust the new activities' order and language according to the current problem's context and compare the two diagrams' structure. Further, the user must also verify the process according to the needs and, if needed, adjust the process. Once the new problem solution is found viable, it can be retained in the case database as a new solution (10). Therefore, the new solution to the current problem is indexed as described in the section storing BPMN diagrams.

6 Discussion

This section analyzes the developed artifacts' usefulness, critique, and possible improvements based on the validation section's answers. The outcome will be a general evaluation of the proposed model.

All interviewees agreed to the usefulness of the theoretical model and its learning from past cases. Although they agreed to the usefulness, they gave feedback to different parts of the model. It was seen positively that the model would work faster in finding old, similar cases than humans improving the knowledge management's performance.

Especially, the pool of retrieved, similar solutions could lead to a more oriented outcome. Although some concerns were expressed that there can still be no fitting solution, the model retrieves the most similar solutions based on a similarity count system, so a solution is always returned. Hence, there should also be a person checking the new process's viability, which is also part of the model. Further, the interviewees mentioned a certain similarity between the model and its use in consultancy companies on a basic level, comparing it to their knowledge management learning from old cases since they do not always start from scratch for every project.

These statements show that learning from past cases has positive aspects and is already used in some professional environments, confirming that CBR used in the developed model is useful.

However, there has also been criticism of the model of different aspects. Some interviewees would have preferred more practical examples and data tests, such as company data, which goes beyond the Thesis's scope. As mentioned in previous sections, the purpose of this Thesis is to develop a theoretical model. If the model is deemed valid, and useful future research involves its practical implications.

Further, it was argued that the BPMN diagram alone is too general to draw similarities and does not provide enough data. Many other parts of the model influence the process's simulation, which is the most crucial part of validating the new process. With this additional information, it is tough for CBR to analyze because every case is very individual. So, if CBR was used, it has to be focused on qualitative analysis. As the beginning of the Thesis mentioned, it has only been focused on the BPMN Diagram for simplicity of the theoretical model.

The recommendations focused mainly on the BPMN details used to find similar cases and to adapt them afterward. If one deletes information and parts of the processes 100, they are not available anymore if any other problems arise with the process where the deleted parts could be useful. It is never practical to delete information 100. Instead, there should be a backup database that stores the deleted parts. As a result, a backup database and practical tests in the adaptation stage should be considered. While the backup database can be incorporated into the adaptation stage, the technical implications are future research.

Further, there should be some kind of validation in every step that verifies why one moves forward. The model's validation steps have been considered, primarily through the final human validation step of the new process's correctness. As mentioned under criticism, just searching for the similarity between the BPMN diagrams is not enough since more detailed information is important to make the process work and verify it through simulations. Thus, it was recommended to continue to find similarities between the BPMN diagrams as proposed. However, instead of just storing and retrieving the diagram, once a similar old diagram has been found, all its additional information should be retrieved with it in order to properly understand the old solution to adapt it correctly to the current problem. Therefore, even the old AS-IS and TO-BE models would be essential to understand the justification. This could be added to the current theoretical model increasing the human validation step's importance. All in all, the theoretical model itself has been evaluated to be useful, although it requires adjustments for additional BPMN information in the adaptation part. It became evident that the human factor in checking and analyzing the new and final process through simulation is significant for its validation. Hence, these points will have to be considered for the theoretical model's practical implications.

7 Conclusion

Based on extensive research in BPM and CBR, together with side topics, such as GED, it can be concluded that a whole CBR life cycle can be applied to BPMN diagram problems with the need for human intervention.

As BPM at its core controls and optimizes the business processes to make them more effective and efficient, it ultimately increases shareholder value. CBR, as part of AI, tries to mimic human learning by analogy. Hence, CBR is adopted to support BPM, making better decisions with existing knowledge when solving process problems. Therefore, BPMN diagrams must be translated first into XPDL files before CBR concepts and transformational analogy can be applied. Using the CBR life cycle helped to structure the solution's steps to build a closed learning process.

During the storing phase, each XPDL file obtains an index with five labels that contain the needed information for efficient comparisons between old solutions and the current problem later on. After the old solutions have been stored in the case database, the retrieving phase starts with semantically analyzing the current problems' index. Further, it will place the current problem in the semantic space with the old solutions that use similar wording, which is expected when dealing with similar solutions. The old solutions the semantic analysis finds closest to the current problem are then extracted for structural analysis. Here the XPDL structure is translated into a graph to apply GED distance to calculate the smallest edit distance to adjust the old solution to the current problem. Thus, the smallest GED distance indicates the final most similar solution. In the final adaptation phase, the transformational analogy is used, fundamentally applying the smallest Graph-Edit Distance to the current problem. Afterward, the new structure of the current problem has to be tested for its viability. If the user finds the current problem's solution successful, its wording must be adjusted before it is stored in the case database starting a new CBR cycle. While the proposed artifact's usefulness is generally approved, it needs improvement in some areas.

This dissertation was developed with the purpose to contribute to the development of learning from past solutions by building an overall system that stores, retrieves, and adapts old BPMN solutions to current BPMN processes. Therefore, this work did not have the objective to solve the whole problem but to contribute to a possible solution by using CBR. Combined with BPMN diagrams each CBR phase needs a different and unique approach to bring the whole CBR life cycle together as one working solution. The solution is constructed on a theoretical level and thus does not provide a practical implementation.

Limitations

This research focused on building a theoretical model; it did not develop the technical side because it is outside the thesis's scope due to time constraints. Further, the developed semantic search engine is a theoretical construct with not yet reached technological requirements. Until now, most text mining tools for natural language processing need excessive amounts of concise text to be trained and to conduct an analysis. Most search engines look for the exact word or phrases of words on other documents limiting the retrieval of other matching knowledge.

Moreover, big BPMN diagrams can still be too big for similarities to be calculated by GED or GES in an A* algorithm. Lastly, adaptation is a challenging part as it is limited to blindly applying the structure without checking if the content of each activity also fits in the same space. There are still many problems in the adaption phase, so the user must intervene to check the outcome, maybe even for the temporary outcomes.

Future Work

Future research has to consider all the additional metadata that describes the diagram's process in detail. However, the model theoretically has the ability to store the additional information and retrieve it together with the BPMN diagram. Additionally, deleted sequences in the adaptation phase should be stored in a backup database for possible later use.

From the technical side, further research needs to determine the semantical model's and GED's capabilities using practical examples to confirm their viability and usefulness.

Hence, future research should extend the model's theoretical possibilities and its technical implications, which will minimize the need for user intervention.

References

1. Aagesen, G., Krogstie, J.: BPMN 2.0 for modeling business processes. In: Handbook on Business Process Management 1: Introduction. Methods, and Information Systems, pp. 219–250. Springer, Heidelberg (2015)
2. Aamodt, A., Plaza, E.: Case-based reasoning: foundational issues, methodological variations, and system approaches. AI Commun. **7**(1), 39–59 (1994)
3. Abu-Aisheh, Z., Raveaux, R., Ramel, J.Y., Martineau, P.: A parallel graph edit distance algorithm. Expert Syst. Appl. **94**, 41–57 (2018)
4. Blumenthal, D.B., Boria, N., Gamper, J., Bougleux, S., Brun, L.: Comparing heuristics for graph edit distance computation. VLDB J. **29**, 419–458 (2020)
5. Blumenthal, D.B., Gamper, J.: On the exact computation of the graph edit distance. Pattern Recogn. Lett. **134**, 46–57 (2020)
6. Carbonell, J.G.: Derivational analogy: a theory of reconstructive problem solving and expertise acquisition. Technical report, Carnegie Mellon University, Pittsburgh (1985)
7. Cheng, R., Sadiq, S., Indulska, M.: Framework for business process and rule integration: a case of BPMN and SBVR. In: Lecture Notes in Business Information Processing. LNBIP, vol. 87, pp. 13–24. Springer, Heidelberg (2011)
8. Dijkman, R., Dumas, M., García-Bañuelos, L.: Graph matching algorithms for business process model similarity search. In: Lecture Notes in Computer Science (including subseries Lecture Notes in Artificial Intelligence and Lecture Notes in Bioinformatics). LNCS, vol. 5701, pp 48–63. Springer, Heidelberg (2009)
9. Dumas, M., La Rosa, M., Mendling, J., Reijers, H.A.: Fundamentals of Business Process Management. Springer, Heidelberg (2013)
10. Gregor, S., Hevner, A.R.: Positioning and presenting design science research for maximum impact. MIS Q.: Manag. Inf. Syst. **37**(2), 337–355 (2013)
11. Grida, M., Soliman, H., Hassan, M.: Short text mining: state of the art and research opportunities. J. Comput. Sci. **15**(10), 1450–1460 (2019)
12. Hammer, M.: What is business process management? In: Handbook on Business Process Management 1: Introduction. Methods, and Information Systems, pp. 3–16. Springer, Heidelberg (2015)
13. Kolodner, J.: Promoting Transfer through Case-Based Reasoning: Rituals and Practices in the Learning by Design Classroom and Evidence of Transfer. Morgan Kaufmann, San Mateo (1995)

14. Kuchibatla, V., Muñoz-Avila, H.: An analysis on transformational analogy: general framework and complexity. In: Roth-Berghofer, T.R., Göker, M.H., Güvenir, H.A. (eds.) ECCBR 2006. LNCS (LNAI), vol. 4106, pp. 458–473. Springer, Heidelberg (2006). https://doi.org/10.1007/11805816_34
15. Pantic, M.: Introduction to machine learning & case-based reasoning. Technical report, Imperial College London (2006)
16. Sciglar, P.: What Is Artificial Intelligence? Understanding 3 Basic AI Concepts - Robotics Business Review. Robotics Business Review (2018)
17. Pérez-Castillo, R., Piattini, M.G.: Uncovering Essential Software Artifacts Through Business Process Archeology. IGI Global (2013)
18. Pichler, A.: Flexibilität in Business process management Systemen durch case-based reasoning. In: Wirtschaftsinformatik Proceedings, pp. 589–597 (2011)
19. Sanfeliu, A., Sanfeliu, A., Fu, K.S.: A distance measure between attributed relational graphs for pattern recognition. IEEE Trans. Syst. Man Cybern. **SMC-13**(3), 353–362 (1983)
20. Van Der Aalst, W.M., La Rosa, M., Santoro, F.M.: Business process management: don't forget to improve the process!. Bus. Inf. Syst. Eng. **58**(1), 1–6 (2016)
21. WfMC: XPDL - Workflow Management Coalition (2019). https://www.wfmc.org/standards/xpdl. Accessed Jan 2021
22. White, S.A.: XPDL and BPMN. Technical report, Future Strategies Inc., WfMC (2003)
23. Xu, J., et al.: Self-Taught convolutional neural networks for short text clustering. Neural Netw. **88**, 22–31 (2017)
24. Yang, S., Huang, G., Cai, B.: Discovering topic representative terms for short text clustering. IEEE Access **7**, 92037–92047 (2019)

Cognitive Expert Assessment

S. A. Kudzh[1] and V. Ya Tsvetkov[1,2(✉)]

[1] Russian Technological University (RTU MIREA), Vernadsky Prospekt, 78, 119454 Moscow, Russia

[2] Research and Design Institute of Design Informati on, Automation and Communication on Railway Transport, 27, bldg 1 Nizhegorodskaya Str, Moscow, Russia

Abstract. The article examines the theory and practice of expert methods and assessments. The article reveals the content of expert methods and expert assessments. The article examines the influence of cognitive factors that exist in expert assessment. Seven models of cognitive information interaction in expert assessment are described. The content of individual expert assessment is considered. Group expert assessment is investigated. The structure of expert assessment model by the subject is given. Models analysis of group expert assessment is given. The value of both cognitive and informational interaction in group expert assessment is shown. The article investigates the measurement scales for finding invariants in expert assessment. The article shows the importance of the information situation as an additional tool in expert assessment. #CSOC1120

Keywords: Expert assessment · Expert assessment method · Information construction · Information cognitive construction · Information situation · Cognitive assessment factors · Individual expert assessment · Group expert assessment

1 Introduction

In expert assessment [1], there is always expert assessment object (OAE), which must be assessed by an expert as a trinitarian [2] or cognitive [3, 4] system. Expert assessment is inextricably linked with the cognitive space of the expert. Expert assessment is carried out to prepare information when making decisions [5, 6] for a decision maker (DM). Therefore, the results of expert assessment are inextricably linked with the cognitive space of the decision maker. Such a double cognitive connection makes it relevant to study cognitive factors in expert assessment. The use of expert assessment is effective in conditions of fuzzy and poorly structured information when the measurement methods and quantitative information processing do not provide the necessary information. Therefore, expert assessment should be considered in the context of the relationship "expert - decision maker" or relationship (E-decision maker). This relationship is cognitive. Different decision makers need different cognitive models and forms of presentation of expert assessments. Cognitive models of information presentation are the first cognitive

© The Author(s), under exclusive license to Springer Nature Switzerland AG 2021
R. Silhavy (Ed.): CSOC 2021, LNNS 229, pp. 742–749, 2021.
https://doi.org/10.1007/978-3-030-77445-5_66

factor of expert assessment. There are methods of expert assessments and expert assessments. Expert assessment methods are methods for obtaining assessments by expert specialists, intelligent systems, cognitive systems using computer or cognitive technology. Expert assessment methods [7] (MEA) are synonymous with expert assessment. Expert judgment methods are procedural cognitive models. Expert assessments (EA) are descriptive models that are usually expressed in quantitative and/or qualitative form and often have a probabilistic meaning. To enhance the significance of assessments and increase their reliability an expert group (EG) is sometimes created which makes a collective decision and forms group expert assessments [8]. The peculiarity of expert assessments is that they transform the implicit knowledge of an expert into an explicit assessment that can be analyzed and available to other subjects. The transformation of implicit knowledge into explicit knowledge often takes the form of a cognitive procedural model. However, most of work in the field of expert assessment does not address the issue of expert assessment cognition. The article aims to fill this gap in expert research.

2 Research Methodology

The research is based on systems analysis, comparative analysis, information analysis. The materials used are publications in the field of expert assessment, representative measurement theory and cognitive modeling.

3 Research Results

3.1 Cognitive Model of Expert Assessment

The cognitive information model [9] is applicable in many areas in which information and cognitive interactions take place. The cognitive model is applicable to the field of expert assessment. Expert assessments are divided into individual and collective (group). Individual assessments are assessments of one expert using: his intelligence, his level of explicit and implicit knowledge, his level of logical constructions, his cognitive space. Any individual expert assessment is formed on the basis of cognitive information interaction with reality. On the basis of this interaction, the expert performs logical constructions in his cognitive space. Figure 1 shows a cognitive model for the formation of individual expert assessment. The model contains two parts: informational situation and expert's cognitive space. The model of the information situation [10] is an important factor in expert assessment. It includes objective information seen by the expert and other subjects. Expert's cognitive space is his personal subjective model, not visible to others. The result of the assessment is an expert situational assessment associated with the situation and the cognitive space of the expert.

In most cases, expert does not consider individual parameters or an object of expert assessment in isolation from the situation in which he is. Therefore, it is necessary to say that expert assessment is conditional since it is carried out in a certain informational situation. In this situation, the expert sees part of the information about assessment object. This visible situational information enters the expert's cognitive space. The expert supplements the situation information with available reference information. It can use reference dictionaries of the database. The available reference information is the "informational support" of expert assessment. In the cognitive space of an expert there are personal stereotypes through which he passes the assessed situation.

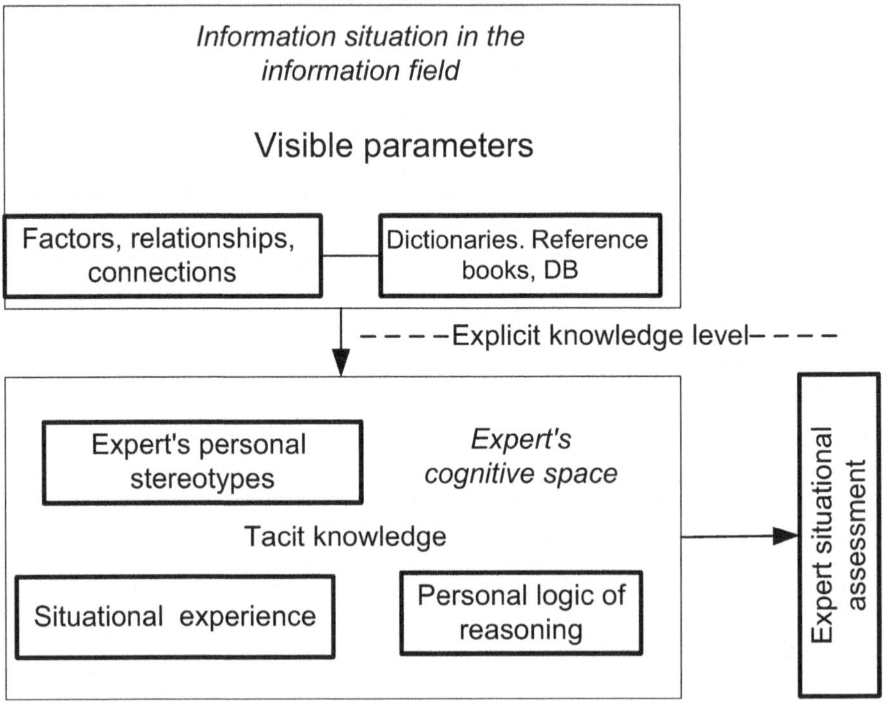

Fig. 1. Cognitive model for the formation of individual expert assessment.

Implicit knowledge exists in the cognitive space of an expert [11, 12]. The implicit knowledge of an expert includes the situational experience of expert cognitive analysis based not only on cause-and-effect but also on associative relationships. The implicit knowledge of an expert also includes personal experience in constructing logical reasoning, logical constructions and logical consequences. Personal experience of constructing logical reasoning depends on the peculiarities of the expert's thinking: right-brain or left-brain. A logical personal construction is a personal scheme of evidence and construction of logical expressions from which a logical conclusion is subsequently made.

3.2 Scenario Expert Assessments

The expert gives his assessment in different ways in relation to other experts. In some cases, the expert (postfix assessment) discusses his assessment with other experts and completes the assessment. In other cases, the expert (prefix score) gives his score independently. However, after that he can discuss it with other experts and all together make a single group assessment. This kind of expert assessment is called point, because it corresponds to a point in space and reflects the individual opinion of an expert.

A more complex assessment is the expert assessment by the scenario method which is usually created by a group of experts and used for expert forecasting. The scenario method is a method of combining expert judgments associated with different time periods. The scenario method describes the options for the development of the situation (scenarios). Moreover, each stage of the scenario and the entire scenario are probabilistic. The concept of a Markov chain [24] is close but not equivalent to scenario method.

It is worth reminding that Markov chain is a sequence of random events in which the probability of a subsequent event occurring depends on the state achieved in the previous event. The concept of "Markov property" is used. The Markov property means that the presence of knowledge about the state of the process at a given moment is sufficient to assess it in the future and other information about its past does not affect its state in any way. In other words, at any moment in time, the distribution of the future states of the process with the given current and past states depends only on the current state and does not depend on the past states. The difference between Markov chain and the scenario is that the stages of the predictive scenario are considered as probabilistic events and relationship between them is searched as causal and not probabilistic as in the Markov chain.

When applying the method of scenarios, it is necessary to carry out three stages of the study: to select significant time intervals; build an exhaustive set of expert assessments for each stage of the scenario (scenes); build a scenario model based on causal relationships between stages. When assembling scenario, the perspective (start → end) and retrospective (end → start) methods of scenario composition are used. Modern scenario development is closely related to the use of situation centers or situation rooms.

Expert assessments can take place in several cycles. They can include one cycle, a fixed number of cycles or an indefinite number of cycles until the desired result is obtained. An analogue is the method of successive approximations [14] in computational mathematics. From a purely formal point of view, the more cycles, the more thorough the analysis of the situation is since experts usually return to the examination of the subject matter many times. However, as in the method of successive approximations, there is a limit to the accuracy of the examination. It is due to information uncertainty or the concept of the level of information certainty. It is impossible to give an expert assessment more precisely than the level of uncertainty of the initial information. Obviously, with a large number of expert assessment cycles, the total time for the expert assessment increases and its cost increases. Here you can draw an analogy with information retrieval [15]. The longer the search, the lower the information utility/cost ratio.

3.3 Influence of Informational and Cognitive Interaction on the Result of Expert Assessment

Consider the advantages and disadvantages of information interaction [16] between experts: lack of interaction - remote interaction - direct interaction with restrictions - direct interaction without restrictions. When there is absence of interaction, the expert makes an assessment without knowing anything about other experts and their assessments. He is completely independent which is both good and bad. Usually this situation corresponds to a one-cycle examination with complex conflicting information when you need to get a radical solution. Remote interaction with anonymity means that the expert gets acquainted with the opinions and arguments of other experts but does not know who exactly expressed this or that position. Consequently, such an examination should provide for at least two cycles. Remote interaction without anonymity corresponds, for example, to the performance of opponents in off-line mode. Absentee expertise options are good because there is no need to bring experts together and you can use specialists located in different geographic locations.

With face-to-face information interactions, experts discuss because they manage to eliminate information uncertainty much more effectively. Communication with constraints is very common, for example, when defending dissertations. This is a meeting following the protocol in which the sequence of speeches is prescribed and there are restrictions on the content of the speeches of opponents, the leader and members of the Dissertation Council. Informational open interaction without restrictions is a free discussion that takes place at seminars and scientific colloquium.

3.4 Representative Measurement Theory in Expert Assessment

The representative measurement theory [17, 18] (RMT) in expert assessment was originally developed as a theory of psychophysical measurements. The founder of RMT, American psychologist S.S. Stevens focused on measurement scales [19]. The two main problems of RMT include: the choice of measurement scale type and the search for expert assessment algorithms the result of which does not change with an acceptable scale transformation. The main requirement of RMT is the invariance of conclusions and expert assessments.

In other words, the result of the algorithm is invariant under the allowed transformation. When considering scales, the cognitive aspect of expert assessment is that the expert answers more correctly (and with fewer errors) questions of a qualitative, for example, comparative, nature than quantitative ones. For example, it is easier for him to say who of people is taller than to indicate their approximate height in centimeters. An example of an invariant algorithm is the use of scaling in factor analysis.

With an acceptable conversion for a given scale, the value of the average for the population may change. But the conclusions for different populations about which set the average is greater and for which it is less, should not change (in accordance with the invariance of the conclusions adopted in the RMT). Let us formulate the corresponding mathematical problem of finding the type of mean values, the comparison result of which is stable with respect to admissible scale transformations.

We have $f(X1, X2,..., Xn)$ – average according to Cauchy. Assume that the population average $(Y1, Y2,..., Yn)$ is less than the population average $(Z1, Z2,..., Zn)$. It means that

$$f(Y1,\ Y2, \ldots, Yn) < f(Z1,\ Z2, \ldots, Zn). \tag{1}$$

According to the RMT, for stability of average comparing result, it is necessary that for any admissible transformation g from the group of admissible transformations in the corresponding scale, the inequality

$$f(g(Y1),\ g(Y2), \ldots,\ g(Yn)) < f(g(Z1),\ g(Z2), \ldots,\ g(Zn)), \tag{2}$$

Expression (2) suggests that the average of the transformed values from the first population is also less than the average of the transformed values for the second population. Moreover, the formulated condition must be true for any admissible transformation g. According to RMT, only such averages can be used when analyzing expert opinions and other data measured in the scale under consideration. In this, we can note the factor of expert cognitive assessment - assessment only within the framework of permissible transformations.

For the ordinal scale, there are five axioms of admissible transformations [5]. Three axioms of identification are valid for the nominal and ordinal scales.

1. $(A = B) \oplus (A \neq B)$.
2. $(A = B) \rightarrow (B = A) \rightarrow (A = C)$.
3. $(A = B) \bigwedge (B = C)$.

Additionally, for the ordinal scale, the ordering axioms.

4. $(\geq B) \;\; \rangle \; (B \leq A) \oplus (A \leq B) \rightarrow (B \geq A)$
5. $(> B) \bigwedge (B > C) \rightarrow (A > C)$

Axiom 4A is called the sorting axiom.

Axiom 5A is called the axiom of transitivity. It is one of the keys in many theories, including the theory of hierarchies. The scaling property will emerge from the fifth axiom.

Let

$$\begin{gathered} B = Ak1, k1 < 1; C = Bk2, k2 < 1 \\ (\mathrm{A} > Ak1) \bigwedge (Ak1 > Ak1k2)(A > Ak1k2) \end{gathered} \tag{3}$$

Expression (3) is among the admissible transformations. The fifth axiom has a second option (5*).

5*. $(\mathrm{A} < B) \bigwedge (B < C) \rightarrow (A < C)$

Let

$$\begin{gathered} B = Ap1, p1 > 1; C = Bp2, p2 > 1 \\ (\mathrm{A} < Ap1) \bigwedge (Ap1 < Ap1p2) \rightarrow (A < Ap1p2) \end{gathered} \tag{4}$$

Expression (4) is among the allowed transformations and expressions (3) and (4) describe scaling. Consequently, scaling is included in the number of admissible transformations and is an invariant in RMT. In the naming scale, only fashion is allowed as an average. Of all Cauchy averages in an ordinal scale, terms of the variation series (ordinal statistics), in particular, the median (for an odd sample size), can be used as averages. For an even sample size, one of the two central terms of the variation series should be used — as they are sometimes called, the left median or right median. You cannot use arithmetic mean, geometric mean, etc. In the scale of intervals, of all Kolmogorov averages, only the arithmetic average can be used. On the scale of ratios, of all Kolmogorov averages, only power-law and geometric average are stable relative to comparison. In this, one more factor of cognitive assessment can be noted - the expert must choose acceptable estimates of the average in accordance with the type of scale in which the assessment is made.

4 Conclusion

The first cognitive factor of expert assessment is the need to take into account the relationship “expert - decision maker”. This factor leads to the need to form different forms of presentation of expert assessments for different cognitive types of decision makers. The second cognitive factor of expert assessment is that visible information enters the expert’s cognitive space. When choosing scales, the cognitive aspect of expert assessment is that the expert answers more correctly (and with fewer errors) questions of a qualitative, for example, comparative, nature than quantitative ones. Because of this, it forms assessments on a comparative scale. Another factor of expert cognitive assessment is that the expert must carry out the assessment only within the framework of admissible transformations. Associated with the previous factor of cognitive assessment is that the expert must choose acceptable estimates of the average in accordance with the type of scale in which the assessment is made. The initial factor of expert cognitive assessment is that the expert must intuitively create a model of the information situation in which expert assessment is carried out. During the assessment process, the expert performs cognitive information construction. The information construction model is becoming a common means of conceptual interpretation and description of phenomena in many areas. They describe processes and entities. In the aspect of expert assessment, one should distinguish between the formal model of the information structure and the cognitive model of the information structure. The formal model of information construction is close to the notion of explicit knowledge. It is fully formalized, has a clear description and can be transferred from one person to another. The cognitive model of information construction is close to the concept of implicit knowledge. It is not formalized, it has a description only in expert’s cognitive space and cannot be transferred from one person to another. An exception is the case when it can be described using an infological model. Thus, the real situation of cognitive expert assessment is closely related to cognitive factors. The cognitive methods of constructing cognitive logical constructions and cognitive and statistical methods for assessing the professionalism of an expert remained outside the scope of this article. This presents the research material of the next article.

References

1. Schömig, N., et al.: Checklist for expert evaluation of HMIs of automated vehicles — discussions on its value and adaptions of the method within an expert workshop. Information **11**(4), 233 (2020)
2. Kudzh, S.A., Tsvetkov, V.Y.: Trinitarian systems. Russian Technol. J. **7**(6), 74–88 (2019). https://doi.org/10.32362/2500-316X-2019-7-6-74-88
3. Cox, M., Dannenhauer, D., Kondrakunta, S.: Goal operations for cognitive systems. In: Proceedings of the AAAI Conference on Artificial Intelligence, vol. 31, no. 1 (2017)

4. Jones, A.T., Romero, D., Wuest, T.: Modeling agents as joint cognitive systems in smart manufacturing systems. Manuf. Lett. **17**, 6–8 (2018)
5. Tikhonov, A.N., Tsvetkov, V.Y.: Decision support methods and systems. MAKS Press, Moscow, Russia, p. 312 (2001)
6. Lytvyn, V. et al.: Methods of building intelligent decision support systems based on adaptive ontology 2018, pp. 145–150. In: IEEE Second International Conference on Data Stream Mining & Processing (DSMP). IEEE (2018)
7. Igrunova, S.V., et al.: Development of the procedure of testing with the application of the expert evaluation method in psychophysiology (2019)
8. Sanzhapov, B.K., Sanzhapov, R.B.: Ordering objects on the basis of potential fuzzy relation for group expert evaluation. ARPN J. Eng. Appl. Sci. **11**(13), 8544–8548 (2016)
9. Tsvetkov, V.Y.: Cognitive information models. Life Sci. J. **11**(4), 468–471 (2014)
10. Tsvetkov, V.Y.: Information situation and information position as a management tool. Eur. Res. **12-1**(36), 2166–70 (2012)
11. Sigov, A.S., Tsvetkov, V.Y.: Tacit knowledge: oppositional logical analysis and Typologization. Herald Russian Acad. Sci. **85**(5), 429–433 (2015). https://doi.org/10.1134/S10193316 15040073
12. Tsvetkov, V.Y.: Implicit knowledge and its varieties. Bull. Mordovian Univ. **24**(3), 199–205 (2014)
13. Rakhman, P.A.: Methodology for calculating the operational readiness coefficient of control systems in reliability models based on Markov chains. Econ. Manag. Control Syst. **4**, 90–99 (2018)
14. Beck, A., Ben-Tal, A., Tetruashvili, L.: A sequential parametric convex approximation method with applications to nonconvex truss topology design problems. J. Global Optim. **47**(1), 29–51 (2010)
15. Tsvetkov, V.Y.: Cognitive science of information retrieval. Eur. J. Psychol. Stud. **1**(5), 37–44 (2015)
16. Tsvetkov, V.Y.: Information interaction. Eur. Res. **11-1**(62), 2573–7 (2013)
17. Khitrova, Y.: Representative theory of measurements. Mod. High Technol. **6**, 89–91 (2013)
18. Gobrecht, A., et al.: Combining linear polarization spectroscopy and the representative layer theory to measure the beer – lambert law absorbance of highly scattering materials. Anal. Chim. Acta **853**, 486–494 (2015)
19. Aaboud, M., et al.: Jet energy scale measurements and their systematic uncertainties in proton-proton collisions at s = 13 TeV with the ATLAS detector. Phys. Rev. D **96**(7), 072002 (2017)

A Percentile Firefly Algorithm an Application to the Set Covering Problem

Lorena Jorquera[1], Pamela Valenzuela[1], Leonardo Causa[2], Paola Moraga[1], and Gabriel Villavicencio[1](✉)

[1] Escuela de ingeniería en construcción, Pontificia Universidad Católica de Valparaíso, Valparaíso, Chile
{lorena.jorquera,pamela.valenzuela,paola.moraga, gabriel.villavicencio}@pucv.cl

[2] Facultad de Ingeniería y Negocios, Universidad de le Américas, Santiago, Chile
lcausamo@edu.udla.cl

Abstract. The application of continuous metaheuristic algorithms to combinatorial problems is an area of interest at the engineering level. This interest is related to the fact that many of the problems that are addressed at the industrial level are of the combinatorial type and a not lesser subset of these are of the NP-hard type. In this article, we propose a binarization mechanism for continuous metaheuristics that uses the concept of percentile. This percentile concept is applied to the firefly optimization algorithm to solve the set covering problem (SCP). To determine the importance of the percentile concept, experiments were designed to identify its contribution. Additionally, to see that the proposal is adequate, the efficiency of the algorithm is compared using reference instances. The results indicate that the binary percentile firefly algorithm (BPFA) obtains adequate results when evaluated with a combinatorial problem such as the SCP.

1 Introduction

In recent years, there is a research line of algorithms inspired by natural phenomena focused on solving optimization problems. As examples of these algorithms we have Cuckoo Search [1], Bat Algorithm [2] and Sine Cosine Algorithm [3] among others. Due to the type of phenomena in which these algorithms are inspired, a large part of these algorithms operate naturally in continuous spaces. On the other hand, the combinatorial problems have a great relevance at the scientific and industrial level. We find them in different areas, such as Civil Engineering [4], Bio-Informatics [5], Operational Research [6–9], Big Data [10], resource allocation [11–13], programming problems [14–16], routing problems [17–19], civil engineering, [20–23], among others. If we want to apply continuous algorithms to combinatorial problems, these algorithms must adapt. In the process of adaptation, the mechanisms of exploration and exploitation of the algorithm can be altered, having consequences in the efficiency of the algorithm.

© The Author(s), under exclusive license to Springer Nature Switzerland AG 2021
R. Silhavy (Ed.): CSOC 2021, LNNS 229, pp. 750–759, 2021.
https://doi.org/10.1007/978-3-030-77445-5_67

Several binarization techniques have been developed to address this situation. In a literature search, the main binarization methods used correspond to transfer functions, angle modulation and quantum approach. In this article we present a binarization method that uses the percentile concept to group the solutions and then perform the binarization process. To verify the efficiency of our method, we used the fireflies algorithm, it was based on the idealized behaviour of the flashing characteristics of fireflies. This algorithm was proposed in [24].

To check our binary percentile fireflies optimisation algorithm (BPFA), we use the well-known set covering problem. Experiments were developed using a random operator to validate the contribution of the percentile technique in the binarization process of the fireflies algorithm. In addition, to verify our results, the JPSO algorithm developed by [25] and MDBBH developed in [26] were chosen. The results show that BPFA algorithm obtain competitive results.

2 Set Covering Problem

The set covering problem corresponds to a classical problem in combinatorics, and complexity theory. The problem aims to find subsets of a set. Given a set and its elements, we want to find subsets that completely cover the set at a minimum cost. The SCP is one of the oldest and most studied optimization problems. It is well-known to be NP-hard [27].

SCP is an active problem because medium and large instances often become intractable and cannot be solved any more using exact algorithms. Additionally, SCP due to its large number of instances, is used to verify the behavior of proposed new algorithms. In recent years, SCP has been approached by various continuous Swarm intelligence metaheuristics and using different methods to binarize metaheuristics. In [28], they used the teaching-learning-based optimisation with a specific binarization scheme to solve medium and large size instances. An SCP in fault diagnosis application was developed by [29]. In this article, the Gravitational Search algorithm was used and transfer functions were applied to perform the binarization. A rail scheduling problem application using SCP was developed in [26]. In [25] a Jumping PSO was used to solve SCP and in [14] CS metaheuristics were binarized using a percentile algorithm.

SCP has many practical applications in engineer, e.g., vehicle routing, facility location, railway, and airline crew scheduling [30–33] problems.

The SCP can be formally defined as follows. Let $A = (a_{ij})$, be a $n \times m$ zero-one matrix, where a column j cover a row i if $a_{ij} = 1$, besides a column j is associated with a non-negative real cost c_j. Let $I = \{1, ..., n\}$ and $J = \{1, ...m\}$, be the row and column set of A, respectively. The SCP consists in searching a minimum cost subset $S \subset J$ for which every row $i \in I$ is covered by at least one column $j \in J$, i.e.,:

$$\text{Minimize } f(x) = \sum_{j=1}^{m} c_j x_j \tag{1}$$

$$\text{Subject to} \sum_{j=1}^{m} a_{ij} x_j \geq 1, \forall i \in I, \text{ and } x_j \in \{0,1\}, \ \forall j \in J \tag{2}$$

where $x_j = 1$ if $j \in S$, $x_j = 0$ otherwise.

3 Binary Percentile Firefly Algorithm

The application of statistics and machine learning techniques, are used in many areas such as transports, smart cities, agriculture and computational intelligence, examples of applications are found in [34–37,37,38]. In this article, we explore the percentile concept applied to binarization of continuous metaheuristics. As an algorithm to binarize, firefly algorithm is used and as problem to solve, SCP. The Proposed BPFA has two modules. The first module corresponds to the initialization of the feasible solutions, and is detailed in the Sect. 3.1. Once the initialization of the particles is performed, it is consulted if the maximum if iterations is satisfied. Subsequently if the criterion is not satisfied, The BPFA is then run in conjunction with the percentile operator, this is detailed in Sect. 3.2. Once the transitions of the different solutions are made, we compare the resulting solutions with the best solution previously obtained. In the event that a superior solution is found, this replaces the previous one. The general algorithm scheme is detailed in Fig. 1.

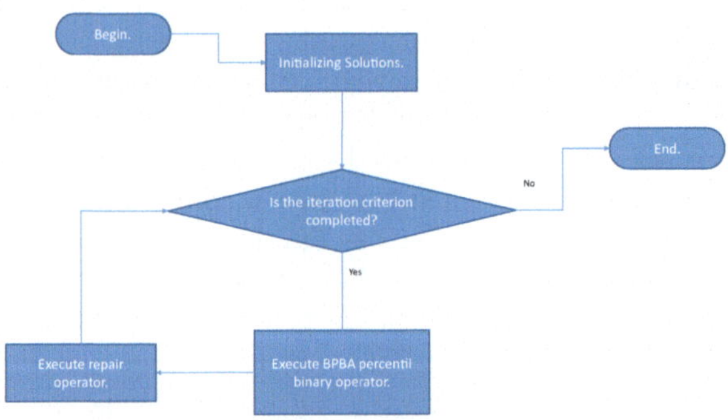

Fig. 1. Flowchart of the BPFA algorithm.

3.1 Initialization

For initialization of a new solution, a column is randomly chosen. It is then queried whether the current solution covers all rows. In the case of the solution does not meet the coverage condition, the heuristic operator is called. This operator aims to select a new column. This heuristic operation is iterated until all rows are covered. Once the coverage condition is met, the solution is optimized.

The optimization consists of eliminating columns where all their rows are covered by more than one column. The detail of the procedure is shown in Algorithm 1.

Algorithm 1. Initialization Operator

Function Initialization()
Input
Output Initialized solution S_{out}
$S \leftarrow$ SRandomColumn()
while All row are not covered **do**
    S.append(Heuristic(S))
end while
$S \leftarrow$ dRepeatedItem(S)
$S_{out} \leftarrow S$
return S_{out}

3.2 Percentile Binary Operator

Since our firefly algorithm is continuous swarm intelligence metaheuristic, it works in an iterative way by updating the position and velocity of the particles in each iteration. As firefly is continuous, the update is done in $\mathbb{R}^n$. In Eq. 3, the position update is written in a general way. The $x(t+1)$ variable represents the x position of the particle at time t+1. This position is obtained from the position x at time t plus a Δ function calculated at time t+1. The function Δ is proper to each metaheuristic and produces values in $\mathbb{R}^n$. For example in Cuckoo Search $\Delta(x) = \alpha \oplus Levy(\lambda)(x)$, in Black Hole $\Delta(x) = \text{rand} \times (x_{bh}(t) - x(t))$ and in the Firefly, Bat and PSO algorithms Δ can be written in simplified form as $\Delta(x) = v(x)$.

$$x(t+1) = x(t) + \Delta_{t+1}(x(t)) \tag{3}$$

In the percentile binary operator, we considering the movements generated by the algorithm in each dimension for all particles. $\Delta^i(x)$ corresponds to the magnitude of the displacement $\Delta(x)$ in the i-th position for the particle x. Subsequently these displacement are grouped using $\Delta^i(x)$, the magnitude of the displacement. This grouping is done using the percentile list. In our case the percentile list used the values {20, 40, 60, 80, 100}.

The percentile operator has as entry the parameters percentile list (pList) and the list of values (vList). Given an iteration, the list of values corresponds to the magnitude $\Delta^i(x)$ of the displacements of the particles in each dimension. As a first step the operator uses the vList and obtains the values of the percentiles given in the pList. Later, each value in the vList is assigned the group of the smallest percentile to which the value belongs. Finally, the list of the percentile to which each value belongs is returned (pGroupValue). The algorithm is shown in Algorithm 2.

A transition probability through the function P_{tr} is assigned to each element of the vList. This assignment is done using the percentile group assigned to each value. For the case of this study, we particularly use the Step function given in Eq. 4.

$$P_{tr}(x^i) = \begin{cases} 0.1, & \text{if } x^i \in \text{group } \{0,1\} \\ 0.5, & \text{if } x^i \in \text{group } \{2,3,4\} \end{cases} \tag{4}$$

Afterwards the transition of each particle is performed. In the case of firefly search the Eq. 5 is used to perform the transition, where $\hat{x}^i$ is the complement of x^i. Finally, each solution is repaired using the heuristic operator.

$$x^i(t+1) := \begin{cases} \hat{x}^i(t), & \text{if } rand < P_{tg}(x^i) \\ x^i(t), & \text{otherwise} \end{cases} \tag{5}$$

Algorithm 2. percentile binary operator

```
1: Function percentileBinary(vList, pList)
2: Input vList, pList
3: Output pGroupValue
4: percentileValue = getPValue(vList, pList)
5: for each value in vList do
6:     pGroupValue = getPGroupValue(pValue,vList)
7: end for
8: return pGroupValue
```

4 Results

4.1 Insight of BPFA Algorithm

In this module we developed experiments that allowed us to study the contribution of the binary percentile operator with respect to the quality of the solutions obtained and the iterations used when solving SCP instances. To perform this evaluation, we used the instances of Balas and Carrera. In the comparison, a random operator was designed which replaces our binary percentile operator. This random operator instead of executing transitions assigning transition probabilities per group, uses a fixed value of 0.5. To compare the distributions of the results of the different experiments we use violin Chart. The horizontal axis X corresponds to the problems, while Y axis uses the measure % - Gap.

Furthermore, a non-parametric test, Wilcoxon signed-rank test is carried out to determine if the results of BPFA with respect to the random algorithm have significant difference or not. The parameter settings and browser ranges are shown in Table 1.

BPFA corresponds to our standard algorithm. *random.0.5* is the random variant. The results are shown in Table 2 and Fig. 2. When we compared the Best Values between BPFA and *rnd.0.5* which are shown in Table 2. In the Average comparison, BPFA outperforms *rnd.0.5* in all problems. The comparison of distributions is shown in Fig. 2. We see the dispersion of the *rnd.0.5* distributions

Table 1. Setting of parameters for Binary percentile firefly Algorithm.

Parameters	Description	Value	Range
N	Number of solutions	25	[20, 25, 30]
G	Number of percentiles	5	[4, 5, 6]
Iteration Number	Maximum iterations	800	[700, 800, 900]

are bigger than the dispersions of BPFA. Therefore, the percentile binary operator, contribute to the precision of the results. Finally, the BPFA distributions are closer to zero than *rnd.0.5* distributions, indicating that BPFA has consistently better results than *random.0.5*. When we evaluate the behaviour of the algorithms through the Wilcoxon test, this test indicates that there is a significant difference between the two algorithms.

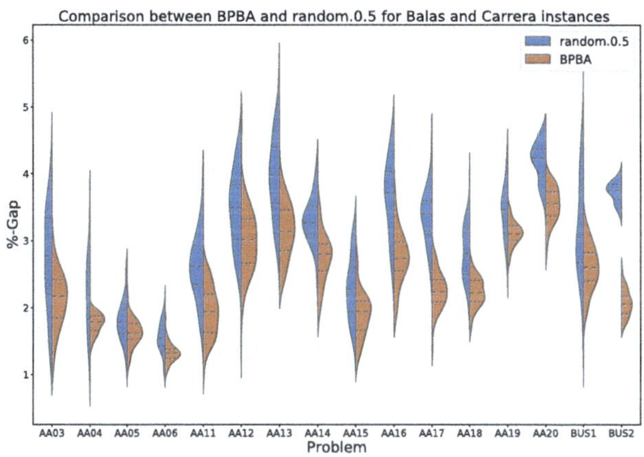

Fig. 2. Evaluation of percentile binary operator.

4.2 BPFA Compared with JPSO and MDBBH

In this section, we detail the comparisons made with the objective of evaluating the performance of our BPFA algorithm. For the evaluation, the larger problems of the OR-library were chosen. To develop the comparison, two algorithms were selected. The first one corresponds to a discretization Particle Swarm Optimization (PSO) technique called Jumping PSO (JPSO), [25]. The second one, is a binarization of the Black Hole technique called Multi Dynamic Binary Black Hole (MDBBH) Algorithm, [26]. JPSO uses a discrete PSO based on the frog jump. JPSO works without the concept of velocity, replacing this one by a component of random jump which allows to perform the movement in the discrete search space. On the other hand, MDBBH uses a binarization mechanism specific to

Table 2. Balas and Carrera instances.

Instance	Row	Col	Density	Best known	BPFA (avg)	Std	(Secs)	*rnd.0.5* (avg)	Std
AA03	106	8661	4.05%	33155	33581.2	127.1	215	33693.5	252.5
AA04	106	8002	4.05%	34573	34871.2	58.5	203	34931	234.1
AA05	105	7435	4.05%	31623	31884.7	65.3	208	31889.2	107.6
AA06	105	6951	4.11%	37464	37659.1	59.1	208	37651.3	93.5
AA11	271	4413	2.53%	35384	35767.9	134.7	201	35901.2	199.5
AA12	272	4208	2.52%	30809	31456.9	153.3	217	31631.3	202.1
AA13	265	4025	2.60%	33211	33999.7	136.5	198	34201.4	236.7
AA14	266	3868	2.50%	33219	33875.1	126	148	33967.3	124.2
AA15	267	3701	2.58%	34409	34827.4	112.9	144	34846.8	168.1
AA16	265	3558	2.63%	32752	33326.7	139.3	184	33601.3	185.4
AA17	264	3425	2.61%	31612	32008.3	89.2	167	32386.2	145.7
AA18	271	3314	2.55%	36782	37293.7	78.7	138	37417.6	146.5
AA19	263	3202	2.63%	32317	33096.1	77.8	143	33109.1	138.3
AA20	269	3095	2.58%	34912	35844.3	93.7	146	35963.1	96.6
BUS1	454	2241	1.88%	27947	28486.2	86.5	192	28756.1	233.4
BUS2	681	9524	0.51%	67760	68545.8	84.5	177	68678.9	105.3
Average p-value				35495.56	36029.9	101.44	176.8	36164.08 1.41e-04	165.11

Table 3. OR-Library benchmarks.

Instance	Row	Col	Density	Best known	JPSO Avg	MDBBH best	Avg	BPFA best	Avg	Time(s)
E.1	500	5000	10%	29	29	29	29	29	30,9	15,1
E.2	500	5000	10%	30	30	31	31,6	30	33,4	18.7
E.3	500	5000	10%	27	27	27	27,4	27	28,7	23,7
E.4	500	5000	10%	28	28	28	29,1	28	30,3	22,4
E.5	500	5000	10%	28	28	28	28	28	30	24,3
F.1	500	5000	20%	14	14	14	14,1	14	14,9	21,2
F.2	500	5000	20%	15	15	15	15,3	15	17,5	27,6
F.3	500	5000	20%	14	14	14	14,8	14	15,5	28,1
F.4	500	5000	20%	14	14	14	14,9	14	15,6	24,6
F.5	500	5000	20%	13	13	14	14,1	14	16,4	28,3
G.1	1000	10000	2%	176	176	177	178,5	177	179,9	217,5
G.2	1000	10000	2%	154	155	157	160,6	156	160,8	218,1
G.3	1000	10000	2%	166	167,2	168	170,4	168	171.4	234,2
G.4	1000	10000	2%	168	168,2	169	170,9	170	172,2	238,1
G.5	1000	10000	2%	168	168	168	169,8	168	171,9	262,1
H.1	1000	10000	5%	63	64	64	64,9	65	67.2	201,8
H.2	1000	10000	5%	63	63	64	64	65	65,7	192,5
H.3	1000	10000	5%	59	59,2	59	60	60	62,2	193.1
H.4	1000	10000	5%	58	58,3	59	60,4	59	62,6	188,1
H.5	1000	10000	5%	55	55	55	56,4	56	57,1	172,3
Average				67.1	67.3	67.7	68.7	67.9	70.86	123.2

the Black Hole algorithm. This is based on the concept of closeness to the Black Hole (BH). BH corresponds to the solution that has obtained the best value. When a solution is close to BH, the transitions are less likely to be performed. In the case of being away from the BH, the probability of transition is greater.

For instances E and F which are of medium size, the results of the MDBBH and BPBA algorithms are similar when comparing their Best Value. Only in instance E.2 BPFA was superior to MDBBH. When analyzing the average, JPSO obtained better results, however, the values of BPFA were close, being MDBBH the one that obtained the worse performance. For the case of problems G and H, BPFA was superior in problems G.1, G.2 and G.3 and MDBBH in problem H.3 when comparing their Best Value. In the case of averages, JPSO was superior in all cases. However, BPFA obtained results quite close, leaving MDBBH behind. In this point we must emphasize that the percentile technique used in BPFA allows binarizing any continuous swarm intelligence algorithm unlike JPSO that is specific for PSO (Table 3).

5 Conclusions

In this work, the BPFA algorithm based on the percentile concept was proposed to perform binarization of firefly metaheuristics. We must emphasize that this percentile concept can be applied in the binarization of any continuous metaheuristics of the swarm intelligence type. To evaluate the performance of Our percentile concept the set covering problem was used together with the firefly algorithm. Experiments were designed to evaluate the contribution of the binary operator percentile of the algorithm. It was found that the operator contributes significantly to improve the accuracy and quality of the solutions, all these visualized through violin graphics. As future works, we believe that the autonomous search for efficient configuration is a little-explored area and it could help to improve the performance of the binarizations both in the quality of the solutions and in the times of convergence. Additionally, it is interesting to see how this percentile technique works with other metaheuristics and other NP-hard problems. Finally, we believe that using machine learning we can generate indicators or obtain correlations that allow us to understand how the percentile algorithm translates the exploration and exploitation properties from the continuous space to the binary space.

References

1. Yang, X.S., Deb, S.: Cuckoo search via lévy flights. In: World Congress on Nature & Biologically Inspired Computing, NaBIC 2009, pp. 210–214. IEEE (2009)
2. Yang, X.-S.: A new metaheuristic bat-inspired algorithm. In: Nature Inspired Cooperative Strategies for Optimization (NICSO 2010), pp. 65–74 (2010)
3. Mirjalili, S.: SCA: a sine cosine algorithm for solving optimization problems. Knowl.-Based Syst. **96**, 120–133 (2016)
4. Khatibinia, M., Yazdani, H.: Accelerated multi-gravitational search algorithm for size optimization of truss structures. Swarm Evol. Comput. **38**, 109–119 (2017)
5. Barman, S., Kwon, Y.-K.: A novel mutual information-based Boolean network inference method from time-series gene expression data. PloS One **12**(2), e0171097 (2017)
6. Crawford, B., et al.: Investigating the efficiency of swarm algorithms for bridge strengthening by conversion to tied-arch: a numerical case study on San Luis bridge. Iranian J. Sci. Technol. Trans. Civ. Eng. 1–13 (2020)

7. Crawford, B., Soto, R., Astorga, G., García, J.: Constructive metaheuristics for the set covering problem. In: International Conference on Bioinspired Methods and Their Applications, pp. 88–99. Springer (2018)
8. García, J., Peña, A.: Robust optimization: concepts and applications. In: Nature-inspired Methods for Stochastic, Robust and Dynamic Optimization. IntechOpen (2018)
9. García, J., et al.: A db-scan binarization algorithm applied to matrix covering problems. Comput. Intell. Neurosci. **2019** (2019)
10. García, J., Altimiras, F., Peña, A., Astorga, G., Peredo, O.: A binary cuckoo search big data algorithm applied to large-scale crew scheduling problems. Complexity **2018** 2018
11. García, J., Maureira, C.: A KNN quantum cuckoo search algorithm applied to the multidimensional knapsack problem. Appl. Soft Comput. **102**, 107077 (2021)
12. García, J., Lalla-Ruiz, E., Voß, S., Droguett, E.L.: Enhancing a machine learning binarization framework by perturbation operators: analysis on the multidimensional knapsack problem. Int. J. Mach. Learn. Cybern. **11**, 1–20 (2020)
13. García, J., Moraga, P., Valenzuela, M., Pinto, H.: A db-scan hybrid algorithm: an application to the multidimensional knapsack problem. Mathematics **8**(4), 507 (2020)
14. García, J., Crawford, B., Soto, R., Astorga, G.: A percentile transition ranking algorithm applied to knapsack problem. In: Proceedings of the Computational Methods in Systems and Software, pp. 126–138. Springer (2017)
15. García, J., Crawford, B., Soto, R., Astorga, G.: A percentile transition ranking algorithm applied to binarization of continuous swarm intelligence metaheuristics. In: International Conference on Soft Computing and Data Mining, pp. 3–13. Springer (2018)
16. García, J., Astorga, G., Yepes, V.: An analysis of a KNN perturbation operator: an application to the binarization of continuous metaheuristics. Mathematics **9**(3), 225 (2021)
17. Tapia, D., et al.: A q-learning hyperheuristic binarization framework to balance exploration and exploitation. In: International Conference on Applied Informatics, pp. 14–28. Springer (2020)
18. García, J., Crawford, B., Soto, R., Astorga, G.: A clustering algorithm applied to the binarization of swarm intelligence continuous metaheuristics. Swarm Evol. Comput. **44**, 646–664 (2019)
19. Garcia, J., Măntoiu, M.: Localization results for zero order pseudodifferential operators. J. Pseudo-Differ. Oper. Appl. **5**(2), 255–276 (2014). https://doi.org/10.1007/s11868-013-0084-y
20. García, J., Yepes, V., Martí, J.V.: A hybrid k-means cuckoo search algorithm applied to the counterfort retaining walls problem. Mathematics **8**(4), 555 (2020)
21. García, J., Martí, J.V., Yepes, V.: The buttressed walls problem: An application of a hybrid clustering particle swarm optimization algorithm. Mathematics **8**(6), 862 (2020)
22. Yepes, V., Martí, J.V., García, J.: Black hole algorithm for sustainable design of counterfort retaining walls. Sustainability **12**(7), 2767 (2020)
23. Martínez-Muñoz, D., Martí, J.V., García, J., Yepes, V.: Embodied energy optimization of buttressed earth-retaining walls with hybrid simulated annealing. Appl. Sci. **11**(4), 1800 (2021)
24. Yang, X.-S.: Firefly algorithm, stochastic test functions and design optimisation. Int. J. Bio-Inspir. Comput. **2**(2), 78–84 (2010)

25. Balaji, S., Revathi, N.: A new approach for solving set covering problem using jumping particle swarm optimization method. Nat. Comput. **15**(3), 503–517 (2016). https://doi.org/10.1007/s11047-015-9509-2
26. García, J., Crawford, B., Soto, R., García, P.: A multi dynamic binary black hole algorithm applied to set covering problem. In: International Conference on Harmony Search Algorithm, pp. 42–51. Springer (2017)
27. Gary, M.R., Johnson, D.S.: Computers and intractability. In: A Guide to the Theory of NP-Completeness (1979)
28. Lu, Y., Vasko, F.J.: An or practitioner's solution approach for the set covering problem. Int. J. Appl. Metaheuristic Comput. (IJAMC) **6**(4), 1–13 (2015)
29. Li, Y., Cai, Z.: Gravity-based heuristic for set covering problems and its application in fault diagnosis. J. Syst. Eng. Electron. **23**(3), 391–398 (2012)
30. Crawford, B., Soto, R., Monfroy, E., Astorga, G., García, J., Cortes, E.: A meta-optimization approach for covering problems in facility location. In: Workshop on Engineering Applications, pp. 565–578. Springer (2017)
31. Kasirzadeh, A., Saddoune, M., Soumis, F.: Airline crew scheduling: models, algorithms, and data sets. EURO J. Transp. Logist. **6**(2), 111–137 (2017)
32. Horváth, M., Kis, T.: Computing strong lower and upper bounds for the integrated multiple-depot vehicle and crew scheduling problem with branch-and-price. Central Eur. J. Oper. Res. **27**, 1–29 (2017). https://doi.org/10.1007/s10100-017-0489-4
33. Stojković, M.: The operational flight and multi-crew scheduling problem. Yugoslav J. Oper. Res. **15**(1) (2016)
34. García, J., Crawford, B., Soto, R., Carlos, C., Paredes, F.: A k-means binarization framework applied to multidimensional knapsack problem. Appl. Intell. **48**, 1–24 (2017). https://doi.org/10.1007/s10489-017-0972-6
35. García, J., Pope, C., Altimiras, F.: A distributed k-means segmentation algorithm applied to Lobesia Botrana recognition. **Complexity**, **2017** (2017)
36. Graells-Garrido, E., García, J.: Visual exploration of urban dynamics using mobile data. In: International Conference on Ubiquitous Computing and Ambient Intelligence, pp. 480–491. Springer (2015)
37. Graells-Garrido, E., Peredo, O., García, J.: Sensing urban patterns with antenna mappings: the case of Santiago, Chile. Sensors **16**(7), 1098 (2016)
38. Peredo, O.F., García, J.A., Stuven, R., Ortiz, J.M.: Urban dynamic estimation using mobile phone logs and locally varying anisotropy. In: Geostatistics Valencia 2016, pp. 949–964. Springer (2017)

Correction to: Numerical Modelling of the Impact of Dredging on Stability of Oum Rabiâ Estuary (Morocco) Using SWAN Model

Hassan Ayad, Mustapha Mouhid, Dalila Loudyi, Laila Mouakkir, and Mohamed Chagdali

Correction to: Chapter "Numerical Modelling of the Impact of Dredging on Stability of Oum Rabiâ Estuary (Morocco) Using SWAN Model" in: R. Silhavy (Ed.): *Artificial Intelligence in Intelligent Systems*, LNNS 229, https://doi.org/10.1007/978-3-030-77445-5_8

In the original version of the book, the following corrections have been incorporated: The author name "Dalila Loudy" has been changed to "Dalila Loudyi" in the Frontmatter, Backmatter and in Chapter 8. The book has been updated with the changes.

The updated original version of this chapter can be found at https://doi.org/10.1007/978-3-030-77445-5_8

© The Author(s), under exclusive license to Springer Nature Switzerland AG 2022
R. Silhavy (Ed.): CSOC 2021, LNNS 229, p. C1, 2022.
https://doi.org/10.1007/978-3-030-77445-5_68

Author Index

© The Editor(s) (if applicable) and The Author(s), under exclusive license
to Springer Nature Switzerland AG 2021

R. Silhavy (Ed.): CSOC 2021, LNNS 229, pp. 761–763, 2021.
https://doi.org/10.1007/978-3-030-77445-5

The manufacturer's authorised representative in the EU is Springer Nature Customer Service Centre GmbH, Europaplatz 3, 69115 Heidelberg, Germany. If you have any concerns regarding our products, please contact ProductSafety@springernature.com

Printed and bound by CPI Group (UK) Ltd, Croydon, CR0 4YY
29/09/2026
02225004-0002